CAMBRIDGE MONOGRAPHS ON PARTICLE PHYSICS, NUCLEAR PHYSICS AND COSMOLOGY

General Editors: T. Ericson, P. V. Landshoff

Available titles in this series:

DYNAMICS OF THE STANDARD MODEL

Describing the fundamental theory of particle physics and its applications, this book provides a detailed account of the Standard Model, focusing on techniques that can produce information about real observed phenomena.

The book begins with a pedagogic account of the Standard Model, introducing essential techniques such as effective field theory and path-integral methods. It then focuses on the use of the Standard Model in the calculation of physical properties of particles. Rigorous methods are emphasized, but other useful models are also described.

This second edition has been updated to include recent theoretical and experimental advances, such as the discovery of the Higgs boson. A new chapter is devoted to the theoretical and experimental understanding of neutrinos, and major advances in *CP* violation and electroweak physics have been given a modern treatment. This book is valuable to graduate students and researchers in particle physics, nuclear physics and related fields. This title, first published in 2014, has been reissued as an Open Access publication on Cambridge Core.

JOHN F. DONOGHUE is Distinguished Professor in the Department of Physics, University of Massachusetts. His research spans particle physics, quantum field theory and general relativity. He is a Fellow of the American Physical Society.

EUGENE GOLOWICH is Emeritus Professor in the Department of Physics, University of Massachusetts. His research has focused on particle theory and phenomenology. He is a Fellow of the American Physical Society and is a recipient of the College Outstanding Teacher award from the University of Massachusetts.

BARRY R. HOLSTEIN is Emeritus Professor in the Department of Physics, University of Massachusetts. His research is in the overlap area of particle and nuclear theory. A Fellow of the American Physical Society, he is also the editor of *Annual Reviews of Nuclear and Particle Science* and is a longtime consulting editor of the *American Journal of Physics*.

DYNAMICS OF THE STANDARD MODEL

SECOND EDITION

JOHN F. DONOGHUE
University of Massachusetts

EUGENE GOLOWICH
University of Massachusetts

BARRY R. HOLSTEIN
University of Massachusetts

CAMBRIDGE
UNIVERSITY PRESS

Shaftesbury Road, Cambridge CB2 8EA, United Kingdom

One Liberty Plaza, 20th Floor, New York, NY 10006, USA

477 Williamstown Road, Port Melbourne, VIC 3207, Australia

314–321, 3rd Floor, Plot 3, Splendor Forum, Jasola District Centre, New Delhi – 110025, India

103 Penang Road, #05–06/07, Visioncrest Commercial, Singapore 238467

Cambridge University Press is part of Cambridge University Press & Assessment, a department of the University of Cambridge.

We share the University's mission to contribute to society through the pursuit of education, learning and research at the highest international levels of excellence.

www.cambridge.org
Information on this title: www.cambridge.org/9781009291002

DOI: 10.1017/9781009291033

First published 2014
Reissued as OA 2022

A catalogue record for this publication is available from the British Library.

ISBN 978-1-009-29100-2 Hardback
ISBN 978-1-009-29101-9 Paperback

To Lincoln Wolfenstein

Contents

Contents

Preface to the second edition

The Standard Model is the basis of our understanding of the fundamental interactions. At the present time, it remains in excellent agreement with experiment. It
is clear that any further progress in the field will need to build on a solid understanding of the Standard Model. Since the first edition was written in 1992 there
have been major discoveries in neutrino physics, in *CP* violation, the discoveries
of the top quark and the Higgs boson, and a dramatic increase in precision in both
electroweak physics and in *QCD*. We feel that the present is a good moment to
update our book, as the Standard Model seems largely complete.

The opportunity to revise our book at this time has also enabled us to survey the
progress since the first edition went to print. Besides the experimental discoveries
that have taken place during these two decades, we have been impressed by the
increase in theoretical sophistication. Many of the topics which were novel at the
time of the first edition have now been extensively developed. Perturbative treatments have progressed to higher orders and new techniques have been developed.
To cover all of these completely would require the expansion of many chapters
into book-length treatments. Indeed, in many cases, entire new books dedicated to
specialized topics have been published.[1] Our revision is meant as a coherent pedagogic introduction to these topics, providing the reader with the basic background
to pursue more detailed studies when appropriate.

There has also been great progress on the possible New Physics which could
emerge beyond the Standard Model – dark matter and dark energy, grand unification, supersymmetry, extra dimensions, etc. We are at a moment where this physics
could emerge in the next round of experiments at the Large Hadron Collider (LHC)
as well as in precision measurements at the intensity frontier. We look forward with
great anticipation to the new discoveries of the next decade.

---

[1] For example, see [BaP 99, Be 00, BiS 00, EISW 03, FuS 04, Gr 04, IoFL 10, La 10, Ma 04, MaW 07, Co 11]

We thank our colleagues and students for feedback about the first edition of this book. A list of errata for the second edition will be maintained at the homepage of John Donoghue at the University of Massachusetts, Amherst. We encourage readers who find any mistakes in this edition to submit them to Professor Donoghue at donoghue@physics.umass.edu.

From the preface to the first edition

The Standard Model lagrangian $\mathcal{L}_{SM}$ embodies our knowledge of the strong and electroweak interactions. It contains as fundamental degrees of freedom the spin one-half quarks and leptons, the spin one gauge bosons, and the spin zero Higgs fields. Symmetry plays the central role in determining its dynamical structure. The lagrangian exhibits invariance under $SU(3)$ gauge transformations for the strong interactions and under $SU(2) \times U(1)$ gauge transformations for the electroweak interactions. Despite the presence of (all too) many input parameters, it is a mathematical construction of considerable predictive power.

There are books available which describe in detail the construction of $\mathcal{L}_{SM}$ and its quantization, and which deal with aspects of symmetry breaking. We felt the need for a book describing the next steps, how $\mathcal{L}_{SM}$ is connected to the observable physics of the real world. There are a considerable variety of techniques, of differing rigor, which are used by particle physicists to accomplish this. We present here those which have become indispensable tools. In addition, we attempt to convey the insights and 'conventional wisdom' which have been developed throughout the field. This book can only be an introduction to the riches contained in the subject, hopefully providing a foundation and a motivation for further exploration by its readers.

In writing the book, we have become all too painfully aware that each topic, indeed each specific reaction, has an extensive literature and phenomenology, and that there is a limitation to the depth that can be presented compactly. We emphasize applications, not fundamentals, of quantum field theory. Proofs of formal topics like renormalizability or the quantization of gauge fields are left to other books, as is the topic of parton phenomenology. In addition, the study by computer of lattice field theory is an extensive and rapidly changing discipline, which we do not attempt to cover. Although it would be tempting to discuss some of the many stimulating ideas, among them supersymmetry, grand unification, and string theory, which attempt to describe physics beyond the Standard Model, limitations of space prevent us from doing so.

Although this book begins gently, we do assume that the reader already has some familiarity with quantum field theory. As an aid to those who lack familiarity with path-integral methods, we include a presentation, in Appendix A, which treats this subject in an introductory manner. In addition, we assume a knowledge of the basic phenomenology of particle physics.

We have constructed the material to be of use to a wide spectrum of readers who are involved with the physics of elementary particles. Certainly it contains material of interest to both theorist and experimentalist alike. Given the trend to incorporate the Standard Model in the study of nuclei, we expect the book to be of use to the nuclear physics community as well. Even the student being trained in the mathematics of string theory would be well advised to learn the role that sigma models play in particle theory.

This is a good place to stress some conventions employed in this book. Chapters are identified with roman numerals. In cross-referencing equations, we include the chapter number if the referenced equation is in a chapter different from the point of citation. The Minkowski metric is $g_{\mu\nu} = \text{diag} \{1, -1, -1, -1\}$. Throughout, we use the natural units $\hbar = c = 1$, and choose $e > 0$ so that the electron has electric charge $-e$. We employ rationalized Heaviside–Lorentz units, and the fine-structure constant is related to the charge via $\alpha = e^2/4\pi$. The coupling-constants for the $SU(3)_c \times SU(2)_L \times U(1)$ gauge structure of the Standard Model are denoted respectively as g_3, g_2, g_1, and we employ coupling-constant phase conventions analogous to electromagnetism for the other abelian and nonabelian covariant derivatives of the Standard Model. The chiral projection operator for left-handed massless spin one-half particles is $(1+\gamma_5)/2$, and in analyzing systems in d dimensions, we employ the parameter $\epsilon \equiv (4-d)/2$. What is meant by the 'Fermi constant' is discussed in Sect. V–2.

Amherst, MA, 2013

I

Inputs to the Standard Model

This book is about the Standard Model of elementary particle physics. If we set the beginning of the modern era of particle physics in 1947, the year the pion was discovered, then the ensuing years of research have revealed the existence of a consistent, self-contained layer of reality. The energy range which defines this layer of reality extends up to about 1 TeV or, in terms of length, down to distances of order 10^{-17} cm. The Standard Model is a field-theoretic description of strong and electroweak interactions at these energies. It requires the input of as many as 28 independent parameters.[1] These parameters are not explained by the Standard Model; their presence implies the need for an understanding of Nature at an even deeper level. Nonetheless, processes described by the Standard Model possess a remarkable insulation from signals of such New Physics. Although the strong interactions remain a calculational challenge, the Standard Model (generalized from its original form to include neutrino mass) would appear to have sufficient content to describe all existing data.[2] Thus far, it is a theoretical structure which has worked splendidly.

I–1 Quarks and leptons

The Standard Model is an $SU(3) \times SU(2) \times U(1)$ gauge theory which is spontaneously broken by the Higgs potential. Table I–1 displays mass determinations [RPP 12] of the Z^0 and $W^\pm$ gauge bosons, the Higgs boson H^0, and the existing mass limit on the photon γ.

In the Standard Model, the fundamental fermionic constituents of matter are the quarks and the leptons. Quarks, but not leptons, engage in the strong interactions as a consequence of their color charge. Each quark and lepton has spin one-half.

---

[1] There are six lepton masses, six quark masses, three gauge coupling constants, three quark-mixing angles and one complex phase, three neutrino-mixing angles and as many as three complex phases, a Higgs mass and quartic coupling constant, and the QCD vacuum angle.

[2] Admittedly, at this time the sources of *dark matter* and of *dark energy* are unknown.

1

Table I–1. *Boson masses.*

Particle	Mass (GeV/c^2)
γ	$< 1 \times 10^{-27}$
$W^\pm$	80.385 ± 0.015
Z^0	91.1876 ± 0.0021
H^0	126.0 ± 0.4

Collectively, they display conventional Fermi–Dirac statistics. No attempt is made in the Standard Model either to explain the variety and number of quarks and leptons or to compute any of their properties. That is, these particles are taken at this level as truly elementary. This is not unreasonable. There is no experimental evidence for quark or lepton compositeness, such as excited states or form factors associated with intrinsic structure.

Quarks

There are six quarks, which fall into two classes according to their electrical charge Q. The u, c, t quarks have $Q = 2e/3$ and the d, s, b quarks have $Q = -e/3$, where e is the electric charge of the proton. The u, c, t and d, s, b quarks are eigenstates of the hamiltonian ('mass eigenstates'). However, because they are believed to be permanently confined entities, some thought must go into properly defining quark mass. Indeed, several distinct definitions are commonly used. We defer a discussion of this issue and simply note that the values in Table I–2 provide

Table I–2. *The quarks.*

Flavor	Mass[a] (GeV/c^2)	Charge	I_3	S	C	B	T
u	$(2.55^{+0.75}_{-1.05}) \times 10^{-3}$	$2e/3$	$1/2$	0	0	0	0
d	$(5.04^{+0.96}_{-1.54}) \times 10^{-3}$	$-e/3$	$-1/2$	0	0	0	0
s	$0.105^{+0.025}_{-0.035}$	$-e/3$	0	-1	0	0	0
c	$1.27^{+0.07}_{-0.11}$	$2e/3$	0	0	1	0	0
b	$4.20^{+0.17}_{-0.07}$	$-e/3$	0	0	0	-1	0
t	173.4 ± 1.6	$2e/3$	0	0	0	0	1

[a]The t-quark mass is inferred from top quark events. All others are determined in $\overline{\text{MS}}$ renormalization (cf. Sect. II–1) at scales $m_{u,d,s}(2 \text{ GeV}/c^2)$, $m_c(m_c)$ and $m_b(m_b)$ respectively.

Table I–3. *The leptons.*

Flavor	Mass (GeV/c^2)	Charge	L_e	L_μ	L_τ
ν_e	$< 0.2 \times 10^{-8}$	0	1	0	0
e	$5.10998928(11) \times 10^{-4}$	$-e$	1	0	0
ν_μ	$< 1.9 \times 10^{-4}$	0	0	1	0
μ	$0.1056583715(35)$	$-e$	0	1	0
ν_τ	< 0.0182	0	0	0	1
τ	$1.77682(16)$	$-e$	0	0	1

an overview of the quark mass spectrum. A useful benchmark for quark masses is the energy scale Λ_{QCD} ($\simeq$ several hundred MeV) associated with the confinement phenomenon. Relative to Λ_{QCD}, the u, d, s quarks are light, the b, t quarks are heavy, and the c quark has intermediate mass. The dynamical behavior of light quarks is described by the chiral symmetry of massless particles (cf. Chap. VI) whereas heavy quarks are constrained by the so-called Heavy Quark Effective Theory (cf. Sect. XIII–3). Each quark is said to constitute a separate *flavor*, i.e. six quark flavors exist in Nature. The s, c, b, t quarks carry respectively the quantum numbers of strangeness (S), charm (C), bottomness (B), and topness (T). The u, d quarks obey an $SU(2)$ symmetry (isospin) and are distinguished by the three-component of isospin (I_3). The flavor quantum numbers of each quark are displayed in Table I–2.

Leptons

There are six leptons which fall into two categories according to their electrical charge. The charged leptons e, μ, τ have $Q = -e$ and the neutrinos ν_e, ν_μ, ν_τ have $Q = 0$. Leptons are also classified in terms of three lepton types: electron (ν_e, e), muon (ν_μ, μ), and tau (ν_τ, τ). This follows from the structure of the charged weak interactions (cf. Sect. II–3) in which these charged-lepton/neutrino pairs are coupled to $W^\pm$ gauge bosons. Associated with each lepton type is a lepton number L_e, L_μ, L_τ. Table I–3 summarizes lepton properties.

At this time, there is only incomplete knowledge of neutrino masses. Information on the mass parameters $m_{\nu_e}, m_{\nu_\mu}, m_{\nu_\tau}$ is obtained from their presence in various weak transition amplitudes. For example, the single beta decay experiment $^3\text{H} \rightarrow {}^3\text{He} + e^- + \bar{\nu}_e$ is sensitive to the mass m_{ν_e}. In like manner, one constrains the masses m_{ν_μ} and m_{ν_τ} in processes such as $\pi^+ \rightarrow \mu^+ + \nu_\mu$ and $\tau^- \rightarrow 2\pi^- + \pi^+ + \nu_\tau$ respectively. Existing bounds on these masses are displayed in Table I–3.

It is known experimentally that upon creation the neutrinos $\{\nu_\alpha\} \equiv (\nu_e . \nu_\mu, \nu_\tau)$ will not propagate indefinitely but will instead mix with each other. This means that the basis of states $\{\nu_\alpha\}$ cannot be eigenstates of the hamiltonian. Diagonalization of the leptonic hamiltonian is carried out in Sect. VI–2 and yields the basis $\{\nu_i\} \equiv \{\nu_1, \nu_2, \nu_3\}$ of mass eigenstates. Information on the neutrino mass eigenvalues m_1, m_2, m_3 is obtained from neutrino oscillation experiments and cosmological studies. Oscillation experiments (cf. Sects. VI–3,VI–4) are sensitive to squared-mass differences.[3] Throughout the book, we adhere to the following relations,

$$\text{definition: } \Delta m_{ij}^2 \equiv m_i^2 - m_j^2, \qquad \text{convention: } m_2 > m_1. \qquad (1.1)$$

From the compilation in [RPP 12], the squared-mass difference $|\Delta m_{32}^2|$ deduced from the study of atmospheric and accelerator neutrinos gives

$$|\Delta m_{32}^2| = 2.32^{+0.12}_{-0.08} \times 10^{-3} \text{ eV}^2, \qquad (1.2a)$$

whereas data from solar and reactor neutrinos imply a squared-mass difference roughly 31 times smaller,

$$\Delta m_{21}^2 = (7.50 \pm 0.20) \times 10^{-5} \text{ eV}^2. \qquad (1.2b)$$

Thus the neutrinos ν_1 and ν_2 form a quasi-doublet. One speaks of a *normal* or *inverted* neutrino mass spectrum, respectively, for the cases[4]

$$\text{normal: } m_3 > m_{1,2}, \qquad \text{inverted: } m_{1,2} > m_3. \qquad (1.2c)$$

Since the largest neutrino mass m_{lgst}, be it m_2 or m_3, cannot be lighter than the mass splitting of Eq. (1.2), we have the bound $m_{\text{lgst}} > 0.049$ eV. Finally, a combination of cosmological inputs can be employed to bound the neutrino mass sum $\sum_{i=1}^{3} m_i$, the precise bound depending on the chosen input data set. In one example [deP *et al.* 12], photometric redshifts measured from a large galaxy sample, cosmic microwave background (CMB) data and a recent determination of the Hubble parameter are used to obtain the bound

$$m_1 + m_2 + m_3 < 0.26 \text{ eV}, \qquad (1.3a)$$

whereas data from the CMB combined with that from baryon acoustic oscillations yields [Ad *et al.* (Planck collab.) 13]

$$m_1 + m_2 + m_3 < 0.23 \text{ eV}. \qquad (1.3b)$$

A further discussion of the neutrino mass spectrum appears in Sect. VI–4.

---

[3] Only two of the mass differences can be independent, so $\Delta m_{12}^2 + \Delta m_{23}^2 + \Delta m_{31}^2 = 0$.

[4] There is also the possibility of a *quasi-degenerate* neutrino mass spectrum ($m_1 \simeq m_2 \simeq m_3$), which can be thought of as a limiting case of both the normal and inverted cases in which the individual masses are sufficiently large to dwarf the $|\Delta m_{32}^2|$ splitting.

Quark and lepton numbers

Individual quark and lepton numbers are known to be not conserved, but for different reasons and with different levels of nonconservation. Individual quark number is not conserved in the Standard Model due to the charged weak interactions (cf. Sect. II–3). Indeed, quark transitions of the type $q_i \rightarrow q_j + W^{\pm}$ induce the decays of most meson and baryon states and have led to the phenomenology of *Flavor Physics*. Individual lepton number is not conserved, as evidenced by the observed $\nu_\alpha \leftrightarrow \nu_\beta$ ($\alpha, \beta = e, \mu, \tau$) oscillations. This source of this phenomenon is associated with nonzero neutrino masses. There is currently no additional evidence for the violation of individual lepton number despite increasingly sensitive limits such as the branching fraction $B_{\mu^- \rightarrow e^- e^- e^+} < 1.0 \times 10^{-12}$.

Existing data are consistent with conservation of total quark and total lepton number, e.g. the proton lifetime bound $\tau_p > 2.1 \times 10^{29}$ yr [RPP 12] and the nuclear half-life limit $t_{1/2}^{0\nu\beta\beta}[^{136}\text{Xe}] > 1.6 \times 10^{25}$ yr [Ac *et al.* (EXO-200 collab.) 11]. These conservation laws are empirical. They are not required as a consequence of any known dynamical principle and in fact are expected to be violated by certain nonperturbative effects within the Standard Model (associated with quantum tunneling between topologically inequivalent vacua – see Sect. III–6).

I–2 Chiral fermions

Consider a world in which quarks and leptons have no mass at all. At first, this would appear to be a surprising supposition. To an experimentalist, mass is the most palpable property a particle has. It is why, say, a muon behaves differently from an electron in the laboratory. Nonetheless, the massless limit is where the Standard Model begins.

The massless limit

Let $\psi(x)$ be a solution to the Dirac equation for a massless particle,

$$i \slashed{\partial} \psi = 0. \tag{2.1}$$

We can multiply this equation from the left by γ_5 and use the anticommutativity of γ_5 with γ^μ to obtain another solution,

$$i \slashed{\partial} \gamma_5 \psi = 0. \tag{2.2}$$

We superpose these solutions to form the combinations

$$\psi_L = \frac{1}{2}(1 + \gamma_5)\psi, \qquad \psi_R = \frac{1}{2}(1 - \gamma_5)\psi, \tag{2.3}$$

where '1' represents the unit 4×4 matrix. The quantities ψ_L and ψ_R are solutions of definite *chirality* (i.e. handedness). For a massless particle moving with precise momentum, these solutions correspond respectively to the spin being anti-aligned (left-handed) and aligned (right-handed) relative to the momentum. In other words, chirality coincides with helicity for zero-mass particles. The matrices $\Gamma_{\substack{L\\R}} = (1 \pm \gamma_5)/2$ are chirality projection operators. They obey the usual projection operator conditions under addition,

$$\Gamma_L + \Gamma_R = 1, \tag{2.4}$$

and under multiplication,

$$\Gamma_L \Gamma_L = \Gamma_L, \qquad \Gamma_R \Gamma_R = \Gamma_R, \qquad \Gamma_L \Gamma_R = \Gamma_R \Gamma_L = 0. \tag{2.5}$$

In the massless limit, a particle's chirality is a Lorentz-invariant concept. For example, a particle which is left-handed to one observer will appear left-handed to all observers. Thus chirality is a natural label to use for massless fermions, and a collection of such particles may be characterized according to the separate numbers of left-handed and right-handed particles.

It is simple to incorporate chirality into a lagrangian formalism. The lagrangian for a massless noninteracting fermion is

$$\mathcal{L} = i\overline{\psi}\,\partial\!\!\!/\,\psi, \tag{2.6}$$

or in terms of chiral fields,

$$\mathcal{L} = \mathcal{L}_L + \mathcal{L}_R, \tag{2.7}$$

where

$$\mathcal{L}_{L,R} = i\overline{\psi}_{L,R}\partial\!\!\!/\,\psi_{L,R}. \tag{2.8}$$

The lagrangians $\mathcal{L}_{L,R}$ are invariant under the global chiral phase transformations

$$\psi_{L,R}(x) \rightarrow \exp(-i\alpha_{L,R})\psi_{L,R}(x), \tag{2.9}$$

where the phases $\alpha_{L,R}$ are constant and real-valued but otherwise arbitrary. Anticipating the discussion of Noether's theorem in Sect. I–4, we can associate conserved particle-number current densities $J^{\mu}_{L,R}$,

$$J^{\mu}_{L,R} = \overline{\psi}_{L,R}\gamma^{\mu}\psi_{L,R} \qquad (\partial_{\mu}J^{\mu}_{L,R} = 0), \tag{2.10}$$

with this invariance. From these chiral current densities, we can construct the vector current $V^{\mu}(x)$,

$$V^{\mu} = J^{\mu}_L + J^{\mu}_R \tag{2.11}$$

and the axial-vector current $A^{\mu}(x)$,

$$A^\mu = J_L^\mu - J_R^\mu. \tag{2.12}$$

Chiral charges $Q_{L,R}$ are defined as spatial integrals of the chiral charge densities,

$$Q_{L,R}(t) = \int d^3x \, J_{L,R}^0(x), \tag{2.13}$$

and represent the number operators for the chiral fields $\psi_{L,R}$. They are time-independent if the chiral currents are conserved. One can similarly define the vector charge Q and the axial-vector charge Q_5,

$$Q(t) = \int d^3x \, V^0(x), \qquad Q_5(t) = \int d^3x \, A^0(x). \tag{2.14}$$

The vector charge Q is the total number operator,

$$Q = Q_R + Q_L, \tag{2.15}$$

whereas the axial-vector charge is the number operator for the difference

$$Q_5 = Q_L - Q_R. \tag{2.16}$$

The vector charge Q and axial-vector charge Q_5 simply count the sum and difference, respectively, of the left-handed and right-handed particles.

Parity, time reversal, and charge conjugation

The field transformations of Eq. (2.9) involve parameters $\alpha_{L,R}$ which can take on a continuum of values. In addition to such continuous field mappings, one often encounters a variety of *discrete* transformations as well. Let us consider the operations of parity

$$x = (x^0, \mathbf{x}) \to x_P = (x^0, -\mathbf{x}), \tag{2.17}$$

and of time reversal

$$x = (x^0, \mathbf{x}) \to x_T = (-x^0, \mathbf{x}), \tag{2.18}$$

as defined by their effects on spacetime coordinates. The effect of discrete transformations on a fermion field $\psi(x)$ will be implemented by a unitary operator P for parity and an antiunitary operator T for time reversal. In the representation of Dirac matrices used in this book, we have

$$P\psi(x)P^{-1} = \gamma^0 \psi(x_P), \qquad T\psi(x)T^{-1} = i\gamma^1\gamma^3 \psi(x_T). \tag{2.19}$$

An additional operation typically considered in conjunction with parity and time reversal is that of charge conjugation, the mapping of matter into antimatter,

$$C\psi(x)C^{-1} = i\gamma^2\gamma^0 \overline{\psi}^T(x), \tag{2.20}$$

Table I–4. *Response of Dirac bilinears to discrete mappings.*

C	P	T
$S(x)$	$S(x_P)$	$S(x_T)$
$P(x)$	$-P(x_P)$	$-P(x_T)$
$-J^\mu(x)$	$J_\mu(x_P)$	$J_\mu(x_T)$
$J_5^\mu(x)$	$-J_{5\mu}(x_P)$	$J_{5\mu}(x_T)$
$-T^{\mu\nu}(x)$	$T_{\mu\nu}(x_P)$	$-T_{\mu\nu}(x_T)$

where $\overline{\psi}_\beta^T \equiv \psi_\alpha^\dagger \gamma_{\alpha\beta}^0$ ($\alpha, \beta = 1, \ldots, 4$). The spacetime coordinates of field $\psi(x)$ are unaffected by charge conjugation.

In the study of discrete transformations, the response of the normal-ordered Dirac bilinears

$$S(x) = \: \overline{\psi}(x)\psi(x) \: \qquad P(x) = \: \overline{\psi}(x)\gamma_5\psi(x) \:$$
$$J^\mu(x) = \: \overline{\psi}(x)\gamma^\mu\psi(x) \: \qquad J_5^\mu(x) = \: \overline{\psi}(x)\gamma^\mu\gamma_5\psi(x) \: \qquad (2.21)$$
$$T^{\mu\nu}(x) = \: \overline{\psi}(x)\sigma^{\mu\nu}\psi(x) \:$$

is of special importance to physical applications. Their transformation properties appear in Table I–4. Close attention should be paid there to the location of the indices in these relations. Another example of a field's response to these discrete transformations is that of the photon $A^\mu(x)$,

$$C \, A^\mu(x) \, C^{-1} \, c = -A^\mu(x), \qquad P \, A^\mu(x) \, P^{-1} = A_\mu(x_P),$$
$$T \, A^\mu(x) \, T^{-1} \, c = A_\mu(x_T). \qquad (2.22)$$

Beginning with the discussion of Noether's theorem in Sect. 1–4, we shall explore the topic of invariance throughout much of this book. It suffices to note here that the Standard Model, being a theory whose dynamical content is expressed in terms of hermitian, Lorentz-invariant lagrangians of local quantum fields, is guaranteed to be invariant under the combined operation *CPT*. Interestingly, however, these discrete transformations are individually symmetry operations only of the strong and electromagnetic interactions, but not of the full electroweak sector. We see already the possibility for such behavior in the occurrence of chiral fermions $\psi_{L,R}$, since parity maps the fields $\psi_{L,R}$ into each other,

$$\psi_{L,R} \to P \, \psi_{L,R}(x) \, P^{-1} = \gamma^0 \psi_{R,L}(x_P). \qquad (2.23)$$

Thus any effect, like the weak interaction, which treats left-handed and right-handed fermions differently, will lead inevitably to parity-violating phenomena.

I–3 Fermion mass

Although the discussion of chiral fermions is cast in the limit of zero mass, fermions in Nature do in fact have nonzero mass and we must account for this. In a lagrangian, a mass term will appear as a hermitian, Lorentz-invariant bilinear in the fields. For fermion fields, these conditions allow realizations referred to as *Dirac* mass and *Majorana* mass.[5]

Dirac mass

The Dirac mass term for fermion fields $\psi_{L,R}$ involves the bilinear coupling of fields with opposite chirality

$$-\mathcal{L}_D = m_D[\,\overline{\psi}_L\psi_R + \overline{\psi}_R\psi_L\,] = m_D\,\overline{\psi}\psi \tag{3.1}$$

where $\psi \equiv \psi_L + \psi_R$ and m_D is the Dirac mass. The Dirac mass term is invariant under the phase transformation $\psi(x) \to \exp(-i\alpha)\psi(x)$ and thus does not upset conservation of the vector current $V^\mu = \overline{\psi}\gamma^\mu\psi$ and the corresponding number fermion operator Q of Eq. (2.15). All fields in the Standard Model, save possibly for the neutrinos, have Dirac masses obtained from their interaction with the Higgs field (cf. Sects. II–3, II–4). Although right-handed neutrinos have no couplings to the Standard Model gauge bosons, there is no principle prohibiting their interaction with the Higgs field and thus generating neutrino Dirac masses in the same manner as the other particles.

Majorana mass

A Majorana mass term is one which violates fermion number by coupling two fermions (or two antifermions). In the Majorana construction, use is made of the *charge-conjugate* fields,

$$\psi^c \equiv C\gamma^0\psi^*, \qquad (\psi_{L,R})^c = (\Gamma_{L,R}\psi)^c, \tag{3.2}$$

where C is the charge-conjugation operator, obeying

$$C = -C^{-1} = -C^\dagger = -C^T. \tag{3.3}$$

In the Dirac representation of gamma matrices (cf. App. C), one has $C = i\gamma^2\gamma^0$. Some useful identities involving ψ^c include

---

[5] We suppress spacetime dependence of the fields in this section.

$$\overline{(\psi_i^c)}\,\psi_j = \psi_i^T\,C\,\psi_j, \qquad \overline{\psi_i}\,\psi_j^c = -\psi_i^{*T}\,C\,\psi_j^*,$$

$$\left(\overline{(\psi_i^c)}\,\psi_j\right)^\dagger = \overline{\psi_j}\,\psi_i^c, \qquad \left(\psi_i^T\,C\,\psi_j\right)^\dagger = -\psi_j^{*T}\,C\,\psi_i^*,$$

$$\overline{(\psi_i^c)}\,\psi_j^c = \overline{\psi_j}\,\psi_i, \qquad \overline{(\psi_i^c)}\,\gamma^\mu\,\psi_j^c = -\overline{\psi_j}\,\gamma^\mu\,\psi_i, \qquad (3.4)$$

$$\overline{(\psi_R^c)}\,\psi_L = 0, \qquad \overline{(\psi_R^c)}\,\gamma^\mu\,\psi_R = 0.$$

The two identities in the bottom line follow from $\Gamma_R C \Gamma_L = 0$.

The possibility of a Majorana mass term follows from the fact that a combination of two fermion fields $\psi^T C \psi$ is an invariant under Lorentz transformations. Two equivalent expressions for a Majorana mass term involving chiral fields $\psi_{L,R}$ are[6]

$$\begin{aligned}
-\mathcal{L}_M &= \frac{m_{L,R}}{2}\left[\overline{(\psi_{L,R})^c}\,\psi_{L,R} + \overline{\psi_{L,R}}\,(\psi_{L,R})^c\right]\\
&= \frac{m_{L,R}}{2}\left[(\psi_{L,R})^T C \psi_{L,R} - (\psi_{L,R}^*)^T C \psi_{L,R}^*\right].
\end{aligned} \qquad (3.5)$$

Because the cross combination $(\psi_R)^T C \psi_L = 0$, the Majorana mass terms involves either two left-chiral fields or two right-chiral fields, and the left-chiral and right-chiral masses are independent. Treating ψ and ψ^* as independent variables, the resulting equations of motion are

$$i\slashed{\partial}\,\psi_R - m_R\,\psi_R^c = 0, \qquad i\slashed{\partial}\,\psi_R^c - m_R\,\psi_R = 0, \qquad (3.6)$$

with a similar set of equations for ψ_L. Iteration of these coupled equations shows that m_R indeed behaves as a mass.

A Majorana mass term clearly does not conserve fermion number and mixes the particle with its antiparticle. Indeed, a Majorana fermion can be identified with its own antiparticle. This can be seen, using ψ_R as an example, by rewriting the lagrangian in terms of the self-conjugate field

$$\psi_M = \frac{1}{\sqrt{2}}\left[\psi_R + \psi_R^c\right], \qquad (3.7)$$

which, given the equations of motion above, will clearly satisfy the Dirac equation. The total Majorana lagrangian can be simply rewritten in terms of this self-conjugate field as

$$\begin{aligned}
\mathcal{L}_{KE}^{(R)} + \mathcal{L}_M^{(R)} &= \overline{\psi}_R i\slashed{\partial}\,\psi_R - \frac{m_R}{2}\left[\overline{(\psi_R)^c}\,\psi_R + \overline{\psi_R}\,(\psi_R)^c\right]\\
&= \overline{\psi}_M i\slashed{\partial}\,\psi_M - m_R\overline{\psi}_M\,\psi_M\\
&= \psi_M^T C i\slashed{\partial}\,\psi_M - m_R\psi_M^T C\,\psi_M,
\end{aligned} \qquad (3.8)$$

[6] The factor of $1/2$ with the Majorana mass parameters $m_{L,R}^{(M)}$ compensates for a factor of 2 encountered in taking the matrix element of the Majorana mass term.

where again the identity $\Gamma_R\, C\Gamma_L = 0$ plays a role in this construction. To avoid the possibility of nonconservation of charge, any Majorana mass term must be restricted to a field with is neutral under the gauge charges. Thus, among the particles of the Standard Model, we will see that only right-handed neutrinos satisfy this condition.

Finally, we note that a Dirac field can be written as two Majorana fields with opposite masses via the construction

$$\psi_a = \frac{1}{\sqrt{2}}\left[\psi_R + \psi_L^c\right], \qquad \psi_b = \frac{1}{\sqrt{2}}\left[\psi_L - \psi_R^c\right], \qquad (3.9)$$

in which case we find

$$\begin{aligned}
-\mathcal{L}_D &= m_D\left[\overline{\psi_L}\,\psi_R + \overline{\psi_R}\,\psi_L\right] \\
&= \frac{m_D}{2}\,\overline{(\psi_a^c)}\,\psi_a - \frac{m_D}{2}\,\overline{(\psi_b)^c}\,\psi_b + \text{h.c.}
\end{aligned} \qquad (3.10)$$

The apparent violation of lepton number that looks like it would arise from this framework does not actually occur, because the effects proportional to the mass of these fields will cancel due to the minus sign between the two mass terms. To make matters look even more puzzling, we can flip the sign on the mass term for the second field, by the field redefinition $\psi_b \to i\,\psi_b$ in which case both masses appear positive. However in this case, the weak current would pick up an unusual factor of i, since the left-handed field would then become

$$\psi_L \to \frac{1}{\sqrt{2}}\left[i\,\psi_b + \psi_a^c\right]. \qquad (3.11)$$

In this case, potential lepton-number violating processes would cancel between the two fields because of the occurrence of a factor of $i^2 = -1$ from the application of the weak currents. These algebraic gymnastics become more physically relevant when we combine both Dirac and Majorana mass terms in Chap. VI.

I–4 Symmetries and near symmetries

A symmetry is said to arise in Nature whenever some change in the variables of a system leaves the essential physics unchanged. In field theory, the dynamical variables are the fields, and symmetries describe invariances under transformations of the fields. For example, one associates with the spacetime translation $x_\mu \to x_\mu + a_\mu$ a transformation of the field $\psi(x)$ to $\psi(x + a)$. In turn, the 'essential physics' is best described by an action, at least in classical physics. If the action is invariant, the equations of motion, and hence the classical physics, will be unchanged. The invariances of quantum physics are identified by consideration of matrix elements or, equivalently, of the path integral. We begin the study of symmetries here by exploring several lagrangians which have invariances and by considering some of the consequences of these symmetries.

Noether currents

The classical analysis of symmetry focusses on the lagrangian, which in general is a Lorentz-scalar function of several fields, denoted by φ_i, and their first derivatives $\partial_\mu \varphi_i$, i.e. $\mathcal{L} = \mathcal{L}(\varphi_i, \partial_\mu \varphi_i)$. *Noether's theorem* states that for any invariance of the action under a continuous transformation of the fields, there exists a classical charge Q which is time-independent ($\dot{Q} = 0$) and is associated with a conserved current, $\partial_\mu J^\mu = 0$. This theorem covers both internal and spacetime symmetries. For most[7] internal symmetries, the lagrangian is itself invariant. Given a continuous field transformation, one can always consider an infinitesimal transformation

$$\varphi_i'(x) = \varphi_i(x) + \epsilon f_i(\varphi), \tag{4.1}$$

where ϵ is an infinitesimal parameter and $f_i(\varphi)$ is a function of the fields in the theory. The procedure for constructing the Noether current of an internal symmetry is to temporarily let ϵ become a function of x and to define the quantity

$$\hat{\varphi}_i(x) = \varphi_i(x) + \epsilon(x) f_i(\varphi), \tag{4.2}$$

such that in the restriction back to constant ϵ, $\mathcal{L}$ becomes invariant and $\hat{\varphi}_i(x) \to \varphi_i'(x)$. For an internal symmetry, the Noether current is then defined by

$$J^\mu(x) \equiv \frac{\partial}{\partial(\partial_\mu \epsilon(x))} \mathcal{L}(\hat{\varphi}, \partial\hat{\varphi}). \tag{4.3}$$

Use of the equation of motion together with the invariance of the lagrangian under the transformation in Eq. (4.1) yields $\partial_\mu J^\mu = \partial \mathcal{L}/\partial \epsilon(x) = 0$ as desired. The Noether charge $Q = \int d^3x\, J_0$ is time-independent if the current vanishes sufficiently rapidly at spatial infinity, i.e.

$$\frac{dQ}{dt} = \int d^3x\, \partial_0 J_0 = -\int d^3x\, \nabla \cdot \mathbf{J} = 0. \tag{4.4}$$

We refer the reader to field theory textbooks for further discussion, including the analogous procedure for constructing Noether currents of spacetime symmetries.

Identifying the current does not exhaust all the consequences of a symmetry but is merely the first step towards the implementation of symmetry relations. Notice that we have been careful to use the word 'classical' several times. This is because the invariance of the action is not generally sufficient to identify symmetries of a quantum theory. We shall return to this point.

---

[7] An exception occurs for the so-called topological gauge symmetries.

Examples of Noether currents

Let us now consider some explicit field theory models in order to get practice in constructing Noether currents.

(i) *Isospin symmetry*: $SU(2)$ isospin invariance of the nucleon–pion system provides a standard and uncomplicated means for studying symmetry currents. Consider a doublet of nucleon fields

$$\psi = \begin{pmatrix} p \\ n \end{pmatrix}, \tag{4.5}$$

and a triplet of pion fields $\boldsymbol{\pi} = \{\pi^i\}$ ($i = 1, 2, 3$) with lagrangian

$$\mathcal{L} = \bar{\psi}\,(i\partial\!\!\!/ - \mathbf{m})\,\psi + \frac{1}{2}\left[\partial_\mu \boldsymbol{\pi} \cdot \partial^\mu \boldsymbol{\pi} - m_\pi^2 \boldsymbol{\pi} \cdot \boldsymbol{\pi}\right] + ig\bar{\psi}\boldsymbol{\tau} \cdot \boldsymbol{\pi}\gamma_5\psi - \frac{\lambda}{4}(\boldsymbol{\pi} \cdot \boldsymbol{\pi})^2 \tag{4.6}$$

where $\mathbf{m}$ is the nucleon mass matrix

$$\mathbf{m} = \begin{pmatrix} m & 0 \\ 0 & m \end{pmatrix}$$

and $\boldsymbol{\tau} = \{\tau^i\}$ ($i = 1, 2, 3$) are the three Pauli matrices. This lagrangian is invariant under the *global* $SU(2)$ rotation of the fields

$$\psi \to \psi' = U\,\psi, \qquad U = \exp\left(-i\boldsymbol{\tau} \cdot \boldsymbol{\alpha}/2\right) \tag{4.7}$$

for any α^i, ($i = 1, 2, 3$) provided the pion fields are transformed as

$$\boldsymbol{\tau} \cdot \boldsymbol{\pi} \to \boldsymbol{\tau} \cdot \boldsymbol{\pi}' = U\boldsymbol{\tau} \cdot \boldsymbol{\pi}U^\dagger. \tag{4.8}$$

In proving this, it is useful to employ the identity

$$\boldsymbol{\pi} \cdot \boldsymbol{\pi} = \frac{1}{2}\,\text{Tr}\,(\boldsymbol{\tau} \cdot \boldsymbol{\pi}\boldsymbol{\tau} \cdot \boldsymbol{\pi}), \tag{4.9}$$

from which we easily see that $\pi^i\pi^i$ is invariant under the transformation of Eq. (4.8). The response of the individual pion components to an isospin transformation can be found from multiplying Eq. (4.8) by τ^i and taking the trace,

$$\pi'^{\,i} = R^{ij}(\boldsymbol{\alpha})\pi^j, \qquad R^{ij}(\boldsymbol{\alpha}) = \frac{1}{2}\,\text{Tr}\left(\tau^i U\tau^j U^\dagger\right). \tag{4.10}$$

To determine the isospin current, one considers the spacetime-dependent transformation with $\boldsymbol{\alpha}$ now *infinitesimal*,

$$\hat{\psi} = (1 - i\boldsymbol{\tau} \cdot \boldsymbol{\alpha}(x)/2)\,\psi, \qquad \hat{\pi}^i = \pi^i - \epsilon^{ijk}\pi^j\alpha^k(x). \tag{4.11}$$

Performing this transformation on the lagrangian gives

$$\mathcal{L}(\hat{\psi}, \hat{\pi}) = \mathcal{L}(\psi, \pi) + \frac{1}{2}\bar{\psi}\gamma^{\mu}\boldsymbol{\tau} \cdot \partial_{\mu}\boldsymbol{\alpha}\psi - \epsilon^{ijk}(\partial_{\mu}\pi^{i})\pi^{j}\partial^{\mu}\alpha_{k}, \tag{4.12}$$

and applying our expression Eq. (4.3) for the current yields the triplet of currents (one for each α_i)

$$V_{\mu}^{i} = \bar{\psi}\gamma_{\mu}\frac{\tau^{i}}{2}\psi + \epsilon^{ijk}\pi^{j}\partial_{\mu}\pi^{k}. \tag{4.13}$$

By use of the equations of motion for ψ and π, it is straightforward to verify that this current is conserved.

(ii) *The linear sigma model*: With a few modifications the above example becomes one of the most instructive of all field theory models, the sigma model [GeL 60]. One adds to the lagrangian of Eq. (4.6) a scalar field σ with judiciously chosen couplings, and removes the bare nucleon mass,

$$\begin{aligned}
\mathcal{L} &= \bar{\psi}i\partial\!\!\!/\psi + \frac{1}{2}\partial_{\mu}\boldsymbol{\pi} \cdot \partial^{\mu}\boldsymbol{\pi} + \frac{1}{2}\partial_{\mu}\sigma\partial^{\mu}\sigma \\
&\quad - g\bar{\psi}\left(\sigma - i\boldsymbol{\tau} \cdot \boldsymbol{\pi}\gamma_5\right)\psi + \frac{\mu^2}{2}\left(\sigma^2 + \boldsymbol{\pi}^2\right) - \frac{\lambda}{4}\left(\sigma^2 + \boldsymbol{\pi}^2\right)^2.
\end{aligned} \tag{4.14}$$

For $\mu^2 > 0$, the model exhibits the phenomenon of spontaneous symmetry breaking (cf. Sect. I–6). In describing the symmetries of this lagrangian, it is useful to rewrite the mesons in terms of a matrix field

$$\Sigma \equiv \sigma + i\boldsymbol{\tau} \cdot \boldsymbol{\pi}, \tag{4.15}$$

such that

$$\sigma^2 + \boldsymbol{\pi}^2 = \frac{1}{2}\operatorname{Tr}\left(\Sigma^{\dagger}\Sigma\right). \tag{4.16}$$

Then we obtain

$$\begin{aligned}
\mathcal{L} &= \bar{\psi}_{L}i\partial\!\!\!/\psi_{L} + \bar{\psi}_{R}i\partial\!\!\!/\psi_{R} + \frac{1}{4}\operatorname{Tr}\left(\partial_{\mu}\Sigma\partial^{\mu}\Sigma^{\dagger}\right) \\
&\quad + \frac{1}{4}\mu^2\operatorname{Tr}\left(\Sigma^{\dagger}\Sigma\right) - \frac{\lambda}{16}\operatorname{Tr}^2\left(\Sigma^{\dagger}\Sigma\right) - g\left(\bar{\psi}_{L}\Sigma\psi_{R} + \bar{\psi}_{R}\Sigma^{\dagger}\psi_{L}\right),
\end{aligned} \tag{4.17}$$

where $\psi_{L,R}$ are chiral fields (cf. Eq. (2.3)). The left-handed and right-handed fermion fields are coupled together only in the interaction with the Σ field. The purely mesonic portion of the lagrangian is obviously invariant under rotations among the σ, π fields. The full lagrangian has separate 'left' and 'right' invariances, i.e. $SU(2)_{L} \times SU(2)_{R}$,

$$\psi_{L,R} \to \psi'_{L,R} = U_{L,R}\psi_{L,R}, \qquad \Sigma \to \Sigma' = U_{L}\Sigma U_{R}^{\dagger}, \tag{4.18}$$

with U_L and U_R being arbitrary $SU(2)$ matrices,

$$U_{L,R} = \exp\left(-i\boldsymbol{\alpha}_{L,R} \cdot \boldsymbol{\tau}/2\right). \tag{4.19}$$

The fermion portions of the transformation clearly involve just the $SU(2)$ isospin rotations on the left-handed and right-handed fermions. However, the mesons involve a combination of a pure isospin rotation among the π fields together with a transformation between the σ and π fields

$$\sigma \to \sigma' = \frac{1}{2} \operatorname{Tr}\left(U_L U_R^\dagger\right)\sigma + \frac{i}{2} \operatorname{Tr}\left(U_L \tau^k U_R^\dagger\right)\pi^k$$

$$\simeq \sigma + \frac{1}{2}\left(\boldsymbol{\alpha}_L - \boldsymbol{\alpha}_R\right) \cdot \boldsymbol{\pi},$$

$$\pi^k \to \pi'^k = -\frac{i}{2} \operatorname{Tr}\left(\tau^k U_L U_R^\dagger\right)\sigma + \frac{1}{2} \operatorname{Tr}\left(\tau^k U_L \tau^\ell U_R^\dagger\right)\pi^\ell$$

$$\simeq \pi^k - \frac{1}{2}\left(\alpha_L^k - \alpha_R^k\right)\sigma - \frac{1}{2}\epsilon^{k\ell m}\pi^\ell\left(\alpha_L^m + \alpha_R^m\right), \tag{4.20}$$

where the second form in each case is for infinitesimal $\boldsymbol{\alpha}_L$, $\boldsymbol{\alpha}_R$. For each invariance there is a separate conserved current

$$J_{L\mu}^k = \bar{\psi}_L \gamma_\mu \frac{\tau^k}{2}\psi_L - \frac{i}{8} \operatorname{Tr}\left(\tau^k\left(\Sigma \partial_\mu \Sigma^\dagger - \partial_\mu \Sigma \, \Sigma^\dagger\right)\right)$$

$$= \bar{\psi}_L \gamma_\mu \frac{\tau^k}{2}\psi_L - \frac{1}{2}\left(\sigma \partial_\mu \pi^k - \pi^k \partial_\mu \sigma\right) + \frac{1}{2}\epsilon^{k\ell m}\pi^\ell \partial_\mu \pi^m,$$

$$J_{R\mu}^k = \bar{\psi}_R \gamma_\mu \frac{\tau^k}{2}\psi_R + \frac{i}{8} \operatorname{Tr}\left(\tau^k\left(\partial_\mu \Sigma^\dagger \Sigma - \Sigma^\dagger \partial_\mu \Sigma\right)\right)$$

$$= \bar{\psi}_R \gamma_\mu \frac{\tau^k}{2}\psi_R + \frac{1}{2}\left(\sigma \partial_\mu \pi^k - \pi^k \partial_\mu \sigma\right) + \frac{1}{2}\epsilon^{k\ell m}\pi^\ell \partial_\mu \pi^m. \tag{4.21}$$

These can be formed into a conserved vector current

$$V_\mu^k = J_{L\mu}^k + J_{R\mu}^k = \bar{\psi}\gamma_\mu \frac{\tau^k}{2}\psi + \epsilon^{k\ell m}\pi^\ell \partial_\mu \pi^m, \tag{4.22}$$

which is just the isospin current derived previously, and a conserved axial-vector current

$$A_\mu^k = J_{L\mu}^k - J_{R\mu}^k = \bar{\psi}\gamma_\mu \gamma_5 \frac{\tau^k}{2}\psi + \pi^k \partial_\mu \sigma - \sigma \partial_\mu \pi^k. \tag{4.23}$$

(iii) *Scale invariance*: Our third example illustrates the case of a spacetime transformation in which the lagrangian changes by a total derivative. Consider classical electrodynamics (cf. Sect. II–1) but with a massless electron,

$$\mathcal{L} = -\frac{1}{4}F_{\mu\nu}F^{\mu\nu} + \bar{\psi}\, i \slashed{D}\psi, \tag{4.24}$$

where ψ is the electron field, $D_\mu \psi = (\partial_\mu + ieA_\mu)\psi$ is the covariant derivative of ψ, A_μ is the photon field, and $F_{\mu\nu}$ is the electromagnetic field strength. We shall describe the construction of both $D_\mu \psi$ and $F_{\mu\nu}$ in the next section. Since there are no dimensional parameters in this lagrangian, we are motivated to consider the effect of a change in coordinate scale $x \rightarrow x' = \lambda x$ together with the field transformations

$$\psi(x) \rightarrow \psi'(x) = \lambda^{3/2} \psi(\lambda x), \qquad A_\mu(x) \rightarrow A'_\mu(x) = \lambda A_\mu(\lambda x). \qquad (4.25)$$

Although the lagrangian itself is not invariant,

$$\mathcal{L}(x) \rightarrow \mathcal{L}'(x) = \lambda^4 \mathcal{L}(\lambda x), \qquad (4.26)$$

with a change of variable the action is easily seen to be unchanged,

$$S = \int d^4x \, \mathcal{L}(x) \rightarrow \int d^4x \, \lambda^4 \mathcal{L}(\lambda x) = \int d^4x' \, \mathcal{L}(x') = S. \qquad (4.27)$$

There is nothing in this classical theory which depends on how length is scaled. The Noether current associated with the change of scale is

$$J^\mu_{\text{scale}} \equiv x_\nu \theta^{\mu\nu}, \qquad (4.28)$$

where $\theta^{\mu\nu}$ is the energy-momentum tensor of the theory,

$$\theta^{\mu\nu} = -g^{\mu\nu} \left[-\frac{1}{4} F^{\lambda\sigma} F_{\lambda\sigma} + \bar{\psi} i \slashed{D} \psi \right] - F^{\mu\lambda} F^\nu{}_\lambda + A^\nu \partial_\lambda F^{\mu\lambda} + \frac{i}{2} \bar{\psi} \gamma^\mu \overleftrightarrow{\partial^\nu} \psi. \tag{4.29}$$

Since the energy-momentum tensor is itself conserved, $\partial_\mu \theta^{\mu\nu} = 0$, the conservation of scale current is equivalent to the vanishing of the trace of the energy-momentum tensor,

$$\partial_\mu J^\mu_{\text{scale}} = \theta_\mu{}^\mu = 0. \qquad (4.30)$$

This trace property may be easily verified using the equations of motion.

Approximate symmetry

Thus far, we have been describing exact symmetries. Symmetry considerations are equally useful in situations where there is 'almost' a symmetry. The very phrase 'approximate symmetry' seems self-contradictory and needs explanation. Quite often a lagrangian would have an invariance if certain of the parameters in it were set equal to zero. In that limit the invariance would have a set of physical consequences which, with the said parameters being nonzero, would no longer obtain. Yet, if the parameters are in some sense 'small', the predicted consequences are still approximately valid. In fact, when the interaction which breaks the symmetry has a well-defined behavior under the symmetry transformation, its effect can generally be analyzed in terms of the basis of unperturbed particle states by using the

Wigner–Eckart theorem. The precise sense in which the symmetry-breaking terms can be deemed small depends on the problem under consideration. In practice, the utility of an approximate symmetry is rarely known *a priori*, but is only evident after its predictions have been checked experimentally.

If a symmetry is not exact, the associated currents and charges will no longer be conserved. For example, in the linear sigma model, the symmetry is partially broken if we add to the lagrangian a term of the form

$$\mathcal{L}' = a\,\sigma = \frac{a}{2}\,\mathrm{Tr}\,\Sigma, \tag{4.31}$$

where Σ is the matrix defined in Eq. (4.15). With this addition, the vector isospin $SU(2)$ symmetry remains exact but the axial $SU(2)$ transformation is no longer an invariance. The axial-current divergence becomes

$$\partial^\mu A^i_\mu = a\pi^i, \tag{4.32}$$

and the charge is time-dependent,

$$\frac{dQ^i_5}{dt} = a \int d^3x\, \pi^i. \tag{4.33}$$

In the linear sigma model, if the parameters g, λ are of order unity it is clear that the perturbation is small provided $1 \gg a/\mu^3$, as μ is the only other mass scale in the theory. However, if either g or λ happens to be anomalously large or small, the condition appropriate for a 'small' perturbation is not *a priori* evident.

In our example (iii) of scale invariance in massless fermion electrodynamics, the addition of an electron mass

$$\mathcal{L}_{\mathrm{mass}} = -m\bar{\psi}\psi \tag{4.34}$$

would explicitly break the symmetry and the trace would no longer vanish,

$$\theta_\mu{}^\mu = m\bar{\psi}\psi \neq 0. \tag{4.35}$$

This is in fact what occurs in practice. Fermion mass is typically *not* a small parameter in *QED* and cannot be treated as a perturbation in most applications.

I-5 Gauge symmetry

In our discussion of chiral symmetry, we considered the effect of global phase transformations, $\psi_{L,R}(x) \to \exp(-i\alpha_{L,R})\psi_{L,R}(x)$. Global phase transformations are those which are constant throughout all spacetime. Let us reconsider the system of chiral fermions, but now insist that the phase transformations be *local*. Each transformation is then labeled by a spacetime-dependent phase $\alpha_{L,R}(x)$,

$$\psi_{L,R}(x) \to \exp(-i\alpha_{L,R}(x))\,\psi_{L,R}(x). \tag{5.1}$$

Such local mappings are referred to as *gauge transformations*. The free massless lagrangian of Eq. (2.1) is not invariant under the gauge transformation

$$i\overline{\psi}_{L,R}(x)\slashed{\partial}\psi_{L,R}(x) \rightarrow i\overline{\psi}_{L,R}(x)\slashed{\partial}\,\psi_{L,R}(x) + \overline{\psi}_{L,R}(x)\gamma^{\mu}\psi_{L,R}(x)\cdot\partial_{\mu}\alpha_{L,R}(x),$$

(5.2)

because of the spacetime dependence of $\alpha_{L,R}$. In order for such a local transformation to be an invariance of the lagrangian, we need an extended kind of derivative D_{μ}, such that

$$D_{\mu}\psi_{L,R}(x) \rightarrow \exp(-i\alpha_{L,R}(x))D_{\mu}\psi_{L,R}(x)$$

(5.3)

under the local transformation of Eq. (5.1). The quantity D_{μ} is a *covariant derivative,* so called because it responds covariantly, as in Eq. (5.3), to a gauge transformation.

Abelian case

Before proceeding with the construction of a covariant derivative, we broaden the context of our discussion. Let $\Theta(x)$ now represent a boson or fermion field of any spin and arbitrary mass. We consider transformations

$$\Theta \rightarrow U(\alpha)\Theta$$

(5.4)

$$D_{\mu}\Theta \rightarrow U(\alpha)D_{\mu}\Theta,$$

(5.5)

with a spacetime-dependent parameter, $\alpha = \alpha(x)$. Suppose these gauge transformations form an abelian group, e.g., as do the set of phase transformations of Eq. (5.1).[8] It is sufficient to consider transformations with just one parameter as in Eqs. (5.4)–(5.5) since we can use direct products of these to construct arbitrary abelian groups.

One can obtain a covariant derivative by introducing a vector field $A_{\mu}(x)$, called a *gauge field*, by means of the relation

$$D_{\mu}\Theta = (\partial_{\mu} + ifA_{\mu})\Theta,$$

(5.6)

where f is a real-valued coupling constant whose numerical magnitude depends in part on the field Θ. For example, in electrodynamics f becomes the electric charge of Θ. The problem is then to determine how A_{μ} must transform under a gauge transformation in order to give Eq. (5.5). This can be done by inspection, and we find

$$A_{\mu} \rightarrow A_{\mu} + \frac{i}{f}\partial_{\mu}U(\alpha)\cdot U^{-1}(\alpha).$$

(5.7)

---

[8] An *abelian* group is one whose elements commute. A *nonabelian* group is one which is not abelian.

The gauge field A_μ must itself have a kinetic contribution to the lagrangian. This is written in terms of a field strength, $F_{\mu\nu}$, which is antisymmetric in its indices. A general method for constructing such an antisymmetric second rank tensor is to use the commutator of covariant derivatives,

$$[D_\mu, D_\nu]\,\Theta \equiv if\,F_{\mu\nu}\Theta. \tag{5.8}$$

By direct substitution we find

$$F_{\mu\nu} = \partial_\mu A_\nu - \partial_\nu A_\mu. \tag{5.9}$$

It follows from Eq. (5.7) and Eq. (5.9) that the field strength $F_{\mu\nu}$ is invariant under gauge transformations. A gauge-invariant lagrangian containing a complex scalar field φ and a spin one-half field ψ, chiral or otherwise, has the form

$$\mathcal{L} = -\frac{1}{4} F_{\mu\nu} F^{\mu\nu} + (D^\mu \varphi)^\dagger D_\mu \varphi + i\overline{\psi}\,\slashed{D}\psi + \cdots, \tag{5.10}$$

where the ellipses stand for possible mass terms and nongauge field interactions. There is no contribution corresponding to a gauge-boson mass. Such a term would be proportional to $A^\mu A_\mu$, which is not invariant under the gauge transformation, Eq. (5.7).

Nonabelian case

The above reasoning can be generalized to nonabelian groups [YaM 54]. First, we need a nonabelian group of gauge transformations and a set of fields which forms a representation of the gauge group. Then, we must construct an appropriate covariant derivative to act on the fields. This step involves introducing a set of gauge bosons and specifying their behavior under the gauge transformations. Finally, the gauge field strength is obtained from the commutator of covariant derivatives, at which point we can write down a gauge-invariant lagrangian.

Consider fields $\Theta = \{\Theta_i\}$ $(i = 1, \ldots, r)$, which form an r-dimensional representation of a nonabelian gauge group $\mathcal{G}$. The Θ_i can be boson or fermion fields of any spin. In the following it will be helpful to think of Θ as an r-component column vector, and operations acting on Θ as $r \times r$ matrices. We take group $\mathcal{G}$ to have a Lie algebra of dimension n, so that the numbers of group generators, group parameters, gauge fields, and components of the gauge field strength are each n. We write the spacetime-dependent group parameters as the n-dimensional vector $\vec{\alpha} = \{\alpha_a(x)\}$ $(a = 1, \ldots, n)$. A gauge transformation on Θ is

$$\Theta' = \mathbf{U}(\vec{\alpha})\Theta, \tag{5.11}$$

where the $r \times r$ matrix $\mathbf{U}$ is an element of group $\mathcal{G}$. For those elements of $\mathcal{G}$ which are connected continuously to the identity operator, we can write

$$\mathbf{U}(\vec{\alpha}) = \exp(-i\alpha^a \mathbf{G}^a), \tag{5.12}$$

where $\vec{\mathbf{G}} = \{\mathbf{G}_a\}$ $(a = 1, \ldots, n)$ are the *generators* of the group $\mathcal{G}$ expressed as hermitian $r \times r$ matrices. The set of generators obeys the Lie algebra

$$[\mathbf{G}^a, \mathbf{G}^b] = ic^{abc}\mathbf{G}^c \qquad (a, b, c = 1, \ldots, n), \tag{5.13}$$

where $\{c^{abc}\}$ are the structure constants of the algebra. We construct the covariant derivative $\mathbf{D}_\mu \Theta$ in terms of gauge fields $\vec{B}_\mu = \{B_\mu^a\}$ $(a = 1, \ldots, n)$ as

$$\mathbf{D}_\mu \Theta = (\mathbf{I}\partial_\mu + ig\mathbf{B}_\mu)\Theta, \tag{5.14}$$

where g is a coupling constant analogous to f in Eq. (5.6). In Eq. (5.14), $\mathbf{I}$ is the $r \times r$ unit matrix, and

$$\mathbf{B}_\mu \equiv \mathbf{G}^a B_\mu^a. \tag{5.15}$$

Realizing that the covariant derivative must transform as

$$(\mathbf{D}_\mu \Theta)' = \mathbf{U}(\vec{\alpha})(\mathbf{D}_\mu \Theta), \tag{5.16}$$

we infer from Eqs. (5.12)–(5.14) the response, in matrix form, of the gauge fields,

$$\mathbf{B}_\mu' = \mathbf{U}(\vec{\alpha})\mathbf{B}_\mu \mathbf{U}^{-1}(\vec{\alpha}) + \frac{i}{g}\partial_\mu \mathbf{U}(\vec{\alpha}) \cdot \mathbf{U}^{-1}(\vec{\alpha}). \tag{5.17}$$

The field strength matrix $\mathbf{F}_{\mu\nu}$ is found, as before, from the commutator of covariant derivatives,

$$[\mathbf{D}_\mu, \mathbf{D}_\nu]\Theta \equiv ig\mathbf{F}_{\mu\nu}\Theta, \tag{5.18}$$

implying

$$\mathbf{F}_{\mu\nu} = \partial_\mu \mathbf{B}_\nu - \partial_\nu \mathbf{B}_\mu + ig[\mathbf{B}_\mu, \mathbf{B}_\nu]. \tag{5.19}$$

Eqs. (5.17) and (5.19) provide the field strength transformation property,

$$\mathbf{F}_{\mu\nu}' = \mathbf{U}(\vec{\alpha})\mathbf{F}_{\mu\nu}\mathbf{U}^{-1}(\vec{\alpha}). \tag{5.20}$$

Unlike its abelian counterpart, the nonabelian field strength is not gauge invariant. Finally, we write down the gauge-invariant lagrangian

$$\mathcal{L} = -\frac{1}{2}\mathrm{Tr}\,(\mathbf{F}^{\mu\nu}\mathbf{F}_{\mu\nu}) + (\mathbf{D}^\mu \Phi)^* \mathbf{D}_\mu \Phi + i\overline{\Psi}\,\displaystyle{\not}\mathbf{D}\Psi + \cdots, \tag{5.21}$$

where Φ and Ψ are distinct multiplets of scalar and spin one-half fields and the ellipses represent possible mass terms and nongauge interactions. Analogously to the abelian case, there is no gauge-boson mass term.

The most convenient approach for demonstrating the theory's formal gauge structure is the matrix notation. However, in specific calculations it is sometimes more convenient to work with individual fields. To cast the matrix equations into component form, we employ a normalization of group generators consistent with Eq. (5.21),

$$\text{Tr}(\mathbf{G}^a\mathbf{G}^b) = \frac{1}{2}\delta^{ab} \qquad (a, b = 1, \ldots, n). \tag{5.22}$$

To obtain the ath component of the field strength $\vec{F}_{\mu\nu} = \{F^a_{\mu\nu}\}$ $(a = 1, \ldots, n)$, we matrix multiply Eq. (5.19) from the left by $\mathbf{G}^a$ and take the trace to find

$$F^a_{\mu\nu} = \partial_\mu B^a_\nu - \partial_\nu B^a_\mu - gc^{abc}B^b_\mu B^c_\nu \qquad (a, b, c = 1, \ldots, n). \tag{5.23}$$

The lagrangian Eq. (5.21) can likewise be rewritten in component form,

$$\mathcal{L} = -\frac{1}{4}F^{a\mu\nu}F^a_{\mu\nu} + (D^\mu_{km}\varphi_m)^\dagger(D_\mu)_{kn}\varphi_n + i\overline{\psi}_i(\not{D})_{ij}\psi_j + \cdots, \tag{5.24}$$

where $a = 1, \ldots, n$ and the remaining indices cover the dimensionalities of their respective multiplets.

Mixed case

In the Standard Model, it is a combination of abelian and nonabelian gauge groups which actually occurs. To deal with this circumstance, let us consider one abelian gauge group $\mathcal{G}$ and one nonabelian gauge group $\mathcal{G}'$ having gauge fields A^μ and $\vec{B}^\mu = \{B^\mu_a\}$ $(a = 1, \ldots, n)$, respectively. Further assume that $\mathcal{G}$ and $\mathcal{G}'$ commute and that components of the generic matter field Θ transform as an r-dimensional multiplet under $\mathcal{G}'$. The key construction involves the generalized covariant derivative, written as an $r \times r$ matrix,

$$\mathbf{D}_\mu\Theta = \left((\partial_\mu + ifA_\mu)\mathbf{I} + ig\vec{B}_\mu \cdot \vec{\mathbf{G}}\right)\Theta, \tag{5.25}$$

where $\mathbf{I}$ is the unit matrix and f, g are distinct real-valued constants. Given this, much of the rest of the previous analysis goes through unchanged. The field strengths associated with the abelian and nonabelian gauge fields have the forms given earlier. So does the gauge-invariant lagrangian, except now the extended covariant derivative of Eq. (5.25) appears, and both the abelian and nonabelian field strengths must be included. For the theory with distinct multiplets of complex scalar fields Φ and spin one-half fields Ψ, the general form of the gauge-invariant lagrangian is

$$\mathcal{L} = -\frac{1}{4}F^{\mu\nu}F_{\mu\nu} - \frac{1}{2}\text{Tr}\,(\mathbf{F}^{\mu\nu}\mathbf{F}_{\mu\nu}) + i\overline{\Psi}\not{D}\Psi - \overline{\Psi}\mathbf{m}\Psi$$

$$+ (\mathbf{D}^{\mu}\Phi)^{\dagger}\mathbf{D}_{\mu}\Phi - V(|\Phi|^2) + \mathcal{L}(\Psi, \Phi), \tag{5.26}$$

where $\mathbf{m}$ is the fermion mass matrix, $V(|\Phi|^2)$ contains the Φ mass matrix and any polynomial self-interaction terms, and $\mathcal{L}(\Psi, \Phi)$ describes the coupling between the spin one-half and spin zero fields.

I–6 On the fate of symmetries

Depending on the dynamics of the theory, a given symmetry of the lagrangian can be manifested physically in a variety of ways. Apparently all such realizations are utilized by Nature. Here we list the various possibilities.

(1) The symmetry may remain exact. The electromagnetic gauge $U(1)$ symmetry, the $SU(3)$ color symmetry of *QCD*, and the global 'baryon-number minus lepton-number' $(B - L)$ symmetry are examples in this class.
(2) The apparent symmetry may have an anomaly. In this case it is not really a true symmetry. Within the Standard Model the global axial $U(1)$ symmetry is thus affected. Our discussion of anomalies is given in Sect. III–3.
(3) The symmetry may be explicitly broken by terms (perhaps small) in the lagrangian which are not invariant under the symmetry. Isospin symmetry, broken by electromagnetism and light-quark mass difference, is an example.
(4) The symmetry may be 'hidden' in the sense that it is an invariance of the lagrangian but not of the ground state, and thus one does not 'see' the symmetry in the spectrum of physical states. This can be produced by different physical mechanisms.
   (a) The acquiring of vacuum expectation values by one or more scalar fields in the theory gives rise to a *spontaneously broken* symmetry, as in the breaking of $SU(2)_L$ invariance by Higgs fields in the electroweak interactions.
   (b) Even in the absence of scalar fields, quantum effects can lead to the *dynamical breaking* of a symmetry. Such is the fate of chiral $SU(2)_L \times SU(2)_R$ symmetry in the strong interactions.

The various forms of symmetry breaking in the above are quite different. In particular, the reader should be warned that the word 'broken' is used with very different meanings in case (3) and the cases in (4). The meaning in (3) is literal – what would have been a symmetry in the absence of the offending terms in the lagrangian is *not* a symmetry of the lagrangian (nor of the physical world). Although the usage in (4) is quite common, it is really a malapropism because the symmetry is not actually broken. Rather, it is realized in a special way, one which turns out to have important consequences for a number of physical processes. The situation is somewhat subtle and requires more explanation, so we shall describe its presence in a magnetic system and in the sigma model.

Hidden symmetry

The phenomenon of hidden symmetry occurs when the ground state of the theory does not have the full symmetry of the lagrangian. Let Q be a symmetry charge as inferred from Noether's theorem, and consider a global symmetry transformation of the vacuum state

$$|0\rangle \to e^{i\alpha Q}|0\rangle, \tag{6.1}$$

where α is a continuous parameter. Invariance of the vacuum,

$$e^{i\alpha Q}|0\rangle = |0\rangle \qquad \text{(all } \alpha), \tag{6.2a}$$

implies that

$$Q|0\rangle = 0. \tag{6.2b}$$

In this circumstance, the vacuum is unique and the symmetry manifests itself in the 'normal' fashion of mass degeneracies and coupling constant identities. Such is the case for the isospin symmetric model of nucleons and pions discussed in Sect. I–4, where the lagrangian of Eq. (4.6) implies the relations

$$m_n = m_p, \qquad m_{\pi^+} = m_{\pi^0} = m_{\pi^-},$$
$$g(pp\pi^0) = -g(nn\pi^0) = g(pn\pi^+)/\sqrt{2} = g(np\pi^-)/\sqrt{2}, \tag{6.3}$$

with $\pi^\pm = (\pi_1 \mp i\pi_2)/\sqrt{2}$.

Alternatively, if *new* states $|\alpha\rangle \neq |0\rangle$ are reached via the transformations of Eq. (6.1), we must have

$$Q|0\rangle \neq 0. \tag{6.4}$$

Since, by Noether's theorem, the symmetry charge is time-independent,

$$\dot{Q} = i[H, Q] = 0, \tag{6.5}$$

all of the new states $|\alpha\rangle$ must have the same energy as $|0\rangle$. That is, if E_0 is the energy of the vacuum state, $H|0\rangle \equiv E_0|0\rangle$, then we have

$$H|\alpha\rangle = H e^{i\alpha Q}|0\rangle = e^{i\alpha Q} H|0\rangle = E_0|\alpha\rangle. \tag{6.6}$$

Because the symmetry transformation is continuous, there must occur a continuous family of degenerate states.

Can one visualize these new states in a physical setting? It is helpful to refer to a ferromagnet, which consists of separate domains of aligned spins. Let us focus on one such domain in its ground state. It is invariant only under rotations about the direction of spin alignment, and hence does not share the full rotational invariance of the hamiltonian. In this context, the degenerate states mentioned above are just the different possible orientations available to the lattice spins in a domain. Since space is rotationally invariant, there is no preferred direction along which a domain must be oriented. By performing rotations, one transfers from one orientation to another, each having the same energy.

Let us try to interpret, from the point of view of quantum field theory, the states which are obtained from the vacuum by a continuous symmetry transformation and which share the energy of the vacuum state. In a quantum field theory any excitation about the ground state becomes quantized and is interpreted as a particle. The minimum excitation energy is the particle's mass. Thus the zero-energy excitations generated from symmetry transformations must be described by massless particles whose quantum numbers can be taken as those of the symmetry charge(s). Thus we are led to Goldstone's theorem [Go 61, GoSW 62] – if a theory has a continuous symmetry of the lagrangian, which is not a symmetry of the vacuum, there must exist one or more massless bosons (*Goldstone bosons*). That is, spontaneous or dynamical breaking of a continuous symmetry will entail massless particles in the spectrum.

This phenomenon can be seen in the magnet analogy, where the excitation is a spin-wave quantum. When the wavelength becomes very large, the spin configuration begins to resemble a uniform rotation of *all* the spins. This is one of the other possible domain alignments discussed above, and to reach it does not cost any energy. Thus, in the limit of infinite wavelength ($\lambda \to \infty$), the excitation energy vanishes ($E \to 0$), yielding a Goldstone boson.[9]

Spontaneous symmetry breaking in the sigma model

We proceed to a more quantitative analysis of hidden symmetry by returning to the sigma model of Sect. I–4. Let us begin by inferring from the sigma model lagrangian of Eq. (4.14) the potential energy

$$V(\sigma, \boldsymbol{\pi}) = -\frac{\mu^2}{2}\left(\sigma^2 + \boldsymbol{\pi}^2\right) + \frac{\lambda}{4}\left(\sigma^2 + \boldsymbol{\pi}^2\right)^2. \tag{6.7}$$

---

[9] In the ferromagnet case, the spin waves actually have $E \propto \mathbf{p}^2 \sim \lambda^{-2}$ for low momentum. In Lorentz-invariant theories, the form $E \propto |\mathbf{p}|$ is the only possible behavior for massless single particle states. For a more complete discussion, see [An 84].

With μ^2 negative, minimization of $V(\sigma, \pi)$ occurs for the unique configuration $\sigma = \pi = 0$. Hidden symmetry occurs for μ^2 positive, where minimization of $V(\sigma, \pi)$ reveals the set of degenerate ground states to be those with

$$\sigma^2 + \pi^2 = \frac{\mu^2}{\lambda}. \tag{6.8}$$

Let us study the particular ground state,

$$\langle \sigma \rangle_0 = \sqrt{\frac{\mu^2}{\lambda}} \equiv v, \qquad \langle \pi \rangle_0 = 0. \tag{6.9}$$

Other choices yield the same physics, but require a relabeling of the fields. For this case, field fluctuations in the pionic direction do not require any energy, so that the pions are the Goldstone bosons. Defining

$$\tilde{\sigma} = \sigma - v, \tag{6.10}$$

we then have for the full sigma model lagrangian

$$\mathcal{L} = \bar{\psi} \left(i \partial\!\!\!/ - g v \right) \psi + \frac{1}{2} \left[\partial_\mu \tilde{\sigma} \partial^\mu \tilde{\sigma} - 2\mu^2 \tilde{\sigma}^2 \right] + \frac{1}{2} \partial_\mu \pi \cdot \partial^\mu \pi$$
$$- g \bar{\psi} \left(\tilde{\sigma} - i \tau \cdot \pi \gamma_5 \right) \psi - \lambda v \tilde{\sigma} \left(\tilde{\sigma}^2 + \pi^2 \right) - \frac{\lambda}{4} \left[\left(\tilde{\sigma}^2 + \pi^2 \right)^2 - v^4 \right]. \tag{6.11}$$

Observe that the pion is massless, while the $\tilde{\sigma}$ and nucleon fields are massive. Thus, at least part of the original symmetry in the sigma model lagrangian of Eq. (4.14) appears to have been lost. Certainly, the mass degeneracy $m_\sigma = m_\pi$ is no longer present, although the normal pattern of isospin invariance survives. However, the full set of original symmetry currents remain conserved. In particular, the axial current of Eq. (4.23), which now appears as

$$A_\mu^i = \bar{\psi} \gamma_\mu \gamma_5 \frac{\tau^i}{2} \psi - v \partial_\mu \pi^i + \pi^i \partial_\mu \tilde{\sigma} - \tilde{\sigma} \partial_\mu \pi^i, \tag{6.12}$$

still has a vanishing divergence, $\partial^\mu A_\mu^i = 0$. We warn the reader that to demonstrate this involves a complicated set of cancellations.

For a normal symmetry, particles fall into mass-degenerate multiplets and have couplings which are related by the symmetry. The isospin relations in Eq. (6.3) are an example of this. In a certain sense, a hidden symmetry likewise gives rise to degenerate states whose couplings are related by the symmetry. The degeneracy consists of a state taken alone or accompanied by an arbitrary number of Goldstone bosons. For example, in the sigma model it can be a nucleon and the same nucleon accompanied by a zero-energy massless pion, which are degenerate. Moreover, the couplings of such configurations are restricted by the symmetry. Historically, predictions of chiral symmetry were originally formulated in terms of *soft-pion theorems* (cf. App. B–3) relating the couplings of the N states to those of the degenerate πN states.

$$\lim_{q_\mu \to 0} \langle \pi^k(q)N'|O|N \rangle = -\frac{i}{F_\pi} \langle N'|[Q_5^k, O]|N \rangle, \tag{6.13}$$

where O is some local operator and N, N' are nucleons or other states. This captures intuitively the nature of symmetry predictions for a hidden symmetry. In this book, we will explore such chiral relations using the more modern techniques of effective lagrangians.

To summarize, if a symmetry of the theory exists but is not apparent in the single-particle spectrum, it still can have a great deal of importance in restricting particle behavior. What happens is actually quite remarkable – in essence, symmetry becomes dynamics. One obtains information about the excitation or annihilation of particles from symmetry considerations. In this regard, hidden symmetries are neither less 'real' nor less useful than normal symmetries – they simply yield a different pattern of predictions.

Problems

(1) **The Poincaré algebra**

   (a) Consider the spacetime (Poincaré) transformations, $x^\mu \to \Lambda^\mu_{\ \nu} x^\nu + a^\mu$, where $\Lambda^\mu_{\ \sigma} \Lambda^{\sigma\nu} = g^{\mu\nu}$. Associated with each coordinate transformation (a, Λ) is the *unitary* operator $U(a, \Lambda) = \exp(ia_\mu P^\mu - \frac{i}{2}\epsilon_{\mu\nu} M^{\mu\nu})$. For two consecutive Poincaré transformations there is a closure property, $U(a', \Lambda')$ $U(a, \Lambda) = U(\ldots)$. Fill in the dots.

   (b) Prove that $U(a^{-1}, 0)U(a', 0)U(a, 0) = U(a', 0)$, and by taking a'_μ, a_μ infinitesimal, determine $[P^\mu, P^\nu]$.

   (c) Demonstrate that $(\Lambda^{-1})_{\lambda\nu} = \Lambda_{\nu\lambda}$, and then show that $U(0, \Lambda^{-1})U(a', \Lambda')U(0, \Lambda) = U(\Lambda^{-1}a', \Lambda^{-1}\Lambda'\Lambda)$.

   (d) For infinitesimal transformations we write $\Lambda^\mu_{\ \lambda} \simeq g^\mu_{\ \lambda} + \epsilon^\mu_{\ \lambda}$. Prove that $\epsilon_{\sigma\lambda} = -\epsilon_{\lambda\sigma}$ and hence $M_{\sigma\lambda} = -M_{\lambda\sigma}$. Upon taking primed quantities in (c) to be infinitesimal, prove $U(0, \Lambda^{-1})P^\mu U(0, \Lambda) = \Lambda^\mu_{\ \nu} P^\nu$ and $U(0, \Lambda^{-1})$ $M^{\mu\nu}U(0, \Lambda) = \Lambda^\mu_{\ \alpha}\Lambda^\nu_{\ \beta} M^{\alpha\beta}$. Finally, letting unprimed quantities be infinitesimal as well, determine $[M^{\alpha\beta}, P^\mu]$ and $[M^{\alpha\beta}, M^{\mu\nu}]$.

(2) **The Meissner effect in gauge theory** [Sh 81]

   The lagrangian for the electrodynamics of a charged scalar field is

$$\mathcal{L}_0 = -\frac{1}{4}F_{\mu\nu}F^{\mu\nu} + (D_\mu\varphi)^*(D^\mu\varphi) - V(\varphi)$$

   with covariant derivative $D_\mu \equiv \partial_\mu + ieA_\mu$ and potential energy,

$$V(\varphi) = \frac{m^2}{2}\varphi^*\varphi + \frac{\lambda}{4}(\varphi^*\varphi)^2 \qquad (\lambda > 0).$$

(a) Identify the electromagnetic current of the φ field.

(b) For $m^2 > 0$, show that the ground state is $\varphi = 0$, $A_\mu = 0$. In this case, the theory is that of normal electrodynamics.

(c) For $m^2 < 0$ ($m^2 \to -\mu^2$ with $\mu^2 > 0$), we enter a different phase of the system. Show that the ground state is now $\varphi = \text{const.} \equiv v$, $A_\mu = 0$. What is the photon mass in this phase? Calculate the potential between two static point charges each of value Q. What sets the scale of the screening length?

(d) Let us now add an external field to the system,

$$\mathcal{L}_0 \to \mathcal{L}_0 + \frac{1}{2}F_{\mu\nu}F^{\mu\nu}_{\text{ext}}.$$

To see that $F^{\mu\nu}_{\text{ext}}$ indeed acts like an applied field, show that if one disregards the field φ the equations of motion require $F^{\mu\nu} = F^{\mu\nu}_{\text{ext}}$.

(e) Demonstrate that there are two simple solutions to the equations of motion in the presence of a *constant* applied field,

$$\varphi = \begin{cases} 0 & (F^{\mu\nu} = F^{\mu\nu}_{\text{ext}}), \\ v & (F^{\mu\nu} = 0). \end{cases}$$

Again, these correspond to unscreened and screened phases of the electromagnetic field.

(f) Calculate the energy of the two phases if $F^{\mu\nu}_{\text{ext}}$ describes a constant magnetic field. Show that the phase in part (e) with $\varphi = 0$ has the lower energy for $B > B_{\text{critical}}$ whereas for $B < B_{\text{critical}}$ it is the phase with $\varphi = v$ which has the lower energy. Discuss the similarity of this result to the Meissner effect.

II

Interactions of the Standard Model

A gauge theory involves two kinds of particles, those which carry 'charge' and those which 'mediate' interactions between currents by coupling directly to charge. In the former class are the fundamental fermions and nonabelian gauge bosons, whereas the latter consists solely of gauge bosons, both abelian and nonabelian. The physical nature of charge depends on the specific theory. Three such kinds of charge, called *color*, *weak isospin*, and *weak hypercharge*, appear in the Standard Model. The values of these charges are not predicted from the gauge symmetry, but must rather be determined experimentally for each particle. The strength of coupling between a gauge boson and a particle is determined by the particle's charge, e.g., the electron–photon coupling constant is $-e$, whereas the u-quark and photon couple with strength $2e/3$. Because nonabelian gauge bosons are both charge carriers and mediators, they undergo self-interactions. These produce substantial nonlinearities and make the solution of nonabelian gauge theories a formidable mathematical problem. Gauge symmetry does not generally determine particle masses. Although gauge-boson mass would seem to be at odds with the principle of gauge symmetry, the Weinberg–Salam model contains a dynamical procedure, the *Higgs mechanism*, for generating mass for both gauge bosons and fermions alike.

II–1 Quantum Electrodynamics

Historically, the first of the gauge field theories was electrodynamics. Its modern version, Quantum Electrodynamics (*QED*), is the most thoroughly verified physical theory yet constructed. *QED* represents the best introduction to the Standard Model, which both incorporates and extends it.

U(1) gauge symmetry

Consider a spin one-half, positively charged fermion represented by field ψ. The classical lagrangian which describes its electromagnetic properties is

28

$$\mathcal{L}_{em} = -\frac{1}{4} F^2 + \overline{\psi} \, (i\,\slashed{D} - m) \, \psi. \tag{1.1}$$

Here, the covariant derivative is $D_\mu \psi \equiv (\partial_\mu + ieA_\mu)\psi$, m and e are, respectively, the mass and electric charge for ψ, A_μ is the gauge field for electromagnetism, $F^{\mu\nu}$ is the gauge-invariant field strength (cf. Eqs. (I–5.8), (I–5.9)), and $F^2 \equiv F^{\mu\nu} F_{\mu\nu}$. This lagrangian is invariant under the local $U(1)$ transformations

$$\psi(x) \to e^{-i\alpha(x)} \psi(x) \,, \tag{1.2}$$

$$A_\mu(x) \to A_\mu(x) + e^{-1} \partial_\mu \alpha(x). \tag{1.3}$$

The associated equations of motion are the Dirac equation

$$(i\,\slashed{\partial} - m - e\,\slashed{A}\,)\psi = 0, \tag{1.4}$$

and the Maxwell equation

$$\partial_\mu F^{\mu\nu} = e\overline{\psi}\gamma^\nu \psi. \tag{1.5}$$

It is worthwhile to consider in more detail the important subject of $U(1)$ gauge invariance, addressing both its extent and its limitations.

(i) *Universality of electric charge*: The deflection of atomic and molecular beams by electric fields establishes that the fractional difference in the magnitude of electron and proton charge is no larger than $\mathcal{O}(10^{-20})$. Likewise, there is no evidence of any difference between the electric charges of the leptons e, μ, τ. Whatever the source of this charge universality may be, it is not the $U(1)$ invariance of electrodynamics. For example assume that in addition to ψ, there exists a second charged fermion field ψ' with charge parameter βe. It is easy to see that gauge invariance alone does not imply $\beta = 1$. The electromagnetic lagrangian for the extended system is

$$\mathcal{L}_{em} = -\frac{1}{4} F^2 + \overline{\psi} \, (i\,\slashed{D} - m) \, \psi + \overline{\psi}' \, (i\,\slashed{D}' - m') \, \psi', \tag{1.6}$$

where $D_\mu' \psi' \equiv (\partial_\mu + i\beta e A_\mu(x))\psi'$. The above lagrangian is invariant under the extended set of gauge transformations

$$\psi(x) \to e^{-i\alpha(x)} \psi(x), \qquad \psi'(x) \to e^{-i\beta\alpha(x)} \psi'(x),$$
$$A_\mu(x) \to A_\mu(x) + e^{-1} \partial_\mu \alpha(x). \tag{1.7}$$

This demonstration of gauge invariance is valid for arbitrary β, and thus says nothing about its value. The $U(1)$ symmetry is compatible with, but does not explain, the observed equality between the magnitudes of the electron and proton charges. We shall return to the issue of charge quantization in Sect. II–3 when we consider how weak hypercharge is assigned in the Weinberg–Salam model.

(ii) *A candidate quantum lagrangian*: The quantum version of $\mathcal{L}_{em}$ is in fact the most general Lorentz-invariant, hermitian, and renormalizable lagrangian which is $U(1)$ invariant. Consider the seemingly more general structure

$$\mathcal{L}_{gen} = -\frac{1}{4}ZF^2 + iZ_R\overline{\psi}_R \slashed{D}\psi_R + iZ_L\overline{\psi}_L \slashed{D}\psi_L - M\overline{\psi}_R\psi_L - M^*\overline{\psi}_L\psi_R, \quad (1.8)$$

where Z, $Z_{R,L}$ are constants, $\slashed{D}$ is the covariant derivative of Eq. (1.1), and M can be complex-valued. This lagrangian not only apparently differs from $\mathcal{L}_{em}$, but seemingly is CP-violating due to the complex mass term. However, under the rescalings

$$A'_\mu = Z^{1/2}A_\mu, \qquad e' = Z^{-1/2}e, \qquad \psi'_{R,L} = Z_{R,L}^{1/2}\psi_{R,L}, \quad (1.9)$$

we obtain

$$\mathcal{L}'_{gen} = -\frac{1}{4}F'^2 + i\overline{\psi}'\slashed{D}'\psi' - M'\overline{\psi}'_R\psi'_L - M'^*\overline{\psi}'_L\psi'_R, \quad (1.10)$$

where $M' = (Z_R Z_L)^{-1/2}M$. A subsequent global chiral change of variable

$$\psi''_{L,R} = e^{-i\alpha\gamma_5}\psi'_{L,R} \qquad (\alpha = \text{constant}) \quad (1.11)$$

does not affect the covariant derivative term but modifies the mass terms,

$$\mathcal{L}''_{gen} = -\frac{1}{4}F'^2 + i\overline{\psi}''\slashed{D}'\psi'' - M'e^{2i\alpha}\overline{\psi}''_R\psi''_L - (M'e^{2i\alpha})^*\overline{\psi}''_L\psi''_R. \quad (1.12)$$

Choosing the parameter α so that $\text{Im}\,(M'e^{2i\alpha}) = 0$ and defining $m \equiv \text{Re}(M'e^{2i\alpha})$, we see that $\mathcal{L}''_{gen}$ reduces to $\mathcal{L}_{em}$ which appears in Eq. (1.1).

(iii) *Renormalizability and $U(1)$*: Renormalizability plays a role in the preceding discussion because $U(1)$ symmetry by itself would admit a larger set of interaction terms. In principle, $U(1)$ invariant terms like $\overline{\psi}\sigma^{\mu\nu}\psi F_{\mu\nu}$, $\overline{\psi}\psi F^{\mu\nu}F_{\mu\nu}$, $\overline{\psi}\gamma^\mu\gamma^\nu \gamma^\alpha\gamma^\beta\psi F_{\mu\nu}F_{\alpha\beta}$, etc. could appear in the *QED* lagrangian. However, they do not because the condition of renormalizability admits only those contributions which have dimension $d \leq 4$. As discussed in App. C–3, the canonical dimension of boson and fermion fields is $d = 1$, $3/2$ respectively, and each derivative adds a unit of dimension. Accordingly, the above candidate operators have $d = 5$, 7, 7 and thus are ruled out. There remains an operator, $F_{\mu\nu}\tilde{F}^{\mu\nu}$, which *is* gauge-invariant and has dimension 4.[1] A noteworthy aspect of this quantity is that, unlike the other operators encountered thus far, it is odd under CP. This follows from writing it as $-4\mathbf{E} \cdot \mathbf{B}$ and realizing that under CP, $\mathbf{E} \to \mathbf{E}$ and $\mathbf{B} \to -\mathbf{B}$. However, a simple exercise shows that we can identify this operator as a four-divergence $F_{\mu\nu}\tilde{F}^{\mu\nu} = \partial_\mu K^\mu$, where $K^\mu \equiv 2\epsilon^{\mu\nu\alpha\beta}A_\nu\partial_\alpha A_\beta$. Thus, a contribution proportional to $F_{\mu\nu}\tilde{F}^{\mu\nu}$

---

[1] We define the tensor $\tilde{F}^{\mu\nu}$ which is dual to $F^{\mu\nu}$ as $\tilde{F}^{\mu\nu} \equiv \epsilon^{\mu\nu\alpha\beta}F_{\alpha\beta}/2$.

can be of no physical consequence. Upon integration over spacetime, it becomes a surface term evaluated at infinity. There is nothing in the structure of *QED* which would cause such a surface term to be anything but zero.

QED to one loop

The perturbative expansion of *QED* is carried out about the free field limit, and is interpreted in terms of Feynman diagrams. Two distinct phenomena are involved, scattering and renormalization. The latter encompasses both an additive mass shift for the fermion (but not for the photon) and rescalings of the charge parameter and of the quantum fields. To carry out the calculational program requires a quantum lagrangian $\mathcal{L}_{QED}$ to establish the Feynman rules, a regularization procedure to interpret divergent loop integrals, and a renormalization scheme.

One can develop *QED* using either canonical or path-integral methods. In either case a proper treatment necessitates modification of the classical lagrangian. As we have seen, the $U(1)$ gauge symmetry implies a certain freedom in defining the $A^\mu(x)$ field. Regardless of the quantization procedure adopted, this freedom can cause problems. For canonical quantization, the procedure of selecting a complete set of coordinates and their conjugate momenta is upset by the freedom to gauge transform away a coordinate at any given time. For path integration, the integration over gauge copies of specific field configurations gives rise to specious divergences (cf. App. A–6). In either case, superfluous gauge degrees of freedom can be eliminated by introducing an auxiliary condition which constrains the gauge freedom. There are a variety of ways to accomplish this. The one adopted here is to employ the following *gauge-fixed* lagrangian,

$$\mathcal{L}_{QED} = -\frac{1}{4}F^2 - \frac{1}{2\xi_0}(\partial \cdot A)^2 + \overline{\psi}\,(i\slashed{\partial} - e_0\slashed{A} - m_0)\psi, \qquad (1.13)$$

where e_0 and m_0 are, respectively, the fermion charge and mass parameters. The quantity ξ_0 is a real-valued, arbitrary constant appearing in the gauge-fixing term. This term is Lorentz-invariant but not $U(1)$-invariant. One of its effects is to make the photon propagator explicitly dependent on ξ_0. The value $\xi_0 = 1$ corresponds to *Feynman* gauge, whereas the limit $\xi_0 \to 0$ defines the *Landau* gauge.

The zero subscripts on the mass, charge, and gauge-fixing parameters denote that these *bare* quantities will be subject to renormalizations, as will the quantum fields. This process is characterized in terms of quantities Z_i and δm,

$$\psi = Z_2^{1/2}\psi^r, \qquad A_\mu = Z_3^{1/2}A^r_\mu,$$
$$e_0 = Z_1 Z_2^{-1} Z_3^{-1/2}e, \quad m_0 = m - \delta m,$$
$$\xi_0 = Z_3\xi, \qquad\qquad\qquad (1.14)$$

where the superscript 'r' labels renormalized fields. The renormalization constants Z_1, Z_2, and Z_3 (associated respectively with the fermion–photon vertex, the fermion wavefunction, and the photon wavefunction) and the fermion mass shift δm are chosen order by order to cancel the divergences occurring in loop integrals. For vanishing bare charge $e_0 = 0$, they reduce to $Z_{1,2,3} = 1$, $\delta m = 0$.

The Feynman rules for *QED* are:

fermion–photon vertex:

$$-i\, e_0\, (\gamma_\mu)_{\alpha\beta}$$

$$\beta \xrightarrow{\hspace{2cm}}^{\mu} \alpha$$

$$(1.15)$$

fermion propagator $i\, S_{\alpha\beta}(p)$:

$$\frac{i\,(\not{p} + m_0)_{\alpha\beta}}{p^2 - m_0^2 + i\epsilon} \qquad\qquad \beta \xrightarrow{\hspace{2cm}}^{p} \alpha$$

$$(1.16)$$

photon propagator $i\, D^{\mu\nu}(q)$:

$$\frac{i}{q^2 + i\epsilon}\left(-g^{\mu\nu} + (1 - \xi_0)\frac{q^\mu q^\nu}{q^2 + i\epsilon}\right) \qquad \nu \,\rule{0pt}{0pt}\!\!\!\sim\!\!\!\sim\!\!\!\sim^{q}\,\mu$$

$$(1.17)$$

In the above ϵ is an infinitesimal positive number.

The remainder of this section is devoted to a discussion of the one-loop radiative correction experienced by the photon propagator.[2] Throughout, we shall work in Feynman gauge.

Let us define a *proper* or *one-particle irreducible* (*1PI*) Feynman graph such that there is no point at which only a single internal line separates one part of the diagram from another part. The proper contributions to photon and to fermion propagators are called *self-energies*. The point of finding the photon self-energy is that the full propagator $i\, D'_{\mu\nu}$ can be constructed via iteration as in Fig. II–1. Performing a summation over self-energies, we obtain

Fig. II–1 The full photon propagator as an iteration.

2 We shall leave calculation of the fermion self-energy to Prob. II–3 and analysis of the photon-fermion vertex to Sect. V–1.

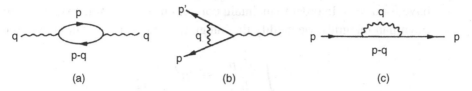

Fig. II–2 One-loop corrections to (a) photon propagator, (b) fermion-photon vertex, and (c) fermion propagator.

$$i D'_{\mu\nu} = i D_{\mu\nu} + i D_{\mu\alpha}(i \Pi^{\alpha\beta}) i D_{\beta\nu} + \cdots$$

$$= \frac{-i}{q^2} \left[\frac{1}{1 + \Pi(q)} \left(g_{\mu\nu} - \frac{q_\mu q_\nu}{q^2} \right) + \xi_0 \frac{q_\mu q_\nu}{q^2} \right], \qquad (1.18)$$

where the proper contribution

$$i \Pi^{\alpha\beta}(q) = (q^\alpha q^\beta - q^2 g^{\alpha\beta}) i \Pi(q) \qquad (1.19)$$

is called the *vacuum polarization tensor*. It is depicted in Fig. II–2(a) (along with corrections to the photon-fermion vertex and fermion propagator in Figs. II–2(b)–(c)), and is given to lowest order by

$$i \Pi^{\alpha\beta}(q) = -(-ie_0)^2 \int \frac{d^4 p}{(2\pi)^4} \, \mathrm{Tr} \left[\gamma^\alpha \frac{i}{\not{p} - m + i\epsilon} \gamma^\beta \frac{i}{\not{p} \not{q} - m + i\epsilon} \right]. \qquad (1.20)$$

This integral is quadratically divergent due to singular high-momentum behavior. To interpret it and other divergent integrals, we shall employ the method of *dimensional regularization* [BoG 72, 'tHV 72, Le 75].

Accordingly, we consider $\Pi^{\alpha\beta}(q)$ as the four-dimensional limit of a function defined in d spacetime dimensions. Various mathematical operations, such as summing over Lorentz indices or evaluating loop integrals, are carried out in d dimensions and the results are continued back to $d = 4$, generally expressed as an expansion in the variable[3]

$$\epsilon \equiv \frac{4 - d}{2}. \qquad (1.21a)$$

Formulae relevant to this procedure are collected in App. C–5. For all theories described in this book, we shall define the process of dimensional regularization such that all parameters of the theory (such as e^2) retain the dimensionality they

---

[3] We shall follow standard convention is using the symbol ϵ for both the infinitesimal employed in Feynman integrals and the variable for continuation away from the dimension of physical spacetime.

have for $d = 4$. In order to maintain correct units while dimensionally regularizing Feynman integrals, we modify the integration measure over momentum to

$$\int \frac{d^4 p}{(2\pi)^4} \rightarrow \mu^{2\epsilon} \int \frac{d^d p}{(2\pi)^d}. \tag{1.21b}$$

The parameter μ is an arbitrary auxiliary quantity having the dimension of a mass. It appears in the intermediate parts of a calculation, but cannot ultimately influence relations between physical observables. Indeed, there exist in the literature a number of variations of the extension to $d \neq 4$ dimensions. These are able to yield consistent results because one is ultimately interested in only the physical limit of $d = 4$. Let us now return to the photon self-energy calculation to see how the dimensional regularization is implemented.

The self-energy of Eq. (1.20), now expressed as an integral in d dimensions, is

$$\Pi^{\alpha\beta}(q) = 4ie_0^2\mu^{2\epsilon} \int \frac{d^d p}{(2\pi)^d} \frac{p^\alpha(p-q)^\beta + p^\beta(p-q)^\alpha + g^{\alpha\beta}(m^2 - p \cdot (p-q))}{[p^2 - m^2 + i\epsilon][(p-q)^2 - m^2 + i\epsilon]}, \tag{1.22}$$

where we retain the same notation $\Pi^{\alpha\beta}(q)$ as for $d = 4$ and we have already computed the trace. Upon introducing the Feynman parameterization, Dirac relations, and integral identities of App. C–5, we can perform the integration over momentum to obtain

$$\Pi^{\alpha\beta}(q) = (q^\alpha q^\beta - q^2 g^{\alpha\beta}) \frac{e_0^2}{2\pi^2} \frac{\Gamma(\epsilon)}{(4\pi)^{-\epsilon}} \mu^\epsilon \int_0^1 dx \frac{x(1-x)}{(m^2 - q^2 x(1-x))^\epsilon}. \tag{1.23}$$

We next expand $\Pi^{\alpha\beta}(q)$ in powers of ϵ and then pass to the limit $\epsilon \to 0$ of physical spacetime. In doing so, we use the familiar

$$a^\epsilon = e^{\ln a^\epsilon} = e^{\epsilon \ln a} = 1 + \epsilon \ln a + \cdots, \tag{1.24}$$

and take note of the combination

$$\frac{\Gamma(\epsilon)}{(4\pi)^{-\epsilon}} = \frac{1}{\epsilon} + \ln(4\pi) - \gamma + \mathcal{O}(\epsilon), \tag{1.25}$$

where $\gamma = 0.57221\ldots$ is the Euler constant. The presence of ϵ^{-1} makes it necessary to expand *all* the other ϵ-dependent factors in Eq. (1.23) and to take care in collecting quantities to a given order of ϵ. To order e^2, the vacuum polarization in Feynman gauge is then found to be

$$\Pi(q) = \frac{e_0^2}{12\pi^2} \left[\frac{1}{\epsilon} + \ln(4\pi) - \gamma \right.$$

$$\left. - 6 \int_0^1 dx \, x(1-x) \ln\left(\frac{m^2 - q^2 x(1-x)}{\mu^2}\right) + \mathcal{O}(\epsilon) \right]$$

$$= \frac{e_0^2}{12\pi^2} \begin{cases} \frac{1}{\epsilon} + \ln(4\pi) - \gamma + \frac{5}{3} - \ln\frac{-q^2}{\mu^2} + \cdots & (|q^2| \gg m^2), \\ \frac{1}{\epsilon} + \ln(4\pi) - \gamma - \ln\frac{m^2}{\mu^2} + \frac{q^2}{5m^2} + \cdots & (m^2 \gg |q^2|). \end{cases} \qquad (1.26)$$

The above expression is an example of the general property in dimensional regularization that divergences from loop integrals take the form of poles in ϵ. These poles are absorbed by judiciously choosing the renormalization constants. Renormalization constants can also have finite parts whose specification depends on the particular renormalization scheme employed. One generally adopts a scheme which is tailored to facilitate comparison of theory with some set of physical amplitudes. In the *minimal subtraction* (MS) renormalization, the Z_i subtract off only the ϵ-poles, and thus have the very simple form,

$$Z_i^{(\text{MS})} - 1 = \sum_{n=1}^{\infty} \frac{c_{i,n}}{\epsilon^n} \qquad (i = 1, 2, 3). \qquad (1.27)$$

Because the $\{Z_i^{(\text{MS})} - 1\}$ have no finite parts, they are sensitive only to the ultraviolet behavior of the loop integrals, and the $c_{i,n}$ are independent of mass. The simple appearance of the MS scheme is somewhat deceptive since further (finite) renormalizations are required if the mass and coupling parameters of the theory are to be asociated with physical masses and couplings. A related renormalization scheme is the *modified minimal subtraction* ($\overline{\text{MS}}$) in which renormalization constants are chosen to subtract off not only the ϵ-poles but also the omnipresent term $\ln(4\pi) - \gamma$ of Eq. (1.25). Minimal subtraction schemes are typically used in *QCD* where, due to the confinement phenomenon (cf. Sect. II–2), there is no renormalization scale that could naturally be associated with the mass of a freely propagating quark. Yet another approach is the *on-shell* (o-s) renormalization, where the renormalized mass and coupling parameters of the theory are arranged to coincide with their physical counterparts.

On-shell renormalization of the electric charge

The renormalization scale for electric charge is set by experimental determinations typically involving solid-state devices like Josephson junctions. These refer to probes of the electromagnetic vertex $-e\Gamma_\nu(p_2, p_1)$ of Fig. II–2(b) with on-shell electrons ($p_2^2 = p_1^2 = m_e^2$) and with $q^2 = (p_1 - p_2)^2 \simeq 0$. The value of the electromagnetic fine-structure constant $\alpha \equiv e^2/4\pi$ obtained under such conditions is given in rationalized units by

$$\alpha^{-1} = 137.035999074(44). \tag{1.28}$$

To interpret this in the context of the theoretical analysis performed thus far, recall from Eq. (1.18) how the photon propagator is modified by radiative corrections,

$$i e_0^2 D_{\mu\nu} = -\frac{i}{q^2} e_0^2 \, g_{\mu\nu} \;\rightarrow\; i e^2 D'_{\mu\nu} = -\frac{i}{q^2} \frac{e_0^2}{1 + \Pi(q)} g_{\mu\nu}. \tag{1.29}$$

We display only the $g_{\mu\nu}$ piece since, in view of current conservation, only it can contribute to the full amplitude upon coupling the propagator to electromagnetic vertices. The above suggests that we associate the physical, renormalized charge e with the bare charge parameter e_0 by

$$e^2 = \frac{e_0^2}{1 + \Pi(0)} \simeq e_0^2[1 - \Pi(0)]. \tag{1.30}$$

In this on-shell renormalization prescription, the $g_{\mu\nu}$ part of the photon propagator $i D'_{\mu\nu}(q)$ is seen to assume its unrenormalized form in the physical limit $q^2 \to 0$. The appellation 'on-shell' means that the physical kinematic point $q^2 = 0$ is used to implement the renormalization condition, and, by absorbing the singular vacuum polarization in the electric charge, one ensures that the photon has zero mass. Likewise, in the on-shell renormalization approach fermion propagators have poles at their physical masses.

Next, we show how to infer the form of the renormalization constant $Z_3^{(\text{o-s})}$ in the on-shell scheme. There is a relation, called the Ward identity, that implies $Z_1 = Z_2$ as a consequence of the gauge symmetry of the theory. From Eq. (1.14), this gives

$$e = \sqrt{Z_3^{(\text{o-s})}} \, e_0. \tag{1.31}$$

Use of the relation $e^2 \equiv Z_3^{(\text{o-s})} e_0^2$ then specifies the on-shell renormalization constant to be

$$Z_3^{(\text{o-s})} = 1 - \frac{e^2}{12\pi^2} \left[\frac{1}{\epsilon} + \ln(4\pi) - \gamma - \ln\left(\frac{m^2}{\mu^2}\right) + \mathcal{O}(\epsilon) \right]. \tag{1.32}$$

One can similarly absorb the ϵ-pole in either the MS or $\overline{\text{MS}}$ schemes by adopting

$$Z_3^{(\text{MS})} = 1 - \frac{e^2}{12\pi^2} \frac{1}{\epsilon} + \mathcal{O}(e^4),$$

$$Z_3^{(\overline{\text{MS}})} = 1 - \frac{e^2}{12\pi^2} \left(\frac{1}{\epsilon} - \gamma + \ln(4\pi) \right) + \mathcal{O}(e^4). \tag{1.33}$$

Fig. II–3 Virtual pair production in the vicinity of a charge.

Eqs. (1.32), (1.33) display how the various renormalization constants differ by finite amounts. The ϵ-poles in the fermion self-energy and the fermion-photon vertex can be dealt with in the same manner and we find, e.g., in MS renormalization (cf. Prob. II–3 and Sect. V–1),

$$Z_1^{(MS)} = Z_2^{(MS)} = 1 - \frac{e^2}{16\pi^2} \frac{1}{\epsilon} + \mathcal{O}(e^4), \qquad (1.34)$$

$$\delta m^{(MS)} = \frac{3e^2}{16\pi^2} m \frac{1}{\epsilon} + \mathcal{O}(e^4). \qquad (1.35)$$

Electric charge as a running coupling constant

The concept of electric charge as a 'running' coupling constant is motivated by the following consideration. In the perturbative Feynman expansion for a given theory, the hope is that corrections to the lowest-order amplitudes will be small. However, potentially large corrections of the form $\ln q^2/q_0^2$ can arise if the theory is renormalized at scale q_0^2 but then applied at a very different scale q^2. It is convenient to deal with this problem by absorbing such logarithms into scale-dependent or 'running' renormalized coupling constants and masses.

To see why scale-dependent charge is not an unreasonable concept, consider the vacuum polarization process of Fig. II–3, which depicts virtual production of a fermion of charge $Q_i e$ together with its antiparticle near a charge source. Due to the source, each such vacuum fluctuation is polarized, and thus the source becomes screened. All charged fermion species contribute to the screening, and the larger the mass of the virtual pair, the closer they lie to the source. The effect is somewhat akin to concentric onion skins, with each virtual pair forming a layer, resulting in an effectively scale-dependent source charge.

Let us seek a method for specifying a running fine structure constant $\alpha(q)$ for nonzero momentum transfers, with $\alpha(0)$ to be identified with the α of Eq. (1.28). The interpretation of $e_0^2/(1 + \text{Re } \Pi(q))$ as a running charge is appealing since it would maintain the simple $-i/q^2$ structure of the lowest-order photon exchange amplitude. The fact that $\Pi(q)$ is divergent (see Eq. (1.26)) can be circumvented by subtracting off its value at $q^2 = 0$ to define a finite quantity $\overline{\Pi}(q) \equiv \Pi(q) - \Pi(0)$ and defining

$$e^2(q) \equiv \frac{e^2}{1 + \text{Re } \overline{\Pi}(q)} \simeq e^2[1 - \text{Re } \overline{\Pi}(q)], \tag{1.36}$$

so that $\alpha(q) = e^2(q)/4\pi$. It is not difficult to deduce the behavior of $\overline{\Pi}(q)$ from the integral representation of Eq. (1.26), and we find

$$\overline{\Pi}(q) = \frac{\alpha}{3\pi} \begin{cases} \dfrac{5}{3} - \ln \dfrac{|q|^2}{m^2} + i\pi\theta(q^2) + \cdots & (|q^2| \gg m^2), \\[2mm] \dfrac{q^2}{5m^2} + \cdots & (m^2 \gg q^2). \end{cases} \tag{1.37}$$

Observe that the arbitrary energy scale μ is absent from $\overline{\Pi}(q)$, as would be expected since $\overline{\Pi}(q)$ is a physically measurable quantity.

The above formulae correspond to the loop correction of one fermion of mass m. Generally, loops from *all* available fermions must be included, although contributions of heavy $(m^2 \gg q^2)$ fermions are seen to be suppressed. Important modern applications of the Standard Model engender phenomena at scales provided by the gauge-boson masses M_W, M_Z. To obtain an estimate for $\alpha(M_Z^2)$, we can apply Eq. (1.37) to find

$$\alpha^{-1}(M_Z^2) = \alpha^{-1}\left[1 - \frac{\alpha}{3\pi} \sum_i Q_i^2 \left(\ln \frac{M_Z^2}{m_i^2} - \frac{5}{3} \right) + \cdots \right]. \tag{1.38}$$

If a sum over quark-loops (each being accompanied by the color factor $N_c = 3$) and lepton-loops is performed, then the mass values in Tables I–2, I–3 yield the approximate determination $\alpha^{-1}(M_Z^2) \simeq 130$. The main uncertainty in this approach arises from quarks. It is possible to perform a more accurate evaluation of $\alpha(M_Z^2)$ (cf. Sect. XVI–6) which avoids this difficulty.

Let us return to the question of how to define a momentum-dependent coupling. To emphasize the fact that a 'running fine-structure constant' is after all a matter of definition, let us consider a somewhat different derivation (and definition) of $\alpha(q^2)$. One is able to renormalize the electric charge in a mass-independent scheme [We 73] by calculating renormalization constants with $m = 0$. If we return to the vacuum polarization diagram, but with $m = 0$, we find

$$\Pi(q^2) = \frac{e_0^2}{12\pi^2} \left(\frac{\mu^2}{-q^2} \right)^\epsilon \left[\frac{1}{\epsilon} + \ln(4\pi) - \gamma + \frac{5}{3} + \mathcal{O}(\epsilon) \right]$$

$$= \frac{e_0^2}{12\pi^2} \left[\frac{1}{\epsilon} + \ln(4\pi) - \gamma + \frac{5}{3} - \ln\left(\frac{-q^2}{\mu^2} \right) + \mathcal{O}(\epsilon) \right]. \tag{1.39}$$

In order to apply the renormalization program, we must specify the value of the coupling at some renormalization point,[4] which we choose to be $q^2 = -\mu_R^2$, identifying

$$e^2(\mu_R^2) = \frac{e_0^2}{1 + \Pi(q^2)\big|_{-q^2 = \mu_R^2}} \simeq e_0^2 \left[1 - \frac{e_0^2}{12\pi^2}\left(\frac{1}{\epsilon} - \ln\frac{\mu_R^2}{\mu^2} + \cdots\right)\right]. \quad (1.40)$$

However, if we had chosen a different renormalization point $\mu_R^{2\,\prime}$, we would have obtained a different value,

$$e^2(\mu_R^{2\,\prime}) = e^2(\mu_R^2) + \frac{e_0^4}{12\pi^2} \ln\frac{\mu_R^{2\,\prime}}{\mu_R^2}. \quad (1.41)$$

The functional dependence of the charge on the renormalization scale is embodied in the so-called *beta* function of electrodynamics [GeL 54],

$$\beta_{QED}(e) \equiv \mu_R \frac{\partial e}{\partial \mu_R} = \frac{e^3}{12\pi^2} + \mathcal{O}(e^5). \quad (1.42)$$

It can be shown [Po 74] that the leading and next-to-leading terms in a perturbative expansion of β_{QED} are independent of both renormalization and gauge choices.

The quantity $e^2(\mu_R^2)$ defined by integrating the beta function,

$$\frac{de}{\beta_{QED}(e)} = \frac{d\mu_R}{\mu_R}, \quad (1.43)$$

is not exactly the same quantity as the running coupling constant defined in Eq. (1.36), differing by a (small) finite renormalization. For example, the electron contribution to the running coupling in the range $m_e^2 \leq \mu_R^2 \leq M_Z^2$ is

$$\alpha^{-1}(\mu_R^2)\big|_{\mu_R^2 = m_e^2} - \alpha^{-1}(\mu_R^2)\big|_{\mu_R^2 = M_Z^2} = \frac{1}{3\pi} \ln\frac{M_Z^2}{m_e^2}, \quad (1.44)$$

which contains the dominant logarithmic dependence, but differs from Eq. (1.38) by a small additive term. However, complete calculations of *all* corrections to physical observables using the two schemes will yield the same answer. Since the running coupling constant is but a bookkeeping device, one's choice is a matter of taste or of convenience. Regardless of the specific definition employed for $\alpha(q^2)$, we see that as the energy scale is increased (or as distance is decreased), the running electric charge grows. This is anticipated from the screening of a test charge due to vacuum polarization (recall our explanation of Fig. II–3). As the momentum transfer of a photon probe is increased, the screening is penetrated and the effective charge increases.

---

[4] Note that the renormalization point μ_R and the scale factor μ in dimensional regularization need not be identical. They are sometimes confused in the literature, and hence we use a different notation for the two quantities.

The use of a mass-independent scheme is convenient for identifying the high-energy scaling behavior of gauge theories. One useful feature is in the calculation of the one-loop beta function. Dimensional analysis requires that the one-loop charge renormalization be of the form,

$$g = g_0 \left[1 - g_0^2 b \left(\frac{\mu^2}{-q^2} \right)^\epsilon \left(\frac{1}{2\epsilon} + \text{finite terms} \right) + \mathcal{O}(g_0^4) \right], \tag{1.45}$$

where g is the 'charge' associated with the gauge theory being considered. Choosing the renormalization point as $q^2 = -\mu_R^2$ and forming the beta function as in Eq. (1.42), we see that $\beta = bg^3$. This allows the beta function to be simply identified with the coefficient of ϵ^{-1} to this order.

II–2 Quantum Chromodynamics

Chromodynamics, the nonabelian gauge description of the strong interactions, contains quarks and gluons instead of electrons and photons as its basic degrees of freedom [FrG 72, Co 11]. A hallmark of Quantum Chromodynamics (*QCD*) is asymptotic freedom [GrW 73a,b, Po 73], which reveals that only in the short-distance limit can perturbative methods be legitimately employed. The necessity to employ approaches alternative to perturbation theory for long-distance processes motivates much of the analysis in this book.

SU(3) gauge symmetry

Chromodynamics is the $SU(3)$ nonabelian gauge theory of color charge. The fermions which carry color charge are the *quarks*, each with field $\psi_j^{(\alpha)}$, where $\alpha = u, d, s, \ldots$ is the flavor label and $j = 1, 2, 3$ is the color index. The gauge bosons, which also carry color, are the *gluons*, each with field A_μ^a, $a = 1, \ldots, 8$.[5] Classical chromodynamics is defined by the lagrangian

$$\mathcal{L}_{\text{color}} = -\frac{1}{4} F^{a\mu\nu} F_{\mu\nu}^a + \sum_\alpha \overline{\psi}_j^{(\alpha)} (i \slashed{D}_{jk} - m^{(\alpha)} \delta_{jk}) \psi_k^{(\alpha)}, \tag{2.1}$$

where the repeated color indices are summed over. The gauge field strength tensor is

$$F_{\mu\nu}^a = \partial_\mu A_\nu^a - \partial_\nu A_\mu^a - g_3 f^{abc} A_\mu^b A_\nu^c, \tag{2.2a}$$

---

[5] In this section, it will be particularly important to explicitly display color indices. We shall reserve indices which begin the alphabet for gluon color indices (e.g., $a, b, c = 1, \ldots, 8$), use mid-alphabetic letters for quark color indices (e.g., $j, k, l = 1, 2, 3$), and employ greek symbols for flavor indices.

where g_3 is the $SU(3)$ gauge coupling parameter, and the quark covariant derivative is

$$\mathbf{D}_\mu \psi = \left(\mathbf{I}\partial_\mu + ig_3 A_\mu^a \frac{\lambda_a}{2} \right) \psi. \qquad (2.2b)$$

The lagrangian of Eq. (2.1) is invariant under local $SU(3)$ transformations of the color degree of freedom, under which the quark and gluon fields transform as given earlier in Eqs. (I–5.11), (I–5.17). Equations of motion for the quark and gluon fields are

$$(i\slashed{D} - m^{(\alpha)})\psi^{(\alpha)} = 0,$$

$$D^\mu F_{\mu\nu}^a = g_3 \sum_\alpha \overline{\psi}^{(\alpha)} \frac{\lambda_a}{2} \gamma_\nu \psi^{(\alpha)}. \qquad (2.3)$$

In its quantum version, the $g_3 \to 0$ limit of $\mathcal{L}_{\text{color}}$ describes an exceedingly simple world. There exist only free massless spin one gluons and massive spin one-half quarks. However, the full theory is quite formidable. In particular, accelerator experiments reveal a particle spectrum which bears no resemblance to that of the noninteracting theory.

The group $SU(3)$ has an infinite number of irreducible representations R. The first several are $R = \mathbf{1}, \mathbf{3}, \mathbf{3}^*, \mathbf{6}, \mathbf{6}^* \mathbf{8}, \mathbf{10}, \mathbf{10}^*, \dots$, where we label an irreducible representation in terms of its dimensionality. Quarks, antiquarks, and gluons are assigned to the representations $\mathbf{3}, \mathbf{3}^*, \mathbf{8}$ respectively. We denote the group generators for representation R by $\{\mathbf{F}_a(R)\}$ $(a = 1, \dots, 8)$. The quantities $\lambda/2$ are group generators for the $d = 3$ *fundamental* representation, i.e., $\mathbf{F}(3) = \lambda/2$. They have the matrix representation

$$\lambda_1 = \begin{pmatrix} 0 & 1 & 0 \\ 1 & 0 & 0 \\ 0 & 0 & 0 \end{pmatrix} \qquad \lambda_4 = \begin{pmatrix} 0 & 0 & 1 \\ 0 & 0 & 0 \\ 1 & 0 & 0 \end{pmatrix} \qquad \lambda_7 = \begin{pmatrix} 0 & 0 & 0 \\ 0 & 0 & -i \\ 0 & i & 0 \end{pmatrix}$$

$$\lambda_2 = \begin{pmatrix} 0 & -i & 0 \\ i & 0 & 0 \\ 0 & 0 & 0 \end{pmatrix} \qquad \lambda_5 = \begin{pmatrix} 0 & 0 & -i \\ 0 & 0 & 0 \\ i & 0 & 0 \end{pmatrix} \qquad \lambda_8 = \begin{pmatrix} \frac{1}{\sqrt{3}} & 0 & 0 \\ 0 & \frac{1}{\sqrt{3}} & 0 \\ 0 & 0 & \frac{-2}{\sqrt{3}} \end{pmatrix}$$

$$\lambda_3 = \begin{pmatrix} 1 & 0 & 0 \\ 0 & -1 & 0 \\ 0 & 0 & 0 \end{pmatrix} \qquad \lambda_6 = \begin{pmatrix} 0 & 0 & 0 \\ 0 & 0 & 1 \\ 0 & 1 & 0 \end{pmatrix}. \qquad (2.4)$$

As generators, they obey the commutation relations

$$[\lambda_a, \lambda_b] = 2if_{abc}\lambda_c \qquad (a, b, c = 1, \dots, 8) \qquad (2.5a)$$

Table II–1. *Nonvanishing f, d coefficients.*

abc	f_{abc}	abc	d_{abc}	abc	d_{abc}
123	1	118	$1/\sqrt{3}$	355	1/2
147	1/2	146	1/2	366	−1/2
156	−1/2	157	1/2	377	−1/2
246	1/2	228	$1/\sqrt{3}$	448	$-1/2\sqrt{3}$
257	1/2	247	−1/2	558	$-1/2\sqrt{3}$
345	1/2	256	1/2	668	$-1/2\sqrt{3}$
367	−1/2	338	$1/\sqrt{3}$	778	$-1/2\sqrt{3}$
458	$\sqrt{3}/2$	344	1/2	888	$-1/\sqrt{3}$
678	$\sqrt{3}/2$				

where the f-coefficients are totally antisymmetric structure constants of $SU(3)$. There exist corresponding anticommutation relations

$$\{\lambda_a, \lambda_b\} = \frac{4}{3}\delta_{ab}\,\mathbf{I} + 2d_{abc}\lambda_c \qquad (a, b, c = 1, \ldots, 8) \qquad (2.5b)$$

with d-coefficients which are totally symmetric. Values for f_{abc} and d_{abc} are given in Table II–1.

Useful trace relations obeyed by the $\{\lambda_a\}$ are

$$\mathrm{Tr}\,\lambda_a = 0 \qquad (a = 1, \ldots, 8) \qquad (2.6)$$

from Eq. (2.4) and

$$\mathrm{Tr}\,\lambda_a\lambda_b = 2\delta_{ab} \qquad (a, b = 1, \ldots, 8) \qquad (2.7)$$

from Eq. (2.5). The statement of completeness takes the form,

$$\lambda_{ij}^a\lambda_{kl}^a = -\frac{2}{3}\delta_{ij}\delta_{kl} + 2\delta_{il}\delta_{jk} \qquad (i, j, k, l = 1, 2, 3), \qquad (2.8)$$

where $a = 1, \ldots, 8$ is summed over. Useful labels for the irreducible representations of $SU(3)$ are provided by the *Casimir invariants*. For any representation R, the quadratic Casimir invariant $C_2(R)$ is defined by squaring and summing the group generators $\{\mathbf{F}_a(R)\}$,

$$C_2(R)\mathbf{I} \equiv \sum_{a=1}^{8} \mathbf{F}_a^2(R). \qquad (2.9)$$

There is also a third-order Casimir invariant,

$$C_3(R)\mathbf{I} \equiv \sum_{a,b,c=1}^{8} d_{abc}\mathbf{F}_a(R)\mathbf{F}_b(R)\mathbf{F}_c(R). \qquad (2.10)$$

The quark and antiquark states form the bases for the smallest nontrivial irreducible representations of $SU(3)$. It is possible to use products of them, say p factors of quarks and q factors of antiquarks, to construct all other irreducible tensors in $SU(3)$. Each irreducible representation R is then characterized by the pair (p, q). For example, we have the correspondences $\mathbf{1} \sim (0, 0)$, $\mathbf{3} \sim (1, 0)$, $\mathbf{3^*} \sim (0, 1)$, $\mathbf{8} \sim (1, 1)$, $\mathbf{10} \sim (3, 0)$, etc. The (p, q) labeling scheme provides useful expressions for the dimension of a representation,

$$d(p, q) = (p + 1)(q + 1)(p + q + 2)/2, \tag{2.11}$$

and of the two Casimir invariants,

$$C_2(p, q) = (3p + 3q + p^2 + pq + q^2)/3,$$
$$C_3(p, q) = (p - q)(2p + q + 3)(2q + p + 3)/18. \tag{2.12}$$

From Eq. (2.12) we find $C_2(\mathbf{3}) = C_2(\mathbf{3^*}) = 4/3$ for the quark and antiquark representations. Equivalently, upon setting $j = k$ and summing in Eq. (2.8) we obtain

$$\lambda_{ij}^a \lambda_{jl}^a = \frac{16}{3} \delta_{il} = 4C_2(\mathbf{3}) \delta_{il}. \tag{2.13}$$

Generators for the $d = 8$ *regular* (or *adjoint*) representation are determined from the structure constants themselves,

$$(F^a(\mathbf{8}))_{bc} = -if_{abc} \qquad (a, b, c = 1, \ldots, 8). \tag{2.14}$$

It follows directly from Eq. (2.14) and from using Eq. (2.12) to compute $C_2(\mathbf{8}) = 3$ that

$$f_{acd} f_{bcd} = C_2(\mathbf{8}) \, \delta_{ab} = 3 \, \delta_{ab}. \tag{2.15}$$

This result, in turn, enables us to determine

$$f_{abc}\lambda_b\lambda_c = \frac{1}{2} f_{abc}[\lambda_b, \lambda_c] = if_{abc} f_{bcd}\lambda_d = iC_2(\mathbf{8})\lambda_a. \tag{2.16}$$

As a final example involving $SU(3)$, we evaluate the quantity

$$\lambda^b\lambda^a\lambda^b = \frac{1}{2} \left(\lambda^b[\lambda^a, \lambda^b] - [\lambda^a, \lambda^b]\lambda^b + \lambda^b\lambda^b\lambda^a + \lambda^a\lambda^b\lambda^b \right)$$
$$= 4C_2(\mathbf{3})\lambda^a + if_{abc}[\lambda^b, \lambda^c] = 4\left(C_2(\mathbf{3}) - \frac{1}{2}C_2(\mathbf{8}) \right)\lambda^a. \tag{2.17}$$

Shortly, we shall see how such combinations of color factors arise in various radiative corrections.

Including only gauge-invariant and renormalizable terms, we can write the most general form for a chromodynamic lagrangian as

$$\mathcal{L}_{\text{gen}} = -\frac{1}{4} Z F_a^{\mu\nu} F_{\mu\nu}^a + \overline{\psi}_L^\alpha Z_L^{\alpha\beta} i\slashed{D} \psi_L^\beta + \overline{\psi}_R^\alpha Z_R^{\alpha\beta} i\slashed{D} \psi_R^\beta - \overline{\psi}_L^\alpha M^{\alpha\beta} \psi_R^\beta$$

$$- \overline{\psi}_R^\alpha M^{\dagger\alpha\beta} \psi_L^\beta + \frac{g_3^2}{64\pi^2} \theta \epsilon^{\mu\nu\lambda\sigma} F_{\mu\nu}^a F_{\lambda\sigma}^a, \tag{2.18}$$

where the flavor matrices $Z_{L,R}$ are hermitian, color and flavor indices are as before, except that for simplicity we suppress quark color notation. The final contribution to Eq. (2.18) is called the θ-*term*. We can reduce $\mathcal{L}_{\text{gen}}$ to the form of $\mathcal{L}_{\text{color}}$ by first rescaling,

$$A_\mu'^a = Z^{1/2} A_\mu^a, \qquad g_3' = Z^{-1/2} g_3, \tag{2.19}$$

and then diagonalizing and rescaling with respect to quark flavors,

$$\psi_{L,R}' = U_{L,R} \psi_{L,R}, \qquad U_{L,R} Z_{L,R} U_{L,R}^\dagger = \Lambda_{L,R}, \qquad \psi_{L,R}'' = \Lambda_{L,R}^{1/2} \psi', \tag{2.20}$$

where $\Lambda_{L,R}$ are diagonal. Finally we diagonalize the mass terms

$$\mathcal{L}_{\text{mass}} = -\overline{\psi}_L''^\alpha M'^{\alpha\beta} \psi_R''^\beta - \overline{\psi}_R''^\alpha M'^{\dagger\alpha\beta} \psi_L''^\beta, \tag{2.21}$$

where $M' = \Lambda_L^{-1/2} U_L M U_R^\dagger \Lambda_R^{-1/2}$, by means of yet another set of unitary transformations on the quark fields. Aside from the θ-term, this results in the canonical expression for $\mathcal{L}_{\text{color}}$ of Eq. (2.1).

We shall demonstrate later in Sect. IX–4 that the above quark mass diagonalization procedure induces a modification in the θ-parameter,

$$\theta \to \overline{\theta} = \theta + \arg \det M'. \tag{2.22}$$

This does not imply $\overline{\theta} = 0$ because both θ and the original quark mass matrices are arbitrary from the viewpoint of renormalizability and $SU(3)$ gauge invariance. In fact, the θ-term cannot be ruled out by any of the tenets which underlie the Standard Model. Moreover, although the θ-term can be expressed as a four-divergence

$$\mathcal{L}_\theta = \frac{g_3^2}{32\pi^2} \overline{\theta} \, \partial_\mu K^\mu, \tag{2.23}$$

$$K^\mu = \epsilon^{\mu\nu\lambda\sigma} A_\nu^a \left(F_{\lambda\sigma}^a + \frac{g_3}{3} f_{abc} A_\lambda^b A_\sigma^c \right), \tag{2.24}$$

analysis demonstrates that K^μ is a singular operator and that its divergence cannot be summarily discarded as was done in electrodynamics. This is a curious situation because the θ-term is *CP*-violating. Thus, one is faced with the specter of large *CP*-violating signals in the strong interactions. Yet such effects are not observed.

Indeed, it has been estimated that the θ-term generates a nonzero value for the neutron electric dipole moment $d_e(n) \simeq 5 \times 10^{-16}\,\overline{\theta}$ e-cm, but to date no signal has been observed experimentally, $d_e(n) < 2.9 \times 10^{-26}$ e-cm at C.L. 90% [RPP 12]. This provides the upper bound $\overline{\theta} < 5.8 \times 10^{-11}$. Perhaps Nature has dictated $\overline{\theta} \equiv 0$, albeit for reasons not yet understood.

QCD to one loop

To develop Feynman rules for *QCD*, we must first obtain an effective lagrangian which properly addresses the issue of $SU(3)$ gauge freedom. For the $U(1)$ gauge invariance of *QED*, this was accomplished by adding a gauge-fixing term to the classical lagrangian. The situation for $SU(3)$ is analogous, but somewhat more complicated due to its nonabelian structure. If we continue to use a Lorentz-invariant gauge-fixing procedure, the effective *QCD* lagrangian (for simplicity, consider just one quark flavor) can be expressed as

$$\mathcal{L}_{QCD} = -\frac{1}{4} F_{\mu\nu}^a F^{a\mu\nu} + \overline{\psi}_j (i\,\slashed{D} - m_0 \mathbf{I})_{jk} \psi_k - \frac{1}{2\xi_0} (\partial_\mu A_a^\mu)^2$$
$$+ \partial_\mu \overline{c}_a \partial^\mu c_a + g_{3,0} f_{abe} A_a^\mu (\partial_\mu \overline{c}_b) c_e. \tag{2.25}$$

Bare quantities carry the subscript '0' and the field strengths and covariant derivative are defined as in Eqs. (2.2a), (2.2b). The quantities $\{c_a(x)\}$ $(a = 1, \ldots, 8)$ are called *ghost fields*. As explained in App. A–5, they are anticommuting c-number quantities (i.e., *Grassmann variables*) which couple only to gluons. Ghosts occur only within loops, and never appear as asymptotic states. Each ghost-field loop contribution must be accompanied by an extra minus sign, analogous to that of a fermion–antifermion loop. Their presence is a consequence of the Lorentz-invariant gauge-fixing procedure. In alternative schemes such as axial or temporal gauge, ghost fields do not appear, but compensating unphysical singularities occur in Feynman integrals instead.

The Feynman rules for *QCD* are

three-gluon vertex:

$-g_{3,0} f_{abc} [g_{\mu\nu} (p - q)_\lambda + g_{\nu\lambda} (q - r)_\mu$
$\quad + g_{\lambda\mu} (r - p)_\nu]$

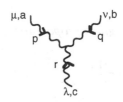

$$\tag{2.26}$$

quark–gluon vertex:

$$-ig_{3,0}(\gamma_\mu)_{\alpha\beta}\left(\frac{\lambda^a}{2}\right)_{jk}$$

(2.27)

four-gluon vertex:

$$-ig_{3,0}^2\left[(f_{abe}f_{cde}(g_{\mu\lambda}g_{\nu\sigma}-g_{\mu\sigma}g_{\nu\lambda})\right.$$
$$+f_{ace}f_{bde}(g_{\mu\nu}g_{\lambda\sigma}-g_{\mu\sigma}g_{\nu\lambda})$$
$$\left.+f_{ade}f_{cbe}(g_{\mu\lambda}g_{\nu\sigma}-g_{\mu\nu}g_{\lambda\sigma})\right]$$

(2.28)

ghost–gluon vertex:

$$-g_{3,0}f_{abc}r_\mu$$

(2.29)

quark propagator $iS_{\alpha\beta}^{jk}(p)$:

$$\frac{i\delta_{jk}\,(\not p+m_0)_{\alpha\beta}}{p^2-m_0^2+i\epsilon}$$

(2.30)

gluon propagator $iD_{\mu\nu}^{ab}(q)$:

$$\frac{i\delta_{ab}}{q^2+i\epsilon}\left(-g^{\mu\nu}+(1-\xi_0)\frac{q^\mu q^\nu}{q^2+i\epsilon}\right)$$

(2.31)

ghost propagator:

$$\frac{i\delta_{ab}}{p^2+i\epsilon}$$

(2.32)

The above rules involve a total of four distinct interaction vertices. Of these, the three-gluon and four-gluon self-vertices, and the ghost–gluon coupling have no counterpart in *QED*. That all four vertices are scaled by a single coupling strength g_3 is a consequence of gauge invariance. Also, chromodynamics exhibits a certain coupling-constant universality, called *flavor independence*, in the quark–gluon sector. All fields which transform according to a given representation of the $SU(3)$ of color have the same interaction structure, e.g., all triplets couple alike, all octets couple alike but differently from triplets, etc. Quarks are assigned solely to the color triplet representation. Thus, the quark–gluon interaction is independent of flavor.

The renormalization constants of *QCD* are

$$
\begin{aligned}
A_\mu^a &= Z_3^{1/2}(A_\mu^a)^r, & g_{3,0} &= Z_1 Z_3^{-3/2} g_3, \\
\psi &= Z_2^{1/2}\psi^r, & &= Z_4^{1/2} Z_3^{-1} g_3, \\
c^a &= \overline{Z}_3^{1/2}(c^a)^r, & &= Z_{1F} Z_2^{-1} Z_3^{-1/2} g_3, \\
\xi_0 &= Z_3\xi, & &= \overline{Z}_1 \overline{Z}_3^{-1} Z_3^{-1/2} g_3, \\
m_0 &= m - \delta m,
\end{aligned}
\tag{2.33}
$$

where the quantities $Z_1, \overline{Z}_1, Z_{1F}$, and Z_4 are defined by the above coupling, constant relations and can be determined from Z_2, Z_3, and $\overline{Z}_3$. In the following, working in $\xi_0 = 1$ gauge we shall compute the one-loop contributions to the gluon self-energy and to the quark–gluon vertex, and, by absorbing the ϵ-poles, thereby obtain expressions for Z_3 and Z_{1F} to leading order. Determination of the remaining renormalization constants, which can be computed from loop corrections to the quark and ghost propagators and the three-gluon, four–gluon, and ghost–gluon vertices will be left as exercises. However, it is clear from the definition of $g_{3,0}$ in Eq. (2.33) that the relations

$$
\frac{Z_4}{Z_1} = \frac{Z_1}{Z_3} = \frac{\overline{Z}_1}{\overline{Z}_3}
\tag{2.34}
$$

must hold in any consistent renormalization scheme. These are the analogs of the Ward identities in *QED*. Physically, they ensure that the coupling-constant relations which appear in the *QCD* lagrangian (as a consequence of gauge invariance) are maintained in the full theory.

The *QCD* one-loop contribution to the quark–antiquark vacuum polarization amplitude of Fig. II–4(a),[6]

---

[6] To avoid notational clutter, we shall not put subscripts on the bare coupling for the remainder of this subsection.

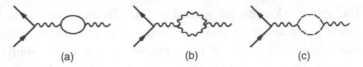

Fig. II–4 One-loop corrections to the gluon propagator: (a) quark–antiquark pair,
(b) gluon pair, and (c) ghosts.

$$i\Pi^{ab}_{\alpha\beta}(q)\big|_{\text{quark}} = -\left(\frac{-ig_3}{2}\right)^2 \int \frac{d^4p}{(2\pi)^4}$$

$$\times \text{Tr}\left[\gamma_\alpha(\lambda^a)_{kj}\frac{i}{\not{p}-m+i\epsilon}\gamma_\beta(\lambda^b)_{jk}\frac{i}{\not{p}-\not{q}-m+i\epsilon}\right], \quad (2.35)$$

differs from the *QED* self-energy only by the group factor $(\lambda^a)_{jk}(\lambda^b)_{kj} = \text{Tr}\,(\lambda^a\lambda^b) = 2\delta^{ab}$ (cf. Eq. (2.7)). Comparing with Eq. (1.39), we obtain

$$i\Pi^{ab}_{\alpha\beta}(q)\big|_{\text{quark}} = i\delta^{ab}(q_\alpha q_\beta - g_{\alpha\beta}q^2)\left(\frac{\mu^2}{-q^2}\right)^\epsilon\left[\frac{g_3^2}{24\pi^2}\frac{1}{\epsilon}+\cdots\right]. \quad (2.36)$$

This must be multiplied by the number of quark flavors n_f which contribute in the
vacuum polarization loops.

The contribution from the gluon–gluon intermediate state of Fig. II–4(b) can be
written

$$i\Pi^{ab}_{\alpha\beta}(q)\big|_{\text{gluon}} = \frac{1}{2}(-i)^2 \int \frac{d^4k}{(2\pi)^4}\frac{N^{ab}_{\alpha\beta}}{[k^2+i\epsilon][(q-k)^2+i\epsilon]} \quad (2.37)$$

with

$$N^{ab}_{\alpha\beta} = g_3 f^{bcd}[-g_{\beta\mu}(q+k)_\nu + g_{\mu\nu}(2k-q)_\beta + g_{\nu\beta}(2q-k)_\mu]$$
$$\times g_3 f^{acd}[g'_\alpha{}^\mu(q+k)^\nu + g^{\mu\nu}(q-2k)_\alpha + g^\nu_\alpha(k-2q)^\mu]. \quad (2.38)$$

The prefactor $1/2$ in Eq. (2.37) is a Feynman symmetry factor associated with the
identical intermediate-state gluons. To arrive at this expression, special care must
be exercised with momentum flow in the three-gluon vertices. Upon extending the
integration to d dimensions and using Eq. (2.15) to evaluate the color factor, we
obtain

$$i\Pi^{ab}_{\alpha\beta}(q)\big|_{\text{gluon}} = -\frac{1}{2}C_2(8)\delta^{ab}g_3^2\mu^{2\epsilon}\int \frac{d^dk}{(2\pi)^d}\frac{N_{\alpha\beta}}{[k^2+i\epsilon][(q-k)^2+i\epsilon]} \quad (2.39)$$

with

$$N_{\alpha\beta} = (-5q^2 + 2q\cdot k - 2k^2)g_{\alpha\beta} + (6-d)q_\alpha q_\beta$$
$$+ (2d-3)(q_\alpha k_\beta + q_\beta k_\alpha) + (6-4d)k_\alpha k_\beta. \quad (2.40)$$

Integration of Eq. (2.39) yields

$$i\,\Pi_{\alpha\beta}^{ab}(q)\big|_{\text{gluon}} = -i\,\frac{g_3^2}{16\pi^2}C_2(\mathbf{8})\delta^{ab}\left(\frac{\mu^2}{-q^2}\right)^\epsilon\left[\frac{11}{3}q_\alpha q_\beta - \frac{19}{6}g_{\alpha\beta}q^2\right]\frac{1}{2\epsilon} + \cdots.$$

(2.41)

The final contribution to the gluon propagator is the ghost-loop amplitude of Fig. II–4(c),

$$i\,\Pi_{\alpha\beta}^{ab}(q)\big|_{\text{ghost}} = -\int\frac{d^4k}{(2\pi)^4}\frac{i}{(k-q)^2 + i\epsilon}$$
$$\times\,[g_3 f^{bdc}(k-q)_\beta]\,\frac{i}{k^2 + i\epsilon}\,[g_3 f^{acd}k_\alpha].$$

(2.42)

The bracketed quantities arise from the gluon–ghost vertices, and the minus prefactor must accompany any ghost loop. Following the standard steps to a d-dimensional form, we arrive at

$$i\,\Pi_{\alpha\beta}^{ab}(q)\big|_{\text{ghost}} = -g_3^2 C_2(\mathbf{8})\delta^{ab}\,\mu^{2\epsilon}\int\frac{d^dk}{(2\pi)^d}\frac{k_\alpha(k-q)_\beta}{[(k-q)^2 + i\epsilon][k^2 + i\epsilon]},$$

(2.43)

which becomes to leading order in ϵ,

$$i\,\Pi_{\alpha\beta}^{ab}(q)\big|_{\text{ghost}} = i\delta^{ab}\frac{g_3^2}{16\pi^2}C_2(\mathbf{8})\left(\frac{\mu^2}{-q^2}\right)^\epsilon\left[\frac{1}{3}q_\alpha q_\beta + \frac{1}{6}g_{\alpha\beta}q^2\right]\frac{1}{2\epsilon}.$$

(2.44)

The sum of gluon and ghost contributions takes the gauge-invariant form

$$i\,\Pi_{\alpha\beta}^{ab}(q)\big|_{\text{gl+gh}} = -i\delta^{ab}\frac{g_3^2}{8\pi^2}\,C_2(\mathbf{8})\,\frac{5}{3}\left(\frac{\mu^2}{-q^2}\right)^\epsilon\left[q_\alpha q_\beta - g_{\alpha\beta}q^2\right]\frac{1}{2\epsilon} + \cdots.$$

(2.45)

Finally, adding the quark contribution for n_f flavors gives the total result

$$\Pi_{\alpha\beta}^{ab}(q) = i\delta^{ab}(q_\alpha q_\beta - g_{\alpha\beta}q^2)\frac{g_3^2}{8\pi^2}\left(\frac{\mu^2}{-q^2}\right)^\epsilon\left[\frac{2n_f}{3} - \frac{5}{3}C_2(\mathbf{8})\right]\frac{1}{2\epsilon} + \cdots.$$

(2.46)

Renormalizing at $q^2 = -\mu_R^2$, we find[7]

$$Z_3 = 1 - \frac{g_3^2}{8\pi^2}\left(\frac{\mu}{\mu_R}\right)^{2\epsilon}\left[\frac{2n_f}{3} - \frac{5}{3}C_2(\mathbf{8})\right]\frac{1}{2\epsilon} + \mathcal{O}(g_3^4).$$

(2.47)

Proceeding next to the quark–gluon vertex, written through first order as

$$-i\frac{g_3}{2}(\Gamma_\nu^a)_{ji}(p_2, p_1) = -i\frac{g_3}{2}\gamma_\nu(\lambda^a)_{ji} - ig_3(\Lambda_\nu^a)_{ji}(p_2, p_1) + \cdots,$$

(2.48)

[7] For notational simplicity, we discontinue displaying the superscript (MS) on renormalization constants.

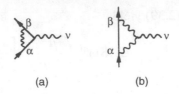

(a)   (b)

Fig. II–5 One-loop corrections to the quark–gluon vertex.

we see from Fig. II–5 that there are radiative corrections from both quark and gluon intermediate states. The quark contribution is

$$
-ig_3[\Lambda_\nu^a(p_2, p_1)]_{ji}\Big|_{\text{quark}} = \left(\frac{-ig_3}{2}\right)^3 \int \frac{d^4k}{(2\pi)^4} \frac{-ig^{\alpha\beta}}{k^2 + i\epsilon} (\lambda^b)_{jn}\gamma_\alpha
$$

$$
\times \frac{i}{\not{p}_2 - \not{k} - m + i\epsilon}(\lambda^a)_{nl}\gamma_\nu \frac{i}{\not{p}_1 - \not{k} - m + i\epsilon} (\lambda^b)_{li}\gamma_\beta. \quad (2.49)
$$

Aside from the replacement $e \to g_3$ and a color factor $\lambda^b\lambda^a\lambda^b/8$, which is evaluated in Eq. (2.17), the remaining expression is the *QED* vertex, which will be analyzed in detail in Sect. V–1. Thus we anticipate from Eq. (V–1.19) that at $p_1 = p_2 = p$ and $|p|^2 \gg m^2$,

$$
[\Lambda_\nu^a(p, p)]_{ji}\Big|_{\text{quark}} = (C_2(3) - \frac{1}{2}C_2(8)) \frac{g_3^2}{8\pi^2} \frac{1}{2\epsilon} \left(\frac{\mu^2}{-p^2}\right)^\epsilon (\lambda^a/2)_{ji}\gamma_\nu + \cdots .
$$
$$(2.50)$$

The two-gluon intermediate state, which has no counterpart in *QED*, has the form

$$
-ig_3(\Lambda_\nu^a(p_2, p_1))_{ji}\Big|_{\text{gluon}} = if_{abc}(\lambda^c\lambda^b)_{ji} \frac{g_3^3}{4} \int \frac{d^4k}{(2\pi)^4} \gamma^\beta(\not{k} + m)\gamma^\alpha
$$

$$
\times \frac{g_{\nu\beta}(2p_2 - k - p_1)_\alpha + g_{\beta\alpha}(2k - p_1 - p_2)_\nu + g_{\alpha\nu}(2p_1 - k - p_2)_\beta}{[k^2 - m^2 + i\epsilon][(p_1 - k)^2 + i\epsilon]^2[(p_2 - k)^2 + i\epsilon]}. \quad (2.51)
$$

By a now-standard set of steps, it is not difficult to extract the ϵ-pole from the extension of the above to d dimensions, and we find

$$
(\Lambda_\nu^a(p, p))_{ji}\Big|_{\text{gluon}} = (\lambda^a/2)_{ji}\gamma_\nu \frac{3}{2}C_2(8) \frac{g_3^2}{8\pi^2} \left(\frac{\mu^2}{-p^2}\right)^\epsilon \frac{1}{2\epsilon} + \cdots , \quad (2.52)
$$

implying a total vertex correction of the form,

$$
(\Lambda_\nu^a(p, p))_{ji}\Big|_{\text{tot}} = (\lambda^a/2)_{ji}\gamma_\nu [C_2(3) + C_2(8)] \frac{g_3^2}{8\pi^2} \left(\frac{\mu^2}{-p^2}\right)^\epsilon \frac{1}{2\epsilon} + \cdots .
$$
$$(2.53)$$

We thus determine the renormalization constant for the quark–gluon vertex at $p_i^2 = -\mu_R^2$ to be

$$Z_{1F} = 1 - [C_2(3) + C_2(8)] \frac{g_3^2}{8\pi^2} \left(\frac{\mu}{\mu_R}\right)^{2\epsilon} \frac{1}{2\epsilon} + \cdots . \tag{2.54}$$

There remains the task of determining Z_2. We shall leave this for an exercise (cf. Prob. II–3) and simply quote the result

$$Z_2 = 1 - C_2(3) \frac{g_3^2}{8\pi^2} \left(\frac{\mu}{\mu_R}\right)^{2\epsilon} \frac{1}{2\epsilon} + \cdots . \tag{2.55}$$

Asymptotic freedom and renormalization group

A striking property of *QCD* is *asymptotic freedom* [GrW 73a,b, Po 73]. This is the statement that, unlike the electric charge, the coupling constant $g_3(\mu_R)$ of color decreases as the scale of renormalization μ_R is increased. To demonstrate this, we first combine our results for Z_1, Z_2 and Z_3 to obtain the coupling renormalization constant Z_g,

$$g_{3,0} = Z_{1F} Z_2^{-1} Z_3^{-1/2} g_3 \equiv Z_g g_3,$$

$$Z_g = 1 - \frac{\alpha_s}{4\pi} \left(11 - \frac{2n_f}{3}\right) \left(\frac{\mu}{\mu_R}\right)^{2\epsilon} \frac{1}{2\epsilon} + \cdots , \tag{2.56}$$

where $\alpha_s \equiv g_3^2/(4\pi)$. From the ϵ^{-1} coefficient of Z_g, we learn that

$$\mu_R \frac{\partial g_3}{\partial \mu_R} = -\left[\frac{11}{3} C_2(8) - \frac{n_f}{2} C_2(3)\right] \frac{g_3^3}{16\pi^2} + \mathcal{O}(g_3^5), \tag{2.57a}$$

or equivalently,

$$\beta_{QCD} = -\left(11 - \frac{2n_f}{3}\right) \frac{g_3^3}{16\pi^2} + \mathcal{O}(g_3^5) \equiv -\beta_0 \frac{g_3^3}{16\pi^2} + \mathcal{O}(g_3^5). \tag{2.57b}$$

The sign of the leading term in β_{QCD} is negative for the six-flavor world $n_f = 6$, becoming positive only if the number of quark flavors exceeds 16. As we have already seen, the *QED* vacuum acts as a dielectric medium with dielectric constant $\epsilon_{QED} > 1$ because spontaneous creation of charged fermion–antifermion pairs results in screening (i.e., vacuum polarization) of electric charge. The dielectric property $\epsilon_{QED} > 1$ means that the *QED* vacuum has magnetic susceptibility $\mu_{QED} < 1$, and thus is a diamagnetic medium. The *QCD* vacuum is the recipient of similar effects from virtual quark–antiquark pairs, but *these are overwhelmed by contributions from virtual gluons*. As a result, the *QCD* vacuum is a paramagnetic medium ($\mu_{QCD} > 1$) and antiscreens ($\epsilon_{QCD} < 1$) color charge [Hu 81].

The effect of asymptotic freedom can be displayed most clearly by performing a renormalization group (RG) analysis on the $1PI$ amplitudes of the theory. A connected[8] renormalized Green's function is defined in coordinate space as

$$G^{(n_F, n_B)}(\{x\}) = \langle 0|T\left(\overline{\psi}^r(x_1) \dots A^r(x_n)\right)|0\rangle_{\text{conn}} \qquad (2.58)$$

where the numbers of quark and gluon fields are n_F, n_B, respectively, and for convenience we suppress color and Lorentz indices. We employ the same symbol $G^{(n_F, n_B)}$ for the momentum Green's function

$$(2\pi)^4 \delta^4(p_1 + \dots + p_n) G^{(n_F, n_B)}(\{p\}) = \int \prod_{j=1}^{n} (d^4x_j \, e^{-ip_j \cdot x_j}) \, G^{(n_F, n_B)}(\{x\})$$

$$(2.59)$$

where $n = n_F + n_B$. The $1PI$ amplitudes $\Gamma^{(n_F, n_B)}$ are obtained by removing the external-leg propagators from $G_{1PI}^{(n_F, n_B)}$,

$$G_{1PI}^{(n_F, n_B)} = \prod_{i'} D(p_{i'}) \prod_{j'} S(p_{j'}) \, \Gamma^{(n_F, n_B)}(\{p\}) \prod_{i} D(p_i) \prod_{j} S(p_j), \qquad (2.60)$$

where unprimed (primed) momenta represent initial (final) states. The relations of Eq. (2.33) imply for any renormalization scheme, which we need not specify yet, that

$$G^{(n_F, n_B)} = Z_2^{-n_F/2} Z_3^{-n_B/2} \, G_0^{(n_F, n_B)},$$
$$D = Z_3^{-1} \, D_0, \qquad S = Z_2^{-1} \, S_0, \qquad (2.61)$$

where the zero subscript denotes unrenormalized quantities. From this, we have

$$\Gamma^{(n_F, n_B)} = Z_2^{n_F/2} Z_3^{n_B/2} \Gamma_0^{(n_F, n_B)}, \qquad (2.62)$$

and the combination of terms

$$Z_2^{-n_F/2}(\mu_R) \, Z_3^{-n_B/2}(\mu_R) \, \Gamma^{(n_F, n_B)}(\{p\}, g_3(\mu_R), m(\mu_R), \xi(\mu_R); \mu_R) \qquad (2.63)$$

is therefore independent of the renormalization scale μ_R.

Let us now ascertain the behavior of $\Gamma^{(n_F, n_B)}$ in the deep Euclidean kinematic limit where all momenta $\{p\}$ are both spacelike (in order to avoid singularities) and very large compared to any other mass scale in the theory. To keep the situation as simple as possible, we omit the dependence of $\Gamma^{(n_F, n_B)}$ on both the quark-mass $m(\mu_R)$ and gauge $\xi(\mu_R)$ parameters.[9] Then from Eq. (2.59) we find in response to a scale transformation $p \to \lambda p$ that

---

[8] All the fields participating in a *connected* Green's function are affected by interactions; in a *disconnected* Green's function, one or more of the field quanta propagate freely.

[9] We shall define a 'running mass parameter' later, in Chap. XIV.

$$G^{(n_F,n_B)}(\{\lambda p\}, g_3(\mu_R); \mu_R) = \lambda^{4-n_B-3n_F/2}\, G^{(n_F,n_B)}(\{p\}, g_3(\mu_R); \mu_R/\lambda). \quad (2.64)$$

This behavior is almost that of a homogeneous function occurring in a scale-invariant theory. Canonical dimensions of the fields appear in the exponent of the scaling factor along with an additive factor of four arising from the four-momentum delta function in Eq. (2.59). However, in $G^{(n_F,n_B)}$, there is also an implicit dependence on λ due to the presence of the renormalization scale μ_R. The corresponding scaling property of the $1PI$ amplitude is found from Eqs. (2.63), (2.64) to be

$$\Gamma^{(n_F,n_B)}(\{\lambda p\}, g_3(\mu_R); \mu_R) = \lambda^{4-n_B-3n_F/2}\Gamma^{(n_F,n_B)}(\{p\}, g_3(\mu_R); \mu_R/\lambda) \quad (2.65)$$

or

$$\Gamma^{(n_F,n_B)}(\{\lambda p\}, g_3(\mu_R); \mu_R) = \lambda^{4-n_B-3n_F/2}\left(\frac{Z_3(\lambda\mu_R)}{Z_3(\mu_R)}\right)^{-n_B/2}$$
$$\times \left(\frac{Z_2(\lambda\mu_R)}{Z_2(\mu_R)}\right)^{-n_F/2} \Gamma^{(n_F,n_B)}(\{p\}, g_3(\lambda\mu_R); \mu_R). \quad (2.66)$$

This functional relationship can be converted to a differential RG equation by taking the λ-derivative of both sides and then setting $\lambda = 1$,

$$\left(\sum_i^n p_i \frac{\partial}{\partial p_i} + n_B(1 + \gamma_B) + n_F\left(\frac{3}{2} + \gamma_F\right) - 4 - \beta_{QCD}\frac{\partial}{\partial g_3}\right)\Gamma^{(n_F,n_B)} = 0,$$
$$(2.67)$$

where

$$\gamma_F = \mu_R\frac{\partial}{\partial \mu_R}\ln Z_2^{1/2}, \qquad \gamma_B = \mu_R\frac{\partial}{\partial \mu_R}\ln Z_3^{1/2} \qquad (2.68)$$

are called the *anomalous dimensions* of the respective fields and β_{QCD} is as in Eq. (2.57).

Let us now see how to obtain leading-order estimates for the above anomalous dimensions. To this order, the result for β_{QCD} is both gauge and renormalization scheme-independent. To start, we can use the result of Eq. (2.55) to determine γ_F,

$$\gamma_F = \frac{1}{2}\mu_R\frac{\partial \ln Z_2}{\partial \mu_R} = \frac{g_3^2}{16\pi^2}C_2(3) + \mathcal{O}(g_3^4), \qquad (2.69)$$

and analogously for γ_B. To solve the RG equation, we employ the variable $t = \ln \lambda$, where λ is the scaling parameter appearing in Eqs. (2.64)–(2.66), and introduce the *running coupling constant* $\overline{g}_3(t)$,

$$\frac{\partial \overline{g}_3}{\partial t} = \beta(\overline{g}_3), \qquad \overline{g}_3(0) = g_3. \qquad (2.70)$$

Then it is straightforward to verify that the solution to Eq. (2.67) is

$$\Gamma(\{e^t p\}, g_3(\mu_R); \mu_R) = e^{t(4 - n_B - 3n_F/2)} \mathcal{D}(t) \Gamma(\{p\}, \bar{g}_3(t); \mu_R), \qquad (2.71)$$

where

$$\mathcal{D}(t) = \exp\left(-\int_0^t dt' \left[n_B \gamma_B \left(\bar{g}_3(t')\right) + n_F \gamma_F \left(\bar{g}_3(t')\right)\right]\right) \qquad (2.72)$$

is the anomalous dimension factor. The scaling behavior of the $1PI$ amplitude is seen to have field dimensions with anomalous contributions in addition to the canonical values.

Despite naive expectations, the interaction strength at the scaled momentum is not the constant g_3, but rather the running coupling constant $\bar{g}_3$ whose magnitude decreases as the momentum is increased. Employing the lowest-order contribution for β_{QCD} in Eq. (2.57b), we can integrate Eq. (2.70) over the interval $t_1 < t < t_2$ to obtain

$$(\bar{g}_3(t_2))^{-2} - (\bar{g}_3(t_1))^{-2} = 2\left(11 - 2n_f/3\right)(t_2 - t_1)/16\pi^2, \qquad (2.73)$$

where n_f is the number of quark flavors having mass less than $\sqrt{t_2}$. It is conventional to express this relation in a somewhat different form. Defining a scale Λ at which $\bar{g}_3$ diverges and letting $\alpha_s(q^2) \equiv \bar{g}_3^2(q^2)/4\pi$, we have to lowest order,

$$\alpha_s(q^2) = \frac{4\pi}{(11 - 2n_f/3)} \frac{1}{\ln(q^2/\Lambda^2)} + \cdots, \qquad (2.74)$$

where n_f is the number of quark flavors with mass less than $\sqrt{q^2}$. Higher order contributions are discussed at the end of this section.

If $\alpha_s(q^2)$ continues to grow as q^2 is lowered, any perturbative representation of β_{QCD} ultimately becomes a poor approximation, and we can no longer integrate Eq. (2.70) with confidence. Although unproven, a popular working hypothesis is that the QCD coupling indeed continues to grow as the energy is lowered, leading to the phenomenon of quark confinement. In QED, the free parameter $\alpha(q \simeq 0) \simeq 1/137$ is quite small and expansions in powers of α converge rapidly. However QCD behaves differently. In particular, it is clear from Eq. (2.74) that α_s is not really a free parameter, but is instead inexorably related to some mass scale, e.g., Λ. This phenomenon, called *dimensional transmutation*, means that an energy such as Λ can effectively serve to replace the dimensionless quantity α_s in the formulae of QCD. Specifying QCD operationally requires not only a lagrangian but also a value for Λ. For example, QCD perturbation theory is useful only if 'large' mass scales M (i.e. those with $(\Lambda/M)^2 \ll 1$) are probed. Because the complexity of low-energy QCD has thus far prevented direct analytic solution of the theory, there have been substantial efforts to develop alternative approaches. These include

Table II–2. *Determinations of* $\alpha_s(M_Z)$.

Experiment	q [GeV]	$\alpha_s(q^2)$	$\alpha_s(M_Z)$
τ decays	1.777	0.330 ± 0.014	0.1197 ± 0.0016
DIS [F_2]	$2 \to 15$	$- - -$	0.1142 ± 0.0023
DIS [$e + p \to$ jets]	$6 \to 100$	$- - -$	0.1198 ± 0.0032
$Q\bar{Q}$ states	7.5	0.1923 ± 0.0024	0.1183 ± 0.0008
Υ decays	9.46	$0.184^{+0.015}_{-0.014}$	$0.1190^{+0.006}_{-0.005}$
e^+e^- jets & shapes	$14 \to 44$	$- - -$	0.1172 ± 0.0051
e^+e^- [ew]	91.17	0.1193 ± 0.0028	0.1193 ± 0.0028
e^+e^- jets & shapes	$91 \to 208$	$- - -$	0.1224 ± 0.0039

attempts to solve *QCD* numerically (lattice-gauge theory), phenomenological study of various theoretical constructs (potential, bag, Skyrme models), exploitation of the invariances contained in $\mathcal{L}_{QCD}$ (notably chiral and flavor symmetries), and consideration of the infinite color limit $N_c \to \infty$ as a first approximation to *QCD* (N_c^{-1} expansion). The first of these topics is beyond the scope of this book (e.g. see [GaL 10, DeD 10]), but the others will form the basis for much of our discussion.

Attempts to infer $\alpha_s(q^2)$ from experimental data are typically carried out under kinematic conditions for which a perturbative analysis of *QCD* presumably makes sense. Systems commonly used for this purpose include decays of the τ lepton, deep-inelastic scattering (DIS) structure functions, Υ decay, and hadronic event shapes and jet production in e^+e^- annihilation. Suppose, as is generally the case, a given process is computed to some order in *QCD* perturbation theory and regularized in the $\overline{\text{MS}}$ scheme. If such a theoretical expression is then used to fit the data with a q-value characteristic of the given process employed, an expression such as Eq. (2.74) can be used to determine Λ and $\alpha_s(q^2)$ can be evolved to different q. Since this operation depends on both the regularization procedure and the number of quark flavors n_f used in Eq. (2.74), a notation like $\Lambda^{(n_f)}_{\overline{\text{MS}}}$ would be precise. Unfortunately there is no uniformity in the rate of convergence of *QCD* perturbation theory from process to process. Thus, determinations of $\alpha_s(q^2)$ are affected by both theoretical and experimental uncertainties, and a scatter of quoted values results. Nonetheless, an impressive consistency now exists between determinations carried out for a variety of conditions. Table II–2 lists values of $\alpha_s(M_Z)$ as inferred from a diverse set of experimental inputs [Be 09], and Figure II–6, which has attained the status of a *QCD* icon, displays the same. The current world average at the Z-boson mass scale is [Be *et al.* 11]

$$\alpha_s(M_Z) = 0.1184 \pm 0.0007, \qquad (2.75)$$

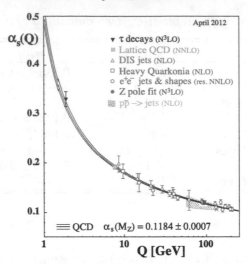

Fig. II–6 Energy dependence of $\alpha_s(Q)$, from [RPP 12] (used with permission).

which implies the value $\Lambda_{\overline{\text{MS}}}^{(5)} = 213 \pm 8$ MeV for the five-flavor sector of *QCD*. Determinations of $\alpha_s(q^2)$ have been found to be qualitatively in accord consistent with the predicted q^2 dependence of *QCD*. Taken over the full range of available data, values in the range $0.2 \leq \Lambda(\text{GeV}) \leq 0.4$ are not uncommon, e.g., $\Lambda_{\overline{\text{MS}}}^{(3)} = 339 \pm 10$ MeV and $\Lambda_{\overline{\text{MS}}}^{(4)} = 296 \pm 10$ MeV as cited in [Be *et al.* 11].

To conclude this section, we briefly comment on the status of higher-order contributions to the running of the strong fine structure constant. To date, analytic calculations on $\alpha_s(\mu)$ have been performed up to the four-loop level,

$$\mu^2 \frac{\partial}{\partial \mu^2} a_s = -\beta_0 a_s^2 - \beta_1 a_s^3 - \beta_2 a_s^4 - \beta_3 a_s^5 + \cdots, \tag{2.76}$$

in which $a_s \equiv \alpha_s/(4\pi)$ is the expansion parameter and exact expressions for the coefficients β_0, β_1, β_2 and β_3 appear in [RiVL 97]. The following useful approximations are also provided,

$$
\begin{aligned}
\beta_0 &\simeq 11 - 0.66667 n_f \\
\beta_1 &\simeq 102 - 12.6667 n_f \\
\beta_2 &\simeq 1428.50 - 279.61 n_f + 6.01852 n_f^2 \\
\beta_3 &\simeq 29243.0 - 6946.30 n_f + 405.089 n_f^2 + 1.49931 n_f^3,
\end{aligned}
\tag{2.77}
$$

where as usual n_f denotes the number of active flavors. The four-loop running of α_s can then be expressed as [ChKS 98],

$$\alpha_s(\mu_R^2) \simeq \frac{4\pi}{\beta_0 t}\left[1 - \frac{\beta_1}{\beta_0^2}\frac{\ln t}{t} + \frac{\beta_1^2(\ln^2 t - \ln t - 1) + \beta_0\beta_2}{\beta_4 t^2}\right.$$
$$\left.\frac{\beta_1^3\left(\ln^3 t - 2.5\ln^2 t - 2\ln t + 0.5\right) + 3\beta_0\beta_1\beta_2\ln t - 0.5\beta_0^2\beta_3}{\beta_0^6 t^3}\right],$$

$$\tag{2.78}$$

where $t \equiv \ln(\mu_R^2/\Lambda^2)$. As an example, let us use this (taking $n_f = 5$) to determine $\alpha_s(\mu)$ at three mass scales involving respectively the b quark, the Z boson, and the Higgs boson, i.e., $\mu_b = 4.18$ GeV, $\mu_Z = 91.1876$ GeV and $\mu_H = 125.5$ GeV,

$$\alpha_s(m_b) \simeq 0.2266, \qquad \alpha_s(M_Z) \simeq 0.1184, \qquad \alpha_s(M_H) \simeq 0.1129. \tag{2.79}$$

These values reflect the behavior expected from asymptotic freedom, as discussed earlier.[10] They will later be of use in discussing running quark mass (Chap. XIV) and Higgs-boson phenomenology (Chap. XV).

II–3 Electroweak interactions

The Weinberg–Salam–Glashow model [Gl 61, We 67b, Sa 69] is a gauge theory of the electroweak interactions whose input fermionic degrees of freedom are massless spin one-half chiral particles. It has the group structure $SU(2)_L \times U(1)$, where the $SU(2)_L$, $U(1)$ represent *weak isospin* and *weak hypercharge* respectively. The subscript 'L' on $SU(2)_L$ indicates that, among fermions, only left-handed states transform nontrivially under weak isospin.

Weak isospin and weak hypercharge assignments

First, we shall discuss how the fermionic weak isospin (T_w, T_{w3}) and weak hypercharge (Y_w) quantum numbers are assigned. The fermion generations are taken to obey a 'template' pattern – we assume that each succeeding generation differs from the first only in mass. Thus, it will suffice to consider just the lightest fermions for the remainder of this section. The first-generation electroweak assignments are displayed in Table II–3.

For weak isospin, experience gained from charged weak current interactions such as nuclear beta decay dictates that left-handed fermions belong to weak isodoublets while right-handed fermions be placed in weak isosinglets, as in

---

[10] Using the exact relations for $\beta_0, \ldots, \beta_4$ yields the same results to the stated level of accuracy.

Table II–3. $SU(2)_L \times U(1)$ *fermion assignments*

Particle	T_w	T_{w3}	Y_w
$\nu_{e,L}$	1/2	1/2	−1
e_L	1/2	−1/2	−1
$\nu_{e,R}$	0	0	0
e_R	0	0	−2
u_L	1/2	1/2	1/3
d_L	1/2	−1/2	1/3
u_R	0	0	4/3
d_R	0	0	−2/3

$$\text{leptons}: \quad \ell_L \equiv \begin{pmatrix} \nu_e \\ e \end{pmatrix}_L \quad \nu_{e,R} \quad e_R,$$

$$\text{quarks}: \quad q_L \equiv \begin{pmatrix} u \\ d \end{pmatrix}_L \quad u_R \quad d_R. \tag{3.1}$$

In view of nonzero neutrino mass, we include a right-handed neutrino. Each of the degrees of freedom displayed above must be assigned a weak hypercharge. There are *a priori* six in all,[11]

$$Y(q_L) \equiv Y_q, \quad Y(u_R) \equiv Y_u, \quad Y(d_R) \equiv Y_d,$$
$$Y(\ell_L) \equiv Y_\ell, \quad Y(e_R) \equiv Y_e, \quad Y(\nu_R) \equiv Y_\nu. \tag{3.2}$$

In the Standard Model one identifies the electromagnetic current, following spontaneous symmetry breaking in the electroweak sector, by its coupling to the linear combination of neutral gauge bosons having zero mass. The electric charge Q carried by a particle is thus *linearly* related to the $SU(2)_L \times U(1)_Y$ quantum numbers T_{w3} and Y_w,

$$aQ = T_{w3} + bY_w, \tag{3.3}$$

where a, b are constants. We can use the freedom in assigning the scale of the electric charge Q to choose $a = 1$. At this point, let us not assume any knowledge of the fermion electric charge values. Ultimately, however, the left-handed and right-handed components of the charged chiral fermions must unite to form the physical states themselves. Consistency demands that the electric charges of the chiral components of each such charged fermion be the same, whatever value that charge might have. Using Eq. (3.3), we find

---

[11] The reason that weak hypercharge engenders so many free parameters in contrast to weak isospin lies in the difference between an abelian gauge structure (like weak hypercharge) and one which is nonabelian (like weak isospin). Thus all doublets have the same weak isospin properties irrespective of their other properties, analogous to flavor independence in *QCD*. For the abelian group of weak hypercharge, the group structure by itself provides no guidelines for assigning the weak hypercharge quantum number. Like the electric charge, it is *a priori* an arbitrary quantity.

$$Y_q = Y_u - \frac{1}{2b} = Y_d + \frac{1}{2b}, \quad Y_\ell = Y_e + \frac{1}{2b} = Y_\nu - \frac{1}{2b}. \tag{3.4}$$

Additional information is contained in axial anomaly cancelation conditions, to be discussed in detail in Sect. III–3 (see especially Eq. (III–3.60b) and subsequent discussion). In particular, the cancelation requirement implies the conditions

$$\mathrm{Tr}\, F_3{}^2 Y_\mathrm{w} = 0, \tag{3.5a}$$

$$\mathrm{Tr}\, T_\mathrm{w3}^2 Y_\mathrm{w} = 0, \tag{3.5b}$$

$$\mathrm{Tr}\, Y_\mathrm{w}^3 = 0, \tag{3.5c}$$

where 'Tr' represents a sum over fermions and in Eq. (3.5a) F_3 is the third generator of the octet of color charges. These constraints imply

$$2Y_q - Y_u - Y_d = 0, \tag{3.6a}$$

$$3Y_q + Y_\ell = 0, \tag{3.6b}$$

$$2(3Y_q{}^3 + Y_\ell{}^3) - 3(Y_u{}^3 + Y_d{}^3) - Y_e^3 - Y_\nu^3 = 0, \tag{3.6c}$$

where the factors of '3' are color related and the minus signs arise from chirality dependence of the anomalies. Then, insertion of Eq. (3.4) into Eqs. (3.6b), (3.6c) yields

$$\left(bY_\ell + \frac{1}{2}\right)^3 - Y_\nu^3 = 0. \tag{3.7}$$

If neutrinos are Majorana particles (i.e. identical to their antiparticles), then they cannot carry electric charge and by Eq. (3.3), one has $Y_\nu = 0$. If so, Eq. (3.7) implies $bY_\ell = -1/2$, which fixes the remaining Y_i via Eq. (3.4). Thus, provided neutrinos are Majorana particles, once the weak isospin is chosen as in Eqs. (3.1), (3.2) and all possible chiral anomalies are arranged to cancel, one obtains a prediction for the fermion electric charge. We also learn that any attempt to determine weak hypercharge values from the known fermion electric charges is affected by an arbitrariness associated with the value of 'b'. This accounts for the variety of conventions seen in the literature. For definiteness, we have taken $b = 1/2$ in Eq. (3.3) and thus the relationship among the various quantum numbers in Table II–3 is

$$Y_\mathrm{w} = 2(Q - T_\mathrm{w3}). \tag{3.8}$$

On the other hand, if neutrinos are Dirac particles, it follows from Eq. (3.4) that Eq. (3.7) becomes a trivality and we learn nothing of weak hypercharge assignments from anomaly cancelation arguments. In this instance, one assigns the weak hypercharge by inserting the observed fermion electric charges into Eq. (3.8). The ability to *predict* $\{Q_i\}$ values has been lost.

$SU(2)_L \times U(1)_Y$ gauge-invariant lagrangian

Having assigned quantum numbers, we turn next to the electroweak interactions. The Weinberg–Salam lagrangian divides naturally into three additive parts, gauge (G), fermion (F), and Higgs (H),

$$\mathcal{L}_{WS} = \mathcal{L}_G + \mathcal{L}_F + \mathcal{L}_H. \tag{3.9}$$

Throughout this section we shall concentrate on establishing the general form of the electroweak sector, referring at times to only a few tree-level amplitudes. We shall return in Chap. V to the subject of electromagnetic radiative corrections, and present the electroweak Feynman rules along with various radiative corrections in Chap. XVI.

The gauge-boson fields, which couple to the weak isospin and weak hypercharge are, respectively, $\vec{W}_\mu = (W_\mu^1, \ W_\mu^2, \ W_\mu^3)$ and B_μ. These contribute to the purely gauge part of the lagrangian as

$$\mathcal{L}_G = -\frac{1}{4} F_i^{\mu\nu} F_{\mu\nu}^i - \frac{1}{4} B^{\mu\nu} B_{\mu\nu}, \tag{3.10}$$

where $F_{\mu\nu}^i$ ($i = 1,2,3$) is the $SU(2)$ field strength,

$$F_{\mu\nu}^i = \partial_\mu W_\nu^i - \partial_\nu W_\mu^i - g_2 \epsilon^{ijk} W_\mu^j W_\nu^k, \tag{3.11}$$

and $B_{\mu\nu}$ is the $U(1)$ field strength,

$$B_{\mu\nu} = \partial_\mu B_\nu - \partial_\nu B_\mu. \tag{3.12}$$

The fermionic sector of the lagrangian density includes both the left-handed and right-handed chiralities. Summing over left-handed weak isodoublets ψ_L and right-handed weak isosinglets ψ_R, we have

$$\mathcal{L}_F = \sum_{\psi_L} \overline{\psi}_L \, i\not{D} \, \psi_L + \sum_{\psi_R} \overline{\psi}_R \, i\not{D} \, \psi_R. \tag{3.13}$$

Since a right-handed chiral fermion does not couple to weak isospin, its covariant derivative has the simple form

$$D_\mu \psi_R = (\partial_\mu + i\frac{g_1}{2} Y_{\mathrm{w}} B_\mu)\psi_R. \tag{3.14}$$

This expression serves to define the $U(1)$ coupling g_1. Its normalization is dictated by our convention for weak hypercharge Y_{w}. The corresponding covariant derivative for the $SU(2)_L$ doublet ψ_L is

$$D_\mu \psi_L = \left(\mathbf{I} \left(\partial_\mu + i\frac{g_1}{2} Y_{\mathrm{w}} B_\mu \right) + i g_2 \frac{\vec{\tau}}{2} \cdot \vec{W}_\mu \right) \psi_L, \tag{3.15}$$

given in terms of the $SU(2)$ gauge coupling constant g_2 and the 2×2 matrices $\mathbf{I}$, $\vec{\tau}$.

We shall not display the quark color degree of freedom in this section for reasons of notational simplicity. However, it is understood that all situations in which quark internal degrees of freedom are summed over, as in Eq. (3.13), must include a color sum. Similarly, relations like Eq. (3.14) or Eq. (3.15) hold for each distinct internal color state when applied to quark fields.

The above equations define a mathematically consistent gauge theory of weak isospin and weak hypercharge. However, it is not a physically acceptable electroweak theory of Nature because the fermions and gauge bosons are massless. A Higgs sector must be added to the above lagrangians to arrive at the full Weinberg–Salam model. Thus, we introduce into the theory a complex doublet

$$\Phi = \begin{pmatrix} \varphi^+ \\ \varphi^0 \end{pmatrix} \tag{3.16}$$

of spin-zero Higgs fields with electric charge assignments as indicated. The quanta of these fields then each carry one unit of weak hypercharge. The Higgs lagrangian $\mathcal{L}_H$ is the sum of two kinds of terms, $\mathcal{L}_{HG}$ and $\mathcal{L}_{HF}$, which contain the Higgs–gauge and Higgs–fermion couplings respectively. The former is written as

$$\mathcal{L}_{HG} = (D^\mu \Phi)^* D_\mu \Phi - V(\Phi), \tag{3.17}$$

where

$$D_\mu \Phi = \left(\mathbf{1}\left[\partial_\mu + i\frac{g_1}{2} B_\mu \right] + ig_2 \frac{\vec{\tau}}{2} \cdot \vec{W}_\mu \right) \Phi, \tag{3.18}$$

and V is the Higgs self-interaction,

$$V(\Phi) = -\mu^2 \Phi^\dagger \Phi + \lambda (\Phi^\dagger \Phi)^2. \tag{3.19}$$

The parameters μ^2 and λ are positive but otherwise arbitrary. For simplicity, we write the Higgs–fermion interaction in this section for just the first generation of fermions. Denoting the left-handed quark and lepton doublets respectively as q_L and ℓ_L, we have

$$\mathcal{L}_{HF} = -g_u \bar{q}_L \widetilde{\Phi} u_R - g_d \bar{q}_L \Phi d_R - g_\nu \bar{\ell}_L \widetilde{\Phi} \nu_{e,R} - g_e \bar{\ell}_L \Phi e_R + \text{h.c.}, \tag{3.20}$$

where the coupling constants g_u, g_d, g_e and g_ν are arbitrary and we employ the charge conjugate to Φ,

$$\widetilde{\Phi} = i\tau_2 \Phi^*. \tag{3.21}$$

In a sense the Higgs potential V and Higgs–fermion coupling $\mathcal{L}_{HF}$ lie outside our guiding principle of gauge invariance because neither contains a gauge field. However, there is no principle which forbids such contributions, and their presence is phenomenologically required. Moreover, note that each is written in $SU(2)_L \times U(1)$ invariant form.

Spontaneous symmetry breaking

Mass generation for fermions and gauge bosons proceeds by means of spontaneous breaking of the $SU(2)_L \times U(1)$ symmetry. To begin, we obtain the ground-state Higgs configuration by minimizing the potential V to give

$$\Phi(-\mu^2 + 2\lambda\Phi^\dagger\Phi) = 0. \tag{3.22}$$

We interpret this ground-state relation in terms of vacuum expectation values, denoted by a zero subscript. Eq. (3.22) has two solutions, the trivial solution $\langle\Phi\rangle_0 = 0$ and the nontrivial solution,

$$\langle\Phi^\dagger\Phi\rangle_0 = \frac{v^2}{2}, \tag{3.23}$$

with

$$v \equiv \sqrt{\frac{\mu^2}{\lambda}}. \tag{3.24}$$

Let us consider the latter alternative. A nontrivial vacuum Higgs configuration, which obeys the constraint Eq. (3.23), respects conservation of electric charge, and describes the spontaneous symmetry breaking of the original $SU(2)_L \times U(1)$ symmetry is

$$\langle\Phi\rangle_0 = \begin{pmatrix} 0 \\ v/\sqrt{2} \end{pmatrix}. \tag{3.25}$$

In one interpretation, it is the order parameter for the Weinberg–Salam model, playing a role analogous to the magnetization in a ferromagnet. Group theoretically, it is seen to transform as a component of a weak isodoublet. The energy scale, v, of the effect is not predicted by the model and must be inferred from experiment.

The fermion and gauge-boson masses are determined by employing Eq. (3.25) for the Higgs field everywhere in the lagrangian $\mathcal{L}_H$. We first define charged fields $W_\mu^\pm$,

$$W_\mu^{\pm} = \sqrt{\frac{1}{2}}(W_\mu^1 \mp i W_\mu^2). \tag{3.26}$$

corresponding to the gauge bosons $W^\pm$. By substitution, we find for the mass contribution to the lagrangian

$$\mathcal{L}_{\text{mass}} = -\frac{v}{\sqrt{2}}(g_u\bar{u}u + g_d\bar{d}d + g_e\bar{e}e) + \left(\frac{vg_2}{2}\right)^2 W_\mu^+ W_-^\mu$$

$$+ \frac{v^2}{8}(W_\mu^3 \ B_\mu) \begin{pmatrix} g_2^2 & -g_1g_2 \\ -g_1g_2 & g_1^2 \end{pmatrix} \begin{pmatrix} W_3^\mu \\ B^\mu \end{pmatrix}. \tag{3.27}$$

The fermion masses are given by

$$m_f = \frac{v}{\sqrt{2}} g_f \quad (f = u, \, d, \, e, \, \ldots). \tag{3.28}$$

Although the theory can accommodate fermions of any mass, it does not predict the mass values. Instead, the measured fermion masses are used to fix the arbitrary Higgs–fermion couplings. The charged W-boson masses can be read off directly from Eq. (3.27),

$$M_W = \frac{v}{2} g_2, \tag{3.29}$$

but the symmetry breaking induces the neutral gauge bosons to undergo mixing. Their mass matrix is not diagonal in the basis of W^3, B states. Diagonalization occurs in the basis

$$Z_\mu = \cos \theta_{\rm w} \, W_\mu^3 - \sin \theta_{\rm w} \, B_\mu,$$
$$A_\mu = \sin \theta_{\rm w} \, W_\mu^3 + \cos \theta_{\rm w} \, B_\mu, \tag{3.30}$$

where the weak mixing angle (or *Weinberg angle*) $\theta_{\rm w}$ is defined by

$$\tan \theta_{\rm w} = \frac{g_1}{g_2}. \tag{3.31}$$

The neutral gauge-boson masses are found to be

$$M_\gamma = 0, \qquad M_Z = \frac{v}{2}\sqrt{g_1^2 + g_2^2}, \tag{3.32}$$

and the fields A_μ and Z_μ correspond to the massless photon and massive Z^0-boson, respectively. Observe that the $W^\pm$-to-Z^0 mass ratio is fixed by

$$\frac{M_W}{M_Z} = \cos \theta_{\rm w}. \tag{3.33}$$

Electroweak currents

Now that we have determined the mass spectrum of the theory in terms of the input parameters, we must next study the various gauge–fermion interactions. The traditional description of electromagnetic and low-energy charged weak interactions of spin one-half particles is expressed as

$$\mathcal{L}_{\rm int} = -e A_\mu J_{\rm em}^\mu - \frac{G_F}{\sqrt{2}} J_{\rm ch}^{\mu\dagger} J_\mu^{\rm ch}, \tag{3.34}$$

where $J_{\rm em}^\mu$ is the electromagnetic current

$$J_{\rm em}^\mu = -\bar{e}\gamma^\mu e + \frac{2}{3}\bar{u}\gamma^\mu u - \frac{1}{3}\bar{d}\gamma^\mu d + \cdots, \tag{3.35}$$

$J_{\rm ch}^\mu$ is the charged weak current (ignoring quark mixing)

$$J_{\text{ch}}^{\mu} = \overline{\nu}_e \gamma^{\mu}(1 + \gamma_5)e + \overline{u}\gamma^{\mu}(1 + \gamma_5)d + \cdots, \tag{3.36}$$

and $G_F \simeq 1.166 \times 10^{-5}$ GeV^{-2} is the Fermi constant (cf. Sect. V–2).

Alternatively, we can use Eqs. (3.13)–(3.15) to obtain the charged and neutral interactions in the $SU(2)_L \times U(1)_Y$ description,

$$\mathcal{L}_{\text{int}}' = -\frac{g_2}{\sqrt{8}}\left(W_{\mu}^+ J_{\text{ch}}^{\mu} + W_{\mu}^- J_{\text{ch}}^{\mu\dagger}\right) - g_2 W_{\mu}^3 J_{\text{w}3}^{\mu} - g_1 B_{\mu}(J_{\text{em}}^{\mu} - J_{\text{w}3}^{\mu}), \tag{3.37}$$

where $J_{\text{w}3}^{\mu}$ is the third component of the weak isospin current,

$$\vec{J}_{\text{w}}^{\mu} = \sum_{\psi_L} \overline{\psi}_L \gamma^{\mu} \frac{\vec{\tau}}{2} \psi_L, \tag{3.38}$$

summed over all left-handed fermion weak isodoublets. Substituting for B_{μ} and W_{μ}^3 in Eq. (3.37) in terms of A_{μ} and Z_{μ} yields

$$\mathcal{L}_{\text{int}}' = -\frac{g_2}{\sqrt{8}}\left(W_{\mu}^+ J_{\text{ch}}^{\mu} + W_{\mu}^- J_{\text{ch}}^{\mu\dagger}\right) - g_1\cos\theta_{\text{w}} A_{\mu} J_{\text{em}}^{\mu} + \mathcal{L}_{\text{ntl}-\text{wk}}, \tag{3.39}$$

where $J_{\text{ch}}^{\mu} = 2J_{\text{w},1+i2}^{\mu}$ is given in Eq. (3.36) and the neutral weak interaction $\mathcal{L}_{\text{ntl}-\text{wk}}$ for fermion f is[12]

$$\mathcal{L}_{\text{ntl}-\text{wk}}^{(f)} = -\frac{g_2}{2\cos\theta_{\text{w}}} Z^{\mu} \bar{f}(g_{\text{v}}^{(f)}\gamma_{\mu} + g_{\text{a}}^{(f)}\gamma_{\mu}\gamma_5)f,$$

$$g_{\text{v}}^{(f)} \equiv T_{\text{w}3}^{(f)} - 2\sin^2\theta_{\text{w}}\, Q_{\text{el}}^{(f)}, \qquad g_{\text{a}}^{(f)} \equiv T_{\text{w}3}^{(f)}. \tag{3.40}$$

Specifically, we have for the vector and axial-vector couplings

$$
\begin{aligned}
g_{\text{v}}^{(e,\mu,\tau)} &= -\frac{1}{2} + 2\sin^2\theta_{\text{w}}, & g_{\text{a}}^{(e,\mu,\tau)} &= -\frac{1}{2}, \\
g_{\text{v}}^{(u,c,t)} &= \frac{1}{2} - \frac{4}{3}\sin^2\theta_{\text{w}}, & g_{\text{a}}^{(u,c,t)} &= \frac{1}{2}, \\
g_{\text{v}}^{(d,s,b)} &= -\frac{1}{2} + \frac{2}{3}\sin^2\theta_{\text{w}}, & g_{\text{a}}^{(d,s,b)} &= -\frac{1}{2}, \\
g_{\text{v}}^{(\nu_e,\nu_{\mu},\nu_{\tau})} &= \frac{1}{2}, & g_{\text{a}}^{(\nu_e,\nu_{\mu},\nu_{\tau})} &= \frac{1}{2}.
\end{aligned}
\tag{3.41}
$$

Observe the structure of the neutral weak couplings $g_{\text{v,a}}^{(f)}$. If θ_{w} were to vanish, neutral weak interactions would be given strictly in terms of $T_{\text{w}3}$, the third component of weak isospin. However in the real world, phenomena like low-energy neutrino

---

[12] One should be careful not to confuse Eq. (3.40) with the alternate form

$$\mathcal{L}_{\text{ntl}-\text{wk}}^{(f)} = -eZ^{\mu}\bar{\psi}_f(v_f\gamma_{\mu} + a_f\gamma_{\mu}\gamma_5)\psi_f,$$

$$v_f = \frac{T_{\text{w}3}^{(f)} - 2\sin^2\theta_{\text{w}}\, Q_{\text{el}}^{(f)}}{2\sin\theta_{\text{w}}\cos\theta_{\text{w}}}, \qquad a_f = \frac{T_{\text{w}3}^{(f)}}{2\sin\theta_{\text{w}}\cos\theta_{\text{w}}},$$

which also appears in the literature.

interactions, M_W/M_Z, deep inelastic lepton scattering data, etc. all depend on the value of θ_w. In addition, we note that because $\sin^2 \theta_w \simeq 0.23$, the leptonic vector coupling constants $g_v^{(e,\mu,\tau)}$ are substantially suppressed relative to the axial-vector couplings.

Comparison of Eq. (3.34) with Eq. (3.39) yields

$$e = g_1 \cos \theta_w = g_2 \sin \theta_w. \tag{3.42}$$

The Fermi interaction of Eq. (3.34) corresponds in the Weinberg–Salam model to a second-order interaction mediated by W-exchange and evaluated in the limit of small momentum transfer $(1 \gg q^2/M_W^2)$,

$$\frac{G_F}{\sqrt{2}} = \frac{g_2^2}{8M_W^2}. \tag{3.43}$$

Together, these relations provide a tree-level expression for the W-boson mass,

$$M_W^2 = \frac{1}{\sin^2 \theta_w} \frac{\pi \alpha}{\sqrt{2}G_F} \simeq \left(\frac{37.281 \text{ GeV}}{\sin \theta_w} \right)^2. \tag{3.44}$$

Also, Eqs. (3.29), (3.43) imply

$$v = 2^{-1/4} G_F^{-1/2} \simeq 246.221(2) \text{ GeV}. \tag{3.45}$$

It is the quantity v which sets the scale of spontaneous symmetry breaking in the $SU(2)_L \times U(1)$ theory, and all masses in the Standard Model are proportional to it, although with widely differing coefficients.

We shall resume in Chaps. XV, XVI discussion of a number of topics introduced in this section, among them the Higgs scalar, the $W^\pm$ and Z^0 gauge bosons, and phenomenology of the neutral weak current. Also included will be a description of quantization procedures for the electroweak sector, including the issue of radiative corrections. First, however, in the intervening chapters we shall encounter a number of applications involving light fermions undergoing electroweak interactions at very modest energies and momentum transfers. For these it will suffice to work with just tree-level $W^\pm$ and/or Z^0 exchange, and to consider only photonic or gluonic radiative corrections. We shall also neglect the gauge-dependent longitudinal polarization contributions to the gauge-boson propagators (analogous to the $q^\mu q^\nu$ term in the photon propagator in Eq. (1.18)), as well as effects of the Higgs degrees of freedom. For photon propagators, the $q^\mu q^\nu$ terms do not contribute to physical amplitudes because of current conservation. Although current conservation is generally not present for the weak interactions, both the $q^\mu q^\nu$ propagator terms and Higgs contributions are suppressed by powers of $(m_f/M_W)^2$ for an external fermion of mass m_f.

II–4 Fermion mixing

In our discussion of the Weinberg–Salam model, we limited the number of fermion generations to one. We now lift that restriction and consider the implication of having n generations. Although the existing experimental situation supports the value $n = 3$, we shall take n arbitrary in our initial analysis.

Diagonalization of mass matrices

To begin, it is necessary to generalize the Higgs–fermion lagrangian $\mathcal{L}_{HF}$ of Eq. (3.20) to

$$-\mathcal{L}_{HF} = g_u^{\alpha\beta}\, \bar{q}'_{L,\alpha}\widetilde{\Phi}u'_{R,\beta} + g_d^{\alpha\beta}\, \bar{q}'_{L,\alpha}\Phi d'_{R,\beta} + g_v^{\alpha\beta}\, \bar{\ell}'_{L,\alpha}\widetilde{\Phi}v'_{R,\beta}$$
$$+ g_e^{\alpha\beta}\, \bar{\ell}'_{L,\alpha}\Phi e'_{R,\beta} + \text{h.c.}, \tag{4.1}$$

where we employ the summation convention $\alpha, \beta = 1, \dots, n$, and adopt the notation

$$\vec{u}' = (u', c', t', \dots),$$
$$\vec{d}' = (d', s', b', \dots),$$
$$\vec{v}' = (v_e, v_\mu, v_\tau, \dots),$$
$$\vec{e}' = (e', \mu', \tau', \dots),$$
$$\vec{q}' = \left(\begin{pmatrix} u' \\ d' \end{pmatrix}, \begin{pmatrix} c' \\ s' \end{pmatrix}, \begin{pmatrix} t' \\ b' \end{pmatrix}, \dots\right),$$
$$\vec{\ell}' = \left(\begin{pmatrix} v_e \\ e' \end{pmatrix}, \begin{pmatrix} v_\mu \\ \mu' \end{pmatrix}, \begin{pmatrix} v_\tau \\ \tau' \end{pmatrix}, \dots\right). \tag{4.2}$$

Observe that we denote the individual neutrino flavor eigenstates as v_e, v_μ, v_τ, with no primes. The states which appear in the original gauge-invariant lagrangian are generally *not* the mass eigenstates. That is, there is no reason why the $n \times n$ generational coupling matrices $\mathbf{g}_u, \mathbf{g}_d, \mathbf{g}_v, \mathbf{g}_e$ should be diagonal. Following spontaneous symmetry breaking, we obtain the generally nondiagonal $n \times n$ mass matrices $\mathbf{m}'_u, \mathbf{m}'_d, \mathbf{m}'_v, \mathbf{m}'_e$ from the analog of Eq. (3.28),

$$\mathbf{m}'_f = \frac{v}{\sqrt{2}}\,\mathbf{g}_f \qquad (f = u, d, v, e). \tag{4.3}$$

Although not diagonal in the *flavor basis*, these matrices can be brought to diagonal form in the *mass basis*. The transformation from flavor eigenstates to mass eigenstates is accomplished by means of the steps

$$-\mathcal{L}_{F,\,\text{mass}} = \vec{\bar{u}}_L' \, \mathbf{m}_u' \vec{u}_R' + \vec{\bar{d}}_L' \, \mathbf{m}_d' \vec{d}_R' + \vec{\bar{v}}_L' \, \mathbf{m}_v' \vec{v}_R' + \vec{\bar{e}}_L' \, \mathbf{m}_e' \vec{e}_R' + \text{h.c.},$$

$$= \vec{\bar{u}}_L' \, \mathbf{S}_L^u \mathbf{S}_L^{u\dagger} \mathbf{m}_u' \mathbf{S}_R^u \mathbf{S}_R^{u\dagger} \, \vec{u}_R' + \vec{\bar{d}}_L' \, \mathbf{S}_L^d \mathbf{S}_L^{d\dagger} \mathbf{m}_d' \mathbf{S}_R^d \mathbf{S}_R^{d\dagger} \, \vec{d}_R'$$

$$+ \vec{\bar{v}}_L' \, \mathbf{S}_L^v \mathbf{S}_L^{v\dagger} \mathbf{m}_v' \mathbf{S}_R^v \mathbf{S}_R^{v\dagger} \, \vec{v}_R' + \vec{\bar{e}}_L' \, \mathbf{S}_L^e \mathbf{S}_L^{e\dagger} \mathbf{m}_e' \mathbf{S}_R^e \mathbf{S}_R^{e\dagger} \, \vec{e}_R' + \text{h.c.}$$

$$= \vec{\bar{u}}_L \, \mathbf{m}_u \, \vec{u}_R + \vec{\bar{d}}_L \, \mathbf{m}_d \, \vec{d}_R + \vec{\bar{v}}_L \, \mathbf{m}_v \, \vec{v}_R + \vec{\bar{e}}_L \, \mathbf{m}_e \, \vec{e}_R + \text{h.c.}$$

$$= \vec{\bar{u}} \, \mathbf{m}_u \, \vec{u} + \vec{\bar{d}} \, \mathbf{m}_d \, \vec{d} + \vec{\bar{v}} \, \mathbf{m}_v \, \vec{v} + \vec{\bar{e}} \, \mathbf{m}_e \, \vec{e}. \tag{4.4}$$

The $n \times n$ unitary matrices $S_{L,R}^\alpha$ $(\alpha = u,\ d,\ v,\ e)$ relate the basis states,

$$\vec{u}_L' = \mathbf{S}_L^u \, \vec{u}_L, \quad \vec{d}_L' = \mathbf{S}_L^d \, \vec{d}_L, \quad \vec{v}_L' = \mathbf{S}_L^v \, \vec{v}_L, \quad \vec{e}_L' = \mathbf{S}_L^e \, \vec{e}_L,$$

$$\vec{u}_R' = \mathbf{S}_R^u \, \vec{u}_R, \quad \vec{d}_R' = \mathbf{S}_R^d \, \vec{d}_R, \quad \vec{v}_R' = \mathbf{S}_R^v \, \vec{v}_R, \quad \vec{e}_R' = \mathbf{S}_R^e \, \vec{e}_R, \tag{4.5}$$

and induce the biunitary diagonalizations

$$\mathbf{m}_\alpha' = \mathbf{S}_L^\alpha \, \mathbf{m}_\alpha \, \mathbf{S}_R^{\alpha\dagger}, \qquad (\alpha = u, d, v, e), \tag{4.6}$$

thus yielding the diagonal quark mass matrices

$$\mathbf{m}_u = \begin{pmatrix} m_u & 0 & 0 & \dots \\ 0 & m_c & 0 & \dots \\ 0 & 0 & m_t & \dots \\ \vdots & \vdots & \vdots & \ddots \end{pmatrix}, \quad \mathbf{m}_d = \begin{pmatrix} m_d & 0 & 0 & \dots \\ 0 & m_s & 0 & \dots \\ 0 & 0 & m_b & \dots \\ \vdots & \vdots & \vdots & \ddots \end{pmatrix}, \tag{4.7a}$$

and the diagonal lepton mass matrices

$$\mathbf{m}_v = \begin{pmatrix} m_1 & 0 & 0 & \dots \\ 0 & m_2 & 0 & \dots \\ 0 & 0 & m_3 & \dots \\ \vdots & \vdots & \vdots & \ddots \end{pmatrix}, \quad \mathbf{m}_e = \begin{pmatrix} m_e & 0 & 0 & \dots \\ 0 & m_\mu & 0 & \dots \\ 0 & 0 & m_\tau & \dots \\ \vdots & \vdots & \vdots & \ddots \end{pmatrix}. \tag{4.7b}$$

Although the Weinberg–Salam model is first written down in terms of the flavor basis states, actual calculations which confront theory with experiment are performed using the mass basis states. We must then transform from one to the other. This turns out to have no effect on the structure of the electromagnetic and neutral weak currents. One simply omits writing the primes, which would otherwise appear. The reason is that (aside from mass) each generation is a replica of the others, and products of the unitary transformation matrices always give rise to the unit matrix in flavor space. Thus, at the lagrangian level, there are no *flavor-changing neutral currents* in the theory.

As an example of this, consider the leptonic contribution to the electromagnetic current,

$$
\begin{aligned}
J_{\text{em}}^{\mu}(\text{lept}) &= -\bar{e}_{\alpha}'\gamma^{\mu}e_{\alpha}' = -\bar{e}_{L,\alpha}'\gamma^{\mu}e_{L,\alpha}' - \bar{e}_{R,\alpha}'\gamma^{\mu}e_{R,\alpha}' \\
&= -(\bar{e}_L \mathbf{S}_L^{e\dagger})_{\alpha}\gamma^{\mu}(\mathbf{S}_L^e e_L)_{\alpha} - (\bar{e}_R \mathbf{S}_R^{e\dagger})_{\alpha}\gamma^{\mu}(\mathbf{S}_R^e e_R)_{\alpha} \\
&= -\bar{e}_{L,\alpha}\gamma^{\mu}e_{L,\alpha} - \bar{e}_{R,\alpha}\gamma^{\mu}e_{R,\alpha} = -\bar{e}_{\alpha}\gamma^{\mu}e_{\alpha},
\end{aligned}
\tag{4.8}
$$

where we sum over family index $\alpha = 1, \ldots, n$ and invoke the unitarity of matrices $\mathbf{S}_{L,R}^e$. Note that there is no difficulty in passing the $\mathbf{S}_{L,R}^e$ through γ^{μ} because the former matrices act in flavor space whereas the latter matrix acts in spin space.

Quark mixing

Thus far, the distinction between flavor basis states and mass eigenstates has been seen to have no apparent effect. However, mixing between generations does manifest itself in the system of quark charged weak currents,

$$
J_{\text{ch}}^{\mu}(\text{qk}) = 2\bar{u}_{L,\alpha}'\gamma^{\mu}d_{L,\alpha}' = 2\bar{u}_{L,\alpha}\gamma^{\mu}\mathbf{V}_{\alpha\beta}d_{L,\beta},
\tag{4.9}
$$

where

$$
\mathbf{V} \equiv \mathbf{S}_L^{u\dagger}\mathbf{S}_L^d.
\tag{4.10}
$$

The quark-mixing matrix $\mathbf{V}$, being the product of two unitary matrices, is itself unitary. The Standard Model does not predict the content of $\mathbf{V}$. Rather, its matrix elements must be phenomenologically extracted from data. For the two-generation case, $\mathbf{V}$ is called the *Cabibbo* matrix [Ca 63]. For three generations, it has been referred to as the *Kobayashi–Maskawa* (KM) matrix [KoM 73] after its originators, but is now usually denoted by the abbreviation 'CKM'. We shall analyze properties of such mixing matrices for the remainder of this section.

An $n \times n$ unitary matrix is characterized by n^2 real-valued parameters. Of these, $n(n-1)/2$ are angles and $n(n+1)/2$ are phases. Not all the phases have physical significance, because $2n - 1$ of them can be removed by *quark rephasing*. The effect of quark rephasing

$$
u_{L,\alpha} \to e^{i\theta_{\alpha}^u}u_{L,\alpha}, \qquad d_{L,\alpha} \to e^{i\theta_{\alpha}^d}d_{L,\alpha} \qquad (\alpha = 1, \ldots, n)
\tag{4.11}
$$

on an element of the mixing matrix is

$$
\mathbf{V}_{\alpha\beta} \to \mathbf{V}_{\alpha\beta}e^{i(\theta_{\beta}^d - \theta_{\alpha}^u)} \qquad (\alpha, \beta = 1, \ldots, n).
\tag{4.12}
$$

Since an overall common rephasing does not affect $\mathbf{V}$, only the $2n - 1$ remaining transformations of the type in Eq. (4.11) are effective in removing complex phases. This leaves $\mathbf{V}$ with $(n-1)(n-2)/2$ such phases. One must be careful to also transform the right-chirality fields of a given flavor in like manner to keep masses real. If so, all terms in the lagrangian other than $\mathbf{V}$ are unaffected by this procedure.

For two generations, there are no complex phases. The only parameter is commonly taken to be the *Cabibbo angle* θ_C and we write

$$\mathbf{V} = \begin{pmatrix} \cos\theta_C & \sin\theta_C \\ -\sin\theta_C & \cos\theta_C \end{pmatrix}. \tag{4.13}$$

A common notation for the $n = 2$ mixed states is

$$\begin{pmatrix} d_C \\ s_C \end{pmatrix} \equiv \mathbf{V}\begin{pmatrix} d \\ s \end{pmatrix}. \tag{4.14}$$

Within the two-generation approximation, weak interaction decay data imply the numerical value, $\sin\theta_C \simeq 0.226$.

The three-generation case involves the 3×3 matrix

$$\mathbf{V} = \begin{pmatrix} V_{ud} & V_{us} & V_{ub} \\ V_{cd} & V_{cs} & V_{cb} \\ V_{td} & V_{ts} & V_{tb} \end{pmatrix}, \tag{4.15}$$

which is a form that emphasizes the physical significance of each matrix element. The $n = 3$ mixing matrix can be expressed in terms of four parameters, of which one is a complex phase. The presence of a complex phase is highly significant because it signals the existence of *CP* violation in the theory. We shall return to this point shortly. The KM representation employs three mixing angles $\theta_{12}, \theta_{13}, \theta_{23}$ and a complex phase δ. It can be viewed as the following Eulerian construction of three matrices,

$$\mathbf{V} = \begin{pmatrix} 1 & 0 & 0 \\ 0 & c_{23} & s_{23} \\ 0 & -s_{23} & c_{23} \end{pmatrix}\begin{pmatrix} c_{13} & 0 & s_{13}e^{-i\delta} \\ 0 & 1 & 0 \\ -s_{13}e^{i\delta} & 0 & c_{13} \end{pmatrix}\begin{pmatrix} c_{12} & s_{12} & 0 \\ -s_{12} & c_{12} & 0 \\ 0 & 0 & 1 \end{pmatrix}, \tag{4.16}$$

where $s_{\alpha\beta} \equiv \sin\theta_{\alpha\beta}$, $c_{\alpha\beta} \equiv \cos\theta_{\alpha\beta}$ ($\alpha, \beta = 1, 2, 3$). In combined form this becomes

$$\mathbf{V} = \begin{pmatrix} c_{12}c_{13} & s_{12}c_{13} & s_{13}e^{-i\delta} \\ -s_{12}c_{23} - c_{12}s_{23}s_{13}e^{i\delta} & c_{12}c_{23} - s_{12}s_{23}s_{13}e^{i\delta} & s_{23}c_{13} \\ s_{12}s_{23} - c_{12}c_{23}s_{13}e^{i\delta} & -s_{23}c_{12} - s_{12}c_{23}s_{13}e^{i\delta} & c_{23}c_{13} \end{pmatrix}. \tag{4.17}$$

By means of quark rephasing, it can be arranged that the angles $\{\theta_{\alpha\beta}\}$ all lie in the first quadrant. In the limit $\theta_{23} = \theta_{13} = 0$, KM mixing reduces to Cabibbo mixing with the identification $\theta_{12} = \theta_C$.

An alternative approach for describing the quark mixing matrix, the *Wolfenstein parameterization* [Wo 83], expresses the mixing matrix as the unit 3×3 matrix together with a perturbative hierarchical structure organized by a smallness parameter λ. In the updated version, the Wolfenstein representation contains four parameters λ, A, ρ, η defined by

$$s_{12} \equiv \lambda, \quad s_{23} \equiv A\lambda^2, \quad s_{13}e^{-i\delta} \equiv A\lambda^3(\rho - i\eta). \tag{4.18}$$

These definitions hold to all orders in λ. Since many phenomenological applications require accuracy to the level of order λ^5, we write

$$\mathbf{V} = \begin{pmatrix} 1 - \dfrac{\lambda^2}{2} - \dfrac{\lambda^4}{4} & \lambda & \lambda^3 A(\rho - i\eta) \\ -\lambda + \dfrac{A^2\lambda^5}{2}(1 - 2(\rho + i\eta)) & 1 - \dfrac{\lambda^2}{2} - \dfrac{\lambda^4}{8}(1 + 4A^2) & \lambda^2 A \\ \lambda^3 A(1 - \bar{\rho} - i\bar{\eta}) & -\lambda^2 A + \dfrac{A\lambda^4}{2}(1 - 2(\rho + i\eta)) & 1 - \dfrac{A^2\lambda^4}{2} \end{pmatrix}.$$

$$\tag{4.19}$$

Observe that the matrix element V_{td} is expressed in terms of $\bar{\rho} \equiv \rho(1 - \lambda^2/2)$ and $\bar{\eta} \equiv \eta(1 - \lambda^2/2)$. These quantities, which are useful in generalizing the so-called unitarity triangle (cf. Sect. XIV–5) beyond leading order, are directly cited in modern fits of the CKM matrix.

Attempts to theoretically predict the content of the CKM matrix have not borne fruit. The CKM matrix elements have come to be thought of as basic quantities, much like particle masses and interaction coupling constants. As such, each matrix element must be determined experimentally (with several experiments per matrix element). This endeavor, which has been a preoccupation of 'Flavor Physics' for many years, has finally reached an acceptable level of sensitivity, particularly with the operation of several B-factories (cf. Chap. XIV). Current values [RPP 12] for the Wolfenstein parameters are

$$\lambda = 0.2257^{+0.0008}_{-0.0010}, \qquad A = 0.814^{+0.021}_{-0.022},$$
$$\bar{\rho} = 0.135^{+0.031}_{-0.016}, \qquad \bar{\eta} = 0.349^{+0.015}_{-0.017}. \tag{4.20a}$$

Alternatively, we have for the original parameter set,

$$s_{12} = 0.2257^{+0.0008}_{-0.0010}, \qquad s_{23} = 0.0415^{+0.0014}_{-0.0015},$$
$$s_{13} = 0.0036^{+0.0004}_{-0.0003}, \qquad \delta = \left(68.9^{+3.0}_{-5.4}\right)^o. \tag{4.20b}$$

Neutrino mixing

Flavor mixing affects not only the quarks, but also the leptons, in the form of neutrino mixing. Just as the 3×3 quark mixing matrix $\mathbf{V}$ is associated with the acronym 'CKM', there will be a 3×3 lepton mixing matrix $\mathbf{U}$ for neutrinos. Its standard acronym 'PMNS', acknowledges the early work of Pontecorvo [Po 68] and of Maki, Nakagawa, and Sakata [MaNS 62]. As we will show in Sect. VI–2, when we include the possibility of a Majorana nature of neutrino mass, lepton mixing has a form very similar to quark mixing,

$$\mathbf{U} = \mathbf{V}^{(\nu)} \mathcal{P}_\nu, \tag{4.21}$$

where $\mathbf{V}^{(\nu)}$ has the same mathematical content as the quark mixing matrix $\mathbf{V}$ of Eq. (4.17) except that the mixing angles $\{\theta_{ij}\}$ and phase δ now pertain to the neutrino sector and

$$\mathcal{P}_\nu = \begin{pmatrix} 1 & 0 & 0 \\ 0 & e^{i\alpha_1/2} & 0 \\ 0 & 0 & e^{i\alpha_2/2} \end{pmatrix}. \tag{4.22}$$

Here, the $\{\alpha_i\}$ are so-called Majorana phases. They are physical, i.e., observable, if the Majorana neutrino option is chosen by Nature. Although not contributing to neutrino oscillations, they will occur in the neutrinoless double beta decay (cf. Sect. VI–5) of certain nuclei.

Thus far, information about the lepton mixing matrix has come from fits to neutrino oscillation data (although highly anticipated searches for neutrinoless double beta decay are underway, cf. Sect. VI–5). There is no evidence at this time for *CP* violation in the lepton sector, so one cannot yet distinguish between the Dirac and Majorana cases described above. For either, the leptonic mixing angles are measured to be [RPP 12][13]

$$\sin^2(2\theta_{12}) = 0.857 \pm 0.024,$$
$$\sin^2(2\theta_{23}) \geq 0.95 \ \text{(at 90\%C.L.)},$$
$$\sin^2(2\theta_{13}) = 0.098 \pm 0.013, \tag{4.23}$$

which translates into angles (here we take the value for θ_{23} from Table 13.7 of [RPP 12])

$$\theta_{12} = (33.9 \pm 1.0)^\circ, \quad \theta_{23} = \left(40.4^{+4.6}_{-1.8}\right)^\circ, \quad \theta_{13} = (9.1 \pm 0.6)^\circ. \tag{4.24}$$

These values are quite different from the quark mixing angles inferred from Eq. (4.20b). Whereas the quark mixing matrix is a 'zeroth-order' unit 3×3 matrix modified by perturbative entries proportional to powers of the smallness parameter λ, current data for lepton mixing are consistent with the 'zeroth-order' representation[14]

---

[13] See also [FoTV 12] and [FoLMMPR 12].
[14] This form, referred to as tri-bimaximal mixing [HaPS 02], is used here as simply a numerical convenience.

$$\mathbf{V}_0^{(\nu)} = \begin{pmatrix} \frac{2}{\sqrt{6}} & \frac{1}{\sqrt{3}} & 0 \\ -\frac{1}{\sqrt{6}} & \frac{1}{\sqrt{3}} & \frac{1}{\sqrt{2}} \\ \frac{1}{\sqrt{6}} & -\frac{1}{\sqrt{3}} & \frac{1}{\sqrt{2}} \end{pmatrix}. \tag{4.25}$$

Perturbative modifications of this can be introduced as

$$\sin\theta_{13} = \frac{r}{\sqrt{2}}, \qquad \sin\theta_{12} = \frac{1+s}{\sqrt{3}}, \qquad \sin\theta_{23} = \frac{1+a}{\sqrt{2}}, \tag{4.26}$$

where the r, s, a parameters are sensitive, in part, to reactor, solar, and atmospheric data, yielding $\mathbf{V}_0^{(\nu)} \rightarrow \mathbf{V}^{(\nu)}$, where [KiL 13]

$$\mathbf{V}^{(\nu)} = \begin{pmatrix} \frac{2}{\sqrt{6}}\left(1 - \frac{s}{2}\right) & \frac{1}{\sqrt{3}}(1+s) & \frac{r}{\sqrt{2}}e^{-i\delta} \\ -\frac{1}{\sqrt{6}}\left(1+s-a+re^{i\delta}\right) & \frac{1}{\sqrt{3}}\left(1 - \frac{s}{2} - a - \frac{1}{2}re^{i\delta}\right) & \frac{1}{\sqrt{2}}(1+a) \\ \frac{1}{\sqrt{6}}\left(1+s+a-re^{i\delta}\right) & -\frac{1}{\sqrt{3}}\left(1 + \frac{s}{2} + a + \frac{r}{2}e^{i\delta}\right) & \frac{1}{\sqrt{2}}(1-a) \end{pmatrix}. \tag{4.27}$$

In the above δ is the phase parameter which reflects the possibility of *CP* violation in the lepton sector but for which there is, as of yet, no evidence. For the others, the current limits

$$r = 0.22 \pm 0.01, \qquad s = -0.03 \pm 0.03, \qquad a = 0.10 \pm 0.05, \tag{4.28}$$

imply that we are not very far from this tri-bimaximal form.

Quark CP violation and rephasing invariants

There is no unique parameterization for three-generation quark mixing. Any scheme which is convenient to the situation at hand may be employed as long as it is used consistently and adheres to the underlying principles. There is, however, a somewhat different logical position to adopt, that of working solely with *rephasing invariants*. After all, only those functions of $\mathbf{V}$ which are invariant under the rephasing operation in Eq. (4.18) can be observable. An obvious set of quadratic invariants are the squared moduli $\Delta_{ij}^{(2)} \equiv |V_{ij}|^2$ where $i, j = 1, 2, 3$. The unitarity conditions $\mathbf{V}^\dagger\mathbf{V} = \mathbf{V}\mathbf{V}^\dagger = \mathbf{I}$ constrain the number of independent squared-moduli to four. They are of course all real-valued. In addition there are quartic functions $\Delta_{ab}^{(4)} \equiv V_{ij}V_{kl}V_{il}^*V_{kj}^*$, where we suspend the summation convention for repeated indices and, to avoid redundant factors of squared-moduli, take a, i, k (b, j, l) cyclic. There are yet higher-order invariants, but they are all expressible in terms of the quadratic and quartic functions. The nine quantities $\Delta_{ab}^{(4)}$ are generally complex-valued. A unique measure of *CP* violations for three generations is provided by the rephasing invariant $\text{Im}\Delta^{(4)}_{ab}$,

$$\text{Im}\,[V_{ij}V_{kl}V^*_{il}V^*_{kj}] = J\sum_{m,n}\epsilon_{ikm}\,\epsilon_{jln}, \tag{4.29}$$

where J is the so-called Jarlskog invariant [Ja 85],

$$J = c_{12}c^2_{13}c_{23}s_{12}s_{13}s_{23}s_\delta = \lambda^6 A^2\bar{\eta} + \mathcal{O}(\lambda^8) = (2.96^{+0.20}_{-0.16}) \times 10^{-5}. \tag{4.30}$$

This combination of quark mixing parameters will always appear in calculations of *CP*-violating phenomena. To have nonzero *CP* violating effects, the KM angles must avoid the values $\theta_{ij} = 0$, $\pi/2$, and $\delta = 0$, π. The *CP*-violating invariant J achieves its maximum value for $c_{13} = 2/\sqrt{3}$, $c_{12} = c_{23} = 1/\sqrt{2}$, $s_\delta = 1$ at which it equals $1/6\sqrt{3}$. This set of circumstances is very unlike the real-world value in Eq. (4.30).

The consideration of rephasing invariants need not involve just the mixing matrix **V**, but can also be applied to the $Q = 2/3$, $-1/3$ nondiagonal mass matrices $\mathbf{m}'_u$, $\mathbf{m}'_d$ themselves. In particular, the determinant of their commutator is found to provide an invariant measure of *CP* violations [Ja 85]. If, for simplicity, we work in a basis where $\mathbf{m}_u'$, $\mathbf{m}'_d$ are hermitian, it can be shown that $\mathbf{S}^{u,d}_L = \mathbf{S}^{u,d}_R \equiv \mathbf{S}^{u,d}$. Thus we have

$$[\mathbf{m}'_u, \mathbf{m}'_d] = \mathbf{S}^u\,[\mathbf{m}_u, \mathbf{V}\mathbf{m}_d\mathbf{V}^\dagger]\,\mathbf{S}^{u\dagger} = \mathbf{S}^u\mathbf{V}\,[\mathbf{V}^\dagger\mathbf{m}_u\mathbf{V}, \mathbf{m}_d]\,\mathbf{V}^\dagger\mathbf{S}^{u\dagger}, \tag{4.31}$$

from which it follows that

$$\det\,[\mathbf{m}'_u, \mathbf{m}'_d] = \det\,[\mathbf{m}_u, \mathbf{V}\mathbf{m}_d\mathbf{V}^\dagger] = \det\,[\mathbf{V}^\dagger\mathbf{m}_u\mathbf{V}, \mathbf{m}_d]. \tag{4.32}$$

The two commutators on the right-hand sides of this relation are skew-hermitian and each of their matrix elements is multiplied by a $Q = 2/3, -1/3$ quark mass difference, respectively. The determinant is thus proportional to the product of all $Q = 2/3$, $-1/3$ quark mass differences, and explicit evaluation reveals

$$\det\,[\mathbf{m}'_u, \mathbf{m}'_d] = 2i\,\text{Im}\Delta^{(4)}\prod_{\alpha>\beta}(m_{u,\alpha} - m_{u,\beta})(m_{d,\alpha} - m_{d,\beta}). \tag{4.33}$$

This provides a more extensive list of necessary conditions for *CP* violations to be present. Not only are the mixing angles constrained as discussed above, but also the quark masses within a given charge sector must not exhibit degeneracies.

Our discussion of the $n = 2, 3$ generation cases suggests how larger systems $n = 4, \dots$ can be addressed, although the number of parameters becomes formidable, e.g., four generations require six mixing angles and three complex phases. However, existing data indicate the existence of just three fermion generations, e.g., measurements from Z^0-decay fail to see additional neutrinos, and there is no evidence from e^+e^- or $\overline{p}p$ collisions for additional quarks or leptons lighter than roughly 50 GeV. Moreover, if there were more than three quark generations the full quark-mixing matrix would be unitary, but one would expect to see violations of unitarity in any submatrix. Yet to the present level of sensitivity, the 3×3 KM mixing matrix obeys the unitarity constraint. The most accurate data occur in the $(V^\dagger V)_{11} = 1$ sector. Here, the contribution from V_{ub} is negligible and one finds [Ma 11, HaT 10]

$$(V^\dagger V)_{11} = |V_{ud}|^2 + |V_{us}|^2 + |V_{ub}|^2 = 0.9999(6). \qquad (4.34)$$

Problems

(1) **SU(3)**

    (a) Starting from the general form $\lambda_{ij}^a \lambda_{kl}^a = A\delta_{ij}\delta_{kl} + B\delta_{il}\delta_{jk} + C\delta_{ik}\delta_{jl}$ ($a = 1, \ldots, 8$ is summed), determine A, B, C by using the trace relations of Eqs. (II–2.5a, 2.5b), etc.

    (b) Determine $\mathrm{Tr}\, \lambda^a \lambda^b \lambda^c$.

    (c) Determine $\epsilon_{ijk}\epsilon_{lmn}\lambda_{mp}^a \lambda_{pk}^b \lambda_{ni}^a \lambda_{lj}^b$.

    (d) Consider the 8×8 matrices $(F_a)_{bc} = -if_{abc}$, where the $\{f_{abc}\}$ are $SU(3)$ structure constants ($a, b, c = 1, \ldots, 8$). Show that these matrices (the *regular* or *adjoint* representation) obey the Lie algebra of $SU(3)$, and determine $\mathrm{Tr}\, F_a F_b$.

(2) **Gauge invariance and the QCD interaction vertices**

    (a) Define constants f, g such that the covariant derivative of quark q is $D_\mu q = (\partial_\mu + if A_\mu)q$ and the QCD gauge transformations are $A_\mu \to UA_\mu U^{-1} + ig^{-1}U\partial_\mu U^{-1}$ and $q \to Uq$, where A_μ are the gauge fields in matrix form (cf. Eq. (I–5.15)). Show that $\bar{q}\,\slashed{D}\,q$ is invariant under a gauge transformation only if $f = g$.

    (b) Define a constant h such that the QCD field strength is $F_{\mu\nu} = \partial_\mu A_\nu - \partial_\nu A_\mu + ih[A_\mu, A_\nu]$. Let the gauge transformation for A_μ be as in (a). Show that $F_{\mu\nu}$ transforms as $F_{\mu\nu} \to UF_{\mu\nu}U^{-1}$ only if $h = g$.

(3) **Fermion self-energy in QED and QCD**

    (a) Express the fermion QED self-energy, $-i\Sigma(p)$, of Fig. II–2(b) as a Feynman integral and use dimensional regularization to verify the forms of $Z_2^{(\mathrm{MS})}$, $\delta m^{(\mathrm{MS})}$ appearing in Eqs. (1.34), (1.35).

    (b) Proceed analogously to determine Z_2 for QCD and thus verify Eq. (2.55).

(4) **Gravity as a gauge theory**

    The only force which remains outside of the present Standard Model is gravity. General relativity is also a gauge theory, being invariant under local-coordinate transformations. The full theory is too complex for presentation here, but we can study the weak field limit. In general relativity the metric tensor becomes a function of spacetime, with the weak field limit being an expansion around flat space, $g^{\mu\nu}(x) = g^{\mu\nu} + h^{\mu\nu}(x)$, with $1 \gg h^{\mu\nu}$. Let us consider weak field gravity coupled to a scalar field, with lagrangian $\mathcal{L} = \mathcal{L}_{\text{grav}} + \mathcal{L}_{\text{matter}}$ defined as

$$\mathcal{L}_{\text{grav}} = \frac{1}{64\pi G_N} \left[\partial_\lambda h_{\mu\nu} \partial^\lambda \bar{h}^{\mu\nu} - 2\partial^\lambda \bar{h}_{\mu\lambda} \partial_\sigma \bar{h}^{\mu\sigma} \right],$$

$$\mathcal{L}_{\text{matter}} = \frac{1}{2} \left(1 - \frac{1}{2} h_\lambda^\lambda \right) \left[(g^{\mu\nu} + h^{\mu\nu}) \partial_\mu \varphi \partial_\nu \varphi - m^2 \varphi^2 \right],$$

where $\bar{h}^{\mu\nu} \equiv h^{\mu\nu} - g^{\mu\nu} h_\lambda^\lambda / 2$, all indices are raised and lowered with the flat space metric $g^{\mu\nu}$, and G_N is the Cavendish constant.

(a) Show that the action is invariant under the action of an infinitesimal coordinate translation, $x^\mu \to x'^\mu = x^\mu + \epsilon^\mu(x)$ $(1 \gg \epsilon^\mu(x))$, together with a gauge change on $h^{\mu\nu}$,

$$\varphi(x) \to \varphi'(x') = \varphi(x),$$
$$h^{\mu\nu}(x) \to h'^{\mu\nu}(x') = h^{\mu\nu}(x) + \partial^\mu \epsilon^\nu(x) + \partial^\nu \epsilon^\mu(x).$$

Note: both ϵ^μ and $h^{\mu\nu}$ are infinitesimal and should be treated to first order only.

(b) Obtain the equations of motion for φ and $h^{\mu\nu}$. The source term for $h^{\mu\nu}$ is $T^{\mu\nu}$, the energy-momentum tensor for φ. Use the equation of motion for $h^{\mu\nu}$ to show that $T^{\mu\nu}$ is conserved. Simplify the equations with the choice of 'harmonic gauge', $\partial_\nu \bar{h}^{\mu\nu} = 0$.

(c) Solve for $h^{\mu\nu}$ near a point mass at rest, corresponding to $T^{00} = M\delta^{(3)}(x)$ and $T^{0i} = T^{ij} = 0$. Perform a nonrelativistic reduction for φ, i.e., $\varphi(x, t) = e^{-imt} \tilde{\varphi}(x, t)$, in order to obtain a Schrödinger equation for $\tilde{\varphi}$ in the gravitational field. Verify that Newtonian gravity is reproduced.

III

Symmetries and anomalies

Application of the concept of symmetry leads to some of the most powerful techniques in particle physics. The most familiar example is the use of gauge symmetry to generate the lagrangian of the Standard Model. Symmetry methods are also valuable in extracting and organizing the physical predictions of the Standard Model. Very often when dealing with hadronic physics, perturbation theory is not applicable to the calculation of quantities of physical interest. One turns in these cases to symmetries and approximate symmetries. It is impressive how successful these methods have been. Moreover, even if one could solve the theory exactly, symmetry considerations would still be needed to organize the results and to make them comprehensible. The identification of symmetries and near symmetries has been considered in Chap. I. This chapter is devoted to their further study, both in general and as applied to the Standard Model, with the intent of providing the foundation for later applications.

III–1 Symmetries of the Standard Model

The treatment of symmetry in Sects. I–4, I–6 was carried out primarily in a general context. In practice, however, we are most interested in the symmetries relevant to the Standard Model. Let us briefly list these, reserving for some a much more detailed study in later sections.

Gauge symmetries: As discussed in Chap. II, these are the $SU(3)_c \times SU(2)_L \times U(1)_Y$ gauge invariances. It is interesting to compare their differing realizations. $SU(3)_c$ is unbroken but evidently confined, whereas $SU(2)_L \times U(1)_Y$ undergoes spontaneous symmetry breaking, induced by the Higgs fields, leaving an unbroken $U(1)_{em}$ gauge invariance.

Fermion-number symmetries: There exist global vector symmetries corresponding to both lepton and quark number. These are of the form

$$\psi_\alpha \to e^{-iQ_\alpha\theta}\psi_\alpha \qquad (1.1)$$

for fields of each chirality. The index α refers to either the set of all leptons or the set of all quarks, and the conserved charges Q_α are just the total number of quarks minus antiquarks and the total number of leptons minus antileptons.[1] Conservation of baryon number B is violated due to an anomaly in the electroweak sector, but $B - L$ remains exact.

Global vectorial symmetries of QCD: If the quarks were all massless, there would be a very high degree of symmetry associated with QCD. Even if $m \neq 0$, symmetries are possible if two or more quark masses are equal. Three of the quarks (c, b, t) are heavy compared to the confinement scale Λ_{QCD} and widely spaced in mass, so they cannot be accommodated into a global symmetry scheme.[2] However, the u, d, and s quarks are light enough that their associated symmetries are useful. The best of these is the *isospin* invariance, which consists of field transformations

$$\psi = \begin{pmatrix} u \\ d \end{pmatrix} \to \psi' = \exp(-i\boldsymbol{\tau} \cdot \boldsymbol{\theta})\psi \qquad (1.2)$$

where $\{\tau^i\}$ $(i = 1, 2, 3)$ are $SU(2)$ Pauli matrices and $\{\theta^i\}$ are the components of an arbitrary constant vector. Associated with the $SU(2)$-flavor invariance are the three Noether currents

$$J_\mu^{(i)} = \bar{\psi}\gamma_\mu \frac{\tau^i}{2}\psi. \qquad (1.3)$$

Isospin symmetry is broken by the up–down mass difference,

$$\mathcal{L}_{mass} = -\frac{m_u + m_d}{2}(\bar{u}u + \bar{d}d) - \frac{m_u - m_d}{2}(\bar{u}u - \bar{d}d) \qquad (1.4)$$

and by electromagnetic and weak interactions. Inclusion of the strange quark extends isospin to $SU(3)$-flavor transformations

$$\psi = \begin{pmatrix} u \\ d \\ s \end{pmatrix} \to \psi' = \exp(-i\boldsymbol{\theta} \cdot \boldsymbol{\lambda})\psi, \qquad (1.5)$$

where $\{\lambda^a\}$ $(a = 1, 2, \ldots, 8)$ are the $SU(3)$ Gell-Mann matrices. The $SU(3)$-flavor symmetry is broken significantly by the strange quark mass, and to a lesser extent by other effects. Predictions of isospin symmetry work at the 1% level, whereas $SU(3)$ predictions hold only to about 30%. It is occasionally convenient to employ

---

[1] In Chap. VI, we return to the study of lepton-number violation through possible Majorana mass terms
[2] See, however, the discussion of the dynamical heavy-quark symmetries in Chap. XII–3

a particular $SU(2)$ subgroup of $SU(3)$, called *U-spin*, which corresponds to the transformations

$$\begin{pmatrix} d \\ s \end{pmatrix} \rightarrow \exp(-i\boldsymbol{\tau} \cdot \boldsymbol{\theta}) \begin{pmatrix} d \\ s \end{pmatrix}. \tag{1.6}$$

U spin is also a symmetry of the electromagnetic interaction, since its generators commute with the electric-charge operator. The U-spin symmetry is broken by the large d-quark, s-quark mass difference.

Approximate chiral symmetries of QCD: The vectorial symmetries are valid if quark masses are equal. If the masses vanish, there are additional *chiral* symmetries, because in this limit the left-handed and right-handed components of the fields are decoupled (cf. Sect. I–3),

$$\mathcal{L}_{QCD}\Big|_{m=0} = -\frac{1}{4}F_{\mu\nu}^a F^{a\mu\nu} + \bar{\psi}_L \not{D} \psi_L + \bar{\psi}_R \not{D} \psi_R, \tag{1.7}$$

i.e., the left-handed and right-handed fields have *separate* invariances. For massless up-and-down chiral quarks, the symmetry operations are

$$\psi_L \rightarrow \exp(-i\boldsymbol{\theta}_L \cdot \boldsymbol{\tau})\psi_L \equiv L\psi_L, \quad \psi_R \rightarrow \exp(-i\boldsymbol{\theta}_R \cdot \boldsymbol{\tau})\psi_R \equiv R\psi_R \tag{1.8}$$

where $\psi_{L,R}$ are chiral projections of the ψ doublet in Eq. (1.2). These can also be expressed as vector and axial-vector isospin transformations,

$$\psi \rightarrow \exp(-i\boldsymbol{\theta}_V \cdot \boldsymbol{\tau})\psi, \qquad \psi \rightarrow \exp(-i\boldsymbol{\theta}_A \cdot \boldsymbol{\tau}\gamma_5)\psi \tag{1.9}$$

with $\boldsymbol{\theta}_V = (\boldsymbol{\theta}_L + \boldsymbol{\theta}_R)/2$, and $\boldsymbol{\theta}_A = (\boldsymbol{\theta}_L - \boldsymbol{\theta}_R)/2$. This invariance is variously referred to as chiral-$SU(2)$, $SU(2)_L \times SU(2)_R$, or $SU(2)_V \times SU(2)_A$. In QCD, it is broken by quark mass terms,

$$\mathcal{L}_{\text{mass}} = -m_u \bar{u}u - m_d \bar{d}d = -m_u(\bar{u}_L u_R + \bar{u}_R u_L) - m_d(\bar{d}_L d_R + \bar{d}_R d_L). \tag{1.10}$$

Thus, if $m_u = m_d \neq 0$, separate left-handed and right-handed invariances no longer exist, but rather only the vector isospin symmetry. The generalization to three massless quarks defines chiral $SU(3)$ (or $SU(3)_L \times SU(3)_R$) and is a straightforward extension of the above ideas.

Discrete symmetries: Since the Standard Model is a hermitian and Lorentz-invariant local quantum field theory, it is invariant under the combined set of transformations *CPT*. Both *QCD* (given the absence of the θ-term) and *QED* conserve P, C, and T separately. By contrast, the electroweak interactions have maximal violation of P and C in the charged-current sector. If a nonzero phase resides in the quark-mixing matrix, there will exist a breaking of *CP*, or equivalently of T, invariance. Otherwise the weak interactions are invariant under the product *CP*.

In addition to the above exact or approximate symmetries of the Standard Model, there are some important 'non-symmetries' of *QCD*. By these we mean invariances of the underlying lagrangian, which might naively be expected to appear as symmetries of Nature but which, for a variety of reasons, do not. These include the following.

Axial $U(1)$: The *QCD* lagrangian would have an axial $U(1)$ invariance of the form

$$\psi = \begin{pmatrix} u \\ d \\ s \end{pmatrix} \rightarrow \psi' = e^{-i\theta\gamma_5}\psi, \tag{1.11}$$

if the u, d, s quarks were massless. However, this turns out not to be even an approximately valid symmetry, as it has an anomaly. We shall return to this point in Sect. III–3.

Scale transformations: If quarks were massless, the *QCD* lagrangian would contain no dimensional parameters. The lagrangian would therefore be invariant under the scale transformations

$$\psi(x) \rightarrow \lambda^{3/2}\psi(\lambda x), \qquad A_\mu^a(x) \rightarrow \lambda A_\mu^a(\lambda x), \tag{1.12}$$

where ψ and A_μ^a are respectively the quark and gluon fields. This invariance is also destroyed by anomalies (see Sect. III–4).

'Flavor symmetry': Because the gluon couplings are independent of the quark flavor, one often finds reference in the literature to a flavor symmetry of *QCD*. Unless the specific application is reducible to one of the above true symmetries, one should not be misled into thinking that such a symmetry exists. For example, flavor symmetry is often used in this context to relate properties of the pseudoscalar mesons $\eta(549)$ and $\eta'(960)$ (or analogous particles in other nonets). However, the result is rarely a symmetry prediction. Rather, this approach typically pertains to specific assumptions about the way quarks behave, and is dressed up by incorrectly being called a symmetry. In group theoretic language, this may arise by assuming that *QCD* has a $U(3)$ symmetry rather than just that of $SU(3)$.

III–2 Path integrals and symmetries

The transition from classical physics to quantum physics is in many ways most transparent in the path-integral formalism. In this chapter we use these techniques to provide a quantum description of symmetries, complementing the treatment at the classical level of Sects. I–4, I–6. A brief pedagogical introduction to those path-integral techniques which are important for the Standard Model is provided in App. A.

The generating functional

In order to implement a quantum description of currents and current matrix elements, one studies the generating functional, Z, of the theory. For a generic field φ, we have

$$Z[j] = e^{iW[j]} = \int [d\varphi] \exp i \int d^4x \, (\mathcal{L}(\varphi, \partial\varphi) - j\varphi), \qquad (2.1)$$

where $j(x)$ is an arbitrary classical source field whose presence allows us to probe the theory by studying its response to the source. The symbol $[d\varphi]$ indicates that at *each* point of spacetime one integrates over all possible values of the field $\varphi(x)$. All the matrix elements needed to describe physical processes in the theory can be obtained from $\ln Z[j]$ by functional derivations, i.e.,

$$\langle 0 \, | \, T \left(\varphi(x_k) \ldots \varphi(x_p) \right) | \, 0 \rangle = (i)^n \frac{\delta^n \ln Z[j]}{\delta j(x_k) \ldots \delta j(x_p)} \bigg|_{j=0}, \qquad (2.2)$$

where n is the number of fields in the matrix element. If there is more than one field, i.e., the set $\{\varphi_i\}$, a separate source is introduced for each field.

If one wants to study a given current J^μ (not to be confused with the source j) associated with some classical symmetry, one simply adds an extra classical source field v_μ, which is coupled to that current,

$$Z[j, v_\mu] = \int [d\varphi] \exp i \int d^4x \, (\mathcal{L} - j\varphi - v_\mu J^\mu). \qquad (2.3)$$

In this case all matrix elements involving J^μ can be obtained by functional derivation with respect to v_μ,

$$\bar{J}^\mu(x) = i \frac{\delta \ln Z}{\delta v_\mu(x)} \bigg|_{v_\mu=0}, \qquad (2.4)$$

where the bar in $\bar{J}^\mu$ indicates that it is a functional describing matrix elements of the current J^μ. Specific matrix elements are obtained by further derivatives, as in

$$\langle 0 \, | \, T \left(J^\mu(x)\varphi(x_1)\varphi(x_2) \right) | \, 0 \rangle = (i)^2 \frac{\delta^2}{\delta j(x_1)\delta j(x_2)} \bar{J}^\mu(x) \bigg|_{j=0}. \qquad (2.5)$$

This device allows one to discuss all possible matrix elements of the current J^μ.

As an example, consider the vector and axial-vector currents of *QED*. We define

$$Z[v_\mu, a_\mu] \equiv \int [d\psi][d\bar{\psi}][dA_\mu] e^{i \int d^4x \left(\mathcal{L}_{QED} - v_\mu \bar{\psi}\gamma^\mu\psi - a_\mu \bar{\psi}\gamma^\mu\gamma_5\psi \right)}. \qquad (2.6)$$

A three-current (connected) matrix element is obtained then as

$$T_{\mu\alpha\beta}(x, y, z)_{\text{conn}} \equiv \langle 0 \,|\, T\left(\bar{\psi}(x)\gamma_\mu\gamma_5\psi(x)\, \bar{\psi}(y)\gamma_\alpha\psi(y)\, \bar{\psi}(z)\gamma_\beta\psi(z)\right)\,|\, 0\rangle$$

$$= (i)^3 \left[\frac{\delta^2}{\delta v^\alpha(y)\delta v^\beta(z)} \frac{\delta}{\delta a^\mu(x)} \ln Z \right]_{\substack{v^\mu=0 \\ a^\mu=0}}$$

$$= (i)^2 \frac{\delta^2}{\delta v^\alpha(y)\delta v^\beta(z)} \bar{J}_{5\mu}(x) \tag{2.7}$$

where the axial-vector quantity $\bar{J}_{5\mu}$ is defined in analogy with Eq. (2.4).

Noether's theorem and path integrals

Returning to the general case, let us consider an infinitesimal transformation of a set of fields $\{\varphi_i\}$ (cf. Eq. (I–3.1))

$$\varphi_i \rightarrow \varphi_i' = \varphi_i + \epsilon(x) f_i(\varphi) \tag{2.8}$$

such that the current under discussion is

$$J^\mu(x) = \frac{\partial \mathcal{L}'}{\partial(\partial_\mu \epsilon)}. \tag{2.9}$$

If this is a symmetry transformation, one has up to a total derivative,

$$\mathcal{L}' = \mathcal{L}\left(\varphi', \partial\varphi'\right) = \mathcal{L}\left(\varphi, \partial\varphi\right) + J^\mu \partial_\mu \epsilon. \tag{2.10}$$

If $\epsilon(x)$ is a constant, the lagrangian is invariant under the transformation. This is the statement of the classical symmetry condition. In order to study the consequences of this situation, we rewrite our previous definition of the current matrix elements

$$\bar{J}^\mu(x) = i \frac{\delta}{\delta v_\mu(x)} \ln Z[v_\nu] \tag{2.11}$$

in integral form by noting

$$\delta \ln Z[v_\mu] = \ln Z[v_\mu + \delta v_\mu] - \ln Z[v_\mu] \equiv -i \int d^4x\, \bar{J}^\mu(x)\delta v_\mu(x), \tag{2.12}$$

which is just the inverse of Eq. (2.11). Now choosing the particular form for δv_μ,

$$\delta v_\mu(x) = -\partial_\mu \epsilon(x), \tag{2.13}$$

we have

$$\delta_\epsilon \ln Z[v_\mu] \equiv \ln Z[v_\mu - \partial_\mu \epsilon] - \ln Z[v_\mu]$$

$$= i \int d^4x\, \bar{J}^\mu(x)\partial_\mu \epsilon(x) = -i \int d^4x\, \epsilon(x)\partial_\mu \bar{J}^\mu(x). \tag{2.14}$$

With this procedure we can isolate a divergence condition for $\bar{J}^\mu$. If $Z[v_\mu - \partial_\mu \epsilon] = Z[v_\mu]$, then $\partial_\mu \bar{J}^\mu(x) = 0$. To check this, consider

$$Z[v_\mu - \partial_\mu \epsilon] = \int [d\varphi_i] \exp i \int d^4x \left(\mathcal{L}(\varphi_i, \partial \varphi_i) - (v_\mu - \partial_\mu \epsilon) J^\mu \right). \quad (2.15)$$

If we can change integration variables so that

$$\int [d\varphi_i] = \int [d\varphi_i'] \quad (2.16)$$

with φ_i' given by Eq. (2.8), then we obtain

$$Z[v_\mu - \partial_\mu \epsilon] = \int [d\varphi_i'] \exp i \int d^4x \left(\mathcal{L}(\varphi_i', \partial \varphi_i') + v_\mu J^\mu \right) = Z[v_\mu], \quad (2.17)$$

and therefore

$$\partial_\mu \bar{J}^\mu(x) = 0. \quad (2.18)$$

This change of variables seems reasonable and in most cases is perfectly legitimate. After all, the symbol $[d\varphi_i(x)]$ means that we integrate over all values of the field φ_i separately at each point in spacetime. Shifting the origin of integration at point x by a constant, $\varphi_i(x) \equiv \varphi_i'(x) - \epsilon(x) f_i$, and then integrating over all values of φ_i' should amount to the original integration. Given this shift, we have obtained in Eq. (2.18) by Noether's theorem a quantum conservation law involving matrix elements. The expression $\partial_\mu \bar{J}^\mu(x) = 0$ means that all matrix elements of J^μ, obtained via further functional derivatives (as in Eq. (2.5)), satisfy a divergenceless condition, i.e., of the current J^μ is conserved in all matrix elements.

It was Fujikawa who first pointed out the consequences if the change of variables, Eq. (2.16), is not a valid operation in a path integral [Fu 79]. Certainly, many procedures involving path integrals need to be examined carefully in order to see if they are well defined. We shall explicitly study some examples in which the change of variable is nontrivial and can be calculated. In such cases one finds $\partial_\mu \bar{J}^\mu(x) \neq 0$, which implies that the classical symmetry is *not* a quantum symmetry. In these situations it is said that there exists an *anomaly*.

III–3 The U(1) axial anomaly

For massless quarks $m_u = m_d = m_s = 0$, the QCD lagrangian contains an invariance $\mathcal{L}_{QCD} \to \mathcal{L}_{QCD}$ under the global $U(1)$ axial transformations

$$\psi = \begin{pmatrix} u \\ d \\ s \end{pmatrix} \to \psi' = e^{-i\theta\gamma_5}\psi. \quad (3.1)$$

In this limit, which we shall adopt until near the end of this chapter, Noether's theorem can be applied to identify the classically conserved axial current,

$$J_{5\mu}^{(0)} = \bar{u}\gamma_\mu\gamma_5 u + \bar{d}\gamma_\mu\gamma_5 d + \bar{s}\gamma_\mu\gamma_5 s, \qquad \partial^\mu J_{5\mu}^{(0)} = 0, \qquad (3.2)$$

where the superscript on $J_{5\mu}^{(0)}$ denotes an $SU(3)$ *singlet* current. We shall see that this is not an approximate symmetry of the full quantum theory because the current divergence has an anomaly. This can be demonstrated in various ways. For a direct 'hands-on' demonstration, the early discussion [Ad 69, BeJ 67, Ad 70] of Adler and of Bell and Jackiw, which we recount below, has still not been improved upon. However, for a deeper understanding, Fujikawa's path-integral treatment [Fu 79], also described below, seems to us to be the most illuminating. The effect of an anomaly is simply stated, although one must go through some subtle calculations to be convinced that the effect is inescapable. An anomaly is said to occur when a symmetry of the classical action is not a true symmetry of the full quantum theory. The Noether current is no longer divergenceless, but receives a contribution arising from quantum corrections. It is this contribution which is often loosely referred to as the anomaly. The Ward identities which relate matrix elements no longer hold, but rather are replaced by a set of *anomalous* Ward identities, which take into account the correct current divergence.

There are two applications of the axial anomaly which have proved to be of particular importance to the Standard Model. One is in connection with the $SU(3)$ singlet axial current described above. Here the anomaly will end up telling us that the current is not conserved in the chiral limit, but rather that

$$\partial^\mu J_{\mu 5}^{(0)} = \frac{3\alpha_s}{4\pi} F_{\mu\nu}^a \tilde{F}^{a\mu\nu} \qquad \left(\tilde{F}_{\mu\nu}^a \equiv \frac{1}{2}\epsilon^{\mu\nu\alpha\beta} F_{\alpha\beta}^a \right). \qquad (3.3)$$

This will serve to keep the ninth pseudoscalar meson, the η', from being a pseudo-Goldstone boson.

The other application is in the decay $\pi^0 \to \gamma\gamma$, which is historically the process wherein the anomaly was discovered. The quantity of interest here is an isovector axial current $J_{5\mu}^{(3)}$ which transforms as the third component of an $SU(3)$-flavor octet,

$$J_{5\mu}^{(3)} = \bar{u}\gamma_\mu\gamma_5 u - \bar{d}\gamma_\mu\gamma_5 d. \qquad (3.4)$$

Without the anomaly, one would expect that the current $J_{5\mu}^{(3)}$ would be conserved in the chiral $SU(2)$ limit even in the presence of electromagnetism. This follows from the apparently correct procedure

$$\partial^\mu J_{5\mu}^{(3)} = \bar{u} \left[\left(\overleftarrow{\not{\partial}} - i Q \not{A} \right) \gamma_5 - \gamma_5 \left(\not{\partial} + i Q \not{A} \right) \right] u$$
$$- \bar{d} \left[\left(\overleftarrow{\not{\partial}} - i Q \not{A} \right) \gamma_5 - \gamma_5 \left(\not{\partial} + i Q \not{A} \right) \right] d = 0. \tag{3.5}$$

However, explicit calculation shows that the current has an anomaly, such that

$$\partial^\mu J_{5\mu}^{(3)} = 2i \left(m_u \bar{u} \gamma_5 u - m_d \bar{d} \gamma_5 d \right) + \frac{\alpha N_c}{6\pi} F_{\mu\nu} \tilde{F}^{\mu\nu}, \tag{3.6}$$

where $F_{\mu\nu}$ is the electromagnetic field strength. This will be important in predicting the $\pi^0 \to \gamma\gamma$ and $\eta^0 \to \gamma\gamma$ rates and serves as a test for the number of quark colors.

Diagrammatic analysis

To review the work of Adler and of Bell and Jackiw, we first consider the Ward identities for the coupling of the $U(1)$ axial current to two gluons. We define

$$T_{\mu\alpha\beta}^{ab}(k, q) \equiv i \int d^4x \, d^4y \, e^{ik \cdot x} e^{iq \cdot y} \left\langle 0 \left| T \left(J_{5\mu}^{(0)}(x) J_\alpha^a(y) J_\beta^b(0) \right) \right| 0 \right\rangle, \tag{3.7}$$

where J_α^a is a flavor-singlet (color-octet) vector current coupled to gluons

$$J_\alpha^a = \sum_{q=u,d,s} \bar{q} \gamma_\alpha \frac{\lambda^a}{2} q. \tag{3.8}$$

It is important to understand that the $SU(3)$ matrices pertain here to the *color* degree of freedom and should not be confused with analogous matrices which operate in flavor space. The amplitude $T_{\mu\alpha\beta}^{ab}$ is related to the vacuum-to-digluon matrix element by

$$\langle G^a(\lambda_1, q) \, G^b(\lambda_2, -k-q) | J_{5\mu}^{(0)} | 0 \rangle = i g_3^2 \epsilon_1^{\dagger\alpha} \epsilon_2^{\dagger\beta} T_{\mu\alpha\beta}^{ab}(k, q). \tag{3.9}$$

There are two Ward identities, representing the conservation of axial and vector currents. The vector Ward identity, corresponding to color current conservation, $\partial^\alpha J_\alpha^a = 0$, is

$$q^\alpha T_{\mu\alpha\beta}^{ab}(k, q) = 0. \tag{3.10}$$

The axial Ward identity is derived in a similar fashion using the assumed conservation of the $U(1)$ axial current in the massless limit,

$$\partial^\mu J_{5\mu}^{(0)}(x) = 0, \tag{3.11}$$

to yield

$$k^\mu T_{\mu\alpha\beta}^{ab}(k, q) = 0. \tag{3.12}$$

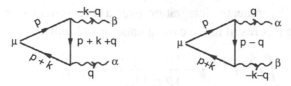

Fig. III–1 Triangle diagrams associated with the axial anomaly.

In order to reveal the anomalous behavior of this coupling, we calculate the vertex in lowest-order perturbation theory via the triangle diagrams of Fig. III–1. With the momenta as labeled in the figures, this produces the amplitude

$$
T^{ab}_{\mu\alpha\beta} = -3 \int \frac{d^4 p}{(2\pi)^4} \left[\text{Tr} \left(\gamma_\mu \gamma_5 \frac{1}{\not p + \not k} \gamma_\beta \frac{\lambda^b}{2} \frac{1}{\not p - \not q} \gamma_\alpha \frac{\lambda^a}{2} \frac{1}{\not p} \right) \right.
$$
$$
\left. + \text{Tr} \left(\gamma_\mu \gamma_5 \frac{1}{\not p + \not k} \gamma_\alpha \frac{\lambda^a}{2} \frac{1}{\not p + \not k + \not q} \gamma_\beta \frac{\lambda^b}{2} \frac{1}{\not p} \right) \right], \quad (3.13)
$$

where the prefactor of 3 arises from the three massless quarks, each of which contributes equally.

Observe that these integrals are linearly divergent, and so may not be well defined. In particular, there exists an ambiguity corresponding to the different possible ways to label the loop momentum. An example will prove instructive, so we consider the integral

$$
I_\gamma = \int d^4 p \left[\frac{p_\gamma}{p^4} - \frac{(p - \ell)_\gamma}{(p - \ell)^4} \right]. \quad (3.14)
$$

This is evaluated by transforming to Euclidean space, where $p_0 = i p_4$ and $p^2 = -p_4^2 - \mathbf{p}^2 \equiv -p_E^2$. In order to perform the integration, one may note that for a general function, $F(p)$, whose four-dimensional integral is linearly divergent (i.e., one with $p^3 F(p) \neq 0$, but $p^3 F'(p) = p^3 F''(p) = \ldots = 0$ for $p \to \infty$), one finds by Taylor expanding and using Gauss' theorem that

$$
\int d^4 p_E \left[F(p) - F(p - \ell) \right] = \int d^4 p_E \left[\ell^\mu \partial_\mu F(p) - \frac{1}{2} \ell^\mu \ell^\nu \partial_\mu \partial_\nu F(p) + \cdots \right]
$$
$$
= \ell^\mu \int d^3 S_\mu \left[F(p) - \frac{1}{2} \ell^\nu \partial_\nu F(p) + \cdots \right]_{p \to \infty}
$$
$$
= \ell^\mu \int d^3 S_\mu F(p) \Big|_{p \to \infty} \quad (3.15)
$$

where $d^3 S_\mu$ indicates integration over a three-dimensional surface at $p \to \infty$.[3]
Applying this result to the case at hand, we obtain a surface integral

$$I_\gamma = i \int d^4 p_E \left(\frac{p_\gamma}{p^4} - \frac{(p - \ell)_\gamma}{(p - \ell)^4} \right) = i\ell^\mu \int d^3 S_\mu \frac{p_\gamma}{p^4} = i\ell^\mu \int d^3 S \frac{p_\mu}{p} \frac{p_\gamma}{p^4}.$$

(3.16)

Note that from euclidean covariance we can replace $p_\mu p_\gamma$ by $\delta_{\mu\gamma} p^2/4$, to yield

$$I_\gamma = i \frac{\ell_\gamma}{4} \int d^3 S \frac{1}{p^3} = i \frac{\pi^2 \ell_\gamma}{2},$$

(3.17)

where the last step uses the surface area of a three-dimensional surface in four-dimensional euclidean space, $S_4 = 2\pi^2 R^3$.

In the case of $T^{ab}_{\mu\alpha\beta}$, consider the effect of shifting the integration variable of the first term in Eq. (3.13) from p to $p + b_1 q + b_2(-k - q)$. In order to maintain the Bose symmetry of $T^{ab}_{\mu\alpha\beta}$ (i.e., symmetry under the interchange $\alpha \leftrightarrow \beta$ at the same time as $q \leftrightarrow (-k - q)$) we must shift the second integration from p to $p + b_1(-k - q) + b_2 q$. Use of Eqs. (3.14)–(3.17) then yields the change in $T^{ab}_{\mu\alpha\beta}$

$$\Delta T^{ab}_{\mu\alpha\beta} = \frac{6i \delta^{ab}}{(2\pi)^4} \epsilon_{\mu\alpha\beta\gamma} [I^\gamma (b_1 q + b_2(-q - k)) - I^\gamma (b_1(-q - k) + b_2 q)]$$

$$= -\frac{3\delta^{ab}}{16\pi^2} (b_1 - b_2) \epsilon_{\mu\alpha\beta\gamma} (2q + k)^\gamma,$$

(3.18)

induced by the shift of the original integration variable p^μ. This is an indication that there may be trouble in the calculation of this diagram, but it is not yet proof of any violation of the Ward identities.

Let us now check the Ward identities. In both cases, use can be made of identities similar to $q^\alpha = p^\alpha - (p^\alpha - q^\alpha)$ in order to change the result into a difference of integrals. We find for the vector Ward identity

$$q^\alpha T^{ab}_{\mu\alpha\beta}(k, q)$$

$$= -\frac{3\delta^{ab}}{2} \int \frac{d^4 p}{(2\pi)^4} \text{Tr} \left[\gamma_\mu \gamma_5 \frac{1}{\not{p} + \not{k}} \gamma_\beta \frac{1}{\not{p} - \not{q}} - \gamma_\mu \gamma_5 \frac{1}{\not{p} + \not{q} + \not{k}} \gamma_\beta \frac{1}{\not{p}} \right]$$

$$= -6i \delta^{ab} \epsilon_{\mu\beta\rho\sigma} \int \frac{d^4 p}{(2\pi)^4} \left[\frac{(p + k)^\rho (p - q)^\sigma}{(p + k)^2 (p - q)^2} - \frac{(p + k + q)^\rho p^\sigma}{(p + k + q)^2 p^2} \right],$$

(3.19)

---

[3] Note that this is just the four-dimensional generalization of the one-dimensional formula

$$\int_{-\infty}^{\infty} dx \, [f(x + y) - f(x)] = \int_{-\infty}^{\infty} dx \left[y f'(x) + \frac{1}{2} y^2 f''(x) + \cdots \right] = y \, [f(\infty) - f(-\infty)],$$

valid for $f(\pm\infty) \neq 0$ but $f'(\pm\infty) = f''(\pm\infty) = \ldots = 0$.

while for the axial-vector case,

$$k^\mu T^{ab}_{\mu\alpha\beta}(k, q) = \frac{3\delta^{ab}}{2} \int \frac{d^4 p}{(2\pi)^4} \, \text{Tr}\left[\gamma_5\gamma_\beta \frac{1}{\not p - \not q}\gamma_\alpha\frac{1}{\not p} + \gamma_5 \frac{1}{\not p + \not k}\gamma_\beta \frac{1}{\not p - \not q}\gamma_\alpha\right.$$

$$\left. + \gamma_5\gamma_\alpha \frac{1}{\not p + \not k + \not q}\gamma_\beta \frac{1}{\not p} + \gamma_5 \frac{1}{\not p + \not k}\gamma_\alpha \frac{1}{\not p + \not k + \not q}\gamma_\beta\right]$$

$$= -6i\delta^{ab}\epsilon_{\alpha\beta\rho\sigma} \int \frac{d^4 p}{(2\pi)^4}\left[\frac{(p + k + q)^\rho p^\sigma}{(p + k + q)^2 p^2} - \frac{(p + k)^\rho (p - q)^\sigma}{(p + k)^2 (p - q)^2}\right.$$

$$\left. + \frac{(p + k)^\rho (p + k + q)^\sigma}{(p + k)^2 (p + k + q)^2} - \frac{(p - q)^\rho p^\sigma}{(p - q)^2 p^2}\right]. \tag{3.20}$$

It is easy to see that if one could freely shift the integration variable, each expression would separately vanish. However, direct calculation using Eqs. (3.14)–(3.17) yields

$$q^\alpha T^{ab}_{\mu\alpha\beta}(k, q) = -\frac{3\delta^{ab}}{16\pi^2}\epsilon_{\mu\beta\rho\sigma}k^\rho q^\sigma \quad \text{and} \quad k^\mu T^{ab}_{\mu\alpha\beta}(k, q) = \frac{3\delta^{ab}}{8\pi^2}\epsilon_{\alpha\beta\rho\sigma}k^\rho q^\sigma. \tag{3.21}$$

If, on the other hand, the original integration variable were shifted as in Eq. (3.18) one would obtain

$$q^\alpha T^{ab}_{\mu\alpha\beta}(k, q) = -\frac{3\delta^{ab}}{16\pi^2}(1 + b_1 - b_2)\,\epsilon_{\mu\beta\rho\sigma}k^\rho q^\sigma,$$
$$k^\mu T^{ab}_{\mu\alpha\beta}(k, q) = \frac{3\delta^{ab}}{8\pi^2}(1 - b_1 + b_2)\,\epsilon_{\alpha\beta\rho\sigma}k^\rho q^\sigma. \tag{3.22}$$

Thus, either *one* of the original Ward identities may be regained by a particular choice of $b_1 - b_2$, but both expressions cannot vanish simultaneously.

Our discussion of the manipulations of Feynman diagrams should not obscure the main physical fact illustrated above, i.e., despite the claim of Noether's theorem that there are *two* sets of conserved currents (vector $SU(3)$ of color and axial-vector $U(1)$), one-loop calculations indicate that only one can in fact be conserved. On physical grounds, we know that in Nature the vector current is conserved, as its charge corresponds to *QCD* color charge. Thus, it must be the axial current which is not conserved. This phenomenon is at first sight quite surprising and it deserves the name 'anomaly' by which it has come to be called. Noether's theorem has misled us, and it is only by direct calculation of the quantum corrections that the true symmetry structure of the theory has been exposed. Note that the situation is not the same as spontaneous symmetry breaking, where the symmetry is hidden by dynamical effects. There the currents remain conserved, as demonstrated in Sect. I–6. Here, current conservation has been violated. In particular, the calculation described above (with $b_1 - b_2 = 1$) is consistent through use of Eq. (3.9) with the operator relation of Eq. (3.3),

$$\partial^\mu J_{5\mu}^{(0)} = \frac{3\alpha_s}{4\pi} F_{\mu\nu}^a \tilde{F}^{a\mu\nu}. \tag{3.23}$$

Both sides of this equation have the same two-gluon matrix elements. It is clear from this that the apparent $U(1)$ symmetry predicted by Noether's theorem is not a symmetry of the quantum theory after all.

Path-integral analysis

In a path-integral treatment [Fu 79], the symmetry of the theory can be tested by considering the generating functional, as described in Sect. III–2. In particular, if we consider a functional of the gluon field A_μ^b and an axial current source a_μ,

$$Z\left[a_\mu, A_\lambda^c\right] = \int [d\psi][d\bar{\psi}] \exp i \int d^4x \left(\mathcal{L}_{QCD}(\psi, \bar{\psi}, A_\lambda^c) - a_\mu J_5^{(0)\mu}\right) \tag{3.24}$$

then the steps leading to Eq. (2.14) produce

$$-i \int d^4x \, \beta(x) \partial^\mu \bar{J}_{5\mu}^{(0)}(x) = \ln Z\left[a_\mu - \partial_\mu \beta, A_\mu^b\right] - \ln Z\left[a_\mu, A_\mu^b\right], \tag{3.25}$$

where $\bar{J}_{5\mu}^{(0)}(x)$ denotes the matrix elements of the current $J_5^{(0)}$,

$$\bar{J}_{5\mu}^{(0)}(x) = i \frac{\delta}{\delta a^\mu(x)} \ln Z\left[a_\nu, A_\lambda^b\right]\Big|_{a_\nu=0}. \tag{3.26}$$

In particular, the two-gluon matrix described above is given by

$$T_{\mu\alpha\beta}^{ab}(x, y, z) = (i)^2 \left[\frac{\delta^2}{\delta A_a^\alpha(y)\delta A_b^\beta(z)} \bar{J}_{5\mu}^{(0)}(x)\right]\Bigg|_{\substack{A_\lambda^c=0 \\ a_\nu=0}}. \tag{3.27}$$

In order to solve for $\partial^\mu \bar{J}_{5\mu}^{(0)}$, we note that the $\partial_\mu \beta$ term can be absorbed into a redefinition of the fermion fields. This can be seen from the identity (for infinitesimal β),

$$\bar{\psi} i \partial\!\!\!/ \psi + \partial_\mu \beta \, \bar{\psi} \gamma^\mu \gamma_5 \psi = \bar{\psi} (1 - i\beta\gamma_5) i \partial\!\!\!/ (1 - i\beta\gamma_5) \psi. \tag{3.28}$$

The following quantities are invariant under this transformation:

$$\begin{aligned}
\bar{\psi} A\!\!\!/^a \lambda^a \psi &= \bar{\psi}(1 - i\beta\gamma_5) A\!\!\!/^a \lambda^a (1 - i\beta\gamma_5)\psi, \\
J_\mu &= \bar{\psi} \gamma_\mu \psi = \bar{\psi}(1 - i\beta\gamma_5)\gamma_\mu(1 - i\beta\gamma_5)\psi.
\end{aligned} \tag{3.29}$$

Mass terms would not be invariant, but we are presently working in the massless limit. Therefore, if we define

$$\psi' = (1 - i\beta\gamma_5)\psi = e^{-i\beta\gamma_5}\psi + \mathcal{O}(\beta^2),$$
$$\bar{\psi}' = \bar{\psi}(1 - i\beta\gamma_5) = \bar{\psi}e^{-i\beta\gamma_5} + \mathcal{O}(\beta^2),$$

(3.30)

we see that the lagrangian can be written in terms of ψ',

$$\mathcal{L}_{QCD}(\psi, \bar{\psi}, A_\mu^a) + \partial^\mu\beta \, J_{5\mu}^{(0)} = \mathcal{L}_{QCD}(\psi', \bar{\psi}', A_\mu^a).$$

(3.31)

Furthermore, we would like to change from ψ to ψ' in the path integration. To be general, we allow for the possibility of a jacobian $\mathcal{J}$ accompanying this change of variables, viz.,

$$\int [d\psi][d\bar{\psi}] \equiv \int [d\psi'][d\bar{\psi}']\mathcal{J}.$$

(3.32)

If, as will be shown later, the jacobian $\mathcal{J}$ is independent of ψ and $\bar{\psi}$, it can be taken to the outside of the path integral, resulting in

$$Z\left[a_\mu - \partial_\mu\beta, A_\mu^a\right] = \int [d\psi'][d\bar{\psi}']\mathcal{J}e^{i\int d^4x(\mathcal{L}_{QCD}(\psi', \bar{\psi}', A_\mu^a) - a_\mu J_5^\mu)}$$
$$= \mathcal{J} \, Z\left[a_\mu, A_\mu^a\right].$$

(3.33)

Thus, the test for the symmetry, Eq. (3.25), depends entirely on $\mathcal{J}$,

$$\ln \mathcal{J} = -i \int d^4x \, \beta(x)\partial^\mu \bar{J}_{5\mu}^{(0)}(x).$$

(3.34)

The lesson learned is that if the lagrangian *and* the path-integral measure are invariant under the $U(1)$ transformation, then there exists a $U(1)$ symmetry in the theory, with $\partial^\mu \bar{J}_{5\mu}^{(0)} = 0$. However, if the lagrangian is invariant, as it is in this case, but the path integral is not (i.e. $\mathcal{J} \neq 1$), then the $U(1)$ transformation is not a symmetry of the theory, i.e., $\partial^\mu J_{5\mu}^{(0)} \neq 0$.

We shall show below that the jacobian, when properly regularized, has the form

$$\mathcal{J} = \exp\left(-2i \, \text{tr} \, \beta\gamma_5\right) = \exp\left[-i \int d^4x \, \beta(x)\frac{3\alpha_s}{4\pi} F_{\mu\nu}^a \tilde{F}^{a\mu\nu}\right],$$

(3.35)

so that the current divergence has the form given in Eq. (3.3),

$$\partial^\mu \bar{J}_{5\mu}^{(0)} = \frac{3\alpha_s}{4\pi} F_{\mu\nu}^a \tilde{F}^{a\mu\nu}.$$

Functional differentiation using Eq. (3.27) yields the same result for $q^\mu T_{\mu\alpha\beta}^{ab}$ as obtained in ordinary perturbation theory. The nontrivial transformation of the path-integral measure has prevented the axial $U(1)$ transformation from being a symmetry of the theory. We now turn to the calculation of the jacobian.

The jacobian in fact diverges, and a regularization is needed in order to make it finite. In Fujikawa's original calculation the regularizer was introduced early into the procedure, allowing each step to be well defined. We will be slightly less rigorous by introducing the regularizer somewhat later. In order to calculate the jacobian we need to review the properties of integration over Grassmann numbers (which are described in more detail in App. A–5). The anticommuting nature of the variables requires that any function constructed from them terminates after linear order in each variable. Thus, a function of two Grassman numbers z_1, z_2 ($z_1 z_2 = -z_2 z_1$, $z_i^2 = 0$) becomes

$$f(z_1, z_2) = f_0 + f_1 z_1 + f_2 z_2 + f_{12} z_1 z_2, \tag{3.36}$$

where f_0, f_1, f_2, f_{12} are real numbers. The primary property of an integral to be transferred to Grassmann numbers is completeness, i.e.,

$$\int dz \, f(z) = \int dz \, f(z + z'), \tag{3.37}$$

where z' is a constant Grassmann number. Expanding both sides we have

$$\int dz \, (f_0 + f_1 z) = \int dz \, (f_0 + f_1 z + f_1 z') \tag{3.38}$$

For this to be true, the condition

$$\int dz = 0 \tag{3.39}$$

is required. Now consider a change of variables

$$z_1 = c_{11} z_1' + c_{12} z_2', \qquad z_2 = c_{21} z_1' + c_{22} z_2', \tag{3.40}$$

involving a matrix of coefficients $\mathbf{C}$. The jacobian is defined by

$$\int dz_1 \, dz_2 \, f(z) = \mathcal{J} \int dz_1' \, dz_2' \, f(\mathbf{C} z'). \tag{3.41}$$

Application of Eq. (3.36) leads to the consideration of only the f_{12} term,

$$f_{12} \int dz_1 \, dz_2 \, z_1 z_2 = \mathcal{J} f_{12} \int dz_1' \, dz_2' \, (c_{11} z_1' + c_{12} z_2')(c_{21} z_1' + c_{22} z_2')$$

$$= \mathcal{J} f_{12}(c_{11} c_{22} - c_{12} c_{21}) \int dz_1' \, dz_2' \, z_1' z_2', \tag{3.42}$$

and hence the identification of the jacobian,

$$\mathcal{J} = [\det \mathbf{C}]^{-1}. \tag{3.43}$$

Although derived in the simple 2×2 case, Eq. (3.43) generalizes to arbitrary dimension. Note that, due to the Grassmann nature of the variables, this result is the inverse of what would be expected with normal commuting variables.

Turning now to the path integral, we temporarily consider $\psi(x)$ as a finite number of Grassmann variables corresponding to four Dirac indices at each point of spacetime (i.e., imagine that the spacetime label is discrete and finite). At each point, the transformation is from $\psi \to \psi'$

$$\psi(x) = e^{i\beta(x)\gamma_5}\psi'(x), \qquad \bar\psi(x) = \bar\psi'(x)\, e^{i\beta(x)\gamma_5}, \tag{3.44}$$

so that the overall jacobian has the form

$$\mathcal{J} = \left[\det\left(e^{i\beta\gamma_5}\right)\right]^{-1}\left[\det\left(e^{i\beta\gamma_5}\right)\right]^{-1} \tag{3.45}$$

with one factor from each of the ψ and $\bar\psi$ variables. The determinant runs over the 4×4 Dirac indices, the three flavors, colors, and also the spacetime indices. This is a rather formal object, but can be made more explicit by using

$$\det \mathbf{C} = e^{\operatorname{tr}\ln\mathbf{C}}, \tag{3.46}$$

valid for finite matrices, to write

$$\mathcal{J} = e^{-2i\operatorname{tr}\beta\gamma_5}. \tag{3.47}$$

The symbol tr denotes a trace acting over spacetime indices plus Dirac indices, flavors, and colors,

$$\operatorname{tr}\beta\gamma_5 = \operatorname{Tr}'\int d^4x\,\langle x|\beta\gamma_5|x\rangle, \tag{3.48}$$

with Tr' indicating the Dirac, color, and flavor trace. This will become clearer through direct calculation below.

The jacobian still is not regulated. Fujikawa suggested the removal of high-energy eigenmodes of the Dirac field in a gauge-invariant way. Consider, for example, the simple extension

$$\mathcal{J} = \lim_{M\to\infty}\exp\left[-2i\operatorname{tr}\left(\beta\gamma_5\,e^{-(\slashed{D}/M)^2}\right)\right], \tag{3.49}$$

where $\slashed{D}$ is the *QCD* covariant derivative. The insertion of a complete set of eigenfunctions of $\slashed{D}$ exponentially removes those with large eigenvalues. There has been an extensive literature demonstrating that other regularization methods produce the same results as Fujikawa's, provided that the regulator preserves the vector gauge invariance.

In order to complete the calculation we employ the following identity:

$$\slashed{D}\slashed{D} = \frac{1}{2}\{\gamma_\mu, \gamma_\nu\}D^\mu D^\nu + \frac{1}{2}[\gamma_\mu, \gamma_\nu]D^\mu D^\nu$$

$$= D_\mu D^\mu + \frac{1}{4}[\gamma_\mu, \gamma_\nu][D^\mu, D^\nu] \tag{3.50}$$

$$= D_\mu D^\mu + \frac{g_3 \lambda^a}{4}\sigma^{\mu\nu} F_{\mu\nu}^a.$$

In this case the expression

$$\langle x| \exp -(\slashed{D}/M)^2 |x\rangle \tag{3.51}$$

has the same form as given in Eqs. (B–1.1), (B–1.9), (B–1.17–18) with the identifications

$$d_\mu = D_\mu, \qquad \sigma = \frac{g_3}{4}\sigma^{\mu\nu}\lambda^a F_{\mu\nu}^a, \qquad \tau = \frac{1}{M^2}. \tag{3.52}$$

Applying the calculation done there to our present situation yields

$$\mathcal{J} = \lim_{M\to\infty} e^{-2i \int d^4x \,\mathrm{Tr}\,(\beta(x)\gamma_5 H(x,M^{-2}))}$$

$$= \lim_{M\to\infty} e^{\frac{1}{8\pi^2} \int d^4x \,\mathrm{Tr}\,(\beta(x)\gamma_5[M^4 a_0 + M^2 a_1 + a_2 + \mathcal{O}(M^{-2})])}. \tag{3.53}$$

The notation is defined in App. B–1. The first two traces vanish, leaving only the factor with two $\sigma^{\mu\nu}$ matrices in a_2. From the result

$$\mathrm{Tr}\,(\gamma_5 \sigma^{\mu\nu}\sigma^{\alpha\beta}) = -\,\mathrm{Tr}\,\gamma_5 \gamma^\mu \gamma^\nu \gamma^\alpha \gamma^\beta = -4i\epsilon^{\mu\nu\alpha\beta}, \tag{3.54}$$

it is easy to calculate

$$\mathcal{J} = \exp\left(\frac{1}{16\pi^2}\int d^4x\,\beta(x)\mathrm{Tr}'\left(\gamma_5 \frac{g_3^2 \lambda^a \lambda^b}{16}\sigma^{\mu\nu} F_{\mu\nu}^a \sigma^{\alpha\beta} F_{\alpha\beta}^b\right)\right)$$

$$= \exp\left(\frac{-1}{16\pi^2}\int d^4x\,\beta(x)\,3\cdot 2\delta^{ab}\cdot 4\,i\epsilon^{\mu\nu\alpha\beta}\frac{g_3^2}{16}F_{\mu\nu}^a F_{\alpha\beta}^b\right) \tag{3.55}$$

$$= \exp\left(-i\int d^4x\,\beta(x)\frac{3\alpha_s}{4\pi}F_{\mu\nu}^a \tilde{F}^{a\mu\nu}\right)$$

where the trace Tr' has produced factors for three flavors, color, and the Dirac trace.

Although the calculation of the jacobian has been somewhat involved, we have succeeded in making sense out of what seemed to be a rather abstract object. The fact that it is not unity is an indication that the $U(1)$ transformation is not a symmetry of the theory. Applying Eq. (3.34) we see that

$$\ln \mathcal{J} = -i\int d^4x\,\beta(x)\partial^\mu \bar{J}_{5\mu}^{(0)}(x) = -i\int d^4x\,\beta(x)\frac{3\alpha_s}{4\pi}F_{\mu\nu}^a \tilde{F}^{a\mu\nu}, \tag{3.56}$$

or once again

$$\partial^\mu \bar{J}^{(0)}_{5\mu} = \frac{3\alpha_s}{4\pi} F^a_{\mu\nu} \tilde{F}^{a\mu\nu}. \tag{3.57}$$

The choice of a regulator which preserves the vector $SU(3)$ gauge symmetry is important. Whereas in the Feynman diagram approach, we had the apparent freedom to shift the integration variable to preserve either the vector or axial-vector symmetries, the corresponding freedom in the path-integral case is in the choice of regularization.

If quark masses are included, the operator relation becomes

$$\partial_\mu J^{(0)}_{5\mu}(x) = 2i(m_u \bar{u}\gamma_5 u + m_d \bar{d}\gamma_5 d + m_s \bar{s}\gamma_5 s) + \frac{3\alpha_s}{4\pi} F^a_{\mu\nu} \tilde{F}^{a\mu\nu}. \tag{3.58}$$

Masses do not modify the coefficient of the anomaly, basically because it arises from the ultraviolet divergent parts of the theory, which are insensitive to masses.

One does not have to go through these lengthy calculations for each new application of the anomaly. The anomalous coupling for currents

$$V^{(b)}_\mu = \bar{\psi}\gamma_\mu T^{(b)}_v \psi, \qquad A^{(b)}_\mu = \bar{\psi}\gamma_\mu\gamma_5 T^{(b)}_a \psi, \tag{3.59}$$

where $T^{(b)}_v$, $T^{(b)}_a$ are matrices in the space of quark flavors, is of the form

$$\partial^\mu A^{(b)}_\mu = \frac{D^{bcd}}{16\pi^2} \epsilon^{\mu\nu\alpha\beta} F^c_{\mu\nu} F^d_{\alpha\beta} + \text{mass terms}, \tag{3.60a}$$

$$D^{bcd} \equiv \frac{N_c}{2} \text{Tr}\left(T^{(b)}_a \left\{ T^{(c)}_v, T^{(d)}_v \right\} \right), \tag{3.60b}$$

where N_c is the number of colors. In particular, for the electromagnetic coupling to the isovector axial current we have

$$\begin{aligned} J^{(3)}_{5\mu} &= \bar{u}\gamma_\mu\gamma_5 u - \bar{d}\gamma_\mu\gamma_5 d, \\ D^{bcd} &= e^2 N_c \text{Tr}\, \tau_3 Q^2 = \frac{N_c}{3} e^2, \end{aligned} \tag{3.61}$$

leading to the result already quoted in Eq. (3.6).

The full content of the anomaly was given by Bardeen [Ba 69]. Consider a fermion with η internal degrees of freedom (flavor or color) coupled to vector and axial-vector currents v_μ, a_μ,

$$\mathcal{L} = \overline{\psi} \left(i\partial\!\!\!/ - v\!\!\!/ - a\!\!\!/\gamma_5 \right) \psi. \tag{3.62}$$

These currents are in an $\eta \times \eta$ representation

$$v_\mu = v^0_\mu I + v^k_\mu \lambda^k, \qquad a_\mu = a^0_\mu I + a^k_\mu \lambda^k. \tag{3.63}$$

Thus, the axial current is $J_{5\mu}^{(k)} = \overline{\psi}\gamma_\mu\gamma_5\lambda^k\psi$, and the anomaly equation becomes

$$
\begin{aligned}
\partial^\mu J_{5\mu}^{(k)} = \frac{1}{4\pi^2}\epsilon^{\mu\nu\alpha\beta}\,\text{Tr}\,&\Bigg[\lambda^k\Bigg(\frac{1}{4}v_{\mu\nu}v_{\alpha\beta} + \frac{1}{12}a_{\mu\nu}a_{\alpha\beta} \\
&- \frac{2i}{3}a_\mu a_\nu v_{\alpha\beta} - \frac{2i}{3}v_{\mu\nu}a_\alpha a_\beta - \frac{2i}{3}a_\mu v_{\nu\alpha}a_\beta - \frac{8}{3}a_\mu a_\nu a_\alpha a_\beta\Bigg)\Bigg],
\end{aligned}
$$

$$
\begin{aligned}
v_{\mu\nu} &= \partial_\mu v_\nu - \partial_\nu v_\mu + i[v_\mu, v_\nu] + i[a_\mu, a_\nu], \\
a_{\mu\nu} &= \partial_\mu a_\nu - \partial_\nu a_\mu + i[v_\mu, a_\nu] - i[v_\nu, a_\mu].
\end{aligned}
\tag{3.64}
$$

This may also be expressed in terms of the left-handed and right-handed field tensors $\ell_{\mu\nu}$ and $r_{\mu\nu}$ by using the identities,

$$
\begin{aligned}
\ell_{\mu\nu} &\equiv \partial_\mu\ell_\nu - \partial_\nu\ell_\mu + i[\ell_\mu, \ell_\nu] = v_{\mu\nu} + a_{\mu\nu}, \\
r_{\mu\nu} &\equiv \partial_\mu r_\nu - \partial_\nu r_\mu + i[r_\mu, r_\nu] = v_{\mu\nu} - a_{\mu\nu}, \\
\frac{1}{4}v_{\mu\nu}v_{\alpha\beta} + \frac{1}{12}a_{\mu\nu}a_{\alpha\beta} &= \frac{1}{12}\left(\ell_{\mu\nu}\ell^{\mu\nu} + r_{\mu\nu}r^{\mu\nu}\right) + \frac{1}{24}\left(\ell_{\mu\nu}r^{\mu\nu} + r_{\mu\nu}\ell^{\mu\nu}\right).
\end{aligned}
\tag{3.65}
$$

In the language of Feynman diagrams, one encounters the anomaly contributions not only in the triangle diagram, but also in square and pentagon diagrams (e.g. from the $a_\mu a_\nu a_\alpha a_\beta$ term). Our previous result, Eq. (3.57), is obtained for $a_\mu = 0$, $v_\mu = g_3 A_\mu^k \lambda^k/2$, with three flavors and three colors of quarks.

We have seen that symmetries of the classical lagrangian are not always symmetries of the full quantum theory. This is the general situation when there are anomalies. These appear in perturbation theory and are associated with divergent Feynman diagrams. This sometimes gives the mistaken impression that the dynamics has 'broken' the symmetry, and hence one might expect a massless particle through the application of Goldstone's theorem. In the path-integral framework the impression is different. There the symmetry never exists in the first place, as the calculation performed above is simply the path-integral test for a symmetry, generalizing Noether's theorem. Hence there is in general no expectation for a Goldstone boson.

Can anomalies cause problems? When the anomaly occurs in a global symmetry, such as the above $U(1)$ example, the answer is, 'no'. They just need to be taken properly into account, e.g., as in Eq. (3.61). Given the specific form of the anomaly operator relation, there exist 'anomalous Ward identities' which contain terms attributable to the anomaly [Cr 78]. These anomalies can even be associated with a variety of specific phenomena. For example, in Sect. VII–6 we shall see how the decay $\pi^0 \to \gamma\gamma$ is attributed to the axial anomaly.

The presence of anomalies in gauge theories is far more serious because they destroy the gauge invariance of the theory and wreak havoc with renormalizability. Thus, one attempts to employ only those gauge theories which have no anomalies.

In some cases this can be arranged by ensuring, through the group or particle content of the theory, that the coefficient D^{bcd} of Eq. (3.60b) vanishes. For example, in the Standard Model it must be checked that this occurs for all combinations of the $SU(3)_c \times SU(2)_L \times U(1)_Y$ generators. These were already compiled in Eqs. (II–3.5a–c) and were seen to lead to the quantized fermion charge values observed in Nature.

III–4 Classical scale invariance and the trace anomaly

If the fermion masses were zero in either *QED* or *QCD*, these theories would contain no dimensional parameters in the lagrangian, and they would exhibit a classical scale invariance. The associated quark and gluon scale transformations would be $\psi(x) \rightarrow \lambda^{3/2} \psi(\lambda x)$ and $A_\mu^a(x) \rightarrow \lambda A_\mu^a(\lambda x)$ for arbitrary λ. We saw in Sect. I–4 that this leads to a traceless energy-momentum tensor, with conserved dilation current J_{scale}^μ,

$$J_{\text{scale}}^\mu = x_\nu \theta^{\mu\nu}, \qquad \partial_\mu J_{\text{scale}}^\mu = \theta^\nu{}_\nu = 0, \qquad (4.1)$$

where $\theta^{\mu\nu}$ is the energy-momentum tensor. Such a situation would have drastic consequences on the theory, since all single particle states would be massless. This can be seen as follows. For any hadron H, the matrix element of the energy-momentum tensor at zero-momentum transfer is

$$\langle H(\mathbf{k}) | \theta^{\mu\nu} | H(\mathbf{k}) \rangle = 2k^\mu k^\nu, \qquad (4.2)$$

where the normalization of states is chosen in accordance with the conventions defined in App. C–3. A vanishing trace would imply zero mass, i.e.,

$$\langle H(\mathbf{k}) | \theta^\mu{}_\mu | H(\mathbf{k}) \rangle = 0 = 2M_H^2. \qquad (4.3)$$

This is most obviously a problem in *QCD* where the quark masses are small compared to most composite particle masses.[4] We would not expect the proton mass to vanish if the quark masses were set equal to zero yet the scale-invariance argument implies that it must.

A resolution is suggested by the method which is used to renormalize the theory. In practice, renormalization prescriptions introduce dimensional scales into the theory. Most commonly, there is the momentum scale at which one specifies the running coupling constant to have a particular value, e.g., $\alpha_s(91 \text{ GeV}) \simeq 0.12$. This in turn defines a scale Λ which enters the formula for the running coupling constant, Eq. (II–2.74). Thus, to fully specify *QCD* one needs to specify not only the lagrangian, but also a scale parameter, and the full quantum theory is *not* scale

---

[4] As can be justified, we neglect here the existence of very heavy quarks, c, b, and t.

invariant. Although this argument does not, at first sight, seem to nullify the reasoning based on Noether's theorem, it turns out that the trace of the energy-momentum tensor has an anomaly [Cr 72, ChE 72, CoDJ 77], and the specification of a scale and the coefficient of the anomaly are in fact related.

In the following, let us start directly with the path-integral treatment [Fu 81], again in the framework of *QCD*, concentrating on the effect of a single quark. We can introduce an external source coupled to θ^μ_μ into the generating functional

$$Z[h, A^a_\mu] = \int d\psi \, d\bar\psi e^{i \int d^4x [\mathcal{L}_{QCD}(\psi, A^a_\mu) + h(x)\theta^\mu_\mu]}, \tag{4.4}$$

where

$$\theta^{\mu\nu} = \frac{i}{2} \bar\psi \gamma^\mu \overleftrightarrow{D}^\nu \psi. \tag{4.5}$$

As in the case of the chiral anomaly, we can use this as a starting point to explore the nature of the trace θ^μ_μ. The key is that if one makes the change of variables

$$\psi(x) = e^{-\alpha(x)/2} \psi'(x), \tag{4.6}$$

one obtains for infinitesimal α

$$\int d^4x \left[\mathcal{L}_{QCD}\left(\psi, A^a_\mu(x)\right) + \alpha(x)\theta^\mu_\mu \right]$$
$$= \int d^4x \left[\mathcal{L}_{QCD}\left(\psi', A^a_\mu\right) + \alpha(x)m\bar\psi'\psi' + i\bar\psi'\gamma_\mu\psi'\partial^\mu\alpha \right]. \tag{4.7}$$

The last term vanishes after an integration by parts. The focus of our calculation can thus be shifted to a jacobian $\mathcal{J}$ by a change of variable,

$$Z\left[h + \alpha, A^a_\mu\right] = \int d\psi \, d\bar\psi \, e^{i \int d^4x [\mathcal{L}_{QCD}(\psi, A^a_\mu) + (h+\alpha)\theta^\mu_\mu]}$$
$$= \int d\psi \, d\bar\psi \, e^{i \int d^4x [\mathcal{L}_{QCD}(\psi', A^a_\mu) + h\theta^\mu_\mu + \alpha m\bar\psi'\psi']}$$
$$= \int d\psi' \, d\bar\psi' \, \mathcal{J} \, e^{i \int d^4x [\mathcal{L}_{QCD}(\psi', A^a_\mu) + h\theta^\mu_\mu + \alpha m\bar\psi'\psi']}. \tag{4.8}$$

Thus, we obtain the identity

$$i \int d^4x \, \theta^\mu_\mu \alpha(x) = \ln \mathcal{J} + i \int d^4x \, m\bar\psi\psi \, \alpha(x). \tag{4.9}$$

The form of the jacobian which follows from the work done in Sect. III–3 is

$$\mathcal{J} = \left[\det \left(e^{-\alpha/2} \right) \right]^{-2} = \lim_{M \to \infty} e^{\mathrm{Tr}' \int d^4x \, \langle x | \alpha \exp -(\slashed{D}/M)^2 | x \rangle}, \tag{4.10}$$

where we have adopted the same regulator as used previously.

The final result is easily obtained from the general heat-kernel calculation of App. B–1, again using the identities of Eqs. (B–1.17), (B–1.18). After some algebra this becomes

$$
\mathrm{Tr}'\langle x \left| \exp -(\not{D}/M)^2 \right| x \rangle
$$
$$
= \frac{iM^4}{16\pi^2} \mathrm{Tr}' \left[1 - \frac{g_3^2 \lambda^a \lambda^b}{32M^4} \sigma^{\mu\nu} \sigma^{\alpha\beta} F_{\mu\nu}^a F_{\alpha\beta}^b + \frac{[D^\mu, D^\nu][D_\mu, D_\nu]}{12M^4} + \cdots \right]
$$
$$
= \frac{3iM^4}{4\pi^2} + \frac{ig_3^2}{48\pi^2} F_{\mu\nu}^a F_a^{\mu\nu} + \cdots. \tag{4.11}
$$

Here we have found both a term which is a divergent constant, and one which involves two-gluon field strengths. The divergent constant corresponds to the infinite zero-point energy of the vacuum. This can be seen by noting that if the zero-point energy is defined by the vacuum matrix element

$$
\langle 0 | \mathcal{H}(x) | 0 \rangle = \frac{E_0}{V} = \langle 0 | \theta^{00}(x) | 0 \rangle, \tag{4.12}
$$

then Lorentz covariance requires a nonzero trace

$$
\langle 0 | \theta^{\mu\nu}(x) | 0 \rangle = \frac{E_0}{V} g^{\mu\nu} \implies \langle 0 | \theta^\mu_\mu(x) | 0 \rangle = 4\frac{E_0}{V}. \tag{4.13}
$$

Thus, a constant in the vacuum matrix element of the trace is just four times the zero-point energy density. It is standard practice to subtract off this zero-point energy, and we shall do so by dropping the constant term. This is similar to the procedure of normal ordering the energy-momentum tensor.

If we now combine these results using Eq. (4.9), we obtain

$$
i \int d^4x \, \theta^\mu_\mu \alpha(x) = i \int d^4x \left[\frac{g_3^2}{48\pi^2} F_{\mu\nu}^a F^{a\mu\nu} + m\bar{\psi}\psi \right] \alpha(x), \tag{4.14}
$$

which is equivalent to the operator relation

$$
\theta^\mu_\mu = \frac{\alpha_s}{12\pi} F_{\mu\nu}^a F^{a\mu\nu} + m\bar{\psi}\psi. \tag{4.15}
$$

One may also derive the trace anomaly via the calculation of Feynman diagrams, the triangle diagrams of Fig. III–1, but with the axial current replaced by the energy-momentum tensor. The trace anomaly is different from the chiral anomaly in that it receives contributions also from gluons. In the Feynman diagram approach, this arises from the replacement of quark lines by gluons, while in the path-integral context it occurs when one considers scale transformations of the gluon field. A full calculation yields

$$
\theta^\mu_\mu = \frac{\beta_{QCD}}{2g_3} F_{\mu\nu}^a F^{a\mu\nu} + m_u \bar{u}u + m_d \bar{d}d + m_s \bar{s}s + \cdots, \tag{4.16}
$$

where β_{QCD} is the beta function of QCD (cf. Eq. (II–2.57b)). The result of our previous calculation, Eq. (4.15), corresponds to the lowest order contribution of a single quark to the beta function.

A feeling of why the beta function enters can be obtained from an extremely simple, but heuristic, derivation of the trace anomaly. Let us rescale the gluon field to $\overline{A}^a_\mu \equiv g_3 A^a_\mu$, such that the massless action becomes

$$\mathcal{L} = -\frac{1}{4g_3^2} \overline{F}^a_{\mu\nu} \overline{F}^{a\mu\nu} + i\overline{\psi}\gamma^\mu \overline{D}_\mu \psi. \tag{4.17}$$

The coupling constant g_3 now enters only as an overall factor in the first term. However in renormalizing the coupling constant, we need to introduce a renormalization scale. If we interpret this coupling as a *running* parameter, the action is no longer invariant under scale transformations. Instead, taking $\lambda = 1 + \delta\lambda$, we find

$$\frac{\delta S}{\delta \lambda} = \int d^4x \, \frac{\partial}{\partial \lambda} \left(-\frac{1}{4g_3^2(\lambda)} \right) \overline{F}^a_{\mu\nu} \overline{F}^{a\mu\nu} = \int d^4x \, \frac{\beta_{QCD}(g_3)}{2g_3} F^a_{\mu\nu} F^{a\mu\nu}, \tag{4.18}$$

where we have changed back to the standard normalization of A^a_μ in the final term. By Noether's theorem, the scale current is no longer conserved, and Eq. (4.16) is reproduced. The need to specify a scale in defining the coupling constant has removed the scale invariance of the theory.

The trace anomaly occupies a significant place in the phenomenology of hadrons because it is the signal for the generation of hadronic masses. Returning to the discussion of masses which began this section, we see that the mass of a state is expressible as a matrix element of the energy-momentum trace. For example, we find for the nucleon state that

$$m_N \bar{u}(\mathbf{p})u(\mathbf{p}) = \langle N(\mathbf{p})|\theta^\mu_\mu|N(\mathbf{p})\rangle$$

$$= \langle N(\mathbf{p})|\frac{\beta_{QCD}}{2g_3} F^a_{\mu\nu} F^{a\mu\nu} + m_s \bar{s}s + m_u \bar{u}u + m_d \bar{d}d|N(\mathbf{p})\rangle. \tag{4.19}$$

The terms containing the light quark masses m_u, m_d are expected to be small, and indeed the 'σ-term' determined in πN scattering (cf. Sect. III–3) implies that they contribute about only 45 MeV. This leaves the bulk of the nucleon's mass to the gluon and s-quark terms in Eq. (4.19), of which the $F^a_{\mu\nu} F^{a\mu\nu}$ part is expected to be dominant. Although this presents a conceptual problem for the naive quark model interpretation of the proton as a composite of three light quarks, it is nevertheless a central result of QCD.

III–5 Chiral anomalies and vacuum structure

There is a fascinating connection between the axial anomaly described previously in this chapter and the vacuum of QCD. This has important phenomenological consequences for both the η' mass and the strong *CP* problem. Here we present an introductory account of this topic [Pe 89].

The θ vacuum

One is used to considering the effect on gluon fields of 'small' gauge transformations, i.e., those which are connected to the identity operator in a continuous manner. There also exist 'large' gauge transformations which change the color gauge fields in a more drastic fashion. For example the gauge transformation [JaR 76] generated by

$$\Lambda_1(\mathbf{x}) \equiv \frac{\mathbf{x}^2 - d^2}{\mathbf{x}^2 + d^2} + \frac{2id\boldsymbol{\tau} \cdot \mathbf{x}}{\mathbf{x}^2 + d^2}, \tag{5.1}$$

where d is an arbitrary parameter and $\boldsymbol{\tau}$ is an $SU(2)$ Pauli matrix in any $SU(2)$ subgroup of $SU(3)$, transforms the null potential $\mathbf{A}(\mathbf{x}) = 0$ into

$$\mathbf{A}_j^{(1)}(\mathbf{x}) = -\frac{i}{g_3} \left(\nabla_j \Lambda_1(\mathbf{x}) \right) \Lambda_1^{-1}(\mathbf{x})$$

$$= -\frac{2d}{g_3 \left(\mathbf{x}^2 + d^2\right)^2} \left[\boldsymbol{\tau}_j (d^2 - \mathbf{x}^2) + 2\mathbf{x}_j (\boldsymbol{\tau} \cdot \mathbf{x}) - 2d(\mathbf{x} \times \boldsymbol{\tau})_j \right]. \tag{5.2}$$

Here, we are using the matrix notation

$$\mathbf{A}_\mu \equiv A_\mu^a \frac{\lambda^a}{2}. \tag{5.3}$$

This potential lies in an $SU(2)$ subgroup of the full color $SU(3)$ group, and is 'large' in the sense that it cannot be brought continuously into the identity. The $\boldsymbol{\tau} \cdot \mathbf{x}$ factor couples the internal color indices to the spatial position such that a path in coordinate space implies a corresponding path in the $SU(2)$ color subspace. All gauge potentials $\mathbf{A}_\mu$ carry a conserved topological charge called the *winding number*,

$$n = \frac{ig_3^3}{24\pi^2} \int d^3x \, \mathrm{Tr} \, \left(\mathbf{A}_i(x) \mathbf{A}_j(x) \mathbf{A}_k(x) \right) \epsilon^{ijk}. \tag{5.4}$$

As can be demonstrated by direct substitution, the gauge field of Eq. (5.2) corresponds to the value $n = 1$. Fields with any integer value of the winding number n can be obtained by repeated applications of $\Lambda_1(x)$, viz.,

$$\Lambda_n(\mathbf{x}) = [\Lambda_1(\mathbf{x})]^n. \tag{5.5}$$

All gauge potentials can be classified into disjoint sectors labeled by their winding number.

The existence of these distinct classes has interesting consequences. For example, consider a configuration of the gluon field that starts off at $t = -\infty$ as the zero potential $\mathbf{A}(\mathbf{x}) = 0$, has some interpolating $\mathbf{A}(x, t)$ for intermediate times, and ends up at $t = +\infty$ lying in the gauge-equivalent configuration $\mathbf{A}(x) = \mathbf{A}^{(1)}(x)$.[5] Then the following integral can be shown to be nonvanishing:

$$\frac{g_3^2}{32\pi^2} \int d^4x \, F_{\mu\nu}^a \tilde{F}^{a\mu\nu} \qquad (\tilde{F}^{a\mu\nu} \equiv \frac{1}{2} \epsilon^{\mu\nu\alpha\beta} F_{\alpha\beta}^a). \tag{5.6}$$

This is surprising because the integrand is a total divergence. As noted previously in Eq. (II–2.23), $F\tilde{F}$ can be written as

$$F_{\mu\nu}^a \tilde{F}^{a\mu\nu} = \partial_\mu K^\mu, \quad K^\mu = \epsilon^{\mu\nu\lambda\sigma}[A_\nu^a F_{\lambda\sigma}^a + \frac{1}{3} g_3 f_{abc} A_\nu^a A_\lambda^b A_\sigma^c], \tag{5.7}$$

and thus the integral can be written as a surface integral at $t = \pm\infty$. For the field configuration under consideration, this reduces to the winding-number integral

$$\begin{aligned}
\frac{g_3^2}{32\pi^2} \int d^4x \, F_{\mu\nu}^a \tilde{F}^{a\mu\nu} &= \frac{g_3^2}{32\pi^2} \int d^4x \, \partial_\mu K^\mu \\
&= \frac{g_3^2}{32\pi^2} \int d^3x \, K_0 \Big|_{t=-\infty}^{t=\infty} \\
&= \frac{g_3^3}{24\pi^2} i \int d^3x \, \epsilon^{ijk} \operatorname{Tr} \left(\mathbf{A}_i^{(1)}(x) \mathbf{A}_j^{(1)}(x) \mathbf{A}_k^{(1)}(x) \right) \\
&= 1.
\end{aligned} \tag{5.8}$$

More generally, the integral of $F\tilde{F}$ gives the *change* in the winding number

$$\frac{g_3^2}{32\pi^2} \int d^4x \, F_{\mu\nu}^a \tilde{F}^{a\mu\nu} = \frac{g_3^2}{32\pi^2} \int d^3x \, K^0 \Big|_{t=-\infty}^{t=\infty} = n_+ - n_- \tag{5.9}$$

between asymptotic gauge-field configurations.

Thus, the vacuum state vector will be characterized by configurations of gluon fields, which fall into classes labeled by the winding number. Moreover, there will exist a corespondence between the gauge transformations $\{\Lambda_n\}$ and unitary operators $\{U_n\}$, which transform the state vectors. For example, a vacuum state dominated by field configurations in the zero winding class ('near' to $A_\mu = 0$) would be transformed by U_1 into configurations with a dominance of $n = 1$ configurations, or more generally,

$$U_1|n\rangle = |n+1\rangle. \tag{5.10}$$

---

[5] Such configurations are known to exist [Co 85].

This implies that a gauge-invariant vacuum state requires contributions from *all* classes, such as the coherent superposition

$$|\theta\rangle = \sum_n e^{-in\theta} |n\rangle, \tag{5.11}$$

where θ is an arbitrary parameter. It follows from Eq. (5.10) that this θ-*vacuum* is gauge-invariant up to an overall phase

$$U_1|\theta\rangle = e^{i\theta}|\theta\rangle. \tag{5.12}$$

The *QCD* vacuum must contain contributions from all topological classes.

The θ term

Given this nontrivial vacuum structure, one requires three ingredients to completely specify QCD: (1) the *QCD* lagrangian, (2) the coupling constant (i.e. Λ_{QCD}), and (3) the vacuum label θ. How can we account for the different vacua corresponding to different choices of θ? In a path-integral representation, the $\theta = 0$ vacuum would imply generic transition elements of the form

$$_{\text{out}}\langle\theta = 0|X|\theta = 0\rangle_{\text{in}} = \int [dA_\mu][d\psi]\,[d\bar\psi]\, X e^{iS_{QCD}} = \sum_{n,m}\,_{\text{out}}\langle m|X|n\rangle_{\text{in}}. \tag{5.13}$$

The presence of a nonzero θ leads to an extra phase,

$$_{\text{out}}\langle\theta|X|\theta\rangle_{\text{in}} = \sum_{n,m} e^{i(m-n)\theta}\,_{\text{out}}\langle m|X|n\rangle_{\text{in}}. \tag{5.14}$$

However, this phase can be accounted for in the path integral by the addition of a new term to S_{QCD}. In particular we have, through the use of Eq. (5.9),

$$\begin{aligned}
_{\text{out}}\langle\theta|X|\theta\rangle_{\text{in}} &= \int [dA_\mu][d\psi][d\bar\psi]\, X\, e^{iS_{QCD}+i\frac{g_3^2}{32\pi^2}\theta\int d^4x\, F_{\mu\nu}^a \tilde{F}^{a\mu\nu}} \\
&= \sum_{n,m} e^{i(m-n)\theta}\,_{\text{out}}\langle m|X|n\rangle_{\text{in}},
\end{aligned} \tag{5.15}$$

where X is some operator. We see that the quantity $(m - n)$ given by the winding-number difference of the fields contributing to the path integral is equivalent to a new exponential factor containing $F_{\mu\nu}^a \tilde{F}^{a\mu\nu}$. Thus, a correct procedure for doing calculations involving θ vacua is to follow the ordinary path-integral methods but with a *QCD* lagrangian containing the new term

$$\mathcal{L}_{QCD} = \mathcal{L}_{QCD}^{(\theta=0)} + \theta\frac{g_3^2}{32\pi^2} F_{\mu\nu}^a \tilde{F}^{a\mu\nu}. \tag{5.16}$$

The parameter θ is to be considered a coupling constant. Since the operator $F\tilde{F}$ is P-odd and T-odd, a nonzero θ can induce measurable T violation. In Sect. IX–4, we shall show how to connect θ to physical observables. There is an important distinction between the various θ vacua of QCD and the many possible vacuum states of a spontaneously broken symmetry such as the Higgs sector of the electroweak theory. In the latter case, the various possible vacuum expectation values of the Higgs field label different states within the same theory. In contrast, each value of θ corresponds to a different theory, just as each value of Λ_{QCD} would label a different theory. Specifying θ and Λ_{QCD} then specifies the content of the version of QCD used by Nature.

Connection with chiral rotations

There is a connection between the axial anomaly and the presence of a θ vacuum ['tH 76a,b]. It involves the matrix element of $F\tilde{F}$ as follows. Consider the limit of N_f massless quarks. The $U(1)$ axial current

$$J_{5\mu}^{(0)} = \sum_{j=1}^{N_f} \bar{\psi}_j \gamma_\mu \gamma_5 \psi_j \tag{5.17}$$

is not conserved due to the anomaly,

$$\partial^\mu J_{5\mu}^{(0)} = \frac{N_f \alpha_s}{4\pi} F_{\mu\nu}^a \tilde{F}^{a\mu\nu}. \tag{5.18}$$

However, because of the fact that $F\tilde{F}$ is a total divergence, one can define a new conserved current

$$\tilde{J}_{5\mu} = J_{5\mu}^{(0)} - \frac{N_f \alpha_s}{4\pi} K_\mu. \tag{5.19}$$

While $\tilde{J}_{5\mu}$ does form a conserved charge,

$$\tilde{Q}_5 = \int d^3x \, \tilde{J}_{5,0}(x), \tag{5.20}$$

neither $\tilde{Q}_5$ nor $\tilde{J}_{5\mu}$ is gauge-invariant. In fact, under the gauge transformation Λ_1 of Eq. (5.1), it follows from Eq. (5.8) that the operator $\tilde{Q}_5$ changes by a c-number integer

$$U_1 \tilde{Q}_5 U_1^{-1} = \tilde{Q}_5 - 2N_f. \tag{5.21}$$

This tells us that in the world of massless quarks, the different θ-vacua are related by a chiral $U(1)$ transformation,

$$U_1 e^{i\alpha\tilde{Q}_5}|\theta\rangle = U_1 e^{i\alpha\tilde{Q}_5} U_1^{-1} U_1|\theta\rangle = e^{i(\theta - 2N_f\alpha)} e^{i\alpha\tilde{Q}_5}|\theta\rangle, \tag{5.22}$$

or, from Eq. (5.12),

$$e^{i\alpha \tilde{Q}_5}|\theta\rangle = |\theta - 2N_f\alpha\rangle, \tag{5.23}$$

where α is a constant. Therefore, in the limit of massless quarks, when $\tilde{Q}_5$ is a conserved quantity, all of the θ vacua are equivalent and one can transform away the θ dependence by a chiral $U(1)$ transformation. The same is not true if quarks have mass, as the mass terms in $\mathcal{L}_{QCD}$ are not invariant under a chiral transformation. We shall return to this topic in Sect. IX–4.

To summarize, one finds that the existence of topologically nontrivial gauge transformations, and of field configurations which make transitions between the different topological sectors of the theory, leads to the existence of nonvanishing effects from a new term in the QCD action. Chiral rotations can change the value of θ, allowing it to be rotated away if any of the quarks are massless. However, for massive quarks, the net effect is a measurable CP-violating term in the QCD lagrangian.

III–6 Baryon- and lepton-number violation in the Standard Model

An even more dramatic effect arises from an anomaly in the current for the total baryon plus lepton number $(B + L)$. Baryon number appears to be a conserved quantity when Noether's theorem is applied to the lagrangian of the Standard Model, as is total lepton number.[6] The invariances are

$$q \to e^{i\varphi_B}q, \quad \ell \to e^{i\varphi_L}\ell \tag{6.1}$$

for all quarks q and leptons ℓ. The corresponding currents involve the sum over all quarks and leptons

$$J_B^\mu = \frac{1}{3}(\bar{u}\gamma^\mu u + \bar{d}\gamma^\mu d + \cdots)$$
$$J_L^\mu = \bar{e}\gamma^\mu e + \bar{v}_{eL}\gamma^\mu v_{eL} + \cdots, \tag{6.2}$$

where the normalization of the baryon current is chosen to give a baryon a charge of $+1$.

The baryon current is vectorial, and naively might not be expected to have an anomaly. However, the coupling of the quarks to the $SU(2)_L$ and $U(1)_Y$ gauge bosons violates parity, so that there are VVA triangle diagrams involving the baryon current with two gauge currents. For example, the triangle diagram

---

[6] If there are neutrino Majorana masses, lepton number will be violated. However, this is independent of the anomaly effect discussed in this section. Majorana masses will be discussed in Chap. VI.

involving the baryon current with the $U(1)_Y$ hypercharge current has a VVA triangle involving the quantum number sum

$$\text{Tr}(B(Y_L + Y_R)(Y_L - Y_R)) = -2 \tag{6.3}$$

where $B = 1/3$ for quarks and $B = 0$ for leptons. These diagrams then yield an anomaly. Because the axial current of this triangle is a gauge current, any gauge-invariant regularization of the triangle diagram will place the anomaly in global baryon-number current even though it is vectorial (see the discussion surrounding Eq. (3.22)). Similar anomalies occur in the lepton number current.[7] The anomalies cancel if we take the difference of the baryon and lepton currents, with the resulting anomaly equations

$$\partial_\mu (J_B^\mu - J_L^\mu) = 0$$
$$\partial_\mu (J_B^\mu + J_L^\mu) = \frac{3}{32\pi^2} \left(g_2^2 F_{\mu\nu}^i \tilde{F}_i^{\mu\nu} - g_1^2 B_{\mu\nu} \tilde{B}^{\mu\nu} \right). \tag{6.4}$$

Here we see that, because of the anomaly, baryon number is in fact not conserved in the Standard Model, although $B - L$ is.

However, the baryon-number violation due to the anomaly is unmeasurably small at low temperature. Any transition that would change baryon number is non-perturbative in nature, as it is not seen in the usual perturbative Feynman rules. In weakly coupled field theory, such nonperturbative phenomena are suppressed in rate by a factor ['tH 76b]

$$[e^{-8\pi^2/g_2^2}]^2 \sim 10^{-160}, \tag{6.5}$$

so that such transitions are unobservable.

At high temperatures the situation is different [KuRS 85]. The classical solution mediating a transition which changes baryon number, a *sphaleron* [KlM 84], is known in the limit $\theta_w \to 0$ and the corrections due to a nonzero θ_w can be estimated. The solution has an energy around $E_{\text{sph}} \sim 10$ TeV, taking into account the measured Higgs-boson mass. At high temperature, thermal effects can cause transitions with a Boltzmann factor $e^{-E_{\text{sph}}/T}$, and at very high temperatures all suppressions disappear and the rate per unit volume scales with the temperature $\Gamma/V \sim T^4$.

This has an important consequence – at equilibrium in the early Universe an initial excess of baryons can disappear. More precisely, the equilibrium value of $B + L$ is zero at high temperature. However, $B - L$ is still conserved, so that an initial excess of $B - L$ will be preserved.

---

[7] Because possible right-handed neutrinos have no gauge couplings, their presence would not modify the anomaly.

It is natural to ask if a sufficiently large baryon asymmetry in the Universe can be generated by out-of-equilibrium processes near the electroweak phase transition, using only Standard Model interactions. The answer appears to be negative [GaHOP 94], as the necessary *CP* violation within the Standard Model is too small and the phase transition is not strong enough. New interactions near the weak scale could provide the needed extra physics. Alternatively, the residual baryon asymmetry may arise from a net $B - L$ generated in the Universe before the electroweak epoch. Within the context of the Standard Model interactions, the simplest such possibility is *leptogenesis* involving heavy right-handed neutrinos with Majorana masses. This mechanism will be discussed in Sect. VI–6.

Problems

(1) **Currents and anomalies**

    (a) Verify that all currents coupled to gauge bosons in the Standard Model are anomaly free.

    (b) Find the relative strength of the anomaly coupling of the baryon number current to the $SU(2)_L$ and $U(1)_R$ gauge bosons.

(2) **Trace anomaly in** *QED*

    In d dimensions, the trace of the energy-momentum tensor does not vanish classically, except at $d = 4$. For example, in massless *QED* the energy-momentum tensor,

$$\theta^{\mu\nu} = -F^{\mu}_{\lambda} F^{\lambda\nu} + \frac{1}{4} g^{\mu\nu} F^{\lambda\sigma} F_{\lambda\sigma} + \frac{i}{2} \bar{\psi} \gamma^{\mu} \overleftrightarrow{D}^{\nu} \psi,$$

    has trace $\theta^{\mu}_{\mu} = \frac{d-4}{4} F^{\lambda\sigma} F_{\lambda\sigma}$. In the renormalization of the operator $F^{\lambda\sigma} F_{\lambda\sigma}$, one encounters a renormalization constant which diverges as $d \to 4$. Use this feature to calculate the *QED* trace anomaly using dimensional regularization.

IV

Introduction to effective field theory

The purpose of an effective field theory is to represent in a simple way the dynamical content of a theory in the low-energy limit. One uses only those light degrees of freedom that are active at low energy, and treats their interactions in a full field-theoretic framework. The effective field theory is often technically non-renormalizable, yet loop diagrams are included and renormalization of the physical parameters is readily accomplished.

Effective field theory is used in all aspects of the Standard Model and beyond, from *QED* to superstrings. Perhaps the best setting for learning about the topic is that of chiral symmetry. Besides being historically important in the development of effective field theory techniques, chiral symmetry is a rather subtle subject, which can be used to illustrate all aspects of the method, viz., the low-energy expansion, non-leading behavior, loops, renormalization and symmetry breaking. In addition, the results can be tested directly by experiment since the chiral effective field theory provides a framework for understanding the very low-energy limit of *QCD*.

In this chapter we introduce effective field theory by a study of the linear sigma model, and discuss the generalization of these techniques to other settings.

IV–1 Effective lagrangians and the sigma model

The linear sigma model, introduced in Sects. I–4, I–6, provides a 'user friendly' introduction to effective field theory because all the relevant manipulations can be explicitly demonstrated. The Goldstone boson fields, the pions, are present at all stages of the calculation. It also introduces many concepts which are relevant for the low enegy limit of *QCD*. However, low-energy *QCD* is far less transparent, involving a transference from the quark and gluon degrees of freedom of the original lagrangian to the pions of the physical spectrum. Nevertheless, the low-energy properties of the two theories have many similarities.

106

The first topic that we need to describe is that of an 'effective lagrangian'. First, let us illustrate this concept by simply quoting the result to be derived below. Recall the sigma model of Eq. (I–4.14),

$$\mathcal{L} = \bar{\psi} i \partial\!\!\!/ \psi + \frac{1}{2} \partial_\mu \boldsymbol{\pi} \cdot \partial^\mu \boldsymbol{\pi} + \frac{1}{2} \partial_\mu \sigma \partial^\mu \sigma$$
$$- g \bar{\psi} \left(\sigma - i \boldsymbol{\tau} \cdot \boldsymbol{\pi} \gamma_5 \right) \psi + \frac{1}{2} \mu^2 \left(\sigma^2 + \boldsymbol{\pi}^2 \right) - \frac{\lambda}{4} \left(\sigma^2 + \boldsymbol{\pi}^2 \right)^2 . \quad (1.1)$$

This is a renormalizable field theory of pions, and from it one can calculate any desired pion amplitude. Alternatively, if one works at low-energy ($E \ll \mu$), then it turns out that all matrix elements of pions are contained in the rather different looking 'effective lagrangian'

$$\mathcal{L}_{\text{eff}} = \frac{F^2}{4} \text{Tr} \left(\partial_\mu U \partial^\mu U^\dagger \right), \qquad U = \exp i \boldsymbol{\tau} \cdot \boldsymbol{\pi} / F, \quad (1.2)$$

where $F = v = \sqrt{\mu^2/\lambda}$ at tree level (cf. Eq. (I–6.9)). This effective lagrangian is to be used by expanding in powers of the pion field

$$\mathcal{L}_{\text{eff}} = \frac{1}{2} \partial_\mu \boldsymbol{\pi} \cdot \partial^\mu \boldsymbol{\pi} + \frac{1}{6F^2} \left[\left(\boldsymbol{\pi} \cdot \partial_\mu \boldsymbol{\pi} \right)^2 - \boldsymbol{\pi}^2 \left(\partial_\mu \boldsymbol{\pi} \cdot \partial^\mu \boldsymbol{\pi} \right) \right] + \cdots, \quad (1.3)$$

and taking tree-level matrix elements. This procedure is a relatively simple way of encoding all the low-energy predictions of the theory. Moreover, with this effective lagrangian is the starting point of a full effective field theory treatment including loops, which we will develop in Sect. IV–3.

Representations of the sigma model

In order to embark on the path to the effective field theory approach, let us rewrite the sigma model lagrangian as

$$\mathcal{L} = \frac{1}{4} \text{Tr} \left(\partial_\mu \Sigma \partial^\mu \Sigma^\dagger \right) + \frac{\mu^2}{4} \text{Tr} \left(\Sigma^\dagger \Sigma \right) - \frac{\lambda}{16} \left[\text{Tr} \, \Sigma^\dagger \Sigma \right]^2$$
$$+ \bar{\psi}_L i \partial\!\!\!/ \psi_L + \bar{\psi}_R i \partial\!\!\!/ \psi_R - g \left(\bar{\psi}_L \Sigma \psi_R + \bar{\psi}_R \Sigma^\dagger \psi_L \right), \quad (1.4)$$

with $\Sigma = \sigma + i \boldsymbol{\tau} \cdot \boldsymbol{\pi}$. The model is invariant under the $SU(2)_L \times SU(2)_R$ transformations

$$\psi_L \to L \psi_L, \qquad \psi_R \to R \psi_R, \qquad \Sigma \to L \Sigma R^\dagger \quad (1.5)$$

for L, R in $SU(2)$. This is the linear representation.[1]

---

[1] A number of distinct 2 × 2 matrix notations, among them Σ, U, and M, are commonly employed in the literature for either the linear or the nonlinear cases. It is always best to check the definition being employed and to learn to be flexible.

After symmetry breaking and the redefinition of the σ field,

$$\sigma = v + \tilde{\sigma}, \qquad v = \sqrt{\frac{\mu^2}{\lambda}}, \tag{1.6}$$

the lagrangian reads[2]

$$\mathcal{L} = \frac{1}{2}\left(\partial_\mu\tilde{\sigma}\partial^\mu\tilde{\sigma} - 2\mu^2\tilde{\sigma}^2\right) + \frac{1}{2}\partial_\mu\boldsymbol{\pi}\cdot\partial^\mu\boldsymbol{\pi} - \lambda v\tilde{\sigma}\left(\tilde{\sigma}^2 + \boldsymbol{\pi}^2\right)$$
$$- \frac{\lambda}{4}\left(\tilde{\sigma}^2 + \boldsymbol{\pi}^2\right)^2 + \bar{\psi}\left(i\not{\partial} - gv\right)\psi - g\bar{\psi}\left(\tilde{\sigma} - i\boldsymbol{\tau}\cdot\boldsymbol{\pi}\gamma_5\right)\psi, \tag{1.7}$$

indicating massless pions and a nucleon of mass gv. All the interactions in the model are simple nonderivative polynomial couplings.

There are other ways to display the content of the sigma model besides the above linear representation. For example, instead of $\tilde{\sigma}$ and $\boldsymbol{\pi}$ one could define fields S and $\boldsymbol{\varphi}$,

$$S \equiv \sqrt{(\tilde{\sigma} + v)^2 + \boldsymbol{\pi}^2} - v = \tilde{\sigma} + \cdots, \qquad \boldsymbol{\varphi} \equiv \frac{v\boldsymbol{\pi}}{\sqrt{(\tilde{\sigma} + v)^2 + \boldsymbol{\pi}^2}} = \boldsymbol{\pi} + \cdots, \tag{1.8}$$

where one expands in inverse powers of v. For lack of a better name, we can call this the *square-root* representation. The lagrangian can be rewritten in terms of the variables S and $\boldsymbol{\varphi}$ as

$$\mathcal{L} = \frac{1}{2}\left[(\partial_\mu S)^2 - 2\mu^2 S^2\right] + \frac{1}{2}\left(\frac{v+S}{v}\right)^2\left[(\partial_\mu\boldsymbol{\varphi})^2 + \frac{(\boldsymbol{\varphi}\cdot\partial_\mu\boldsymbol{\varphi})^2}{v^2 - \boldsymbol{\varphi}^2}\right]$$
$$- \lambda v S^3 - \frac{\lambda}{4}S^4 + \bar{\psi}i\not{\partial}\psi - g\left(\frac{v+S}{v}\right)\bar{\psi}\left[\left(v^2 - \boldsymbol{\varphi}^2\right)^{1/2} - i\boldsymbol{\varphi}\cdot\boldsymbol{\tau}\gamma_5\right]\psi. \tag{1.9}$$

Although this looks a bit forbidding, no longer having simple polynomial interactions, it is nothing more than a renaming of the fields. This form has several interesting features. The pion-like fields, still massless, no longer occur in the potential part of the lagrangian, but instead appear with derivative interactions. For vanishing S, this is called the *nonlinear sigma model*.

Another nonlinear form, the *exponential* parameterization, will prove to be of importance to us. Here the fields are written as

$$\Sigma = \sigma + i\boldsymbol{\tau}\cdot\boldsymbol{\pi} = (v+S)U, \qquad U = \exp\left(i\boldsymbol{\tau}\cdot\boldsymbol{\pi}'/v\right) \tag{1.10}$$

such that $\boldsymbol{\pi}' = \boldsymbol{\pi} + \cdots$. Using this form, we find

---

[2] Here, and in subsequent expressions for $\mathcal{L}$, we drop all additive constant terms.

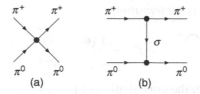

Fig. IV–1 Contributions to $\pi^+\pi^0$ elastic scattering.

$$\mathcal{L} = \frac{1}{2}\left[(\partial_\mu S)^2 - 2\mu^2 S^2\right] + \frac{(v+S)^2}{4}\,\text{Tr}\,(\partial_\mu U \partial^\mu U^\dagger)$$

$$- \lambda v S^3 - \frac{\lambda}{4}S^4 + \bar\psi i \partial\!\!\!/ \psi - g(v+S)\left(\bar\psi_L U \psi_R + \bar\psi_R U^\dagger \psi_L\right). \qquad (1.11)$$

The quantity U transforms under $SU(2)_L \times SU(2)_R$ in the same way as does Σ, i.e.,

$$U \to L U R^\dagger. \qquad (1.12)$$

This lagrangian is reasonably compact and also has only derivative couplings for pions.

Representation independence

We have introduced three sets of interactions with very different appearances. They are all nonlinearly related. In each of these forms the free-particle sector, found by looking at terms bilinear in the field variables, has the same masses and normalizations. To compare their dynamical content, let us calculate the scattering of the Goldstone bosons of the theory, specifically $\pi^+\pi^0 \to \pi^+\pi^0$. The diagrams that enter at tree level are displayed in Fig. IV–1. The relevant terms in the lagrangians and their tree-level scattering amplitudes are as follows.

(1) *Linear form*:

$$\mathcal{L}_I = -\frac{\lambda}{4}\left(\pi^2\right)^2 - \lambda v \tilde\sigma \pi^2,$$

$$i\mathcal{M}_{\pi^+\pi^0 \to \pi^+\pi^0} = -2i\lambda + (-2i\lambda v)^2 \frac{i}{q^2 - m_\sigma^2}$$

$$= -2i\lambda \left[1 + \frac{2\lambda v^2}{q^2 - 2\lambda v^2}\right] = \frac{iq^2}{v^2} + \cdots, \qquad (1.13)$$

where $q = p'_+ - p_+ = p_0 - p'_0$ and the relation $m_\sigma^2 = 2\lambda v^2 = 2\mu^2$ has been used. The contributions of Figs. IV–1(a), 1(b) are seen to cancel at $q^2 = 0$. Thus, to leading order, the amplitude is momentum-dependent even though the interaction contains no derivatives. The vanishing of the amplitudes at zero momentum is universal in the limit of exact chiral symmetry.

(2) *Square-root representation*:

$$\mathcal{L}_I = \frac{1}{2} \frac{\left(\boldsymbol{\varphi} \cdot \partial_\mu \boldsymbol{\varphi}\right)^2}{\left(v^2 - \boldsymbol{\varphi}^2\right)} + \frac{S}{v} \partial_\mu \boldsymbol{\varphi} \cdot \partial^\mu \boldsymbol{\varphi}. \tag{1.14}$$

For this case, the contribution of Fig. IV–1(b) involves four factors of momentum, two at each vertex, and so may be dropped at low-energy. For Fig. IV–1(a) we find

$$\mathcal{L}_I = \frac{1}{2v^2} \left(\varphi^0 \partial_\mu \varphi^0 + \varphi^+ \partial_\mu \varphi^- + \varphi^- \partial_\mu \varphi^+\right)^2,$$

$$i\mathcal{M}_{\varphi^+ \varphi^0 \to \varphi^+ \varphi^0} = \frac{i\left(p'_+ - p_+\right)^2}{v^2} = \frac{i\,q^2}{v^2} + \cdots. \tag{1.15}$$

(3) *Exponential representation*:

$$\mathcal{L}_I = \frac{(v + S)^2}{4} \operatorname{Tr}\left(\partial_\mu U \partial^\mu U^\dagger\right) + \cdots. \tag{1.16}$$

Again Fig. IV–1(b) has a higher-order $(\mathcal{O}(p^4))$ contribution, leaving only Fig. IV–1(a),

$$\mathcal{L}_I = \frac{1}{6v^2} \left[\left(\boldsymbol{\pi}' \cdot \partial_\mu \boldsymbol{\pi}'\right)^2 - \boldsymbol{\pi}'^2 \left(\partial_\mu \boldsymbol{\pi}' \cdot \partial^\mu \boldsymbol{\pi}'\right)\right],$$

$$i\mathcal{M}_{\pi^+ \pi^0 \to \pi^+ \pi^0} = \frac{i\left(p'_+ - p_+\right)^2}{v^2} + \cdots. \tag{1.17}$$

The lesson to be learned is that all three representations give the *same* answer despite very different forms and even different Feynman diagrams. A similar conclusion would follow for any other observable that one might wish to calculate.

The above analysis demonstrates a powerful field-theoretic theorem, proved first by R. Haag [Ha 58, CoWZ 69, CaCWZ 69], on representation independence. It states that if two fields are related nonlinearly, e.g., $\varphi = \chi F(\chi)$ with $F(0) = 1$, then the same experimental observables result if one calculates with the field φ using $\mathcal{L}(\varphi)$ or instead with χ using $\mathcal{L}(\chi F(\chi))$. The proof consists basically of demonstrating that (i) two S-matrices are equivalent if they have the same single particle singularities, and (ii) since $F(0) = 1$, φ and χ have the same free field behavior and single-particle singularities. This result can be made plausible if we think of the scattering in non-mathematical terms. If the free particles are isolated, they have the same mass and charge and experiment cannot tell the φ particle from the χ particle. At this level they are in fact the same particles, due to $F(0) = 1$. The scattering experiment is then performed by colliding the particles. The results cannot depend on whether a theorist has chosen to calculate the amplitude using the φ or the χ names. That is, the physics cannot depend on a labeling convention.

This result is quite useful as it lets us employ nonlinear representations in situations where they can simplify the calculation. The linear sigma model is a good example. We have seen that the amplitudes of this theory are momentum-dependent. Such behavior is obtained naturally when one uses the nonlinear representations, whereas for the linear representation more complicated calculations involving assorted cancelations of constant terms are required to produce the correct momentum dependence. In addition, the nonlinear representations allow one to display the low-energy results of the theory without explicitly including the massive $\tilde{\sigma}$ (or S) and ψ fields.

IV–2 Integrating out heavy fields

When one is studying physics at some energy scale E, one must explicitly take into account all the particles which can be produced at that energy. What is the effect of fields whose quanta are too heavy to be directly produced? They may still be felt through virtual effects. When using an effective low-energy theory, one does not include the heavy fields in the lagrangian, but their virtual effects are represented by various couplings between light fields. The process of removing heavy fields from the lagrangian is called *integrating out* the fields. Here, we shall explore this process.

The decoupling theorem

There is a general result in field theory, called the decoupling theorem, which describes how the heavy particles must enter into the low-energy theory [ApC 75, OvS 80]. The theorem states that *if the remaining low-energy theory is renormalizable, then all effects of the heavy-particle appear either as a renormalization of the coupling constants in the theory or else are suppressed by powers of the heavy-particle mass.* We shall not display the formal proof. However, the result is in accord with physical expectations. If the heavy particle's mass becomes infinite, one would indeed expect the influence of the particle to disappear. Any shift in the coupling constants is not directly observable because the values of these couplings are always determined from experiment. Inverse powers of heavy-particle mass arise from propagators involving virtual exchange of the heavy particle.

In the Standard Model, the most obvious example of this is the role played in low-energy physics by the $W^{\pm}$ and Z gauge bosons. For example, while $W^{\pm}$-loops can contribute to the renormalization of the electric charge, the effect cannot be isolated at low energies. Also, the residual form of $W^{\pm}$-exchange amplitudes is that of a local product of two weak currents (Fermi interaction) with coupling strength G_F. Its effect is suppressed because $G_F \propto M_W^{-2}$.

However, in the Standard Model there is an example where the heavy-particle effects do *not* decouple. For a heavy top quark, there are many loop diagrams which do not vanish as $m_t \to \infty$, but instead behave as m_t^2 or $\ln(m_t^2)$. This can occur because the electroweak theory with the t quark removed violates the $SU(2)_L$ symmetry, as the full $\binom{t}{b}$ doublet is no longer present. Without the constraint of weak-isospin symmetry, the theory is not renormalizable and new divergences can occur in flavor-changing processes. These would-be divergences are cut off in the real theory by the mass m_t. Note that at the same time as $m_t \to \infty$, the top quark Yukawa coupling also goes to infinity, and hence induces strong coupling, which can also lead to a violation of decoupling.

In the sigma model, *all* the low-energy couplings of the pions are proportional to powers of $1/v^2 \propto 1/m_\sigma^2$, the simplest example being Eq. (1.9). Hence the effective renormalizable theory is in fact a free field theory, without interactions. The interactions have been suppressed by powers of heavy-particle masses. We shall use the energy expansion of the next section to organize the expansion in powers of the inverse heavy mass.

Integrating out heavy fields at tree level

The name of this procedure comes from the path-integral formalism, where the process of integrating out a heavy field H and leaving behind light fields ℓ_i is defined in terms of an effective action $W_{\text{eff}}[\ell_i]$,

$$Z[\ell_i] = e^{i W_{\text{eff}}[\ell_i]} \equiv \int [dH] \, e^{i \int d^4x \, \mathcal{L}(H(x), \ell_i(x))}. \tag{2.1}$$

However, the procedure is equally familiar from perturbation theory, in which the effect of the path integral is represented by a sum of Feynman diagrams.

Let us proceed with a path-integral example. Consider a linear coupling of H to some combination of fields J, with the lagrangian

$$\mathcal{L} = \frac{1}{2} \left(\partial_\mu H \partial^\mu H - m_H^2 H^2 \right) + JH. \tag{2.2}$$

One way to integrate out H is to 'complete the square', i.e., we write

$$\int d^4x \, \mathcal{L}(H, J) = \int d^4x \left[-\frac{1}{2} H \mathcal{D} H + JH \right]$$

$$= -\frac{1}{2} \int d^4x \left[\left(H - \mathcal{D}^{-1} J \right) \mathcal{D} \left(H - \mathcal{D}^{-1} J \right) - J \mathcal{D}^{-1} J \right]$$

$$= -\frac{1}{2} \int d^4x \left[H' \mathcal{D} H' - J \mathcal{D}^{-1} J \right], \tag{2.3}$$

where we have used the shorthand notations,

$$\mathcal{D} = \Box + m_H^2,$$

$$\mathcal{D}^{-1} J = -\int d^4 y \, \Delta_F(x - y) J(y),$$

$$(\Box_x + m_H^2) \, \Delta_F(x - y) = -\delta^4(x - y),$$

$$H'(x) = H(x) + \int d^4 y \, \Delta_F(x - y) J(y),$$

$$\int d^4 x \, J \mathcal{D}^{-1} J = -\int d^4 x \, d^4 y \, J(x) \Delta_F(x - y) J(y), \qquad (2.4)$$

and have integrated by parts repeatedly. Since we integrate in the path integral over all values of the field at each point of spacetime, we may change variables $[dH] = [dH']$ so

$$Z[J] = e^{i W_{\text{eff}}[J]} = \int [dH] e^{i \int d^4 x \, \mathcal{L}(H, J)}$$

$$= \int [dH'] e^{i \int d^4 x \left[-\frac{1}{2} H' \mathcal{D} H' + \frac{1}{2} J \mathcal{D}^{-1} J \right]}$$

$$= Z[0] \, e^{\frac{i}{2} \int d^4 x \, J \mathcal{D}^{-1} J}, \qquad (2.5)$$

where

$$Z[0] = \int [dH'] e^{i \int d^4 x \left[-\frac{1}{2} H' \mathcal{D} H' \right]}. \qquad (2.6)$$

Here, $Z[0]$ is an overall constant that can be dropped from further consideration. From this result we obtain the effective action

$$W_{\text{eff}}[J] = -\frac{1}{2} \int d^4 x \, d^4 y \, J(x) \, \Delta_F(x - y) J(y). \qquad (2.7a)$$

This action is nonlocal because it includes an integral over the propagator. However, the heavy-particle propagator is peaked at small distances, of order $1/m_H^2$. This allows us to obtain a local lagrangian by Taylor expanding $J(y)$ as

$$J(y) = J(x) + (y - x)^\mu \left[\partial_\mu J(y) \right]_{y=x} + \cdots . \qquad (2.7b)$$

Keeping the leading term and using

$$\int d^4 y \, \Delta_F(x - y) = -\frac{1}{m_H^2}, \qquad (2.8)$$

we obtain

$$W_{\text{eff}}[J] = \int d^4 x \, \frac{1}{2 m_H^2} J(x) \, J(x) + \cdots , \qquad (2.9)$$

where the ellipses denote terms suppressed by additional powers of m_H. Outside of the path-integral context, this result is familiar from W-exchange in the weak interactions.

Matching the sigma model at tree level

We can apply this procedure to the lagrangian for the sigma model, where the scalar field S is heavy with respect to the Goldstone bosons. Thus, considering the theory in the low-energy limit, we may integrate out the field S. Referring to Eq. (1.11) and neglecting the S^2 interactions, it is clear that we should make the identifications $H \rightarrow S$ and $J \rightarrow v \, \mathrm{Tr} \, (\partial_\mu U \partial^\mu U^\dagger)/2$. The effective lagrangian then takes the form

$$\mathcal{L}_{\mathrm{eff}} = \frac{v^2}{4} \, \mathrm{Tr} \, \left(\partial_\mu U \partial^\mu U^\dagger\right) + \frac{v^2}{8m_S^2} \left[\, \mathrm{Tr} \, \left(\partial_\mu U \partial^\mu U^\dagger\right)\right]^2 + \cdots, \qquad (2.10)$$

where the second term in Eq. (2.10) is the result of integrating out the S-field and gives rise to the diagram of Fig. IV–1(b). Additional tree-level diagrams are implied by the sigma model when one includes the S^3 and S^4 interactions. Since these carry more derivatives, the above result is the correct tree-level answer with up to four derivatives.

This calculation is an illustration of the concept of 'matching', here applied at tree level. We match the effective field theory to the full theory in order to reproduce the correct matrix elements. From the starting point of Eq. (1.11), we expect that there will be a low-energy effective lagrangian, which is written as an expansion in powers of $\mathrm{Tr} \, (\partial_\mu U \partial^\mu U^\dagger)$, with coefficients that are initially unknown. In the matching procedure, we choose the coefficients to be those appropriate for the full theory.

In calculating transitions of pions, this is then used by expanding the U matrix in terms of the pion fields and taking matrix elements. At the lowest energies, only the lagrangian with two derivatives is required, justifying the result quoted in Eq. (1.2).[3] Interested readers may verify that the two terms in Eq. (2.10) reproduce the first two terms in the $\pi^+\pi^0$ scattering amplitude previously obtained in Eq. (1.13). However, we have gained a great deal by using the effective lagrangian framework, because now *all* matrix elements of pions can be calculated simply to this order in the energy by simply expanding the effective lagrangian and reading off the answer.

---

[3] We will show that this term is not modified by loop effects, aside from the renomalization of the parameter v.

IV–3 Loops and renormalization

The treatment above has left us with a nonlinear effective lagrangian of the form that is called 'non-renormalizable'. It is also incomplete because loop diagrams have not yet been considered. One might worry that because the effective lagrangian is non-renormalizeable, loops would cause trouble. However, that is not the case. Indeed, this situation helps demonstrate the 'effectiveness' of effective field theory – we will see that the important loop processes are reproduced in a simpler manner using the effective field theory.

Continuing our treatment of the linear sigma model, let us display the precise formal correspondence between the full theory and the effective theory. If we are only considering matrix elements involving the light pions, we can write the path integral defining the theory[4] as

$$Z[\mathbf{j}] = N \int [d\boldsymbol{\pi}(x)] [d\sigma(x)] \exp\left[i \int d^4x \ (\mathcal{L}[\boldsymbol{\pi}(x), \sigma(x)] + \mathbf{j}(x) \cdot \boldsymbol{\pi}(x))\right].$$

$$(3.1)$$

When working at low energies, we can then integrate out the heavy field σ to produce the effective theory

$$Z[\mathbf{j}] = N \int [d\boldsymbol{\pi}(x)] \exp\left[i \int d^4x \ (\mathcal{L}_{\text{eff}}[\boldsymbol{\pi}(x)] + \mathbf{j}(x) \cdot \boldsymbol{\pi}(x))\right]. \qquad (3.2)$$

Because the σ field is heavy, its influence will not propagate far and the resulting effective lagrangian will be local. However, this correspondence emphasizes the fact that one is still left with a full field theory. It is not only at tree level that the effective lagrangian must be applied. Loop processes must also be considered, as is the case in any field theory. The original theory involves both σ and π loops, while the effective theory has only the π loop diagrams. We will demonstrate how to match the effective theory to the full theory through an explicit calculation.

In order to accomplish the renormalization and matching procedure for the effective theory we will need a lagrangian similar to the tree level form, but with initially unknown coefficients that will be chosen later, i.e.,

$$\mathcal{L}_{\text{eff}} = \frac{v^2}{4} \text{Tr} \left(\partial_\mu U \partial^\mu U^\dagger\right)$$
$$+ \ell_1 [\text{Tr} \left(\partial_\mu U \partial^\mu U^\dagger\right)]^2 + \ell_2 \text{Tr} \left(\partial_\mu U \partial_\nu U^\dagger\right) \text{Tr} \left(\partial^\mu U \partial^\nu U^\dagger\right). \qquad (3.3)$$

This is the most general form consistent with the symmetry $U \to L U R^\dagger$, containing up to four derivatives. The first portion of this lagrangian, when expanded in terms of the pion field, yields the usual pion propagator as well as the lowest-order result for the $\pi\pi$ scattering amplitudes.

---

[4] Recall that $\sigma = S$ in some previous formulas.

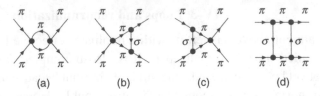

Fig. IV–2 A subset of one-loop diagrams contributing to $\pi^+\pi^0$ elastic scattering.

Let us again consider the process $\pi^+ + \pi^0 \to \pi^+ + \pi^0$, this time to one loop. The full linear sigma model is renormalizeable and will yield finite predictions in terms of the (renormalized) parameters of the theory. The effective theory has been constructed to have the same vertices at the lowest energies, but will have quite different high-energy properties because it is missing the extra high-energy degree of freedom. There will be new divergences present in perturbation theory. However, the low-energy effects will be similar in both calculations.

For example, consider the set of diagrams depicted in Fig. IV–2. In the full theory, all of these diagrams exist, and our previous result of Eq. (1.13) can be used to write the combined amplitudes as

$$
i\mathcal{M}_{\text{full}} = \int \frac{d^4k}{(2\pi)^4} \left[-2i\lambda + (-2i\lambda v)^2 \frac{i}{(k+p_+)^2 - m_\sigma^2} \right] \frac{i}{(k+p_++p_0)^2} \frac{i}{k^2}
$$
$$
\times \left[-2i\lambda + (-2i\lambda v)^2 \frac{i}{(k+p_+')^2 - m_\sigma^2} \right]. \tag{3.4}
$$

The result is a sum of bubble, triangle, and box diagrams. The box in particular is a very complicated function of the kinematic invariants, involving di-logarithms ['tHV 79, DeNS 91, ElZ 08]. The divergence from the bubble diagram goes into the renormalization of the $\lambda\varphi^4$ coupling of the original lagrangian. For the effective theory, in contrast, one uses only pions and considers only the bubble diagram. The low-energy limit of the vertex is employed. Again, drawing from our results of Eq. (1.13), also visible by taking the leading approximation for the vertices in Eq. (3.4), one finds

$$
i\mathcal{M}_{\text{eff}} = \int \frac{d^4k}{(2\pi)^4} \frac{i(k+p_+)^2}{v^2} \frac{i}{(k+p_++p_0)^2} \frac{i}{k^2} \frac{i(k+p_+')^2}{v^2}. \tag{3.5}
$$

This diagram has a different divergence than the full theory. It is also much simpler kinematically, and its dimensional regularized form is easily evaluated as

$$i\mathcal{M}_{\text{eff}} = \frac{i}{96\pi^2 v^4} s(s-u) \left[\frac{2}{4-d} - \gamma + \ln 4\pi - \ln \frac{-s - i\epsilon}{\mu^2} \right]$$

$$+ \frac{i}{288\pi^2 v^4} [2s^2 - 5su], \qquad (3.6)$$

using the usual variables $s = (p_+ + p_0)^2$, $t = (p_+ - p'_+)^2$, $u = (p_0 - p'_+)^2$.

There are various interesting features of this result. Note that the whole amplitude is of order (energy)^4, while the original scattering vertex of Eq. (1.13) was of order (energy)^2. Technically, this follows simply from noting that the loop has factors of $1/v^4$ and that in dimensional regularization the only other dimensional factors are the external energies. On a more profound level it is an example of the energy expansion of the effective theory – loops produce results that are suppressed by higher powers of the momenta at low-energy. Because of this kinematic dependence, one can also readily see that the divergence cannot be absorbed into the renormalization of the original $\mathcal{O}(E^2)$ effective lagrangian. In fact we know that this divergence is spurious. It was generated because the effective theory had the wrong high-energy behavior compared to the full theory. This is to be expected in an effective theory – it does not pretend to know the content of the theory at all energies. However, the divergence will disappear in the matching of the two theories through the renormalization of a term in the $\mathcal{O}(E^4)$ lagrangian – this will be demonstrated below.

Even more interesting from the physics point of view is that the $s(s-u)\ln -s$ behavior is exactly what is found by taking the low-energy limit of the complicated result from the full theory and expanding it to this order in the momenta. This occurs because the $\ln -s$ factor comes from the low-energy regions of the loop momenta, of order $k \sim s$, so that the logarithm represents long-distance propagation.[5] Indeed, the imaginary part of the amplitude arising from $\ln(-s - i\epsilon) = \ln(s) - i\pi$ (for $s > 0$) comes from the on-shell intermediate state of two pions. This logarithm could never be represented by a local effective lagrangian and is a distinctive feature of long-distance (low-energy) quantum loops. These features match in the two calculations because when the loop momenta are small the effective field theory approximation for the vertex is valid. Overall, the effective field theory has an incorrect high-energy behavior but does capture the correct low-energy dynamics.

The comparison of the full theory and the effective theory can be carried out directly for this reaction. The dimensionally regularized result for the full theory is given in [MaM 08], but is too complicated to be reproduced here. However the

---

[5] Short distance pieces from higher values of k would be analytic functions able to be Taylor expanded around $s = 0$.

expansion of the full theory at low-energy in terms of renormalized parameters is relatively simple [GaL 84]

$$
\mathcal{M}_{\text{full}} = \frac{t}{v^2} + \left[\frac{1}{m_\sigma^2 v^2} - \frac{11}{96\pi^2 v^4} \right] t^2
$$
$$
- \frac{1}{144\pi^2 v^4} [s(s-u) + u(u-s)]
$$
$$
- \frac{1}{96\pi^2 v^4} \left[3t^2 \ln \frac{-t}{m_\sigma^2} + s(s-u) \ln \frac{-s}{m_\sigma^2} + u(u-s) \ln \frac{-u}{m_\sigma^2} \right]. \tag{3.7}
$$

The effective theory result [Le 72, GaL 84] has a very similar form but does not know about the existence of the σ,

$$
\mathcal{M}_{\text{eff}} = \frac{t}{v^2} + \left[8\ell_1^r + 2\ell_2^r + \frac{5}{192\pi^2} \right] \frac{t^2}{v^4}
$$
$$
+ \left[2\ell_2^r + \frac{7}{576\pi^2} \right] [s(s-u) + u(u-s)]/v^4 \tag{3.8}
$$
$$
- \frac{1}{96\pi^2 v^4} \left[3t^2 \ln \frac{-t}{\mu^2} + s(s-u) \ln \frac{-s}{\mu^2} + u(u-s) \ln \frac{-u}{\mu^2} \right],
$$

where we have defined[6]

$$
\ell_1^r = \ell_1 + \frac{1}{384\pi^2} \left[\frac{2}{4-d} - \gamma + \ln 4\pi \right]
$$
$$
\ell_2^r = \ell_2 + \frac{1}{192\pi^2} \left[\frac{2}{4-d} - \gamma + \ln 4\pi \right]. \tag{3.9}
$$

At this stage we can match the two theories, providing identical scattering amplitudes to this order, through the choice

$$
\ell_1^r = \frac{v^2}{8m_\sigma^2} + \frac{1}{192\pi^2} \left[\ln \frac{m_\sigma^2}{\mu^2} - \frac{35}{6} \right]
$$
$$
\ell_2^r = \frac{1}{384\pi^2} \left[\ln \frac{m_\sigma^2}{\mu^2} - \frac{11}{6} \right]. \tag{3.10}
$$

The reader is invited to compare this result with the tree-level matching, Eq. (2.10). We have not only obtained a more precise matching, we also have generated important kinematic dependence, particularly the logarithms, in the scattering amplitude.

We have seen that the predictions of the full theory can be reproduced even when using only the light degrees of freedom, as long as one chooses the coefficient of the effective lagrangian appropriately. This holds for *all* observables. Once

---

[6] Readers who compare with [GaL 84] should be aware that our normalization of the ℓ_i coefficients differs by a factor of four.

the matching is done, other processes can be calculated using the effective theory without the need to match again for each process.[7] The total effect of the heavy particle, both tree diagrams and loops, has been reduced to a few numbers in the lagrangian which we have deduced from matching conditions to a given order in an expansion in the energy.

In this example we match to a known calculable theory. In other realizations of effective field theory, the full theory may be unknown (for example, in the case of gravity [Do 94]) or very difficult to calculate (as we will discuss for *QCD*). In cases where direct matching is not possible, the renormalized coefficients in the lagrangian could be determined through measurement. Measuring the value of the coefficients in one reaction would allow them to be used by the effective theory in other processes.

IV–4 General features of effective field theory

After this explicit example, let us think more generally about effective field theories. In quantum mechanics and quantum field theory, we face what appears to be an impossible situation. Intermediate states in perturbation theory and in loop diagrams include all energies, even beyond those which have been probed experimentally. Yet we expect more new particles and new interactions to be present eventually at higher energies. How can we then reliably perform *any* calculation without knowing the particles and interactions at all energies which enter in our calculations?

The answer essentially comes from the uncertainty principle. Effects from high energy appear local when viewed at low energy. This means that they are equivalent to terms in a local lagrangian. Most often the coefficient of a particular term in a lagrangian – a mass or a coupling constant – is something that we have to measure. So the effects of physics from high energy is contained in the parameters that we measure at low energy.

Effective field theory embraces this fact and uses it to perform calculations at low energy. In theories where the high-energy limit is known, such as our linear sigma model example above, the coefficients of the effective lagrangian can be determined by matching. In theories where the high-energy physics is not known, we still know that its effect is local, so that we parameterize it by the most general local lagrangian.

The decoupling theorem tells us that the high-energy effect appears in renormalized couplings or in terms suppressed by powers of the heavy scale. In this sense, all of our theories can be viewed as effective field theories. The class of renormalizable

---

[7] As part of our treatment of *QCD*, we show the universality of the renormalization in Appendix B–2.

field theories is a subset of effective field theories in which the power-suppressed lagrangians have not yet been needed.

Effective lagrangians and symmetries

What would happen if, instead of having a straightforward known theory like the linear sigma model, we were dealing with an unknown or unsolvable theory with the same $SU(2)_L \times SU(2)_R$ chiral symmetry? In this case there would exist some set of pion interactions which, although not explicitly known, would be greatly restricted by the $SU(2)$ chiral symmetry. Once again we could choose to describe the pion fields in terms of the exponential parameterization U, with a symmetry transformation

$$U \to L U R^\dagger \qquad (4.1)$$

for L, R in $SU(2)$. Not having an explicit prescription, we would proceed to write out the most general effective lagrangian consistent with the chiral symmetry. In view of the infinite number of possible terms contained in such a description, this would appear to be a daunting process. However, the energy expansion allows it to be manageable.

It is not difficult to generate candidate interactions which are invariant under chiral $SU(2)$ transformations. For the purpose of illustration, we list the following two-derivative, four-derivative, and six-derivative terms in the exponential parameterization,

$$\mathrm{Tr}\left(\partial_\mu U \partial^\mu U^\dagger\right), \quad \mathrm{Tr}\left(\partial_\mu U \partial_\nu U^\dagger\right) \cdot \mathrm{Tr}\left(\partial^\mu U \partial^\nu U^\dagger\right),$$
$$\mathrm{Tr}\left(\partial_\mu U \partial^\mu U^\dagger\right) \cdot \mathrm{Tr}\left(\partial_\nu U \Box \partial^\nu U^\dagger\right). \qquad (4.2)$$

There can be no derivative-free terms in a list such as this because $\mathrm{Tr}\left(U U^\dagger\right) = 2$ is a constant. It is clear that one can generate innumerable similar terms with arbitrary numbers of derivatives. The general lagrangian can be organized by the dimensionality of the operators,

$$\mathcal{L} = \mathcal{L}_2 + \mathcal{L}_4 + \mathcal{L}_6 + \mathcal{L}_8 + \cdots$$
$$= \frac{F^2}{4} \mathrm{Tr}\left(\partial_\mu U \partial^\mu U^\dagger\right) + \ell_1 [\mathrm{Tr}\left(\partial_\mu U \partial^\mu U^\dagger\right)]^2$$
$$+ \ell_2 \mathrm{Tr}\left(\partial_\mu U \partial_\nu U^\dagger\right) \cdot \mathrm{Tr}\left(\partial^\mu U \partial^\nu U^\dagger\right) + \cdots. \qquad (4.3)$$

The important point is that, at sufficiently low energies, the matrix elements of most of these terms are very small since each derivative becomes a factor of the momentum q when matrix elements are taken. It follows from dimensional analysis that the coefficient of an operator with n derivatives behaves as $1/M^{n-4}$, where

M is a mass scale which depends on the specific theory. Therefore, the effect of an n-derivative vertex is of order E^n / M^{n-4}, and, at an energy small compared to M, large-n terms have a very small effect. At the lowest energy, only a single lagrangian, the one in Eq. (3.3) with two derivatives, is required. We shall call this an '$\mathcal{O}(E^2)$' contribution in subsequent discussions. The most important corrections to this involve four derivatives, and are therefore '$\mathcal{O}(E^4)$'. In practice then, the infinity of possible contributions is reduced to only a small number. The coefficients of these terms are not generally known, and must thus be determined phenomenologically. However, once fixed by experiment (or by matching to the full theory if possible) they can be used to allow predictions to be made for a variety of reactions.

Power counting and loops

It would appear that loop diagrams could upset the dimensional counting described above. This might happen in the calculation of a given loop diagram if, for example, two of the momentum factors from an $\mathcal{O}(E^4)$ lagrangian are involved in the loop and are thus proportional to the loop momentum. Integrating over the loop momentum apparently leaves only two factors of the 'low' energy variable. It would therefore seem that for certain loop diagrams, an $\mathcal{O}(E^4)$ lagrangian could behave as if it were $\mathcal{O}(E^2)$. If this happened, it would be a disaster because arbitrarily high order lagrangians would contribute at $\mathcal{O}(E^2)$ when loops were calculated. As we shall show, this does not occur. In fact, the reverse happens. When $\mathcal{O}(E^2)$ lagrangians are used in loops, they contribute to $\mathcal{O}(E^4)$ or higher.

Before we give the formal proof of this result, let us note that we saw this effect in the linear sigma model calculation above. We started by using the order E^2 lagrangian in the loop diagram and the result was the renormalization of a lagrangian at order E^4. It is also straightforward to demonstrate why this occurs. Consider a pion loop diagram, as in Fig. IV–2. From the explicit form displayed in Eq. (3.5), we see that

$$\mathcal{M}^{(\text{loop})}_{\pi^+ \pi^0 \to \pi^+ \pi^0} \equiv \frac{1}{v^4} I(p_+, p_0, p'_+), \tag{4.4}$$

where I is the loop integral with the factor v^{-4} extracted. Counting powers of energy factors is most easily done in dimensional regularization. The loop integral contains no dimensional factors other than p_+, p_0, and p'_+. Since, in four dimensions it has the overall energy unit E^4, it must therefore be expressible as fourth order in momentum. Despite the loop integration, the end result is expressed only in terms of the external momenta. These momenta are small, and hence all the energy factors involved in power counting are taken at low-energy. In dimensional regularization, there can also be a dependence on the arbitrary scale μ,

$$\int d^4\ell \to \mu^{4-d} \int d^d\ell, \tag{4.5}$$

but in the limit $d \to 4$ this occurs only in dimensionless logarithms such as $\ln(E^2/\mu^2)$. Thus, the order of momentum can be found by counting the factors of $1/v^2$ which occur for every vertex from the lowest-order lagrangians. Each factor of $1/v^2$ must be accompanied by momenta in the numerator in order to produce a dimensionless amplitude. Each vertex in a diagram contributes powers of $1/v^2$, and higher-order loop diagrams require more vertices. Thus, every time a loop is formed, the overall momentum power of the amplitude must *increase* rather than decrease.

We have also seen that any divergences present can be handled in the usual way, by renormalizations of the parameters in the theory. Again, the uncertainty principle comes into play – the divergences come from the extreme high-energy part of the calculation and thus they must look like some term in a local lagrangain. If the original effective lagrangian which we have written down is indeed the most general one consistent with the given symmetry, then it must have enough parameters of the right form to encompass any divergences which occur. In particular, our power-counting argument tells us that when $\mathcal{L}_2$ is used in one-loop diagrams, the divergences are of order E^4 and should be capable of being absorbed into the parameters of that order. Since the parameters are generally unknown and are to be determined phenomenologically, the only difference this makes is to cast physical results in terms of the renormalized parameters instead of the bare ones.

Weinberg's power-counting theorem

To prove this result [We 79b], consider some diagram with a total of N_V vertices. Then letting N_n be the number of vertices arising from the subset of effective lagrangians which contain n derivatives (e.g. N_4 is the number of vertices coming from four-derivative lagrangians), we have $N_V = \Sigma_n N_n$. The overall energy dimensionality of the coupling constants is thus M^{N_C} with

$$N_C = \sum_n N_n(4 - n), \tag{4.6}$$

where M is a mass scale entering into the coefficients of the effective lagrangian (e.g., the quantity v in the sigma model). Each pion field comes with a factor of $1/v$, so that associated with N_E external pions and N_I internal pion lines is an energy factor $(1/M)^{2N_I+N_E}$. (Recall that two pions must be contracted to form an internal line.) However, the number of internal lines can be eliminated in terms of the number of vertices and loops (N_L),

$$N_I = N_L + N_V - 1 = N_L + \sum_n N_n - 1. \tag{4.7}$$

Any remaining dimensional factors must be made up of powers of the energy E times a dimensionless factor of E/μ where μ is the scale employed for renormalizing the coupling constants. (When using dimensional regularization, these factors of E/μ enter only in logarithms.) Thus the overall matrix element is composed of energy factors

$$\mathcal{M} \sim (M)^{\sum_n N_n (n-4)} \frac{1}{M^{N_E + 2N_L + 2\sum_n N_n - 2}} E^D F(E/\mu)$$
$$\sim (\text{mass or energy})^{4-N_E}, \tag{4.8}$$

where the second line is the overall dimension of an amplitude with N_E external bosons. The renormalization scale μ can be chosen of the order of E so no large factors are present in $F(E/\mu)$. Overall the energy dimension is then

$$D = 2 + \sum_n N_n(n-2) + 2N_L. \tag{4.9}$$

A diagram containing N_L loops contributes at a power E^{2N_L} higher than the tree diagrams. This theorem is of great practical consequence. At low energy, it allows one to work with only small numbers of loops. In particular, at $\mathcal{O}(E^4)$ only one-loop diagrams generated from $\mathcal{L}_2$ need to be considered.

The end result is a very simple rule for counting the order of the energy expansion. The lowest-order (E^2) behavior is given by the two-derivative lagrangians treated at tree level. There are two sources at the next order (E^4): (i) the $\mathcal{O}(E^2)$ one-loop amplitudes, and (ii) the tree-level $\mathcal{O}(E^4)$ amplitudes. When the coefficients of the E^4 lagrangians are renormalized, finite predictions result. Other effective field theories will have power-counting rules analogous to this one appropriate for chiral theories.

The limits of an effective field theory

The effective field theory of the linear sigma model is valid for energies well below the mass of the scalar particle in the theory, the σ or S. Once there is enough energy to directly excite the S particle, it is clear that the effective theory is inadequate. This energy scale is visible even within the effective theory itself. Scattering matrix elements are an expansion in the energy, with a schematic form

$$\mathcal{M} \sim \frac{q^2}{v^2} \left[1 + \frac{q^2}{m_\sigma^2} + \cdots \right] \tag{4.10}$$

and the scale of the energy dependence is determined largely by the scalar mass. As the energy increases the corrections to the lowest-order result grow and eventually all terms in the energy expansion become equally important and the effective theory breaks down. Thus, the effective theory reveals its own limits.

In more general effective field theories, there is always a separation of the heavy degrees of freedom, which are integrated out from the theory, and the light degrees of freedom, which are treated dynamically. In many instances, the natural separation scale is set by a particle's mass, as in the linear sigma model. We will see that in the case of *QCD*, the meson resonances such as the $\rho(770)$ do not appear explicitly in the low-energy effective theory. Therefore, these have been integrated out and help define the limits of the effective field theory. In other cases, we could integrate some of the high-momentum modes of certain fields, while still keeping the low-momentum modes of these same fields as active dynamical participants in the low-energy theory. This is done for the effective hamiltonian for weak decays, where we integrate out the high-energy modes of the gluonic fields. In these cases, the scale that we have used to separate high and low energy defines the limit of validity of the effective field theory.

Let us also address a rather subtle point concerning the energy scale of the effective theory. While we regularly use this idea of an energy scale defining the limit of validity of the effective theory, there are times that we do not apply this separation fully. In loop diagrams, if we wanted to only include loop effects below a certain energy scale, we would need to employ a cut-off in the loop integral. This is often inconvenient and if done carelessly could upset some of the symmetries of the theory. Moreover, the presence of an additional dimensional factor in loop diagrams would upset some of the power-counting arguments described above. Most often, practical calculations are performed using dimensional regularization. This regulator has no knowledge of the energy scale of the theory and thus loop diagrams will in general include effects from energies where the effective theory is not valid. However, again the uncertainty principle comes to our rescue. Even if these spurious high-energy contributions are not correct, we know that their effect is equivalent to a local term in the effective lagrangian. Any mistakes made in the loop can be corrected by modifying the coefficients of the terms in the effective lagrangian. Careful application of the procedures for matching or measuring the parameters will return the the same physical predictions independent of the choice of regularization scheme.

IV–5 Symmetry breaking

Effective lagrangians can be used not only in the limit of exact symmetry but also to analyze the effect of small symmetry breaking. Let us first return to the sigma model for an illustration of the method, and then consider the general technique.

The $SU(2)_L \times SU(2)_R$ symmetry of the sigma model is explicitly broken if the potential $V(\sigma, \pi)$ is made slightly asymmetric, e.g., by the addition of the term

$$\mathcal{L}_{\text{breaking}} = a\sigma = \frac{a}{4} \text{Tr} \left(\Sigma + \Sigma^+ \right) \tag{5.1}$$

to the basic lagrangian of Eq. (1.4). To first order in the quantity a, this shifts the minimum of the potential to

$$v = \sqrt{\frac{\mu^2}{\lambda} + \frac{a}{2\mu^2}}, \tag{5.2}$$

and produces a pion mass

$$m_\pi^2 = \frac{a}{v}. \tag{5.3}$$

Although the latter result can be found by using the linear representation and expanding the fields about their vacuum expectation values, it is easier to use the exponential representation,

$$\mathcal{L}_{\text{breaking}} = \frac{a}{4}(v + S) \, \text{Tr} \left(U + U^\dagger \right) = \frac{a}{4}(v + S) \, \text{Tr} \left(2 - \left(\frac{\tau \cdot \pi}{v} \right)^2 + \cdots \right)$$

$$= a(v + S) - \frac{a}{2v} \pi \cdot \pi + \cdots = a(v + S) - \frac{m_\pi^2}{2} \pi \cdot \pi + \cdots . \tag{5.4}$$

The chiral $SU(2)$ symmetry is seen to be slightly broken, but the vectorial $SU(2)$ isospin symmetry remains exact.

As we have seen, the $\mathcal{O}(E^2)$ lagrangian is obtained by setting $S = 0$,

$$\mathcal{L}_2 = \frac{v^2}{4} \text{Tr} \left(\partial_\mu U \partial^\mu U^\dagger \right) + \frac{m_\pi^2}{4} v^2 \, \text{Tr} \left(U + U^\dagger \right). \tag{5.5}$$

Higher-order terms will contain products like

$$\left[m_\pi^2 \, \text{Tr} \left(U + U^\dagger \right) \right]^2, \quad m_\pi^2 \, \text{Tr} \left(U + U^\dagger \right) \cdot \text{Tr} \left(\partial_\mu U \partial^\mu U^\dagger \right), \quad \ldots, \tag{5.6}$$

and can be obtained by integrating out the field S as was done in Sect. IV–2. It is important to realize that the symmetry-breaking sector also has a low-energy expansion, with each factor of m_π^2 being equivalent to two derivatives. If m_π^2 is small, *the expansion is a dual expansion in both the energy and the mass.*

If we encounter a theory more general than the sigma model, the effect of a small pion mass can be similarly expressed in low orders by,

$$\mathcal{L}_{\text{breaking}} = a_1 m_\pi^2 \, \text{Tr} \left(U + U^\dagger \right) + a_2 \left[m_\pi^2 \, \text{Tr} \left(U + U^\dagger \right) \right]^2$$

$$+ a_3 m_\pi^2 \, \text{Tr} \left(U + U^\dagger \right) \text{Tr} \left(\partial_\mu U \partial^\mu U^\dagger \right) + a_4 m_\pi^2 \, \text{Tr} \left[(U + U^\dagger) \partial_\mu U \partial^\mu U^\dagger \right], \tag{5.7}$$

with coefficients that are generally not known. An important consideration is the symmetry-transformation property of the perturbation. The symmetry-breaking term of Eq. (5.1) is not invariant under separate left-handed and right-handed transformations but only under those with $L = R$. All the terms in Eq. (5.7) have this property.

Other symmetry breakings can be analyzed in a manner analogous to the treatment just given of the mass term. One identifies the symmetry-transformation property of the perturbing effect and writes the most general effective lagrangian with that property. Most often the perturbation is treated to only first order, but higher-order behavior can also be studied.

IV–6 Matrix elements of currents

There is an elegant technique which allows one, at a minimal increase in complexity, to calculate matrix elements of currents from a chiral effective lagrangian [GaL 84, 85a]. The idea is to add to the lagrangian terms containing external sources coupled to the currents in question. Construction of the effective lagrangian, including source terms, then allows the current matrix elements to be easily identified. We shall explain this technique here, and use it extensively in our discussion of QCD in subsequent chapters.

First, consider how current matrix elements are identified in a path-integral framework. We have seen in Chap. III (see also App. A) that by adding a source coupled to the desired current, matrix elements can be obtained from differentiation of the path integral, e.g., Eqs. (III–2.2), (III–2.4). For example, we can modify three-flavor QCD by adding sources to obtain

$$\mathcal{L} = -\frac{1}{4} F^a_{\mu\nu} F^{\mu\nu}_a + \bar{\psi} i \not{D} \psi - \bar{\psi} \gamma_\mu \frac{1 + \gamma_5}{2} \ell^\mu \psi - \bar{\psi} \gamma_\mu \frac{1 - \gamma_5}{2} r^\mu \psi$$
$$- \bar{\psi}_L (s + ip) \psi_R - \bar{\psi}_R (s - ip) \psi_L, \tag{6.1}$$

where ℓ_μ, r_μ, s, p are 3×3 matrix source functions expressible as

$$\ell_\mu = \ell^0_\mu + \ell^a_\mu \lambda^a, \quad r_\mu = r^0_\mu + r^a_\mu \lambda^a, \quad s = s^0 + s^a \lambda^a, \quad p = p^0 + p^a \lambda^a, \tag{6.2}$$

with $a = 1, \ldots, 8$. The lagrangian in Eq. (6.1) reduces to the usual QCD lagrangian in the limit $\ell_\mu = r_\mu = p = 0, s = \mathbf{m}$, where $\mathbf{m}$ is the 3×3 quark mass matrix. The electromagnetic coupling can be obtained with the choice $\ell_\mu = r_\mu = eQA_\mu$, where A_μ is the photon field and Q is the electric charge operator defined in units of e. Various currents can be read off from the lagrangian, such as the left-handed current

$$J_{L\mu}^k(x) = -\frac{\partial \mathcal{L}}{\partial \ell_k^\mu(x)} = \bar{\psi}(x)\gamma_\mu \frac{1+\gamma_5}{2}\lambda^k \psi(x) \qquad (6.3)$$

or the scalar density

$$\bar{\psi}(x)\psi(x) = -\frac{\partial \mathcal{L}}{\partial s^0(x)}. \qquad (6.4)$$

Moreover, matrix elements of these currents can be formed from the path integral by taking functional derivatives. The simplest example is

$$\langle 0 | \bar{\psi}(x)\psi(x) | 0 \rangle = i\frac{\delta \ln Z}{\delta s^0(x)}\Bigg|_{\substack{\ell=r=p=0 \\ s=m}}, \qquad (6.5)$$

while other examples appear in Sect. III–2.

Matrix elements and the effective action

A low-energy effective action for the Goldstone bosons of *QCD* will be a functional of the external sources. One way to define the connection of the effective action with *QCD* is to consider the effect of the sources,

$$e^{iW(\ell_\mu, r_\mu, s, p)} = \int [d\psi][d\bar{\psi}][dA_\mu^a] \, e^{i\int d^4x \, \mathcal{L}_{QCD}(\psi, \bar{\psi}, A_\mu^a, \ell_\mu, r_\mu, s, p)}. \qquad (6.6)$$

At low-energy, all heavy degrees of freedom can be integrated out and absorbed into coefficients in the effective action W. However, the Goldstone bosons propagate at low-energy, and they must be explicitly taken into account. One then writes a representation of the form

$$e^{iW(\ell_\mu, r_\mu, s, p)} = \int [dU] \, e^{i\int d^4x \, \mathcal{L}_{\text{eff}}(U, \ell_\mu, r_\mu, s, p)}, \qquad (6.7)$$

where as usual U contains the Goldstone fields. This form then allows inclusion of all low-energy effects while maintaining the symmetries of *QCD*.

The lagrangian of Eq. (6.1) has an exact *local* chiral $SU(3)$ invariance if we have the external fields transform in the same way as gauge fields. In particular, the transformations

$$\psi_L \to L(x)\psi_L, \qquad \psi_R \to R(x)\psi_R,$$
$$\ell_\mu \to L(x)\ell_\mu L^\dagger(x) + i\partial_\mu L(x)L^\dagger(x),$$
$$r_\mu \to R(x)r_\mu R^\dagger(x) + i\partial_\mu R(x)R^\dagger(x), \qquad (6.8)$$
$$(s+ip) \to L(x)(s+ip)R^\dagger(x)$$

provide an invariance for any $L(x)$, $R(x)$ in $SU(3)$.

In constructing the effective action, these invariances must be included. This is easy to do if ℓ_μ and r_μ enter in the same way as gauge fields. In particular, upon defining a covariant derivative

$$D_\mu U = \partial_\mu U + i\ell_\mu U - iU r_\mu, \tag{6.9}$$

and field-strength tensors

$$L_{\mu\nu} = \partial_\mu \ell_\nu - \partial_\nu \ell_\mu + i[\ell_\mu, \ell_\nu],$$
$$R_{\mu\nu} = \partial_\mu r_\nu - \partial_\nu r_\mu + i[r_\mu, r_\nu], \tag{6.10}$$

we obtain the following covariant responses to local transformations:

$$
\begin{aligned}
U &\to L(x)UR^\dagger(x), & D_\mu U &\to L(x)D_\mu U R^\dagger(x), \\
L_{\mu\nu} &\to L(x)L_{\mu\nu}L^\dagger(x), & R_{\mu\nu} &\to R(x)R_{\mu\nu}R^\dagger(x).
\end{aligned} \tag{6.11}
$$

The effective action is then expressed in terms of these quantities. At order E^2, there are only two terms in the effective lagrangian,

$$\mathcal{L}_2 = \frac{F_\pi^2}{4}\, \mathrm{Tr}\left(D_\mu U D^\mu U^\dagger\right) + \frac{F_\pi^2}{4}\, \mathrm{Tr}\left(\chi U^\dagger + U \chi^\dagger\right), \tag{6.12}$$

where

$$\chi \equiv 2B_0(s + ip) \tag{6.13}$$

and B_0 is a constant with the dimension of mass. In the limit $\ell_\mu = r_\mu = p = 0$, $s = m$, this is the same effective lagrangian with which we have been dealing in the $SU(2)$ examples, with the identification $m_\pi^2 = (m_u + m_d)B_0$. Note that this usage requires B_0 to be positive.

Having constructed the effective action, we can obtain a number of interesting matrix elements. For example, use of Eq. (6.5) provides the identification of the vacuum scalar-density matrix element as

$$\langle 0 |\bar{\psi}_i \psi_j| 0 \rangle = -F_\pi^2 B_0 \delta_{ij} \tag{6.14}$$

to this order in the effective lagrangian. Similarly, use of Eq. (6.3) reveals the left-handed current to be

$$L_\mu^k = -i\frac{F_\pi^2}{2}\, \mathrm{Tr}\left(\lambda^k U \partial_\mu U^\dagger\right). \tag{6.15}$$

One other advantage of the source method is to allow the use of the equations of motion. The standard Noether procedure for identifying currents does not work if the equations of motion are employed in the lagrangian. To become convinced of this, one can consider the following exercise. We examine the response of the two trial lagrangians,

$$\mathcal{L}_1 = \varphi^* \Box \varphi, \quad \mathcal{L}_2 = -m^2 \varphi^* \varphi \qquad (6.16)$$

to a phase transformation $\varphi \to e^{i\alpha} \varphi$. The first contributes to the Noether current while the second does not. However, these two forms are identical on-shell if φ satisfies the Klein–Gordon equation. In an effective lagrangian which is meant to be used always on-shell it is often convenient to drop terms which vanish by virtue of the equations of motion. The use of source fields as described above avoids this problem.

IV–7 Effective field theory of regions of a single field

In our presentation earlier in this chapter, the construction of an effective field theory was described by the integrating out of heavy particles, while leaving the light particles as dynamical degrees of freedom. However, often one can make an effective field theory from a single particle. In this case, certain energy regions of the field are treated as heavy and others are light, and one retains the light regions in the effective field theory. Indeed, sometimes there are multiple regions that are 'light' in some sense, and one splits the original single field into multiple fields. This section provides some of the background for such decompositions.

The simplest example of the division of a single field into 'heavy' and 'light' is in the nonrelativistic reduction. When the energy is small, the antiparticle degrees of freedom are heavy and can be removed from the theory, leaving a nonrelativistic particle description. For example, if one redefines a four-component Dirac field ψ into upper and lower two-component fields, ψ_u and ψ_ℓ by factoring out the leading energy dependence at low-energy via

$$\psi(x, t) = e^{-imt} \begin{pmatrix} \psi_u(x, t) \\ \psi_\ell(x, t) \end{pmatrix}, \qquad (7.1)$$

ψ_u will behave as a nonrelativistic field and ψ_ℓ will account for the two heavy degrees of freedom. The free Dirac lagrangian shows this separation,

$$\begin{aligned} \mathcal{L} &= \bar{\psi}(i\slashed{\partial} - m)\psi \\ &= \psi_u^* i \partial_t \psi_u + \psi_\ell^*[i\partial_t + 2m]\psi_\ell + \psi_u^* i\sigma \cdot \nabla \psi_\ell + \psi_\ell^* i\sigma \cdot \nabla \psi_u. \end{aligned} \qquad (7.2)$$

While no approximation has yet been made by this redefinition, the nonrelativistic limit is taken by assuming that the residual energy dependence is small compared to the mass (i.e., one neglects ∂_t compared to $2m$). One can then integrate out the lower component through its equation of motion,

$$(i\partial_t + 2m)\psi_\ell \approx 2m\,\psi_\ell = i\sigma \cdot \nabla\psi_u, \tag{7.3}$$

leaving the upper component as the active non-relativistic degree of freedom.

$$\mathcal{L} = \psi_u^* i\partial_t \psi_u - \frac{(\nabla\psi_u^*) \cdot \nabla\psi_u}{2m}. \tag{7.4}$$

With inclusion of the interactions, this can lead to a full nontrivial effective field theory. A well-developed example of this is the Non-Relativistic *QCD* (*NRQCD*) effective field theory [CaL 86]. We will also return to this procedure in more generality in the discussion of Heavy Quark Effective Theory (*HQET*) in Chap. XIII.

A second common way of splitting up a single field is to integrate out the high momentum portions of a field. This logic is often called Wilsonian [Wi 69]. Imagine splitting the momenta in a problem into those above an energy scale Λ and those below this scale. By first performing the calculation of the high-energy portion, one is left with an effective field theory. The operators defining that theory will carry factors, the Wilson coefficients, that depend on the scale Λ. This means that one obtains a set of new operators $\mathcal{O}_n$ in the lagrangian

$$\mathcal{L} = \cdots + \sum_n C_n(\Lambda)\mathcal{O}_n, \tag{7.5}$$

where $C_n(\Lambda)$ are the Wilson coefficients and the series is infinite. The operators are local because they capture high-energy physics, and their matrix elements will depend on the separation scale, $\langle\mathcal{O}_n\rangle = \langle\mathcal{O}_n(\Lambda)\rangle$. One regularly uses the renormalization group to describe the running of the Wilson coefficients with changes of scale. The low-energy theory remains a full field theory and one must calculate the full quantum effects in the matrix elements of $\mathcal{O}_n$ up to the scale Λ. When the high-energy physics in C_n and the low-energy physics in the matrix elements of $\mathcal{O}_n$ are properly matched, in the end the separation scale Λ will disappear from the description. Nevertheless, this separation is often useful. For example, in *QCD* the high-energy behavior may be reliably calculated in perturbation theory, while the low-energy behavior may be best accomplished with lattice calculations. Examples appearing in this book include the Wilson coefficients of the non-leptonic weak hamiltionian, cf. Sect. VIII–3, and those used in *QCD* sum rules, cf. Sect. XI–5.

In practice, however, we most often do *not* use a Wilsonian separation scale Λ, but instead employ dimensional regularization. Dimensional regularized loop integrals do not carry information about any particular scale, and therefore extend over all energies. The extension to $d < 4$ damps the high-energy divergences in a scale-independent way. Nonetheless, this procedure works for logarithmically running Wilson coefficients. Aside from the momenta, the only scale in a dimensionally regularized integral is the $\mu^{2\epsilon}$ inserted in front of the loop integral. This ends up

Fig. IV–3 The scalar vertex diagram analysed in the text.

appearing in the final answer as $\ln \mu^2$ when expanded close to $d = 4$. The fact that cut-off regularization and dimensional regularization have the equivalence

$$\ln \Lambda^2 \Leftrightarrow \frac{1}{\epsilon} + \ln \mu^2 \tag{7.6}$$

allows the scale μ to be a proxy for the separation scale Λ. However, the correspondence of μ with a Wilsonian separation scale does not hold for Wilson coefficients with power-law running [CiDG 00].

For a yet more subtle example, consider the interaction of a high-energy massless particle in the vertex diagram of Fig. IV–3. For the purposes of this example, let us consider these as scalars and the current vertex as $J = \varphi^2/2$. We can analyse the resultant scalar vertex integral,

$$I = \mu^{4-d} \int \frac{d^d k}{(2\pi)^d} \frac{1}{(p+k)^2} \frac{1}{k^2} \frac{1}{(p'+k)^2}, \tag{7.7}$$

in the limit where $p^2 \sim p'^2 \ll Q^2 = (p - p')^2$. The only scales in this problem are Q^2, which is treated as a large scale, and $p^2 \sim p'^2$, which is the small scale. The relative size is labeled $\lambda^2 \sim p^2/Q^2 \sim p'^2/Q^2$.

This integral can be analyzed by the *method of regions* [BeS 98, Sm 02].[8] In this technique, one identifies all the important momentum regions of the loop integral, and makes appropriate approximations within each region. A portion of the integral will have all the components of the loop momenta of order Q and higher. This will be called the *hard* region. A region labeled *soft* has all the components much smaller than Q. In addition, there will be regions where the momentum is of order Q in the direction of p or p'. In these *collinear* regions, some invariant products can be smaller than Q^2.

In order to quantify this one takes light-like reference four-vectors

$$n^\mu = (1, 0, 0, 1), \quad \bar{n}^\mu = (1, 0, 0, -1), \quad n^2 = \bar{n}^2 = 0, \quad n \cdot \bar{n} = 2. \tag{7.8}$$

For an arbitrary four-vector expressed using these and a transverse component,

$$V^\mu = n \cdot V \frac{\bar{n}^\mu}{2} + \bar{n} \cdot V \frac{n^\mu}{2} + V_\perp^\mu \equiv V_+ \frac{\bar{n}^\mu}{2} + V_- \frac{n^\mu}{2} + V_\perp^\mu, \tag{7.9}$$

the invariant product is

---

[8] This example and the treatment of it follows the lectures of [Be 10].

$$V^2 = (n \cdot V)(\bar{n} \cdot V) + V_\perp^2 = V_+ V_- + V_\perp^2.$$

$$A_\mu B^\mu = \frac{1}{2}(A_+ B_- + A_- B_+) + A_\perp \cdot B_\perp \tag{7.10}$$

These are useful because we can choose a frame with p along n and with p' along $\bar{n}$, and we can refer to the n direction as 'right' and the $\bar{n}$ direction as 'left'. This allows us to classify the different regions. Of the original momenta, we have

$$(V_+, V_-, V_\perp)$$
$$p \sim (\lambda^2, \ 1, \ 0) \, Q$$
$$p' \sim (1, \ \lambda^2, \ 0) \, Q. \tag{7.11}$$
$$Q \sim (1, \ 1, \ 0) \, Q$$

Q is a hard momentum because it takes a hard interaction to change an energetic right-moving particle into one moving left. Using this decomposition, one can identify the regions of the loop momentum

$$(k_+, k_-, k_\perp)$$
$$k \sim (1, \ 1, \ 1) \, Q \quad \text{hard}$$
$$k \sim (\lambda^2, \ 1, \ \lambda) \, Q \quad \text{collinear R.} \tag{7.12}$$
$$k \sim (1, \ \lambda^2, \ \lambda) \, Q \quad \text{collinear L}$$
$$k \sim (\lambda^2, \lambda^2, \lambda^2) \, Q \quad \text{soft}$$

In each region, one can drop small momentum components in terms of large ones. For example, when k is in the hard region, one can drop p^2, p'^2, $k_- p_+$, $k_+ p'_-$, which are all of order λ^2, in order to obtain[9]

$$I_{\text{hard}} = \mu^{4-d} \int \frac{d^d k}{(2\pi)^d} \frac{1}{(k^2 + i\epsilon)(k^2 + k_- p_+ + i\epsilon)(k^2 + k_+ p'_- + i\epsilon)}$$
$$= \frac{i\Gamma(1+\epsilon)}{(4\pi)^{d/2} Q^2} \left[\frac{1}{\epsilon^2} + \frac{1}{\epsilon} \ln \frac{\mu^2}{-Q^2} + \frac{1}{2} \ln^2 \frac{\mu^2}{-Q^2} - \frac{\pi^2}{6} \right]. \tag{7.13}$$

Similarly, in the right collinear region, one can expand $(k + p')^2 = k_- p'_+ + \mathcal{O}(\lambda^2)$, such that

$$I_{\text{col}-R} = \mu^{4-d} \int \frac{d^d k}{(2\pi)^d} \frac{1}{(k^2 + i\epsilon)((k + p)^2 + i\epsilon)(k_- p'_+ + i\epsilon)}$$
$$= \frac{i\Gamma(1+\epsilon)}{(4\pi)^{d/2} Q^2} \left[-\frac{1}{\epsilon^2} - \frac{1}{\epsilon} \ln \frac{\mu^2}{-p'^2} + \frac{1}{2} \ln^2 \frac{\mu^2}{-p'^2} + \frac{\pi^2}{6} \right]. \tag{7.14}$$

An observation that will be relevant for the eventual construction of an effective theory is that when the exchanged propagator carrying momentum k is in the right

[9] The integrals of this section are displayed in the useful appendix of [Sm 02]. See also [Sm 12].

collinear region, the other propagator on the p side is also collinear, but the third propagator on the p' side is hard. A similar result is obviously found when k is in the left collinear region, obtained by replacing p by p'. Finally, in the soft region, one keeps only terms of order λ^2, finding

$$
\begin{aligned}
I_{\text{soft}} &= \mu^{4-d} \int \frac{d^d k}{(2\pi)^d} \frac{1}{(k^2 + i\epsilon)(k_- p_+ + p^2 + i\epsilon)(k_+ p'_- + p'^2 + i\epsilon)} \\
&= \frac{i\Gamma(1+\epsilon)}{(4\pi)^{d/2} Q^2} \left[\frac{1}{\epsilon^2} + \frac{1}{\epsilon} \ln \frac{\mu^2 Q^2}{-p^2 p'^2} + \frac{1}{2} \ln^2 \frac{\mu^2 Q^2}{-p^2 p'^2} + \frac{\pi^2}{6} \right].
\end{aligned} \tag{7.15}
$$

If one tries to identify other regions besides these and makes the corresponding simplifications of the loop integral, one ends up with a scale-less integral which vanishes within dimensional regularization. For example, if one considers the region where k scales as $k \sim (\lambda^2, \lambda^2, \lambda)^{10}$, one would use $k^2 \sim k_\perp^2$ and keep terms of order λ^2 in each propagator

$$
\begin{aligned}
I' &= \int \frac{d^d k}{(2\pi)^d} \frac{1}{(k_\perp^2 + i\epsilon)(k_- p_+ + p^2 + k_\perp^2 + i\epsilon)(k_+ p'_- + p'^2 + k_\perp^2 + i\epsilon)} \\
&= \frac{1}{p_+ p'_-} \int \frac{d^d k'}{(2\pi)^d} \frac{1}{(k_\perp^2 + i\epsilon)(k'_- + i\epsilon)(k'_+ + i\epsilon)} \\
&= 0,
\end{aligned} \tag{7.16}
$$

where in the second line we have defined shifted variables $k'_- = k_- + (p^2 + k_\perp^2)/p_+$ and $k'_+ = k_+ + (p'^2 + k_\perp^2)/p'_-$, with the result being an integral without any scale. Such integrals are set to zero within dimensional regularization.

The sum of the four subregions yields the correct total integral,

$$
I = \frac{i}{16\pi^2 Q^2} \left[\ln \frac{Q^2}{p^2} \ln \frac{Q^2}{p'^2} + \frac{\pi^2}{3} \right], \tag{7.17}
$$

up to terms suppressed by powers of λ. As expected, this result is finite, even though the integrals from the individual regions are not. The approximations that we made lead to infrared divergences in the hard integral, and ultraviolet divergences in the others. However, these cancel when added together.

The other interesting feature of this procedure is that we have not restricted the integration ranges when calculating the integrals for the different regions. The full integration range is used in each case. The reason that this does not amount to double counting within dimensional regularization is that if there is a single unique scale within the integral, as has been deliberately constructed in each region, the integral is determined by momenta around that scale. This is the key observation

---

[10] This region is referred to as the *Glauber* region. The treatment of the integral given in the text appears adequate for this example, although the understanding of the Glauber region is still evolving [BaLO 11].

that allows the method of regions to work. By constructing approximations that scale in unique fashions, one can isolate the physics of that region alone.[11] That this actually happens in these integrals can be seen from the above integral where the factors of Q^2, p^2, and $p^2 p'^2 / Q^2$ all signal the dominant scale in the respective diagrams, showing that the effects come from different regions of the momentum integration.

One can convert the analysis of the method of regions into an effective field theory whose applicability extends beyond this particular example. The initial field can be divided up into new effective fields for each of the important regions. The goal is to choose these fields and their interactions to yield the same results as the method of regions analysis outlined above. The hard-momentum region can be integrated out completely and replaced by effective operators of the light fields. These operators will come with Wilson coefficients to ensure the matching with the full calculation. However, the dynamical light fields need to come in three varieties for the different light-momentum regions. Thus, the original scalar field $\varphi(x)$ now comes in three varieties, $\varphi(x) = \varphi_{cR}(x) + \varphi_{cL}(x) + \varphi_s(x)$. The interactions of the light fields among themselves is relatively simple to construct. If the interaction vertex of the original theory was a simple φ^3 vertex, we expand that to include the possible interaction between the light fields,

$$-\mathcal{L} = \frac{g}{3}\varphi^3 \to \frac{g}{3}\varphi_{cR}^3 + \frac{g}{3}\varphi_{cL}^3 + \frac{g}{3}\varphi_s^3 + g\varphi_{cR}^2\varphi_s + g\varphi_{cL}^2\varphi_s. \tag{7.18}$$

Vertices not listed above, such as $\varphi_{cR}\varphi_s^2$, are ones which cannot occur due to momentum conservation (e.g., a collinear particle cannot split into two soft particles).

It is somewhat more subtle to choose the other effective operators and their Wilson coefficients. For the scalar example shown above, the 'current' carrying the momentum Q in the full theory is $J = \varphi^2/2$. Since it transfers this large momentum it can connect φ_{cR} to φ_{cL} such that we expect a vertex $J \sim \varphi_{cR}\varphi_{cL}$. However, in addition we need to recall that we have integrated out the hard scalars. This leads to additional vertices. For example, in the diagram of Fig. IV–4(a) the propagator is hard because it carries the momenta of both left-moving and right-moving fields, which couple to it at the lower vertex. When the other fields are light, this propagator shrinks to a point vertex as in Fig. IV–4(b). This, then, is a new contribution to the current operator, and we expect that the current has the form

$$J = C_2\varphi_{cR}\varphi_{cL} + C_3\varphi_{cR}^2\varphi_{cL} + C_3'\varphi_{cR}\varphi_{cL}^2 + \cdots, \tag{7.19}$$

---

[11] In cases where regions are defined which have overlapping contributions there are also methods for cleanly separating the regions [MaS 07].

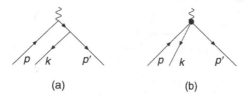

(a) (b)

Fig. IV–4 (a) An interaction of collinear particles through a hard propagator; (b) the effective local vertex representing this interaction at low-energy.

where C_2 and C_3 are the Wilson coefficients. Calculation from the original theory shows that to this order

$$C_2 = 1 + g^2 I_{\text{hard}}, \qquad C_3 = \frac{2g}{k_- - p'_+ - i\epsilon}, \qquad C'_3 = \frac{2g}{k_+ + p_- - i\epsilon}, \qquad (7.20)$$

where I_{hard} refers back to Eq. (7.13).

At this stage, we can reproduce the original vertex calculation using the effective theory by the calculation of the diagrams of Fig. IV–5. The diagrams of Fig. IV–5 (a),(b),(c) refer to the new vertices given in Eq. (7.20), while Fig. IV–5 (d) refers to the soft contribution of Eq. (7.15). By construction, one can see how all four of the regions of the original diagram are reproduced. We note how the hard propagators that occur when k is in one of the collinear regions have been accounted for by a new local vertex in the current operator, with the Wilson coefficient describing the effect of the hard propagator.

The reader may object that the construction of the effective theory was more trouble than evaluating the original diagram. However, once we have developed the effective theory, we can apply it in multiple new contexts. The example above is analogous to the Soft Collinear Effective Theory (*SCET*) of *QCD* [BaFPS 01]. Similar techniques are used in the various realizations of *NRQCD* [CaL 86, PiS 98, BrPSV 05]. Outside of the Standard Model, related methods are applied in the classical effective field theory of General Relativity [GoR 06], which has been used to systematize the classical treatment of gravitational radiation from binary systems [PoRR 11]. Further development of the method of regions and threshold expansions can be found in [BeS 98, Sm 02].

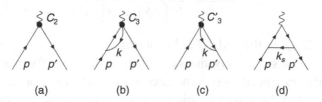

(a) (b) (c) (d)

Fig. IV–5 The diagrams involving the light fields reconstructing the scalar vertex.

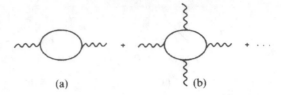

(a)   (b)

Fig. IV–6 Photon amplitudes containing a single fermion loop.

IV–8 Effective lagrangians in *QED*

We have explored in some detail the structure of effective field theory by using chiral symmetry as an example. However, this is not meant to imply that effective lagrangians are useful only in that one context. In fact, they can be applied to a wide variety of situations. Here, we apply the technique to *QED*.

Consider situations in which the photon's four-momentum is small compared to the electron mass. In such cases, the electron and other fermions cannot be produced directly, but instead influence the physics of photons only through virtual processes. The lowest-order diagrams, i.e., those which contain a single electron loop, with increasing numbers of external photon legs, are shown in Fig. IV–6. Note that the one-loop diagram containing three photons, or indeed any odd number of photons, vanishes by virtue of charge-conjugation invariance. This is true to all orders in the coupling *e*, and is refered to as *Furry's theorem*. Diagrams like those in Fig. IV–6 have effects at low-energy which are typically calculated in perturbation theory. The associated amplitudes have coefficients which scale as some power of the inverse electron mass. They can be generated by means of an effective lagrangian, as we shall now discuss.

Let us seek a description which eliminates the electron degrees of freedom. That is, we wish to write a lagrangian which involves only photons, but nevertheless includes effects like the ones in Fig. IV–6. The result must of course be gauge-invariant. The procedure may be defined by the path-integral relation

$$
\int [dA_\mu] \, \exp\left[i \int d^4x \, \mathcal{L}_{\text{eff}}(A_\mu)\right]
$$
$$
\equiv \frac{\int [dA_\mu][d\psi][d\bar\psi] \, \exp\left[i \int d^4x \, \mathcal{L}_{QED}(A_\mu, \psi, \bar\psi)\right]}{\int [d\psi][d\bar\psi] \, \exp\left[i \int d^4x \, \mathcal{L}_0(\psi, \bar\psi)\right]}, \tag{8.1}
$$

where $\mathcal{L}_{QED}$ is the full *QED* lagrangian, and $\mathcal{L}_0$ is the free fermion lagrangian. Thus $\mathcal{L}_{\text{eff}}$ has precisely the same matrix elements for photons as does the full *QED* theory. Specifically, it includes the virtual effects of electrons. The techniques described in App. A–5 enable us to formally express the content of Eq. (8.1) as [Sc 51]

$$\int d^4x \, \mathcal{L}_{\text{eff}}(A_\mu) = -\frac{1}{4} \int d^4x \, F_{\mu\nu}F^{\mu\nu} - i \, \text{Tr} \, \ln\left[\frac{i\slashed{D} - m}{i\slashed{\partial} - m}\right]. \tag{8.2}$$

This form, although formally correct, does not readily lend itself to physical interpretation. However, we can determine various interesting effects directly from perturbation theory. For example, the vacuum polarization of Fig. IV–6(a) modifies the photon propagator, i.e., the two-point function. From Eqs. (II–1.26), (II–1.29), we determine the result for a photon of momentum q to be

$$i\hat{\Pi}_{\mu\nu}(q) = i\frac{\alpha}{15\pi}\left(q_\mu q_\nu - g_{\mu\nu}q^2\right)\frac{q^2}{m^2} + \cdots . \tag{8.3}$$

The essence of the effective lagrangian approach is to represent such information as the matrix element of a local lagrangian. In the present example, we find that the term in Eq. (8.3) corresponds to the interaction

$$\mathcal{L}_{\text{eff}} = \frac{\alpha}{60\pi m^2}F_{\mu\nu}\square F^{\mu\nu}, \tag{8.4}$$

where $\square \equiv \partial_\mu \partial^\mu$.

The calculation of Fig. IV–6(b) is a somewhat more difficult, but still straightforward, exercise in perturbation theory. We shall lead the reader through a calculation using path integrals in a problem at the end of this chapter. It too can be represented as a local lagrangian, and is usually named after Euler and Heisenberg [ItZ 80]. One finds the full result to one-loop order to be

$$\mathcal{L}_{\text{eff}}(A_\mu) = -\frac{1}{4}F_{\mu\nu}F^{\mu\nu} + \frac{\alpha}{60\pi m^2}F_{\mu\nu}\square F^{\mu\nu}$$
$$+ \frac{\alpha^2}{90m^4}\left[\left(F_{\mu\nu}F^{\mu\nu}\right)^2 + \frac{7}{4}\left(F_{\mu\nu}\tilde{F}^{\mu\nu}\right)^2\right] + \cdots, \tag{8.5}$$

where $\tilde{F}_{\mu\nu} \equiv \frac{1}{2}\epsilon_{\mu\nu\alpha\beta}F^{\alpha\beta}$. Corrections to this effective lagrangian can be of two forms: (i) even at one loop there are additional terms of higher dimension

$$F_{\mu\nu}\frac{\square^2}{m^4}F^{\mu\nu}, \quad F_{\mu\nu}\frac{\square}{m^6}F^{\mu\nu}F_{\alpha\beta}F^{\alpha\beta}, \quad \frac{1}{m^8}\left(F_{\mu\nu}F^{\mu\nu}\right)^3, \ldots, \tag{8.6}$$

involving either more fields or more derivatives; or (ii) the coefficients of these operators can receive corrections of higher order in α through multi-loop diagrams. We see here an example of the energy expansion, which we have discussed at length earlier in this chapter. In this case it is an expansion in powers of q^2/m^2. The effective lagrangian of Eq. (8.5) can be used to compute aspects of low-energy photon physics such as the low-energy contribution of the vacuum polarization process or the matrix element for photon–photon scattering.

IV–9 Effective lagrangians as probes of New Physics

One of the most common and important uses of effective lagrangians is to parameterize how *new* physics at high energy may influence low-energy observables. The general procedure can be abstracted from our earlier discussion. Remember that one is trying to represent the low-energy effects from a 'heavy' sector of the theory. This is accomplished by employing an effective lagrangian

$$\mathcal{L}_{\text{eff}} = \sum_n \mathcal{C}_n \, O_n, \tag{9.1}$$

where the $\{O_n\}$ are local operators having the symmetries of the theory and are constructed from fields that describe physics at low-energy. There need be no restriction to renormalizable combinations of fields. Most often the operators can be organized by dimension. The lagrangian itself has mass dimension 4, so that if an operator has dimension d_i the coefficient must have mass dimension

$$\mathcal{C}_n \sim M^{4-d_n}. \tag{9.2}$$

The mass scale M is associated with the heavy sector of the theory. It is clear that operators of high dimension will be suppressed by powers of the heavy mass. To leading order, this allows one to keep a small set of operators.

Some applications will involve phenomena for which the dynamics is well understood. If so, the coefficients of the effective lagrangian can be determined through direct calculation as in the preceding sections. Other examples occur in the theory of weak nonleptonic interactions (cf. Sect. VIII–3) and in the interactions of W-bosons (cf. Sect. XVI–3). Even more generally, effective lagrangians can also be used to describe the effects of new types of interactions. In these cases, dimensional analysis supplies an estimate for the magnitude of the energy scales of possible New Physics. We shall conclude this section by using effective lagrangians to characterize the size of possible violations of some of the symmetries of the Standard Model.

Given certain input parameters, the Standard Model is a closed, self-consistent description of physics up to at least the mass of the Z^0, and is described by the most general renormalizable lagrangian consistent with the underlying gauge symmetries. What would happen if there were new interactions having an intrinsic energy scale of several TeV or beyond? In general, such new theories would be expected to modify predictions of the Standard Model. The modifications would be described by non-renormalizable interactions, organized by dimension in an effective lagrangian description as

$$\mathcal{L}_{\text{eff}} = \mathcal{L}_{\text{SM}} + \frac{1}{\Lambda}\mathcal{L}_5 + \frac{1}{\Lambda^2}\mathcal{L}_6 + \cdots , \qquad (9.3)$$

where $\mathcal{L}_n$ has mass dimension n and Λ is the energy scale of the new interaction.

There is a single operator of dimension 5 which will be displayed in the neutrino chapter. At dimension 6, there are 80 distinct operators consistent with the gauge symmetries of the Standard Model [BuW 86]. These can generate a variety of effects which deviate from the Standard Model. For example, the operator

$$\mathcal{L}_6\left(c'\right) \equiv \frac{c'}{\Lambda^2}(\Phi^\dagger\Phi)\mathbf{W}_{\mu\nu}\cdot\mathbf{W}^{\mu\nu}, \qquad (9.4)$$

containing the Higgs field Φ and the field tensor $\mathbf{W}_{\mu\nu}$ of $SU(2)$ gauge bosons, produces a deviation from unity in the rho-parameter,[12]

$$\rho \equiv \frac{M_W^2}{M_Z^2\cos^2\theta_{\text{w}}} = 1 - c'\frac{v^2}{\Lambda^2} + \cdots . \qquad (9.5)$$

The current level of precision, $|\rho - 1| \leq 0.0029$ (at 2σ), requires $\Lambda > 4.5\,\text{TeV}$ for $c' = 1$. Yet another possibility concerns the violation of flavor symmetries in the Standard Model. The operator,

$$\mathcal{L}_6(c'') \equiv \frac{c''}{\Lambda^2}\bar{e}\gamma_\mu(1 + \gamma_5)\mu\,\bar{s}\gamma^\mu(1 + \gamma_5)d + \text{h.c.}, \qquad (9.6)$$

conserves generational or family number, but violates the separate lepton-number symmetries. It leads to the transition $K_L \to e^-\mu^+$ such that

$$\frac{\Gamma_{K_L^0 \to \mu^+ e^-}}{\Gamma_{K^+ \to \mu^+\nu_\mu}} = \left(\frac{c''}{V_{\text{us}}G_F\Lambda^2}\right)^2 . \qquad (9.7)$$

The present bound, $\text{Br}_{K_L^0 \to \mu e} < 4.7 \times 10^{-12}$ at 90% confidence level, requires $\Lambda > 1700\,\text{TeV}$ for $c'' \simeq 1$. In a similar manner, constraints on other physical processes imply bounds on their corresponding energy scales Λ, generally in the range $5 \to 5000\,\text{TeV}$.

Dimension-six contact interactions also are searched for at the highest energies of the Large Hadron Collider (LHC). The effect of the contact interaction becomes relatively more pronounced at high energy when competing with background processes which fall off due to propagator effects. For example, an operator leading to $q\bar{q} \to \mu^+\mu^-$,

$$\mathcal{L}_6(g) \equiv \frac{g^2}{2\Lambda^2}\bar{q}_L\gamma_\nu q_L\bar{\mu}_L\gamma^\nu\mu_L, \qquad (9.8)$$

---

[12] More precisely the comparison is with a form of the rho-parameter after radiative corrections.

becomes increasingly visible over the Drell–Yan process at high energy. Early LHC results [Aa *et al.* (ATLAS collab.) 11] bound this interaction with $\Lambda > 4.5$ TeV at 95% confidence for $g^2/4\pi = 1$; such limits will clearly improve in the future. Interestingly, some operators are better bounded by low-energy precision experiments and others are better probed at high energy [Bh *et al.* 12], demonstrating the value of both lines of research.

Of course, if there is new physics in the TeV energy range, it need not generate all 80 possible effective interactions. The ones actually appearing would depend on the couplings and symmetries of the new theory. In addition, the coefficients of contributing operators could contain small coupling constants or mixing angles, diminishing their effects at low-energy. However, the effective lagrangian analysis indicates that the continued success of the Standard Model is quite nontrivial and places meaningful bounds on possible new dynamical structures occurring at TeV, and even higher, energy scales.

Problems

(1) **$U(1)$ effective lagrangian**

Consider a theory with a complex scalar field φ with a $U(1)$ global symmetry $\varphi \to \varphi' = \exp(i\theta)\,\varphi$. The lagrangian will be

$$\mathcal{L} = \partial_\mu \varphi^* \partial^\mu \varphi + \mu^2 \varphi^* \varphi - \lambda (\varphi^* \varphi)^2.$$

(a) Minimize the potential to find the ground state and write out the lagrangian in the basis

$$\varphi = \frac{1}{\sqrt{2}}(v + \varphi_1(x) + i\varphi_2(x))$$

Show that φ_2 is the Goldstone boson.

(b) Use this lagrangian to calculate the low-energy scattering of $\varphi_2 + \varphi_2 \to \varphi_2 + \varphi_2$. Show that despite the nonderivative interactions of the lagrangian, cancelations occur such that leading scattering amplitude starts at order p^4.

(c) Instead of the basis above, express the lagrangian using an exponential basis,

$$\varphi = \frac{1}{\sqrt{2}}(v + \Phi(x))e^{i\chi(x)/v}.$$

Show that in this basis a 'shift symmetry' $\chi \to \chi + c$ is manifest.

(d) Calculate the same scattering amplitude using this basis and show that the results agree. Note that the fact that the amplitude is of order p^4 is more readily apparent in this basis.

(2) **Path integrals and the Fermi effective lagrangian**

Consider the path integral $Z_W = \int [dW^+][dW^-] \exp[i \int d^4x \mathcal{L}_W(x)]$, where $\mathcal{L}_W$ is the $W^\pm$-boson lagrangian $\mathcal{L}_W = \mathcal{L}_W^{(\text{free})} + \mathcal{L}_W^{(\text{int})}$, with

$$\mathcal{L}_W^{(\text{free})} = -\frac{1}{2} F_{\mu\nu}^+ F^{\mu\nu}_- + M_W^2 W_\mu^+ W_-^\mu, \quad \mathcal{L}_W^{(\text{int})} = -\frac{g_2}{\sqrt{8}} \left(W_\mu^+ J_{\text{ch}}^\mu + \text{h.c.} \right).$$

Integrating out the heavy $W^\pm$ fields in Z_W leads to an effective interaction between charged weak currents called the Fermi model.

(a) Show that, upon discarding a total derivative term, one can write the free field contribution in Z_W as

$$\int d^4x \, \mathcal{L}_W^{(\text{free})} = \int d^4x \, d^4y \, W_\mu^+ \alpha K^{\mu\nu}(x, y) W_\nu^-(y),$$

where $K_{\mu\nu}(x, y) \equiv \delta^{(4)}(x - y) \left[g_{\mu\nu} \left(\partial^2 + M_W^2 \right) - \partial_\mu \partial_\nu \right]$.

(b) Further steps allow the path integral to be expressed as

$$Z_W = \exp \left[-i \frac{g_2^2}{8} \int d^4x \, d^4y \, J_{\text{ch}}^{\mu\dagger}(x) \Delta_{\mu\nu}(x, y) J_{\text{ch}}^\nu(y) \right],$$

where $\Delta_{\mu\nu}(x, y)$ is the Fourier transform of the $W^\pm$ propagator $\Delta_{\mu\nu}(k) = -\left(g_{\mu\nu} - k_\mu k_\nu / M_W^2 \right)$. Upon expanding this form of Z_W in powers of M_W^{-2}, show that to lowest order,

$$\mathcal{L}_W^{(\text{eff})}(x) = -\frac{G_F}{\sqrt{2}} J_{\text{ch}}^{\mu\dagger}(x) J_\mu^{\text{ch}}(x) \qquad \text{(Fermi model)},$$

where the Fermi constant obeys $G_F/\sqrt{2} \equiv g_2^2/(8M_W^2)$.

(3) **The Euler–Heisenberg lagrangian: constant magnetic field**

Consider a charged scalar field φ interacting with a constant external magnetic field $\mathbf{B} = B\hat{\mathbf{k}}$. The corresponding Klein–Gordon equation is $(D^2 + m^2)\varphi(x) = 0$, where $D_\mu = \partial_\mu + ieA_\mu$ is the covariant derivative, and the effective action is then given by

$$e^{i S_{\text{eff}}(B)} = \frac{\int [d\varphi(x)][d\varphi^*(x)] e^{i \int d^4x \, \varphi^*(x)(D^2 + m^2)\varphi(x)}}{\int [d\varphi(x)][d\varphi^*(x)] e^{i \int d^4x \, \varphi^*(x)(\Box^2 + m^2)\varphi(x)}}$$

$$= \det(\Box^2 + m^2)/\det(D^2 + m^2),$$

$$S_{\text{eff}}(B) = i \, \text{Tr} \ln \frac{D^2 + m^2}{\Box + m^2}.$$

The operation 'Tr ln' applied to a differential operator is not a trivial one and the purpose of this problem is to evaluate this quantity for the case at hand.

(a) Demonstrate that

$$S_{\text{eff}}(B) = i \operatorname{Tr} \int_0^\infty e^{-m^2 s} \left(e^{-\Box s} - e^{-D^2 s} \right).$$

(b) In order to evaluate the trace we require a complete set of solutions to the equations

$$D^2 \bar{\varphi}_n(x, y, z, t) = \lambda_n \bar{\varphi}_n(x, y, z, t),$$
$$\Box \varphi_n(x, y, z, t) = \kappa_n \varphi_n(x, y, z, t),$$

so that we may write

$$S_{\text{eff}}(B) = i \sum_n \int_0^\infty \frac{ds}{s} e^{-m^2 s} \left(e^{-\kappa_n s} - e^{-\lambda_n s} \right).$$

(c) With the gauge choice $A_\mu = (0, B x \hat{\mathbf{j}})$ show that the eigenstates are

$$\varphi(x, y, z, t) = e^{i(k_x x + k_y y + k_z z - k_t t)},$$
$$\bar{\varphi}(x, y, z, t) = e^{i(k_z z + k_y y - k_t t)} \psi_n(x - k_y / eB),$$

where $\psi_n(x)$ is an eigenstate of the harmonic-oscillator hamiltonian, and the eigenvalues are $\kappa_n = -k_t^2 + k_x^2 + k_y^2 + k_z^2$, $\lambda_n = -k_t^2 + k_z^2 + eB(2n+1)$.

(d) Rotate to euclidean space and evaluate the trace using box quantization. Taking a box with sides L_1, L_2, L_3 and a time interval T, we have

$$\kappa : \sum_n \to L_1 L_2 L_3 T \int_{-\infty}^\infty \frac{d^4 k}{(2\pi)^4},$$

$$\lambda : \sum_n \to L_2 L_3 T \int_0^{eBL_1} dk_y \int_{-\infty}^\infty \frac{dk_0 dk_z}{(2\pi)^2} \sum_{n=0}^\infty,$$

where the integration on k_y is over all values with $x' = x - k_y / eB$ positive.

(e) Evaluate the effective action

$$S_{\text{eff}}(B) = L_1 L_2 L_3 T \int_0^\infty \frac{ds}{s} \int_{-\infty}^\infty \frac{dk_0 dk_z}{(2\pi)^2} e^{-(m^2 + k_0^2 + k_z^2)s}$$

$$\times \left[\frac{eB}{2\pi} \sum_{n=0}^\infty e^{-eB(2n+1)s} - \int_{-\infty}^\infty \frac{dk_x dk_y}{(2\pi)^2} e^{-(k_x^2 + k_y^2)s} \right]$$

and show that

$$S_{\text{eff}}(B) = L_1 L_2 L_3 T \frac{1}{16\pi^2} \int_0^\infty \frac{ds}{s^3} e^{-m^2 s} \left(\frac{eBs}{\sinh eBs} - 1 \right).$$

Expand this in powers of B, finding the (divergent) wavefunction renormalization and the B^4 piece of the Euler–Heisenberg lagrangian.

(f) Show that the corresponding result of a constant *electric* field can be found by the substitution $B \rightarrow iE$ so that

$$S_{\text{eff}}(E) = L_1 L_2 L_3 T \frac{1}{16\pi^2} \int_0^\infty \frac{ds}{s^3} e^{-m^2 s} \left(\frac{eEs}{\sin eEs} - 1 \right).$$

(g) Demonstrate that, although Im $S_{\text{eff}}(B) = 0$, one nonetheless obtains

$$\text{Im } S_{\text{eff}}(E) = L_1 L_2 L_3 T \frac{e^2 E^2}{16\pi^3} \sum_{n=1}^\infty \frac{(-)^n}{n^2} e^{-n\pi m^2 / eE},$$

and discuss the meaning of this result [Sc 51, Sc 54].

V

Charged leptons

From the viewpoint of probing the basic structure of the Standard Model, the charged leptons constitute an attractive starting point. Since effects of the strong interaction are generally either absent or else play a secondary role, the theoretical analysis is relatively clean. Moreover, a great deal of high-quality data has been amassed involving these particles. Thus, charged leptons serve as an ideal system for defining our renormalization prescription, and for investigating the effects of various radiative corrections.

V–1 The electron

Some of the most precise tests of the Standard Model (or more exactly of QED) occur within the elementary electron–proton system. The renormalization program for the theory has been introduced in Sect. II–1, where it was shown how ultraviolet divergent contributions to such calculations can be removed by means of subtraction from a finite number of suitably constructed counterterms. Here we examine the finite pieces which remain after such subtractions and compare theory with experiment.

Breit–Fermi interaction

The electromagnetic properties of the electron are studied by use of a photon probe. To lowest order, the $\bar{e}e\gamma$ vertex has the structure

$$\langle e(\mathbf{p}'_e, \lambda'_e)|J^\mu_{\text{em}}|e(\mathbf{p}_e, \lambda_e)\rangle = -e\,\bar{u}(\mathbf{p}'_e, \lambda'_e)\gamma^\mu u(\mathbf{p}_e, \lambda_e), \qquad (1.1)$$

and the interaction between two charged particles is governed by the exchange of a single virtual photon. An important example is the electron–proton interaction, which has the invariant amplitude[1]

---

[1] We work temporarily with an 'ideal' proton – a point particle having no anomalous magnetic moment.

$$\mathcal{M}_{eP} = e^2 \bar{u}(\mathbf{p}'_e, \lambda'_e) \gamma^\mu u(\mathbf{p}_e, \lambda_e) \frac{1}{q^2} \bar{u}(\mathbf{p}'_p, \lambda'_p) \gamma_\mu u(\mathbf{p}_p, \lambda_p), \tag{1.2}$$

where $\mathbf{p}_e$, $\mathbf{p}'_e$ and $\mathbf{p}_p$, $\mathbf{p}'_p$ are respectively electron and proton momenta and $q = p_e - p'_e$ is the four-momentum transfer. In the following, we shall demonstrate how the above single-photon exchange amplitude is associated with well-known contributions in atomic physics. Denoting proton two-spinors with tildes, we begin by reducing the amplitude of Eq. (1.2) in the small-momentum limit to

$$\mathcal{M}_{eP} \simeq -\frac{e^2}{\mathbf{q}^2} \left[1 - \frac{\mathbf{p}_e^2 + \mathbf{p}'^2_e}{8m_e^2}\right] \left[1 - \frac{\mathbf{p}_p^2 + \mathbf{p}'^2_p}{8m_p^2}\right]$$
$$\times \left[\tilde{\chi}'^\dagger \left[1 + \frac{\mathbf{p}'_p \cdot \mathbf{p}_p + i\boldsymbol{\sigma} \cdot \mathbf{p}'_p \times \mathbf{p}_p}{4m_p^2}\right] \tilde{\chi}\, \chi'^\dagger \left[1 + \frac{\mathbf{p}'_e \cdot \mathbf{p}_e + i\boldsymbol{\sigma} \cdot \mathbf{p}'_e \times \mathbf{p}_e}{4m_e^2}\right] \chi \right.$$
$$\left. - \tilde{\chi}'^\dagger \frac{\mathbf{p}_p + \mathbf{p}'_p - i\boldsymbol{\sigma} \times (\mathbf{p}_p - \mathbf{p}'_p)}{2m_p} \tilde{\chi} \cdot \chi'^\dagger \frac{\mathbf{p}_e + \mathbf{p}'_e - i\boldsymbol{\sigma} \times (\mathbf{p}_e - \mathbf{p}'_e)}{2m_e} \chi\right], \tag{1.3}$$

where m_e, m_p are, respectively, the electron and proton masses. The various terms in the above expression can be interpreted physically by recalling that in Born approximation the transition amplitude and interaction potential are Fourier transforms of each other,

$$V_{eP}(\mathbf{r}) = \int \frac{d^3q}{(2\pi)^3} e^{-i\mathbf{q}\cdot\mathbf{r}} \mathcal{M}_{eP}, \tag{1.4}$$

where $\mathbf{r} = \mathbf{r}_e - \mathbf{r}_p$. From the relation

$$\int \frac{d^3q}{(2\pi)^3} e^{-i\mathbf{q}\cdot\mathbf{r}} \frac{1}{\mathbf{q}^2} = \frac{1}{4\pi r}, \tag{1.5}$$

we recognize the leading (velocity-independent) term,

$$V_{\mathrm{Coul}} = -\frac{e^2}{4\pi r} \tilde{\chi}'^\dagger \tilde{\chi}\, \chi'^\dagger \chi, \tag{1.6}$$

as the Coulomb interaction between electron and proton. The identity

$$-\int \frac{d^3q}{(2\pi)^3} \frac{e^2}{\mathbf{q}^2} \frac{i\boldsymbol{\sigma} \cdot \mathbf{p}'_e \times \mathbf{p}_e}{4m_e^2} e^{-i\mathbf{q}\cdot\mathbf{r}} = \frac{e^2}{4m_e^2} \frac{\boldsymbol{\sigma} \cdot \mathbf{r} \times \mathbf{p}_e}{4\pi r^3} \tag{1.7}$$

allows us to recognize an additional piece of Eq. (1.3) as the spin–orbit potential, which is often expressed as

$$V_{s-o} = \frac{1}{2m_e^2} \frac{1}{r} \frac{dV_0}{dr} \tilde{\chi}'^\dagger \tilde{\chi}\, \chi'^\dagger \frac{\boldsymbol{\sigma}}{2} \cdot \mathbf{r} \times \mathbf{p}_e\, \chi, \tag{1.8}$$

but evaluated in this instance with $V_0 = -e^2/4\pi r$. Combining the remaining $\mathcal{O}(\mathbf{p}^2/m_e^2)$ terms in Eq. (1.3), we can cancel the $\mathbf{q}^2$ term in the denominator to obtain the so-called Darwin potential,

$$V_D = \frac{e^2}{8m_e^2}\delta^{(3)}(r)\, \tilde{\chi}'^{\dagger}\tilde{\chi}\, \chi'^{\dagger}\chi. \tag{1.9}$$

This term has its origin in the electric interaction between the particles, and by employing the Gauss' law relation,

$$\nabla \cdot \mathbf{E}_{\text{Coul}} = e\delta^{(3)}(r), \tag{1.10}$$

it can be re-expressed in the equivalent form

$$V_D = \frac{e}{8m_e^2}\nabla \cdot \mathbf{E}_{\text{Coul}}\, \tilde{\chi}'^{\dagger}\tilde{\chi}\, \chi'^{\dagger}\chi. \tag{1.11}$$

The spin–orbit and Darwin potentials, together with the $\mathcal{O}(\mathbf{p}^2/m_e^2)$ relativistic corrections to the electron kinetic energy, give rise to atomic *fine structure* energy effects.

The remaining terms in the photon exchange interaction of Eq. (1.3) are effects produced by electron and proton current densities, the terms $(\mathbf{p}_e + \mathbf{p}'_e)/2m_e$ and $-i\boldsymbol{\sigma} \times (\mathbf{p}_e - \mathbf{p}'_e)/2m_e$ representing convection and magnetization contributions, respectively. In particular, the interaction between magnetization densities is equivalent to the dipole–dipole potential

$$V_{\text{dple–dple}} = -\frac{e}{m_e}\, \chi'^{\dagger}\frac{\boldsymbol{\sigma}}{2}\chi \cdot \nabla \times \left(\frac{e}{m_p}\, \tilde{\chi}'^{\dagger}\frac{\boldsymbol{\sigma}}{2}\tilde{\chi} \times \nabla\frac{1}{4\pi r}\right). \tag{1.12}$$

Recognizing that the magnetic field produced by the magnetic dipole moment of a (point) proton is

$$\mathbf{B}_{\text{proton}} = \nabla \times \left(\frac{e}{m_p}\, \tilde{\chi}'^{\dagger}\frac{\boldsymbol{\sigma}}{2}\tilde{\chi} \times \nabla\frac{1}{4\pi r}\right), \tag{1.13}$$

we can interpret the hyperfine energy as the interaction between the electron magnetic moment and the spin-induced proton magnetic field. Upon dropping the proton and electron spinors and using the identity

$$\nabla_i\nabla_j\frac{1}{r} = 3\frac{x_i x_j}{r^5} - \delta_{ij}\left[\frac{1}{r^3} + \frac{4\pi}{3}\delta^{(3)}(\mathbf{r})\right], \tag{1.14}$$

the dipole–dipole interaction may be written as a sum of *hyperfine* and *tensor* terms,

Table V–1. *Precision tests of QED.*

	Experiment	Theory
ν_{hyp}^b	1420.405 751 767(1)	1420.403(1)
a_e^a	1159.65218076(27)	1159.65218178(77)
a_μ^a	1165.92089(54)(33)	1165.91802(2)(42)(26)
$\Delta E_{n=2}^{(\text{Lamb})b}$	1057845.0(9.0)	1057844.4(1.8)

$^a \times 10^{-6}$
^{b}In units of kHz.

$$V_{\text{dple–dple}} = V_{\text{hyp}} + V_{\text{tensor}},$$

$$V_{\text{hyp}} = \frac{8\pi\alpha}{3m_e m_p} \mathbf{s}_e \cdot \mathbf{s}_p \delta^{(3)}(\mathbf{r}), \tag{1.15}$$

$$V_{\text{tensor}} = \frac{\alpha}{m_e m_p r^3} \left[3(\mathbf{s}_e \cdot \hat{\mathbf{r}})(\mathbf{s}_p \cdot \hat{\mathbf{r}}) - \mathbf{s}_e \cdot \mathbf{s}_p\right].$$

Denoting the total electron–proton spin as $\mathbf{s}_{\text{tot}} = \mathbf{s}_e + \mathbf{s}_p$, it follows that the hyperfine interaction splits the hydrogen atom ground state into components with $s_{\text{tot}} = 1$ and $s_{\text{tot}} = 0$. The frequency associated with this splitting is one of the most precisely measured constants in physics and is the source of the famous 21-cm radiation of radioastronomy. As seen in Table V–1, the experimental determination is about six orders of magnitude more precise than the theoretical value. Precision in the latter is limited by the nuclear force contribution (about 3 parts in 10^5).

Let us gather all the terms discussed thus far. In addition, we treat the proton and electron on an equal footing, since it will prove instructive when we discuss models of quark interactions in Chaps. XI–XIII. We then obtain the full one-photon exchange potential (Breit–Fermi interaction) for the electron–proton system,

$$
\begin{aligned}
V_{\text{one–photon}} = &-\frac{\alpha}{r} + \frac{8\pi\alpha}{3m_e m_p}\delta^{(3)}(\mathbf{r})\mathbf{s}_e \cdot \mathbf{s}_p + \frac{\pi\alpha}{2}\delta^{(3)}(\mathbf{r})\left[\frac{1}{m_e^2} + \frac{1}{m_p^2}\right] \\
&+ \frac{\alpha}{m_e m_p r^3}\left[3(\mathbf{s}_e \cdot \hat{\mathbf{r}})(\mathbf{s}_p \cdot \hat{\mathbf{r}}) - \mathbf{s}_e \cdot \mathbf{s}_p\right] \\
&+ \frac{\alpha}{r^3}\left[\frac{\mathbf{s}_e \cdot \mathbf{r} \times \mathbf{p}_e}{2m_e^2} - \frac{\mathbf{s}_p \cdot \mathbf{r} \times \mathbf{p}_p}{2m_p^2} + \frac{\mathbf{s}_p \cdot \mathbf{r} \times \mathbf{p}_e - \mathbf{s}_e \cdot \mathbf{r} \times \mathbf{p}_p}{m_e m_p}\right] \\
&+ \frac{\alpha}{2m_e m_p r}\left[\mathbf{p}_e \cdot \mathbf{p}_p + \hat{\mathbf{r}}(\hat{\mathbf{r}} \cdot \mathbf{p}_e) \cdot \mathbf{p}_p\right], \tag{1.16}
\end{aligned}
$$

where we recall $\mathbf{r} \equiv \mathbf{r}_e - \mathbf{r}_p$ and note that a spin-independent *orbit–orbit* interaction has been included as the final term. The single-photon exchange interaction is

seen to include a remarkable range of effects, all of which are necessary to understand details of atomic spectra.

QED corrections

Also important in precision tests of atomic systems are the higher order *QED*
corrections. We have just demonstrated how the simple q^2 piece of the photon
propagator leads to the Breit–Fermi interaction between electron and proton. The
vacuum polarization correction discussed in Sect. II–1 produces an additional component of the $e-P$ interaction called the *Uehling* potential. From Eq. (II–1.37), we
recall that in the on-shell renormalization scheme the subtracted vacuum polarization $\overline{\Pi}$ behaves in the small-momentum limit $m_e^2 \gg q^2$ as

$$\overline{\Pi}(q) = \frac{e^2}{60\pi^2}\frac{q^2}{m_e^2} + \mathcal{O}\left(\frac{q^4}{m_e^4}\right).$$

By the process of Fig. II–3, this yields the contribution

$$V_{\text{Uehling}}(r) = \int \frac{d^3q}{(2\pi)^3} e^{-i\mathbf{q}\cdot\mathbf{r}} \frac{e^2}{q^2} \times \frac{e^2}{60\pi^2}\frac{q^2}{m_e^2} = \frac{4}{15}\frac{\alpha^2}{m_e^2}\delta^{(3)}(\mathbf{r}). \tag{1.17}$$

The presence of the delta function implies that S-wave states of the hydrogen atom
are shifted by this potential while other partial waves are not. Contributions from
the Uehling potential have been observed in scattering experiments despite its $\mathcal{O}(\alpha)$
suppression relative to the dominant Coulomb scattering [Ve *et al.* 89].

The photon-electron vertex is also affected by radiative corrections. Let us write
the proper ($1PI$) electron–photon vertex through first order in α as

$$ie\Gamma_v(p'_e, p_e) = ie\gamma_v + ie\Lambda_v(p'_e, p_e) + \cdots, \tag{1.18}$$

where, referring to Fig. II–2(b), we have in Feynman gauge

$$ie\Lambda_v(p'_e, p_e) = (ie)^3 \int \frac{d^4k}{(2\pi)^4} \frac{-ig^{\alpha\beta}}{k^2 - \lambda^2 + i\epsilon}$$

$$\times \gamma_\alpha \frac{i}{\not{p}'_e - \not{k} - m_e + i\epsilon}\gamma_v \frac{i}{\not{p}_e - \not{k} - m_e + i\epsilon}\gamma_\beta. \tag{1.19}$$

Note that a small photon mass λ has been inserted to act as a cut-off in the small-
momentum domain, and we take both incoming and outgoing electrons to obey
$p_e^2 = p'^2_e = m_e^2$. With a modest effort, the integral in Eq. (1.19) can be continued
to d spacetime dimensions,

$$ie\Lambda_v\left(p'_e,\,p_e\right)=(ie)^3i\mu^{2\epsilon}\int_0^1 dx\int\frac{d^dk}{(2\pi)^d}$$

$$\times\frac{(2\epsilon-2)\not k\gamma_v\not k+4\not k\left(p_e+p'_e\right)_v-4m_ek_v-4\left(p_e+p'_e\right)\cdot k\gamma_v+4p_e\cdot p'_e\gamma_v}{\left[k^2-\lambda^2+i\epsilon\right]\left[(k-p_x)^2-p_x^2+i\epsilon\right]^2},$$

$$(1.20)$$

where $p_x\equiv xp_e+(1-x)p'_e$, and the result of performing the k-integration can be expressed as

$$ie\Lambda_v(p'_e,\,p_e)=(I_1)_v+(I_2)_v,\qquad(1.21)$$

where $(I_1)_v$ is singular in the $\epsilon\to 0$ limit,

$$(I_1)_v=i\gamma_v\frac{e^3}{(4\pi)^2}\frac{\Gamma(\epsilon)}{(4\pi)^{-\epsilon}}\frac{(2\epsilon-2)^2\mu^{2\epsilon}}{2}\int_0^1 dx\int_0^1 dy\frac{y}{\left[y^2p_x^2+\lambda^2(1-y)\right]^\epsilon},$$

$$(1.22)$$

and $(I_2)_v$ is not,

$$(I_2)_v=i\frac{e^3}{(4\pi)^2}\frac{\mu^{2\epsilon}\Gamma(1+\epsilon)}{(4\pi)^{-\epsilon}}\int_0^1 dx\int_0^1 dy\frac{N_v}{\left[y^2p_x^2+\lambda^2(1-y)\right]^{\epsilon+1}}$$

$$N_v=y^3(2\epsilon-2)\not p_x\gamma_v\not p_x+4y^2\left[(p_e+p'_e)_v\not p_x-m_e(p_x)_v-(p_e+p'_e)\cdot p_x\gamma_v\right]$$

$$+4yp_e\cdot p'_e\gamma_v.\qquad(1.23)$$

The singular term $(I_1)_v$, which arises from the $\not k\gamma_v\not k$ term in Eq. (1.20), is infrared-finite, and thus the photon mass λ can be dropped from it. Upon expanding $(I_1)_v$ in powers of ϵ and performing the y-integral, we obtain

$$(I_1)_v=ie\gamma_v\frac{e^2}{16\pi^2}\left[\frac{1}{\epsilon}+\ln(4\pi)-\gamma-1-\int_0^1 dx\ln\left(\frac{p_x^2}{\mu^2}\right)+\mathcal{O}(\epsilon)\right].\quad(1.24)$$

Because $(I_2)_v$ is not multiplied by any quantity which is singular in ϵ, we can immediately take the $\epsilon\to 0$ limit to cast it in the form

$$(I_2)_v=-i\frac{e^3}{16\pi^2}\int_0^1 dx\int_0^1 dy\frac{N_v}{y^2p_x^2+\lambda^2(1-y)},$$

$$N_v=-2y^3\not p_x\gamma_v\not p_x+4yp_e\cdot p'_e\gamma_v\qquad(1.25)$$

$$+4y^2\left[(p_e+p'_e)_v\not p_x-m_e(p_x)_v-(p_e+p'_e)\cdot p_x\gamma_v\right].$$

The photon mass λ can be dropped from the terms in N_v proportional to y^2 and y^3 since they are nonsingular even if $\lambda=0$. Performing the y-integration then yields the result,

$$(I_2)_\nu = -i\frac{e^3}{16\pi^2}\int_0^1 dx\; p_x^{-2}\left(-\not{p}_x\gamma_\nu\not{p}_x + 2p_e\cdot p_e'\gamma_\nu\ln(p_x^2/\lambda^2)\right.$$
$$\left.+4\left[(p_e+p_e')_\nu\not{p}_x - m_e(p_x)_\nu - (p_e+p_e')\cdot p_x\gamma_\nu\right]\right). \tag{1.26}$$

The identities

$$\not{p}_x\gamma_\nu\not{p}_x = 2m_e(p_x)_\nu - p_x^2\gamma_\nu, \qquad (p_e+p_e')\cdot p_x = 2m_e^2 - q^2/2, \tag{1.27}$$
$$p_x^2 = m_e^2 - q^2 x(1-x)$$

allow $(I_2)_\nu$ to be expressed in terms of $q^2 = (p_e - p_e')^2$, and the dependence of p_x^2 on the symmetric combination $x(1-x)$ implies

$$\int_0^1 dx\; (p_x)_\nu f(p_x^2) = \frac{1}{2}\left(p_e+p_e'\right)_\nu \int_0^1 dx\; f(p_x^2). \tag{1.28}$$

These steps, plus use of the Gordon decomposition of Eq. (C–2.8) finally lead to the expression

$$i e\Gamma_\nu(p_e', p_e) = ie\gamma_\nu\left[1 + \frac{e^2}{4\pi^2}\left\{\frac{1}{4\epsilon} - \frac{2+\gamma-\ln(4\pi)}{4}\right.\right.$$
$$-\frac{1}{4}\int_0^1 dx\;\ln\left(\frac{m_e^2 - q^2 x(1-x)}{\mu^2}\right) + \frac{1}{2}\int_0^1 dx\;\frac{3m_e^2 - q^2}{m_e^2 - q^2 x(1-x)}$$
$$\left.\left.-\frac{2m_e^2 - q^2}{4}\int_0^1 dx\;\frac{1}{m_e^2 - q^2 x(1-x)}\ln\left(\frac{m_e^2 - q^2 x(1-x)}{\lambda^2}\right)\right\}\right]$$
$$-ie\left[-\frac{i\sigma_{\nu\beta}q^\beta}{2m_e}\frac{e^2}{8\pi^2}\int_0^1 dx\;\frac{m_e^2}{m_e^2 - q^2 x(1-x)}\right]. \tag{1.29}$$

In the on-shell renormalization program, the electron–photon vertex

$$i e\Gamma_\nu^{(o-s)}(p_e', p_e) \equiv ieZ_1^{(o-s)}\Gamma_\nu(p_e', p_e) \tag{1.30}$$

is constrained to obey

$$\lim_{q\to 0} i e\Gamma_\nu^{(o-s)}(p_e', p_e) = ie\gamma_\nu, \tag{1.31}$$

so that

$$Z_1^{(o-s)} = 1 - \frac{e^2}{4\pi^2}\left[\frac{1}{4\epsilon} + 1 - \frac{\gamma-\ln(4\pi)}{4} - \frac{1}{4}\ln\left(\frac{m_e^2}{\mu^2}\right) - \frac{1}{2}\ln\left(\frac{m_e^2}{\lambda^2}\right)\right]$$
$$= Z_1^{(MS)} - \frac{e^2}{4\pi^2}\left[1 - \frac{\gamma-\ln(4\pi)}{4} - \frac{1}{4}\ln\left(\frac{m_e^2}{\mu^2}\right) - \frac{1}{2}\ln\left(\frac{m_e^2}{\lambda^2}\right)\right]. \tag{1.32}$$

The on-shell renormalized vertex is thus given by

$$i e\Gamma_\nu^{(o-s)}\left(p_e', p_e\right) = ie\left(\gamma_\nu F_1\left(q^2\right) - \frac{i\sigma_{\nu\beta}q^\beta}{2m_e}F_2\left(q^2\right)\right), \tag{1.33}$$

where $F_1(q^2)$ is given by a complicated expression which we do not reproduce here, and

$$F_2\left(q^2\right) = \frac{e^2}{8\pi^2} \int_0^1 dx \, \frac{m_e^2}{m_e^2 - q^2 x(1-x)}. \tag{1.34}$$

In addition to its original spin structure γ_ν, the electromagnetic vertex is seen in Eq. (1.33) to have picked up a contribution proportional to $\sigma_{\nu\beta}q^\beta$. The γ_ν and $\sigma_{\nu\beta}q^\beta$ contributions are called the *Dirac* and *Pauli* terms respectively, and $F_1(q^2)$ and $F_2(q^2)$ are the Dirac and Pauli *form factors* of the electron. The vertex correction turns out to have several important experimental consequences.

Consider the interaction of an electron with a classical electromagnetic field for very small q^2. Using Eq. (1.33) and the Gordon identity, we have

$$\begin{aligned}
\mathcal{H}_{\text{int}} &= eA_\nu(x)\langle e(\mathbf{p}'_e)|J_{\text{em}}^\nu(x)|e(\mathbf{p}_e)\rangle \\
&= -eA_\nu(x)\bar{u}(\mathbf{p}'_e)\left[\gamma^\nu - \frac{i\sigma^{\nu\beta}q_\beta}{2m_e}\frac{e^2}{8\pi^2}\right]u(\mathbf{p}_e)e^{-iq\cdot x} + \mathcal{O}\left(q^2\right) \\
&= -eA_\nu(x)\bar{u}(\mathbf{p}'_e)\left[\frac{(p_e + p'_e)^\nu}{2m_e} - \frac{i\sigma^{\nu\beta}q_\beta}{2m_e}\left(1 + \frac{e^2}{8\pi^2}\right)\right]u(\mathbf{p}_e)e^{-iq\cdot x} + \mathcal{O}\left(q^2\right).
\end{aligned} \tag{1.35}$$

Precision tests of QED

Some of the most severe tests of the Standard Model have come from comparing theory and experiment in ever more precise determinations of electromagnetic particle properties [MoNT 12]. Among these, the topic of lepton magnetic moments has achieved a deserved prominence, and we turn to this now by continuing the discussion of the previous section.

The first term in Eq. (1.35) describes the coupling of the photon to the convective current of electron, but it is the second term which interests us here. Ignoring the convective term and integrating by parts, we obtain to lowest order in q,

$$\begin{aligned}
eA_\nu(x)\langle p'_e|J_{\text{em}}^\nu(x)|p_e\rangle &= -e\bar{u}\left(p'_e\right)\frac{\sigma^{\beta\nu}\partial_\beta A_\nu(x)}{2m_e}\left(1 + \frac{e^2}{8\pi^2}\right)u(p_e)e^{-iq\cdot x} \\
&= -e\bar{u}\left(p'_e\right)\frac{\sigma^{\beta\nu}F_{\beta\nu}(x)}{4m_e}\left(1 + \frac{e^2}{8\pi^2}\right)u(p_e)e^{-iq\cdot x}. \tag{1.36}
\end{aligned}$$

Noting that in the nonrelativistic limit $\sigma^{\beta\nu}F_{\beta\nu}/2 \to -\boldsymbol{\sigma}\cdot\mathbf{B}$, we see that this is the coupling of a magnetic field to the electron magnetic moment. The result is usually expressed in terms of the gyromagnetic ratio g_{el}, where $\boldsymbol{\mu}_e \equiv -eg_{\text{el}}\mathbf{s}_e/2m_e$, and to order e^2 we have

$$a_e \equiv \frac{g_{\text{el}} - 2}{2} = \frac{\alpha}{2\pi} + \mathcal{O}\left(\frac{\alpha^2}{\pi^2}\right) \simeq 0.00116\ldots \qquad (1.37)$$

Clearly, the radiative corrections have modified the Dirac equation value, $g_{\text{el}}^{(\text{Dirac})} = 2$. The factor $\alpha/2\pi$, which arises from the Pauli term, is but the first of the anomalous *QED* contributions.

For definiteness, let us now focus on theoretical corrections to the *muon* magnetic moment.[2] The *QED* component, whose first nontrivial term is shown in Eq. (1.37), encompasses Feynman diagrams with multiple photon exchanges as well as charged lepton loops. It is expressible as a series in powers of α/π,

$$a_\mu^{(\text{QED})} = \frac{\alpha}{2\pi} + 0.765857410(27)\left(\frac{\alpha}{\pi}\right)^2 + \cdots . \qquad (1.38a)$$

Contributions through $(\alpha/\pi)^5$ have, in fact, been calculated.

There is a smaller electroweak (EW) sector with diagrams comprising virtual $W^\pm$, Z^0 and Higgs-boson exchanges. The leading order term is given by

$$a_\mu^{(\text{EW})} = \frac{G_\mu m_\mu^2}{8\sqrt{2}\pi^2}\left[\frac{5}{3} + \frac{1}{3}\left(1 - 4\sin^2\theta_{\text{w}}\right)^2 + \mathcal{O}\left(\frac{m_\mu^2}{M_W^2}\right) + \mathcal{O}\left(\frac{m_\mu^2}{M_H^2}\right) + \cdots\right], \qquad (1.38b)$$

where G_μ and θ_{w} are respectively the muon decay constant and the Weinberg angle. We will discuss each of these later, G_μ in Sect. V–2 and θ_{w} in Sect. XVI–2.

Finally, there are important corrections from the strong interactions. It can be shown that these are largely influenced by effects of relatively low-energy hadronic physics. For example, at the *QCD* level the lowest order correction amounts to a quark–antiquark vacuum polarization, expressible in terms of either e^+e^- cross sections or vector spectral functions from τ decay (see Sect. V–3),

$$a_\mu^{(\text{Had})}[\text{LO}] \times 10^{11} = 6\,923(42)(3) \qquad [\sigma(e^+e^- \to \text{hadrons}]$$
$$= 7\,015(42)(19)(3) \qquad [\tau \text{ decay}].$$

Upon using the e^+e^- cross-section data, one finds for the total,

$$a_\mu^{(\text{SM})} = a_\mu^{(\text{QED})} + a_\mu^{(\text{EW})} + a_\mu^{(\text{Had})}$$
$$= [116\,584\,718.09(0.15) + 154.(1)(2) + 6\,923(42)(3)] \times 10^{-11}$$
$$= 116591802(2)(42)(26) \times 10^{-11}. \qquad (1.39a)$$

This amounts to a difference between experiment and theory of

$$\Delta a_\mu \equiv a_\mu^{(\text{expt})} - a_\mu^{(\text{thy})} = 287(63)(49) \times 10^{-11} \qquad (1.39b)$$

[2] We follow the treatment of Hoecker and Marciano [RPP 12], which lists many references.

or 3.6 times the corresponding one-sigma error. If instead tau decay data are used, the discrepancy between experiment and theory is 2.4σ.

The theoretical values [AoHKN 12] of the magnetic moments a_e and a_μ for both electron and muon are displayed in Table V–1. There is at present no consensus about whether the theoretical predictions are in accord with the experimental determinations, and work continues on this subject. At any rate, these represent an even more stringent test of QED than the hyperfine frequency in hydrogen because theory is far less influenced by hadronic effects, and is thus about a factor of 10^4 more precise.

Radiative corrections also modify the form of the Dirac coupling. One effect of this vertex correction is to contribute to the *Lamb shift* which lifts the degeneracy between the $2S_{1/2}$ and $2P_{1/2}$ states of the hydrogen atom. Recall that the fine-structure corrections, computed as perturbations of the atomic hamiltonian, give a total energy contribution

$$(\Delta E)_{\text{fine str}} = (\Delta E)_{\text{Darwin}} + (\Delta E)_{\text{spin-orbit}} + (\Delta E)_{\text{rel kin en}}$$

$$= -\frac{7.245 \times 10^{-4}\text{eV}}{n^3}\left(\frac{1}{j+1/2} - \frac{3}{4n}\right), \tag{1.40}$$

which depends only upon the quantum numbers n and j. Thus, the $2S_{1/2}$ and $2P_{1/2}$ atomic levels are degenerate to this order, and in fact to all orders. However, the vertex radiative correction breaks the degeneracy, lowering the $2P_{1/2}$ level with respect to the $2S_{1/2}$ level by 1010 MHz. When the anomalous magnetic moment coupling ($+68$ MHz), the Uehling vacuum polarization potential (-27 MHz), and effects of higher order in α/π are added to this, the result agrees with the experimental value (cf. Table V–1). Since the entire Lamb shift arises from field-theoretic radiative corrections, one must regard the agreement with experiment as strong confirmation for the validity of QED and of the renormalization prescription.

The infrared problem

Viewed collectively, the results of this section point to a remarkable success for QED. Yet there remains an apparent blemish – the theory still contains an infinity. When the photon 'mass' λ is set equal to zero, the vertex modification of Eq. (1.29) diverges logarithmically due to the presence of terms logarithmic in λ^2. The resolution of this difficulty lies in realizing that any electromagnetic scattering process is unavoidably accompanied by a background of events containing one or more *soft* photons whose energy is too small to be detected. For example, consider Coulomb scattering of electrons from a heavy point source of charge Ze. The spin-averaged cross section for the scattering of unpolarized electrons in the absence of electromagnetic corrections is

$$\frac{d\sigma^{(0)}}{d\Omega} = \frac{Z^2\alpha^2}{4} \cdot \frac{1 - \beta^2 \sin^2\frac{\theta}{2}}{|\mathbf{p}_e|^2\beta^2 \sin^4\frac{\theta}{2}}, \tag{1.41}$$

where $\beta = |\mathbf{p}_e|/E$ is the electron speed. Radiative corrections modify this result. Using the on-shell subtraction prescription and neglecting the anomalous magnetic-moment contribution, one has in the limit $m_e^2 \gg q^2$,

$$\frac{d\sigma}{d\Omega} = \frac{d\sigma^{(0)}}{d\Omega}\left[1 + \frac{2\alpha}{3\pi}\frac{q^2}{m_e^2}\left(\ln\left(\frac{m_e}{\lambda}\right) - \frac{3}{8}\right) + \cdots\right] \tag{1.42}$$

from the *QED* vertex correction. This diverges if we attempt to take $\lambda \to 0$.

However, we must also consider the bremsstrahlung process, in which the scattering amplitude is accompanied by emission of a soft photon of infinitesimal mass λ and four-momentum k^μ. For k_0 sufficiently small, the inelastic bremsstrahlung process cannot be experimentally distinguished from the radiatively corrected elastic scattering of Eq. (1.42). To lowest order in the photon momentum k the invariant amplitude $\mathcal{M}_B$ for bremsstrahlung is[3]

$$\begin{aligned}
\mathcal{M}_B &= \frac{Ze}{\mathbf{q}^2}\bar{u}(p'_e)\left[(-ie\slashed{\epsilon})\frac{i}{\slashed{p}'_e + \slashed{k} - m_e}(-ie\gamma_0)\right]u(p_e) \\
&\quad + \bar{u}(p'_e)\left[(-ie\gamma_0)\frac{i}{\slashed{p}_e - \slashed{k} - m_e}(-ie\slashed{\epsilon})\right]u(p_e) \\
&= \mathcal{M}^{(0)} \times e\left(\frac{p'_e \cdot \epsilon}{p'_e \cdot k} - \frac{p_e \cdot \epsilon}{p_e \cdot k}\right), \tag{1.43}
\end{aligned}$$

and has the corresponding cross section

$$d\sigma_\gamma = d\sigma^{(0)}e^2\int{}'\frac{d^3k}{(2\pi)^3}\frac{1}{2k_0}\sum_{\text{pol}}\left(\frac{p'_e \cdot \epsilon}{p'_e \cdot k} - \frac{p_e \cdot \epsilon}{p_e \cdot k}\right)^2. \tag{1.44}$$

The prime on the integral sign denotes limiting the range of photon energy, $\lambda \leq k_0 \leq \Delta E$, where ΔE is the detector energy resolution. The polarization sum in Eq. (1.44) is performed with the aid of the completeness relation for massive spin-one photons

$$\sum_{\text{pol}} \epsilon_\mu(k)\epsilon_\nu(k) = -g_{\mu\nu} + \frac{k_\mu k_\nu}{\lambda^2}, \tag{1.45}$$

to yield

$$\frac{d\sigma_\gamma}{d\Omega} = \frac{d\sigma^{(0)}}{d\Omega}e^2\int{}'\frac{d^3k}{(2\pi)^3}\frac{1}{2k_0}\left(\frac{2p_e \cdot p'_e}{p'_e \cdot k \, p_e \cdot k} - \frac{m_e^2}{(p_e \cdot k)^2} - \frac{m_e^2}{(p'_e \cdot k)^2}\right). \tag{1.46}$$

---

[3] For simplicity, we shall take the photon as massless in this amplitude, and at the end indicate the effect of this omission.

Performing the angular integration in Eq. (1.46) with the aid of

$$\int d\Omega \, \frac{m_e^2}{(p \cdot k)^2} = \frac{4\pi}{k_0^2} + \mathcal{O}(\lambda^2),$$

$$\int d\Omega \, \frac{2p_e \cdot p_e'}{p_e' \cdot k \, p_e \cdot k} = \int_0^1 dx \int d\Omega \, \frac{2p_e \cdot p_e'}{(k \cdot p_x)^2} + \mathcal{O}(\lambda^2) \qquad (1.47)$$

$$= \frac{4\pi}{k_0^2} \left(2 - \frac{q^2}{m_e^2}\right) \int_0^1 dx \left(1 - \frac{q^2}{m_e^2} x(1-x)\right)^{-1} + \mathcal{O}(\lambda^2),$$

we find

$$\frac{d\sigma_\gamma}{d\Omega} = -\frac{d\sigma^{(0)}}{d\Omega} \left[\frac{2\alpha}{3\pi} \frac{q^2}{m_e^2} \left(\ln\left(\frac{2(\Delta E)}{\lambda}\right) - 1\right) + \mathcal{O}\left(\frac{q^4}{m_e^4}\right)\right]. \qquad (1.48)$$

Adding this to the nonradiative cross section of Eq. (1.42), we obtain the finite result,

$$\frac{d\sigma}{d\Omega} + \frac{d\sigma_\gamma}{d\Omega} = \frac{d\sigma^{(0)}}{d\Omega} \left[1 + \frac{2\alpha}{3\pi} \frac{q^2}{m_e^2} \left(\ln\left(\frac{m_e}{2(\Delta E)}\right) + \frac{5}{8}\right)\right]. \qquad (1.49)$$

Thus, the net effect of soft-photon emission is to replace the photon mass λ by the detector resolution $2\Delta E$, leaving a finite result.[4]

V–2 The muon

The analysis just presented for the electron can just as well be repeated for the muon. However, the muon has the additional property of being an unstable particle, and in the following we shall focus entirely on this aspect. The subject of muon decay is important because it provides a direct test of the spin structure of the charged weak current. It is also important to be familiar with the calculation of photonic corrections to muon decay, as they are part of the process whereby the Fermi constant G_μ is determined from experiment.

Muon decay at tree level

Muon decay does not proceed like the $2p \to 1s + \gamma$ transition in atomic hydrogen because the radiative process $\mu \to e + \gamma$ would conserve neither muon nor electron number and is predicted to be highly suppressed in the Standard Model. Indeed the current bound [Ad *et al.* (MEG collab.) 13] for this mode is extremely tiny, $Br_{\mu \to e+\gamma} < 5.7 \times 10^{-13}$ at 90% confidence level.[5]

---

[4] As anticipated, the result quoted in Eq. (1.49) is not quite correct, since although we have given the photon an effective mass λ we have not consistently included it, as in Eq. (1.43). In a more careful evaluation the constant $\frac{5}{8}$ is replaced by the value $\frac{11}{24}$.

[5] See Problem V–1 for a further discussion of $\mu \to e + \gamma$.

In fact, it is the weak transition $\mu(\mathbf{p}_1, \mathbf{s}) \rightarrow \nu_\mu(\mathbf{p}_2) + e(\mathbf{p}_3) + \bar{\nu}_e(\mathbf{p}_4)$ which is the dominant decay mode of the muon. In the Standard Model, this process occurs through W-boson exchange between the leptons. However, since the momentum transfer is small compared to the W-boson mass, it is possible to express muon decay in terms of the local *Fermi* interaction,

$$\mathcal{L}_{\text{Fermi}} = -\frac{G_\mu}{\sqrt{2}} \, \overline{\psi}^{(\nu_\mu)} \gamma^\alpha (1 + \gamma_5) \psi^{(\mu)} \, \overline{\psi}^{(e)} \gamma_\alpha (1 + \gamma_5) \psi^{(\nu_e)} \tag{2.1}$$

$$= -\frac{G_\mu}{\sqrt{2}} \, \overline{\psi}_{(e)} \gamma^\alpha (1 + \gamma_5) \psi_{(\mu)} \, \overline{\psi}_{(\nu_\mu)} \gamma_\alpha (1 + \gamma_5) \psi_{(\nu_e)}, \tag{2.2}$$

where the coupling constant G_μ is to be considered a phenomenological quantity determined from the muon lifetime. At tree level, G_μ is related to basic Standard Model parameters as in Eq. (II–3.43). The orderings in Eqs. (2.1)–(2.2) are called, respectively, the *charge-exchange* and *charge-retention* forms of the interaction, and are related by the Fierz transformation of Eq. (C–2.11).

Let us consider the decay of a polarized muon, with rest-frame spin vector $\hat{\mathbf{s}}$, into final states in which spin is not detected. For simplicity, we set the electron mass to zero. The muon decay width is given in terms of a three-body phase space integral by

$$\Gamma_{\mu \rightarrow e\nu_\mu \bar{\nu}_e} = \frac{1}{(2\pi)^5} \frac{1}{2E_1} \int \prod_{j=2}^4 \frac{d^3 p_j}{2E_j} \, \delta^{(4)}(p_1 - p_2 - p_3 - p_4) \sum_{s_2, s_3, s_4} |\mathcal{M}|^2, \tag{2.3}$$

where in charge-exchange form,

$$\mathcal{M} = \frac{G_\mu}{\sqrt{2}} \, \overline{u}(\mathbf{p}_2, s_2) \gamma^\alpha (1 + \gamma_5) u(\mathbf{p}_1, s_1) \, \overline{u}(\mathbf{p}_3, s_3) \gamma_\alpha (1 + \gamma_5) v(\mathbf{p}_4, s_4). \tag{2.4}$$

The muon polarization is described by a four-vector s^μ, which equals $(0, \hat{\mathbf{s}})$ in the muon rest frame. In computing the squared matrix element, we employ

$$u_\beta(\mathbf{p}_1, s_1) \overline{u}_\alpha(\mathbf{p}_1, s_1) = \frac{1}{2} \left[(m_\mu + \not{p}_1)(1 - \gamma_5 \not{s}) \right]_{\beta\alpha} \tag{2.5}$$

to obtain

$$\sum_{s_2, s_3, s_4} |\mathcal{M}|^2 = 64 G_\mu^2 \, (p_1 \cdot p_4 \, p_2 \cdot p_3 - m_\mu p_4 \cdot s \, p_2 \cdot p_3). \tag{2.6}$$

The neutrino phase space integral is easily found to be

$$\int \frac{d^3 p_2}{2E_2} \frac{d^3 p_4}{2E_4} \, \delta^{(4)}(Q - p_2 - p_4) \, p_2^\alpha p_4^\beta = \frac{\pi}{24} \left(g^{\alpha\beta} Q^2 + 2 Q^\alpha Q^\beta \right), \tag{2.7}$$

where $Q = p_1 - p_3$. For the electron phase space, it is convenient to define a reduced electron energy $x = E_e / W$, where $W = m_\mu / 2$ is the maximum electron energy in the limit of zero electron mass. The standard notation for the electron spectrum involves the so-called *Michel parameters* ρ, δ, ξ whose values depend on the tensorial nature of the beta decay interaction,

$$d^2\Gamma_{\mu \to e\nu_\mu \bar{\nu}_e} = \frac{G_\mu^2 m_\mu^5}{192\pi^3} \left[6(1-x) + 4\rho \left(\frac{4x}{3} - 1 \right) \right. $$
$$\left. -2\xi \cos\theta \left(1 - x + 2\delta \left(\frac{4x}{3} - 1 \right) \right) \right] x^2 dx \sin\theta d\theta. \quad (2.8)$$

For the $V-A$ chiral structure of the Fermi model, we predict

$$\rho = \delta = 0.75, \qquad \xi = 1.0, \quad (2.9a)$$

in good agreement with the current experimental values [RPP 12],

$$\rho = 0.74979 \pm 0.00026, \qquad \delta = 0.75047 \pm 0.00034, \qquad \xi P_\mu \delta / \rho = 1.0018^{+0.0016}_{-0.0007} \quad (2.9b)$$

where P_μ is the longitudinal muon polarization from pion decay ($P_\mu = P_\nu / E_\nu = 1$ in $V-A$ theory). In making comparisons between Eq. (2.9a) and Eq. (2.9b), one shoud first subtract from the data corrections due to radiative effects. Upon integration over the electron phase space, Eq. (2.8) gives rise to the well-known formula,

$$\Gamma_{\mu \to e\nu_\mu \bar{\nu}_e} = \frac{1}{\tau_{\mu \to e\nu_\mu \bar{\nu}_e}} = \frac{G_\mu^2 m_\mu^5}{192\pi^3}. \quad (2.10)$$

This relation has been used to provide an order-of-magnitude estimate for decay rates of heavy leptons and quarks.

Precise determination of G_μ

Thus far, we have worked to lowest order in the local Fermi interaction and have assumed massless final state particles. This is not sufficient to describe results from modern experiments, e.g., the recent measurement of $\tau_{\mu \to e\nu_\mu \bar{\nu}_e}$ by Webber *et al.* [We *et al.* (MuLan collab.) 11] is 15 times as precise as any previous determination and provides a value of G_μ with an uncertainty of only 0.6 ppm.

Including corrections to Eq. (2.10) yields

$$\Gamma_{\mu \to e\nu_\mu \bar{\nu}_e} = \frac{G_\mu^2 m_\mu^5}{192\pi^3} f(x) r_\gamma \left(\hat{\alpha}(m_\mu), x \right) r_W(y_\mu), \quad (2.11a)$$

where $f(x)$ is a phase space factor, with $x \equiv m_e^2 / m_\mu^2$,

$$f(x) = 1 - 8x + 8x^3 - x^4 - 12x^2 \ln x \simeq 0.999812961, \quad (2.11b)$$

Table V–2. *Determinations of Fermi-model couplings.*

Factor	Determination
G_μ	Muon decay
$G_\tau^{(\ell)}$	Tau decay into lepton ℓ
G_β	G_μ plus *QED* theory
$G_\beta V_{ud}$	$\begin{cases} \text{Nuclear beta decay} \\ \pi_{\ell 3} \end{cases}$
$G_\beta V_{us}$	$\begin{cases} \text{Hyperon beta decay} \\ K_{\ell 3} \end{cases}$
$G_\beta V_{ud} F_\pi$	Pion beta decay $(\pi_{\ell 2})$
$G_\beta V_{us} F_K$	Kaon beta decay $(K_{\ell 2})$

and $r_W(y_\mu) = 1 + 3 y_\mu/5 + \cdots$ is a W-boson propagator correction, with $y_\mu \equiv m_\mu^2/M_W^2$.[6] The quantity $r_\gamma(\hat{\alpha}(m_\mu), x)$ provides a perturbative expression of the photonic radiative corrections,

$$r_\gamma(\hat{\alpha}, x) = H_1(x) \frac{\hat{\alpha}(m_\mu)}{\pi} + H_2(x) \frac{\hat{\alpha}^2(m_\mu)}{\pi^2} + \cdots, \qquad (2.11c)$$

where $\hat{\alpha}(m_\mu)$ refers to the $\overline{\text{MS}}$ subtracted quantity

$$\hat{\alpha}(m_\mu)^{-1} = \alpha^{-1} + \frac{1}{3\pi} \ln x + \cdots \simeq 135.901. \qquad (2.11d)$$

The functions $H_1(x)$, $H_2(x)$ appear, together with references to original work, in Chap. 10 of [RPP 12]. In the subsection to follow, we will calculate the leading-order contribution to r_γ. The above theoretical relations lead to the determination

$$G_\mu = 1.1663787(6) \times 10^{-5} \text{ GeV}^{-2}. \qquad (2.12)$$

The above analysis serves to define the Fermi constant in the context of muon decay. Fermi couplings $G_\tau^{(\ell)}$ for the weak leptonic transitions $\tau^- \to e^- + \bar{\nu}_e + \nu_\tau$ and $\tau^- \to \mu^- + \bar{\nu}_\mu + \nu_\tau$ can likewise be defined and compared with G_μ (see Sect. V–3). However, for weak semileptonic transitions of hadrons (e.g. nuclear beta decay) the photonic corrections are *not* identical to those in muon decay because quark charges differ from lepton charges. Such processes define instead a quantity called G_β, and we shall present in Sect. VII–1 a calculation of G_β for the case of pion decay (cf. Eq. (VII–1.31)). As seen in Table V–2, determinations involving G_β generally contain quark mixing factors and also meson decay

---

[6] It is possible to study muon decay corrections either within just the Fermi effective theory or with the full Standard Model, For the former choice $r_W(y_\mu)$ is omitted in Eq. (2.11a), whereas for the latter it is included. Which choice is made affects details of higher-order corrections. We have opted to follow the first edition of this book by including $r_W(y_\mu)$. In fact, its effect is numerically tiny, affecting only the final decimal place in the value of G_μ given in Eq. (2.12).

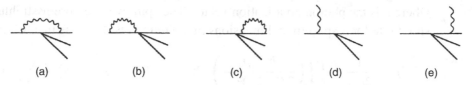

(a) (b) (c) (d) (e)

Fig. V–1 Contributions to muon decay from (a) vertex, (b)–(c) wavefunction renormalization, and (d)–(e) bremsstrahlung amplitudes.

constants. Marciano [Ma 11] has used, among other processes, semileptonic decays (nuclear, kaon and B-meson) and CKM unitarity to determine the Fermi constant without recourse to muon decay. He finds $G_F^{(\text{CKM})} = 1.166309(350) \times 10^{-5}\ \text{GeV}^2$, which is the second most accurate determination after the muon decay value of Eq. (2.12), but relatively far less accurate.

Leading-order photonic correction

Computation of the lowest order electron and W-boson mass corrections appearing in Eq. (2.10) is not difficult, and is left to Prob. V–2. However, the *QED* radiative correction is rather more formidable, and it is to that which we now turn our attention. Rather than attempt a detailed presentation, we summarize the analysis of [GuPR 80]. We shall work in Feynman gauge, and employ the charge-retention ordering for the Fermi interaction. There is an advantage to performing the calculation as if the muon existed in a spacetime of arbitrary dimension d. Working in $d = 4$ dimensions entails factors which are logarithmic in the electron mass and which would forbid the simplifying assumption $m_e = 0$. Dimensional regularization frees one from this restriction, and such potential singularities become displayed as poles in the variable $\epsilon = (4 - d)/2$. Although there would appear to be difficulty in extending the Dirac matrix γ_5 to arbitrary spacetime dimensions, this turns out not to be a problem here. The set of radiative corrections consists of three parts, which are displayed in Fig. V–1, (i) vertex (Fig. V–1(a)), (ii) self-energy (Fig. V–1(b)–(c)), and (iii) bremsstrahlung (Fig. V–1(d)–(e)). We shall begin with the bremsstrahlung part of the calculation and then proceed to the vertex and self-energy contributions.

The amplitude for the bremsstrahlung (B) process $\mu(\mathbf{p}_1) \to \nu_\mu(\mathbf{p}_2) + e(\mathbf{p}_3) + \bar{\nu}_e(\mathbf{p}_4) + \gamma(\mathbf{p}_5)$ is given by

$$\mathcal{M}_B = \frac{eG_\mu}{\sqrt{2}} \bar{u}(\mathbf{p}_2)\gamma^\alpha(1 + \gamma_5)v(\mathbf{p}_4)$$

$$\times \bar{u}(\mathbf{p}_3)\left[\gamma_\alpha(1 + \gamma_5)\frac{1}{\not{p}_1 - \not{p}_5 - m_\mu}\not{\epsilon} + \not{\epsilon}\frac{1}{\not{p}_3 + \not{p}_5 - m_e}\gamma_\alpha(1 + \gamma_5)\right]u(\mathbf{p}_1),$$

$$(2.13)$$

where ϵ is the photon polarization vector. The spin-averaged bremsstrahlung transition rate for d spacetime dimensions in the muon rest frame is then given by[7]

$$\Gamma_B = \frac{1}{2m_\mu} \int \prod_{j=2}^{5} \left(\frac{d^{d-1}p_j}{2E_j(2\pi)^{d-1}} \right) (2\pi)^d \delta^{(d)} \left(Q' - p_2 - p_4 \right) \frac{1}{2} \sum_{\text{spins}} |\mathcal{M}_B|^2, \tag{2.14}$$

where $Q' \equiv Q - p_5 = p_1 - p_3 - p_5$. A lengthy analysis yields a result which can be expanded in powers of $\epsilon = (4 - d)/2$ to read

$$\Gamma_B = \frac{G_\mu^2 m_\mu^5}{192\pi^3} \frac{3\alpha}{4} \left(\frac{m_\mu^3}{32\pi^{3/2}} \right)^{-2\epsilon} \frac{\Gamma(2 - \epsilon)}{\Gamma\left(\frac{3}{2} - \epsilon\right) \Gamma\left(\frac{5}{2} - \epsilon\right) \Gamma(5 - 3\epsilon)}$$

$$\times \left(\frac{6}{\epsilon^2} - \frac{5 - 6\gamma}{\epsilon} - 5\gamma + 3\gamma^2 - \frac{7\pi^2}{2} + \frac{215}{6} + \mathcal{O}(\epsilon) \right). \tag{2.15}$$

Observe that singularities are encountered as $\epsilon \to 0$.

The radiative correction (R) contribution to the muon transition rate is given by

$$\Gamma_R = \frac{1}{2m_\mu} \int \prod_{j=2}^{4} \left(\frac{d^{d-1}p_j}{2E_j(2\pi)^{d-1}} \right) (2\pi)^d \delta^{(d)} \left(Q - p_2 - p_4 \right) \frac{1}{2} \sum_{\text{spins}} |\mathcal{M}|^2_{\text{int}}, \tag{2.16}$$

where $|\mathcal{M}|^2_{\text{int}}$ is the interference term between the Fermi-model amplitudes, which are, respectively, zeroth order ($\mathcal{M}^{(0)}$) and first order ($\mathcal{M}^{(1)}$) in e^2,

$$|\mathcal{M}|^2_{\text{int}} = \mathcal{M}^{(0)*} \mathcal{M}^{(1)} + \mathcal{M}^{(1)*} \mathcal{M}^{(0)}. \tag{2.17}$$

The first-order amplitude can be written as a product of neutrino factors and a term ($\mathcal{M}_R$) containing radiative corrections of the charged leptons,

$$\mathcal{M}^{(1)} = \frac{e^2 G_\mu}{\sqrt{2}} \bar{u}(\mathbf{p}_2) \gamma_\alpha (1 + \gamma_5) v(\mathbf{p}_4) \mathcal{M}_R^\alpha. \tag{2.18}$$

The quantity $\mathcal{M}_R^\alpha$ is itself expressible as the sum of vertex (V) and self-energy (SE) contributions,

$$\mathcal{M}_R^\alpha = \bar{u}(\mathbf{p}_3) \left(\mathcal{M}_V^\alpha + \mathcal{M}_{SE}^\alpha \right) u(\mathbf{p}_1). \tag{2.19}$$

The vertex modification of Fig. V–1(c)

$$\mathcal{M}_V^\alpha = \frac{1}{i} \int \frac{d^4k}{(2\pi)^4} \frac{\gamma^\mu \left(\not{p}_3 - \not{k} \right) \gamma^\alpha (1 + \gamma_5) \left(\not{p}_1 - \not{k} + m_\mu \right) \gamma_\mu}{k^2 (p_3 - k)^2 \left[(p_1 - k)^2 - m_\mu^2 \right]}, \tag{2.20}$$

---

[7] Since the result that we seek is finite and scale independent, in this section we shall suppress the scale parameter μ introduced in Eq. (II–1.21b).

has the same form as the electromagnetic vertex correction for the electron discussed previously (cf. Eq. (1.19)) except that it contains the weak vertex $\gamma^\alpha(1+\gamma_5)$. Upon employing the Feynman parameterization in Eq. (2.20) and using the muon equation of motion, the extension of the vertex amplitude to d dimensions can ultimately be expressed in terms of hypergeometric functions,

$$\mathcal{M}_V^\alpha = 4\Gamma\left(3 - \frac{d}{2}\right) \frac{m_\mu^{-(4-d)}}{(4\pi)^{d/2}} \left[\gamma^\alpha(1 + \gamma_5)A_1 + (1 - \gamma_5)\frac{p_1^\alpha B + p_3^\alpha C}{m_\mu}\right], \tag{2.21}$$

where

$$A_1 = \frac{F\left(3 - \frac{d}{2}, 1; \frac{d}{2}; \xi\right)}{(d-3)(d-2)} - \frac{(d-3)F\left(2 - \frac{d}{2}, 1; \frac{d}{2}; \xi\right)}{(d-4)(d-2)}$$
$$- \frac{(1-\xi)F\left(3 - \frac{d}{2}, 1; \frac{d}{2} - 1, \xi\right)}{(d-4)^2(d-3)}, \tag{2.22}$$

$$B = \frac{F\left(3 - \frac{d}{2}, 1; \frac{d}{2} + 1; \xi\right)}{d},$$

$$C = \frac{2F\left(3 - \frac{d}{2}, 2; \frac{d}{2} + 1; \xi\right)}{d(d-2)} - \frac{2F\left(3 - \frac{d}{2}, 1; \frac{d}{2}; \xi\right)}{(d-2)(d-3)}.$$

For the muon self-energy amplitude of Fig. (V–1(d)), we write

$$\Sigma(p) = -i \int \frac{d^d q}{(2\pi)^d} \frac{\gamma^\lambda(\not{p} - \not{q} + m_\mu)\gamma_\lambda}{q^2\left(q^2 - 2p\cdot q + p^2 - m_\mu^2\right)}, \tag{2.23}$$

remembering that a factor of e^2 has already been extracted in Eq. (2.18). Implementing the Feynman parameterization and integrating over the virtual momentum yields an expression,

$$\Sigma(p) = \frac{\Gamma\left(2 - \frac{d}{2}\right)}{(4\pi)^{d/2}} \frac{1}{m_\mu^{4-d}} \int_0^1 dx\,\left[m_\mu d + x(2 - d)\not{p}\right] \frac{(1-x)^{(4-d)/2}}{\left(1 - x\frac{p^2}{m^2}\right)^{(4-d)/2}}, \tag{2.24}$$

which with the aid of Eq. (C–5.5) in App. C can be written in terms of hypergeometric functions,

$$\Sigma(p) = \frac{\Gamma(2 - \frac{d}{2})}{(4\pi)^{d/2}} \frac{1}{m_\mu^{4-d}} \left[m_\mu \frac{2d}{d-2} F\left(\frac{4-d}{2}, 1; \frac{d}{2}; \frac{p^2}{m_\mu^2}\right) \right.$$
$$\left. -\not{p}\frac{4}{d} F\left(\frac{4-d}{2}, 2; \frac{d+2}{2}; \frac{p^2}{m_\mu^2}\right)\right]. \tag{2.25}$$

When the self-energy is expanded in powers of $\not{p} - m_\mu$, the leading term is just the mass shift, which is removed by mass renormalization. We require the $\not{p}$-derivative of $\Sigma(p)$ evaluated at $\not{p} = m_\mu$. Being careful while carrying out the differentiation to interpret p^2 factors as $\not{p}\not{p}$, we find

$$\left.\frac{\partial \Sigma}{\partial \not{p}}\right|_{\not{p}=m_\mu} = \frac{\Gamma\left(2 - \frac{d}{2}\right)}{(4\pi)^{d/2}} \frac{1}{m_\mu^{4-d}} \frac{1-d}{d-3}. \tag{2.26}$$

It is this quantity multiplied by the vertex $\gamma^\alpha(1 + \gamma_5)$ which ultimately gives rise to $\mathcal{M}_{SE}^\alpha$. However, in addition to mass renormalization there is also wavefunction renormalization, whose effect is to reduce the above quantity by a factor of 2, yielding

$$\mathcal{M}_{SE}^\alpha = \frac{\Gamma\left(3 - \frac{d}{2}\right)}{(4\pi)^{d/2}} \frac{1}{m_\mu^{4-d}} \frac{1-d}{4(4-d)(d-3)} \gamma^\alpha(1 + \gamma_5). \tag{2.27}$$

In principle, there also exists the electron self-energy contribution. As can be verified by direct calculation, this vanishes because the electron is taken as massless. Thus, we conclude that

$$\mathcal{M}_{SE}^\alpha = 4\Gamma\left(3 - \frac{d}{2}\right) \frac{m_\mu^{d-4}}{(4\pi)^{d/2}} A_2 \gamma^\alpha(1 + \gamma_5),$$

$$A_2 = -\frac{1}{4}\frac{d-1}{(4-d)(d-3)}. \tag{2.28}$$

The net effect of the self-energy contribution is to replace A_1 in the vertex amplitude of Eq. (2.21) by $A = A_1 + A_2$.

Insertion of the radiatively corrected amplitudes into Eq. (2.16) leads to a transition rate Γ_R, which expanded to lowest order in $\epsilon = (4 - d)/2$, has the form

$$\Gamma_R = \frac{G_\mu^2 m_\mu^5}{192\pi^3} \frac{3\alpha}{4} \left(\frac{m_\mu^3}{32\pi^{3/2}}\right)^{-2\epsilon} \frac{\Gamma(2-\epsilon)}{\Gamma\left(\frac{3}{2}-\epsilon\right)\Gamma\left(\frac{5}{2}-\epsilon\right)\Gamma(5-3\epsilon)}$$

$$\times \left(-\frac{6}{\epsilon^2} + \frac{5-6\gamma}{\epsilon} + 5\gamma - 3\gamma^2 - \frac{5\pi^2}{2} + \frac{5}{3} + \mathcal{O}(\epsilon)\right). \tag{2.29}$$

Like the bremsstrahlung contribution, the radiatively corrected decay rate is found to be singular in the $\epsilon \to 0$ limit. However, the final result which is obtained by adding the radiative correction of Eq. (2.29) to that of the bremsstrahlung expression of Eq. (2.15) is found to be free of divergences,

$$\delta\Gamma_{\text{muon}}^{(\text{QED})} = \Gamma_R + \Gamma_B = -\frac{G_\mu^2 m_\mu^5}{192\pi^3} \frac{\alpha(m_\mu)}{2\pi}\left(\pi^2 - \frac{25}{4}\right), \tag{2.30}$$

which is the leading-order contribution to the function r_γ of Eq. (2.11c).

V–3 The τ lepton

The heaviest known lepton is $\tau(1777)$, having been discovered in e^+e^- collisions in 1975. There exists also an associated neutrino ν_τ with current mass limit $m_{\nu_\tau} <$ 18.2 MeV. Like the muon, the τ can decay via purely *leptonic* modes,

$$
\tau^- \to \left[\begin{array}{l} \mu^- + \bar{\nu}_\mu + \nu_\tau \\ e^- + \bar{\nu}_e + \nu_\tau \\ \vdots \end{array} \right. . \tag{3.1}
$$

However, a new element exists in τ decay, for numerous *semileptonic* modes are also present,

$$
\tau^- \to \left[\begin{array}{l} \pi^- + \nu_\tau \\ \pi^- + \pi^0 + \nu_\tau \\ \vdots \end{array} \right. . \tag{3.2}
$$

Experiment has revealed the semileptonic sector to be an important component of tau decay, e.g., [Am *et al.* (Heavy Flavor Averaging Group collab.) 12],

$$
R_\tau \equiv \left. \frac{\Gamma_{\text{semileptonic}}}{\Gamma_{\tau \to e\bar{\nu}_e \nu_\tau}} \right|_{\text{expt}} = \frac{1 - \text{Br}_{\tau \to e\bar{\nu}_e \nu_\tau} - \text{Br}_{\tau \to \mu\bar{\nu}_\mu \nu_\tau}}{\text{Br}_{\tau \to e\bar{\nu}_e \nu_\tau}} = 3.6280 \pm 0.0094, \tag{3.3}
$$

where Br denotes branching ratio. It is possible to obtain a simple but naive estimate of R_τ as follows. Because the τ is lighter than any charmed hadron, semileptonic decay amplitudes must involve the quark charged weak current

$$
J_{\text{ch}}^\mu = V_{\text{ud}} \bar{d}\gamma^\mu(1 + \gamma_5)u + V_{\text{us}} \bar{s}\gamma^\mu(1 + \gamma_5)u, \tag{3.4}
$$

where V_{ud} and V_{us} are CKM mixing elements. Neglect of all final state masses and of effects associated with quark hadronization (an assumption only approximately valid at this relatively low energy) implies the estimates

$$
R_\tau^{(\text{naive})} \simeq N_c \left[|V_{\text{ud}}|^2 + |V_{\text{us}}|^2 \right] \simeq N_c \xrightarrow[N_c=3]{} 3.0,
$$

$$
\text{Br}_{\tau \to e\bar{\nu}_e \nu_\tau}^{(\text{naive})} \simeq \text{Br}_{\tau \to \mu\bar{\nu}_\mu \nu_\tau}^{(\text{naive})} \simeq \frac{1}{2 + N_c} \xrightarrow[N_c=3]{} 0.2, \tag{3.5}
$$

where N_c is the number of quark color degrees of freedom. The above analysis although rough, nonetheless yields estimates for R_τ, $\tau \to e\bar{\nu}_e \nu_\tau$ and $\tau \to \mu\bar{\nu}_\mu \nu_\tau$ in approximate accord with the corresponding experimental values. Also, it is not inconsistent with our belief that $N_c = 3$. However, we can and will improve upon this state of affairs (cf. Eq. (3.27b)) and as a bonus will obtain a determination of the strong coupling $\alpha_s(m_\tau)$.

Exclusive leptonic decays

The momentum spectra of the electron and muon modes also probe the nature of τ decay. The Michel parameter ρ of Eq. (2.8) should equal 0.75 for the usual $V - A$ currents, zero for the combination $V + A$ and 0.375 for V or A separately. The observed value $\rho = 0.745 \pm 0.008$ is in accord with the $V - A$ structure.

The τ leptonic decays afford an opportunity to test the principle of *lepton universality*, i.e., the premise that the only physical difference among the charged leptons is that of mass. In particular, all the charged leptons are expected to have identical charged current weak couplings, cf. Eqs. (II–3.36),(II–3.37),

$$g_{2e} = g_{2\mu} = g_{2\tau} \equiv g_2 \qquad \text{(universality condition)}. \tag{3.6}$$

This has been tested in [Am *et al.* (Heavy Flavor Averaging Group collab.) 12], using the following Standard Model description for the leptonic decay mode of a heavy lepton L,

$$\Gamma_{L \to \nu_L \ell \bar{\nu}_\ell} = \frac{G_L G_\ell m_L^5}{192\pi^3} f(x) r_\gamma(m_L) r_W(y_L),$$

$$G_\ell = \frac{g_{2\ell}^2}{4\sqrt{2} M_W^2}, \qquad r_\gamma(m_L) = 1 + \frac{\alpha(m_L)}{2\pi}\left(\frac{25}{4} - \pi^2\right), \tag{3.7}$$

where $f(x)$ is as in Eq. (2.11b) and $r_W(y_L) = 1 + 3m_L^2/(5M_W^2 + \cdots$. They find [Am *et al.* (Heavy Flavor Averaging Group collab.) 12],

$$\frac{g_{2\tau}}{g_{2\mu}} = 1.0006 \pm 0.0021, \quad \frac{g_{2\tau}}{g_{2e}} = 1.0024 \pm 0.0021, \quad \frac{g_{2\mu}}{g_{2e}} = 1.0018 \pm 0.0014, \tag{3.8}$$

consistent with the universality condition of Eq. (3.6).

There are other ways to study the universality principle. Looking forward to Chap. XVI, we shall exhibit in Eq. (XVI–2.6) the result of testing lepton universality with the decays $Z^0 \to \ell\bar{\ell}$ ($\ell = e, \mu, \tau$). Unlike the above example in Eq. (3.8), the Z^0 decay widths are functions of *neutral* weak coupling constants. Yet another approach, which uses charged current couplings, is to compare leptonic and semileptonic decays, like $H \to \mu\bar{\nu}_\mu$ (where H can be a pion, kaon, etc.) with $\tau \to H\nu_\tau$.

Exclusive semileptonic decays

Matters are somewhat more complex for the hadronic final states, due in part to the large number of modes. Still, for many of these we can make detailed confrontation of theoretical predictions with experimental results. We begin by noting that the semileptonic decay amplitude factorizes into purely leptonic and hadronic matrix elements of the weak current,

$$\mathcal{M}_{\text{semilept}} = \frac{G_\mu}{\sqrt{2}} L^\mu H_\mu,$$

$$L^\mu = \langle \nu_\tau(\mathbf{p}') | J^\mu_{\text{lept}} | \tau(\mathbf{p}) \rangle = \bar{\nu}_\tau(\mathbf{p}') \Gamma^\mu_L \tau(\mathbf{p}),$$

$$H_\mu = \left\langle \text{hadron} \left| \left(J^{\text{qk}}_\mu\right)^\dagger \right| 0 \right\rangle = \langle \text{hadron} | V^*_{\text{ud}} \, \bar{d} \Gamma^L_\mu u + V^*_{\text{us}} \, \bar{s} \Gamma^L_\mu u \, | 0 \rangle, \qquad (3.9)$$

where $\Gamma^L_\mu \equiv \gamma_\mu(1 + \gamma_5)$. In the following, we analyze some modes containing a single meson,

$$\tau^- \to \text{meson} + \nu_\tau \qquad \left(\text{meson} = \pi^-, \ K^-, \ \rho^-(770), \ K^{*-}(892)\right). \qquad (3.10)$$

Weak-current matrix elements which connect the vacuum with spin-parity $J^P = 0^-, 1^+$ hadrons are sensitive to only the axial-vector current, whereas $J^P = 0^+, 1^-$ states arise from the vector current. In each case, the vacuum-to-meson matrix element has a form dictated up to a constant by Lorentz invariance,

$$\langle \pi^-(\mathbf{q}) | \left[\bar{d}\gamma_\mu\gamma_5 u\right](0) | 0 \rangle \equiv -i\sqrt{2}F_\pi q_\mu, \qquad (3.11\text{a})$$

$$\langle K^-(\mathbf{q}) | \left[\bar{s}\gamma_\mu\gamma_5 u\right](0) | 0 \rangle \equiv -i\sqrt{2}F_K q_\mu, \qquad (3.11\text{b})$$

$$\langle \rho^-(\mathbf{q}, \lambda) | \left[\bar{d}\gamma_\mu u\right](0) | 0 \rangle \equiv \sqrt{2}g_\rho \epsilon^*_\mu(\mathbf{q}, \lambda), \qquad (3.11\text{c})$$

$$\langle K^{*-}(\mathbf{q}, \lambda) | \left[\bar{s}\gamma_\mu u\right](0) | 0 \rangle \equiv \sqrt{2}g_{K^*} \epsilon^*_\mu(\mathbf{q}, \lambda), \qquad (3.11\text{d})$$

where the quantities g_ρ and g_{K^*} are the vector meson decay constants. These quantities contribute to the transition rates for pseudoscalar (p) and vector (v) emission, and we find from straightforward calculations

$$\Gamma_{\tau \to p\nu_\tau} = \eta_{\text{KM}} G^2_\mu m^3_\tau \frac{F^2_{\text{p}}}{8\pi} \left(1 - \frac{m^2_{\text{p}}}{m^2_\tau}\right)^2,$$

$$\Gamma_{\tau \to v\nu_\tau} = \eta_{\text{KM}} \frac{G^2_\mu}{8\pi} \left(\frac{g_{\text{v}}}{m^2_{\text{v}}}\right)^2 m^3_\tau m^2_{\text{v}} \left(1 - \frac{m^2_{\text{v}}}{m^2_\tau}\right)^2 \left(1 + 2\frac{m^2_{\text{v}}}{m^2_\tau}\right), \qquad (3.12)$$

where $m_{\text{p}}, m_{\text{v}}$ are the meson masses, $\eta = |V_{\text{ud}}|^2$ for $\Delta S = 0$ decay and $\eta = |V_{\text{us}}|^2$ for $\Delta S = 1$ decay.

It is possible to use the above formulae to extract constants such as $F_\pi, \ldots, g_{K^*}$ from tau decay data. However, such quantities are obtained more precisely from other processes and, in practice, one employs them in tau decay to make branching-ratio predictions. Although *QCD*-lattice studies have steadily improved on their predictions of such constants, we shall focus instead on phenomenological determinations. In Chap. VII, we shall show how the values $F_\pi = 92.21$ MeV and $F_K/F_\pi = 1.197$ are found from a careful analysis of pion and kaon leptonic weak

Table V–3. *Some hadronic modes in tau decay.*

Mode	Hadronic input	Br [thy][a]	Br [expt][a]
$\tau^- \to \pi^- + \nu_\tau$	$c_1 F_\pi$	11.4	10.83 ± 0.06
$\tau^- \to K^- + \nu_\tau$	$c_3 s_1 F_K$	0.8	0.70 ± 0.01
$\tau^- \to \rho^- + \nu_\tau$	$c_1 g_\rho$	23.4 ± 0.8	25.52 ± 0.9
$\tau^- \to K^{*-} + \nu_\tau$	$c_3 s_1 g_{K^*}$	1.1 ± 0.1	1.33 ± 0.13
$\tau^- \to \pi^- \pi^- \pi^+ \pi^0 \nu_\tau$	$\sigma(e^+ e^- \to \text{hadr})$	4.9	4.76 ± 0.06
$\tau^- \to \pi^- \pi^0 \pi^0 \pi^0 \nu_\tau$	$\sigma(e^+ e^- \to \text{hadr})$	0.98	1.05 ± 0.07

[a]Branching ratios are given in percent.

decay. Interestingly, the hadronic matrix elements which contribute there are just the conjugates of those appearing in Eqs. (3.11a), (3.11b). By contrast, the quantity g_ρ is obtained not from weak decay data, but rather from an electromagnetic decay such as $\rho^0 \to e\bar{e}$. That the *same* quantity g_ρ should occur in both weak and electromagnetic transitions is a consequence of the isospin structure of quark currents. That is, the electromagnetic current operator is expressed in terms of octet vector current operators by $J_\mu^{\text{em}} = V_\mu^3 + \frac{1}{\sqrt{3}} V_\mu^8$. Since the latter component is an isotopic scalar whereas the ρ meson carries isospin one, it follows from the Wigner–Eckart theorem that

$$\langle 0 | J_\mu^{\text{em}} | \rho^0(\mathbf{p}) \rangle = \langle 0 | V_\mu^3 | \rho^0(\mathbf{p}) \rangle = \frac{1}{\sqrt{2}} \langle 0 | V_\mu^{1+i2} | \rho^-(\mathbf{p}) \rangle = g_\rho \epsilon_\mu(\mathbf{p}). \qquad (3.13)$$

The transition rate for the electromagnetic decay $\rho^0 \to e\bar{e}$ is given, with final state masses neglected, by

$$\Gamma_{\rho^0 \to e\bar{e}} = \frac{4\pi\alpha^2}{3} \left(\frac{g_\rho}{m_\rho^2} \right)^2 m_\rho, \qquad (3.14)$$

from which we find $g_\rho/m_\rho^2 = 0.198 \pm 0.009$. The $K^{*-}\nu_\tau$ mode can be estimated by using the flavor-$SU(3)$ relation $g_{K^*} = g_\rho$. The predictions for single-hadron branching ratios are collected in Table V–3, and are seen to be in satisfactory agreement with the observed values.

A somewhat different approach can be used to obtain predictions for strangeness-conserving modes with $J^P = 1^-$. Since matrix elements of the vector-charged current can be obtained through an isospin rotation from the isovector part of the $e\bar{e}$ annihilation cross section into hadrons, we can write for a given neutral $I = 1$ hadronic final state f^0,

$$\sigma_{e\bar{e} \to f^0}^{(I=1)} = \frac{8\pi^2\alpha^2}{q^2} \Pi_f^0\left(q^2\right), \qquad (3.15)$$

where we have defined

$$\sum_{f^0} (2\pi)^3 \delta^{(4)}(q - p_{f^0}) \langle f^0 | J_\mu^{(3)} | 0 \rangle \langle f^0 | J_\nu^{(3)} | 0 \rangle^* \equiv \Pi_f^0(q^2)\left(-q^2 g_{\mu\nu} + q_\mu q_\nu\right).$$

(3.16)

The τ^- transition into the isotopically related charged state f^-,

$$|f^-\rangle \equiv \frac{1}{\sqrt{2}}(I_1 - iI_2)|f^0\rangle,$$

(3.17)

is governed by the *same* function $\Pi_f(q^2)$ of hadronic final states as occurs in Eq. (3.15). Including the lepton current and relevant constants, and performing the integration over ν_τ phase space yields a decay rate

$$\Gamma_{\tau \to f\nu_\tau} = \frac{G_\mu^2 |V_{\mathrm{ud}}|^2 S_{\mathrm{EW}}}{32\pi^2 m_\tau^3} \int_0^{m_\tau^2} dq^2 \left(m_\tau^2 - q^2\right)^2 \left(m_\tau^2 + 2q^2\right) \Pi_f^-\left(q^2\right),$$

(3.18)

where S_{EW} is given in Eq. (3.25c). The content of Eq. (3.18) is often expressed as a ratio,

$$\frac{\Gamma_{\tau \to f\nu_\tau}}{\Gamma_{\tau \to e\nu_\tau \bar{\nu}_e}} = \frac{3|V_{\mathrm{ud}}|^2 S_{\mathrm{EW}}}{2\pi\alpha^2 m_\tau^8} \int_0^{m_\tau^2} ds \left(m_\tau^2 - s\right)^2 \left(m_\tau^2 + 2s\right) s \, \sigma_{e^+e^- \to f^0}^{(I=1)}(s).$$

(3.19)

Thus, we find, e.g., for 4π final states, the results listed in Table V–3.

There exist numerous additional hadronic decay modes of the τ lepton. Examples include final hadronic states containing $K\bar{K}$, $K\bar{K}\pi$, etc., and it is possible to analyze each of these with various degrees of theoretical confidence. Another interesting use of the τ semileptonic decay has been to confirm by inference the fundamental structure of the weak quark current from the absence of the mode $\tau^- \to \pi^-\eta\nu_\tau$. This mode, proceeding through the vector current, would violate G-parity invariance. Here G-parity refers to the product of charge conjugation and a rotation by π radians about the 2-axis in isospin space,

$$G \equiv Ce^{-i\pi I_2}.$$

(3.20)

A weak current which could induce a $\Delta G \neq 0$ transition is referred to as a *second-class* current. Such currents do not occur naturally within the quark model. The $\pi^-\eta\nu_\tau$ mode has not been detected, with an existing sensitivity [RPP 12] of $\mathrm{Br}_{\tau \to \pi\eta\nu_\tau} < 9.9 \times 10^{-5}$. This result, consistent with the absence of second-class currents, fits securely within the framework of the Standard Model.

Inclusive semileptonic decays

The inclusive semileptonic decay of the tau is denoted as $\tau \to \nu_\tau + X$, where X represents the sum over all kinematically allowed hadronic states. Let us restrict our attention to the Cabibbo-allowed component, i.e., decay into an even or odd number of pions. The decay rate at invariant squared-energy s is

$$\frac{d\Gamma\left[\tau \to \nu_\tau \left(\genfrac{}{}{0pt}{}{\text{even}}{\text{odd}}\right)\right]}{ds}$$

$$= \frac{G_\mu^2 V_{ud}^2}{8\pi m_\tau^3} \left(m_\tau^2 - s\right)^2 \left[\left(m_\tau^2 + 2s\right) \begin{pmatrix} \rho_V(s) \\ \rho_A(s) \end{pmatrix} + m_\tau^2 \begin{pmatrix} 0 \\ \rho_A^{(0)} \end{pmatrix}\right], \qquad (3.21)$$

as expressed in terms of the so-called *vector* and *axial-vector spectral functions*, spin-one $\rho_V(s)$, $\rho_A(s)$ and spin-zero $\rho_A^{(0)}(s)$.

We can gain some physical understanding of the spectral functions by first studying the propagator $i\Delta(x)$ for a free, scalar field $\varphi(x)$ (cf. Eq. (C–2.12)). Its Fourier transform,

$$\Pi(q^2) = \frac{1}{\mu^2 - q^2 - i\epsilon} \qquad \text{with} \qquad \mathcal{I}m\,\Pi(q^2) = \pi\delta\left(q^2 - \mu^2\right), \qquad (3.22)$$

reveals that the free field $\varphi(x)$ excites the vacuum to just the single state with $q^2 = \mu^2$. For the $V - A$ currents which induce the inclusive tau decay, the momentum space propagators are written

$$i \int d^4x \, e^{iq\cdot x} \, \langle 0|T \left(V_3^\mu(x) V_3^\nu(0) - A_3^\mu(x) A_3^\nu(0)\right) |0\rangle$$

$$= \left(q^\mu q^\nu - q^2 g^{\mu\nu}\right) \left(\Pi_{V,3}^{(1)} - \Pi_{A,3}^{(1)}\right) (q^2) - q^\mu q^\nu \Pi_{A,3}^{(0)}(q^2), \qquad (3.23)$$

where $\Pi_{V,3}^{(1)}$, $\Pi_{A,3}^{(1)}$ and $\Pi_{A,3}^{(0)}$ are respectively the spin-one and spin-zero correlators. The spectral functions are proportional to the imaginary parts of the corresponding correlators,

$$\mathcal{I}m\,\Pi_{V/A,3}^{(1)}(s) = \pi\rho_{V/A,3}(s) \qquad \mathcal{I}m\,\Pi_{A,3}^{(0)}(s) = \pi\rho_{A,3}^{(0)}(s). \qquad (3.24)$$

They encode how the isospin vector and axial-vector currents excite various n-pion states at invariant energy $s < m_\tau^2$. Figure V–2 displays the $V - A$ spectral function[8] $(\rho_{V,3} - \rho_{A,3})(s)$ as measured in tau decay from the ALEPH collaboration. The first peak is from the ρ meson, followed by the negative a_1 peak and then the four-pion component, etc.

Some applications of τ decays

Studies of τ decays have also proved valuable in providing a measure of $\alpha_s(m_\tau)$. Such a determination is significant because the tau lepton mass is one of the lowest energy scales (the charm quark mass is another) at which this is possible. The procedure essentially amounts to performing a more careful evaluation of R_τ than

---

[8] Here and henceforth we abbreviate $\rho_{V,3} \to \rho_V$ and $\rho_{A,3} \to \rho_A$.

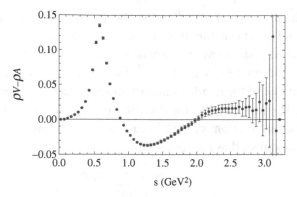

Fig. V–2 Authors' representation of ALEPH data for the $V - A$ spectral functions from tau decay.

the naive reasoning leading to the result in Eq. (3.5). Recall that we have previously displayed in Eq. (3.3) the measured value of R_τ. If we restrict ourselves to the Cabibbo-allowed decays (notationally, $R_\tau \rightarrow R_\tau^{ud}$) by subtracting off the Cabibbo suppressed transitions, then experiment gives [Pi 13]

$$R_\tau^{ud} = 3.4771 \pm 0.0084. \tag{3.25a}$$

On the theory side, a careful analysis of tau decays yields

$$R_\tau^{ud} = N_C |V_{ud}|^2 S_{EW} [1 + \delta_{NP} + \delta_P], \tag{3.25b}$$

where S_{EW} is an electroweak correction,

$$S_{EW} = 1 + \frac{2\alpha(m_\tau)}{\pi} \ln\left(\frac{M_Z}{m_\tau}\right) + \cdots = 1.0201 \pm 0.0003 \tag{3.25c}$$

and $\delta_{NP} \simeq -0.0059 \pm 0.0014$ represents the nonperturbative *QCD* corrections. These two are insignificant compared to δ_P, the perturbative *QCD* correction, whose numerical value is inferred by comparing the above experimental and theoretical relations,

$$\delta_P = 0.2030 \pm 0.0033, \tag{3.25d}$$

amounting to a 20% effect. To derive a theoretical expression for δ_P, one first expresses R_τ^{ud} as the contour integral [BrNP 92]

$$R_\tau^{ud} = 6\pi i \oint_{|s|=m_\tau^2} \frac{ds}{m_\tau^2} \left(1 - \frac{s}{m_\tau^2}\right)^2$$
$$\times \left[\left(1 + 2\frac{s}{m_\tau^2}\right) \Pi^{(1+0)}(s) - 2\frac{s}{m_\tau^2}\Pi^{(0)}(s)\right], \tag{3.26}$$

where the presence of $\Pi^{(1+0)}$ in the above is associated with requiring singularity-free behavior in the complex-q^2 plane upon passing to the chiral limit. We shall not detail the next steps, which involve use of the operator-product expansion (OPE) for $\Pi^{(1+0)}(s)$ on the circle $|s| = m_\tau^2$. However, the physical picture which emerges is akin to that of the quark–antiquark loop in electromagnetic vacuum polarization (cf. Fig. II–2(a)), except now the currents are the weak currents, and the photon-exchange perturbations of the EM case become instead gluon exchanges. The result of this is expressed as the series

$$\delta_P = \sum_{n=1} K_n A^{(n)}(\alpha_s) \tag{3.27a}$$

where $K_1 = 1$, $K_2 = 1.63982$, etc., and

$$A^{(n)}(\alpha_s) = \frac{1}{2\pi i} \oint_{|s|=m_\tau^2} \frac{ds}{s} \left(\frac{\alpha_s(-s)}{\pi} \right)^n \left(1 - 2\frac{s}{m_\tau^2} + 2\frac{s^3}{m_\tau^6} - \frac{s^4}{m_\tau^8} \right). \tag{3.27b}$$

The $\alpha_s(-s)$ dependence is expressible in terms of $\alpha_s(m_\tau)$, which finally can be determined in terms of the experimentally measured R_τ^{ud}. Depending on details of the analysis, some scatter occurs in the value found for $\alpha_s(m_\tau^2)$. The averaging in [Pi 13] (see also [Bo *et al.* 12]) arrives at

$$\alpha_s(m_\tau^2) = 0.334 \pm 0.014, \tag{3.28}$$

having about 4% uncertainty. Renormalization group running of this result up to the Z-boson mass gives $\alpha_s(M_Z) = 0.1204 \pm 0.0016$, in accord with the 2011 world average value $\alpha_s(M_Z) = 0.1183 \pm 0.0010$.

A rather different bit of τ-related physics involves a set of sum rules which contain the $\rho_{V,A}$ spectral functions [We 67a; Das *et al.* 67],

$$\int_{m_\pi^2}^{\infty} ds \, \frac{\rho_V(s) - \rho_A(s)}{s} = -4L_{10}^r(\mu) + \frac{1}{16\pi^2} \ln \left(\frac{m_\pi^2}{\mu^2} \right), \tag{3.29a}$$

where $L_{10}^r(\mu)$ is a chiral coefficient to be defined in Sect. VII–2 and quantified in Table VII–1, and μ is an arbitrary energy scale which cancels between the two terms on the right-hand side,

$$\int_{m_\pi^2}^{\infty} ds \, (\rho_V(s) - \rho_A(s)) = F_\pi^2, \tag{3.29b}$$

$$\int_0^{\infty} ds \, s \, (\rho_V(s) - \rho_A(s)) = 0, \tag{3.29c}$$

$$\int_0^{\infty} ds \, s \, \ell n \left(\frac{s}{\Lambda^2} \right) (\rho_V(s) - \rho_A(s)) = -\frac{16\pi^2 F_\pi^2}{3e^2} (m_{\pi^\pm}^2 - m_{\pi^0}^2), \tag{3.29d}$$

where Λ is an energy scale which is arbitrary by virtue of Eq. (3.29c). Although the sum rules of Eqs. (3.29a,b) hold in the physical world, those of Eqs. (3.29c,d) are derived in the chiral limit of massless u, d quarks, so the quantities F_π and $m_{\pi^\pm}$ are understood to have slightly different numerical values from their physical counterparts. Like the extraction of $\alpha_s(m_\tau)$ from τ decay, these sum rules have been the subject of much study over time. Their convergence is sensitive to the spectral functions in the large s limit. This is known in the $m_u = m_d$ limit to be $\left(\rho_{V,3} - \rho_{A,3}\right)(s) \sim s^{-3}$, which suffices to provide convergence.

Another, perhaps surprising, application of the spectral functions $\rho_{V,3}$ and $\rho_{A,3}$ involves *CP* violation in the kaon system. This is presented in Sect. IX–3, where the association of these spectral functions with ϵ'/ϵ (a measure of direct to indirect *CP* violation) is described.

Problems

(1) **Effective lagrangian for $\mu \to e + \gamma$**

In describing the decay $\mu \to e + \gamma$, one may try to use an effective lagrangian $\mathcal{L}_{3,4}$ which contains terms of dimensions 3 and 4,

$$\mathcal{L}_{3,4} = a_3 \left(\bar{e}\mu + \bar{\mu}e\right) + ia_4 \left(\bar{e}\not{D}\mu + \bar{\mu}\not{D}e\right),$$

where $D_\mu \equiv \partial_\mu + ieQ_{\mathrm{el}}A_\mu$ and a_3, a_4 are constants.

(a) Show by direct calculation that $\mathcal{L}_{3,4}$ does *not* lead to $\mu \to e + \gamma$.

(b) If $\mathcal{L}_{3,4}$ is added to the *QED* lagrangian for muons and electrons, show that one can define new fields μ' and e' to yield a lagrangian which is diagonal in flavor. Thus, even in the presence of $\mathcal{L}_{3,4}$, there are *two* conserved fermion numbers.

(c) At dimension 5, $\mu \to e + \gamma$ can be described by a gauge-invariant effective lagrangian containing constants c, d,

$$\mathcal{L}_5 = \bar{e}\sigma^{\alpha\beta}(c + d\gamma_5)\mu F_{\alpha\beta} + \text{h.c.}$$

Obtain bounds on c, d from the present limit for $\mu \to e + \gamma$.

(2) **Muon decay**

(a) Obtain the leading $\mathcal{O}(m_e^2/m_\mu^2)$ correction to the Fermi-model expression Eq. (2.10) for the muon decay width.

(b) Do the same for the leading $\mathcal{O}(m_\mu^2/M_W^2)$ correction.

(3) **Vacuum polarization and dispersion relations**

The vacuum polarization $\Pi(q^2)$ associated with a loop containing a spin one-half fermion–antifermion pair, each of mass m, can be written as the sum of a term containing an ultraviolet cut-off Λ and a finite contribution $\hat{\Pi}_f(q^2)$,

$$\Pi(q^2) = \frac{\alpha}{\pi} \left[\frac{1}{3} \ln \frac{\Lambda^2}{m^2} - 2 \int_0^1 dx \, x(1-x) \ln \left(1 - \frac{q^2}{m^2} x(1-x) \right) \right]$$

$$\equiv \frac{\alpha}{3\pi} \ln \frac{\Lambda^2}{m^2} + \hat{\Pi}_f(q^2).$$

(a) Show that $\hat{\Pi}_f(q^2)$ is an analytic function of q^2 with branch point at $q^2 = 4m^2$ and with Im $\hat{\Pi}_f(q^2) = \alpha R_f(q^2)/3$, where

$$R_f(q^2) \equiv \sqrt{\frac{q^2 - 4m^2}{q^2}} \frac{2m^2 + q^2}{q^2}$$

is related to the rate for radiative pair creation via

$$\sum_f (2\pi)^3 \delta^{(4)}(q - p_f) \langle f | J_{em}^\nu | 0 \rangle^* \langle f | J_{em}^\mu | 0 \rangle = (-q^2 g^{\mu\nu} + q^\mu q^\nu) \frac{R_f(q^2)}{3}.$$

(b) Use Cauchy's theorem and the result of (a) to express

$$\hat{\Pi}_f(q^2) = \frac{\alpha q^2}{3\pi} \int_{4m^2}^\infty ds \, \frac{R_f(s)}{s(s - q^2 - i\epsilon)}.$$

(c) The form of $\hat{\Pi}_f(q^2)$ given in part (a) can be re-expressed in a dispersion representation. First change variables in (a) to $y = 1 - 2x$ and integrate by parts to obtain

$$\hat{\Pi}_f(q^2) = -\frac{\alpha}{2\pi} \int_0^1 dy \, \ln \left[1 - \frac{q^2(1 - y^2)}{4m^2 - i\epsilon} \right] \frac{d}{dy} \left(y - \frac{1}{3} y^3 \right)$$

$$= \frac{\alpha}{2\pi} \int_0^1 dy \, 2y \left(y - \frac{y^3}{3} \right) \frac{q^2}{4m^2 - q^2(1 - y^2) - i\epsilon}.$$

Then, change variables again to $s = 4m^2/(1 - y^2)$ and demonstrate that the dispersion result of (b) obtains.

VI

Neutrinos

When the Standard Model first emerged, there was no evidence of neutrino mass. Since only left-chiral neutrino fields are coupled to the gauge bosons, the simplest way to accommodate the lack of a neutrino mass was to omit any right-handed counterparts to the neutrino field, in which case masslessness is automatic. Because of the degeneracy of the three massless neutrinos, the charged weak leptonic current can be made diagonal and there exists no lepton analog to the CKM matrix.

In light of evidence for neutrino mass, the most conservative response is to postulate the existence of right-handed neutrinos, similar to the right-handed partners of the other fields. Because the right-handed neutrino carries no gauge charge, its mass may be Dirac or Majorana (or both), and it may be heavy or light. Whether one considers this modification to be an extension beyond the Standard Model or not is largely a matter of semantics. In this chapter, we will describe the rich physics induced by the inclusion of a right-handed neutrino. We note in passing that all fermion fields appearing here will be described as four-component spinors.

VI–1 Neutrino mass

A right-handed neutrino ν_R has no couplings to any of the gauge fields because its Standard Model charges are zero.[1] Nonetheless, it can enter the lagrangian in two ways: there can be a Yukawa coupling to lepton doublet ℓ_L plus a Higgs and there can be a Majorana mass term involving ν_R. Considering only one generation for the moment, these possibilities are[2]

$$\mathcal{L}_{\nu_R} = -g_\nu \bar{\ell}_L \widetilde{\Phi} \nu_R - \frac{m_M}{2} \overline{(\nu_R)^c}\, \nu_R + \text{h.c.} \tag{1.1}$$

---

[1] ν_R is electrically neutral ($Q = 0$) and like all RH particles in the Standard Model is a weak isosinglet ($T_{w3} = 0$), so by Eq. (II–3.8) it has zero weak hypercharge, $Y_w = 0$.

[2] In neutrino physics, a *sterile neutrino* is defined as one which has no interactions whatsoever with Standard Model particles. The right-handed neutrino discussed here is *not* sterile if $g_\nu \neq 0$ because it can then couple to the Higgs field as in Eq. (1.1).

Recall that the left-handed neutrino field is part of an $SU(2)_L$ doublet and so can have no Majorana mass term because the combination $\overline{(\nu_L)^c}\,\nu_L$ is not gauge-invariant. The right-handed Majorana mass could be set equal to zero by the imposition of a discrete symmetry (e.g. lepton number) but this is an additional assumption beyond the Standard Model gauge symmetries.

When the Higgs field picks up a vacuum expectation value, this leads to a mass matrix of the form

$$-2\,\mathcal{L}_{D+M} = \left(\overline{\nu_L}\ \ \overline{(\nu_R)^c}\right) \begin{pmatrix} 0 & m_D \\ m_D & m_M \end{pmatrix} \begin{pmatrix} \nu_L^c \\ \nu_R \end{pmatrix} + \text{h.c.,} \tag{1.2}$$

where the Dirac mass is $m_D = g_\nu v / \sqrt{2}$ and where we have used the fact that $\overline{(\psi_i)^c}\psi_j^c = \overline{(\psi_j)}\psi_i$. The above matrix can be diagonalized by defining fields

$$\nu_a = \cos\theta\,\nu_R + \sin\theta\,\nu_L^c, \qquad \nu_b = \cos\theta\,\nu_L - \sin\theta\,\nu_R^c, \tag{1.3}$$

with $\tan 2\theta = 2m_D/m_M$. The mass terms then become

$$-\mathcal{L}_{D+M} = \frac{m_a}{2}\left[\overline{(\nu_a)^c}\nu_a + \overline{(\nu_a)}\nu_a^c\right] + \frac{m_b}{2}\left[\overline{(\nu_b)^c}\nu_b + \overline{(\nu_b)}\nu_b^c\right], \tag{1.4}$$

with

$$m_a = m_M \cos^2\theta + m_D \sin 2\theta, \qquad m_b = m_M \sin^2\theta - m_D \sin 2\theta. \tag{1.5}$$

The mass matrix of Eq. (1.2) has one negative eigenvalue, and given the mixing angle this can be seen to be m_b. As discussed earlier in Sect. I–3, this is kinematically equivalent to a positive mass, and the eigenvalue can be made positive by the phase change $\nu_b \to i\nu_b$. However, we shall leave the phase unchanged as it would induce an unusual phase in the weak mixing matrix. Finally, inverting Eq. (1.3) yields the following relation between the neutrino field ν_L and the mass eigenstates ν_a^c and ν_b,

$$\nu_L = \cos\theta\,\nu_b + \sin\theta\,\nu_a^c. \tag{1.6}$$

It is this combination of the mass eigenstates which constitutes the neutrino component of charged and neutral weak currents first encountered in Sect. II–3.

There are two obvious limiting cases for the mass matrix of Eq. (1.2). In one limit, the Majorana mass term vanishes, $m_M = 0$, with the result that the neutrino is a Dirac fermion with mass m_D. Here, both particle and antiparticle can have positive helicity (right-handed) or negative helicity (left-handed), so there are four degrees of freedom. As noted in Sect. I–3, despite appearances, Eq. (1.4) reduces to the standard Dirac lagrangian in this limit (with $\theta = \pi/4$ and $m_a = -m_b = m_D$).

The other case is that of a very large Majorana mass m_M. Here, one eigenvalue becomes large and the other small,

$$m_a = m_M , \qquad m_b = -\frac{m_D^2}{m_M}. \tag{1.7}$$

The mixing angle in this case becomes tiny, $\theta = m_D/m_M \ll 1$, so that eigenfunctions are just $\nu_a = \nu_R$, $\nu_b = \nu_L$ up to corrections of order m_D/m_M. Both of these eigenstates are Majorana fields.[3] There are still four degrees of freedom present, viz. left-handed and right-handed helicity states for each of the two self-conjugate neutrinos. This is the famous *seesaw mechanism* [GeRS 79], which has the potential to explain the fact that the neutrinos are much lighter than the quarks and other leptons. As an example, given the mass constraints cited in Chap. I, at least one neutrino must have rest-energy in excess of 0.05 eV, and if we use m_τ for the corresponding Dirac mass, this would be compatible with $m_M \sim 6 \times 10^{10}$ GeV. We see that the light field in this case is a Majorana field of the left-handed neutrino. Even though the direct left-handed Majorana mass term was forbidden by gauge symmetry, after symmetry breaking the left-handed field assumes a Majorana nature. We can understand this feature more directly using effective lagrangian techniques, to which we now turn.

Equivalence of heavy Majorana mass to a dimension-five operator

We have explained in Chap. IV how a heavy field may be integrated out from a theory. Here, we consider a process that involves two applications of the first term in Eq. (1.1), $g_\nu \bar{\ell}_L \tilde{\Phi} \nu_R$, in which the right-handed neutrino is a (self-conjugate!) Majorana fermion. If the Majorana mass is large, ν_R becomes heavy and can accordingly be removed. We then find the following residual interaction [We 79a] involving just the light fields,

$$\mathcal{L}_5 = -\frac{1}{\Lambda_M} \bar{\ell}_L \tilde{\Phi} \tilde{\Phi} \ell_L^c + \text{h.c.}, \tag{1.8}$$

where $1/\Lambda_M \equiv g_\nu^2/(2m_M)$. This interaction is invariant under $SU(2)_L$ gauge interactions because the lepton doublet and the Higgs doublet both transform in the same way. The fields in $\mathcal{L}_5$ carry a total mass dimension of five. Hence, this operator must have a coupling constant with the dimensions of an inverse mass, and so the operator cannot be part of a renormalizable lagrangian. However, in effective field theory, this operator *is* an allowed addition to the lagrangian of the light Standard Model fields, and it is suppressed by a single power of the heavy scale m_M.

---

[3] In the general case, where the mass parameters m_a, m_b are allowed to have arbitrary values, both neutrinos are Majorana. In fact, a Dirac neutrino can itself be interpreted as a pair of degenerate Majorana neutrinos.

Once the Higgs field picks up its vacuum expectation value, this lagrangian turns into a Majorana lagrangian for the left-handed neutrino,

$$\mathcal{L}_5 \rightarrow -\frac{v^2}{2\Lambda_M}\bar{\nu}_L \nu_L^c + \text{h.c.}, \tag{1.9}$$

reproducing the mass eigenvalue and eigenfunction calculated above via diagonalization. So we see that a left-handed Majorana mass term *is* allowed after the electroweak symmetry breaking if we include operators of dimension five. Indeed, although we have just found the above operator by integrating out a particular heavy field, its existence can be more general than this particular calculation. There could be other theories beyond the Standard Model which might generate this operator.

The properties of neutrino mass are suggestive of physics beyond the Standard Model, although they are not conclusive proof of that. We have seen that there is no conflict between the idea of neutrino mass and the symmetries of the Standard Model. Once one allows the possibility of right-handed neutrino fields, both Dirac and Majorana mass terms will occur unless one makes an additional symmetry assumption of lepton-number conservation, which would set the Majorana mass equal to zero. Even if this extra discrete symmetry were imposed, Dirac masses could still account for observations. However, the small magnitude of the observed neutrino masses is puzzling in one way or another. If the Majorana masses are small or zero such that Dirac masses are dominant, one would require the Yukawa couplings to be remarkably small – roughly a billion times smaller than the Yukawas for the charged leptons. On the other hand, if the Majorana mass is large, the neutrino masses are naturally small via the seesaw mechanism, but then one has to understand the large value of the Majorana scale. A Majorana mass in the range $10^6 \rightarrow 10^{14}$ GeV would not match any of the scales of the Standard Model (nor does it match estimates of Grand Unification scales). While the present structure is consistent with the interactions of the Standard Model, we hope that future New Physics will explain the puzzles of the quark and lepton mass scales, which are most dramatic in the case of neutrino masses.

VI–2 Lepton mixing

In the previous section we considered mass diagonalization for a single species of neutrino. In the Standard Model, there are, however, three generations of leptons. This means that both the Dirac and Majorana mass terms will involve 3×3 matrices, $\mathbf{m}_D$ and $\mathbf{m}_M$. The Dirac mass matrix is, in general, complex but not Hermitian, while the Majorana mass matrix will be complex symmetric. The overall mass matrix must be diagonalized, and there will be a resultant weak mixing matrix for the charged weak current. We shall consider lepton mixing in the two limiting cases discussed above, first for a pure Dirac neutrino mass and then in the seesaw limit.

Dirac mass: The biunitary diagonalization of the lepton mass matrices has already been carried out in Eqs. (II–4.1)–(II–4.7b) for the case of a pure Dirac mass. The results for leptons proceed analogously to those for quarks. Mixing between generations occurs in the leptonic charged weak current (recall that the lepton mass eigenstates are $\vec{\nu}_L = \{\nu_1, \nu_2, \nu_3\}_L$ and $\vec{e}_L = \{e, \mu, \tau\}_L$),

$$J_{\text{ch}}^{\mu}(\text{lept}) = 2\vec{\bar{\nu}}_L' \gamma^{\mu} \vec{e}_L' = 2\vec{\bar{\nu}}_L \mathbf{S}_L^{\nu\dagger} \mathbf{S}_L^e \gamma^{\mu} \vec{e}_L \equiv 2\vec{\bar{\nu}}_L \mathbf{V}^{(\nu)} \gamma^{\mu} \vec{e}_L, \qquad (2.1)$$

where

$$\mathbf{V}^{(\nu)} \equiv \mathbf{S}_L^{\nu\dagger} \mathbf{S}_L^e \qquad (2.2)$$

is the Dirac lepton mixing matrix. As an example, the electron's contribution to the charged weak current is given by

$$J_{\text{ch}}^{\mu}(\text{e}) = 2[\bar{\nu}_{L,1} \mathbf{V}_{1e}^{(\nu)} + \bar{\nu}_{L,2} \mathbf{V}_{2e}^{(\nu)} + \bar{\nu}_{L,3} \mathbf{V}_{3e}^{(\nu)}] \gamma^{\mu} e_L \equiv 2\bar{\nu}_{L,e} \gamma^{\mu} e_L, \qquad (2.3)$$

which shows the neutrino $\nu_{L,e}$ created in this process to be a linear combination of the three neutrino mass eigenstates. The lepton mixing matrix $\mathbf{V}^{(\nu)}$ of Eq. (2.2) will have the same structure as the quark mixing matrix of Eq. (II–4.17) with three mixing angles $\{\theta_{ij}\}$ and one *CP*-violating phase δ.

Majorana mass: If the right-handed Majorana mass is very large, or if we invoke the dimension-five operator of the previous section, we see that the light eigenstate is a left-handed Majorana particle with mass

$$\mathbf{m}_L = -\mathbf{m}_D \frac{1}{\mathbf{m}_M} \mathbf{m}_D^T. \qquad (2.4)$$

Here, the factors are themselves 3×3 matrices and we have been careful with the ordering of the elements.

The matrix $\mathbf{m}_L$ is nondiagonal, as are the individual elements $\mathbf{m}_D$ and $\mathbf{m}_M$. The Dirac part is diagonalized as

$$\mathbf{m}_D^{(\text{diag})} = \mathbf{S}_L^{\nu\dagger} \mathbf{m}_D \mathbf{S}_R^{\nu}. \qquad (2.5)$$

Inserting the diagonalized Dirac part into the full mass matrix yields

$$\mathbf{m}_L = \mathbf{S}_L^{\nu} \, \mathcal{C} \, \mathbf{S}_L^{\nu T}, \qquad (2.6)$$

where the central matrix $\mathcal{C}$ is defined as

$$\mathcal{C} \equiv \mathbf{m}_D^{(\text{diag})} \mathbf{S}_R^{\nu\dagger} \frac{1}{\mathbf{m}_M} \mathbf{S}_R^{\nu*} \mathbf{m}_D^{(\text{diag})}. \qquad (2.7)$$

The symmetric (but generally complex-valued) central matrix C can be diagonalized with a unitary matrix $\mathcal{F}$,

$$C = \mathcal{F} \, \mathbf{m}_\nu \, \mathcal{F}^T = \mathcal{F} \begin{pmatrix} m_1 & 0 & 0 \\ 0 & m_2 & 0 \\ 0 & 0 & m_3 \end{pmatrix} \mathcal{F}^T. \tag{2.8}$$

The masses in the diagonal matrix $\mathbf{m}_\nu$ are the physical neutrino masses.

The PMNS matrix involves the rotations that diagonalize the mass matrices of the charged leptons and the neutrinos [Po 68, MaNS 62]. This also includes the rotation that diagonalizes the central matrix. Therefore, in terms of the quantities defined above, the PMNS matrix becomes

$$\mathbf{U} = \mathcal{F}^\dagger \mathbf{S}_L^{\nu\dagger} \mathbf{S}_L^e. \tag{2.9}$$

Like the $n \times n$ Dirac mixing matrix for quarks and leptons, the Majorana mixing matrix has n^2 real-valued parameters, of which $n(n-1)/2$ are angles and $n(n+1)/2$ are phases. However, whereas field redefinitions remove $2n-1$ phases for the Dirac case, only n such phases can be removed (via redefinitions of the charged lepton fields) for the Majorana mixing matrix. The reason is that Majorana fields are self-conjugate (cf. Sect. I–3) and thus not subject to phase redefinitions.[4] Thus, the number of remaining phases in the Majorana mixing matrix is $n(n-1)/2$. For $n = 3$ there are three phases, of which one is identified as the phase δ in the Dirac mixing matrix and two others, α_1, α_2, are commonly called Majorana phases. It can be shown that

$$\mathbf{U} = \mathbf{V}^{(\nu)} \mathcal{P}_\nu \quad \text{with} \quad \mathcal{P}_\nu = \begin{pmatrix} 1 & 0 & 0 \\ 0 & e^{i\alpha_1/2} & 0 \\ 0 & 0 & e^{i\alpha_2/2} \end{pmatrix}, \tag{2.10}$$

where the $\{\alpha_i\}$ are the Majorana phases. For convenience, we give the neutrino mixing matrix $\mathbf{V}^{(\nu)}$,

$$\mathbf{V}^{(\nu)} = \begin{pmatrix} c_{12}c_{13} & s_{12}c_{13} & s_{13}e^{-i\delta} \\ -s_{12}c_{23} - c_{12}s_{23}s_{13}e^{i\delta} & c_{12}c_{23} - s_{12}s_{23}s_{13}e^{i\delta} & s_{23}c_{13} \\ s_{12}s_{23} - c_{12}c_{23}s_{13}e^{i\delta} & -s_{23}c_{12} - s_{12}c_{23}s_{13}e^{i\delta} & c_{23}c_{13} \end{pmatrix}, \tag{2.11}$$

where $s_{\alpha\beta} \equiv \sin\theta_{\alpha\beta}$, $c_{\alpha\beta} \equiv \cos\theta_{\alpha\beta}$ ($\alpha, \beta = 1, 2, 3$).

---

[4] If $\psi \to e^{i\theta}\psi$, then $(\psi)^c \to e^{-i\theta}(\psi)^c$. Maintaining the Majorana condition $\psi = \psi^c$ occurs only for $\theta_n = n\pi$, so θ cannot be arbitrary.

VI–3 Theory of neutrino oscillations

Our current information on neutrino mass and mixing comes via the phenomenon of neutrino oscillations. We review the foundation of this subject in the present section.

Oscillations in vacuum

Suppose that at time $t = 0$ an electron neutrino is produced by a weak process induced by the charged current $J_{ch}^{\mu}(e)$ of Eq. (2.3) and thereafter propagates as an eigenstate of momentum **p**,

$$|\nu_e(0)\rangle \rightarrow |\nu_e(t)\rangle = U_{e1}^*|\nu_1\rangle e^{-iE_1 t} + U_{e2}^*|\nu_2\rangle e^{-iE_2 t} + U_{e3}^*|\nu_3\rangle e^{-iE_3 t}, \qquad (3.1)$$

where $E_i = (\mathbf{p}^2 + m_i^2)^{1/2}$. In this relation, the mixing matrix elements $\{U_{ek}\}$ ($k = 1, 2, 3$) appear as complex conjugates because the neutrino field in the charged current is in the form of a Hermitian conjugate $\bar{\nu}_e$. Actually, as written Eq. (3.1) is theoretically tainted because the superposition cannot be a simultaneous eigenstate of both momentum and energy since $m_1 \neq m_2 \neq m_3$. However, since this simplified description leads to the correct oscillation phase under rather general conditions, we continue to use it here.[5]

To proceed, we take $p \gg m_i$, implying that $E_i \simeq p + m_i^2/(2p) \simeq p + m_i^2/(2E)$. Upon replacing the time by the distance traveled, $t \simeq L$, we obtain from Eq. (3.1),

$$|\nu_e(L)\rangle \simeq e^{-iE_1 L}\left(U_{e1}^*|\nu_1\rangle + U_{e2}^*|\nu_2\rangle \exp\left[-i\frac{m_2^2 - m_1^2}{2E}L\right]\right.$$
$$\left. + U_{e3}^*|\nu_3\rangle \exp\left[-i\frac{m_3^2 - m_1^2}{2E}L\right]\right). \qquad (3.2)$$

Let us now truncate the description to just two neutrino flavors by working in the small θ_{13} limit, evidently a reasonable approximation given that $|U_{e3}/U_{e1}| \simeq 0.16$.

Then, the amplitude $\mathcal{A}_{\nu_e \nu_e}(L)$ and probability $\mathcal{P}_{\nu_e \nu_e}(L)$ for remaining in the original weak eigenstate $\nu_e(0)$ at distance L become

$$\mathcal{A}_{\nu_e \nu_e}(L) = \langle \nu_e(0)|\nu_e(L)\rangle = e^{-iE_1 L}\left(|U_{e1}|^2 + |U_{e2}|^2 \exp\left[-i\frac{m_2^2 - m_1^2}{2E}L\right]\right)$$

$$\mathcal{P}_{\nu_e \nu_e}(L) = |\mathcal{A}_{\nu_e \nu_e}(L)|^2 = c_{12}^4 + s_{12}^4 + 2c_{12}^2 s_{12}^2 \cos\left[\frac{\Delta m_{21}^2}{2E}L\right]. \qquad (3.3)$$

---

[5] Two recent discussions of this point appear in [KaKRV 10] and [CoGL 09], but many others have contributed to the topic. See references cited in [GiK 07] and [RPP 12].

With a bit of algebra, we then obtain for the survival and transition probabilities $\mathcal{P}_{\nu_e \nu_e}(L)$ and $\mathcal{P}_{\nu_e \nu_\mu}(L)$,

$$\mathcal{P}_{\nu_e \nu_e}(L) = 1 - \sin^2 2\theta_{12} \, \sin^2\left[\frac{\Delta m_{21}^2 L}{4E}\right],$$

$$\mathcal{P}_{\nu_e \nu_\mu}(L) = \sin^2 2\theta_{12} \sin^2\left[\frac{\Delta m_{21}^2 L}{4E}\right]. \tag{3.4a}$$

Let us comment on aspects of these important relations. The amplitude of the oscillation factor is $\sin^2 2\theta_{12}$. The oscillation phase $\Phi_{21} \equiv \Delta m_{21}^2 L/(4E)$ informs about the squared mass-difference Δm_{21}^2, given that the energy (E) and distance (L) are dictated by constraints of Nature and/or by experimental design.[6] An expression useful for numerical work is

$$\Phi_{21} \simeq 1.267 \frac{\Delta m_{21}^2 [\text{eV}^2] \, L[m]}{E[\text{MeV}]}. \tag{3.4b}$$

Another involves defining an oscillation *length* $L_{\text{osc}}^{(21)}$,

$$\sin^2 \Phi_{21} = \frac{1}{2}\left(1 - \cos\left[2\pi L/L_{\text{osc}}^{(21)}\right]\right),$$

$$L_{\text{osc}}^{(21)} \equiv \frac{4\pi E}{\Delta m_{21}^2}, \qquad L_{\text{osc}}^{(21)}[m] \simeq 2.48 \frac{E[\text{MeV}]}{\Delta m_{21}^2 [\text{eV}^2]}, \tag{3.4c}$$

which is the length for obtaining a half-cycle of oscillation. If conditions are such that $2\pi L \ll L_{\text{osc}}^{(21)}$, oscillations will not have had a chance to occur because the oscillation phase is too small. Finally, we stress that Eq. (3.4a) is a result of the two-flavor restriction. Although 'three-flavor' phenomenology was already advocated shortly after the discovery of the τ lepton [DeLMPP 80] and is currently used in precise analyses of neutrino data, e.g., [FoTV 12], it can happen that the two-flavor approach is a valid approximation in certain circumstances (see Prob. 2 at the end of this chapter). For example, it is often used to describe both solar mixing $(\theta_{12} \to \theta_\odot, \Delta m_{21}^2 \to \Delta m_\odot^2)$ and atmospheric mixing $(\theta_{23} \to \theta_A, |\Delta m_{32}^2| \to |\Delta m_A^2|)$.

We have been considering the vacuum propagation of neutrinos. The *vacuum evolution equation* for the relativistic energy eigenstates ν_1 and ν_2 as expressed in the energy basis is $i d\nu_E/dx = \mathbf{H}_E \nu_E$, where

$$\nu_E \equiv \begin{pmatrix} \nu_1 \\ \nu_2 \end{pmatrix} \qquad \text{and} \qquad \mathbf{H}_E = \begin{pmatrix} E_1 & 0 \\ 0 & E_2 \end{pmatrix} \to \begin{pmatrix} \frac{m_1^2}{2E} & 0 \\ 0 & \frac{m_2^2}{2E} \end{pmatrix}. \tag{3.5}$$

---

[6] As will be discussed in Sect. VI–4, the predicted oscillation pattern of Eq. (3.4a) was first observed in 2002 (for electron antineutrinos) by the KamLAND collaboration.

The right-most matrix form in Eq. (3.5) has been obtained by expanding the energy in powers of the momentum, followed by the phase transformation $\nu_E \rightarrow \exp(-ipx)\nu_E$. Proceeding to the weak basis ν_W,

$$\nu_W \equiv \begin{pmatrix} \nu_e \\ \nu_\mu \end{pmatrix} = \mathbf{U}\nu_E \quad \text{and} \quad \mathbf{U} = \begin{pmatrix} \cos\theta_{12} & \sin\theta_{12} \\ -\sin\theta_{12} & \cos\theta_{12} \end{pmatrix}, \quad (3.6a)$$

the evolution equation can be written $i\,d\nu_W/dx = \mathbf{H}_W\nu_W$, where

$$\mathbf{H}_W = \mathbf{U}\mathbf{H}_E\mathbf{U}^{-1} = \begin{pmatrix} -\frac{\Delta m_{21}^2}{4E}\cos 2\theta_{12} & \frac{\Delta m_{21}^2}{4E}\sin 2\theta_{12} \\ \frac{\Delta m_{21}^2}{4E}\sin 2\theta_{12} & \frac{\Delta m_{21}^2}{4E}\cos 2\theta_{12} \end{pmatrix}. \quad (3.6b)$$

As shown earlier in this section, the evolution in Eq. (3.6b) describes $\nu_e \leftrightarrow \nu_\mu$ vacuum oscillations. Using the current PDG value for θ_{12} (see Eq. (II–4.24)), we have from Eq. (3.6a) the numerical expressions

$$|\nu_1\rangle = 0.83|\nu_e\rangle - 0.56|\nu_\mu\rangle, \qquad |\nu_2\rangle = 0.56|\nu_e\rangle + 0.83|\nu_\mu\rangle. \quad (3.6c)$$

The dominant component of $|\nu_2\rangle$ resides in $|\nu_\mu\rangle$, a fact we will refer to in the next section.

Oscillations in matter: MSW effect

Neutrino propagation in matter is a problem of intrinsic theoretical interest. It is also of practical importance because many past and present experiments involve, in part, neutrinos traveling in the interiors of the Sun and of the Earth. In the following, we consider a neutrino moving radially with position coordinate r and continue to employ the two-flavor description.

For neutrino propagation in *matter*, a key difference with the vacuum description is that the neutrinos will undergo $W^\pm$ and Z^0 exchange scattering from atomic electrons and quarks confined within protons and neutrons. Only elastic scattering in the forward direction maintains the coherence of the initial mixed ν_e–ν_μ state. In particular, the quark contributions cancel and it is $W^\pm$ exchange in the ν_e–e interaction which produces a potential difference between electron and muon neutrinos,

$$\Delta V \equiv V(\nu_e) - V(\nu_\mu) = \sqrt{2}G_F N_e(r), \quad (3.7)$$

where $N_e(r)$ is the electron number density at distance r from the origin. That neutrinos in matter experience this potential energy was first pointed out by Wolfenstein [Wo 78], who cited a well-known analogous effect in K^0–$\bar{K}^0$ mixing as neutral kaons move through nuclear matter. To properly account for the Wolfenstein effect, we must alter the diagonal matrix elements in Eq. (3.6b) to

$$\mathbf{H}^{(M)}(r) \equiv \begin{pmatrix} \sqrt{\frac{1}{2}} G_F N_e(r) - \dfrac{\Delta m_{21}^2}{4E} \cos 2\theta_{12} & \dfrac{\Delta m_{12}^2}{4E} \sin 2\theta_{12} \\ \dfrac{\Delta m_{12}^2}{4E} \sin 2\theta_{12} & -\sqrt{\frac{1}{2}} G_F N_e(r) + \dfrac{\Delta m_{21}^2}{4E} \cos 2\theta_{12} \end{pmatrix},$$

$$(3.8)$$

where the superscript in $\mathbf{H}^{(M)}$ refers to 'matter'. Since the electron number density is generally spatially dependent, the above matrix $\mathbf{H}^{(M)}$ will have spatially dependent eigenvalues $E_{\pm}(r)$,

$$E_{\pm}(r) = \pm \frac{1}{4} \left[\left(4 \mathbf{H}_{11}^{(M)}(r) \right)^2 + \left(\frac{\Delta m_{21}^2}{E} \sin 2\theta_{12} \right)^2 \right]^{1/2}. \qquad (3.9)$$

In the discussion to follow, we shall consider neutrino propagation in the Sun. At the solar core $r = 0$, the potential energy of Eq. (3.7) becomes $\Delta V^{(\mathrm{core})} \simeq 7.6 \times 10^{-12}$ eV upon taking $\sqrt{2} G_F \simeq 7.63 \times 10^{-14}$eV-cm$^3/N_A$ and $N_e^{(\mathrm{core})} \simeq 100 N_A$ cm$^{-3} \simeq 6.0 \times 10^{25}$ cm^{-3}. Let us next make two working hypotheses:

(1) We assume that the electron number density $N_e^{(\mathrm{core})}$ is sufficiently large to ensure that $\mathbf{H}_{11}^{(M)}(0) > 0$ at the core. Using the value of $N_e^{(\mathrm{core})}$ just given above and adopting the current PDG values for Δm_{21}^2 and θ_{12}, this will be valid for neutrinos with energy above $E \sim 2$ MeV. This energy is, however, not precisely fixed since the core is a region and not just a point.

   If indeed $\mathbf{H}_{11}^{(M)}$ is positive at the solar core, it becomes negative before reaching the surface (since N_e vanishes at the surface) and vice versa for $\mathbf{H}_{22}^{(M)}$. The matrix elements $\mathbf{H}_{11}^{(M)}$ and $\mathbf{H}_{22}^{(M)}$ thus cross at the point where each vanishes. In the limit of neglecting the off-diagonals of $\mathbf{H}^{(M)}$, the diagonals become the eigenvalues and we have the phenomenon of *level crossing*, familiar from atomic and nuclear physics. In reality, the off-diagonals do not vanish and so the level crossing is avoided.

(2) We assume that propagation of an electron neutrino in the solar matter is *adiabatic*, i.e. the fractional change in the electron density of the matter is small per neutrino oscillation cycle. If so, a neutrino that starts in one of the energy eigenstates will not experience a transition as it passes through the solar medium. This is akin to a particle in an eigenstate of the one-dimensional infinite well maintaining its quantum state as the wall separation changes sufficiently slowly.

Let us now follow the behavior of $E_{\pm}(r)$ from the solar core at $r = 0$ to the solar surface at $r = R_\odot$. As we move outward from the core, $N_e(r)$ will decrease[7] until a point $r = r_{\mathrm{res}}$ is reached at which $\mathbf{H}_{11}^{(M)}(r_{\mathrm{res}}) = \mathbf{H}_{22}^{(M)}(r_{\mathrm{res}}) = 0$, with

---

[7] A popular model for the number density profile is $N_e(r) = N_e(0) e^{-r/r_0}$ with $r_0 \simeq R_\odot/10$.

Table VI–1. *Evolution of* $|\nu_2^M(r)\rangle$.

r	$\theta(r)$	$|\nu_2^M(r)\rangle$
0	$\pi/2$	$|\nu_e\rangle$
r_{res}	$\pi/4$	$\left(|\nu_e\rangle + |\nu_\mu\rangle\right)/\sqrt{2}$
$R_\odot$	θ_{12}	$\sin\theta_{12}|\nu_e\rangle + \cos\theta_{12}|\nu_\mu\rangle$

$$N_e^{(\text{res})} \equiv \frac{\Delta m_{12}^2 \cos 2\theta_{12}}{2\sqrt{2}G_F E}, \qquad (3.10)$$

after which $E_+(r)$ grows until the surface is reached. Similarly, E_- will increase from $r = 0$ to $r = r_{\text{res}}$ and decrease thereafter. The label *res* used here stands for 'resonance', as will be explained shortly.

The eigenstates $|\nu^M(r)\rangle$ of the matrix $\mathbf{H}^{(M)}$ are likewise spatially dependent,

$$\begin{aligned} |\nu_1^M(r)\rangle &= \cos\theta(r)|\nu_e\rangle + \sin\theta(r)|\nu_\mu\rangle, \\ |\nu_2^M(r)\rangle &= -\sin\theta(r)|\nu_e\rangle + \cos\theta(r)|\nu_\mu\rangle, \end{aligned} \qquad (3.11)$$

as is also the associated mixing angle $\theta(r)$ which, after some algebra, can be written as

$$\sin 2\theta(r) = \frac{\sin 2\theta_{12}}{\left[\left(N_e(r)/N_e^{(\text{res})} - 1\right)^2 \cos^2 2\theta_{12} + \sin^2 2\theta_{12}\right]^{1/2}}. \qquad (3.12)$$

The square of this relation has the profile of a Lorentzian distribution, indicating the presence of a resonance [MiS 85].

Suppose an electron neutrino ν_e is created at the solar core $r = 0$ under the assumption $N_e(0) \gg N_e^{(\text{res})}$. Its evolution to the solar surface $r = R_\odot$ is briefly summarized in Table VI–1 and explained as follows. The condition $N_e(0) \gg N_e^{(\text{res})}$ implies from Eq. (3.12) that $\theta(0) \simeq \pi/2$, and so from Eq. (3.11) that $|\nu_2^M\rangle \simeq |\nu_e\rangle$ at the core. Thus, a newly created electron neutrino will reside in the energy eigenstate $|\nu_2^M\rangle$ as it begins its journey to the solar surface. If the matter eigenstates undergo adiabatic flow through the resonance, then $|\nu_2^M\rangle$ suffers no transitions. As the surface is eventually approached, the electron number density decreases to zero, $N_e(R_\odot) = 0$ and so, from Eq. (3.12), $\theta(r) \to \theta(R_\odot) \simeq \theta_{12}$, the vacuum mixing angle. What is new and exciting is that electron neutrinos of sufficiently high energies, which are created by nuclear reactions at the solar core, have an appreciable probability for conversion to muon neutrinos by the time the solar surface is reached. This is, in essence, the phenomenon known as the Mikheyev–Smirnov–Wolfenstein (MSW) effect [Wo 78, MiS 85]. The mixing between electron and muon type neutrinos has occurred within the Sun and since $|\nu_2\rangle$ is an energy eigenstate, no further mixing occurs en route to Earth. Measurement at a detector on Earth will yield either ν_e or ν_μ according to the quantum state $|\nu_2\rangle$ of Eq. (3.6c).

Not all neutrinos created in the solar core will experience MSW mixing. As shown earlier, it may be that the neutrino energy is too small (roughly $E_{\nu_e} <$ 2 MeV) for level crossing to take place. Or the neutrino flow to the solar surface may not be adiabatic. The quantitative condition for adiabaticity is most stringent at the resonant point $r = r_{\text{res}}$,

$$\frac{\sin^2 2\theta_{12}}{\cos 2\theta_{12}} \frac{\Delta m_{21}^2}{2E} \left| \frac{N_e^{(\text{res})}}{N_e'(r_{\text{res}})} \right| \gg 1, \qquad (3.13)$$

where $N_e'(r_{\text{res}})$ is the density gradient, $N_e' \equiv dN_e/dr$, evaluated at the resonant point. Thus, adiabaticity will occur provided the solar electron number density does not change too rapidly with position. The relation in Eq. (3.13) amounts to demanding that the splitting between the energy eigenvalues $E_\pm(r)$ of $\mathbf{H}^{(M)}$ (cf. Eq. (3.9)), which is minimal at the resonant point, nonetheless be much larger than the off-diagonal matrix elements of $\mathbf{H}^{(M)}$ (which would produce transitions between the energy eigenstates). We return to this subject in Sect. VI–4, where we further discuss solar neutrinos.

CP violation

The *CP*-violating phase in the PMNS matrix has physical implications in neutrino oscillations, relating the oscillations of neutrinos to those of antineutrinos. It is reasonably straightforward to use the general form of the oscillation formula to calculate the difference of the oscillation probabilities,

$$A_{ij} \equiv P_{\nu_i \to \nu_j} - P_{\bar{\nu}_i \to \bar{\nu}_j} = 4 \sum_{k>\ell} \text{Im}(U_{ik} U_{jk}^* U_{jl} U_{il}^*) \sin\left(\frac{\Delta m_{kl}^2 L}{2E}\right). \qquad (3.14)$$

It is less straightforward to measure this. We note that A_{ij} vanishes unless all three flavors of neutrinos are involved. This can be found from direct calculation but is easy to understand on general principles, as the *CP*-violating phase can be removed from any 2×2 submatrix by redefining the fields. Moreover, the numerator in this asymmetry is the same for any $i \neq j$,

$$A_{ij} = \sin\delta \cos\theta_{13} \sin 2\theta_{13} \sin 2\theta_{23} \sin 2\theta_{12}$$
$$\times \left[\sin\left(\frac{\Delta m_{21}^2 L}{2E}\right) + \sin\left(\frac{\Delta m_{13}^2 L}{2E}\right) + \sin\left(\frac{\Delta m_{32}^2 L}{2E}\right) \right], \qquad (3.15)$$

where we have used $\Delta m_{13}^2 = -\Delta m_{31}^2$. We see that two independent mass differences, e.g., Δm_{21}^2 and Δm_{13}^2, contribute. In addition, the asymmetry will produce small corrections to the oscillations with the largest amplitudes and will be most visible for oscillations where the *CP*-even transitions are the smallest, such as $\nu_e \leftrightarrow \nu_\mu$. Although uncovering *CP* violation in oscillations will be an experimental challenge, the rewards of such a measurement will be considerable. For example, lepton *CP* violation is a necessary ingredient for leptogenesis (cf. Sect. VI–6).

VI–4 Neutrino phenomenology

Determination of the set of mixing parameters $\{\theta_{ij}\}$ and $\{\Delta m_{ij}^2\}$ has taken years of careful experimentation. This has involved a variety of neutrino sources, including our Sun, the Earth's atmosphere, nuclear reactors and particle accelerators. Many references and detailed accounts exist in the literature.[8]

Solar and reactor neutrinos: θ_{12} and Δm_{21}^2

Solar neutrinos: The current evaluations [RPP 12] of the parameters $\sin^2 2\theta_{12}$ and Δm_{21}^2 from a three-neutrino fit give

$$\sin^2 2\theta_{12} = 0.857 \pm 0.024 \qquad \Delta m_{21}^2 = (7.50 \pm 0.20) \times 10^{-5}\ \text{eV}^2. \qquad (4.1)$$

This represents an uncertainty of under 3%, which is one indication of how successful the search for these basic parameters has turned out. The earliest progress in this area involved the detection of solar neutrinos. It was Davis [Da 64] who used a chlorine detector to probe solar neutrinos and Bahcall [Ba 64] who provided the theoretical basis for such an ambitious undertaking. An important conceptual contribution came from Pontecorvo, who suggested testing whether leptonic charge was conserved, and who wrote 'From the point of view of detection possibilities, an ideal object is the Sun' [Po 68].

The initial intent of the chlorine experiment was actually to test physics at the core of the Sun. A significant achievement of solar neutrino studies has been to demonstrate that stars are, indeed, powered by nuclear fusion reactions. Energy produced by the Sun arises from thermonuclear reactions in the solar core and the underlying theoretical description is called the Standard Solar Model (SSM). Solar burning utilizes all three types of Standard Model reactions – strong, weak, and electromagnetic – as well as using gravity to provide the required high density. The primary ingredients of the SSM are:

(1) The Sun evolves in hydrostatic equilibrium, balancing the gravitational force and the pressure gradient. The equation of state is specified as a function of temperature, density, and solar composition.

---

[8] Some recent examples include [AnMPS 12], [Ba 90], [BaH 13], [FoTV 12], [FoLMMPR 12], [GiK 07], [GoMSS 12], [HaRS 12], [KiL 13], [MoA *et al.* 07], and Chap. XIII in [RPP 12] by Nakamura and Petkov.

(2) Energy proceeds through the solar medium by radiation and convection. While the solar envelope is convective, radiative transport dominates the core region where the thermonuclear reactions take place.

(3) The primary thermonuclear chain involves the conversion $4p \rightarrow {}^4\text{He} + 2e^+ + 2\nu_e$. This pp chain produces 26.7 MeV per cycle, and the associated neutrino production rate is firmly tied to the amount of energy production. The core temperature and electron number density of the Sun are respectively $T_c \sim 1.5 \times 10^7$ K and $N_e \sim 6 \times 10^{25}$ cm^{-3}.

(4) The model is constrained to produce the observed solar radius, mass and luminosity. The initial ${}^4\text{He}/\text{H}$ ratio is adjusted to reproduce the luminosity at the Sun's current 4.57 Giga-year age.

The dominant process is the 'pp chain', occurring in stages I $\rightarrow$ IV:

$$
\begin{pmatrix}
\text{Stage} & \text{Nuclear reaction} & \text{Br (\%)} \\
\text{I} & p + p \rightarrow {}^2\text{H} + e^+ + \nu_e & 99.75 \\
& p + e + p \rightarrow {}^2\text{H} + \nu_e & 0.25 \\
\text{II} & {}^2\text{H} + p \rightarrow {}^3\text{He} + \gamma & 100.00 \\
\text{III} & {}^3\text{He} + {}^3\text{He} \rightarrow {}^4\text{He} + 2p & 86.00 \\
\text{or} & {}^3\text{He} + {}^4\text{He} \rightarrow {}^7\text{Be} + \gamma & 14.00 \\
\text{IV} & {}^7\text{Be} + e^- \rightarrow {}^7\text{Li} + \nu_e & 99.89 \\
& {}^7\text{Li} + p \rightarrow {}^4\text{He} + {}^4\text{He} & \\
\text{or} & {}^7\text{Be} + p \rightarrow {}^8\text{B} + \gamma & 0.11 \\
& {}^8\text{B} \rightarrow {}^8\text{Be}^* + e^+ + \nu_e &
\end{pmatrix}
$$

Let us isolate those processes which produce neutrinos and order them according to increasing maximum neutrino energy:

Label	Reaction	E^ν_{max}(MeV)
pp	$p + p \rightarrow {}^2\text{H} + e^+ + \nu_e$	0.42
${}^7\text{Be}$	${}^7\text{Be} + e^- \rightarrow {}^7\text{Li} + \nu_e$	0.86
pep	$p + e + p \rightarrow {}^2\text{H} + \nu_e$	1.44
${}^8\text{B}$	${}^8\text{B} \rightarrow {}^8\text{Be}^* + e^+ + \nu_e$	14.06
hep	${}^3\text{He} + p \rightarrow {}^4\text{H} + e^+ + \nu_e$	18.47

The energy spectra of the pp, ${}^8\text{B}$ and hep neutrinos are continuous whereas the pep and ${}^7\text{Be}$ neutrinos are monoenergetic. Within this general framework, there is, however, still a degree of theoretical uncertainty and work continues to this day on solar modeling. Table VI–2 (taken from [HaRS 12] and [AnMPS 12]) lists SSN flux predictions according to two sets, labelled GS98 and AGSS09, and taken respectively from [GrS 98] and [AsBFS 09]. Note the marked decrease in flux with increasing neutrino energy.

Table VI–2. *Neutrino flux^a in the pp chain.*

Label	GS98	AGSS09	Solar data
pp	$5.98(1 \pm 0.006) \times 10^{-1}$	$6.03(1 \pm 0.006) \times 10^{-1}$	$6.05(1^{+0.003}_{-0.011})$
^{7}Be	$5.00(1 \pm 0.07) \times 10^{-1}$	$4.561(1 \pm 0.07) \times 10^{-1}$	$4.82(1^{+0.05}_{-0.04}) \times 10^{-1}$
pep	$1.44(1 \pm 0.012) \times 10^{-1}$	$1.47(1 \pm 0.012) \times 10^{-1}$	$1.46(1^{+0.010}_{-0.014}) \times 10^{-2}$
^{8}B	$5.58(1 \pm 0.13) \times 10^{-1}$	$4.59(1 \pm 0.13) \times 10^{-1}$	$5.25(1 \pm 0.038) \times 10^{-4}$
hep	$8.04(1 \pm 0.30) \times 10^{-1}$	$8.31(1 \pm 0.30) \times 10^{-1}$	—

^{a}Expressed in units of 10^{10}cm^{-2}s^{-1}.

On the basis of such flux predictions, results from various solar neutrino experiments could be compared with the SSM. The following compilation, taken from [AnMPS 12], summarizes early results for the ratio of observed-to-predicted electron neutrino flux,

$$\text{Homestake} \quad 0.34 \pm 0.03, \qquad \text{Super-K} \quad 0.46 \pm 0.02,$$
$$\text{SAGE} \quad 0.59 \pm 0.06, \qquad \text{Gallex,GNO} \quad 0.58 \pm 0.05.$$

We now know that this spread of values arises from the interplay between the range of neutrino energies and the influence of the MSW effect. At the time, however, it was unclear whether the SSM itself was at fault. The issue was resolved by a series of experiments which probed flavor mixing of solar neutrinos while simultaneously testing the SSN prediction for the total solar flux. This was carried out by the SNO collaboration; for a summary see [Ah *et al.* (SNO collab.) 11]. Since the SNO detector employed heavy water, there was sensitivity to the three reactions:

charged current (CC): $\nu_e + d \rightarrow p + p + e^-$

neutral current (NC): $\nu_x + d \rightarrow p + n + \nu_x \quad (x = e, \mu, \tau)$

elastic scattering (ES): $\nu_x + e^- \rightarrow \nu_x + e^- \quad (x = e, \mu, \tau).$ (4.2)

In the Standard Model, only ν_e contributes to the CC reaction, but all neutrino flavors contribute, with equal rates, to the the NC reactions (and also to the ES, but with ν_e having about six times the rate of ν_μ and ν_τ). Early CC measurements found $f_{\nu_e}^{(CC)} = (1.76 \pm 0.11) \times 10^6$ cm^{-2} s^{-1}, much less than the (then) predicted total flux $f_{\nu_e}^{tot} = (5.05 \pm 0.91) \times 10^6$ cm^{-2} s^{-1}. Then, NC measurements obtained $f_{\nu_e}^{(NC)} = (5.09 \pm 0.62) \times 10^6$ cm^{-2} s^{-1}, consistent with the $f_{\nu_e}^{tot}$ prediction. Within errors, the only reasonable explanation is that the conversion of $\nu_e \rightarrow \nu_\mu, \nu_\tau$ must be occurring. A more recent determination of the total flux from the ^{8}B reaction reduces the uncertainty,

$$\Phi_{^8\text{B}}^{(tot)} = \left(5.25 \pm 0.16 \text{ (stat)} ^{+0.11}_{-0.13} \text{ (syst)}\right) \times 10^6 \text{ cm}^{-2}\text{s}^{-1}, \qquad (4.3)$$

consistent with but having smaller uncertainty than the SSN predictions of Table VI–2.

In summary, the versatility of solar neutrino experiments is that they are sensitive to various nuclear reactions in the Sun through the measurement of different energy neutrinos. The survival probability for electron neutrinos to reach the Earth will depend on the neutrino energy E and will in part be affected by the solar MSW effect. The survival probability in the two-flavor description can be expressed as [Pa 86]

$$\overline{\mathcal{P}}_{\nu_e \to \nu_e} = \frac{1}{2} + \left(\frac{1}{2} - \mathcal{P}_{\text{non-adbtc}} \right) \cos 2\theta(0) \cos 2\theta_{12}. \tag{4.4a}$$

In the above, $\theta(0)$ represents the matter mixing angle at the point of neutrino production (taken here at $r = 0$), averaging of oscillatory behavior has been carried out, and $\mathcal{P}_{\text{non-adbtc}}$ describes the nonadiabatic mixing (which is sensitive to the electron number density $N_e(r)$) as in Eq. (3.13).

Let us explore Eq. (4.4a) in the limits of low-energy and high-energy neutrino energy, while assuming just adiabatic transitions ($\mathcal{P}_{\text{non-adbtc}} = 0$). For very low-energy neutrinos, as explained previously, there is no MSW resonance and the the situation reduces to simple vacuum mixing,

$$\overline{\mathcal{P}}_{\nu_e \to \nu_e} = 1 - \frac{1}{2} \sin^2 2\theta_{12} \simeq 0.57, \tag{4.4b}$$

whereas for very energetic neutrinos, we have $\theta(0) \simeq \pi/2$ and so

$$\overline{\mathcal{P}}_{\nu_e \to \nu_e} = \sin^2 2\theta_{12} \simeq 0.31. \tag{4.4c}$$

For intermediate neutrino energy, the average survival probability interpolates smoothly between these two limits. The overall pattern is as depicted in Fig. VI–1. The recent experiment [Be *et al.* (Borexino collab.) 12a] on *pep* neutrinos, whose energy $E = 1.44$ MeV is at the low end of the spectrum, finds a survival probability $\overline{\mathcal{P}}_{\nu_e \to \nu_e} = 0.62 \pm 0.17$, which is in accord with the above analysis.

The relations in Eqs. (4.4a–c) pertain to neutrino propagation directly from the Sun to the Earth. This is referred to as 'daytime' detection, sometimes denoted by $\overline{\mathcal{P}}^{(D)}_{\nu_e \to \nu_e}$. The 'nighttime' probability $\overline{\mathcal{P}}^{(N)}_{\nu_e \to \nu_e}$ would be sensitive to matter effects from passage through the Earth. Letting R_D and R_N represent the day and night counting rates, the 'day–night' asymmetry,

$$\mathcal{A}_{D-N} \equiv 2 \frac{R_N - R_D}{R_N + R_D}, \tag{4.5}$$

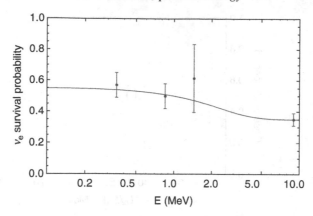

Fig. VI–1 Average survival probability of solar neutrinos vs. neutrino energy. Data points represent (from left to right) *pp*, ^{7}Be, *pep*, and ^{8}B neutrinos.

is an observable which isolates the effect of Earth matter on neutrino propagation. This is in distinction to the MSW effect in the Sun, which cannot be turned off. Several experiments, the SNO and Super-K experiments (with ^{8}B) and Borexino (with ^{7}Be) have studied the day–night effect. For example, the results [Be *et al.* (Borexino collab.) 12b],

$$\mathcal{A}_{D-N} = - (0.1 \pm 1.2 \pm 0.7) \% \qquad [\text{Borexino}]$$
$$= - (4.0 \pm 1.3 \pm 0.8) \% \qquad [\text{Super-K}], \qquad (4.6)$$

are consistent with the theory predictions $|\mathcal{A}_{D-N}| \lesssim 0.1\%$ (Borexino) and $\mathcal{A}_{D-N} \simeq -3\%$ (Super-K), although the latter is also 2.6σ from zero.

Reactor antineutrinos: The KamLAND experiment was able to observe oscillations of $\bar{\nu}_e$ antineutrinos under laboratory conditions. The $\bar{\nu}_e$ beam originates from nuclear beta decays from several nuclear reactors and detection is obtained via the inverse beta-decay process

$$\bar{\nu}_e + p \rightarrow e^+ + n. \qquad (4.7)$$

In the KamLAND experiment, the average baseline between sources and detector is $L_0 \sim 180$ km and the antineutrino energy spectrum covers the approximate range $1 \leq E_{\bar{\nu}_e} \leq 7$ MeV. The $\bar{\nu}_e$ survival formula, as in Eq. (3.4), suggests plotting the data as a function of $L_0/E_{\bar{\nu}_e}$. The result, shown in Fig. VI–2, clearly exhibits the oscillation pattern. This important observation yielded the most accurate determination of Δm_{21}^2 at the time and continues to be a significant contributor to the current database.[9]

---

[9] Since properties of electron antineutrinos are being studied, it is necessary to assume the validity of *CPT* invariance to compare the KamLAND results with those from solar neutrino studies (and any other experiment using neutrinos and not antineutrinos).

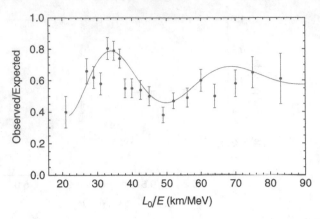

Fig. VI–2 Authors' representation of the KamLAND observation of neutrino oscillations. The curve represents a fit to the oscillation hypothesis.

Atmospheric and accelerator neutrinos: θ_{23} and $|\Delta m^2_{32}|$

Since 1996, the Super-Kamiokande experiment has utilized a 50-kiloton Cherenkov detector to study oscillations of so-called 'atmospheric' muon neutrinos. When high-energy cosmic rays strike the Earth's atmosphere a multitude of secondary particles are produced, most of which travel at nearly the speed of light in the same direction as the incident cosmic ray. Many of the secondaries are pions and kaons, which decay into electrons, muons, and their neutrinos. Using known cross sections and decay rates, one expects about twice as many muon neutrinos as electron neutrinos from the cosmic-ray events. For example, a π^+ decays predominantly as

$$\pi^+ \to \mu^+ + \nu_\mu \to e^+ + \nu_e + \bar{\nu}_\mu + \nu_\mu,$$

i.e., two muon-type neutrinos are produced but only one that is electron-type. Detection of these atmospheric neutrinos yielded evidence for oscillations, to wit, a deficit of muon-type neutrinos, but *no* such deficit for the electron neutrinos. This has since been augmented with data containing dependence on the azimuthal angle (and hence distance from the source) and on the neutrino energy. Because the deficit is of just muon neutrinos, the hypothesis is that these oscillations involve the conversion $\nu_\mu \to \nu_\tau$. Any ν_τ thus generated is not energetic enough to react via the charged current to produce a τ.

Accelerator-based efforts to probe the same oscillation parameters include the K2K and MINOS experiments. In particular, since 2005 MINOS has studied muon neutrinos originating from Fermilab and traveling 735 km through the Earth to a detector at the Soudan mine in the state of Minnesota. At Fermilab, an injector beam of protons strikes a target, producing copious numbers of pions, whose decay is the source of muon neutrinos.

Data from both the Super-Kamiokande and MINOS experiments support the $\nu_\mu \to \nu_\tau$ scenario with mixing angle and mass difference given by [RPP 12]:

$$\sin^2 2\theta_{23} > 0.95 \qquad |\Delta m^2_{32}| = 0.00232^{+0.0012}_{-0.00008} \text{ eV}^2. \qquad (4.8)$$

Moreover, each experiment has also studied muon *antineutrino* oscillations, finding mixing parameters consistent with these values, although less precisely determined.

Finally, the T2K collaboration announcement in 2012 of the first evidence for ν_e appearance in a ν_μ beam has been confirmed in a recent update [Ab *et al.* (T2K collab.) 13]. The ν_e appearance probability at oscillation maximum is

$$P_{\nu_\mu \to \nu_e} \simeq 4c^2_{13}s^2_{13}s^2_{23} \left[1 + \frac{2a}{\Delta m^2_{31}}\right] - 8c^2_{13}c_{12}c_{23}s_{12}s_{13}s_{23}\Phi_{21}\sin\delta, \qquad (4.9)$$

where $\Phi_{21} \equiv \Delta m^2_{21}L/(4E)$ and $a \equiv 2\sqrt{2}G_F n_e E$. In particular, the value of $\sin^2(2\theta_{13})$ inferred from the data depends on whether a normal or inverted neutrino mass hierarchy is assumed. This can, in turn, be compared to reactor values for $\sin^2(2\theta_{13})$. Thus, the importance of this type of experiment lies in its sensitivity both to the hierarchy issue and to detection of a *CP*-violating signal ($\delta \neq 0$). Future data from the T2K and NOνA experiments have the potential for significant progress in our understanding of neutrino physics.

Short-baseline studies: θ_{13}

The last of the neutrino oscillation angles to be determined with precision is θ_{13}. Initial fits to mixing data indicated their smallness. This led to the concern that signals of neutrino *CP* violation, i.e., determination of the *CP*-violating phase δ, might be experimentally inaccessible. For example, recall from Eq. (3.15) that the *CP*-violating asymmetry A_{ij} is linear in both $\sin\delta$ and $\sin\theta_{13}$. Hence, the attempt to measure θ_{13} took on a certain urgency.

Following a growing number of indications that indeed $\theta_{13} \neq 0$, it was several reactor short-baseline experiments which provided the needed precision. A key point is that in $\bar\nu_e$ disappearance experiments, with a relatively short baseline of roughly 1 km, the influence of $\sin^2(2\theta_{12})$ and Δm^2_{21} on the survival probability $P^{(\mathrm{surv})}_{\bar\nu_e}$ for electron antineutrinos can safely be neglected. We then have (see Prob. VI–2),

$$P^{(\mathrm{surv})}_{\bar\nu_e} \simeq 1 - 2|U_{13}|^2 \left(1 - |U_{13}|^2\right) \left(1 - \cos\left[\frac{\Delta m^2_{31}}{2E}L\right]\right)$$
$$= 1 - \sin^2(2\theta_{13})\sin^2\left(1.267\Delta m^2_{31}L/E\right). \qquad (4.10)$$

Based on data from the collaborations Daya Bay [An *et al.* 12], RENO [Ahn *et al.* 12] and Double Chooz [Abe *et al.* 12], the current RPP listing gives [RPP 12]

$$\sin^2 2\theta_{13} = 0.098 \pm 0.013. \tag{4.11}$$

Finally, the future of short-baseline experiments has the potential for additional interesting findings. In particular, the inverse relation between L and Δm^2 in the neutrino oscillation relations implies that a very short-baseline study (say, with $L \sim 5 \rightarrow 50$ m) would be sensitive to much larger values of squared mass difference (say, having order of magnitude $\Delta m^2 \sim 1$ eV^2) than those observed for Δm^2_{21} and Δm^2_{32}. Such a large neutrino mass difference evidently occurred in the LSND experiment [Ag *et al.* 01 (LSND collab.)], which found evidence at 3.5σ for $\bar{\nu}_\mu \rightarrow \bar{\nu}_e$ oscillations with $\Delta m^2 > 0.2$ eV^2. We shall not discuss this experiment further, except to note that, if validated, it would represent effects (e.g., one or more sterile neutrinos) beyond the Standard Model.

VI–5 Testing for the Majorana nature of neutrinos

In order to determine if the neutrino mass has a Majorana component, one can use the fact that Majorana masses violate lepton-number conservation. A sensitive measure occurs in the process of neutrinoless double beta decay. There are many situations in Nature where one has a nucleus which is kinematically forbidden to decay via ordinary beta decay,

$$^Z\!A \not\rightarrow {}^{Z-1}\!A + e^- + \bar{\nu}_e, \tag{5.1}$$

but which is allowed to decay via emission of two lepton pairs ($2\nu\beta\beta$),

$$^Z\!A \rightarrow {}^{Z-2}\!A + e^- + e^- + \bar{\nu}_e + \bar{\nu}_e. \tag{5.2}$$

This $2\nu\beta\beta$ process is attributable to the pairing force in nuclei and occurs only for even–even nuclei. It is produced at order G_F^2 through the exchange of two W bosons. When $2\nu\beta\beta$ can occur, it is also kinematically possible to have a *neutrinoless* double beta decay ($0\nu\beta\beta$),

$$^Z\!A \rightarrow {}^{Z-2}\!A + e^- + e^-. \tag{5.3}$$

However, this latter process violates lepton-number conservation by two units and would be forbidden if the neutrino possessed a standard Dirac mass. We will see that this becomes a sensitive test of the Majorana nature of neutrino mass.

First, consider $2\nu\beta\beta$ decay. Because this involves five-body phase space as well as two factors of the weak coupling constant G_F this process is very rare, with

Table VI–3. *Half-lives of some two-neutrino double beta emitters.*

Nucleus	$T_{1/2}^{2\nu}(\text{yr})$
^{96}Zr	$(2.0 \pm 0.3 \pm 0.2) \times 10^{19}$
^{76}Ge	$(1.7 \pm 0.2) \times 10^{21}$
^{136}Xe	$(2.23 \pm 0.017 \pm 0.22) \times 10^{21}$
^{76}Ge	$(1.7 \pm 0.2) \times 10^{21}$

lifetimes of order 10^{20} years. Even so, it has been observed in many nuclei, and some examples are cited in Table VI–3. The rate for such processes is

$$\Gamma_{2\nu} \sim m_e^{11} F_2 \left(Q/m_e \right) \left| g_a^2 M_{GT} - g_v^2 M_F \right|^2 \cdot \frac{\mathcal{F}(Z)}{E_i - \langle E_n \rangle - \frac{1}{2} E_0}, \qquad (5.4)$$

where $\mathcal{F}(Z)$ is a Fermi function, $F_2 \left(Q/m_e \right)$ is a kinematic factor,

$$F_2(x) = x^7 \left(1 + \frac{x}{2} + \frac{x^2}{9} + \frac{x^3}{90} + \frac{x^4}{1980} \right), \qquad (5.5)$$

and M_F, M_{GT} are, respectively, the Fermi and Gamow–Teller matrix elements,

$$M_F = \langle f | \frac{1}{2} \sum_{ij} \tau_i^+ \tau_j^- | i \rangle, \qquad M_{GT} = \langle f | \frac{1}{2} \sum_{ij} \tau_i^+ \tau_j^- \sigma_i \cdot \sigma_j | i \rangle. \qquad (5.6)$$

In Eq. (5.4), the closure approximation has been made to represent a sum over intermediate states via an average excitation energy $\langle E_n \rangle$. The experimental $2\nu\beta\beta$ decay rates then determine these matrix elements, which unfortunately are extremely difficult to determine theoretically.

If the neutrino has a Majorana mass component, then neutrinoless double beta decay is possible. The basic weak process underlying $0\nu\beta\beta$ decay involves the transition $W^- W^- \rightarrow e^- e^-$ through the Feynman diagram of Fig. VI–3. Let us initially treat this process by considering only one generation and invoking the mass diagonalization framework of Sect. VI–2. Because the charged weak current couples to

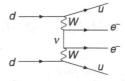

Fig. VI–3 The basic weak process underlying $0\nu\beta\beta$ decay.

the combination (cf. Eqs. (1.3), (1.6)) $v_L = \cos\theta \; v_b + \sin\theta \; v_a^c$, the exchange of the two neutrino eigenstates leads to a matrix element which is proportional to

$$\frac{\sin^2\theta \; m_a}{Q^2 - m_a^2} + \frac{\cos^2\theta \; m_b}{Q^2 - m_b^2},\tag{5.7}$$

noting that the $\not{q}$ portion of the propagator numerator vanishes due to the chiral relation $\Gamma_L \not{q} \Gamma_L = 0$. If the Majorana mass term is vanishingly small compared to the Dirac mass, this reaction also vanishes since in this case $\theta = \pi/4$ (so that $\sin\theta = \cos\theta$) and $m_a = -m_b = m_D$. Despite the apparent existence of two Majorana fields, the fermion-number violating transition vanishes because the two contributions are equal and opposite.[10]

However, if the Majorana mass term does not vanish, the transition can occur. Let us consider the case of the seesaw mechanism, in which $m_a \simeq m_M \gg M_W$ and $\theta \simeq m_D/m_M \ll 1$. Then the contribution of the first propagator becomes tiny, and a nonvanishing transition occurs due to the second propagator. The process is now directly proportional to the light Majorana mass m_b. The momentum flowing in this propagator is of order the electron mass, so we can neglect the mass dependence m_b^2 in the denominator. This leaves the transition proportional to m_b/Q^2. In this scenario the light neutrino acts like a pure Majorana fermion.

When all three generations of neutrinos can contribute the result depends also on the lepton mixing matrix. If one is dealing with Majorana neutrinos, and neglects the neutrino mass in the denominator of the neutrino propagator, then the figure of merit is the averaged Majorana mass $\langle m_v \rangle$ obtained by summing over all neutrino species,

$$\langle m_v \rangle \equiv \sum_{i=1}^{3} U_{ie}^2 m_i \; .\tag{5.8}$$

Note that it is the *square UU* of the PMNS matrix, not the usual combination $U^\dagger U$, that enters the reaction. This is because both weak currents lead to e^- emission in the final state. It is this feature which allows the Majorana phases $\alpha_{1,2}$ to contribute to $\langle m_v \rangle$. The decay rate for such a neutrinoless decay has a form analogous to that in Eq. (5.4),

$$\Gamma_{0v} \sim m_e^7 F_0(Q/m_e) |g_a \tilde{M}_{GT} - g_v^2 \tilde{M}_F|^2 \frac{\langle m_v \rangle^2}{m_e^2},\tag{5.9}$$

---

[10] If one had chosen to redefine m_b to be positive via a phase redefinition, as described in Sect. VI–1, there would be an extra phase in the weak current of v_b such that the cancelation would still occur due to a factor of $i^2 = -1$ in the double beta decay matrix element.

except now with the phase space factor $F_0(x)$,

$$F_0(x) = x \left(1 + 2x + \frac{4x^2}{3} + \frac{x^3}{3} + \frac{x^4}{30} \right), \tag{5.10}$$

and nuclear matrix elements

$$\widetilde{M}_F = \langle f | \frac{1}{2} \sum_{ij} \tau_i^+ \tau_j^- \frac{1}{r_{ij}} | i \rangle, \qquad \widetilde{M}_{GT} = \langle f | \frac{1}{2} \sum_{ij} \tau_i^+ \tau_j^- \sigma_i \cdot \sigma_j \frac{1}{r_{ij}} | i \rangle. \tag{5.11}$$

The factor of $1/r_{ij}$ in Eq. (5.11) comes from spatial dependence associated with the neutrino propagator in the limit that one neglects the neutrino mass in the denominator of Eq. (5.7).

Neutrinoless double beta decay, $0\nu\beta\beta$, is a topic of considerable theoretical importance and is currently under investigation experimentally. As of yet no such mode has been observed. Present limits are $\langle m_\nu \rangle < 140 \to 380$ meV [Ac *et al.* (EXO-200 collab.) 11] and $\langle m_\nu \rangle < 260 \to 540$ meV [Ga *et al.* (KamLAND–ZEN collab.) 12]. There are a number of planned experiments which aim to lower these bounds.

VI–6 Leptogenesis

The material Universe is mostly comprised of matter – protons, neutrons and electrons – rather than their antiparticles. The net baryon asymmetry is described by the ratio

$$\eta_B = \frac{N_B - N_{\bar{B}}}{N_\gamma} \sim 6 \times 10^{-10}. \tag{6.1}$$

Because baryon number and other symmetries are violated in the Standard Model and in most of its extensions, it is plausible that this asymmetry was generated dynamically in the early Universe. Such a dynamical mechanism requires a process which is out of thermal equilibrium and which violates both baryon number and *CP* conservation [Sa 67].

If heavy right-handed Majorana neutrinos exist, as allowed by the general mass analysis of the Standard Model, they can generate the net baryon asymmetry. The basic point is that the heavy Majorana particles can decay differently to leptons and antileptons as they fall out of equilibrium in the early Universe through the *CP* violation that is present in the PMNS matrix, and this lepton number asymmetry can be reprocessed into a baryon-number asymmetry through the $B + L$ anomaly of the Standard Model.

The decay of Majorana particles need not conserve lepton number, as the Majorana mass itself violates this quantity. However, to violate *CP* symmetry requires a

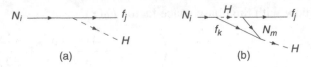

Fig. VI–4 Violating *CP* symmetry in the lepton sector.

specific dynamical mechanism. There can be an interference between the phases of the PMNS matrix and phases generated by unitarity effects for a given final state. To see this, consider the decay diagrams depicted in Fig. VI–4. The tree-level diagrams are proportional to the Yukawa couplings, which are in general complex. However, this is not enough, as an overall phase leads to an unobservable effect when calculating decay rates. But loop diagrams with on-shell intermediate states, like those in Fig. VI–4, pick up extra imaginary parts from on-shell rescattering. Computationally, this comes from the $i\epsilon$ in Feynman propagators. In addition, loop amplitudes have different PMNS phases because they sum over all the particles in the loop. Schematically, this is manifest in decay amplitudes as

$$A_{N_i \to H f_j} = g_{ij} + \sum_{k,m} |L_{km}| e^{i\delta_{km}} g_{ik} g^*_{km} g_{mj},$$

$$A_{N_i \to \bar{H} \bar{f}_j} = g^*_{ij} + \sum_{k,m} |L_{km}| e^{i\delta_{km}} g^*_{ik} g_{km} g^*_{mj}, \tag{6.2}$$

where the loop diagram is represented by $|L_{km}| e^{i\delta_{km}}$ with a rescattering phase due to on-shell intermediate states δ_{km}. The weak phases in the Yukawa couplings g_{ij} change sign under the change from particle to antiparticle, but the rescattering phase does not. We see that a differential rate develops $|A_{N \to H_j f_i}|^2 - |A_{N \to \bar{H} \bar{f}_i}|^2 \neq 0$ through the interference of tree and loop processes and between the different components of the loop diagram.

Producing a net lepton asymmetry would not be sufficient to explain the observed matter asymmetry unless some of the leptons could be transformed to baryons. This can be accomplished through the baryon anomaly described earlier in Chap. III. In the early Universe, with temperatures above the weak scale, processes which change baryon number, but conserve $B - L$, can occur rapidly. This transfers some of the initial lepton excess into a net number of baryons.

The detailed prediction of the amount of baryon production depends on the size of the *CP*-violating phases, as well as the masses of the heavy Majorana particles. While a unique set of parameters is not available, in general one needs heavy particles of at least 10^9 GeV in order to reproduce the observed asymmetry. This fits well with the observed size of the light neutrino masses, as described earlier in Sect. VI–1.

VI–7 Number of light neutrino species

It might seem that the subject of this section, the number of light neutrino species, is a non-issue. After all, the very structure of the Standard Model has each charged fermion paired with its own neutrino in a weak isospin doublet. Since three charged fermions are known to exist, so there must be three neutrinos. Let us, however, view this purely as an issue of *experimental* physics. In particular, data from Z^0-decay [Sc *et al.* 06] and the cosmic microwave background (CMB) [Hi *et al.* (WMAP collab.) 13] have been used to obtain independent determinations of the number of 'light' neutrino species N_ν. We discuss these two approaches in turn.

Studies at the Z^0 peak

Since final-state neutrinos are the only Standard Model particles *not* detected in Z^0 decay, they contribute to the so-called invisible width Γ_{inv}. In the Standard Model, this is predicted to be

$$\Gamma_{inv} = \Gamma_Z - (\Gamma_{had} + \Gamma_{ee} + \Gamma_{\mu\mu} + \Gamma_{\tau\tau}) = (497.4 \pm 2.5) \text{ MeV}, \qquad (7.1)$$

where Γ_Z is the total Z^0 width and Γ_{had}, Γ_{ee}, $\Gamma_{\mu\mu}$, $\Gamma_{\tau\tau}$ are the hadronic and leptonic widths. But is this what is actually found experimentally?

Several approaches have been explored using the invisible width to determine N_ν, but the one cited here has the advantage of minimizing experimental uncertainties. The trick is to work with the ratio of measured quantities $\Gamma_{inv}/\Gamma_{\ell\bar{\ell}}$,

$$\frac{\Gamma_{inv}}{\Gamma_{\ell\bar{\ell}}} = N_\nu \left(\frac{\Gamma_{\nu\bar{\nu}}}{\Gamma_{\ell\bar{\ell}}}\right)_{SM}. \qquad (7.2)$$

The interpretation of this relation is clear, that the measured invisible width is the product of the number of light neutrino species N_ν and the decay width into a single neutrino–antineutrino pair. Using data collected from the collection of LEP and SLD experiments,[11] one finds [Sc *et al.* 06]

$$N_\nu = 2.984 \pm 0.008, \qquad (7.3)$$

which is consistent with the Standard Model value of $N_\nu = 3$.

Astrophysical data

Astrophysics supplies an independent determination of N_ν which, although currently much less precise, is nonetheless of value. The issue of interest to us is

---

[11] Certain assumptions are made, among them that lepton universality is valid, and that the top-quark and Higgs masses are respectively $m_t = 178.0$ GeV and $M_H = 150$ GeV. See [Sc *et al.* 06] for additional discussion.

that the CMB has sensitivity, in part, to the neutrino number. Some insight can be gained by considering the thermal content of an expanding Universe. We take the radiation energy density ρ_r as

$$\rho_r = \rho_\gamma + \rho_\nu, \tag{7.4}$$

referring respectively to photons (ρ_γ) and relativistic neutrinos (ρ_ν). The photon and neutrino components obey the well-known thermal relations,

$$\rho_\gamma = \frac{\pi^2}{15} T_\gamma^4, \qquad \rho_\nu = \frac{\pi^2}{15} T_\nu^4 \cdot \frac{7}{8} N_\nu. \tag{7.5}$$

For a temperature somewhat in excess of 10 MeV, the Universe is pervaded by an $e^\pm$, ν, γ plasma in thermal equilibrium via the electroweak interactions (so that $T_\nu = T_\gamma$). As the temperature drops to about 10 MeV, the expansion rate of the Universe starts to exceed the rate of weak interactions, causing the neutrinos to begin decoupling from the plasma. Still later, the process of $e^\pm$ annihilation releases entropy to the photons, increasing their temperature relative to the neutrinos. In fact, $T_\nu = T_\gamma \cdot (4/11)^{1/3}$ provided the neutrino decoupling is complete by the annihilation era. Since this is not quite true and to account for any hypothetical 'extra radiation species' (er), one introduces the *effective number of relativistic species* N_{eff} and writes instead

$$\rho_\nu + \rho_{\text{er}} \equiv \frac{\pi^2}{15} T_\nu^4 \cdot \frac{7}{8} N_{\text{eff}}. \tag{7.6}$$

Altogether, the radiation density can be written as

$$\rho_r = \rho_\gamma \left[1 + \frac{7}{8} \left(\frac{4}{11} \right)^{4/3} N_{\text{eff}} \right]. \tag{7.7}$$

Finally, modern experiments have probed with increasing precision the CMB radiation density, which reveals conditions at the epoch of photon decoupling (redshift $z \simeq 1090$). Because of its contribution to ρ_r, N_{eff} affects various properties of the CMB [Hi *et al.* (WMAP collab.) 13], among them the peak locations of the baryon acoustic oscillations (BAO). The current fit with minimum uncertainty is found from combining data from BAO and CMB measurements [Ad *et al.* (Planck collab.) 13],

$$N_{\text{eff}} = 3.30 \pm 0.27, \tag{7.8}$$

consistent with the Standard Model determination $N_{\text{eff}} = 3.046$ [MaMPPPS 05].

Problems

(1) **Three-generation neutrino mixing**

In three-generation mixing, the flavor ($\alpha = e, \nu, \tau$) and energy ($j = 1, 2, 3$) eigenstates are related by $|\nu_\alpha\rangle = U^*_{\alpha j}|\nu_j\rangle = U^\dagger_{j\alpha}|\nu_j\rangle$, as in Eq. (3.1).

(a) Show that the amplitude connecting initial and final flavor states $|\nu_\alpha\rangle$ and $|\nu_\beta\rangle$ is $\mathcal{A}_{\alpha\beta} = U_{\beta j}\mathcal{D}_j U^\dagger_{j\alpha}$, given that $\mathcal{D}$ is a phase factor (temporarily unspecified) describing the neutrino's propagation.

(b) Show that the transition probability is $\mathcal{P}_{\alpha\beta} = |\mathcal{A}_{\alpha\beta}|^2$ is expressible as

$$\mathcal{P}_{\alpha\beta} = \sum_{j=1}^{3} |U_{\beta j}|^2 |U_{\alpha j}|^2 + 2\sum_{j>k} U_{\beta j} U^*_{\beta k} U_{\alpha k} U^*_{\alpha j} \mathcal{D}_j \mathcal{D}^*_k.$$

Hint: Partition the double sum as $\sum_{j,k=1}^{3} = \sum_{j=1}^{3} + 2\sum_{j>k}$.

(c) Determine $\sum_{\beta=1}^{3} \mathcal{P}_{\alpha\beta}$.

(d) Assume that the neutrino propagation factor can be expressed as $\mathcal{D}_j\mathcal{D}^*_k = e^{-i\Delta m^2_{kj}L/2E}$, where L is the source-detector separation, E is the (relativistic) neutrino energy and, as in the text, $\Delta m^2_{kj} \equiv m^2_k - m^2_j$. Then show

$$\mathcal{P}_{\alpha\beta} = \sum_{j=1}^{3} |U_{\beta j}|^2 |U_{\alpha j}|^2 + 2\sum_{j>k} \cos\left[\frac{\Delta m^2_{kj}}{2E}L - \varphi_{\beta,\alpha;j,k}\right] |U_{\beta j} U^*_{\beta k} U_{\alpha k} U^*_{\alpha j}|,$$

where $\varphi_{\beta,\alpha;j,k}$ is the (*CP*-violating phase) of the U_{jk} factors.

(2) **Two-generation $1 \leftrightarrow 3$ neutrino mixing**

The aim is to obtain a simple expression for the survival probability $\mathcal{P}_{ee}(L)$ for $1 \leftrightarrow 3$ oscillations starting from the general relation derived above. We shall ignore *CP*-violating effects (and thus set $\varphi_{\beta,\alpha;j,k} = 0$) and use $|\Delta m^2_{21}| \ll |\Delta m^2_{31}| \simeq |\Delta m^2_{32}|$, which is already known from the text. Because we wish to observe $1 \leftrightarrow 3$ oscillations, we take $|\Delta m^2_{31}|L/2E \geq 1$ (i.e. $2\pi L/L^{(31)}_{\text{osc}} \geq 1$). We also take $\Delta m^2_{21}L/2E \ll 1$ (i.e. $2\pi L/L^{(21)}_{\text{osc}} \ll 1$) to suppress $1 \leftrightarrow 2$ oscillations. Then show that the survival probability $\mathcal{P}_{ee}(L)$ can be written as

$$\mathcal{P}_{ee}(L) = 1 - 2|U_{e3}|^2 \left(1 - |U_{e3}|^2\right) \left(1 - \cos(2\pi L/L^{(31)}_{\text{osc}})\right).$$

Hint: You will want to make use of the unitarity property of the mixing matrix U.

VII

Effective field theory for low-energy QCD

At the lowest possible energies, the Standard Model involves only photons, electrons, muons, and pions, as these are the lightest particles in the spectrum. As we increase the energy slightly, kaons and etas become active. The light pseudoscalar hadrons would be massless Goldstone bosons in the limit that the u, d, s quark masses vanished. We give a separate discussion of this portion of the theory because it is an important illustration of effective field theory and because it can be treated with a higher level of rigor than most other topics.

VII–1 QCD at low energies

The $SU(2)$ chiral transformations,

$$\psi_{L,R} \equiv \begin{pmatrix} u \\ d \end{pmatrix}_{L,R} \rightarrow \exp\left(-i\boldsymbol{\theta}_{L,R} \cdot \boldsymbol{\tau}\right)\psi_{L,R}, \tag{1.1}$$

almost give rise to an invariance of the QCD lagrangian for small m_u, m_d, but do not appear to induce a left–right symmetry of the particle spectrum. This is because the axial symmetry is dynamically broken (i.e. hidden) with the pion being the (approximate) Goldstone boson. Vectorial isospin symmetry, i.e., simultaneous $SU(2)$ transformations of ψ_L and ψ_R, remains as an approximate symmetry of the spectrum.

Isospin symmetry is seen from the near equality of masses in the multiplets $(\pi^{\pm}, \pi^0)$, (K^+, K^0), (p, n), etc. In the language of group theory, we say that $SU(2)_L \times SU(2)_R$ has been dynamically broken to $SU(2)_V$. What is the evidence that such a scenario is correct? Ultimately it comes from the predictions which result, such as those which we detail in the remainder of this chapter.

The effective lagrangian for pions at very low-energy has already been developed in Chap. IV. In particular we recall the formalism of Sect. IV–6 which includes couplings to left-handed (right-handed) currents $\ell_\mu(x)$ ($r_\mu(x)$), and scalar and pseudoscalar densities $s(x)$ and $p(x)$, with the resulting $\mathcal{O}(E^2)$ lagrangian,

$$\mathcal{L}_2 = \frac{F_\pi^2}{4} \operatorname{Tr} \left(D_\mu U D^\mu U^\dagger \right) + \frac{F_\pi^2}{4} \operatorname{Tr} \left(\chi U^\dagger + U \chi^\dagger \right),$$

$$U = \exp(i\boldsymbol{\tau} \cdot \boldsymbol{\pi}/F_\pi), \qquad \chi = 2B_0(s + ip), \tag{1.2}$$

where $D_\mu U \equiv \partial_\mu U + i\ell_\mu U - iU r_\mu$ and B_0 is a constant. QCD in the absence of sources is recovered with $\ell_\mu = r_\mu = p = 0$ and $s = \mathbf{m}$, where $\mathbf{m}$ is the quark mass matrix.

Vacuum expectation values and masses

With a dynamically broken symmetry, the lagrangian is invariant but the vacuum state does not share this symmetry. A useful measure of this noninvariance in QCD is the vacuum expectation value of a scalar bilinear,

$$\langle 0 | \bar{\psi}\psi | 0 \rangle = \langle 0 | \bar{\psi}_L \psi_R | 0 \rangle + \langle 0 | \bar{\psi}_R \psi_L | 0 \rangle. \tag{1.3}$$

Up to small corrections, isospin symmetry implies

$$\langle 0 | \bar{u}u | 0 \rangle = \langle 0 | \bar{d}d | 0 \rangle. \tag{1.4}$$

Such matrix elements, if nonzero, cannot be invariant under separate left-handed or right-handed $SU(2)$ transformations. Indeed, it is evident from Eq. (1.3) that the vacuum expectation value couples together the left-handed and right-handed sectors.

One way that the vacuum expectation values of Eq. (1.4) affect phenomenology is through the pion mass. If the u and d quarks were massless, the pion would be a true Goldstone boson with $m_\pi = 0$. The part of the QCD lagrangian which explicitly violates chiral symmetry is the collection of quark mass terms,

$$\mathcal{H}_{\text{mass}} = -\mathcal{L}_{\text{mass}} = m_u \bar{u}u + m_d \bar{d}d. \tag{1.5}$$

To first order in the symmetry breaking, the pion mass is generated by the expectation value of this hamiltonian,

$$m_\pi^2 = \langle \pi | m_u \bar{u}u + m_d \bar{d}d | \pi \rangle. \tag{1.6}$$

This quantity can be related to the vacuum expectation value by using the chiral lagrangian. Taking both the pion and vacuum matrix elements of Eq. (1.2) and using the notation of Sect. IV–6, we have

$$m_\pi^2 = (m_u + m_d)B_0, \qquad \langle 0 | \bar{q}q | 0 \rangle = -\frac{\partial \mathcal{L}}{\partial s^0} = -F_\pi^2 B_0 = -\frac{F_\pi^2 m_\pi^2}{m_u + m_d}. \tag{1.7}$$

Thus since both B_0 and the quark masses are required to be positive, consistency requires that $\langle 0|\bar{q}q|0\rangle$ be nonzero and negative. However, without a separate determination of the quark masses (the origin of which must lie outside chiral symmetry) we do not know either $\langle 0|\bar{q}q|0\rangle$ or $m_u + m_d$ independently.

As an aside, we note that for Goldstone bosons there is a clear answer to the perennial question of whether one should treat symmetry breaking in terms of a linear or quadratic formula in the meson mass. For states of appreciable mass, the two procedures are equivalent to first order in the symmetry breaking since

$$\delta(m^2) \equiv (m_0 + \delta m)^2 - m_0^2 = 2m_0\,\delta m + \cdots. \tag{1.8}$$

However, when the symmetry expansion is about a *massless* limit, the m vs m^2 distinction becomes important. Because pions are bosonic fields we require their effective lagrangian to have the properly normalized form,

$$\mathcal{L} = \frac{1}{2}\left(\partial_\mu \boldsymbol{\pi} \cdot \partial^\mu \boldsymbol{\pi} - m_\pi^2 \boldsymbol{\pi} \cdot \boldsymbol{\pi}\right) + \cdots. \tag{1.9}$$

The prediction for the pion mass must then have the form,

$$m_\pi^2 = (m_u + m_d)B_0 + (m_u + m_d)^2 C_0 + (m_u - m_d)^2 D_0 + \cdots. \tag{1.10}$$

In principle, Nature could decide in favor of either $m_\pi^2 \propto m_q$ or $m_\pi^2 \propto m_q^2$ depending on whether the renormalized parameter B_0 vanishes or not. However, the choice $B_0 = 0$ is not 'natural' in that there is no symmetry constraint to force this value. Since one generally expects a nonzero value for B_0, the squared pion mass is linear in the symmetry-breaking parameter m_q. There is every indication that $B_0 \neq 0$ in QCD.

Quark mass ratios

The addition of an extra quark adds to the number of possible hadrons. If the strange quark mass is not too large, there are additional low-mass particles associated with the breaking of chiral symmetry. Including the quark mass terms,

$$\mathcal{L}_{\text{mass}} = \bar{\psi}_L \mathbf{m}\psi_R + \bar{\psi}_R \mathbf{m}\psi_L, \qquad \mathbf{m} = \begin{pmatrix} m_u & 0 & 0 \\ 0 & m_d & 0 \\ 0 & 0 & m_s \end{pmatrix}, \tag{1.11}$$

the QCD lagrangian has an approximate $SU(3)_L \times SU(3)_R$ global symmetry. If the u, d, s quarks were massless, the dynamical breaking of $SU(3)_L \times SU(3)_R$ to vector $SU(3)$ would produce eight Goldstone bosons, one for each generator of $SU(3)$. These would be the three pions $\pi^\pm, \pi^0$, four kaons $K^\pm, K^0, \bar{K}^0$, and one neutral particle η_8 with the quantum numbers of the eighth component of the octet.

Due to nonzero quark masses, these mesons are not actually massless, but should be light if the quark masses are not 'too large'.

What should the K, η_8 masses be? Unfortunately, *QCD* is unable to answer this question, even if we were able to solve the theory precisely. This is because the quark masses are free parameters in *QCD*, and thus must be determined from experiment. This means that the π, K, and η_8 masses can be used to determine the quark masses rather than vice versa. The discussion is somewhat more subtle than this simple statement would indicate. Quark masses need to be renormalized, and hence to specify their values one has to specify the renormalization prescription and the scale at which they are renormalized. Under changes of scale, the mass values change, i.e., they 'run.' However, quark mass *ratios* are rather simpler. The *QCD* renormalization is flavor-independent, at least to lowest order in the masses. In this situation, mass ratios are independent of the renormalization. There can be some residual scheme dependence through higher-order dependence of the renormalization constants on the quark masses. However, to first order, we can be confident that the mass ratio determined by the π, K, η_8 masses is the *same* ratio as found from the mass parameters of the *QCD* lagrangian.

The content of chiral $SU(3)$ is contained in an effective lagrangian expressed in terms of $U = \exp[i(\lambda \cdot \varphi)/F]$ and having the same form as Eq. (1.2). The matrix field $\lambda \cdot \varphi$ contained in U has the explicit representation,

$$\frac{1}{\sqrt{2}} \sum_{a=1}^{8} \lambda^a \varphi^a = \begin{pmatrix} \frac{1}{\sqrt{2}}\pi^0 + \frac{1}{\sqrt{6}}\eta_8 & \pi^+ & K^+ \\ \pi^- & -\frac{1}{\sqrt{2}}\pi^0 + \frac{1}{\sqrt{6}}\eta_8 & K^0 \\ K^- & \overline{K}^0 & -\frac{2}{\sqrt{6}}\eta_8 \end{pmatrix}, \tag{1.12}$$

as expressed in terms of the pseudoscalar meson fields. If we choose the parameters in Eq. (1.2) to correspond to *QCD* without external sources, viz.,

$$s = \mathbf{m}, \qquad p = 0, \qquad D_\mu U = \partial_\mu U, \tag{1.13}$$

the meson masses obtained by expanding to order φ^2 are

$$m_\pi^2 = B_0(m_u + m_d), \qquad m_{K^\pm}^2 = B_0(m_s + m_u),$$

$$m_{K^0}^2 = B_0(m_s + m_d), \qquad m_{\eta_8}^2 = \frac{1}{3}B_0(4m_s + m_u + m_d). \tag{1.14}$$

Defining $m_K^2 = \frac{1}{2}(m_{K^\pm}^2 + m_{K^0}^2)$, we obtain from Eq. (1.14) the mass relations,

$$\frac{\hat{m}}{m_s} = \frac{m_\pi^2}{2m_K^2 - m_\pi^2} \simeq \frac{1}{26}, \tag{1.15a}$$

$$m_{\eta_8}^2 = \frac{1}{3}\left(4m_K^2 - m_\pi^2\right), \tag{1.15b}$$

where $\hat{m} \equiv (m_u + m_d)/2$. Eq. (1.15a) demonstrates the extreme lightness of the u, d quark masses. Most estimates of the strange quark mass place it at around $m_s(2\ \text{GeV}) \sim 100\ \text{MeV}$ [RPP 12], so that $\hat{m} \sim 4\ \text{MeV}$, i.e., significantly smaller than the scale of QCD, Λ_{QCD}. Of course, the existence of very light quarks in the Standard Model is no more (or less) a mystery than is the existence of very heavy quarks. Both are determined by the Yukawa couplings of fermions to the Higgs boson, which are unconstrained (and not understood) input parameters of the theory. In any case, the small values of the u, d masses are responsible in QCD for the light pion, and for the usefulness of chiral symmetry techniques.

The mass relation of Eq. (1.15b) is the Gell-Mann–Okubo formula as applied to the octet of Goldstone bosons [GeOR 68]. It predicts $m_{\eta_8} = 566\ \text{MeV}$, not far from the mass of the $\eta(549)$. The small difference between these mass values can be accounted for by second-order effects in the mass expansion. In particular, mixing of the η_8 with an $SU(3)$ singlet pseudoscalar produces a mass shift of order $(m_s - \hat{m})^2$. The difference between the predicted and physical masses is then an estimate of accuracy of the lowest-order predictions.

The use of the full pseudoscalar octet allows us to be sensitive to isospin breaking due to quark mass differences in a way not possible using only pions. This is because, to first order, the $\Delta I = 2$ mass difference $m_{\pi^{\pm}} - m_{\pi^0}$ is independent of the $\Delta I = 1$ mass difference $m_d - m_u$. In contrast, the kaons experience a mass splitting of first order in $m_d - m_u$. In particular, the quark mass contribution to the kaon mass difference is

$$\left(m^2_{K^0} - m^2_{K^+}\right)_{\text{qk-mass}} = (m_d - m_u)\, B_0 = \left[\frac{m_d - m_u}{m_s - \hat{m}}\right]\left(m^2_K - m^2_\pi\right). \qquad (1.16)$$

In addition, there are electromagnetic contributions of the form

$$\left(m^2_{K^0} - m^2_{K^+}\right)_{\text{em}} = m^2_{\pi^0} - m^2_{\pi^+}. \qquad (1.17)$$

This result, called Dashen's theorem [Da 69], follows in an effective lagrangian framework from (i) the vanishing of the electromagnetic self-energies of neutral mesons at lowest order in the energy expansion, and (ii) the fact that K^+ and π^+ fall in the same U-spin multiplet and hence are treated identically by the electromagnetic interaction, itself a U-spin singlet.[1] By isolating the quark mass and electromagnetic contributions to the kaon mass difference, we can write a sum rule,

$$\begin{aligned}
\left[\frac{m_d - m_u}{m_s - \hat{m}}\right]\left(m^2_K - m^2_\pi\right) &= \left[\frac{m_d - m_u}{m_d + m_u}\right] m^2_\pi \\
&= \left(m^2_{K^0} - m^2_{K^+}\right) - \left(m^2_{\pi^0} - m^2_{\pi^+}\right),
\end{aligned} \qquad (1.18)$$

---

[1] Recall that U-spin is the $SU(2)$ subgroup of $SU(3)$ under which the d and s quarks are transformed.

which yields

$$\frac{m_d - m_u}{m_s - \hat{m}} = 0.023, \qquad \frac{m_d - m_u}{m_d + m_u} = 0.29. \tag{1.19}$$

The u quark is seen to be lighter than the d quark, with $m_u/m_d \simeq 0.55$. The reason why this large deviation from unity does not play a major role in low-energy physics is that *both m_u and m_d are small compared to the confinement scale of QCD*. This is, in fact, the origin of isospin symmetry, which in terms of quark mass is simply the statement that neither m_u nor m_d plays a major physical role, aside from the crucial fact that $m_\pi \neq 0$. Why these two masses lie so close to zero is a question which the Standard Model does not answer.

Pion leptonic decay, radiative corrections, and F_π

Throughout our previous discussion of chiral lagrangians, the pion decay constant F_π has played an important role. It is defined by the relation

$$\langle 0 | A_\mu^j(0) | \pi^k(\mathbf{p}) \rangle = i F_\pi p_\mu \delta^{jk}, \tag{1.20}$$

where the axial-vector current A_μ^j is expressible in terms of the quark fields $\psi \equiv \begin{pmatrix} u \\ d \end{pmatrix}$ as

$$A_\mu^j = \bar{\psi} \gamma_\mu \gamma_5 \frac{\tau^j}{2} \psi. \tag{1.21}$$

This amplitude gives us the opportunity to display the way that the electroweak interactions are matched on to the low-energy strong interactions, and so we treat this topic in some detail.

The pion matrix element is probed experimentally in the decays $\pi \to \bar{e}\nu_e$ and $\pi \to \bar{\mu}\nu_\mu$, which are induced by the weak hamiltonian,

$$\mathcal{H}_w = \frac{G_F}{\sqrt{2}} V_{ud} \bar{\psi}_d \gamma_\lambda (1 + \gamma_5) \psi_u \left[\bar{\psi}_{\nu_e} \gamma^\lambda (1 + \gamma_5) \psi_e + \bar{\psi}_{\nu_\mu} \gamma^\lambda (1 + \gamma_5) \psi_\mu \right]. \tag{1.22}$$

The decay $\pi^+ \to \mu^+ \nu_\mu$ has invariant amplitude,

$$\begin{aligned}
\mathcal{M}_{\pi^+ \to \mu^+ \nu_\mu} &= \frac{G_F}{\sqrt{2}} V_{ud} \sqrt{2} F_\pi p_\lambda \bar{u}_\nu \gamma^\lambda (1 + \gamma_5) v_\mu \\
&= -G_F V_{ud} F_\pi m_\mu \bar{u}_\nu (1 - \gamma_5) v_\mu,
\end{aligned} \tag{1.23}$$

where the Dirac equation has been used to obtain the second line. An analogous expression holds for $\pi^+ \to e^+ \nu_e$. We see here the well-known *helicity suppression* phenomenon. That is, the weak interaction current contains the left-handed chiral projection operator $(1 + \gamma_5)$, which in the *massless* limit produces only left-handed particles and right-handed antiparticles. However, such a configuration is forbidden in the decay of a spin-zero particle to massless $\bar{\mu}\nu_\mu$ or $\bar{e}\nu_e$ because the leptons would be required to have combined angular momentum $J_z = 1$ along the decay axis. Thus the amplitudes for $\pi \rightarrow \bar{\mu}\nu_\mu$, $\bar{e}\nu_e$ must vanish in the limit $m_\mu = m_e = 0$. Since the neutrino is always left-handed, the μ^+, e^+ in pion decay must have right-handed helicity to conserve angular momentum. It is helicity flip which introduces the factors of m_μ, m_e. The decay rate is found to be

$$\Gamma_{\pi^+ \rightarrow \mu^+ \nu_\mu} = \frac{G_F^2}{4\pi} F_\pi^2 m_\mu^2 m_\pi |V_{\text{ud}}|^2 \left(1 - \frac{m_\mu^2}{m_\pi^2}\right)^2. \tag{1.24}$$

However, before using this expression to extract the pion decay constant, one must include radiative corrections. We shall do this in some detail because it illustrates the way to match electroweak loops onto hadronic calculations. Since a complete analysis would be overly lengthy, we present a simplified argument which stresses the underlying physics.

In Chap. V we found that the radiative correction to the muon lifetime is ultraviolet finite even in the approximation of a strictly local weak interaction. However, this is *not* the case for semileptonic transitions, as can be easily demonstrated. Consider the photon loop diagrams shown in Fig. VII–1. We divide the photon integration into hard and soft components. The former, which determine the ultraviolet properties of the diagrams, have short wavelengths $\lambda \ll R$, where R is a typical hadronic size, and are sensitive to the weak interaction at the quark level. In Landau gauge (i.e. $\xi = 0$), the ultraviolet divergences arising from the wavefunction renormalization and vertex renormalization diagrams depicted in Fig. VII–1 vanish. For example, the vertex term is

$$I_{\text{vertex}}^{(\text{u.v.})} \sim \frac{iG_F}{\sqrt{2}} e^2 Q_4 Q_3 \int \frac{d^4k}{(2\pi)^4} \frac{1}{k^2} \left(-g_{\mu\nu} + \frac{k_\mu k_\nu}{k^2}\right)$$
$$\times \bar{u}_4 \gamma^\mu \frac{\not{k}}{k^2} \gamma_\lambda (1 + \gamma_5) \frac{\not{k}}{k^2} \gamma^\nu u_3 \bar{u}_2 \gamma^\lambda (1 + \gamma_5) u_1$$
$$\sim \frac{iG_F}{\sqrt{2}} e^2 Q_4 Q_3 \int \frac{d^4k}{(2\pi)^4} \bar{u}_4 \left[\frac{2\not{k}\gamma_\lambda \not{k}}{k^6} + \frac{\gamma_\lambda}{k^4}\right] (1 + \gamma_5) u_3 \bar{u}_2 \gamma^\lambda (1 + \gamma_5) u_1,$$
$$\tag{1.25}$$

where $Q_i e$ is the electric charge of the ith particle. Using

$$\int \frac{d^4k}{(2\pi)^4} \frac{k_\mu k_\nu}{k^6} = \frac{g_{\mu\nu}}{4} \int \frac{d^4k}{(2\pi)^4} \frac{1}{k^4} \sim \frac{ig_{\mu\nu}}{32\pi^2} \ln \Lambda, \tag{1.26}$$

we find that $I_{\text{vertex}}^{(\text{u.v.})} = 0$ as claimed. It is clear, employing a Fierz transformation, that photon exchange between particles 4,1 and 2,3 is also ultraviolet-finite.

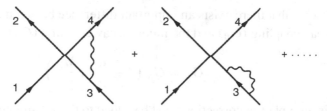

Fig. VII–1 Photonic radiative corrections to the weak quark–quark interaction.

This result simply represents the nonrenormalization of the vertex of a conserved current found in Chap. V. The only ultraviolet divergences then arise from final-state and initial-state interactions, i.e., photon exchange between particles 2,4, and 1,3, for which

$$I_{\text{fsi}}^{(\text{u.v.})} \sim \frac{-iG_F}{\sqrt{2}} e^2 Q_4 Q_2 \int \frac{d^4k}{(2\pi)^4} \frac{1}{k^2} \left(-g_{\mu\nu} + \frac{k_\mu k_\nu}{k^2}\right)$$
$$\times \bar{u}_4 \gamma^\mu \frac{\not{k}}{k^2} \gamma^\lambda (1 + \gamma_5) u_3 \bar{u}_2 \gamma^\nu \frac{\not{k}}{k^2} \gamma_\lambda (1 + \gamma_5) u_3 + (2, 4 \to 1, 3)$$
$$\sim -\frac{G_F}{\sqrt{2}} \frac{e^2}{32\pi^2} Q_4 Q_2 \ln \Lambda [\bar{u}_4 \gamma_\mu \gamma_\alpha \gamma_\lambda (1 + \gamma_5) u_3 \bar{u}_2 \gamma^\mu \gamma^\alpha \gamma^\lambda (1 + \gamma_5) u_1$$
$$- 4\bar{u}_4 \gamma_\lambda (1 + \gamma_5) u_3 \bar{u}_2 \gamma^\lambda (1 + \gamma_5) u_1] + (2, 4 \to 1, 3). \tag{1.27}$$

Using the identity in Eq. (C–2.5) for reducing the product of three gamma matrices, Eq. (1.27) becomes

$$I_{\text{fsi}}^{(\text{u.v.})} = -\mathcal{M}^{(0)} \times \frac{3\alpha}{2\pi} (Q_4 Q_2 + Q_3 Q_1) \ln(\Lambda/\mu_L), \tag{1.28}$$

where $\mathcal{M}^{(0)}$ is the lowest-order vertex. However, the full calculation of the radiative corrections must include the propagator for the W boson as well. When the contact weak interaction is replaced by the W-exchange diagram and is added to that with the photon-exchange replaced by Z-exchange, one obtains a finite result at the ultraviolet end with $\Lambda = m_Z$. The integral is cut off at the lower end at some point $\mu_L \sim m_\ell$ below which the full hadronic structure must be considered. In the case of muon decay we have

$$Q_e Q_{\nu_\mu} + Q_{\nu_e} Q_\mu = 0. \tag{1.29}$$

Thus, as found in Chap. V, there is no divergence. On the other hand, for beta decay we obtain

$$Q_e Q_u + Q_{\nu_e} Q_{d,s} = -\frac{2}{3}. \tag{1.30}$$

We observe that there exists an important difference between the beta-decay effective weak coupling (G_β) and the muon-decay coupling (G_μ)

$$G_\beta = G_\mu \left(1 + \frac{\alpha}{\pi} \ln \frac{M_Z}{\mu_L} \right). \tag{1.31}$$

This hard-photon correction must be added to the soft-photon component, which can be found be evaluating the radiative corrections to a structureless ('point') pion with a high-energy cut-off μ_H. These were calculated long ago with the result [Be 58, KiS 59],

$$\frac{\Gamma_{\pi^+ \to \mu^+ \nu_\mu}}{\Gamma^{(0)}_{\pi^+ \to \mu^+ \nu_\mu}} = 1 + \frac{\alpha}{2\pi} \left(B(x) + 3 \ln \frac{\mu_H}{m_\pi} - 6 \ln \frac{\mu_H}{m_\mu} \right), \tag{1.32}$$

where

$$B(x) = 4 \left[\frac{x^2 + 1}{x^2 - 1} \ln x - 1 \right] \left[\ln(x^2 - 1) - 2 \ln x - \frac{3}{4} \right]$$
$$+ 4 \frac{x^2 + 1}{x^2 - 1} L(1 - x^{-2}) - \ln x - \frac{3}{4} + \frac{10x^2 - 7}{(x^2 - 1)^2} \ln x + \frac{15x^2 - 21}{4(x^2 - 1)}, \tag{1.33}$$

with $L(z) = \int_0^z \frac{dt}{t} \ln(1 - t)$ being the Spence function and $x = m_\pi/m_\mu$. Adding the hard- and soft-photon contributions with $\mu_H = \mu_L \simeq m_\rho$, we find the full radiative correction,

$$\Gamma_{\pi^+ \to \mu^+ \nu_\mu} \simeq \Gamma^{(0)} \left[1 + \frac{\alpha}{2\pi} \left(B(x) + 3 \ln \frac{M_Z}{m_\pi} + \ln \frac{M_Z}{m_\rho} - 6 \ln \frac{m_\rho}{m_\mu} \right) \right]. \tag{1.34}$$

Taking V_{ud} from Sect. XII–4 and $\Gamma^{(\text{expt})}_{\pi^+ \to \mu^+ \nu_\mu} = 3.841 \times 10^7 \text{ s}^{-1}$, we find

$$F_\pi = 92.2 \pm 0.2 \text{ MeV}, \tag{1.35}$$

where we have appended an uncertainty associated with possible radiative effects $\mathcal{O}(\alpha/2\pi)$ that are not included in Eq. (1.34). For chiral symmetry applications in this book we shall generally employ the value

$$F_\pi \simeq 92 \text{ MeV}. \tag{1.36}$$

A clear indication of the importance of radiative corrections can be seen in the ratio

$$R = \frac{\Gamma_{\pi^+ \to e^+ \nu_e}}{\Gamma_{\pi^+ \to \mu^+ \nu_\mu}}, \tag{1.37}$$

which is strongly suppressed by the helicity mechanism discussed earlier. Application of the lowest-order formula given in Eq. (1.24) leads to a prediction

$$R^{(0)} = \frac{m_e^2}{m_\mu^2} \left(\frac{m_\pi^2 - m_e^2}{m_\pi^2 - m_\mu^2} \right)^2 = 1.283 \times 10^{-4}, \tag{1.38}$$

in disagreement with the measured value

$$R_{\text{expt}} = (1.230 \pm 0.004) \times 10^{-4}. \tag{1.39}$$

However, when the full radiative correction given in Eq. (1.34) is employed, the theoretical prediction is modified to become

$$R_{\text{thy}} = R^{(0)} \left(1 - 3\frac{\alpha}{\pi} \ln \frac{m_\mu}{m_e} + \cdots \right) = (1.2353 \pm 0.0001) \times 10^{-4}, \tag{1.40}$$

which is consistent with the experimental value.

VII–2 Chiral perturbation theory to one loop

Let us summarize the development thus far. Interactions of the Goldstone bosons can be expressed in terms of an effective lagrangian having the correct symmetry properties. To lowest order in the energy expansion, i.e., to order E^2, it suffices to use the minimal lagrangian of Eq. (1.2) at tree level. In the $SU(2)$ theory, this involves just the known constants F_π and m_π. At the next order, one encounters both the general $\mathcal{O}(E^4)$ lagrangian, given below, and also one-loop diagrams [ApB 81, GaL 84, 85a]. The $\mathcal{O}(E^4)$ lagrangian introduces new parameters, which must be determined from experiment. It is also necessary to give a prescription which allows one to handle the loop calculations. The general method is described in this section.

The program is called *chiral perturbation theory*. If one works to order E^4 in the energy expansion, there are typically three ingredients:

(1) the general lagrangian $\mathcal{L}_2$ (of order E^2) which is to be used both in loop diagrams and at tree level,
(2) the general lagrangian $\mathcal{L}_4$ (of order E^4) which is to be used only at tree level,
(3) the renormalization program which describes how to make physical predictions at one-loop level.

The general $\mathcal{O}(E^2)$ lagrangian has already been given in Eq. (1.2). Now we shall turn to the construction of the chiral $SU(n)$ lagrangian to order E^4.

The order E^4 lagrangian

The $\mathcal{O}(E^4)$ lagrangian can involve either four-derivative operators or two-derivative operators together with one factor of the quark mass term, $\chi \sim 2m_q B_0$ (which itself is of order m_π^2 or m_K^2) or products of two quark mass factors. There are four possible chiral-invariant terms with four separate derivatives,

$$\mathrm{Tr}\left(D_\mu U D^\mu U^\dagger D_\nu U D^\nu U^\dagger\right), \quad \mathrm{Tr}\left(D_\mu U D_\nu U^\dagger D^\mu U D^\nu U^\dagger\right),$$
$$\mathrm{Tr}\left(D_\mu U D_\nu U^\dagger\right) \cdot \mathrm{Tr}\left(D^\mu U D^\nu U^\dagger\right), \quad [\mathrm{Tr}\left(D_\mu U D^\mu U^\dagger\right)]^2. \tag{2.1}$$

Other structures, such as

$$\left[\mathrm{Tr}\left(\lambda^a U^\dagger D_\mu U\right) \mathrm{Tr}\left(\lambda^a U^\dagger D^\mu U\right)\right]^2, \tag{2.2}$$

can be expressed in terms of these by using $SU(n)$ matrix identities.

For the case of $SU(3)$, the operators in Eq. (2.1) are not linearly independent. The identities quoted in Eq. (II–2.17) can be used to show that

$$\mathrm{Tr}\left(D_\mu U D_\nu U^\dagger D^\mu U D^\nu U^\dagger\right) = \frac{1}{2}\left[\mathrm{Tr}\left(D_\mu U D^\mu U^\dagger\right)\right]^2$$
$$+ \mathrm{Tr}\left(D_\mu U D_\nu U^\dagger\right) \cdot \mathrm{Tr}\left(D^\mu U D^\nu U^\dagger\right) - 2\,\mathrm{Tr}\left(D_\mu U D^\mu U^\dagger D_\nu U D^\nu U^\dagger\right), \tag{2.3}$$

leaving only three independent operators in this class. In $SU(2)$, a further identity,

$$2\,\mathrm{Tr}\left(D_\mu U D^\mu U^\dagger D_\nu U D^\nu U^\dagger\right) = \left[\mathrm{Tr}\left(D_\mu U D^\mu U^\dagger\right)\right]^2, \tag{2.4}$$

leaves us with only two independent $\mathcal{O}(E^4)$ terms.

Another conceivable class of operators could have at least two derivatives acting on a single chiral matrix, such as

$$\mathrm{Tr}\left(D_\mu U D^\mu U^\dagger\right) \cdot \mathrm{Tr}\left(U^\dagger D_\nu D^\nu U\right). \tag{2.5}$$

However, since the E^4 lagrangian is to be used only at tree level, all states to which it is applied obey the equation of motion,

$$D^\mu\left(U^\dagger D_\mu U\right) + \frac{1}{2}\left(\chi^\dagger U - U^\dagger \chi\right) = 0. \tag{2.6}$$

This can be used to eliminate all the double-derivative operators in favor of those involving four single derivatives or with factors of χ.

The remaining operators are reasonably straightforward to determine, and the *most general* $\mathcal{O}(E^4)$ $SU(3)$ chiral lagrangian is,[2]

$$
\mathcal{L}_4 = \sum_{i=1}^{10} L_i O_i
$$

$$
\begin{aligned}
&= L_1 \left[\text{Tr} \left(D_\mu U D^\mu U^\dagger \right) \right]^2 + L_2 \, \text{Tr} \left(D_\mu U D_\nu U^\dagger \right) \cdot \text{Tr} \left(D^\mu U D^\nu U^\dagger \right) \\
&\quad + L_3 \, \text{Tr} \left(D_\mu U D^\mu U^\dagger D_\nu U D^\nu U^\dagger \right) \\
&\quad + L_4 \, \text{Tr} \left(D_\mu U D^\mu U^\dagger \right) \text{Tr} \left(\chi U^\dagger + U \chi^\dagger \right) \\
&\quad + L_5 \, \text{Tr} \left(D_\mu U D^\mu U^\dagger \left(\chi U^\dagger + U \chi^\dagger \right) \right) + L_6 \left[\text{Tr} \left(\chi U^\dagger + U \chi^\dagger \right) \right]^2 \\
&\quad + L_7 \left[\text{Tr} \left(\chi^\dagger U - U \chi^\dagger \right) \right]^2 + L_8 \, \text{Tr} \left(\chi U^\dagger \chi U^\dagger + U \chi^\dagger U \chi^\dagger \right) \\
&\quad + i L_9 \, \text{Tr} \left(L_{\mu\nu} D^\mu U D^\nu U^\dagger + R_{\mu\nu} D^\mu U^\dagger D^\nu U \right) + L_{10} \, \text{Tr} \left(L_{\mu\nu} U R^{\mu\nu} U^\dagger \right),
\end{aligned}
$$

$$(2.7)$$

where $L_{\mu\nu}$, $R_{\mu\nu}$ are the field-strength tensors of external sources given in Eq. (IV–6.10). This is a central result of the effective lagrangian approach to the study of low-energy strong interactions. Much of the discussion in the chapters to follow will concern the above operators and involve a phenomenological determination of the $\{L_i\}$. In chiral $SU(2)$, three operators become redundant.

For completeness, we note that there may also exist two combinations of the external fields,

$$
\mathcal{L}_{\text{ext}} = \beta_1 \, \text{Tr} \left(L_{\mu\nu} L^{\mu\nu} + R_{\mu\nu} R^{\mu\nu} \right) + \beta_2 \, \text{Tr} \left(\chi^\dagger \chi \right),
$$

which are chirally invariant without involving the matrix U. These do not generate any couplings to the Goldstone bosons and hence are not of great phenomenological interest. However, if one were to use the effective lagrangian to describe correlation functions of the external sources, these two operators can generate contact terms.

Finally, we summarize in Table VII–1 a set of values for the low-energy constants $\{L_i\}$ as obtained phenomenologically via a global fit to a range of low-energy data [BiJ 12]. (In this extraction certain assumptions are made also about the size of $\mathcal{O}(p^6)$ chiral coefficients, since they also contribute to observables.) These constants provide a characterization of the low-energy dynamics of QCD.

The renormalization program

The renormalization procedure is as follows. The lagrangian, $\mathcal{L}_2$, when expanded in terms of the meson fields, specifies a set of interaction vertices. These can be

---

[2] We are using the operator basis and notation first set down by Gasser and Leutwyler [GaL 85a].

Table VII–1. *Renormalized coefficients in the chiral
lagrangian $\mathcal{L}_4$ given in units of 10^{-3} and evaluated at
renormalization point $\mu = m_\rho$ [BiJ 12].*

Coefficient	Value	Origin
L_1^r	1.12 ± 0.20	$\pi\pi$ scattering
L_2^r	2.23 ± 0.40	and
L_3^r	-3.98 ± 0.50	$K_{\ell 4}$ decay
L_4^r	1.50 ± 1.01	F_K/F_π
L_5^r	1.21 ± 0.08	F_K/F_π
L_6^r	1.17 ± 0.95	F_K/F_π
L_7^r	-0.36 ± 0.18	Meson masses
L_8^r	0.62 ± 0.16	F_K/F_π
L_9^r	7.0 ± 0.2	Rare pion
L_{10}^r	-5.6 ± 0.2	decays

used to calculate tree-level and one-loop diagrams for any transition of interest.
This result is added to the contribution which comes from the vertices contained
in the $\mathcal{O}(E^4)$ lagrangian $\mathcal{L}_4$, treated at tree level only. At this stage, the result contains both bare parameters and divergent loop integrals. One needs to determine the
parameters from experiment. The first step involves mass and wavefunction renormalization, as well as renormalization of F_π. In addition, the parameters entering
from $\mathcal{L}_4$ need to be determined from data. If the lagrangian is indeed the most general one possible, relations between observables will be *finite* when expressed in
terms of physical quantities. All the divergences will be absorbed into defining a
set of renormalized parameters. This fundamental result is demonstrated explicitly
in App. B–2.

There exists always an ambiguity of what finite constants should be absorbed
into the renormalized parameters L_i^r. This ambiguity does not affect the relationship between observables, but only influences the numerical values quoted for the
low-energy constants. Similarly, the regularization procedure for handling divergent integrals is arbitrary.[3] We use dimensional regularization and the renormalization prescription,

$$L_i^r = L_i - \frac{\gamma_i}{32\pi^2} \left[\frac{2}{d-4} - \ln(4\pi) + \gamma - 1 \right], \qquad (2.8)$$

---

[3] Care must be taken that the regularization procedure does not destroy the chiral symmetry. Dimensional
regularization does not cause any problems. When using other regularization schemes, one sometimes needs
to append an extra contact interaction to maintain chiral invariance [GeJLW 71]. The problem arises due to
the presence of derivative couplings, which imply that the interaction Hamiltonian is not simply the negative
of the interaction lagrangian. The contact interaction vanishes in dimensional regularization.

where the constants γ_i are numbers given in Table B–1. When working to $\mathcal{O}(E^4)$ the following procedure is applied. One first computes the relevant vertices from $\mathcal{L}_2$ and $\mathcal{L}_4$. There are too many possible vertices to make a table of Feynman rules practical. In practice, the needed amplitudes are calculated for each application. The vertices from $\mathcal{L}_2$ are then used in loop diagrams, including mass and wave-function renormalizations. The results may be expressed in terms of the renormalized parameters of Eq. (2.8). If these low-energy constants can be determined from other processes, one has obtained a well-defined result. Including loops does add important physics to the result. The low-energy portion of the loop integrals describes the propagation and rescattering of low-energy Goldstone bosons, as required by the unitarity of the S matrix. One-loop diagrams add the unitarity corrections to the lowest-order amplitudes and in addition contain mass contributions and other effects from low energy.

The effective lagrangian may be used in the context either of chiral $SU(2)$ or of chiral $SU(3)$. Because $SU(2)$ is a subgroup of $SU(3)$, the general $SU(3)$ lagrangian of Eq. (2.7) is also valid for chiral $SU(2)$. However, the $SU(2)$ version has fewer low-energy constants, so that only certain combinations of the L_i^r will appear in pionic processes. If one is dealing with reactions involving only pions at low energy, the kaons and the eta are heavy particles and may be integrated out, such that only pionic effects need to be explicitly considered. This procedure produces a shift in the values of the low-energy renormalized constants L_i^r such that the L_i^r of a purely $SU(2)$ chiral lagrangian and an $SU(3)$ one will differ by a finite calculable amount. In this book, we shall use the $SU(3)$ values as our basic parameter set. The $SU(2)$ coefficients can be found by first performing calculations in the $SU(3)$ limit and then treating m_K^2, m_η^2 as large. Equivalently, all may be calculated at the same time using the background field method [GaL 85a]. The results are

$$2L_1^{(2)r} + L_3^{(2)r} = 2L_1^r + L_3^r - \frac{1}{4}\ell_K, \qquad L_2^{(2)r} = L_2^r - \frac{1}{4}\ell_K,$$

$$2L_4^{(2)r} + L_5^{(2)r} = 2L_4^r + L_5^r - \frac{3}{2}\ell_K, \qquad L_9^{(2)r} = L_9^r - \ell_K,$$

$$2L_6^{(2)r} + L_8^{(2)r} = 2L_6^r + L_8^r - \frac{3}{4}\ell_K - \frac{1}{12}\ell_\eta, \qquad L_{10}^{(2)r} = L_{10}^r + \ell_K,$$

$$L_4^{(2)r} - L_6^{(2)r} - 9L_7^{(2)r} - 3L_8^{(2)r} = L_4^r - L_6^r - 9L_7^r - 3L_8^r + \frac{3}{2}\ell_k$$

$$+ \frac{F_\pi^2}{24m_\eta^2} + \frac{5}{1152\pi^2}\ln\frac{m_\eta^2}{\mu^2}, \qquad (2.9)$$

where we use the superscript (2) to indicate constants in the $SU(2)$ theory and define $\ell_i \equiv \left[\ln(m_i^2/\mu^2) + 1\right]/384\pi^2$. In practice, these shifts are much smaller than the magnitude of the low-energy constants, so that we always simply quote the $SU(3)$ value.

Let us now calculate the mass and wavefunction renormalization constants to $\mathcal{O}(E^4)$ in chiral $SU(2)$. Setting $\chi = (m_u + m_d)B_0 \equiv m_0^2$, we may expand the basic lagrangian as

$$
\mathcal{L}_2 = \frac{1}{2} \left[\partial^\mu \varphi \cdot \partial_\mu \varphi - m_0^2 \varphi \cdot \varphi \right] + \frac{m_0^2}{24 F_0^2} (\varphi \cdot \varphi)^2
$$

$$
+ \frac{1}{6 F_0^2} \left[(\varphi \cdot \partial^\mu \varphi)(\varphi \cdot \partial_\mu \varphi) - (\varphi \cdot \varphi)(\partial^\mu \varphi \cdot \partial_\mu \varphi) \right] + \mathcal{O}(\varphi^6), \quad (2.10)
$$

$$
\mathcal{L}_4 = \frac{m_0^2}{F_0^2} \left[16 L_4^{(2)} + 8 L_5^{(2)} \right] \frac{1}{2} \partial_\mu \varphi \cdot \partial^\mu \varphi
$$

$$
- \frac{m_0^2}{F_0^2} \left[32 L_6^{(2)} + 16 L_8^{(2)} \right] \frac{1}{2} m_0^2 \varphi \cdot \varphi + \mathcal{O}(\varphi^4),
$$

where F_0 denotes the value of F_π prior to loop corrections. When this lagrangian is used in the calculation of the propagator, the terms of $\mathcal{O}(\varphi^4)$ in $\mathcal{L}_2$ will contribute to the self-energy via one-loop diagrams, which involve the following d-dimensional integrals,

$$
\delta_{jk} I(m^2) = i \Delta_{Fjk}(0) = \langle 0|T\varphi_j(x)\varphi_k(x)|0\rangle,
$$

$$
I(m^2) = \mu^{4-d} \int \frac{d^d k}{(2\pi)^d} \frac{i}{k^2 - m^2} = \frac{\mu^{4-d}}{(4\pi)^{d/2}} \Gamma\left(1 - \frac{d}{2}\right)(m^2)^{\frac{d}{2}-1},
$$

$$
\delta_{jk} I_{\mu\nu}(m^2) = -\partial_\mu \partial_\nu i \Delta_{Fjk}(0) = \langle 0|T\partial_\mu\varphi_j(x)\partial_\nu\varphi_k(x)|0\rangle,
$$

$$
I_{\mu\nu}(m^2) = \mu^{4-d} \int \frac{d^d k}{(2\pi)^d} k_\mu k_\nu \frac{i}{k^2 - m^2} = g_{\mu\nu} \frac{m^2}{d} I(m^2). \quad (2.11)
$$

These contributions can be read off from $\mathcal{L}_2$ by considering all possible contractions among the $\mathcal{O}(\varphi^4)$ terms, and result in the quadratic effective lagrangian,

$$
\mathcal{L}_{\text{eff}} = \frac{1}{2} \partial^\mu \varphi \cdot \partial_\mu \varphi - \frac{1}{2} m_0^2 \varphi \cdot \varphi + \frac{5 m_\pi^2}{12 F_\pi^2} I(m_\pi^2) \varphi \cdot \varphi
$$

$$
+ \frac{1}{6 F_\pi^2} \left(\delta_{ik}\delta_{jl} - \delta_{ij}\delta_{kl} \right) I(m_\pi^2) \left(\delta_{ij}\partial^\mu\varphi_k\partial_\mu\varphi_l + \delta_{k\ell}m_\pi^2\varphi_i\varphi_j \right)
$$

$$
+ \frac{1}{2} \partial_\mu \varphi \cdot \partial^\mu \varphi \frac{m_\pi^2}{F_\pi^2} \left[16 L_4^{(2)} + 8 L_5^{(2)} \right] - \frac{1}{2} m_\pi^2 \varphi \cdot \varphi \frac{m_\pi^2}{F_\pi^2} \left[32 L_6^{(2)} + 16 L_8^{(2)} \right]
$$

$$
= \frac{1}{2} \partial^\mu \varphi \cdot \partial_\mu \varphi \left[1 + \left(16 L_4^{(2)} + 8 L_5^{(2)} \right) \frac{m_\pi^2}{F_\pi^2} - \frac{2}{3 F_\pi^2} I(m_\pi^2) \right]
$$

$$
- \frac{1}{2} m_0^2 \varphi \cdot \varphi \left[1 + \left(32 L_6^{(2)} + 16 L_8^{(2)} \right) \frac{m_\pi^2}{F_\pi^2} - \frac{1}{6 F_\pi^2} I(m_\pi^2) \right]. \quad (2.12)
$$

To one-loop order, there are no other contributions to the self energy. Observe that we have changed m_0, F_0 into m_π, F_π in all of the $\mathcal{O}(E^4)$ corrections, as the difference between the two is of yet higher order in the energy expansion. If we expand in powers of $d - 4$ and define the renormalized pion field as $\varphi_r = Z_\pi^{-1/2} \varphi$ with

$$Z_\pi = 1 - \frac{8m_\pi^2}{F_\pi^2}\left(2L_4^{(2)} + L_5^{(2)}\right) + \frac{m_\pi^2}{24\pi^2 F_\pi^2}\left[\frac{2}{d-4} + \gamma - 1 - \ln 4\pi + \ln\frac{m_\pi^2}{\mu^2}\right],$$

(2.13)

then the lagrangian assumes the canonical form

$$\mathcal{L}_{\text{eff}} = \frac{1}{2}\partial_\mu\varphi_r \cdot \partial^\mu\varphi_r - \frac{1}{2}m_\pi^2\varphi_r \cdot \varphi_r.$$

(2.14)

Note that, using the definitions of the renormalized parameters, the physical pion mass is identified as

$$m_\pi^2 = m_0^2\left[1 - \frac{8m_\pi^2}{F_\pi^2}\left[2L_4^{(2)r} + L_5^{(2)r} - 4L_6^{(2)r} - 2L_8^{(2)r}\right] + \frac{m_\pi^2}{32\pi^2 F_\pi^2}\ln\frac{m_\pi^2}{\mu^2}\right].$$

(2.15)

The quantity $\mathcal{L}_{\text{eff}}$ in Eq. (2.14) is the quadratic portion of the one-loop effective lagrangian. Since loop effects have already been accounted for, it is to be used at tree level. This is a simple application of the background field renormalization discussed in App. B–2.

VII–3 The nature of chiral predictions

In order to understand how predictions are made in effective field theory as well as the range of validity of the energy expansion, let us work out several examples. At first, these will seem to be rather obscure processes, but they are the simplest hadronic reactions of QCD. As the bosonic interactions of the Goldstone bosons of the theory, they are the cleanest processes for demonstrating the dynamical content of the symmetries and anomalies of QCD.

The pion form factor

The electromagnetic form factor of charged pions is required by Lorentz invariance and gauge invariance to have the form[4]

$$\langle\pi^+(\mathbf{p}_2)|J_{\text{em}}^\mu|\pi^+(\mathbf{p}_1)\rangle = G_\pi(q^2)(p_1 + p_2)^\mu,$$

(3.1)

---

[4] The neutral pion form factor is required to vanish by charge conjugation invariance.

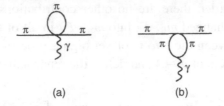

Fig. VII–2 Radiative corrections to the pion form factor.

where $q^\mu = (p_1 - p_2)^\mu$ and $G_\pi(0) = 1$. The electromagnetic current may be identified from the effective lagrangian of Eq. (1.2) by setting $\ell^\mu = r^\mu = eQA^\mu$, $\chi = 2B_0\mathbf{m}$, where Q is the quark charge matrix and $\mathbf{m}$ is the quark mass matrix. To $\mathcal{O}(E^4)$, we then find

$$J^\mu_{em} = -\frac{\partial \mathcal{L}}{\partial(eA_\mu)} = (\boldsymbol{\varphi} \times \partial^\mu\boldsymbol{\varphi})_3 \left[1 - \frac{1}{3F^2}\boldsymbol{\varphi} \cdot \boldsymbol{\varphi} + \mathcal{O}(\varphi^4)\right]$$

$$+ (\boldsymbol{\varphi} \times \partial^\mu\boldsymbol{\varphi})_3 \left[16L_4^{(2)} + 8L_5^{(2)}\right]\frac{m_\pi^2}{F^2} + \frac{4L_9^{(2)}}{F^2}\partial^\nu (\partial^\mu\boldsymbol{\varphi} \times \partial_\nu\boldsymbol{\varphi})_3 + \cdots .$$

$$(3.2)$$

The renormalization of this current involves the Feynman diagrams in Fig. VII–2. That of Fig. VII–2(a) is simply found using the integral previously defined in Eq. (2.11),

$$J^\mu_{em}\bigg|_{(2a)} = -\frac{5}{3F_\pi^2}(\boldsymbol{\varphi} \times \partial^\mu\boldsymbol{\varphi})_3 \, I(m_\pi^2).$$

$$(3.3)$$

Evaluation of Fig. VII–2(b) is somewhat more complicated. Using the elastic $\pi^+\pi^-$ scattering amplitude given by $\mathcal{L}_2$,

$$\langle \pi^+(\mathbf{k}_1)\pi^-(\mathbf{k}_2)|\pi^+(\mathbf{p}_1)\pi^-(\mathbf{p}_2)\rangle$$

$$= \frac{i}{3F_0^2}\left(2m_0^2 + p_1^2 + p_2^2 + k_1^2 + k_2^2 - 3(p_1 - k_1)^2\right),$$

$$(3.4)$$

we compute the vertex amplitude to be

$$\langle J^\mu_{em}\rangle_{(2b)} = -\frac{i}{3F_\pi^2}\int \frac{d^4k}{(2\pi)^4}\frac{1}{\left(k + \frac{1}{2}q\right)^2 - m_\pi^2 + i\epsilon}\frac{1}{\left(k - \frac{1}{2}q\right)^2 - m_\pi^2 + i\epsilon}$$

$$\times \left[4m_\pi^2 + \left(k + \frac{q}{2}\right)^2 + \left(k - \frac{q}{2}\right)^2 - 3\left(k + \frac{(p_1 + p_2)}{2}\right)^2\right]2k^\mu.$$

$$(3.5)$$

Upon integration, most terms drop out because of antisymmetry under $k^\mu \to -k^\mu$, and we find

$$\langle J_{\text{em}}^{\mu} \rangle_{(2b)} = \frac{2i}{F_{\pi}^2} \int \frac{d^4k}{(2\pi)^4} \frac{k^{\mu} k \cdot (p_1 + p_2)}{\left(\left(k + \frac{1}{2}q\right)^2 - m_{\pi}^2 \right) \left(\left(k - \frac{1}{2}q\right)^2 - m_{\pi}^2 \right)}. \tag{3.6}$$

We can evaluate this integral using dimensional regularization,

$$\langle J_{\text{em}}^{\mu} \rangle_{(2b)} = -\frac{2}{F_{\pi}^2} \frac{\mu^{4-d}}{(4\pi)^{d/2}} \int dx \left[-\frac{1}{2} \frac{(p_1 + p_2)^{\mu} \Gamma\left(1 - \frac{d}{2}\right)}{\left(m_{\pi}^2 - q^2 x(1-x)\right)^{1-d/2}} \right.$$
$$\left. + q^{\mu} q \cdot (p_1 + p_2) \left(x + \frac{1}{2}\right)^2 \Gamma\left(2 - \frac{d}{2}\right) \left(m_{\pi}^2 - q^2 x(1-x)\right)^{d/2-2} \right], \tag{3.7}$$

where as usual μ is an arbitrary scale introduced in order to maintain the proper dimensions. On-shell, we can disregard the term in q^{μ} since $q \cdot (p_1 + p_2) = m_{\pi}^2 - m_{\pi}^2 = 0$. For the remaining piece, we expand about $d = 4$ to obtain

$$\langle J_{\text{em}}^{\mu} \rangle_{(2b)} = \frac{1}{(4\pi F_{\pi})^2} (p_1 + p_2)^{\mu} \int_0^1 dx \left(m_{\pi}^2 - q^2 x(1-x)\right)$$
$$\times \left[\left(\frac{2}{d-4} + \gamma - 1 - \ln 4\pi\right) + \ln \frac{m_{\pi}^2 - q^2 x(1-x)}{\mu^2} \right], \tag{3.8}$$

and the x-integration then yields

$$\langle J_{\text{em}}^{\mu} \rangle_{(2b)} = \frac{1}{(4\pi F_{\pi})^2} (p_1 + p_2)^{\mu} \left\{ \left(m_{\pi}^2 - \frac{1}{6}q^2\right) \left[\frac{2}{d-4} + \gamma - 1 - \ln 4\pi \right. \right.$$
$$\left. \left. + \ln \frac{m_{\pi}^2}{\mu^2} \right] + \frac{1}{6} (q^2 - 4m_{\pi}^2) H\left(\frac{q^2}{m_{\pi}^2}\right) - \frac{1}{18}q^2 \right\}, \tag{3.9}$$

where

$$H(a) \equiv -\int_0^1 dx \ln (1 - ax(1-x))$$

$$= \begin{cases} 2 - 2\sqrt{\frac{4}{a} - 1} \, \text{ctn}^{-1} \sqrt{\frac{4}{a} - 1} & (0 < a < 4) \\ 2 + \sqrt{1 - \frac{4}{a}} \left[\ln \frac{\sqrt{1-\frac{4}{a}}-1}{\sqrt{1-\frac{4}{a}}+1} + i\pi\theta(a-4) \right] & \text{(otherwise)}. \end{cases} \tag{3.10}$$

Now we add everything together. The tree-level amplitude is modified by wave-function renormalization,

$$Z_{\pi} G_{\pi}^{(\text{tree})}(q^2) = \left[1 - \frac{8m_{\pi}^2}{F_{\pi}^2} (2L_4^{(2)} + L_5^{(2)}) \right.$$
$$\left. + \frac{m_{\pi}^2}{24\pi^2 F_{\pi}^2} \left\{ \frac{2}{d-4} + \gamma - 1 - \ln 4\pi + \ln \frac{m_{\pi}^2}{\mu^2} \right\} \right]$$

$$\times \left[1 + \frac{8m_\pi^2}{F_\pi^2} \left(2L_4^{(2)} + L_5^{(2)} \right) + 2q^2 \frac{L_9^{(2)}}{F_\pi^2} \right]$$

$$= \left[1 + \frac{m_\pi^2}{24\pi^2 F_\pi^2} \left(\frac{2}{d-4} + \gamma - 1 - \ln 4\pi + \ln \frac{m_\pi^2}{\mu^2} \right) + \frac{2L_9^{(2)}}{F_\pi^2} q^2 \right],$$

$$(3.11)$$

while Figs. VII–2(a,b) contribute as

$$G_\pi(q^2) \Big|_{(2a)} = -\frac{5m_\pi^2}{48\pi^2 F_\pi^2} \left\{ \frac{2}{d-4} + \gamma - 1 - \ln 4\pi + \ln \frac{m_\pi^2}{\mu^2} \right\},$$

$$G_\pi(q^2) \Big|_{(2b)} = \frac{1}{16\pi^2 F_\pi^2} \left\{ \left(m_\pi^2 - \frac{1}{6}q^2 \right) \left[\frac{2}{d-4} + \gamma - 1 - \ln 4\pi + \ln \frac{m_\pi^2}{\mu^2} \right] \right.$$

$$\left. + \frac{1}{6} \left(q^2 - 4m_\pi^2 \right) H \left(\frac{q^2}{m_\pi^2} \right) - \frac{1}{18} q^2 \right\},$$

$$(3.12)$$

respectively. Summing Eqs. (3.11), (3.12) we see that all terms independent of q^2 cancel, $L_9^{(2)}$ becomes $L_9^{(2)r}$ and the final result is

$$G_\pi(q^2) = 1 + \frac{2L_9^{(2)r}}{F_\pi^2} q^2 + \frac{1}{96\pi^2 F_\pi^2} \left[\left(q^2 - 4m_\pi^2 \right) H \left(\frac{q^2}{m_\pi^2} \right) - q^2 \ln \frac{m_\pi^2}{\mu^2} - \frac{q^2}{3} \right].$$

$$(3.13)$$

The divergences have been absorbed in $L_9^{(2)r}$, while the imaginary part required by unitarity is contained in $H(q^2/m_\pi^2)$. Note that the loops also induce a non-power-law behavior in $G_\pi(q^2)$. However, numerically this turns out to be small and is unobservable in practice. A simple linear approximation,

$$G_\pi(q^2) = 1 + q^2 \left[\frac{2L_9^{(2)r}}{F_\pi^2} - \frac{1}{96\pi^2 F_\pi^2} \left(\ln \frac{m_\pi^2}{\mu^2} + 1 \right) \right] + \cdots,$$

$$(3.14)$$

is obtained by Taylor expanding about $q^2 = 0$. The corresponding result for chiral $SU(3)$ is

$$G_\pi(q^2) = 1 + q^2 \left[\frac{2L_9^r}{F_\pi^2} - \frac{1}{96\pi^2 F_\pi^2} \left(\ln \frac{m_\pi^2}{\mu^2} + \frac{1}{2} \ln \frac{m_K^2}{\mu^2} + \frac{3}{2} \right) \right] + \cdots.$$

$$(3.15)$$

The pion form factor is generally parameterized in terms of a charge radius,

$$G_\pi(q^2) = 1 + \frac{1}{6} \langle r_\pi^2 \rangle q^2 + \cdots.$$

$$(3.16)$$

Thus, for any given value of the energy scale μ, the parameter L_9^r can be determined from the experimental charge radius. From the present experimental value $\langle r_\pi^2 \rangle = (0.45 \pm 0.01)$ fm^2, we obtain $L_9^r(\mu = m_\rho) = (7.0 \pm 0.2) \times 10^{-3}$.

The scale μ enters the calculation in such a way that, had we used a different scale μ' but kept the physical result invariant, we would have had

$$L_9^r(\mu') = \begin{cases} L_9^r(\mu) - \dfrac{1}{192\pi^2} \ln\left(\dfrac{\mu'^2}{\mu^2}\right) & (SU(2)) \\[3mm] L_9^r(\mu) - \dfrac{1}{128\pi^2} \ln\left(\dfrac{\mu'^2}{\mu^2}\right) & (SU(3)). \end{cases} \tag{3.17}$$

In fact, if we look back to the origin of $\ln \mu$ in the transition from Eq. (3.7) to Eq. (3.8) using

$$\mu^{4-d} \frac{2}{d-4} = \frac{2}{d-4} - \ln \mu^2 + \mathcal{O}(d-4), \tag{3.18}$$

we see that the scale dependence is always tied to the coefficient of the divergence.[5] The general result is then

$$L_i^r(\mu') = L_i^r(\mu) - \frac{\gamma_i}{32\pi^2} \ln\left(\frac{\mu'^2}{\mu^2}\right), \tag{3.19}$$

where $\{\gamma_i\}$ are the constants of Table B–1 of App. B, used in the renormalization condition of Eq. (2.8).

This calculation also nicely illustrates the range of validity of the energy expansion. The pion form factor is well described by a monopole form,

$$G_\pi(q^2) \simeq \frac{1}{1 - q^2/m^2} = 1 + \frac{q^2}{m^2} + \cdots, \tag{3.20}$$

with $m \simeq m_{\rho(770)}$. The energy expansion is then in powers of q^2/m^2. At the other extreme, the pion form factor can also be treated in *QCD* when q^2 is very large [BrL 80].

Rare processes

The calculation above is clearly non-predictive as it contains a free parameter, L_9^r, which must be determined phenomenologically. However, predictions do arise when more reactions are considered because relations exist between amplitudes as a consequence of the underlying chiral symmetry. In particular, there exists a set of reactions which are described in terms of two low-energy constants. These pionic reactions are shown in Table VII–2. With the additional input of F_K/F_π, the kaonic reactions shown there are also predicted. Each case contains hadronic form factors

---

[5] We have chosen to keep the low-energy constants $\{L_i\}$ dimensionless in the extension to d dimensions. In [GaL 85a], the constants have dimension μ^{d-4}. However, the resulting physics is identical in the limit $d \to 4$.

Table VII–2. *The radiative complex of pion and*
kaon transitions.

Pions	Kaons
$\gamma \to \pi^+\pi^-$	$\gamma \to K^-K^+$
$\gamma\pi^+ \to \gamma\pi^+$	$\gamma K^+ \to \gamma K^+$
$\pi^+ \to e^+\nu_e\gamma$	$K^+ \to e^+\nu_e\gamma$
$\pi^+ \to \pi^0 e^+\nu_e$	$K \to \pi e^+\nu_e$
$\pi^+ \to e^+\nu_e e^+e^-$	$K^+ \to e^+\nu_e e^+e^-$
	$K^+ \to \pi^0 e^+\nu_e\gamma$

which need to be calculated. This section briefly describes the procedure for relating such reactions in chiral perturbation theory. All calculations follow the pattern described above, so that we shall only quote the results [GaL 85a, DonH 89].

In the processes involving photons ($\pi^+ \to e^+\nu_e\gamma$, $\pi^+ \to e^+\nu_e e^+e^-$ and $\gamma\pi^+ \to \gamma\pi^+$), there are always Born diagrams where the photon couples to hadrons through the known $\pi\pi\gamma$ coupling. These are shown in Fig. VII–3. In addition, there can be direct contact interactions associated with the structure of the pions. These introduce new form factors. For the decays $\pi^+ \to e^+\nu_e\gamma$, $e^+\nu_e e^+e^-$, the matrix elements are

$$\mathcal{M}_{\pi^+ \to e^+\nu_e\gamma} = -\frac{eG_F}{\sqrt{2}}\cos\theta_1 M_{\mu\nu}(p,q)\epsilon^{\mu*}(q)\bar{u}(p_\nu)\gamma^\nu(1+\gamma_5)v(p_e),$$

$$\mathcal{M}_{\pi^+ \to e^+\nu_e e^+e^-} = \frac{e^2 G_F}{\sqrt{2}}\cos\theta_1 M_{\mu\nu}(p,q)\frac{1}{q^2}$$
$$\times \bar{u}(p_2)\gamma^\mu v(p_1)\bar{u}(p_\nu)\gamma^\nu(1+\gamma_5)v(p_e), \qquad (3.21)$$

where the hadronic part of the quantity $M_{\mu\nu}$ has the general structure

$$M_{\mu\nu}(p,q) = \int d^4x\, e^{iq\cdot x}\langle 0\,|\,T\left(J_\mu^{\mathrm{em}}(x)J_\nu^{1-i2}(0)\right)|\,\pi^+(\mathbf{p})\rangle$$

$$= -\sqrt{2}\,F_\pi\frac{(p-q)_\nu}{(p-q)^2 - m_\pi^2}\langle\pi^+(\mathbf{p}-\mathbf{q})\,|\,J_\mu^{\mathrm{em}}\,|\,\pi^+(\mathbf{p})\rangle + \sqrt{2}\,F_\pi g_{\mu\nu}$$

$$- h_A\left((p-q)_\mu q_\nu - g_{\mu\nu}q\cdot(p-q)\right) - r_A(q_\mu q_\nu - g_{\mu\nu}q^2)$$

$$+ i h_V\epsilon_{\mu\nu\alpha\beta}q^\alpha p^\beta. \qquad (3.22)$$

The first line represents the tree diagram and in subsequent lines the subscripts V and A indicate whether the vector or axial-vector portions of the weak currents are involved. The form factor r_A in Eq. (3.22) can only contribute with virtual photons, i.e., as in $\pi^+ \to e^+\nu_e e^+e^-$.

The $\gamma\pi^+ \to \gamma\pi^+$ reaction is analyzed in terms of the pion's electric and magnetic polarizabilities, α_E and β_M, which describe the response of the pion to electric

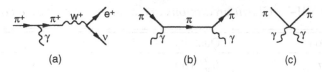

Fig. VII–3 Tree diagrams for (a) $\pi^+ \to e^+ \nu_e \gamma$, $\pi^+ \to e^+ \nu_e e^+ e^-$, and (b)–(c) $\gamma \pi^+ \to \gamma \pi^+$.

and magnetic fields. In the static limit, electromagnetic fields induce the electric and magnetic dipole moments,

$$\mathbf{p}_E = 4\pi \alpha_E \mathbf{E}, \qquad \boldsymbol{\mu}_M = 4\pi \beta_M \mathbf{H}, \tag{3.23}$$

which correspond to an interaction energy

$$V = -2\pi \left(\alpha_E \mathbf{E}^2 + \beta_M \mathbf{H}^2 \right). \tag{3.24}$$

These forms emerge in the non-relativistic limit of the general Compton amplitude

$$-iT_{\mu\nu}(p, p', q_1) = -i \int d^4x \, e^{iq_1 \cdot x} \langle \pi^+(\mathbf{p}') | T \left(J_\mu^{em}(x) J_\nu^{em}(0) \right) | \pi^+(\mathbf{p}) \rangle$$

$$= \frac{(2p' + q_2)_\nu (2p - q_1)_\mu}{(p - q_1)^2 - m_\pi^2} + \frac{(2p' + q_1)_\mu (2p - q_2)_\nu}{(p - q_2)^2 - m_\pi^2} - 2g_{\mu\nu}$$

$$+ \sigma \left(q_{2\mu} q_{1\nu} - g_{\mu\nu} q_1 \cdot q_2 \right) + \cdots, \tag{3.25}$$

where σ is a coefficient proportional to the polarizability and q_1^μ, q_2^μ are the photon momenta, taken as outgoing, with $p = p' + q_1 + q_2$. The first three pieces are the Born and seagull diagrams. The last contains the extra term which emerges from higher-order chiral lagrangians, and the ellipses indicate the presence of other possible gauge-invariant structures, which we shall not need.

The chiral predictions are obtained in the same manner as used for the pion form factor. The results at $q^2 \simeq 0$ are

$$h_V = \frac{N_c}{12\sqrt{2}\,\pi^2 F_\pi} \bigg|_{N_c=3} = 0.027\, m_\pi^{-1}, \qquad \frac{h_A}{h_V} = 32\pi^2 \left(L_9^{(2)r} + L_{10}^{(2)r} \right),$$

$$\frac{r_A}{h_V} = 32\pi^2 \left[L_9^{(2)r} - \frac{1}{192\pi^2} \left(\ln \frac{m_\pi^2}{\mu^2} + 1 \right) \right], \qquad \alpha_E + \beta_M = 0,$$

$$\alpha_E = \frac{\alpha}{2m_\pi} \sigma = \frac{4\alpha}{m_\pi F_\pi^2} \left[L_9^{(2)r} + L_{10}^{(2)r} \right] - \frac{\alpha}{m_\pi} \frac{1}{16\pi^2 F_\pi^2} \left(1 + F(t/m_\pi^2) \right), \tag{3.26}$$

where $t = (q_1 + q_2)^2$ and

$$F(x) \equiv -\frac{4}{x} \sinh^2 \left(\sqrt{x}/2 \right) \xrightarrow[1 \gg x]{} -1 - \frac{x}{12} + \cdots. \tag{3.27}$$

Table VII–3. *Chiral predictions and data in the radiative complex of transitions.*

Reaction	Quantity	Theory	Experiment
$\gamma \to \pi^+\pi^-$	$\langle r_\pi^2 \rangle$ (fm^2)	0.45^a	0.45 ± 0.01
$\gamma \to K^+K^-$	$\langle r_K^2 \rangle$ (fm^2)	0.45	0.31 ± 0.03
$\pi^+ \to e^+\nu_e\gamma$	$h_V(m_\pi^{-1})$	0.027	0.0254 ± 0.0017
	h_A/h_V	0.441^a	0.441 ± 0.004
$K^+ \to e^+\nu_e\gamma$	$(h_V + h_A)(m_K^{-1})$	0.136	0.133 ± 0.008
$\pi^+ \to e^+\nu_e e^+e^-$	r_A/h_V	2.6	2.2 ± 0.3
$\gamma\pi^+ \to \gamma\pi^+$	$(\alpha_E + \beta_M)(10^{-4} \text{ fm})$	0	0.17 ± 0.02
	$(\alpha_E - \beta_M)(10^{-4} \text{ fm})$	5.6	13.6 ± 2.8
$K \to \pi e^+\nu_e$	$\xi = f_-(0)/f_+(0)$	−0.13	−0.17 ± 0.02
	λ_+ (fm^2)	0.067	0.0605 ± 0.001
	λ_0 (fm^2)	0.040	0.0400 ± 0.002

^{a}Used as input.

The prediction for h_V is especially interesting since h_V is related by an isospin rotation to the amplitude for $\pi^0 \to \gamma\gamma$ (cf. Prob. VII–2). As we will show in Sect. VII–6, this is absolutely predicted from the axial anomaly. The presence of L_{10}^r implies that one of the above measurements must be used to determine it. We use the precisely known value for h_A/h_V to yield

$$L_{10}^r(\mu = m_\rho) = -(5.6 \pm 0.2) \times 10^{-3}. \tag{3.28}$$

The results are compared with experiment in Table VII–3.

We see that, with one exception, the chiral predictions are in agreement with experiment. That exception, the electric polarizability in $\gamma\pi^+ \to \gamma\pi^+$, comes from two difficult experiments. One uses a pion beam on a heavy Z atom [An *et al.* 85] and the coulomb exchange in $\pi^+A \to \gamma\pi^+A$ is used to provide the extra photon (this is called the Primakoff effect). The tree diagram must be carefully subtracted off. The second experiment involves the use of a high-energy photon beam and the $p(\gamma, \gamma\pi^+)n$ reaction and extrapolation to the virtual pion pole [Ah *et al.* 05]. In this case there exist a large number of background processes which must be subtracted. We note however that a recent experiment [Fr 12] using the Primakoff effect, not yet included in the averages above, obtains $\alpha_E - \beta_M = 3.8 \pm 1.4 \pm 1.6$. Before being concerned with the possible discrepancy with the chiral prediction, it would be preferable to have the experimental situation clarified. We have also listed the known results on kaonic processes predicted by the same constants. The analyses for $\gamma \to K^+K^-$ and $K \to e^+\nu_e\gamma$ are identical to the above results.

Pion–pion scattering

The elastic scattering of two Goldstone bosons is the purest manifestation of the chiral effective field theory of *QCD*. It is a classic topic with a long history. We use it here as an example of the convergence of the perturbative effective field theory expansion.

The scattering of pions can be classified as *S*-wave, *P*-wave, *D*-wave, etc., with low partial waves dominating at low-energy. The amplitudes also can be decomposed in overall isospin, $I = 0, 1, 2$. Because the pions are spinless bosons, Bose symmetry requires that the even partial waves carry $I = 0, 2$ and odd angular momentum requires $I = 1$. At low energies, the partial wave amplitude can be expanded in terms of a scattering length a_ℓ^I and slope b_ℓ^I, defined by

$$\text{Re}\, T_\ell^I = \left(\frac{q^2}{m_\pi^2}\right)^\ell \left(a_\ell^I + b_\ell^I \frac{q^2}{m_\pi^2} + \cdots\right), \tag{3.29}$$

where $q^2 \equiv \left(s - 4m_\pi^2\right)/4$. Since the chiral expansion is similarly a power series in the energy, a_ℓ^I and b_ℓ^I provide a useful set of quantities to study. In practice, they are extracted from data by using dispersion relations and crossing symmetry to extrapolate some of the higher-energy data down to threshold. The only accurate very low-energy data are those from $K \to \pi\pi e\bar{\nu}_e$. The experimental values are given in Table VII–4.

At lowest order in the energy expansion, the amplitude for $\pi\pi$ scattering [Wc 66] can be obtained from $\mathcal{L}_2$ with the result

$$A(s, t, u) = \frac{s - m_\pi^2}{F_\pi^2}. \tag{3.30}$$

Table VII–4. *The pion scattering lengths and slopes.*

	Experimental	Lowest order[a]	First two orders[a]
a_0^0	0.220 ± 0.005	0.16	0.20
b_0^0	0.25 ± 0.03	0.18	0.26
a_0^2	-0.044 ± 0.001	-0.045	-0.041
b_2^2	-0.082 ± 0.008	-0.089	-0.070
a_1^1	0.038 ± 0.002	0.030	0.036
b_1^1	—	0	0.043
a_2^0	$(17 \pm 3) \times 10^{-4}$	0	20×10^{-4}
a_2^2	$(1.3 \pm 3) \times 10^{-4}$	0	3.5×10^{-4}

[a] Predictions of chiral symmetry.

This produces the scattering lengths and slopes

$$a_0^0 = \frac{7m_\pi^2}{32\pi F_\pi^2}, \quad a_0^2 = -\frac{m_\pi^2}{16\pi F_\pi^2}, \quad a_1^1 = \frac{m_\pi^2}{24\pi F_\pi^2},$$

$$b_0^0 = \frac{m_\pi^2}{4\pi F_\pi^2}, \quad b_0^2 = -\frac{m_\pi^2}{8\pi F_\pi^2}, \tag{3.31}$$

with the numerical values shown in the table. It is remarkable that the lowest-energy form of a scattering process may be determined entirely from symmetry considerations. At next order in the expansion, one considers loop diagrams and the E^4 lagrangian. The convergence towards the experimental values can be seen in the table. Also shown are the best theoretical results which combine dispersive constraints with chiral perturbation theory [CoGL 01].

The nature of the chiral expansion also becomes evident within this process. The lowest-order predictions are real and grow monotonically. As such, they must eventually violate the unitarity constraint at some point. The worst case is the $I = \ell = 0$ amplitude

$$T_0^0 = \frac{1}{32\pi F_\pi^2}(2s - m_\pi^2), \tag{3.32}$$

which violates the simplest consequence of unitarity,

$$\frac{s - 4m_\pi^2}{s}\left|T_\ell^I\right|^2 < 1, \tag{3.33}$$

below $\sqrt{s} = 700$ MeV. In addition, there are no imaginary terms, which must be present due to the unitarity constraint,

$$\text{Im } T_\ell^I = \left(\frac{s - 4m_\pi^2}{s}\right)^{1/2}\left|T_\ell^I\right|^2. \tag{3.34}$$

These drawbacks are remedied order by order in the energy expansion. Note that since $\left|T_\ell^I\right|$ starts at order E^2, Im T_ℓ^I starts at order E^4. When one works to order E^4, loop diagrams generate an imaginary piece given by Eq. (3.34) with the lowest-order predictions for T_ℓ^I inserted on the right-hand side. This process proceeds order by order in the energy expansion.

We have displayed in Table VII–4 how the $\mathcal{O}(E^4)$ predictions modify the scattering lengths. Aside from the renormalization of m_π and F_π, the corrections depend only on the low-energy constants $(2L_1^r + L_3^r)$ and L_2^r. Let us also give a pictorial representation of the result. One may see the order E^4 improvement and the nature of the chiral expansion by considering the $I = 1$, $\ell = 1$ channel, where some of the higher-energy data are shown in Fig. VII–4. The resonance structure visible is the $\rho(770)$. The chiral prediction is

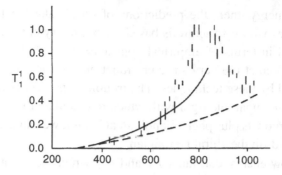

Fig. VII–4 Scattering in the $I = 1$, $\ell = 1$ channel.

$$T_1^1 = \frac{s - 4m^2}{96\pi^2 F^2}\left[1 + 4\left(L_2^{(2)r} - 2L_1^{(2)r} - L_3^{(2)r}\right)\frac{s}{F^2}\right], \qquad (3.35)$$

with loops having a negligible effect. The lowest-order result is given by the dashed line. It clearly does not reflect the presence of the $\rho(770)$ resonance. The solid line represents the result at order E^4 and starts to reproduce the low-energy tail of the $\rho(770)$. It is, of course, impossible to represent a full Breit–Wigner shape by two terms in an energy expansion; all orders are required. The chiral predictions at $\mathcal{O}(E^4)$ may reproduce the first two terms, with the resulting expansion being in powers of q^2/m_ρ^2.

VII–4 The physics behind the QCD chiral lagrangian

For the most part, we have been using chiral lagrangians as our primary tool for making predictions based on the symmetry structure of QCD. In this section, we pause to examine which features of QCD are important in determining the structure of chiral lagrangians. The general strategy can perhaps be appreciated by a comparison of low-energy and high-energy QCD methodology. At high energies, due to the asymptotic freedom of QCD, hard scattering processes can be calculated in a power-series expansion in the strong coupling constant. However, some dependence on 'soft' physics remains in the form of structure functions, fragmentation functions, etc. These are not calculable perturbatively and must be determined phenomenologically from the data. At high energy, then, the predictions of QCD are relations among amplitudes parameterized in terms of various phenomenological structure functions and the strong coupling constant. At very low energies, because of the symmetries of QCD, low-energy scatterings and decays can be calculated in a power series expansion in the energy. However, some dependence on 'harder' physics remains in the form of the constants $\{L_i^r\}$. These are not calculable from the symmetry structure and must be determined phenomenologically from the data.

At low energy, then, the predictions of *QCD* take the form of relations among amplitudes whose structure is based on symmetry constraints but which are parameterized in terms of empirical constants. Nevertheless, *QCD* should in principle also predict the very structure functions and low-energy constants which are employed by these techniques. The trouble at present is that we do not have techniques of comparable rigor with which to calculate these quantities. Nevertheless, by using models plus phenomenological insight we can learn a bit about the physics which leads to the chiral lagrangian.

The low-energy constants F_π and m_π which occur at order E^2 do not reveal much about the structure of the theory. All theories with a slightly broken chiral $SU(2)$ symmetry will have an identical structure at order E^2. The pion decay constant F_π will be sensitive to the mass scale of the underlying theory, while the pion mass m_π will be determined by the amount of symmetry breaking. However, approached phenomenologically, these are basically free parameters and do not differentiate between competing theories.

The situation is different at order E^4. Here, the chiral lagrangian contains many terms, and the pattern of coefficients is a signature of the underlying theory. The linear sigma model without fermions provides us with an example of how one can compare a theory with the real world. In Sects. IV–2,4 we calculated the tree-level terms in $\mathcal{L}_4$ which would be present in the linear sigma model, and obtained a result expressible as

$$2L_1 + L_3 = 2L_4 + L_5 = 8L_6 + 4L_8 = \frac{F_\pi^2}{4m_S^2} = \frac{1}{8\lambda}, \quad L_{2,7,9,10} = 0. \qquad (4.1)$$

This pattern is quite different from the structure obtained phenomenologically. It appears that the linear sigma model is not a good representation of the real world.

Unfortunately, it is harder to theoretically infer the $\{L_i\}$ directly from *QCD*. However, a look at phenomenology indicates that we should consider the effects of vector mesons, in particular the $\rho(770)$. This is the most clear in the pion form factor that shows a dramatic ρ resonance in the timelike region. Indeed, the whole form factor can be well understood in a simple model as being a Breit–Wigner shape due to the ρ resonance

$$G_\pi(q^2) = -\frac{m_\rho^2}{q^2 - m_\rho^2 + im_\rho\Gamma_\rho(q)\theta(q^2 - 4m_\pi^2)}, \qquad (4.2)$$

where the normalization is chosen to enforce the condition $G_\pi(0) = 1$. This works even in the timelike region. Comparison with the chiral lagrangian approach implies that this model would predict

$$L_9 = \frac{F_\pi^2}{2m_\rho^2} = 7.2 \times 10^{-3}, \tag{4.3}$$

in good agreement with the value obtained earlier, $L_9 = (7.0 \pm 0.2) \times 10^{-3}$.

This analysis can be extended to L_{10}. This enters into the $W^+\pi^+\gamma$ vertex, which occurs in $\pi^+ \to e^+\nu_e\gamma$. Here, both vector and axial-vector mesons can generate corrections to the basic couplings. Explicit calculation yields [EcGPR 89]

$$L_{10} = \frac{F_{a_1}^2}{4m_{a_1}^2} - \frac{F_\rho^2}{4m_\rho^2} = -5.8 \times 10^{-3}. \tag{4.4}$$

Here, a_1 refers to the lightest axial-vector meson $a_1(1260)$ (cf. Sect. V–3), and F_{a_1} and F_ρ are the couplings of a_1 and ρ to the W^+ and the photons respectively. Again, the result is close to the empirical value cited in Table VII–1, viz. $L_{10}^r = (-5.6 \pm 0.2) \times 10^{-3}$.

The phenomenological low-energy constants are scale-dependent, and their analysis includes loop effects, while those in Eqs. (4.3), (4.4) are constants, to be used at tree level. Nevertheless, there is some sense in comparing them. The effect of loops in processes involving L_9, L_{10} is small, and the scale dependence only makes a minor change, $L_9^r(\mu = 300 \text{ MeV}) = 7.7 \times 10^{-3}$ vs. $L_9^r(\mu = 1 \text{ GeV}) = 6.5 \times 10^{-3}$. Presumably the appropriate scale is near $\mu = m_{\rho(770)}$. The $\rho(770)$ provides a much more important effect here than any other input.

Finally, it also turns out that the use of vector meson exchange leads to a good description of $\pi\pi$ scattering [DoRV 89, EcGPR 89]. This is not too surprising in light of the need for the chiral lagrangians to reproduce the tail of the $\rho(770)$, as described in Sect. VII–3. As a consequence of crossing symmetry, the $\rho(770)$ must also influence the other scattering channels. To a large extent, the chiral coefficients L_1, L_2, L_3 are dominated by the effect of $\rho(770)$ exchange. We see from these examples that phenomenology indicates that the exchange of light vector particles is the most important physics effect behind the chiral coefficients which we have been discussing.

The idea that vector mesons play an important dynamical role is not new. It predates the Standard Model, originating with Sakurai [Sa 69], in a form called *vector dominance*. The vector dominance idea has never been derived from the Standard Model, but nevertheless enjoys considerable phenomenological support. Put most broadly, vector dominance states that the main dynamical effect at energies less than about 1 GeV is associated with the exchange of vector mesons. The use of a chiral lagrangian with parameters described by ρ-exchange is compatible with this idea and puts it on a firmer footing. These considerations suggest that for chiral lagrangians the prime ingredient of QCD is the spectrum of the theory. The linear sigma model has a quite different spectrum, with a light scalar and no ρ, and hence does not agree with the data. *QCD*, however, seems to predict that deviations from the lowest-order chiral relation must be in such a form as to reproduce the low-energy tails of the light resonances, in particular the ρ. At present, we cannot rigorously prove this connection. However, it remains a useful picture in estimating various effects of chiral lagrangians.

VII–5 The Wess–Zumino–Witten anomaly action

At this stage one must also include the effect of the axial anomaly. The anomaly influences not only processes involving photons, such as $\pi^0 \to \gamma\gamma$, but also purely hadronic processes. For example the reaction $K\bar{K} \to \pi^+\pi^-\pi^0$, allowed by *QCD*, is not present in any of the chiral lagrangians appearing in previous sections. Its absence is easy to understand because the hadronic part of the lagrangian, with external fields set equal to zero, has the discrete symmetry $\varphi^i \to -\varphi^i$ (i.e. $U \leftrightarrow U^\dagger$) which forbids the transition of an even number of mesons to an odd number. However this is *not* a symmetry of *QCD*. More importantly, there are a set of low-energy relations, the Wess–Zumino consistency conditions [WeZ 71], which must be satisfied in the presence of the anomaly and which involve hadronic reactions. The effect of the anomaly was first analyzed by Wess and Zumino who noted that the result could not be expressed as a single local effective lagrangian, and gave a Taylor expansion representation for it.[6] Witten [Wi 83a] subsequently gave an elegant representation of the Wess–Zumino contribution as an integral over a five-dimensional space whose boundary is physical four-dimensional spacetime.

Since the considerations leading to the Wess–Zumino–Witten action can be rather formal, it is best to adopt a direct calculational approach. Fortunately, we are able to employ the familiar sigma model (with fermions) because it contains the same anomaly structure as *QCD*. That is, it is the presence of fermions having the same quantum numbers as quarks which ensures that the anomaly will occur. The absence of gluons in the sigma model is not a problem since, according to the Adler–Bardeen theorem [AdB 69], the inclusion of gluons would not modify the result. Since the sigma model involves coupling between mesons and fermions, we can also observe directly the influence of the anomaly on the Goldstone bosons. Although somewhat technically difficult, our approach will clearly illustrate the connection with treatments of the anomaly based on perturbative calculations.

Consider as a starting point the lagrangian, Eq. (IV–1.11), of the linear sigma model

$$\mathcal{L} = \bar{\psi} i \not{\partial} \psi - gv \left(\bar{\psi}_L U \psi_R + \bar{\psi}_R U^\dagger \psi_L \right) + \cdots . \tag{5.1}$$

---

[6] For a textbook treatment, see [Ge 84].

We have displayed neither the term containing $\mathrm{Tr}\,(\partial_\mu U \partial^\mu U^\dagger)$ nor any term containing the scalar field S. Such contributions are not essential to our study of the anomaly and will be dropped hereafter. In order to simulate the light quarks of QCD, we shall endow each fermion with a color quantum number (letting the number N_c of colors be arbitrary) and assume there are three fermion flavors, each of constituent mass $M = gv$. Although the original linear sigma model has a flavor-$SU(2)$ chiral symmetry, Eq. (5.1) is equally well defined for flavor $SU(3)$.

Our analysis begins by imposing on Eq. (5.1) the change of variable

$$\psi_L'' \equiv \xi^\dagger \psi_L, \qquad \psi_R'' \equiv \xi \psi_R, \qquad \xi\xi = U, \tag{5.2}$$

like that described in App. B–4. This yields

$$\mathcal{L} = \bar\psi''(i\slashed{\mathcal{D}} - M)\psi'', \qquad \mathcal{D}_\mu \equiv \partial_\mu + i\overline{V}_\mu + i\overline{A}_\mu\gamma_5,$$

$$\overline{V}_\mu = -\frac{i}{2}\left(\xi^\dagger\partial_\mu\xi + \xi\partial_\mu\xi^\dagger\right), \qquad \overline{A}_\mu = -\frac{i}{2}\left(\xi^\dagger\partial_\mu\xi - \xi\partial_\mu\xi^\dagger\right). \tag{5.3}$$

For this change of variable the jacobian is not unity, and thus we must write the effective action as

$$e^{i\Gamma(U)} = \int [d\psi][d\bar\psi]\, e^{i\int d^4x \left(\bar\psi i\slashed{\partial}\psi - M(\bar\psi_L U\psi_R + \bar\psi_R U^\dagger\psi_L)\right)}$$

$$= \int [d\psi''][d\bar\psi'']\, \mathcal{J}\, e^{i\int d^4x\, \bar\psi''(i\slashed{\mathcal{D}}-M)\psi''}$$

$$= e^{\ln \mathcal{J}}\, e^{\mathrm{tr}\,\ln(i\slashed{\mathcal{D}}-M)}. \tag{5.4}$$

For large M, it can be shown that the $\mathrm{tr}\,\ln(i\slashed{\mathcal{D}} - M)$ factor does not produce any terms at order E^4 that contain the $\epsilon^{\mu\nu\alpha\beta}$ dependence characteristic of the anomaly.[7] Hence, the effect of the anomaly must lie in the jacobian $\mathcal{J}$, and it is this we must calculate.

It is possible to determine the jacobian by integrating a sequence of infinitesimal transformations. Thus we introduce the extension $\xi \to \xi_\tau$,

$$\xi_\tau \equiv e^{i\frac{\tau\lambda\cdot\bar\varphi}{2F_\pi}} \equiv \exp i\tau\overline{\varphi}, \tag{5.5}$$

where τ is a continuous parameter and $\xi = \xi_{\tau=1}$. Transformations induced by the infinitesimal parameter $\delta\tau$ will give rise to the infinitesimal quantities $\xi_{\delta\tau}$ and $\delta\mathcal{J}$,

---

[7] This can be verified by expanding as

$$\mathrm{tr}\,\ln(i\slashed{\mathcal{D}} - M) = \mathrm{tr}\,\ln(-M(1 - i\slashed{\mathcal{D}}/M)) = \mathrm{tr}\,\ln(-M) - \mathrm{tr}\,(i\slashed{\mathcal{D}})^2/2M^2 + \cdots.$$

The first term can be regularized as in the text and directly calculated using the techniques described in App. B. The remaining terms vanish for large M.

$$\psi = \psi_L + \psi_R \to \psi' = \left[\xi_{\delta\tau}^\dagger \frac{1 + \gamma_5}{2} + \xi_{\delta\tau} \frac{1 - \gamma_5}{2} \right] \psi,$$

$$\int [d\psi][d\bar{\psi}] = \int [d\psi'][d\bar{\psi}']\, e^{\ln \delta \mathcal{J}}. \tag{5.6}$$

From Eqs. (III–3.44), (III–3.47), we find $\delta \mathcal{J}$ to be

$$\delta \mathcal{J} = e^{-2i\delta\tau\, \mathrm{tr}\, (\overline{\varphi}\gamma_5)}, \tag{5.7}$$

or

$$\left. \frac{d \ln \mathcal{J}}{d\tau} \right|_{\tau=0} = -2i\, \mathrm{tr}\, (\overline{\varphi}\, \gamma_5). \tag{5.8}$$

This result should be familiar from our discussion of the axial anomaly in Sect. III–3. There remain two steps, first to calculate the *regularized* representation of $\mathrm{tr}\, (\overline{\varphi}\, \gamma_5)$, and then to integrate with respect to τ.

To regularize the trace, we employ the limiting procedure

$$\mathrm{tr}\, (\overline{\varphi}\gamma_5) = \lim_{\epsilon \to 0} \mathrm{tr}\, (\overline{\varphi}\gamma_5 \exp\left[-\epsilon \mathcal{D}_\tau \mathcal{D}_\tau \right]) \quad (\mathcal{D}_\tau^\mu \equiv \partial^\mu + i\overline{V}_\tau^\mu + i\overline{A}_\tau^\mu \gamma_5), \tag{5.9}$$

with $\overline{A}_\tau^\mu$ and $\overline{V}_\tau^\mu$ as in Eq. (5.3), except now constructed from ξ_τ and $\xi_\tau^\dagger$. For arbitrary τ, we make use of the identities

$$\overline{V}_\tau^{\mu\nu} = \partial^\mu \overline{V}_\tau^\nu - \partial^\nu \overline{V}_\tau^\mu + i\left[\overline{V}_\tau^\mu, \overline{V}_\tau^\nu \right] + i\left[\overline{A}_\tau^\mu, \overline{A}_\tau^\nu \right] = 0,$$

$$\overline{A}_\tau^{\mu\nu} = \partial^\mu \overline{A}_\tau^\nu - \partial^\nu \overline{A}_\tau^\mu + i\left[\overline{V}_\tau^\mu, \overline{A}_\tau^\nu \right] + i\left[\overline{A}_\tau^\mu, \overline{V}_\tau^\nu \right] = 0, \tag{5.10}$$

to express $\mathcal{D}_\tau \mathcal{D}_\tau$ in the form

$$\mathcal{D}_\tau \mathcal{D}_\tau = d_\mu d^\mu + \sigma,$$

$$d_\mu = \partial_\mu + i\overline{V}_{\tau\mu} + \sigma_{\mu\nu} \overline{A}_\tau^\nu \gamma_5 = \partial_\mu + \Gamma_{\tau\mu},$$

$$\sigma = -2\overline{A}_{\tau\mu} \overline{A}_\tau^\mu + i\left[(\partial_\mu + i\overline{V}_{\tau\mu}), \overline{A}_\tau^\mu \right] \gamma_5. \tag{5.11}$$

From the heat-kernel expansion of App. B, we have[8]

$$\mathrm{tr}\, (\overline{\varphi}\gamma_5) \equiv \lim_{\epsilon \to 0} i \int d^4x\, \mathrm{Tr} \left(\frac{\overline{\varphi}\gamma_5}{(4\pi\epsilon)^2} \sum_n \epsilon^n a_n \right)$$

$$= \frac{i}{16\pi^2} \lim_{\epsilon \to 0} \int d^4x\, \mathrm{Tr} \left(\overline{\varphi}\gamma_5 \left[\frac{a_1}{\epsilon} + a_2 + \cdots \right] \right). \tag{5.12}$$

Carrying out the 'Tr' operation, which involves some application of Dirac algebra, yields

---

[8] Note the distinction between 'tr' and 'Tr', as in Eq. (III–3.48).

$$\text{Tr}\,(\gamma_5 \overline{\varphi} a_2) = 2i N_c \,\text{Tr}\,\left(\frac{8}{3}\epsilon_{\mu\nu\alpha\beta}\overline{\varphi}\overline{A}_\tau^\mu\overline{A}_\tau^\nu\overline{A}_\tau^\alpha\overline{A}_\tau^\beta\right) + \cdots ,\tag{5.13}$$

where the ellipses denote contributions not involving $\epsilon_{\mu\nu\alpha\beta}$ and the factor N_c comes from the sum over each fermion color. Combining the above ingredients, we have for the regulated action

$$\Gamma\,(\overline{\varphi}) = -i\ln\mathcal{J} + \cdots$$
$$= \frac{N_c}{4\pi^2}\int_0^1 d\tau \int d^4x \,\text{Tr}\,\left(\frac{8\overline{\varphi}}{3}\epsilon_{\mu\nu\alpha\beta}\overline{A}_\tau^\mu\overline{A}_\tau^\nu\overline{A}_\tau^\alpha\overline{A}_\tau^\beta\right) + \cdots ,\tag{5.14}$$

where we recall that $\overline{\varphi} \equiv \vec{\lambda}\cdot\vec{\varphi}/(2F_\pi)$. This result expresses the effect of the anomaly on the Goldstone bosons.

Unfortunately, there is no simple way to integrate the entire expression of Eq. (5.14) in closed form. In principle, we could represent each of the axial-vector currents therein (e.g. $\overline{A}_\tau^\mu$) as a Taylor series expanded about $\tau = 0$ and perform the integrations to obtain a series of local lagrangians. Alternatively, however, one can simply express Eq. (5.14) as an integral over a five-dimensional space provided we identify τ with a fifth-coordinate x_5 (defined to be timelike). In this case, we use

$$\xi_\tau A_\tau^\mu \xi_\tau^\dagger = -\frac{i}{2}U_\tau\partial^\mu U_\tau^\dagger \equiv -\frac{i}{2}L^\mu,$$
$$\overline{\varphi} = \frac{1}{2}U_\tau\frac{\partial}{\partial\tau}U_\tau^\dagger \equiv \frac{i}{2}L^5,\tag{5.15}$$

plus the cyclic property of the trace to write

$$\Gamma_{WZW}(U) = \frac{iN_c}{240\pi^2}\int d^5x\,\epsilon_{ijklm}\,\text{tr}\,\left(L^i L^j L^k L^l L^m\right),\tag{5.16}$$

where $i,\ldots,m = 5,0,1,2,3$ with $\epsilon^{50123} = +1$. This is Witten's form for the Wess–Zumino anomaly function. The $\tau = 1$ boundary is our physical space-time, and the fifth coordinate is just an integration variable. Since each term in the Taylor expansion can be integrated, the result depends only on the remaining four spacetime variables. Observe that $\Gamma_{WZW}(U)$ vanishes for U in $SU(2)$ due to the properties of Pauli matrices. For chiral $SU(3)$, the process $K^+K^- \to \pi^+\pi^-\pi^0$ is the simplest one described by this action and after expanding Γ_{WZW}, it is given by the lagrangian,

$$\mathcal{L} = \frac{N_c}{240\pi^2 F_\pi^5}\epsilon^{\mu\nu\alpha\beta}\,\text{Tr}\,(\varphi\partial_\mu\varphi\partial_\nu\varphi\partial_\alpha\varphi\partial_\beta\varphi),\tag{5.17}$$

with $\varphi \equiv \lambda\cdot\varphi$.

The above discussion has concerned the impact of the anomaly on the Goldstone modes. We must also determine its proper form in the presence of photons or $W^\pm$ fields. For this purpose, we can obtain the maximal information by generalizing the fermion couplings to include arbitrary left-handed or right-handed currents ℓ_μ, r_μ,

$$\mathcal{L} = \bar{\psi} i \not{D} \psi - M \left(\bar{\psi}_L U \psi_R + \bar{\psi}_R U^\dagger \psi_L \right),$$

$$D_\mu = \partial_\mu + i\ell_\mu \frac{1+\gamma_5}{2} + ir_\mu \frac{1-\gamma_5}{2}. \tag{5.18}$$

The calculation of the jacobian then involves the operator

$$\mathcal{D}_\mu = \partial_\mu + i\bar{\ell}_\mu \frac{1+\gamma_5}{2} + i\bar{r}_\mu \frac{1-\gamma_5}{2},$$

$$\bar{\ell}_\mu = \xi_\tau^\dagger \ell_\mu \xi_\tau - i\xi_\tau^\dagger \partial_\mu \xi_\tau, \qquad \bar{r}_\mu = \xi_\tau r_\mu \xi_\tau^\dagger - i\xi_\tau \partial_\mu \xi_\tau^\dagger, \tag{5.19}$$

which generalizes Eq. (5.3). It is somewhat painful to work out the full answer directly, but fortunately we may invoke Bardeen's result of Eq. (III–3.64) for the general anomaly. Using the identities

$$\bar{\ell}_{\mu\nu} = \xi_\tau^\dagger \ell_{\mu\nu} \xi_\tau, \qquad \bar{v}_{\mu\nu} = \xi_\tau^\dagger \ell_{\mu\nu} \xi_\tau + \xi_\tau r_{\mu\nu} \xi_\tau^\dagger,$$

$$\bar{r}_{\mu\nu} = \xi_\tau r_{\mu\nu} \xi_\tau^\dagger, \qquad \bar{a}_\mu = \left(\bar{\ell}_\mu - \bar{r}_\mu \right)/2, \tag{5.20}$$

where $\ell_{\mu\nu}, r_{\mu\nu}$ are given in Eq. (III–3.65), we obtain

$$\Gamma_{WZW} = -\frac{N_c}{4\pi^2} \int_0^1 d\tau \int d^4x \ \epsilon^{\mu\nu\alpha\beta} \, \mathrm{Tr} \left[\bar{\varphi} \left(-\frac{8}{3} \bar{a}_\mu \bar{a}_\nu \bar{a}_\alpha \bar{a}_\beta \right. \right.$$

$$+ \frac{1}{12} (\bar{\ell}_{\mu\nu} \bar{\ell}_{\alpha\beta} + \bar{r}_{\mu\nu} \bar{r}_{\alpha\beta}) + \frac{1}{24} (\bar{\ell}_{\mu\nu} \bar{r}_{\alpha\beta} + \bar{r}_{\mu\nu} \bar{\ell}_{\alpha\beta})$$

$$\left. \left. - \frac{2i}{3} (\bar{a}_\mu \bar{a}_\nu \bar{v}_{\alpha\beta} + \bar{a}_\mu \bar{v}_{\nu\alpha} \bar{a}_\beta + \bar{v}_{\mu\nu} \bar{a}_\alpha \bar{a}_\beta) \right) \right]. \tag{5.21}$$

Note that the first term corresponds to our previous calculation of Eq. (5.14). The WZW anomaly action contains the full influence of the anomalous low energy couplings of mesons to themselves and to gauge fields. By construction, it is gauge invariant. The τ integration can be explicitly performed for all terms but the first in Eq. (5.21). However, in the general nonabelian case the result is extremely lengthy [PaR 85]. For the simpler but still interesting example of coupling to a photon field A_μ, the result is

$$\Gamma_{WZW}\left(U, A_\mu\right) = \Gamma_{WZW}(U)$$

$$+ \frac{N_c}{48\pi^2} \epsilon^{\mu\nu\alpha\beta} \int d^4x \left[eA_\mu \operatorname{Tr} \left(Q\left(R_\nu R_\alpha R_\beta + L_\nu L_\alpha L_\beta\right)\right) \right.$$

$$\left. - i e^2 F_{\mu\nu} A_\alpha \operatorname{Tr} \left(Q^2 \left(L_\beta + R_\beta\right) + \frac{1}{2}\left(QU^\dagger QUR_\beta + QU\,QU^\dagger L_\beta\right)\right)\right],$$

$$(5.22)$$

where $R_\mu \equiv (\partial_\mu U^\dagger)U$, $L_\mu \equiv U\partial_\mu U^\dagger$.[9]

We have seen here that whereas the anomalous divergence of the axial current represents the response to an infinitesimal anomaly transformation, the WZW lagrangian represents the integration of a series of infinitesimal transformations. In our analysis of the sigma model, the anomaly has forced the occurrence of certain couplings, among them $\pi^0 \to \gamma\gamma$, $\gamma \to 3\pi$ and $K\bar{K} \to 3\pi$. As noted earlier, although these results are based on an instructional model, the result has the same anomaly structure as QCD because the answer must depend on symmetry properties alone. Indeed, such conclusions were originally deduced from anomalous Ward identities [WeZ 71] without any reference to an underlying model. We regard such predictions as among the most profound consequences of the Standard Model.

VII–6 The axial anomaly and $\pi^0 \to \gamma\gamma$

The description of pions and photons presented thus far does not include the decay $\pi^0 \to \gamma\gamma$. This process is important in QCD, because to understand it one must include the anomaly in the axial current. The $\pi^0 \to \gamma\gamma$ amplitude has the general structure

$$\mathcal{M}_{\pi^0 \to \gamma\gamma} = -i A_{\gamma\gamma} \epsilon^{\mu\nu\alpha\beta} \epsilon_\mu^* k_\nu \epsilon_\alpha^{\prime *} k_\beta', \tag{6.1}$$

as required by Lorentz invariance, parity conservation, and gauge invariance, and leads to the decay rate

$$\Gamma_{\pi^0 \to \gamma\gamma} = \frac{m_{\pi^0}^3}{64\pi} \left| A_{\gamma\gamma} \right|^2. \tag{6.2}$$

From the experimental value, $\Gamma = 7.74 \pm 0.37$ eV, we find

$$A_{\gamma\gamma} = 0.0252 \pm 0.0006 \,\mathrm{GeV}^{-1}.$$

We can obtain the lagrangian for $\pi^0 \to \gamma\gamma$ from the WZW action of the previous section, restricting the chiral matrices to $SU(2)$.

---

[9] Witten's original result did not conserve parity, and this was subsequently corrected [PaR 85, KaRS 84].

$$\mathcal{L}_A = \frac{N_c}{48\pi^2}\epsilon^{\mu\nu\alpha\beta}\left[eA_\mu \operatorname{Tr}\left(QL_\nu L_\alpha L_\beta + QR_\nu R_\alpha R_\beta\right) - ie^2 F_{\mu\nu}A_\alpha T_\beta\right], \quad (6.3)$$

with

$$L_\mu \equiv U\partial_\mu U^\dagger, \qquad R_\mu \equiv \partial_\mu U^\dagger U,$$

$$T_\beta = \operatorname{Tr}\left(Q^2 L_\beta + Q^2 R_\beta + \frac{1}{2}QUQU^\dagger L_\beta + \frac{1}{2}QU^\dagger QU R_\beta\right), \quad (6.4)$$

where A_α is the photon field, $F_{\mu\nu}$ is the photon field strength, and $N_c = 3$ is the number of colors. A crucial aspect of this expression is that it has a known coefficient. In this respect, it is unlike other terms in the effective lagrangian, which have free parameters that need to be determined phenomenologically. This is because it is a prediction of the anomaly structure of QCD. A corollary of this is that $\mathcal{L}_A$ must not be renormalized by radiative corrections. This was proven at the quark–gluon level by Adler and Bardeen [AdB 69].

The $\pi^0 \to \gamma\gamma$ amplitude is found by expanding $\mathcal{L}_A$ to first order in the pion field, yielding

$$\mathcal{L}_A = \frac{e^2 N_c}{48\pi^2 F_\pi} 3\operatorname{Tr}\left(Q^2\tau_3\right)\epsilon^{\mu\nu\alpha\beta}F_{\mu\nu}A_\alpha\partial_\beta\pi^0 = \frac{\alpha N_c}{24\pi F_\pi}\epsilon^{\mu\nu\alpha\beta}F_{\mu\nu}F_{\alpha\beta}\pi^0, \quad (6.5)$$

where we have integrated by parts in the second line. This produces a $\pi^0 \to \gamma\gamma$ matrix element of the form

$$A_{\gamma\gamma} = \frac{\alpha N_c}{3\pi F_\pi} \xrightarrow[N_c=3]{} 0.0251\,\text{GeV}^{-1}, \quad (6.6)$$

in excellent agreement with the experimental value. This is widely recognized as an important test of QCD, both as a measurement of the number of colors and also as a reflection of the symmetries and anomalies of the theory. It is a remarkable result.

What would have happened if the axial anomaly were not present? The decay $\pi^0 \to \gamma\gamma$ could still occur, but it would be suppressed. The $\pi^0 \to \gamma\gamma$ transition must be at least of order E^4, as it must involve the dimension-four operator $F\tilde{F}$. The anomaly occurs at this order. However, non-anomalous lagrangians leading to this transition can be constructed at order E^6. This result was first derived by Sutherland and Veltman using a soft-pion technique [Su 67, Ve 67].

At what level would we expect corrections to the anomaly prediction for $\pi^0 \to \gamma\gamma$? It has been checked that $m_\pi^2 \ln m_\pi^2$ corrections, which in principle can occur when meson loops are present, do not in fact modify the lowest-order result when it is expressed in terms of the physical decay constant F_π. However, there are still corrections of order m_π^2/Λ^2, where Λ is the scale in the energy expansion, which amount to modifications of order 2% [GoBH 02].

Problems

(1) **Radiative corrections and $\pi_{\ell 2}$ decay**

To bring $\Gamma_{\pi \to e \nu_e} / \Gamma_{\pi \to \mu \nu_\mu}$ into agreement with experiment requires a radiative correction whose dominant contribution is the so-called *seagull* component $(\sqrt{2} F_\pi g_{\mu\nu})$ of $M_{\mu\nu}$ (cf. Eq. (3.22)).

(a) Verify that gauge invariance requires

$$iq^\mu M_{\mu\nu}(p, q) = \langle 0 | A_\nu^{1-i2} | \pi^+(\mathbf{p}) \rangle,$$

and show that the seagull term is required in this regard to cancel the pion pole contribution.

(b) Use the seagull term in Feynman gauge to calculate the radiative correction to $\pi_{\ell 2}$ decay. Introduce a photon cut-off via

$$\frac{1}{k^2} \to \frac{1}{k^2} \frac{-\Lambda^2}{k^2 - \Lambda^2}$$

so that

$$\mathcal{M}^{\text{rad}} = i e^2 \frac{G_F}{\sqrt{2}} V_{\text{ud}} \int \frac{d^4 k}{(2\pi)^4} \frac{1}{k^2} \frac{-\Lambda^2}{k^2 - \Lambda^2} \sqrt{2} F_\pi g^{\mu\lambda}$$

$$\times \bar{u}(p_\nu) \gamma_\lambda (1 + \gamma_5) \frac{1}{-\not{p}_e + \not{k} - m_e} \gamma_\mu v(p_e),$$

and show that

$$\mathcal{M}^{(0)} \to \mathcal{M}^{(0)} \left(1 - \frac{3\alpha}{2\pi} \ln \frac{\Lambda}{m_\ell} \right),$$

where

$$\mathcal{M}^{(0)} = i \frac{G_F}{\sqrt{2}} V_{\text{ud}} \sqrt{2} F_\pi \, p_\lambda \bar{u}(p_\nu) \gamma_\lambda (1 + \gamma_5) v(p_e)$$

$$= -i \frac{G_F}{\sqrt{2}} V_{\text{ud}} \sqrt{2} F_\pi m_\ell \bar{u}(p_\nu)(1 - \gamma_5) v(p_e)$$

is the lowest order amplitude for the $\pi_{\ell 2}$ process. This then is the origin of the lepton-mass-dependent radiative correction.

(2) **Radiative pion decay and the anomaly**

Writing the π^0 decay amplitude as

$$\mathcal{M}_{\pi^0 \to \gamma\gamma} = -i e^2 \epsilon_1^{\mu*} \epsilon_2^{\nu*} \int d^4 x \, e^{i q_1 \cdot x} \langle 0 | T(V_\mu^{\text{em}}(x) V_\nu^{\text{em}}(0)) | \pi^0(\mathbf{p}) \rangle$$

$$\equiv -i \epsilon_1^{\mu*} \epsilon_2^{\nu*} \epsilon_{\mu\nu\alpha\beta} q_1^\alpha p^\beta A_{\gamma\gamma},$$

and the vector current amplitude in radiative π^+ decay as

$$\mathcal{M}^{(V)}_{\pi^+ \to \ell^+ \nu_\ell \gamma} = -ie\epsilon_1^{\mu*} \int d^4x\, e^{iq_1 \cdot x} \langle 0|T(V_\mu^{em}(x)V_\nu^{1-i2}(0))|\pi^+(\mathbf{p})\rangle$$

$$\equiv -ie\epsilon_1^{\mu*}\epsilon_2^\nu \epsilon_{\mu\nu\alpha\beta} q_1^\alpha p^\beta h_V,$$

demonstrate that isotopic spin invariance requires $\sqrt{2}h_V = A_{\gamma\gamma}$.

(3) **Unitarity and the pion form factor**

(a) Verify that the pion form factor given in Eq. (3.13) obeys the strictures of unitarity, i.e.,

$$2\,\mathrm{Im}\,G_\pi(q^2)(p_1 - p_2)_\mu = \int \frac{d^3q_1 d^3q_2}{(2\pi)^6 2q_1^0 2q_2^0}$$

$$\times (2\pi)^4 \delta^4(p_1 + p_2 - q_1 - q_2)(q_1 - q_2)_\mu \langle \pi^+(\mathbf{q}_1)\pi^-(\mathbf{q}_2)|\pi^+(\mathbf{p}_1)\pi^-(\mathbf{2}_2)\rangle$$

where the matrix element is the two-derivative (tree-level) pion–pion scattering amplitude given in Eq. (3.4).

(b) How does this result change if the K^+K^- intermediate state is added to $\pi^+\pi^-$?

(4) **Other worlds**

Describe changes in the *macroscopic* world if the quark masses were slightly different in the following ways:

(a) $m_u = m_d = 0$,

(b) $m_u > m_d$,

(c) $m_u = 0$, $m_d = m_s$.

VIII

Weak interactions of kaons

The kaon is the lightest hadron having a nonzero strangeness quantum number. It is unstable and decays weakly into states with zero strangeness, containing pions, photons, and/or leptons. We shall consider decays in the leptonic, semileptonic, and hadronic sectors to illustrate aspects of both weak and strong interactions.

VIII–1 Leptonic and semileptonic processes

Leptonic decay

The simplest weak decay of the charged kaon, denoted by the symbol $K_{\ell 2}$, is into purely leptonic channels $K^+ \to \mu^+ \nu_\mu$, $K^+ \to e^+ \nu_e$. Such decays are characterized by the constant F_K,

$$\langle 0 | \bar{s} \gamma_\mu \gamma_5 u | K^+(\mathbf{k}) \rangle = i\sqrt{2} F_K k_\mu. \tag{1.1}$$

As discussed previously, because of $SU(3)$ breaking F_K is about 20% larger than the corresponding pion decay constant F_π. As with the pion, but even more so because of the larger kaon mass, helicity arguments require strong suppression of the electron mode relative to that of the muon. The ratio of $e^+ \nu_e$ to $\mu^+ \nu_\mu$ decay rates, as in pion decay, provides a test of lepton universality [RPP 12],

$$\frac{\Gamma^*_{K^+ \to e^+ \nu_e}}{\Gamma_{K^+ \to \mu^+ \nu_\mu}} \bigg|_{\text{expt}} = (2.488 \pm 0.012) \times 10^{-5}, \tag{1.2}$$

in good agreement with the suppression predicted theoretically [CiR 07],

$$\frac{\Gamma'_{K^+ \to e^+ \nu_e}}{\Gamma_{K^+ \to \mu^+ \nu_\mu}} = \frac{m_e^2}{m_\mu^2} \left[\frac{1 - m_e^2/m_K^2}{1 - m_\mu^2/m_K^2} \right]^2 (1 + \delta) = (2.477 \pm 0.001) \times 10^{-5}, \tag{1.3}$$

237

where $\delta = -0.04$ is the electromagnetic radiative correction including the bremsstrahlung component.[1] The notation Γ' indicates that the experimenters have subtracted off the large structure-dependent components of $K^+ \to \ell^+ \nu_\ell \gamma$ but have included the small bremsstrahlung component.

Kaon beta decay and V_{us}

The kaon beta decay reactions $K^+ \to \pi^0 \ell^+ \nu_\ell$ and $K^0 \to \pi^- \ell^+ \nu_\ell$, called $K^+_{\ell 3}$ and $K^0_{\ell 3}$ respectively, also are important in Standard Model physics. They are each parameterized by two form factors,

$$\langle \pi^-(\mathbf{p}) \left| \bar{s}\gamma_\mu u \right| K^0(\mathbf{k}) \rangle = f^{K^0 \pi^-}_+ (q^2) (k+p)_\mu + f^{K^0 \pi^-}_- (q^2) (k-p)_\mu \,,$$

$$\langle \pi^0(\mathbf{p}) \left| \bar{s}\gamma_\mu u \right| K^+(\mathbf{k}) \rangle = \left[\frac{f^{K^+ \pi^0}_+(q^2)}{\sqrt{2}} (k+p)_\mu + \frac{f^{K^+ \pi^0}_-(q^2)}{\sqrt{2}} (k-p)_\mu \right]. \quad (1.4)$$

Isospin invariance implies $f^{K^0 \pi^-}_\pm = f^{K^+ \pi^0}_\pm \equiv f_\pm$. $SU(3)$ symmetry can be invoked to relate these matrix elements to the strangeness-conserving transition $\pi^+ \to \pi^0 \ell^+ \nu_\ell$, resulting in $f_+(0) = -1$ and $f_-(0) = 0$. The deviation of $f_+(0)$ from unity is predicted to be second order in $SU(3)$ symmetry breaking, i.e., of order $(m_s - \hat{m})^2$. This result, the Ademollo–Gatto [AdG 64] theorem, is proved by considering the commutation of quark vector charges,

$$[Q^{\bar{u}s}, Q^{\bar{s}u}] = Q^{\bar{u}u - \bar{s}s}, \quad (1.5)$$

where

$$Q^{\bar{i}j} \equiv \int d^3x \, \bar{q}^i(x)\gamma_0 q^j(x). \quad (1.6)$$

Taking matrix elements and inserting a complete set of intermediate states gives

$$1 = \sum_n \left(\left| \langle n \left| Q^{\bar{s}u} \right| K^0 \rangle \right|^2 - \left| \langle n \left| Q^{\bar{u}s} \right| K^0 \rangle \right|^2 \right). \quad (1.7)$$

Finally, we isolate the single π^- state from the sum and note that in the $SU(3)$ limit the charge operator can only connect the kaon to another state within the *same* $SU(3)$ multiplet. This implies[2]

$$\langle n \neq \pi^- \left| Q^{\bar{u}s} \right| K^0 \rangle = \mathcal{O}(\epsilon), \quad (1.8)$$

---

[1] The dominant term here is the simple contact contribution $-3(\alpha/\pi) \ln(m_\mu/m_e)$ discussed in Sect. VII–1.
[2] This is easiest to obtain in the limit $\mathbf{p}_K \to \infty$.

where ϵ is a measure of $SU(3)$ breaking, and we thus conclude that

$$1 - \left[f_+^{K^0\pi^-}(0) \right]^2 = \mathcal{O}(\epsilon^2), \tag{1.9}$$

which is the result we were seeking.

It is interesting that the $SU(2)$ mass difference $m_u \neq m_d$ can modify $f_+^{K^+\pi^0}(0)$ in first order despite the Ademollo–Gatto theorem. This can be seen by considering a K^+ in the formulae of Eq. (1.7). Now there exist two intermediate states in the same octet as the kaon, i.e. π^0 and η^0, and it is their *sum* which obeys the Ademollo–Gatto theorem,

$$\frac{1}{4} \left| f_+^{K^+\pi^0}(0) \right|^2 + \frac{3}{4} \left| f_+^{K^+\eta^0}(0) \right|^2 = 1 + \mathcal{O}(\epsilon^2). \tag{1.10}$$

In the isospin limit, each term must separately obey the theorem because of the isospin relation $f_+^{K^+\pi^0} = f_+^{K^0\pi^-}$. However, when $m_u \neq m_d$ each form factor in Eq. (1.10) can separately deviate from unity to first order in $m_u - m_d$ as long as the first order effect cancels in the sum. Indeed this is what happens, yielding (cf. Prob. VIII.1)

$$\frac{f_+^{K^+\pi^0}(0)}{f_+^{K^0\pi^-}(0)} = 1 + \frac{3}{4} \left(\frac{m_d - m_u}{m_s - \hat{m}} \right) + \ell_{K\pi} \simeq 1.021, \tag{1.11}$$

where $\ell_{K\pi} = 0.004$ arises from chiral corrections at $\mathcal{O}(E^4)$ [GaL 85b 85b]. This number can also be easily extracted from experiment by using the ratio of K^+ and K^0 beta decay rates, with the result [CiR 07],

$$\frac{f_+^{K^+\pi^0}(0)}{f_+^{K^0\pi^-}(0)} = 1.027 \pm 0.004, \tag{1.12}$$

in agreement with the prediction.

The prime importance of the $K_{\ell 3}$ process is that it provides the best determination of the weak mixing element V_{us}. Because of the Ademollo-Gatto theorem, the reaction is protected from large symmetry breaking corrections. In addition, the use of chiral perturbation theory allows a reliable treatment of the reaction. The above study of the form factors indicates that the theory is under control within the limits of experimental precision. The value [LeR 84]

$$V_{us} = 0.2253 \pm 0.0013 \tag{1.13}$$

follows from an analysis of the K^0 and K^+ decay rates.

VIII–2 The nonleptonic weak interaction

For leptonic and semileptonic processes, at most one hadronic current is involved. There exist also *nonleptonic* interactions, in which two hadronic charged weak currents are coupled by the exchange of $W^\pm$ gauge bosons,

$$\mathcal{H}_{nl} = \frac{g_2^2}{8} \int d^4x \, D_F^{\mu\nu}(x, M_W) T \left(J_\mu^{\dagger\text{had}}(x/2) \, J_\nu^{\text{had}}(-x/2) \right),$$

$$J_\mu^{\text{had}} = \left(\bar{u} \ \bar{c} \ \bar{t} \right) \mathbf{V} \gamma_\mu (1 + \gamma_5) \begin{pmatrix} d \\ s \\ b \end{pmatrix}, \tag{2.1}$$

with $\mathbf{V}$ being the CKM matrix, given in Eq. (II–4.17). Such interactions are difficult to analyze theoretically because the product of two hadronic currents is a complicated operator. If one imagines inserting a complete set of intermediate states between the currents, all states from zero energy to M_W are important, and the product is singular at short distances. Thus, one needs to have theoretical control over the physics of low-, intermediate-, and high-energy scales in order to make reliable predictions. Because this is not the case at present, our predictive power is substantially limited.

Let us first consider the particular case of $\Delta S = 1$ nonleptonic decays. These are governed by the products of currents

$$\bar{d}\Gamma^\mu u \, \bar{u}\Gamma_\mu s, \qquad \bar{d}\Gamma^\mu c \, \bar{c}\Gamma_\mu s, \qquad \bar{d}\Gamma^\mu t \, \bar{t}\Gamma_\mu s, \tag{2.2}$$

where $\Gamma_\mu \equiv \gamma_\mu(1 + \gamma_5)$ and color labels are suppressed. The first of these would naively be expected to be the most important, because kaons and pions predominantly contain u, d, s quarks. However, the others also contribute through virtual effects. Some properties of the $\Delta S = 1$ nonleptonic interactions can be read off from these currents. The first product contains two flavor-$SU(3)$ octet currents, one carrying $I = 1/2$ and one carrying $I = 1$,

$$SU(3): \qquad (\mathbf{8} \otimes \mathbf{8})_{\text{symm}} = \mathbf{8} \oplus \mathbf{27}, \tag{2.3a}$$

$$\text{isospin}: \qquad 1 \otimes \frac{1}{2} = \frac{1}{2} \oplus \frac{3}{2}, \tag{2.3b}$$

where the symmetric product is taken because the two currents are members of the same octet. The singlet $SU(3)$ representation is excluded from Eq. (2.3a) because a $\Delta S = 1$ interaction changes the $SU(3)$ quantum numbers and hence cannot be an $SU(3)$ singlet. The other two products in Eq. (2.2) are purely $SU(3)$ octet and isospin one-half operators. The currents are also purely left-handed. Thus, the nonleptonic hamiltonian transforms under separate left-handed and right-handed chiral rotations as $(\mathbf{8}_L, \mathbf{1}_R)$ and $(\mathbf{27}_L, \mathbf{1}_R)$. These symmetry properties, valid regardless of the dynamical difficulties occurring in nonleptonic decay, allow one to write down effective chiral lagrangians for the nonleptonic kaon decays. The hamiltonian is a Lorentz scalar, charge neutral, $\Delta S = 1$ operator, and has the above specified chiral properties.

At order E^2, there exist two possible effective lagrangians for the octet part, viz., $\mathcal{L}_{\text{octet}} = \mathcal{L}_8 + \overline{\mathcal{L}}_8$, where in the notation of Sect. IV–6,

$$\mathcal{L}_8 = g_8 \operatorname{Tr} \left(\lambda_6 D_\mu U D^\mu U^\dagger \right), \qquad \overline{\mathcal{L}}_8 = \bar{g}_8 \operatorname{Tr} \left(\lambda_6 \chi U^\dagger \right) + \text{h.c.} \qquad (2.4)$$

It can easily be checked that both $\mathcal{L}_8$ and $\overline{\mathcal{L}}_8$ are singlets under right-handed transformations, but transform as members of an octet for the left-handed transformations. The barred lagrangian in Eq. (2.4) can in fact be removed, so that it does not contribute to physical processes. This is seen in two ways. At the simplest level, direct calculation of $K \to 2\pi$ and $K \to 3\pi$ amplitudes using $\overline{\mathcal{L}}_8$, including all diagrams, yields a vanishing contribution. Alternatively, this can be understood by noting that in QCD the quantity χ appearing in Eq. (2.4) is proportional to the quark mass matrix, $\chi = 2B_0 m_q$. Thus, the effect of $\overline{\mathcal{L}}_8$ is equivalent to a modification of the mass matrix,

$$m_q \to m_q' = m_q + \bar{g}_8 \lambda_6 m_q. \qquad (2.5)$$

This new mass matrix can be diagonalized by a chiral rotation

$$\operatorname{Tr} \left(m_q' U \right) \to \operatorname{Tr} \left(R m_q' L U \right) \equiv \operatorname{Tr} \left(m_D U \right), \qquad (2.6)$$

with m_D diagonal. The transformed theory clearly has conserved quantum numbers, as it is flavor diagonal. This means that the original theory also has conserved quantum numbers, one of which can be called strangeness. When particles are mass eigenstates, even in the presence of $\overline{\mathcal{L}}_8$, the kaon state does not decay. Hence, this $\overline{\mathcal{L}}_8$ can be discarded from considerations, leaving only $\mathcal{L}_8$ as responsible for octet K decays [Cr 67]. This octet operator is necessarily $\Delta I = 1/2$ in character. Another allowed operator, transforming as $(27_L, 1_R)$, contains both $\Delta I = 1/2$ and $\Delta I = 3/2$ portions,

$$\mathcal{L}_{27} = \mathcal{L}_{27}^{(1/2)} + \mathcal{L}_{27}^{(3/2)}, \qquad (2.7)$$

where

$$\mathcal{L}_{27}^{(1/2)} = g_{27}^{(1/2)} C_{ab}^{1/2} \operatorname{Tr} \left(\lambda^a \partial_\mu U U^\dagger \lambda^b \partial_\mu U U^\dagger \right) + \text{h.c.}, \qquad (2.8a)$$

$$\mathcal{L}_{27}^{(3/2)} = g_{27}^{(3/2)} C_{ab}^{3/2} \operatorname{Tr} \left(\lambda^a \partial_\mu U U^\dagger \lambda^b \partial_\mu U U^\dagger \right) + \text{h.c.} \qquad (2.8b)$$

The coefficients are given by

$$C^{1/2}_{6+i7/2,\,3} = 1, \qquad C^{1/2}_{4+i5/2,\,1-i2/2} = -\sqrt{2}, \qquad C^{1/2}_{6+i7/2,\,8} = -\frac{3\sqrt{3}}{2},$$

$$C^{3/2}_{6+i7/2,\,3} = 1, \qquad C^{3/2}_{4+i5/2,\,1-i2/2} = \frac{1}{\sqrt{2}}. \tag{2.9}$$

The complete classification at order E^4 is difficult, but has been obtained [KaMW 90]. We shall apply these lagrangians to the data in Sects. VIII–4, XII–6. There we shall see that $g_8 \gg g^i_{27}$, whereas naive expectations would have octet and 27-plet amplitudes being of comparable strength. This is part of the puzzle of the $\Delta I = 1/2$ rule. The reliable theoretical calculation of the nonleptonic decay amplitudes, which is tantamount to predicting the quantities g_8, $g^{(1/2)}_{27}$ and $g^{(3/2)}_{27}$, is one of the difficult problems mentioned earlier. It has not yet been convincingly accomplished. The best we can do is to describe the theoretical framework of the short distance expansion, to which we now turn.

VIII–3 Matching to *QCD* at short distance

At short distances, the asymptotic freedom property of *QCD* allows a perturbative treatment of the product of currents. The philosophy is to use perturbative *QCD* to treat the strong interactions for energies $M_W \geq E \geq \mu$. The result is an effective lagrangian which depends on the scale μ. Ultimately matrix elements must be taken which include the strong interaction below energy scale μ and the final result should be independent of μ. The subject provides a classic example of the techniques of perturbative matching to effective lagrangians and the use of the renormalization group.

Short-distance operator basis

As introduced in Sect. IV–7, the outcome of the short-distance calculation can be expressed as an effective nonleptonic hamiltonian expanded in a set of local operators with scale-dependent coefficients (*Wilson coefficients*) [Wi 69],

$$\mathcal{H}^{\Delta S=1}_{\rm nl} = \frac{G_F}{2\sqrt{2}} V^*_{\rm ud} V_{\rm us} \sum_i \mathcal{C}_i(\mu)\mathcal{O}_i. \tag{3.1}$$

As in any effective lagrangian, those operators of lowest dimension should be dominant. If the operator $\mathcal{O}_i$ has dimension d, its Wilson coefficient obeys the scaling property $\mathcal{C}_i \sim M_W^{6-d}$. Let us first see how this hamiltonian is generated in perturbation theory. We can later use the renormalization group to sum the

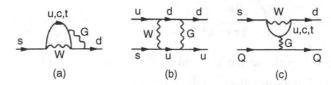

Fig. VIII–1 QCD Radiative corrections to the $\Delta S = 1$ nonleptonic hamiltonian.

leading logarithmic contributions. The lowest-order diagrams renormalizing the current product appear in Fig. VIII–1.

The process in Fig. VIII–1(a) corresponds to a left-handed, gauge-invariant operator of dimension 4,

$$O^{(d=4)} = \bar{d} \not{D} (1 + \gamma_5) s. \tag{3.2}$$

This operator can be removed from consideration by a redefinition of the quark fields (cf. Prob. IV–1). The remaining operators are of dimension 6. Simple W exchange with no gluonic corrections gives rise in the short-distance expansion to the local operator

$$O_A \equiv \bar{d} \gamma_\mu (1 + \gamma_5) u \, \bar{u} \gamma^\mu (1 + \gamma_5) s, \tag{3.3}$$

with a coefficient $C_A = 2$ in the normalization of Eq. (3.1). The gluonic correction of Fig. VIII–1(b) generates an operator of the form

$$\bar{d} \gamma_\mu (1 + \gamma_5) \lambda^a u \, \bar{u} \gamma^\mu (1 + \gamma_5) \lambda^a s, \tag{3.4}$$

where the $\{\lambda^a\}$ are color $SU(3)$ matrices. However, use of the Fierz rearrangement property (see App. C) and the completeness property Eq. (II–2.8) of $SU(3)$ matrices allow this to be rewritten in color-singlet form

$$\bar{d} \gamma_\mu (1 + \gamma_5) \lambda^a u \, \bar{u} \gamma^\mu (1 + \gamma_5) \lambda^a s = -\frac{2}{3} O_A + 2 O_B,$$

where

$$O_B \equiv \bar{u} \gamma_\mu (1 + \gamma_5) u \bar{d} \gamma^\mu (1 + \gamma_5) s. \tag{3.5}$$

The strong radiative correction is seen to generate a new operator O_B.

Perturbative analysis

Consider now the one-loop renormalizations of the four-fermion interaction Fig. VIII–1(b). In calculating Feynman diagrams we typically encounter integrals such as (neglecting quark masses)

$$I(\mu) = \int \frac{d^4k}{(2\pi)^4} \frac{1}{k^4} \frac{1}{k^2 - M_W^2} = -\frac{i}{16\pi^2 M_W^2} \ln \frac{\kappa^2}{\kappa^2 + M_W^2}\bigg|_\mu^\infty, \qquad (3.6)$$

where we evaluate the integral at the lower end using a scale μ. Clearly, M_W presents a natural cut-off in the sense that

$$I(\mu) \simeq \frac{-i}{8\pi^2 M_W^2} \begin{cases} 0 & (\mu \sim M_W), \\ \ln \mu/M_W & (\mu \ll M_W). \end{cases} \qquad (3.7)$$

The modification of the matrix element to first order in *QCD* is then

$$O_A \to O_A - \frac{g_3^2}{16\pi^2} \ln \left(\frac{M_W^2}{\mu^2} \right) (3O_B - O_A), \qquad (3.8)$$

where g_3 is the quark–gluon coupling strength. The gluonic correction to O_B must also be examined, and a similar analysis yields

$$O_B \to O_B - \frac{g_3^2}{16\pi^2} \ln \left(\frac{M_W^2}{\mu^2} \right) (3O_A - O_B). \qquad (3.9)$$

We observe that the operators,

$$O_\pm = \frac{1}{2}(O_A \pm O_B), \qquad (3.10)$$

are form-invariant, $O_\pm \to C_\pm O_\pm$, with coefficients $C_\pm$,

$$C_\pm = 1 + d_\pm \frac{g_3^2}{16\pi^2} \ln \frac{M_W^2}{\mu^2}, \qquad (3.11)$$

where $d_+ = -2$ and $d_- = +4$. The isospin content of the various operators can be determined in various ways. Perhaps the easiest method involves the use of raising and lowering operators [Ca 66],

$$I_+ d = u, \quad I_+ \bar{u} = -\bar{d}, \quad I_- u = d, \quad I_- \bar{d} = -\bar{u}, \qquad (3.12)$$

to show that $I_+ O_- = 0$, implying that O_- is the $I_z = 1/2$ member of an isospin doublet. With repeated use of raising and lowering operators, one can demonstrate that O_- is purely $\Delta I = 1/2$ whereas O_+ is a combination of $\Delta I = 1/2$ and $\Delta I = 3/2$ operators.

From Eq. (3.11), we see that under one-loop corrections the operator O_- is enhanced by the factor

$$C_- = 1 + 4\frac{g_3^2}{16\pi^2} \ln \frac{M_W^2}{\mu^2} \simeq 2.1, \qquad (3.13)$$

where we use $\alpha_s(\mu) \simeq 0.4$ ($\Lambda_{QCD} \simeq 0.2$ GeV) at $\mu \simeq 1$ GeV. Similarly O_+ is accompanied by the suppression factor

$$C_+ = 1 - 2\frac{g_3^2}{16\pi^2} \ln \frac{M_W^2}{\mu^2} \simeq 0.4. \tag{3.14}$$

Renormalization-group analysis

Choosing an even smaller value of μ would lead to an even larger correction. However, maintaining just the lowest-order perturbation in the *QCD* interaction would then be unjustified. It is possible to do better than the lowest-order perturbative estimate by using the renormalization group to sum the logarithmic factors [GaL 74, AlM 74]. In a renormalizable theory physically measurable quantities can be written as functions of couplings which are renormalized at a renormalization scale μ_R. Physical quantities calculated in the theory must be independent of μ_R. Denoting an arbitrary physical quantity by Q, this may be written

$$Q = f\left(g_3(\mu_R), \mu_R\right), \tag{3.15}$$

where f is some function of μ_R and g_3 is the strong coupling constant of QCD. Differentiating with respect to μ_R, we have

$$\mu_R \frac{d}{d\mu_R} f\left(g_3(\mu_R), \mu_R\right) = 0, \tag{3.16}$$

which is the renormalization-group equation. It represents the feature that a change in the renormalization scale must be compensated by a modification of the coupling constants, leaving physical quantities invariant. In order to see how this program can be carried out for the effective weak hamiltonian, consider the following irreducible vertex function which represents a typical weak nonleptonic matrix element,

$$\left\langle 0 \left| T\left(J_\lambda^{had^\dagger}(x) J_{had}^\lambda(0) q_1(p_1) q_2(p_2) \bar{q}_3(p_3) \bar{q}_4(p_4)\right) \right| 0 \right\rangle_{ren}^{irr}$$
$$= \left(\sqrt{Z_2}\right)^4 \left\langle 0 \left| T\left(J_\lambda^{had^\dagger}(x) J_{had}^\lambda(0) q_1(p_1) q_2(p_2) \bar{q}_3(p_3) \bar{q}_4(p_4)\right) \right| 0 \right\rangle_{unren}^{irr}, \tag{3.17}$$

where the $\{q_i\}$ are quark fields carrying momenta $\{p_i\}$. Z_2 is the quark wavefunction renormalization constant for the fermion field, and subscripts 'ren', 'unren' denote renormalized and unrenormalized quantities.

Choosing the subtraction point $p_i^2 = -\mu_R^2$, we require that unrenormalized quantities be independent of μ_R,

$$\mu_R \frac{d}{d\mu_R} \left\langle 0 \left| T\left(J_\lambda^\dagger J^\lambda q_1 q_2 \bar{q}_3 \bar{q}_4\right) \right| 0 \right\rangle_{unren}^{irr} = 0. \tag{3.18}$$

This implies

$$\left(\mu_R \frac{\partial}{\partial \mu_R} + \beta_{QCD} \frac{\partial}{\partial g_{3r}} - 4\gamma_F\right) \left\langle 0 \left| T \left(J_\lambda^\dagger J^\lambda q_1 q_2 \bar{q}_3 \bar{q}_4\right)\right| 0\right\rangle_{\text{ren}}^{\text{irr}} = 0,$$

where g_{3r} is the renormalized strong coupling constant, β_{QCD} is the QCD beta function of Eq. (II–2.57(b)) and γ_F is the quark field anomalous dimension of Eq. (II–2.69). As we have seen, QCD radiative corrections generally mix the local operators appearing in the short-distance expansion,

$$\langle 0 | T (O_n q_1 q_2 \bar{q}_3 \bar{q}_4) | 0\rangle_{\text{ren}}^{\text{irr}} = \sum_{n'} X_{nn'} \langle 0 | T (O_{n'} q_1 q_2 \bar{q}_3 \bar{q}_4) | 0\rangle_{\text{unren}}^{\text{irr}}, \qquad (3.19)$$

and the mixing matrix can be diagonalized to obtain a set of multiplicatively renormalized operators

$$\langle 0 | T (O_k q_1 q_2 \bar{q}_3 \bar{q}_4) | 0\rangle_{\text{ren}}^{\text{irr}} = Z_k \langle 0 | T (O_k q_1 q_2 \bar{q}_3 \bar{q}_4) | 0\rangle_{\text{unren}}^{\text{irr}}. \qquad (3.20)$$

If operator O_k has anomalous dimension γ_k, we can write

$$Z_k \sim 1 + \gamma_k \ln \mu_R + \cdots, \qquad (3.21)$$

and so the coefficient functions $C_k(\mu_R x)$ satisfy

$$\left(\mu_R \frac{\partial}{\partial \mu_R} + \beta_{QCD} \frac{\partial}{\partial g_{3r}} + \gamma_k - 4\gamma_F\right) C_k(\mu_R x) = 0. \qquad (3.22)$$

From the above, we have for the operators $O_\pm$

$$\gamma_\pm - 4\gamma_F \rightarrow \frac{g_3^2}{16\pi^2} d_\pm. \qquad (3.23)$$

We can solve Eq. (3.22) with methods analogous to those employed in Sect. II–2. That is, because QCD is asymptotically free and we are working at large momentum scales, we can use the perturbative result (cf. Eq. (II–2.57(b))),

$$\beta_{QCD}(g_{3r}) = \mu_R \frac{\partial g_{3r}}{\partial \mu_R} = -\frac{g_{3r}^3}{16\pi^2} b + \cdots, \qquad (3.24)$$

where $b = 11 - \frac{2}{3} n_f$, n_f being the number of quark flavors. Upon inserting the leading term in the perturbative expression for α_s (cf. Eq. (II–2.74)),

$$\alpha_s(\mu_R) = \frac{12\pi}{(33 - 2n_f) \ln \mu_R^2 / \Lambda^2}, \qquad (3.25)$$

one can verify that the solution to Eq. (3.22) is given by

$$\frac{\mathcal{C}_\pm(\mu_R)}{\mathcal{C}_\pm(M_W)} = \left(1 + \frac{g_3^2}{16\pi^2}b\ln\frac{M_W^2}{\mu_R^2}\right)^{d_\pm/b}. \qquad (3.26)$$

Note that in the perturbative regime where $\alpha_s \ll 1$, we have

$$\frac{\mathcal{C}_\pm(\mu_R)}{\mathcal{C}_\pm(M_W)} = 1 + d_\pm\frac{g_3^2}{16\pi^2}\ln\frac{M_W^2}{\mu_R^2}, \qquad (3.27)$$

which agrees with our previous result, Eq. (3.11). It is the renormalization group which has allowed us to sum all the 'leading logs'. Of course, at scale M_W one must be able to reproduce the original weak hamiltonian, implying $\mathcal{C}_+(M_W) = \mathcal{C}_-(M_W) = 1$. Taking $\mu_R \simeq 1$ GeV and $\alpha_s = 0.4$ as before, we find

$$\mathcal{H}_{\text{nl}}^{\Delta S=1}(\mu_R) \propto \mathcal{C}_+(\mu_R)O_+ + \mathcal{C}_-(\mu_R)O_-, \qquad (3.28)$$

with

$$\mathcal{C}_-(\mu_R) \simeq 1.5, \qquad \mathcal{C}_+(\mu_R) \simeq 0.8. \qquad (3.29)$$

We observe then a $\Delta I = 1/2$ enhancement of a factor of two or so, which is encouraging but still considerably smaller than the experimental value of $A_0/A_2 \sim 22$ discussed in the next section.

Two additions to the above analysis must now be addressed. One is the proper treatment of heavy-quark thresholds. In reducing the energy scale from M_W down to μ_R, one passes through regions where there are successively six, five, four, or three light quarks, the word 'light' meaning relative to the energy scale μ_R. The beta function changes slightly from region to region. A proper treatment must apply the renormalization group scheme in each sector separately. This is a straightforward generalization of the procedures described above.

The other addition is the inclusion of *penguin* diagrams of Fig. VIII–1(c) [ShVZ 77, ShVZ 79b, BiW 84], whimsically named because of a rough resemblance to this antarctic creature. The gluonic penguin is noteworthy because it is purely $\Delta I = 1/2$, thus helping to build a larger $\Delta I = 1/2$ amplitude, and because it is the main source of CP violation in the $\Delta S = 1$ hamiltonian. The electroweak penguin, wherein the gluon is replaced by a photon or a Z^0 boson, also enters the theory of CP violation. The CP-conserving portion of the penguin diagrams involves a GIM cancelation between the c, u quarks and hence enters significantly at scales below the charmed quark mass. On the other hand, in the CP violating component, the GIM cancelation is between the t, c quarks and thus this piece is short-distance dominated. At lowest order, before renormalization-group enhancement, one obtains the following Hamiltonians for the penguin and electroweak penguin interactions (cf. Fig. VIII–2),

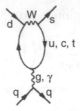

Fig. VIII–2 Penguin diagram.

$$\mathcal{H}_w^{(peng)} = -\frac{G_F \alpha_s}{12\pi\sqrt{2}} \left[V_{ud}^* V_{us} \ln \frac{m_c^2}{\mu_R^2} + V_{td}^* V_{ts} \ln \frac{m_t^2}{m_c^2} \right] \bar{d}\gamma_\mu (1+\gamma_5)\lambda^a s \bar{q}\gamma^\mu \lambda^a q,$$

$$\mathcal{H}_w^{(ewp)} = -\frac{2G_F \alpha}{9\pi\sqrt{2}} \left[V_{ud}^* V_{us} \ln \frac{m_c^2}{\mu_R^2} + V_{td}^* V_{ts} \ln \frac{m_t^2}{m_c^2} \right] \bar{d}\gamma_\mu (1+\gamma_5)d \bar{q}\gamma^\mu Q_q q.$$

(3.30)

We have used a scale μ_R instead of the up quark mass and have quoted only the logarithmic m_t dependence. The quarks $q = u, d, s$ are summed over and Q_q is the charge of quark q. Note that since the vector current can be written as a sum of left-handed and right-handed currents, this is the only place where right-handed currents enter $\mathcal{H}_w$. The gluonic penguin contains the right-handed current in an $SU(3)$ singlet, hence retaining the $(8_L, 1_R)$ property of $\mathcal{H}_w$. However, the electroweak penguin introduces a small $(8_L, 8_R)$ component.

The full result can be described with the four-quark $\Delta S = 1$ operators,

$$O_1 = H_A - H_B, \qquad\qquad O_4 = H_A + H_B - H_C,$$

$$O_2 = H_A + H_B + 2H_C + 2H_D, \quad O_5 = \bar{d}\gamma_\mu (1+\gamma_5)\lambda^a s \, \bar{q}\gamma_\mu (1-\gamma_5)\lambda^a q,$$

$$O_3 = H_A + H_B + 2H_C - 3H_D, \quad O_6 = \bar{d}\gamma_\mu (1+\gamma_5)s \, \bar{q}\gamma_\mu (1-\gamma_5)q,$$

$$O_7 = \frac{3}{2}\bar{s}\gamma_\mu (1+\gamma_5)d \, \bar{q}\gamma^\mu (1-\gamma_5)Q_q q,$$

(3.31)

$$O_8 = -\frac{3}{2}\bar{s}_i \gamma_\mu (1+\gamma_5)d_j \, \bar{q}_j \gamma^\mu (1-\gamma_5)Q_q q_i,$$

where $q = u, d, s$ are summed over in $O_{5,6,7,8}$, i and j are color labels, Q_q is the charge of quark q and

$$H_A = \bar{d}\gamma_\mu (1+\gamma_5)u \, \bar{u}\gamma^\mu (1+\gamma_5)s, \qquad H_C = \bar{d}\gamma_\mu (1+\gamma_5)s \, \bar{d}\gamma^\mu (1+\gamma_5)d,$$

$$H_B = \bar{d}\gamma_\mu (1+\gamma_5)s \, \bar{u}\gamma^\mu (1+\gamma_5)u, \qquad H_D = \bar{d}\gamma_\mu (1+\gamma_5)s \, \bar{s}\gamma^\mu (1+\gamma_5)s.$$

(3.32)

The operators are arranged such that $O_{1,2,5,6}$ have octet and $\Delta I = 1/2$ quantum numbers, $O_3(O_4)$ are in the 27-plet with $\Delta I = 1/2(\Delta I = 3/2)$, while $O_{7,8}$ arise only from the electroweak penguin diagram. The full hamiltonian is

$$\mathcal{H}_{\text{eff}}^{\Delta S=1} = \frac{G_F}{2\sqrt{2}} V_{ud}^* V_{us} \sum_{i=1}^{8} \mathcal{C}_i O_i. \tag{3.33}$$

A renormalization-group analysis of the coefficients [BuBH 90] yields

$$
\begin{aligned}
\mathcal{C}_1 &= 1.90 - 0.62\tau, & \mathcal{C}_5 &= -0.011 - 0.079\tau, \\
\mathcal{C}_2 &= 0.14 - 0.020\tau, & \mathcal{C}_6 &= -0.001 - 0.029\tau, \\
\mathcal{C}_3 &= \mathcal{C}_4/5, & \mathcal{C}_7 &= -0.009 - (0.010 - 0.004\tau)\alpha, \\
\mathcal{C}_4 &= 0.49 - 0.005\tau, & \mathcal{C}_8 &= (0.002 + 0.160\tau)\alpha,
\end{aligned}
\tag{3.34}
$$

with $\Lambda \simeq 0.2$ GeV, $\mu_R \simeq 1$ GeV, $m_t = 150$ GeV, and $\tau = -V_{td}^* V_{ts}/V_{ud}^* V_{us}$. The number multiplying τ has a dependence on m_t whereas (within the leading logarithm approximation) the remainder does not if $m_t > M_W$. The τ dependence in $\mathcal{C}_4$ arises only because of the electroweak penguin diagram. This hamiltonian summarizes the *QCD* short-distance analysis and is the basis for estimates of weak amplitudes. For a treatment of corrections beyond those of leading order, see [BuBL 96].

VIII–4 The $\Delta I = 1/2$ rule

Phenomenology

In the decays $K \to \pi\pi$, the S-wave two-pion final state has a total isospin of either 0 or 2 as a consequence of Bose symmetry. Thus, such decays can be parameterized (ignoring the tiny effect of *CP* violation) as

$$A_{K^0 \to \pi^+\pi^-} = A_0 \, e^{i\delta_0} + \frac{A_2}{\sqrt{2}} e^{i\delta_2},$$

$$A_{K^0 \to \pi^0\pi^0} = A_0 \, e^{i\delta_0} - \sqrt{2} A_2 e^{i\delta_2},$$

$$A_{K^+ \to \pi^+\pi^0} = \frac{3}{2} A_2' \, e^{i\delta_2}, \tag{4.1}$$

where the subscripts $0, 2$ denote the total $\pi\pi$ isospin and the strong interaction S-wave $\pi\pi$ phase shifts δ_I enter as prescribed by Watson's theorem (cf. Eq. (C–3.15)). There are, in principle, two distinct $I = 2$ amplitudes A_2 and A_2'. These are equal if there are no $\Delta I = 5/2$ components in the weak transition, as is the case in the Standard Model if electromagnetic corrections are neglected. Including electromagnetism leads to a small difference between A_2 and A_2' [CiENPP 12], but we will neglect this possibility from now on, and employ just the two isospin amplitudes A_0 and A_2. The experimental decay rates themselves imply

$$|A_{K^0 \to \pi^+\pi^-}| = (2.772 \pm 0.0013) \times 10^{-7} \text{ GeV},$$

$$|A_{K^0 \to \pi^0\pi^0}| = (2.592 \pm 0.0022) \times 10^{-7} \text{ GeV},$$

$$|A_{K^+ \to \pi^+\pi^0}| = (0.1811 \pm 0.0004) \times 10^{-7} \text{ GeV}. \tag{4.2a}$$

The $\pi\pi$ phase difference $\delta_0 - \delta_2$ can be obtained via

$$\cos(\delta_0 - \delta_2) = \frac{\sqrt{3}}{2\sqrt{2}} \cdot \frac{|A_{+-}|^2 - |A_{00}|^2 + 2|A_{+0}|^2/3}{|A_{+0}| \left[2|A_{+-}|^2 + |A_{00}|^2 - 4|A_{+0}|^2/3\right]^{1/2}},$$

$$\delta_0 - \delta_2 = (44.55 \pm 1.04)^0. \tag{4.2b}$$

This is consistent with the phase difference which emerges from the analysis of $\pi\pi$ scattering. The magnitude of the isospin amplitudes can be found to be

$$|A_0| = (2.711 \pm 0.0011) \times 10^{-7} \text{ GeV},$$

$$|A_2| = (1.207 \pm 0.0026) \times 10^{-8} \text{ GeV}. \tag{4.3}$$

The ratio of magnitudes,

$$|A_2/A_0| = 0.0445 \pm 0.0001 \simeq 1/22.47, \tag{4.4}$$

indicates a striking dominance of the $\Delta I = 1/2$ amplitude (which contributes to A_0) over the $\Delta I = 3/2$ amplitude (which contributes only to A_2). This enhancement of A_0 over A_2, together with related manifestations to be discussed later, is called the $\Delta I = 1/2$ rule. As we have seen in previous sections, a naive estimate (and even determinations which are less naive!) do not suggest this much of an enhancement. However, the factor 22.5 dominance of $\Delta I = 1/2$ effects over those with $\Delta I = 3/2$ is common to both kaon and hyperon decay.[3]

A similar enhancement of $\Delta I = 1/2$ is found in the $K \to \pi\pi\pi$ channel. In this case, it is customary to expand the transition amplitude about the center of the Dalitz plot. For the decay amplitude $K(k) \to \pi(p_1) \pi(p_2) \pi(p_3)$, the relevant variables are

$$s_i = (k - p_i)^2 \Big|_{i=1,2,3}, \quad s_0 = \frac{1}{3}(s_1 + s_2 + s_3) = \frac{m_K^2}{3} + m_\pi^2,$$

$$\bar{x} = \frac{s_1 - s_2}{s_0}, \qquad \bar{y} = \frac{s_3 - s_0}{s_0}, \tag{4.5}$$

where s_3 labels the 'odd' pion, i.e. the third pion in each of the final states $\pi^+\pi^-\pi^0$, $\pi^0\pi^0\pi^+$, $\pi^+\pi^+\pi^-$. The large $\Delta I = 1/2$ amplitudes are considered up to quadratic

---

[3] Although our discussion stresses the relative magnitudes of the $\Delta I = 1/2, 3/2$ amplitudes, the relative *phases* of these amplitudes turns out to place important restrictions on the structure of the $|\Delta S| = 1$ nonleptonic hamiltonian [GoH 75].

order in these variables while the $\Delta I = 3/2$ amplitudes contain only constant plus linear terms,

$$\sqrt{2}\, A_{K^0 \to \pi^+\pi^-\pi^0} = a_1 - 2a_3 + (b_1 - 2b_3)\bar{y}e^{i\delta_{M1}} - \frac{2}{3}b_{23}\bar{x}e^{i\delta_{21}}$$
$$+ c\left(\bar{y}^2 + \frac{1}{3}\bar{x}^2\right) + d\left(\bar{y}^2 - \frac{1}{3}\bar{x}^2\right)e^{i\delta_{M1}},$$

$$\sqrt{2}\, A_{K^0 \to \pi^0\pi^0\pi^0} = 3(a_1 - 2a_3) + 3c\left(\bar{y}^2 + \frac{1}{3}\bar{x}^2\right),$$

$$A_{K^+ \to \pi^+\pi^+\pi^-} = 2(a_1 + a_3) - (b_1 + b_3)\bar{y}e^{i\delta_{M1}} + b_{23}\bar{y}e^{i\delta_{21}}$$
$$+ 2c\left(\bar{y}^2 + \frac{1}{3}\bar{x}^2\right) + d\left(\bar{y}^2 - \frac{1}{3}\bar{x}^2\right)e^{i\delta_{M1}},$$

$$A_{K^+ \to \pi^0\pi^0\pi^+} = a_1 + a_3 + (b_1 + b_3)\bar{y}e^{i\delta_{M1}} + b_{23}\bar{y}e^{i\delta_{21}}$$
$$+ c\left(\bar{y}^2 + \frac{1}{3}\bar{x}^2\right) + d\left(\bar{y}^2 - \frac{1}{3}\bar{x}^2\right)e^{i\delta_{M1}}, \qquad (4.6)$$

where a_1, b_1, c, d are $\Delta I = 1/2$ amplitudes, a_3, b_3, b_{23} are $\Delta I = 3/2$ amplitudes, and the phases $\{\delta_I\}$ in δ_{M1} ($\equiv \delta_M - \delta_1$) and δ_{21} ($\equiv \delta_2 - \delta_1$) refer to final-state phase shifts in the $I = 1, 2$ and mixed symmetry $I = 1$ states respectively. Because of the relatively small Q value for such decays ($Q_{\pi\pi\pi} = m_K - 3m_\pi \simeq 75\,\text{MeV}$), such phases are presumably small and are often omitted. Also, this representation in terms of simple energy-independent phase factors is clearly idealistic. Analysis of the available data [BiBD 03] yields (in units of 10^{-7})

$$a_1 = 9.32 \pm 0.04, \quad a_3 = 0.34 \pm 0.03,$$
$$b_1 = 14.2 \pm 0.2, \quad b_3 = -0.6 \pm 0.1, \quad b_{23} = 2.7 \pm 0.3, \qquad (4.7)$$
$$c = -1.1 \pm 0.5, \quad d = -5.0 \pm 0.8.$$

Dominance of the $\Delta I = 1/2$ signal is again clear in magnitude and in slope terms, e.g., we find at the center of the Dalitz plot,

$$|a_3/a_1| \simeq 1/27. \qquad (4.8)$$

In $SU(3)$ language, the dominance of $\Delta I = 1/2$ effects over $\Delta I = 3/2$ implies the dominance of octet transitions over those involving the 27-plet. This is a consequence, within the Standard Model, of the fact that the $\Delta I = 1/2$ 27-plet operator contributes relative to the $\Delta I = 3/2$ 27-plet operator with a fixed strength given by the scale-independent ratio of coefficients $C_3/C_4 \simeq 1/5$ (viz. Eq. (3.34)). The 27-plet operator then gives only a small contribution to the $\Delta I = 1/2$ amplitudes, with the major portion coming from the octet operators. We shall therefore ignore the $\Delta I = 1/2$ 27-plet contribution henceforth.

Chiral lagrangian analysis

The left-handed chiral property of the Standard Model may be directly tested by the use of chiral symmetry to relate the amplitudes in $K \to \pi\pi\pi$ to those in $K \to \pi\pi$. We have already constructed the effective lagrangians for $(8_L, 1_R)$ and $(27_L, 1_R)$ transitions. Dropping $g_{27}^{(1/2)}$, the nonleptonic decays are described by the two parameters g_8 and $g_{27}^{(3/2)}$ at $\mathcal{O}(E^2)$. Let us see how well this parameterization works, and afterwards add $\mathcal{O}(E^4)$ corrections. The two free parameters may be determined from A_0 and A_2 in $K \to \pi\pi$ decays. From the chiral lagrangians of Eqs. (2.4), (2.8b), we find

$$A_0 = \frac{\sqrt{2}\, g_8}{F_\pi^3} \left(m_K^2 - m_\pi^2 \right), \qquad A_2 = \frac{2\, g_{27}^{(3/2)}}{F_\pi^3} \left(m_K^2 - m_\pi^2 \right), \tag{4.9}$$

which yields upon comparison with Eq. (4.3),

$$g_8 \simeq 7.8 \times 10^{-8} F_\pi^2, \qquad g_{27}^{(3/2)} \simeq 0.25 \times 10^{-8} F_\pi^2. \tag{4.10}$$

The $K \to \pi\pi\pi$ amplitude may be predicted from these. Because there are only two factors of the energy, no quadratic terms are present in the predictions,

$$A_{K_L^0 \to \pi^+\pi^-\pi^0}^{(1/2)} = \frac{\sqrt{2} A_0 m_K^2}{6 F_\pi (m_K^2 - m_\pi^2)} \left[1 + \frac{m_K^2 + 3m_\pi^2}{m_K^2} \bar{y} \right],$$

$$A_{K_L^0 \to \pi^+\pi^-\pi^0}^{(3/2)} = -\frac{A_2 m_K^2}{3 F_\pi (m_K^2 - m_\pi^2)} \left[1 - \frac{5}{4} \frac{m_K^2 + 3m_\pi^2}{m_K^2} \bar{y} \right],$$

$$A_{K^+ \to \pi^+\pi^+\pi^-}^{(3/2)} = \frac{A_2 m_K^2}{3 F_\pi (m_K^2 - m_\pi^2)} \left[1 + 4 \frac{m_K^2 + 3m_\pi^2}{m_K^2} \bar{y} \right], \tag{4.11a}$$

which correspond to the numerical values (again in units of 10^{-7}),

$$A_{K_L^0 \to \pi^+\pi^-\pi^0}^{(1/2)} = 7.5 + 9.1\, \bar{y},$$

$$A_{K_L^0 \to \pi^+\pi^-\pi^0}^{(3/2)} = -0.47 + 0.74\, \bar{y},$$

$$A_{K^+ \to \pi^+\pi^+\pi^-}^{(3/2)} = 0.47 + 2.3\, \bar{y}. \tag{4.11b}$$

These are to be compared to the experimental results,

$$A_{K_L^0 \to \pi^+\pi^-\pi^0}^{(1/2)} = 9.32 + 14.2\, \bar{y} - 6.1\, \bar{y}^2 + 1.3\, \bar{x}^2,$$

$$A_{K_L^0 \to \pi^+\pi^-\pi^0}^{(3/2)} = -0.68 + 1.2\, \bar{y},$$

$$A_{K^+ \to \pi^+\pi^+\pi^-}^{(3/2)} = 0.68 + 3.3\, \bar{y}. \tag{4.12}$$

This comparison can be seen in Fig. VIII–3, where a slice across the Dalitz plot is given. Also shown are the extrapolations outside the physical region to the 'soft-pion point' where either $p_+^\mu \to 0$ or $p_0^\mu \to 0$. Predictions at these locations are obtained by using the soft-pion theorem (see Prob. VIII–2).

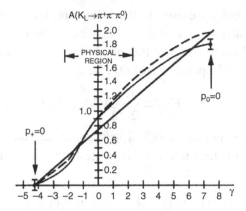

Fig. VIII–3 Dalitz plot.

The chiral relations clearly capture the main features of the amplitude and demonstrate that the $K \to 3\pi$ $\Delta I = 1/2$ enhancement is not independent of that observed in $K \to 2\pi$ decay. However, for the $\Delta I = 1/2$ transitions we may do somewhat better. The kinematic dependence of $\bar{x}^2$ or $\bar{y}^2$ can come only from a chiral lagrangian with four factors of the momentum, and only two combinations are possible:

$$\Lambda_{\text{quad}} = \gamma_1 k \cdot p_0 p_+ \cdot p_- + \gamma_2 \left(k \cdot k_+ p_0 \cdot p_- + k \cdot p_- p_0 \cdot p_+ \right). \tag{4.13}$$

Such behavior can be generated from a variety of chiral lagrangians,

$$\mathcal{L}_{\text{quad}} = g_8' \operatorname{Tr} \left(\lambda_6 \partial_\mu U \partial^\mu U^\dagger \partial_\nu U \partial^\nu U^\dagger \right)$$
$$+ g_8'' \operatorname{Tr} \left(\lambda_6 \partial_\mu U \partial_\nu U^\dagger \partial^\mu U \partial^\nu U^\dagger \right) + \cdots . \tag{4.14}$$

However the predictions in terms of γ_i are unique. Fitting the quadratic terms to determine γ_1, γ_2 yields the full amplitude,

$$\sqrt{2} A^{(1/2)}_{K^0 \to \pi^+ \pi^- \pi^0} = (9.5 \pm 0.7) + (16.0 \pm 0.5)\, \bar{y} - 4.85\, \bar{y} + 0.88\, \bar{x}^2, \tag{4.15}$$

which provides an excellent representation of the data. Final-state interaction effects also provide an important contribution and must be included in a complete analysis [KaMW 90]. Note that in the process of determining the quadratic coefficients, the constant and linear terms have also become improved. This process cannot be repeated for $\Delta I = 3/2$ amplitudes due to a lack of data on quadratic terms.

Vacuum saturation

Direct calculations of nonleptonic amplitudes have proven very difficult to perform. On the whole, no single analytical or numerical method for overcoming the general problem yet exists. In the following, we describe the simplest analytical approach, called *vacuum saturation*, which often serves as a convenient benchmark with which to compare the theory. For convenience we consider only O_1 (the largest $\Delta I = 1/2$ operator) and O_4 (the $\Delta I = 3/2$ operator),

$$\mathcal{H}_W \simeq \frac{G_F}{2\sqrt{2}} V_{ud}^* V_{us} (C_1 O_1 + C_4 O_4), \qquad (4.16)$$

with $C_1 \simeq 1.9$ and $C_4 \simeq 0.5$. The vacuum saturation approximation consists of inserting the vacuum intermediate state between the two currents in all possible ways, e.g.,

$$
\begin{aligned}
&\langle \pi^+(\mathbf{p}_+) \pi^-(\mathbf{p}_-) | \bar{d}\gamma^\mu (1 + \gamma_5) u \bar{u}\gamma^\mu (1 + \gamma_5) s | \bar{K}^0(\mathbf{k}) \rangle \\
&= \langle \pi^-(\mathbf{p}_-) | \bar{d}\gamma^\mu \gamma_5 u | 0 \rangle \langle \pi^+(\mathbf{p}_+) | \bar{u}\gamma^\mu s | \bar{K}^0(\mathbf{k}) \rangle \\
&\quad + \langle \pi^+(\mathbf{p}_+) \pi^-(\mathbf{p}_-) | \bar{u}_\beta \gamma^\mu u_\alpha | 0 \rangle \langle 0 | \bar{d}_\alpha \gamma^\mu \gamma_5 s_\beta | \bar{K}^0(\mathbf{k}) \rangle \\
&= -i\sqrt{2}\, F_\pi f_+ p_-^\mu (k + p_+)_\mu - \frac{i}{3}\sqrt{2}\, F_K f_+ k_\mu (p_- - p_+)^\mu. \qquad (4.17)
\end{aligned}
$$

In obtaining this result the Fierz rearrangement property

$$\bar{d}_\alpha \gamma^\mu (1 + \gamma_5) u_\alpha \bar{u}_\beta \gamma^\mu (1 + \gamma_5) s_\beta = \bar{d}_\alpha \gamma^\mu (1 + \gamma_5) s_\beta \bar{u}_\beta \gamma^\mu (1 + \gamma_5) u_\alpha$$

has been used, where α, β are color indices which are summed over. In addition, the color singlet property of currents is employed,

$$\langle 0 | \bar{d}_\alpha \gamma_\mu \gamma_5 s_\beta | \bar{K}^0(\mathbf{k}) \rangle = i\sqrt{2}\, F_K k_\mu \frac{\delta_{\alpha\beta}}{3}. \qquad (4.18)$$

Within the vacuum saturation approximation, we see that the amplitudes are expressed in terms of known semileptonic decay matrix elements. Putting in all of the constants, we find that

$$A_0 = \frac{G_F}{3} V_{ud}^* V_{us} F_\pi \left(m_K^2 - m_\pi^2 \right) C_1 \simeq 0.34 \times 10^{-7} \text{ GeV},$$

$$A_2 = \frac{2\sqrt{2}\, G_F}{3} V_{ud}^* V_{us} F_\pi \left(m_K^2 - m_\pi^2 \right) C_4 \simeq 2.5 \times 10^{-8} \text{ GeV}. \qquad (4.19)$$

The above estimate of A_2 is seen to work reasonably well, but that of A_0 falls considerably short of the observed $\Delta I = 1/2$ amplitude. This demonstrates that vacuum saturation is not a realistic approximation. However, it does serve to indicate how much additional $\Delta I = 1/2$ enhancement is required to explain the data.

Nonleptonic lattice matrix elements

While historically there have been many attempts to improve on the naive vacuum saturation method, the present state of the art is to rely on lattice calculations.

However, nonleptonic matrix elements have been a particular challenge for lattice methods. The transition from a kaon to two pions requires three external sources to create the mesons involved, as well as a singular four-quark operator for the weak hamiltonian. In addition, there are diagrams where quarks in the hamiltonian form disconnected loops not connected to external states. Recent advances have allowed the extraction of the A_2 amplitude with reasonable precision [Bl *et al.* 12]

$$|A_2| = (1.381 \pm 0.046_{\text{stat}} \pm 0.258_{\text{syst}}) \times 10^{-8} \text{ GeV} \qquad (4.20)$$

consistent with the experimental result of Eq. (4.3). However, the isospin-zero final state in A_0 implies the existence of disconnected diagrams, which make the numerical evaluation difficult, and we do not yet have a reliable lattice calculation of A_0 [Bl *et al.* 11].

VIII–5 Rare kaon decays

Thus far, we have discussed the dominant decay modes of the kaon. There are, however, many additional modes which, despite tiny branching ratios, have been the subject of intense experimental and theoretical activity. We can divide this activity into three main categories.

(1) Forbidden decays – These include tests of the flavor conservation laws of the Standard Model such as $K_L \to e^+\mu^-$. Positive signals would represent evidence for physics beyond our present theory.

(2) Rare decays within the Standard Model – These include decays which occur only at one-loop order. Such processes can be viewed as tests of chiral dynamics as developed in this and preceding chapters (e.g., radiative kaon decays) or as particularly sensitive to short-distance effects, which probe the particle content of the theory.

(3) CP-violation studies – There is now confirmation of CP-violating processes involving kaons and B mesons (and searches of the same for D mesons). Also, the observed baryon–antibaryon asymmetry of the Universe requires the existence of CP violation within the standard cosmological model. There remain, however, interesting opportunities for further studies of CP violation within the subfield of rare kaon decays.

Any of these have the potential to yield exciting physics. We shall content ourselves with discussing only a small sample of the many possibilities. Surveys of rare kaon modes appear in [CiENPP 12] and also in [RPP 12].

Consider first the rare decay $K^+ \to \pi^+\nu\bar{\nu}$, where the neutrino flavor $\nu = \nu_e, \nu_\mu, \nu_\tau$ is summed over. This mode is often called 'K^+ to π^+ plus nothing', in reference to its unique experimental signature. This process can take place only

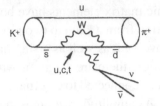

Fig. VIII–4 The decay $K^+ \to \pi^+ \nu_\ell \bar{\nu}_\ell$.

through loop diagrams, such as the ones in Fig. VIII–4. We content ourselves here to show just the effective hamiltonian for the dominant t-quark loop at leading order in *QCD*, [InL 81, HaL 89]

$$\mathcal{H}_{\text{eff}} = \frac{G_F}{\sqrt{2}} \cdot \frac{\alpha}{2\pi \sin^2 \theta_{\text{w}}} V_{\text{ts}}^* V_{\text{td}} X_0(x_t) \sum_\ell \bar{\nu}_\ell \gamma^\mu (1 + \gamma_5) \nu_\ell \, \bar{s} \gamma_\mu (1 + \gamma_5) d, \quad (5.1)$$

where $x_t \equiv m_t^2/m_W^2$ and

$$X_0(x_t) = \frac{x_t}{8} \left(\frac{x_t + 2}{x_t - 1} + \frac{3x_t - 6}{(1 - x_t)^2} \right). \quad (5.2)$$

The overall factor of m_t^2/m_W^2 in $X_0(x_t)$ is associated with the GIM effect; the c-quark and u-quark amplitudes, were they included, would contain analogous mass factors such that in the limit $m_u = m_c = m_t$ the total amplitude would vanish via GIM cancelation. Although the calculation of many hadronic processes in this book are hindered by QCD uncertainties, such is not the case here. The quark matrix element is related by isospin to the known charged current amplitude

$$\langle \pi^+(\mathbf{p}) | \bar{s} \gamma_\mu d | K^+(\mathbf{k}) \rangle = \sqrt{2} \langle \pi^0(\mathbf{p}) | \bar{s} \gamma_\mu u | K^+(\mathbf{k}) \rangle = f_+(q^2) (k + p)_\mu \quad (5.3)$$

with $f_+(0) = -1$. This makes the $K^+ \to \pi^+ \nu_\ell \bar{\nu}_\ell$ example a theoretically 'clean' process and is responsible in large part for all the attention this transition has attracted. Our discussion is relatively brief, and a more careful analysis would include effects like *QCD* perturbative corrections, the c-quark loop contribution (roughly 30%), etc. A recent prediction, [BrGS 11], along with the current experimental result [RPP 12], reads

$$\text{Br}^{(\text{theo})}_{K^+ \to \pi^+ \nu_\ell \bar{\nu}_\ell} = (7.8 \pm 0.8) \times 10^{-11}, \quad \text{Br}^{(\text{expt})}_{K^+ \to \pi^+ \nu_\ell \bar{\nu}_\ell} = (1.7 \pm 1.1) \times 10^{-10}. \quad (5.4)$$

The experimental result, still consistent with zero, is reaching the sensitivity needed to probe the theory prediction and further advances are anticipated. The main source of uncertainty in the theory prediction is from CKM factors and quark mass values, which presumably can and will be improved upon.

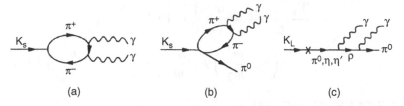

Fig. VIII–5 Long-distance contributions to radiative kaon decays.

A different class of rare decays consists of the radiative processes $K_S \to \gamma\gamma$ and $K_L \to \pi^0\gamma\gamma$. These transitions provide interesting tests of chiral perturbation theory at one-loop order. In this case, the long-distance process, Fig. VIII–5, is dominant. An important feature is that there is no tree-level contribution at order E^2 or E^4 from any of the strong or weak chiral lagrangians because all of the hadrons involved are neutral. Thus, the decays can only come from loop diagrams, or from lagrangians at $\mathcal{O}(E^6)$. There is also an interesting corollary of this result concerning the renormalization behavior of the loops. Since there are no tree-level counterterms at $\mathcal{O}(E^4)$ with which to absorb divergences from the loop diagrams, and recalling that we have proven all divergences can be handled in this fashion, it follows that the sum of the loop diagrams must be finite. This is in fact borne out by direct calculation.

For $K_S \to \gamma\gamma$, the prediction of chiral [D'AE 86, Go 86] loops is given in terms of known quantities such as

$$\Gamma_{K_S \to \gamma\gamma} = \frac{\alpha^2 m_K^2 g_8^2 F_\pi^2}{16\pi^3} \left(1 - \frac{m_\pi^2}{m_K^2}\right)^2 \left| F\left(\frac{m_K^2}{m_\pi^2}\right) \right|^2,$$
$$F(z) = 1 - z \left[\pi^2 - \ln^2 Q(z) - 2\pi i \ln Q(z)\right],$$
$$Q(z) = \frac{1 - \sqrt{1 - 4z}}{1 + \sqrt{1 - 4z}}, \tag{5.5}$$

where g_8 is the nonleptonic coupling defined previously in Eq. (2.4). Comparison of the theoretical one-loop branching ratio and the experimental result,

$$\text{Br}^{(\text{theo})}_{K_S \to \gamma\gamma} = 2.0 \times 10^{-6}, \qquad \text{Br}^{(\text{expt})}_{K_S \to \gamma\gamma} = (2.63 \pm 0.17) \times 10^{-6}, \tag{5.6}$$

shows reasonable agreement but implies the need to consider $\mathcal{O}(E^6)$ corrections. In particular, the 'unitarity correction' $K_S \to \pi^+\pi^- \to \gamma\gamma$ has been shown to provide an improved theoretical prediction [KaH 94].

The case of $K_L \to \pi^0\gamma\gamma$ is also instructive. Again, one-loop contributions are finite and unambiguous [EcPR 88]. Indeed, we know that $K_L \to \pi^0\gamma\gamma$ and $K_L \to \gamma\gamma$ are related by the soft-pion theorem in the limit $p_\pi^\mu \to 0$, yielding

$$\frac{d\Gamma_{K_L \to \pi^0 \gamma \gamma}}{dz} = \frac{\alpha^2 m_K^5}{(4\pi)^5} g_8^2 \left[\lambda \left(z, \frac{m_\pi^2}{m_K^2} \right) \right]^{1/2}$$

$$\times \left| \left(z - \frac{m_\pi^2}{m_K^2} \right) F(z \frac{m_K^2}{m_\pi^2}) + \left(1 - z + \frac{m_\pi^2}{m_K^2} \right) F(z) \right|^2, \qquad (5.7)$$

where $z = m_{\gamma\gamma}^2 / m_K^2$ and

$$\lambda(a, b) \equiv 1 + a^2 + b^2 - 2(a + b + ab). \qquad (5.8)$$

If we compare the theoretical branching ratio based on the above description with the experimental value, we find

$$\mathrm{Br}^{(\mathrm{loop})}_{K_L \to \pi^0 \gamma \gamma} = 0.68 \times 10^{-6}, \qquad \mathrm{Br}_{K_L^0 \to \pi^0 \gamma \gamma} = (1.273 \pm 0.033) \times 10^{-6}. \quad (5.9)$$

This indicates the need for an $\mathcal{O}(E^6)$ correction. Indeed, the most recent data input, from [Ab *et al.* (KTeV collab.) 08], provides evidence for a vector exchange contribution. It is easy to take this into account, viz. the diagram of Fig. VIII–5(c) shows the effect of ρ-exchange.

Problems

(1) **$K_{\ell 3}$ decay**

The ratio $f_+^{K^+\pi^0}(0)/f_+^{K^0\pi^-}(0)$ of semileptonic form factors is a measure of isospin violation. Part of this quantity arises from π^0-η_8^0 mixing.

   (a) By diagonalizing the pseudoscalar meson mass matrix, show that $m_d \neq m_u$ induces the mixing $|\pi^0\rangle = \cos\epsilon \, |\varphi_3\rangle + \sin\epsilon \, |\varphi_8\rangle$ where $\epsilon \simeq \sqrt{3}(m_d - m_u)/ [4(m_s - \hat{m})]$ and $\hat{m} \equiv (m_u + m_d)/2$.

   (b) Demonstrate that this leads to the result (cf. Eq. (1.11))

$$\frac{f_+^{K^+\pi^0}(0)}{f_+^{K^0\pi^-}(0)} = \frac{\cos\epsilon f_+^{K^+\varphi^3}(0) + \sin\epsilon f_+^{K^+\varphi^8}(0)}{f_+^{K^0\pi^-}(0)} \simeq 1 + \sqrt{3}\sin\epsilon.$$

(2) **Soft pions and $K \to 3\pi$ decay**

The results derived in Sect. VIII–4 with effective lagrangians can also be obtained by means of soft pion methods (see App. B–3).

   (a) Using the soft pion theorem, show that the soft-pion limit of the $K \to 3\pi$ transition amplitude is given by

$$\lim_{q_a \to 0} \langle \pi_{q_a}^a \pi_{q_b}^b \pi_{q_c}^c | \mathcal{H}_w(0) | K_k^n \rangle = \frac{-i}{F_\pi} \langle \pi_{q_b}^b \pi_{q_c}^c | [Q_5^a, \mathcal{H}_w(0)] | K_k^n \rangle,$$

where $Q_5^a = \int d^3x \, A_0^a(\mathbf{x}, t)$ is the axial charge.

(b) Demonstrate that this may be also written as

$$\lim_{q_a \to 0} \langle \pi^a_{q_a} \pi^b_{q_b} \pi^c_{q_c} | \mathcal{H}_w(0) | K^n_k \rangle = \frac{-i}{F_\pi} \langle \pi^b_{q_b} \pi^c_{q_c} | [Q^a, \mathcal{H}_w(0)] | K^n_k \rangle,$$

where Q^a is an isotopic spin operator, and hence that

$$\lim_{q_0 \to 0} \langle \pi^+_{q_+} \pi^-_{q_-} \pi^0_{q_0} | \mathcal{H}^I_w(0) | K^0_k \rangle = \frac{-i}{2F_\pi} A^I_{K^0 \to \pi^+ \pi^-},$$

$$\lim_{q_+ \to 0} \langle \pi^+_{q_+} \pi^-_{q_-} \pi^0_{q_0} | \mathcal{H}^I_w(0) | K^0_k \rangle = \frac{-i}{F_\pi} (A^I_{K^0 \to \pi^0 \pi^0} - A^I_{K^0 \to \pi^+ \pi^-}),$$

$$\lim_{q_- \to 0} \langle \pi^+_{q_+} \pi^-_{q_-} \pi^0_{q_0} | \mathcal{H}^I_w(0) | K^n_0 \rangle = \frac{i}{F_\pi} (A^I_{K^0 \to \pi^0 \pi^0},$$

$$- A^I_{K^0 \to \pi^+ \pi^-} + \frac{1}{\sqrt{2}} A^I_{K^+ \to \pi^+ \pi^0}),$$

where $I = 1/2, 3/2$ signifies the isospin component of the quantities in question.

(c) Use a linear expansion of the $K \to 3\pi$ transition amplitude, (i.e. Eq. (4.6) with $c = d = 0$) to reproduce the results of Eq. (4.11), up to corrections of order m^2_π.

IX

Mass mixing and *CP* violation

Aside from a concluding section on the strong *CP* problem, this chapter is about the *CP* violation of kaons. We set up the general framework for meson–antimeson mixing, which is also used in the weak interactions of heavy quarks, treated later in Chap. XIV. In this chapter we apply the formalism to K^0–$\bar{K}^0$ mixing and *CP*-violating processes involving kaons.

IX–1 K^0–$\bar{K}^0$ mixing

It is clear that K^0 and $\bar{K}^0$ *should* mix with each other. In addition to less obvious mechanisms discussed later, the most easily seen source of mixing occurs through their common $\pi\pi$ decays, i.e., $K^0 \leftrightarrow \pi\pi \leftrightarrow \bar{K}^0$. We can use second-order perturbation theory to study the phenomenon of mixing. Writing the wavefunctions in two-component form

$$|\psi(t)\rangle = \begin{pmatrix} a(t) \\ b(t) \end{pmatrix} \equiv a(t)|K^0\rangle + b(t)|\bar{K}^0\rangle, \qquad (1.1)$$

we have the time development

$$i\frac{d}{dt}|\psi(t)\rangle = \left(M - \frac{i}{2}\Gamma\right)|\psi(t)\rangle, \qquad (1.2)$$

where, to second order in perturbation theory, the quantity in parentheses is called the *mass matrix* and is given by[1]

---

[1] The factors $1/2m_K$ are required by the normalization convention of Eq. (C–3.7) for state vectors.

$$\left[M - \frac{i}{2}\Gamma \right]_{ij} \equiv \frac{\langle K_i^0 | \mathcal{H}_{\text{eff}} | K_j^0 \rangle}{2m_K}$$

$$= m_K^{(0)} \delta_{ij} + \frac{\langle K_i^0 | \mathcal{H}_w | K_j^0 \rangle}{2m_K} + \frac{1}{2m_K} \sum_n \frac{\langle K_i^0 | \mathcal{H}_w | n \rangle \langle n | \mathcal{H}_w | K_j^0 \rangle}{m_K^{(0)} - E_n + i\epsilon}.$$

(1.3)

Here, the absorptive piece Γ arises from use of the identity

$$\frac{1}{\omega - E_n + i\epsilon} = P\left(\frac{1}{\omega - E_n} \right) - i\pi \, \delta(E_n - \omega),$$

(1.4)

and hence involves only physical intermediate states

$$\Gamma_{ij} = \frac{1}{2m_K} \sum_n \langle K_i^0 | \mathcal{H}_w | n \rangle \langle n | \mathcal{H}_w | K_j^0 \rangle 2\pi \delta(E_n - m_K).$$

(1.5)

Because M and Γ are hermitian, we have $M_{21} = M_{12}^*$ and $\Gamma_{21} = \Gamma_{12}^*$. The diagonal elements of the mass matrix are required to be equal by *CPT* invariance, leading to a general form

$$M - \frac{i}{2}\Gamma = \begin{pmatrix} A & p^2 \\ q^2 & A \end{pmatrix},$$

(1.6)

where A, p^2, and q^2 can be complex. The states $\bar{K}^0$ and K^0 are related by the unitary *CP* operation,

$$CP|K^0\rangle = \xi_K |\bar{K}^0\rangle$$

(1.7)

with $|\xi_K|^2 = 1$. Our convention will be to choose $\xi_K = -1$. The assumption of *CP* invariance would relate the off-diagonal elements in the mass matrix so as to imply $p = q$,

$$\langle K^0 | \mathcal{H}_{\text{eff}} | \bar{K}^0 \rangle = \langle K^0 | (CP)^{-1} CP \, \mathcal{H}_{\text{eff}} \, (CP)^{-1} CP | \bar{K}^0 \rangle = \langle \bar{K}^0 | \mathcal{H}_{\text{eff}} | K^0 \rangle, \quad (1.8)$$

where $\langle \bar{K}^0 | \mathcal{H}_{\text{eff}} | K^0 \rangle$ is defined in Eq. (1.3). Combined with the hermiticity of M and Γ, this would imply that M_{12} and Γ_{12} are real. In the actual *CP*-noninvariant world, this is not the case and we have instead for the eigenstates of the mass matrix,

$$|K_{\substack{L \\ S}}\rangle = \frac{1}{\sqrt{|p|^2 + |q|^2}} \left[p|K^0\rangle \pm q|\bar{K}^0\rangle \right],$$

(1.9)

where, from the above discussion,

$$\frac{p}{q} = \sqrt{\frac{M_{12} - \frac{i}{2}\Gamma_{12}}{M_{12}^* - \frac{i}{2}\Gamma_{12}^*}}, \quad M_{12} - \frac{i}{2}\Gamma_{12} = \langle K^0 | \mathcal{H} | \bar{K}^0 \rangle.$$

(1.10)

The difference in eigenvalues is given by

$$2qp = (m_L - m_S) - \frac{i}{2}(\Gamma_L - \Gamma_S)$$

$$= 2\left(M_{12} - \frac{i}{2}\Gamma_{12}\right)^{1/2}\left(M_{12}^* - \frac{i}{2}\Gamma_{12}^*\right)^{1/2} \simeq 2\operatorname{Re} M_{12} - i\operatorname{Re}\Gamma_{12}, \quad (1.11)$$

where the final relation is an approximation valid if *CP* violation is small ($1 \gg \operatorname{Im} M_{12}/\operatorname{Re} M_{12}$). The subscripts in K_L and K_S, standing for 'long' and 'short', refer to their respective lifetimes, whose ratio is substantial, $\tau_L/\tau_S \simeq 571$. To understand this large difference, we note that if *CP* were conserved ($p = q$), these states would become *CP* eigenstates $K_\pm^0$ (not to be confused with the charged kaons $K^\pm$!),

$$|K_S\rangle \xrightarrow[p=q]{} |K_+^0\rangle, \qquad\qquad |K_L\rangle \xrightarrow[p=q]{} |K_-^0\rangle,$$
$$|K_\pm^0\rangle \equiv \tfrac{1}{\sqrt{2}}\left[|K^0\rangle \mp |\bar{K}^0\rangle\right], \qquad CP|K_\pm^0\rangle = \pm|K_\pm^0\rangle. \qquad (1.12)$$

In this limit, which well approximates reality, K_S would decay only to *CP*-even final states like $\pi\pi$, whereas K_L would decay only to *CP*-odd final states, e.g., 3π. Since the phase space for the former considerably exceeds that of the latter at the rather low energy of the kaon mass, K_S has much the shorter lifetime. The states $K_{S,L}$, expanded in terms of *CP* eigenstates, are

$$|K_{\,L \atop \,S}\rangle = \frac{1}{\sqrt{1 + |\bar{\epsilon}|^2}}\left[|K_{\mp}^0\rangle + \bar{\epsilon}|K_\pm^0\rangle\right], \qquad \frac{p}{q} = \frac{1 + \bar{\epsilon}}{1 - \bar{\epsilon}},$$

$$\bar{\epsilon} = \frac{p - q}{p + q} \simeq \frac{i}{2}\frac{\operatorname{Im} M_{12} - i\operatorname{Im}\Gamma_{12}/2}{\operatorname{Re} M_{12} - i\operatorname{Re}\Gamma_{12}/2} \simeq \frac{1}{2}\frac{M_{12} - M_{21} - \frac{i}{2}(\Gamma_{12} - \Gamma_{21})}{m_L - m_S - \frac{i}{2}(\Gamma_L - \Gamma_S)}. \quad (1.13)$$

K^0–$\bar{K}^0$ mixing can be observed experimentally from the time development of a state which is produced via a strong interaction process, and therefore starts out at $t = 0$ as either a pure K^0 or $\bar{K}^0$,

$$|K^0(t)\rangle = g_+(t)|K^0\rangle + \frac{q}{p}g_-(t)|\bar{K}^0\rangle,$$

$$|\bar{K}^0(t)\rangle = \frac{p}{q}g_-(t)|K^0\rangle + g_+(t)|\bar{K}^0\rangle,$$

$$g_\pm(t) = \frac{1}{2}e^{-\Gamma_L t/2}e^{-im_L t}\left[1 \pm e^{-\Delta\Gamma t/2}e^{i\Delta m t}\right], \qquad (1.14)$$

where $\Delta\Gamma \equiv \Gamma_S - \Gamma_L$ and $\Delta m \equiv m_L - m_S$, each defined to be a *positive* quantity. From such experiments, the very precise value

$$\Delta m_{\text{expt}} = (3.484 \pm 0.006) \times 10^{-12}\ \text{MeV} \qquad (1.15)$$

has been obtained.

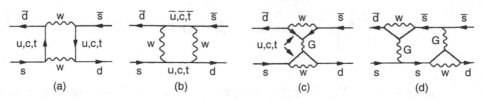

Fig. IX–1 Box (a),(b) and other contributions to *CP* violation.

CP-conserving mixing

There are two main classes of contributions, associated respectively with the short-distance box diagrams of Fig. IX–1(a),(b) and the long-distance contributions like those in Fig. IX–2,

$$\Delta m_{\text{theory}} = (\Delta m)^{\text{SD}}_{\text{theory}} + (\Delta m)^{\text{LD}}_{\text{theory}}. \tag{1.16a}$$

We shall consider the first of these here, the short distance component

$$(\Delta m)^{\text{SD}}_{\text{theory}} = 2\text{Re}\langle K^0 | \mathcal{H}^{\text{box}}_{\text{w}} | \bar{K}^0 \rangle. \tag{1.16b}$$

Determining $(\Delta m)^{\text{SD}}_{\text{theory}}$ has long been, and continues to be, a significant topic in kaon physics. It involves almost all the field theory tools we describe in this book. Our discussion will of necessity include some advanced features in order to present a realistic picture of the current state of the art.

The construction of $\mathcal{H}^{\text{box}}_{\text{w}}$ follows a standard procedure: to a given order of *QCD* perturbation theory, first specify the Wilson coefficient at the scale $\mu = M_W$, then use the renormalization group (RG) to evolve down to a hadronic scale $\mu < m_c$ and finally match onto the effective three-quark (i.e. u, d, s) theory. The result of this is

$$\mathcal{H}^{\text{box}}_{\text{w}} = \mathcal{C}(\mu) O^{\Delta S=2}, \tag{1.17}$$

where $O^{\Delta S=2}$ is the local four-quark operator

$$O^{\Delta S=2} = \bar{d}\gamma_\mu(1 + \gamma_5)s \, \bar{d}\gamma^\mu(1 + \gamma_5)s, \tag{1.18}$$

and $\mathcal{C}(\mu)$ is the corresponding Wilson coefficient,

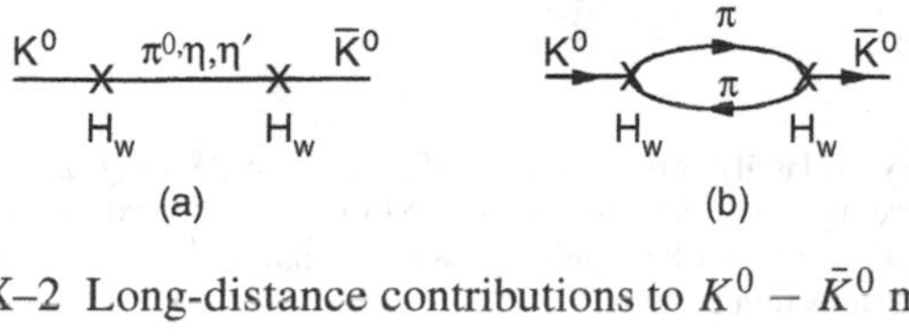

$$
\begin{array}{cc}
K^0 \overset{\times}{\underset{H_w}{}} \overset{\pi^0,\eta,\eta'}{} \overset{\times}{\underset{H_w}{}} \bar{K}^0 & \quad K^0 \underset{H_w}{} \overset{\pi}{\underset{\pi}{}} \underset{H_w}{} \bar{K}^0 \\
\text{(a)} & \text{(b)}
\end{array}
$$

Fig. IX–2 Long-distance contributions to $K^0 - \bar{K}^0$ mixing.

$$C(\mu) = \frac{G_F^2}{16\pi^2}\left[\xi_c^2 H(x_c)m_c^2\eta_{cc} + \xi_t^2 H(x_t)m_t^2\eta_{tt} + 2\xi_c\xi_t\bar{G}(x_c, x_t)m_c^2\eta_{ct}\right]b(\mu) ,$$

$$(1.19)$$

with $\xi_i \equiv V_{id}^*V_{is}$ $(i=c, t)$ and $x_i \equiv m_i^2/M_W^2$. The above expression for $C(\mu)$ is more complicated than the $C_\pm(\mu)$ encountered in our earlier $\Delta S = 1$ discussion (cf. Eq. (VIII–3.11)) because the box amplitude for $\Delta S = 2$ has loop contributions from all the u, c, t quarks. Actually, Eq. (1.19) has already been simplified in that CKM unitarity has allowed removal of ξ_u and the tiny mass of the u quark has been neglected with respect to the heavy-quark masses m_c, m_t. The quantities $H(x_t)$, $H(x_c)$ and $\bar{G}(x_c, x_t)$ in Eq. (1.19) are so-called Inami–Lim functions [InL 81] that describe the quark-level loop amplitudes of Fig. IX–1(a),(b) in the no-QCD limit,

$$H(x) = \left[\frac{1}{4} + \frac{9}{4}\frac{1}{1-x} - \frac{3}{2}\frac{1}{(1-x)^2}\right] - \frac{3}{2}\frac{x^2}{(1-x)^3}\ln x,$$

$$\bar{G}(x, y) = y\left[-\frac{1}{y-x}\left(\frac{1}{4} + \frac{3}{2}\frac{1}{1-x} - \frac{3}{4}\frac{1}{(1-x)^2}\right)\ln x\right.$$
$$\left. + (y \leftrightarrow x) - \frac{3}{4}\frac{1}{(1-x)(1-y)}\right].$$

$$(1.20)$$

This leaves in Eq. (1.19) the factors η_{cc}, η_{tt}, η_{ct}, and $b(\mu)$. These arise from calculating perturbative corrections to $\mathcal{H}_w^{\text{box}}$.[2] Such corrections will contain dependence on both the scale (μ) and renormalization scheme (say, the NDR approach, described in App. C–5). These cannot be present in the full amplitude and must be cancelled by analogous dependence in the matrix element $\langle K^0|O^{\Delta S=2}|\bar{K}^0\rangle$. For convenience, the scale and scheme dependence present in the Wilson coefficient $C(\mu)$ is placed into the factor $b(\mu)$, which for K^0–$\bar{K}^0$ mixing has the perturbative form

$$b(\mu) = \alpha_s(\mu)^{-2/9}\sum_{n=0}^{\infty} J^{(n)}\frac{\alpha_s(\mu)^n}{4\pi} = \alpha_s(\mu)^{-2/9}\left[1 + \frac{\alpha_s(\mu)}{4\pi}J^{(1)} + \cdots\right].$$

$$(1.21)$$

Scheme dependence first appears in $J^{(1)}$ via the anomalous dimension $\gamma^{(1)}$ of operator $O^{\Delta S=2}$,

---

[2] It is customary to classify corrections according to the order of QCD perturbation theory used to determine them, e.g. 'leading' (LO), 'next-to-leading' (NLO), 'next-to-next-to leading' (NNLO) and so on. It is disturbing that $\eta_{cc}^{\text{NNLO}} \simeq 1.87$ is about 36% larger than $\eta_{cc}^{\text{NLO}} \simeq 1.38$. This is an unexpectedly large result, one which warrants further study.

$$J^{(1)} = \frac{\gamma^{(0)}\beta^{(1)}}{2\beta_0^2} - \frac{\gamma^{(1)}}{2\beta^{(0)}} = 12 \cdot \frac{153 - 19n_f}{(33 - 2n_f)^2} - \frac{1}{6} \cdot \frac{4n_f - 63}{33 - 2n_f}, \tag{1.22}$$

shown here for n_f flavors and in NDR renormalization. The above expression for $b(\mu)$ serves at the same time to define the perturbative factors η_{cc}, η_{tt}, and η_{ct}. For completeness, we display the most recent determinations [BrG 12] of the $\{\eta_i\}$ (with perturbative order shown as well),

$$\eta_{cc}^{\text{NNLO}} = 1.87(76) \qquad \eta_{tt}^{\text{NLO}} = 0.5765(65) \qquad \eta_{ct}^{\text{NNLO}} = 0.496(47). \tag{1.23}$$

The determination of $\langle K^0 | O^{\Delta S=2}(\mu) | \bar{K}^0 \rangle$ at a hadronic scale $\mu < m_c$ involves nonperturbative physics, so its evaluation by analytical means is problematic. It has become standard to express this quantity relative to its vacuum saturation value and introduce a parameter B_K as

$$\langle K^0 | O^{\Delta S=2}(\mu) | \bar{K}^0 \rangle = \frac{16}{3} F_K^2 m_K^2 \, B_K(\mu), \tag{1.24}$$

with $F_K = 110.4 \pm 0.6\,\text{MeV}$.[3] There has been substantial progress in the calculation of nonperturbative quantities such as B_K using lattice QCD methods. The scale- and scheme-independent version is defined as

$$\hat{B}_K = b(\mu) B_K(\mu) \tag{1.25}$$

and the value used in [BrG 12] is

$$\hat{B}_K = 0.737 \pm 0.020. \tag{1.26}$$

Finally, given present values for the CKM elements and the t-quark mass, the most important contribution to the *real* part of $\mathcal{H}_w^{\text{box}}$ is found to be from the c quark. In view of this, and noting that $H(x_c) \simeq 1$ (cf. Eq. (1.20)), we then have

$$\text{Re}\,\mathcal{H}_w^{\text{box}} \simeq \frac{G_F^2}{16\pi^2} m_c^2 \, \text{Re}\,\left(V_{cd}^* V_{cs}\right)^2 \, \eta_{cc}^{\text{NNLO}} b(\mu)\, O^{\Delta S=2}. \tag{1.27}$$

At this point, we have the ingredients for determining $(\Delta m)_{\text{theory}}^{\text{SD}}$ and one obtains [BrG 12]

$$(\Delta m)_{\text{theory}}^{\text{SD}} = (3.1 \pm 1.2) \times 10^{-15}\,\text{GeV}, \tag{1.28}$$

which is consistent with the value cited for Δm_{expt} in Eq. (1.15) within the quoted uncertainty.

---

[3] The reader should be wary of occasional notational confusion between F_K and $f_K = \sqrt{2} F_K$.

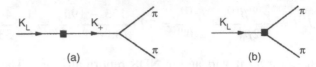

Fig. IX–3 Mechanisms for *CP* violation.

IX–2 The phenomenology of kaon *CP* violation

The $\pi\pi$ final state of kaon decay is even under *CP* provided the strong interactions are invariant under this symmetry. For the $\pi^0\pi^0$ system, this is clear since π^0 is itself a *CP* eigenstate, $CP|\pi^0\rangle = -|\pi^0\rangle$, and the two pions must be in an *S*-wave ($\ell = 0$) state,

$$CP|\pi^0\pi^0\rangle = (-1)^2(-1)^\ell|\pi^0\pi^0\rangle = +|\pi^0\pi^0\rangle. \tag{2.1}$$

The corresponding result for charged pions reflects the fact that π^+ and π^- are *CP*-conjugate partners, $CP|\pi^\pm\rangle = -|\pi^\mp\rangle$. We have seen that if *CP* were conserved, the two neutral kaons would organize themselves into *CP* eigenstates, with only K_S decaying into $\pi\pi$. Alternatively, K_L decays primarily into the $\pi\pi\pi$ final state, which is *CP*-odd if the pions are in relative *S* waves. The observation of *both* neutral kaons decaying into $\pi\pi$ is then a signal of *CP* violation.

There can be two sources of *CP* violation in $K_L \to \pi\pi$ decay. We have already seen that $K^0-\bar{K}^0$ mixing can generate a mixture of the *CP* eigenstates in physical kaons due to *CP* violation in the mass matrix. There also exists the possibility of *direct CP* violation in the weak decay amplitude, such that the *CP*-odd kaon eigenstate $|K^0_-\rangle$ makes a transition to $\pi\pi$. These two mechanisms are pictured in Fig. IX–3. The $K\pi\pi$ decay amplitudes have already been written down in Eq. (VIII–4.1) in terms of real-valued moduli A_0, A_2, and pion–pion scattering phases δ_0, δ_2. This decomposition is a consequence of Watson's theorem, and relies in part upon the assumption of time-reversal invariance. However, if direct *CP* violation occurs, A_0 and A_2 can themselves become complex-valued,

$$A_0 \equiv |A_0|e^{i\xi_0}, \qquad A_2 \equiv |A_2|e^{i\xi_2}, \tag{2.2}$$

with *CP* violation in the decay amplitude being characterized by the phases ξ_0 and ξ_2. Consequently, the $K_0 \to \pi\pi$ and $\bar{K}_0 \to \pi\pi$ decay amplitudes assume the modified form

$$A_{K^0\to\pi^+\pi^-} = |A_0|e^{i\xi_0}e^{i\delta_0} + \frac{|A_2|}{\sqrt{2}}e^{i\xi_2}e^{i\delta_2},$$

$$A_{\bar{K}^0\to\pi^+\pi^-} = -|A_0|e^{-i\xi_0}e^{i\delta_0} - \frac{|A_2|}{\sqrt{2}}e^{-i\xi_2}e^{i\delta_2}. \tag{2.3}$$

Using the definitions of K_L and K_S in Eq. (1.13), a straightforward calculation leads to the following measures of *CP* violation:

$$\frac{\langle \pi^+\pi^-|\mathcal{H}_{\rm w}|K_L\rangle}{\langle \pi^+\pi^-|\mathcal{H}_{\rm w}|K_S\rangle} \equiv \eta_{+-} \equiv \epsilon + \epsilon', \qquad \frac{\langle \pi^0\pi^0|\mathcal{H}_{\rm w}|K_L\rangle}{\langle \pi^0\pi^0|\mathcal{H}_{\rm w}|K_S\rangle} \equiv \eta_{00} \equiv \epsilon - 2\epsilon', \quad (2.4)$$

where

$$\epsilon = \bar{\epsilon} + i\xi_0,$$

$$\epsilon' = \frac{i e^{i(\delta_2-\delta_0)}}{\sqrt{2}}\left|\frac{A_2}{A_0}\right|(\xi_2 - \xi_0) = \frac{i e^{i(\delta_2-\delta_0)}}{\sqrt{2}}\left|\frac{A_2}{A_0}\right|\left(\frac{{\rm Im}\,A_2}{{\rm Re}\,A_2} - \frac{{\rm Im}\,A_0}{{\rm Re}\,A_0}\right). \quad (2.5)$$

The expression for ϵ can be simplified by approximating the numerical value $\Delta m/\Delta\Gamma = 0.475 \pm 0.001$ by $\Delta m/\Delta\Gamma \simeq 1/2$. This yields the approximate relation,

$$\frac{i}{\Delta m + \dfrac{i}{2}\Delta\Gamma} \simeq \frac{e^{i\pi/4}}{\sqrt{2}}\frac{1}{\Delta m}, \quad (2.6)$$

which we shall use repeatedly in the analysis to follow. In addition, since the rate for $K \to \pi\pi$ is much larger than that for $K \to \pi\pi\pi$, and $K^0 \to \pi\pi$ is in turn dominated by the $I = 0$ final state because of the $\Delta I = 1/2$ rule, we have

$$ {\rm Im}\,\Gamma_{12} \simeq \xi_0\Gamma_S \simeq 2\xi_0\Delta m. \quad (2.7)$$

The above relations allow us to write

$$\epsilon = \bar{\epsilon} + i\xi_0 \simeq \frac{e^{i\frac{\pi}{4}}}{\sqrt{2}}\left(\frac{{\rm Im}\,M_{12}}{\Delta m} - i\xi_0\right) + i\xi_0$$

$$= \frac{e^{i\frac{\pi}{4}}}{\sqrt{2}}\left(\frac{{\rm Im}\,M_{12}}{\Delta m} + \xi_0\right) = \frac{e^{i\frac{\pi}{4}}}{\sqrt{2}}\left(\frac{{\rm Im}\,M_{12}}{2{\rm Re}\,M_{12}} + \frac{{\rm Im}\,A_0}{{\rm Re}\,A_0}\right),$$

$$\epsilon' = \frac{i\omega}{\sqrt{2}}e^{i(\delta_2-\delta_0)}(\xi_2 - \xi_0) = \frac{i\omega e^{i(\delta_2-\delta_0)}}{\sqrt{2}}\left(\frac{{\rm Im}\,A_2}{{\rm Re}\,A_2} - \frac{{\rm Im}\,A_0}{{\rm Re}\,A_0}\right), \quad (2.8)$$

where $\omega \equiv {\rm Re}\,A_2/{\rm Re}\,A_0 \simeq 1/22$. All *CP*-violating observables must involve an interference of two amplitudes. In Eq. (2.8), the quantity ϵ expresses the interference of $K^0 \to \pi\pi$ with $K^0 \to \bar{K}^0 \to \pi\pi$, while ϵ' involves interference of the $I = 0$ and $I = 2$ final states.

The formulae for ϵ and ϵ' exhibit an important theoretical property. Since the choice of phase convention for any meson M is arbitrary, its state vector may be modified by the global strangeness transformation $|M\rangle \to e^{i\lambda S}|M\rangle$. For the $\bar{K}^0$ and K^0 states, this becomes

$$|K^0\rangle \to e^{i\lambda}|K^0\rangle, \qquad |\bar{K}^0\rangle \to e^{-i\lambda}|\bar{K}^0\rangle, \quad (2.9)$$

which has the effect,

$$\frac{\operatorname{Im} A_I}{\operatorname{Re} A_I} \to \frac{\operatorname{Im} A_I}{\operatorname{Re} A_I} + \lambda, \qquad \frac{\operatorname{Im} M_{12}}{\operatorname{Re} M_{12}} \to \frac{\operatorname{Im} M_{12}}{\operatorname{Re} M_{12}} - 2\lambda. \qquad (2.10)$$

We see that the values of ϵ and ϵ' are left unchanged. Various phase conventions appear in the literature. In the *Wu–Yang* convention, λ is chosen so that the A_0 amplitude is real-valued. This is always possible to achieve by properly choosing the phase of the kaon state. However, it is inconvenient for the Standard Model, where the A_0 amplitude naturally picks up a *CP*-violating phase. We shall therefore employ the convention in which no such additional phases occur in the definitions of the kaon states.

It was in the $K \to \pi\pi$ system that *CP* violation was first observed. The current status of measurements is

$$|\epsilon| = (2.228 \pm 0.011) \times 10^{-3},$$

$$\operatorname{Re}\left(\frac{\epsilon'}{\epsilon}\right) = (1.66 \pm 0.26) \times 10^{-3},$$

$$\varphi_{+-} \equiv \operatorname{phase}(\eta_{+-}) = (43.51 \pm 0.05)^\circ$$

$$\varphi_{00} \equiv \operatorname{phase}(\eta_{00}) = (43.52 \pm 0.05)^\circ. \qquad (2.11)$$

A violation of *CP* symmetry has also been observed in the semileptonic decays of K_L and K_S. These are related to matrix elements of the weak hadronic currents. Since K^0 must always decay into $e^+\nu_e\pi^-$ while $\bar{K}^0$ goes to $e^-\bar{\nu}_e\pi^+$, we have

$$A_{K_L \to \pi^- e^+ \nu_e} = \frac{1 + \bar{\epsilon}}{\sqrt{2}} A_{K^0 \to \pi^- e^+ \nu_e},$$

$$A_{K_L \to \pi^+ e^- \bar{\nu}_e} = \frac{1 - \bar{\epsilon}}{\sqrt{2}} A_{\bar{K}^0 \to \pi^+ e^- \bar{\nu}_e}. \qquad (2.12)$$

If the semileptonic decays proceed as in the Standard Model, there is no direct *CP* violation in the transition amplitude, so that

$$\frac{\Gamma_{K_L \to \pi^- e^+ \nu_e}}{\Gamma_{K_L \to \pi^+ e^- \bar{\nu}_e}} = \frac{1 + 2\operatorname{Re} \bar{\epsilon}}{1 - 2\operatorname{Re} \bar{\epsilon}} \simeq 1 + 4\operatorname{Re} \bar{\epsilon}. \qquad (2.13)$$

Since $\operatorname{Re} \bar{\epsilon} = \operatorname{Re} \epsilon$, the above asymmetry is sensitive to the same parameter as appears in the $K_L \to \pi\pi$ studies. Here, measurement gives

$$\operatorname{Re} \epsilon = (1.596 \pm 0.013) \times 10^{-3} = |\epsilon| \cos(44.3 \pm 0.8)^\circ, \qquad (2.14)$$

which is consistent with the experimental values from $K \to \pi\pi$.

Finally, precision experiments also probe the *CPT* transformation. For example, two such predictions, involving kaon masses ($m_{K^0} = m_{\bar{K}^0}$) and phases ($\varphi_{+-} = \varphi_{00}$ up to very small corrections from ϵ'), are seen to be consistent with existing data,

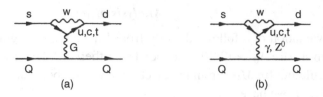

Fig. IX–4 (a) Penguin and (b) electroweak-penguin contributions to CP violation in $\Delta S = 1$ transitions.

$$\frac{|m_{K^0} - m_{\bar{K}^0}|}{m_{K^0}} \leq 8 \times 10^{-19},$$

$$\varphi_{00} - \varphi_{+-} = (0.2 \pm 0.4)°. \tag{2.15}$$

Further study of CPT invariance is left to Prob. IX–2.

IX–3 Kaon *CP* violation in the Standard Model

After diagonalization, there can remain a single phase in the CKM matrix. This phase generates the imaginary parts of amplitudes, which are required for CP violation. It is a physical requirement that results be invariant under rephasing of the quark fields. As a consequence, all observables must be proportional to

$$\text{Im } \Delta^{(4)} = A^2\lambda^6\eta = c_1c_2c_3s_1^2s_2s_3s_\delta, \tag{3.1}$$

written in the notation of Sect. II–4. This shows that all CP-violating signals must vanish if any of the CKM angles vanish. We shall now study the path whereby this phase is transferred from the lagrangian to experimental observables. For kaons, we have seen that the relevant amplitudes are those for K^0–$\bar{K}^0$ mixing ($\Delta S = 2$) and for $K \to \pi\pi$ decays ($\Delta S = 1$). Tree-level amplitudes in kaon decay can never be sensitive to the full rephasing invariant, so that one must consider loops. Typical diagrams are displayed in Fig. IX–4.

Experiment can help in simplifying the theoretical analysis. Note that ϵ' is sensitive to $\Delta S = 1$ physics through the penguin diagram [GiW 79], while ϵ is sensitive to $\Delta S = 2$ mass-matrix physics as well as to $\Delta S = 1$ effects. However, since experiment tells us that $|\epsilon| \gg |\epsilon'|$, it follows that the $\Delta S = 1$ contributions to ϵ must be small. Likewise, the long-distance contributions of Fig. IX–2 and the contribution of Fig. IX–1(d) must both be small because each also involves the $\Delta S = 1$ interaction. This leaves the box diagrams of Fig. IX–1(a),(b) as the dominant component of ϵ. Moreover, since the CKM phase δ is associated with the heavy-quark couplings, only the heavy-quark parts of the box diagrams are needed. Hence ϵ is very clearly short-distance dominated.

Analysis of $|\epsilon|$

The evaluation of ϵ follows directly from Eq. (2.8). To begin, we shall ignore the tiny $\mathrm{Im}\,A_0/\mathrm{Re}\,A_0 \sim \mathcal{O}(10^{-5})$ dependence therein.[4] This leaves us with the issue of calculating $\mathrm{Im}\,M_{12}$. From the discussion of the 'box' hamiltonian $\mathcal{H}_{\mathrm{w}}^{\mathrm{box}}$ given in Sect. IX–1, we have

$$\mathrm{Im}\,M_{12} = \frac{G_F^2}{3\pi^2} F_K^2 m_K \hat{B}_K A^2 \lambda^6 \bar{\eta}$$
$$\times \left[\eta_{cc} m_c^2 H(x_c) - \eta_{tt} m_t^2 H(x_t) A^2 \lambda^4 (1 - \bar{\rho}) - \eta_{ct} m_c^2 \bar{G}(x_c, x_t) \right]. \quad (3.2)$$

Some CKM-related relations and definitions useful in obtaining the above form are

$$\mathrm{Re}\,\xi_c = -\lambda \left(1 - \frac{\lambda^2}{2} \right), \qquad \mathrm{Re}\,\xi_t = -\lambda \left(1 - \frac{\lambda^2}{2} \right) A^2 \lambda^4 (1 - \bar{\rho}),$$

$$\mathrm{Im}\,\xi_c = -\mathrm{Im}\,\xi_t = -\eta A^2 \lambda^5,$$

$$\bar{\rho} \equiv \rho \left(1 - \frac{\lambda^2}{2} \right), \qquad \bar{\eta} \equiv \eta \left(1 - \frac{\lambda^2}{2} \right).$$

From Eq. (2.8) and Eq. (3.2), we obtain the Standard Model prediction,

$$|\epsilon|_{\mathrm{SM}} = \frac{G_F^2}{3\sqrt{2}\pi^2} \frac{F_K^2 m_K \hat{B}_K A^2 \lambda^6 \bar{\eta}}{\Delta m_K}$$
$$\times \left[\eta_{cc} m_c^2 - \eta_{tt} m_t^2 H(x_t) A^2 \lambda^4 (1 - \bar{\rho}) - \eta_{ct} m_c^2 \bar{G}(x_c, x_t) \right], \quad (3.3)$$

roughly in accord with the experimental value, given the uncertainties in several of the above factors, when lattice determinations of the $\hat{B}_K$ parameter are used.

Analysis of $|\epsilon'|$

The importance of ϵ' lies in the fact that it proves that CP violation also occurs in the direct $\Delta S = 1$ weak transition, which is a hallmark of the Standard Model's pattern of CP breaking. For this process, the CP-violating phases from the CKM elements can occur only in loop diagrams, and these appear in the penguin diagram and in the electroweak penguin process in which the gluon is replaced by a photon or a Z^0 boson, as shown in Fig. IX–4. At first sight, it appears surprising that the electroweak penguin plays any significant role, as it is suppressed by a power of α compared to the gluonic penguin. However, recall from Eq. (2.8), that ϵ' measures the relative phase difference of the $K \to \pi\pi$ amplitudes A_0 and A_2

$$|\epsilon'| = \frac{\omega}{\sqrt{2}} \left| \frac{\mathrm{Im}\,A_2}{\mathrm{Re}\,A_2} - \frac{\mathrm{Im}\,A_0}{\mathrm{Re}\,A_0} \right| \simeq 0.032 \left| \frac{\mathrm{Im}\,A_2}{\mathrm{Re}\,A_2} - \frac{\mathrm{Im}\,A_0}{\mathrm{Re}\,A_0} \right|. \quad (3.4)$$

---

[4] Besides, the combination $\mathrm{Im}\,A_0/\mathrm{Re}\,A_0$ also contributes to ϵ' (as seen in Eq. (2.8)) and since $|\epsilon'| \ll |\epsilon|$, the contribution of this ratio to ϵ is presumably ignorable.

The gluonic penguin only contributes an imaginary part to A_0 because its effect is purely $\Delta I = 1/2$. The electroweak penguin involves an extra factor of the electric charge $Q = \frac{1}{2}\lambda_3 + \frac{1}{2\sqrt{3}}\lambda_8$, which means that the corresponding operator has both $\Delta I = 1/2,\ 3/2$ components and can contribute an imaginary part to A_2. Because the real part of A_2 is much smaller than that of A_0, by a factor of $\omega \equiv \mathrm{Re}A_2/\mathrm{Re}A_0 \sim 1/22$, the effect of the electroweak penguin is enhanced by the small denominator. However, while both diagrams make important contributions, it does appear that the gluonic penguin is the larger effect.

The ingredients to ϵ' can be expressed numerically [CiFMRS 95] at the scale $\mu = 2$ GeV in the $\overline{\mathrm{MS}}$–NDR scheme as

$$\frac{\epsilon'}{\epsilon} = 2 \cdot 10^{-3} \left(\frac{\mathrm{Im}\,(V_{td}^* V_{ts})}{1.3 \times 10^{-3}} \right) \left[2.0\ \mathrm{GeV}^{-3} \langle (\pi\pi)_{I=0} | O_6 | K^0 \rangle_{2\ \mathrm{GeV}} (1 - \Omega_{\mathrm{IB}}) \right.$$
$$\left. - 0.5\ \mathrm{GeV}^{-3} \langle (\pi\pi)_{I=2} | O_8 | K^0 \rangle_{2\ \mathrm{GeV}} - 0.06 \right]. \tag{3.5}$$

Here, we see the primary dependence of the gluonic penquin effect in the matrix elements of the penguin operator O_6, while the electroweak-penguin (EWP hereafter) operator is O_8. These operators refer back to the decomposition of Eq. (VIII–3.31). The factor Ω_{IB} describes isospin breaking.

As we mentioned in Sect. VIII–4, present lattice methods are able to calculate the A_2 amplitude with reasonable precision, while the isospin-zero final-state amplitude A_0 remains uncalculable. This means that the EWP contribution can be obtained, with the result [Bl *et al.* 12]

$$\mathrm{Re}\left(\frac{\epsilon'}{\epsilon}\right)_{\mathrm{EWP}} = -(6.25 \pm 0.44_{\mathrm{stat}} \pm 1.19_{\mathrm{syst}}) \times 10^{-4}, \tag{3.6}$$

which has the *opposite* sign from the experimental result and is about one third the magnitude. A chiral analysis that we will describe shortly agrees with this. This implies that the phase due to the gluonic penguin $\mathrm{Im}\,A_0/\mathrm{Re}\,A_0$ must be the larger effect and must have the same sign as the experimental determination. This seems reasonable in estimates which have been made, as discussed in [CiENPP 12]. However, it means that we do not yet have a precise prediction for ϵ' within the Standard Model.

Chiral analysis of $(\epsilon'/\epsilon)_{EWP}$

The chiral symmetry approach to low-energy hadron dynamics emphasized earlier in this book can be used to analyze the electroweak-penguin contribution to ϵ'/ϵ in the chiral limit.

$$\lim_{p=0} \langle (\pi\pi)_{I=2} | O_8 | K^0 \rangle_\mu = -\frac{2}{F_\pi^{(0)}} \left[\frac{1}{3} \langle 0 | \mathcal{Q}_1 | 0 \rangle_\mu + \frac{1}{2} \langle 0 | \mathcal{Q}_8 | 0 \rangle_\mu \right]$$

$$\simeq -\frac{1}{F_\pi^{(0)}} \langle 0 | \mathcal{Q}_8 | 0 \rangle_\mu \qquad (3.7a)$$

where $\mathcal{Q}_8$ is the four-quark operator

$$\mathcal{Q}_8 \equiv \bar{q}\gamma^\mu \lambda^a \frac{\tau_3}{2} q \bar{q}\gamma_\mu \lambda^a \frac{\tau_3}{2} q - \bar{q}\gamma^\mu \gamma_5 \lambda^a \frac{\tau_3}{2} q \bar{q}\gamma_\mu \gamma_5 \lambda^a \frac{\tau_3}{2} q, \qquad (3.7b)$$

and, for notational simplicity, we have suppressed dependence on a second four-quark operator $\langle 0 | \mathcal{Q}_1 | 0 \rangle_\mu$ since $\langle 0 | \mathcal{Q}_8 | 0 \rangle_\mu \gg \langle 0 | \mathcal{Q}_1 | 0 \rangle_\mu$ [CiDGM 01].[5] This relation can be found either by constructing effective lagrangians or by use of the soft-pion theorem of App. B–3.

Thus, a chiral estimate of the EWP part of ϵ'/ϵ amounts to determining the vacuum matrix element of $\mathcal{Q}_8$. It turns out that such information is obtainable from the large Q^2 behavior of $V - A$ correlators measured in τ decay (cf. Sect. V–3),

$$\Delta\Pi(Q^2) \equiv \left(\Pi_{V,3} - \Pi_{A,3} \right)(Q^2). \qquad (3.8)$$

The operator-product expansion (OPE) reveals that $\Delta\Pi(Q^2)$ obeys the asymptotic behavior

$$\Delta\Pi(Q^2) \sim \frac{1}{Q^6} \left[a_6(\mu) + b_6(\mu) \ln \frac{Q^2}{\mu^2} \right] + \mathcal{O}(Q^{-8}), \qquad (3.9a)$$

where, from a two-loop study [CiDGM 01], we have

$$a_6(\mu) = 2\pi \langle 0 | \alpha_s \mathcal{Q}_8 | 0 \rangle_\mu + \frac{25}{4} \langle 0 | \alpha_s^2 \mathcal{Q}_8 | 0 \rangle_\mu + \cdots ,$$

$$b_6(\mu) = -\langle 0 | \alpha_s^2 \mathcal{Q}_8 | 0 \rangle_\mu + \cdots , \qquad (3.9b)$$

where the ellipses represent higher-order terms in the OPE. Thus, the needed information (i.e. $\langle 0 | \mathcal{Q}_8 | 0 \rangle_\mu$) is contained in the large energy component of $\Delta\Pi(Q^2)$, but how can we access it? This problem has been solved in two different papers, which use two alternative approaches. In the first of these [CiDGM 01], one employs sum rules like

$$\langle 0 | \mathcal{Q}_8 | 0 \rangle_\mu = \int_0^\infty ds \, s^2 \frac{\mu^2}{s + \mu^2} \Delta\rho(s) + \cdots , \qquad (3.10)$$

where the ellipses denote contributions from $d > 6$ terms in the OPE. This approach yields a determination $\left[\epsilon'/\epsilon \right]_{\text{EWP}}^{(0)} = (-22 \pm 7) \times 10^{-4}$, having a 32% uncertainty. The superscript indicates working in the chiral limit of massless u, d, s quarks. A second method [CiGM 03], which analyzes tau decay spectral functions by using a finite-energy sum rule (FESR), leads to $\left[\epsilon'/\epsilon \right]_{\text{EWP}}^{(0)} = (-15.0 \pm 2.7) \times 10^{-4}$, having

---

[5] The effect of $\langle 0 | \mathcal{Q}_1 | 0 \rangle_\mu$ is, of course, included in the full analysis of [CiDGM 01].

an 18% percent uncertainty. Upon including chiral corrections, the physical result $\left[\epsilon'/\epsilon\right]_{\text{EWP}} = (-11.0 \pm 3.6) \times 10^{-4}$ is obtained. Together with the lattice evaluation quoted in Eq. (3.6), these evaluations firmly imply that $\left[\epsilon'/\epsilon\right]_{\text{EWP}} < 0$ and that the QCD penguin effect must be large and positive in order to reproduce the experimental value for ϵ'/ϵ of Eq. (2.11).

IX–4 The strong *CP* problem

The possibility of a θ term in the QCD lagrangian raises potential problems (see Sect. III–5). For $\theta \neq 0$, QCD will in general violate parity and, even worse, time-reversal invariance. The strength of T violation (and hence, by the *CPT* theorem, *CP* violation) is known to be small, even by the standards of the weak interaction. This knowledge comes from both the observed $K_L \to 2\pi$ decay and bounds on electric dipole moments. From these it becomes clear that QCD must be T invariant to a very high degree. However, there is nothing within the Standard Model which would force the θ parameter to be small; indeed, it is a free parameter lying in the range $0 \leq \theta \leq 2\pi$. The puzzle of why $\theta \simeq 0$ in Nature is called the *strong CP problem*.

One is tempted to resolve the issue with an easy remedy first. If QCD were the only ingredient in our theory, we could remove the strong *CP* problem by imposing an additional discrete symmetry on the QCD lagrangian, the discrete symmetry being *CP* itself. This wouldn't really explain anything but would at least reduce a continuous problem to a discrete choice. In reality, this will not work for the full Standard Model since, as we have seen, the electroweak sector inherently violates *CP*. It would thus be improper to impose *CP* invariance upon the full lagrangian. Moreover, even if one could set $\theta_{\text{bare}} = 0$ in QCD, electroweak radiative corrections would generate a nonzero value. These turn out to occur only at high orders of perturbation theory, and are expected to be divergent by power-counting arguments, although they have not been explicitly calculated. This divergence is not a fundamental problem because one could simply absorb θ_{bare} plus the divergence into a definition of a renormalized parameter θ_{ren}, which could be inferred from experiment. However, we are then back to an arbitrary value of θ_{ren} and to the problem of why θ_{ren} is small.

The parameter $\bar{\theta}$

The situation is actually worse than this in the full Standard Model, as the quark mass matrix can itself shift the value of θ by an unknown amount. Recall that *CP* violation in the Standard Model arises from the Yukawa couplings between the Higgs doublet and the fermions. When the Higgs field picks up a vacuum expectation value, these couplings produce mass matrices for the quarks, which are neither diagonal nor *CP*-invariant. The mass matrices are diagonalized by separate left-handed and right-handed transformations, and *CP* violation is shifted to the weak mixing matrix. However, because different left-handed and right-handed rotations are generally required, one encounters an axial $U(1)$ rotation in this transformation to the quark mass eigenstates and, as discussed in Sect. III–5, this produces a shift in the value of θ. Let us determine the magnitude of this shift. Denoting by primes the original quark basis, one has the transformation to mass eigenstates given by (cf. Eqs. II–4.5,4.6)

$$\mathbf{m} = S_L^\dagger \mathbf{m}' S_R, \qquad \psi_L = S_L^\dagger \psi_L', \qquad \psi_R = S_R^\dagger \psi_R'. \tag{4.1}$$

Here, we have combined the u and d mass matrices into a single mass matrix. Expressing $S_{L,R}$ as products of $U(1)$ and $SU(N)$ factors,

$$S_L = e^{i\varphi_L} \overline{S}_L, \qquad S_R = e^{i\varphi_R} \overline{S}_R, \tag{4.2}$$

with $\overline{S}_L$, $\overline{S}_R$ in $SU(N)$, one obtains an axial $U(1)$ transformation angle of $\varphi_R - \varphi_L$. From the discussion of Sect. III–5, this is seen to lead to a change in the θ parameter,

$$\theta \to \overline{\theta} = \theta + 2N_f(\varphi_L - \varphi_R), \tag{4.3}$$

where $N_f = 6$ for the three-generation Standard Model. However, noting that the final mass matrix $\mathbf{m}$ is purely real, we have

$$
\begin{aligned}
\arg(\det \mathbf{m}) = 0 &= \arg\left(\det S_L^\dagger \det \mathbf{m}' \det S_R\right) \\
&= \arg\left(\det S_L^\dagger\right) + \arg\left(\det \mathbf{m}'\right) + \arg\left(\det S_R\right) \\
&= 2N(\varphi_R - \varphi_L) + \arg\left(\det \mathbf{m}'\right),
\end{aligned} \tag{4.4}
$$

where we have used the $SU(N)$ property, $\det \overline{S}_R = \det \overline{S}_L = 1$. The resultant θ parameter is then

$$\overline{\theta} = \theta + \arg\left(\det \mathbf{m}'\right), \tag{4.5}$$

with $\mathbf{m}'$ being the original nondiagonal mass matrix. The real strong *CP* problem is to understand why $\overline{\theta}$ is small.

One possible solution to the strong *CP* problem occurs if one of the quark masses vanishes. In this case, the ability to shift θ by an axial transformation would allow one to remove the effect of θ by performing an axial phase transformation on the massless quark. Equivalently stated, any effect of θ must vanish if any quark mass vanishes. Unfortunately, phenomenology does not favor this solution. The u quark is the lightest, but a value $m_u \neq 0$ is favored.

Connections with the neutron electric dipole moment

The $\bar\theta$ term is not the source of the observed *CP* violation in K decays. This can be seen because it occurs in a $\Delta S = 0$ operator, and while this may ultimately generate effects in $\Delta S = 1$ processes, its influence is stronger in the $\Delta S = 0$ sector. In particular, it generates an electric dipole moment d_e for the neutron. Since no such dipole moment has been detected, one can obtain a bound on the magnitude of $\bar\theta$.

To determine the effect of $\bar\theta$, it is most convenient to use a chiral rotation to shift the $\bar\theta$ dependence back into the quark mass matrix. A small axial transformation produces the modified mass matrix

$$\mathcal{L}_{\text{mass}} = \bar\psi \begin{pmatrix} m_u & & \\ & m_d & \\ & & m_s \end{pmatrix} \psi + i\eta\bar\psi T\gamma_5\psi \equiv \bar\psi_L \tilde M \psi_R + \bar\psi_R \tilde M^\dagger \psi_L, \quad (4.6)$$

where η is a small parameter proportional to $\bar\theta$ having units of mass, and T is a 3×3 hermitian matrix. Consistency requires T to be proportional to the unit matrix. If this were not the case, and instead we wrote $T \equiv 1 + \lambda_i T_i/2$, the effective lagrangian would start out with a term linear in the meson fields,

$$\mathcal{L}_{\text{eff}} \sim i\eta \, \text{Tr}\left(TU^\dagger - UT^\dagger\right) = 2\frac{\eta}{F_\pi}\left(T_3\pi_0 + T_8\eta_8 + \cdots\right), \quad (4.7)$$

rather than the usual quadratic dependence. The vacuum would then be unstable because it could lower its energy by producing nonzero values of, say, the π_0 field. Thus, to incorporate θ-dependence without disturbing vacuum stability, one chooses $T = 1$. The act of rotating away any dependence on $\bar\theta$ produces a nonzero value of $\arg(\det \tilde M)$, and also determines η,

$$\bar\theta = \arg(\det \tilde M) = \arg\left[(m_u + i\eta)(m_d + i\eta)(m_s + i\eta)\right],$$
$$\eta \simeq \bar\theta \frac{m_u m_d m_s}{m_u m_d + m_u m_s + m_d m_s} \quad \text{(for small } \eta\text{)}, \quad (4.8)$$

such that the mass terms become

$$\mathcal{L}_{\text{mass}} = m_u \bar u u + m_d \bar d d + m_s \bar s s$$
$$+ i\bar\theta \frac{m_d m_d m_s}{m_u m_d + m_u m_s + m_d m_s}\left(\bar u \gamma_5 u + \bar d \gamma_5 d + \bar s \gamma_5 s\right). \quad (4.9)$$

The last term is the *CP*-violating operator of the *QCD* sector. Note that, as expected, $\bar\theta$ vanishes if any quark is massless.

A nonzero neutron electric dipole moment d_e requires both the action of the above *CP*-odd operator and that of the electromagnetic current,

$$d_e \bar{u}(\mathbf{p}')\sigma_{\mu\nu}q^\nu\gamma_5 u(\mathbf{p}) = \sum_I \langle n(\mathbf{p}') | \mathcal{L}_{\text{mass}}^{CP\text{-}odd} | I \rangle \frac{1}{E_n - E_I} \langle I | J_\mu^{\text{em}} | n(\mathbf{p}) \rangle, \quad (4.10)$$

where $q = p' - p$ and we have inserted a complete set of intermediate states $\{I\}$ in the neutron-to-neutron matrix element. For intermediate baryon states, the matrix elements of $\bar{\psi}\gamma_5\psi$ are dimensionless numbers of order unity and magnetic moment effects are of order the nucleon magneton, μ_n. Thus, we find for d_e,

$$d_e \simeq \bar{\theta} \frac{m_u m_d m_s}{m_u m_d + m_u m_s + m_d m_s} \frac{e\mu_n}{\Delta M}, \quad (4.11)$$

where ΔM is some typical energy denominator. Using $\Delta M = 300$ MeV, we obtain

$$d_e \sim \bar{\theta} \times 10^{-15} \text{ e-cm}. \quad (4.12)$$

Far more sophisticated methods have been used to calculate this, with results that have a spread of values [EnRV 13]. Our simple estimate is near the average. In explicit calculations, some subtlety is required because one must be sure that the evaluation correctly represents the $U(1)_A$ behavior of the theory. However, the precise value is not too important; the significant fact is that bounds on $d_e \lesssim 3 \times 10^{-26}$ e-cm require $\bar{\theta}$ to be tiny, $\bar{\theta} \lesssim 10^{-11}$.

The strong *CP* problem does not have a good resolution within the Standard Model. It would appear that the abnormally small value of $\bar{\theta}$, and of the cosmological constant as well, are indications that more physics is required beyond that contained in the Standard Model.

Problems

(1) **Strangeness gauge invariance**

    (a) Physics must be invariant under a global strangeness transformation $|M\rangle \rightarrow \exp(i\lambda S)|M\rangle$, where λ is arbitrary. Explain why this is the case.

    (b) Demonstrate that such a transformation has the effect

$$\frac{\text{Im } A_I}{\text{Re } A_I} \rightarrow \frac{\text{Im } A_I}{\text{Re } A_I} + \lambda, \quad \frac{\text{Im } M_{12}}{\text{Re } M_{12}} \rightarrow \frac{\text{Im } M_{12}}{\text{Re } M_{12}} - 2\lambda,$$

    as claimed in Eq. (2.10), and that, while unphysical quantities such as $\bar{\epsilon}, \xi_0$ are affected by such a change, physical parameters such as ϵ, ϵ' are not.

(2) **Neutral kaon mass matrices and *CPT* invariance**

Some of the ideas discussed in this chapter can be addressed in terms of simple models of the neutral kaon mass matrix M which appears in Eq. (1.2).

(a) Consider the following *CP*-conserving parameterization as defined in the $(K^0, \bar{K}^0)$ basis:

$$M_0 = \begin{pmatrix} m_0 & \Delta \\ \Delta & m_0 \end{pmatrix},$$

where Δ is real-valued. Determine the basis states (K_-, K_+) in which $M_0 \rightarrow M_\pm$ becomes diagonal and obtain numerical values for m_0, Δ.

(b) Working in the (K_-, K_+) basis, extend the model of (a) to allow for *CP* violation by introducing a real-valued parameter δ,

$$M_\pm = \begin{pmatrix} m_- & 0 \\ 0 & m_+ \end{pmatrix} \rightarrow M_\pm' = \begin{pmatrix} m_- & -i\delta \\ i\delta & m_+ \end{pmatrix},$$

and assume there is no direct *CP* violation. This mass matrix corresponds to the *superweak* (*SW*) model. By expressing $M_\pm'$ in the $(K^0, \bar{K}^0)$ basis, use the analysis of Sects. IX–1,2 to predict $\varphi_\epsilon^{(SW)} \equiv$ phase ϵ and determine δ from the measured value of $|\epsilon|$.

(c) Finally, extend the model in (b) to

$$M_\pm'' = \begin{pmatrix} m_- & \chi \\ \chi^* & m_+ \end{pmatrix},$$

where Re χ is a *T*-conserving, *CP*-violating, and *CPT*-violating parameter. Show that the states which diagonalize $M_\pm''$ are

$$|K_S\rangle \simeq |K_+\rangle - \frac{\chi}{\mathcal{D}}|K_-\rangle,$$

$$|K_L\rangle \simeq |K_-\rangle + \frac{\chi^*}{\mathcal{D}}|K_+\rangle,$$

where $\mathcal{D} \equiv (m_L - m_S)/2 + i\Gamma_S/4$. Then evaluate η_{+-} and η_{00}, allowing for the presence of direct *CP* violation (i.e. $\epsilon' \neq 0$), and derive the following relation between phases,

$$|\chi|\left(\frac{2}{3}\varphi_{+-} + \frac{1}{3}\varphi_{00} - \varphi_\epsilon^{(SW)}\right) = \frac{1}{2m_{K^0}} \cdot \frac{|m_{\bar{K}^0} - m_{K^0}|}{m_L - m_S} \sin \varphi_\epsilon^{(SW)}.$$

The result $|m_{\bar{K}^0} - m_{K^0}|/m_{K^0} < 5 \times 10^{-18}$, which follows from this relation, is the best existing limit on *CPT* invariance.

X

The N_c^{-1} expansion

The N_c^{-1} expansion is an attempt to create a perturbative framework for *QCD* where none exists otherwise. One extrapolates from the physical value for the number of colors, $N_c = 3$, to the limit $N_c \to \infty$ while scaling the *QCD* coupling constant so that $g_3^2 N_c$ is kept fixed ['tH 74]. The amplitudes in the theory are then analyzed in powers of N_c^{-1}. The hope is that the $N_c \to \infty$ world bears sufficient resemblance to the real world to yield significant dynamical insights. There is no magical process which makes the $N_c \to \infty$ theory analytically trivial; nonlinearities of the nonabelian gauge interactions are present, and the theory is still not solvable. However any consistent approximation scheme for *QCD* is welcome, and the large N_c expansion is especially useful for organizing one's thoughts in the analysis of hadronic processes.

X–1 The nature of the large N_c limit

In passing from $SU(3)$ to $SU(N_c)$, the quark and gluon representations, originally **3** and **8**, become $\mathbf{N_c}$ and $\mathbf{N_c^2 - 1}$ respectively. The analysis of Feynman graphs at large N_c is simplified by modifying the notation used to describe gluons. As usual, quarks carry a color label j, with $j = 1, 2, \ldots, N_c$. Gluons can be described by two such labels, i.e.

$$A_\mu^a \; \to \; A_{\mu j}^{\ k} \qquad (A_{\mu j}^{\ j} = 0), \qquad (1.1)$$

where $a = 1, \ldots, N_c^2 - 1$ and $j, k = 1, \ldots, N_c$. In doing so, no approximation is being made. The new notation is simply an embodiment of the group product $\mathbf{N_c} \times \mathbf{N_c} \to (\mathbf{N_c^2 - 1}) \oplus \mathbf{1}$. The quark–gluon coupling is then written

$$g_3 \bar\psi^j \gamma^\mu \psi_k A_{\mu j}^{\ k}, \qquad (1.2)$$

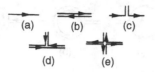

Fig. X–1 Double-line notation: (a) quark and (b) gluon propagators, (c) quark–gluon, (d) three-gluon, and (e) four-gluon vertices.

and the gluon propagator is

$$\int d^4x \, e^{iq \cdot x} \langle 0 \left| T \left(A^i_{\mu j}(x) A^k_{\nu l}(0) \right) \right| 0 \rangle = \left(\delta^i_l \delta^k_j - N_c^{-1} \delta^i_j \delta^k_l \right) i \, D_{\mu\nu}(q). \qquad (1.3)$$

The term proportional to N_c^{-1} must be present to ensure that the color singlet combination vanishes, $A^j_{\mu j} = 0$. However, as long as we avoid the color-singlet channel, this term will be suppressed in the large N_c limit and may be dropped when working to leading order.

Using this new notation, the Feynman diagrams for propagators and vertices are displayed in Fig. X–1. A solid line is drawn for each color index, and each gluon is treated as if it were a quark–antiquark pair (as far as color is concerned). In this double-line notation, certain rules which are obeyed by amplitudes to leading order in $1/N_c$ emerge in an obvious manner. Although general topological arguments exist, we shall review these rules by examining the behavior of specific graphs. The power of Feynman diagrams to build intuition is rather compelling in this case.

We consider first the familiar quark and gluon propagators. The quark propagator, unadorned by higher-order corrections, is $\mathcal{O}(1)$ in the $N_c \to \infty$ limit. Fig. X–2 depicts two radiative corrections. Fig. X–2(a), the one-gluon loop, is $\mathcal{O}(1)$ in powers of N_c because the suppression from the squared coupling g_3^2 is compensated for by the single closed loop, which corresponds to a sum over a free color index and thus contributes a factor of N_c. The graph then is of order $g_3^2 N_c$, which is taken to be constant. The graph Fig. X–2(b) with overlapping gluon loops is $\mathcal{O}(N_c^{-2})$ because, with no free color loops, it is of order $g_3^4 = \left(g_3^2 N_c \right)^2 N_c^{-2} \sim N_c^{-2}$. The terms *planar* and *nonplanar* are used, respectively, to describe Figs. X–2(a),(b), because the latter cannot be drawn in the plane without at least some internal lines crossing each other.

Four distinct contributions to the gluon propagator are exhibited in Fig. X–3. Figs. X–3(a),(b) depict in double line notation the quark–antiquark and two-gluon

Fig. X–2 Radiative corrections to the quark propagator: (a) planar, (b) nonplanar.

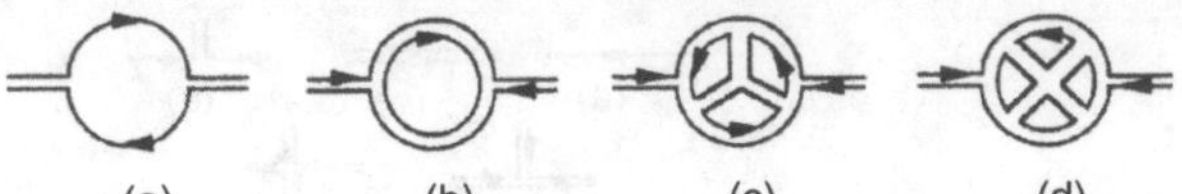

(a)   (b)   (c)   (d)

Fig. X–3 Various radiative corrections to the gluon propagator.

loop contributions. It should be obvious from the above discussion that these are respectively $\mathcal{O}(N_c^{-1})$ and $\mathcal{O}(1)$. A new diagram, involving the three-gluon coupling, appears in Fig. X–3(c). With three color loops and six vertices, it is of order $\left(g_3^2 N_c\right)^3 = \mathcal{O}(1)$. Figure X–3(d) is a nonplanar process with six vertices and one color sum, and is thus $\mathcal{O}(N_c^{-2})$.

The discussion of the gluon propagator indicates why we constrain $g_3^2 N_c$ to be fixed when taking the large N_c limit. The beta function of *QCD* is determined to leading order by Figs. X–3(a),(b). If g_3^2 were held fixed, the beta function would become infinite in the large N_c limit, leading to the immediate onset of asymptotic freedom. The choice $g_3^2 N_c \sim$ constant leads to a running coupling constant and is compatible with the behavior for the realistic case of $N_c = 3$.

To summarize, there are several rules which can be abstracted from examples such as these: (i) the leading-order contributions are planar diagrams containing the minimum number of quark loops; (ii) each internal quark loop is suppressed by a factor of N_c^{-1}; and (iii) nonplanar diagrams are suppressed by factors of N_c^{-2}. The suppressions in rules (ii), (iii) are combinatorial in origin. Quark loops and nonplanarities each limit the number of color-bearing intermediate states, and consequently cost factors of N_c^{-1}.

X–2 Spectroscopy in the large N_c limit

In order for the large N_c limit to be relevant to the real world, it must be assumed that confinement of color-singlet states continues to hold. In this case, we expect the particle spectrum to continue to be divided into mesons and baryons. Let us treat the mesons first.

One can form color-singlet meson contributions from $Q\bar{Q}$ pairs. To form a color singlet, one must sum over the quark colors. In order to produce a properly normalized $Q\bar{Q}$ state one must therefore include a normalization factor of $N_c^{-1/2}$ into each $Q\bar{Q}$ meson wavefunction, such that

$$\left| Q^{(\alpha)} \bar{Q}^{(\beta)} \right\rangle_{\substack{\text{color} \\ \text{singlet}}} \sim \frac{1}{\sqrt{N_c}} b_i^{(\alpha)\dagger} d_i^{(\beta)\dagger} |0\rangle, \qquad (2.1)$$

where α, β are *flavor* labels, $i = 1, \ldots, N_c$ is the *color* label, and $b^\dagger$ ($d^\dagger$) are the quark (antiquark) creation operators. Meson propagators, as represented in

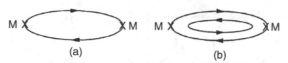

Fig. X–4 Mesons in the double line notation.

Fig. X–4(a), are then $\mathcal{O}(1)$ in N_c since the factors of $(N_c^{-1/2})^2$ from the normalization of the wavefunction are compensated by a factor of N_c from the quark loop. This leads to the prediction that meson masses are of $\mathcal{O}(1)$ in the large N_c limit, i.e., they remain close to their physical values. Multiquark intermediate states, as in Fig. X–4(b), are suppressed by $1/N_c$, indicating a suppression of mixing between $Q\bar{Q}$ and $Q^2\bar{Q}^2$ sectors. That is, large N_c plus confinement implies the existence of $Q\bar{Q}$ mesons which contain an arbitrary amount of glue in their wavefunction, but which do not mix with $Q^2\bar{Q}^2$ states.

The quark content of a given hadron remains an issue of some theoretical and phenomenological interest. Several examples are given in Sect. XIII–4 of hadron states which are thought to be 'nonconventional'. One such is the σ hadron, which is the lightest resonance found in $\pi\pi$ scattering. Analysis has yielded insight as to the N_c dependence of the σ mass (cf. Eq. (XIII–4.7)). We reserve further comment on this interesting topic to Chap. XIII.

What about the decay widths of $Q\bar{Q}$ mesons? The decay amplitude is pictured in Fig. X–5 (other possibilities involve the suppressed quark loops). This diagram contains three meson wavefunctions and one quark loop and hence is of order $(N_c^{-1/2})^3 N_c = N_c^{-1/2}$ in amplitude or N_c^{-1} in rate. The large N_c limit thus involves narrow resonances, i.e., $\Gamma/M \to 0$, where Γ is the meson decay width and M is the meson mass. This is reasonably similar to the real world, where most of the observed resonances have $\Gamma/M \sim 0.1$–0.2 [RPP 12].

Color-singlet gluonic states, called *glueballs*, may also exist. The normalization of a glueball state can be fixed by means of the following argument. Suppose, as will be defined in a gauge-invariant manner in Sect. XIII–4, that a neutral meson can be created from two gluons. Then in normalizing this configuration, one must sum over the N_c^2 gluon color labels. As a consequence, a normalization factor N_c^{-1} is associated with each glueball state. Glueball propagators also emerge as being $\mathcal{O}(1)$. There is no physical distinction between two-gluon states, three-gluon states,

Fig. X–5 Strong interaction decay of a $Q\bar{Q}$ meson.

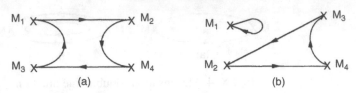

Fig. X–6 Meson–meson scattering.

etc., because all are mixed with each other by the strong interaction. As a result, there need not be any simple association between a specific physical state and gluon number, and thus the concept of a 'constituent gluon' need not be inferred. In glueball decays, however, one must distinguish between glueballs decaying to other glueballs, and those decaying to $Q\bar{Q}$ mesons. Where kinematically allowed, the decay of glueballs to glueballs is $\mathcal{O}(1)$, while that to $Q\bar{Q}$ states is $\mathcal{O}(1/N_c)$. The lowest-lying glueball(s) will then be narrow, while those above the threshold for decay into two glueballs will be of standard, nonsuppressed width.

Meson–meson scattering amplitudes are also restricted by large N_c counting rules. Consider the diagrams of Fig. X–6. That of Fig. X–6(a) is of order $(N_c^{-1/2})^4 N_c \sim N_c^{-1}$, whereas Fig. X–6(b) is $\mathcal{O}(N_c^{-2})$ because of the extra quark loop. The scattering amplitudes thus vanish in the large N_c limit, and the leading contributions are connected, planar diagrams.

The large N_c limit also predicts that neutral mesons (i.e., $Q^{(\alpha)}\bar{Q}^{(\beta)}$ composites with $\alpha = \beta$) do not mix with each other. The possible mixing diagram is given in Fig. X–7, and includes any number of gluons. However, because of the extra quark loop, it is of order N_c^{-1}, and thus vanishes in the infinite color limit. This means that $u\bar{u}$ states do not mix with $d\bar{d}$ or $\bar{s}s$, nor do the latter two mix. The large N_c spectrum thus displays a nonet structure with the $u\bar{u}$ and $d\bar{d}$ states degenerate (to the extent that electromagnetism and the mass difference between the u and d quarks are neglected) and the $s\bar{s}$ states somewhat heavier. This pattern is reflected in Nature, except that the $\bar{u}u$ and $\bar{d}d$ configurations now appear as states of definite isospin, $\bar{u}u \pm \bar{d}d$. For example, let us consider the $J^{PC} = 1^{--}, 2^{++}$ mesons. For the former, $\rho(770)$ and $\omega(783)$ are interpreted as $u\bar{u}, d\bar{d}$ isospin $I = 1$ and $I = 0$ combinations, while $\varphi(1020)$ is the $s\bar{s}$ member of the nonet. Including the $K^*(892)$ doublet as the $\bar{u}s, \bar{d}s$ combinations, a simple additivity in the quark mass would imply

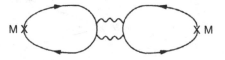

Fig. X–7 Meson–meson mixing.

$$m_{\varphi(1020)} - m_{\rho(770)} = 2\left(m_{K^*(892)} - m_{\rho(770)}\right),\qquad(2.2)$$

which works well. A similar treatment of the 2^{++} mesons, identifying $a_2(1320)$ and $f_2(1270)$ as the corresponding $u\bar{u}$, $d\bar{d}$ states and $f_2'(1525)$ as an $s\bar{s}$ composite, predicts

$$m_{f_2'(1525)} - m_{a_2(1320)} = 2\left(m_{K_2^*(1430)} - m_{a_2(1320)}\right),\qquad(2.3)$$

which is also approximately satisfied. The fact that $\rho(770)$, $\omega(783)$, $f_2(1270)$ and $a_2(1320)$ decay primarily to pions, and $\varphi(1020)$ and $f_2'(1525)$ decay primarily to kaons, reinforces this interpretation.

The world of baryons in the large N_c limit is quite different from that of mesons and glueballs [DaJM 94, Je 98]. In order to form a color singlet, one needs to combine not three quarks but rather N_c quarks in a totally color-antisymmetric fashion. This forces the baryon mass to grow as N_c, i.e., to become infinitely heavy in the $N_c \to \infty$ limit. In an attempt to model this behavior, it has been suggested that baryons can be associated with the soliton solution, called the *Skyrmion*, of a certain chiral lagrangian [Sk 61, Wi 83b]. We shall discuss this idea in Sect. XI–4 in the context of a model with $SU(2)_L \times SU(2)_R$ symmetry.

X–3 Goldstone bosons and the axial anomaly

As stated in the previous section, it must be assumed that color confinement continues to hold in the large N_c limit. Given this behavior, it can be proven under reasonable conditions that chiral symmetry is spontaneously broken [CoW 80]. In this circumstance, the large N_c limit turns out to imply a fascinating unity between the $\eta'(960)$ meson and the octet of Goldstone bosons in massless *QCD* [Wi 79]. The $N_c = \infty$ analog of $\eta'(960)$ is *also* a Goldstone boson if quarks are massless.

In order to see this, let us first consider the pseudoscalar decay constants. Because the axial-current matrix elements

$$\langle P_j(\mathbf{q}) | A_k^\mu(0) | 0 \rangle = -i F_j q^\mu \delta_{jk}\qquad(3.1)$$

involve one meson normalization factor and one quark loop, they are of order $(N_c^{-1/2})N_c \sim N_c^{1/2}$, which then implies $F_j = \mathcal{O}(N_c^{1/2})$. Now consider the current divergence in the limit of massless quarks. In general, we have

$$\langle P_j(\mathbf{q}) | \partial_\mu A_k^\mu(0) | 0 \rangle = F_j m_j^2 \delta_{jk}.\qquad(3.2)$$

For the octet of currents, the divergence vanishes for zero quark mass, and as usual leads to the identification of π, K, η_8 as Goldstone bosons. However, for the singlet current the anomaly is present. Even in the limit of vanishing quark mass, the current divergence has nonzero matrix elements, in particular,

$$\langle \eta^0(\mathbf{q}) \left| \partial_\mu A_0^\mu(0) \right| 0 \rangle = F_{\eta^0} m_{\eta^0}^2 = \left\langle \eta^0(\mathbf{q}) \left| \frac{3g_3^2}{32\pi^2} F_{\mu\nu}^a \tilde{F}^{a\mu\nu} \right| 0 \right\rangle. \tag{3.3}$$

If one repeats the calculation of the anomalous triangle diagram as in Sect. III–3 but now allows N_c to be arbitrary, one sees that it is proportional to $\mathrm{Tr}\,(\lambda^a \lambda^b) = 2\delta^{ab}$ and is therefore independent of N_c. However, by using large N_c-counting rules, the matrix element in Eq. (3.3) is seen to be of order $g_3^2 N_c^{1/2} \sim N_c^{-1/2}$.[1] This implies that the gluonic contribution to the axial anomaly *vanishes* in the large N_c limit. When we take into account the behavior of $F_{\eta'}$, we conclude that $m_{\eta'}^2 \sim 1/N_c \to 0$. The η' is thus massless in the large N_c limit, and we end up with a *nonet* of Goldstone bosons.

To illustrate what happens when the number of colors is treated perturbatively, let us consider the $1/N_c$ corrections to the meson spectrum together with the effects of quark masses. If we first add quark masses, we have, in analogy with the results of Sect. VII–1, the mass matrix

$$m_{ij}^2 = \langle P_i \left| \hat{m}(\bar{u}u + \bar{d}d) + m_s \bar{s}s \right| P_j \rangle, \tag{3.4}$$

where we have taken $m_u = m_d = \hat{m}$. This leads to a squared-mass matrix

$$\mathbf{m}^2 = B_0 \begin{pmatrix} 2\hat{m} & 0 & 0 & 0 \\ 0 & m_s + \hat{m} & 0 & 0 \\ 0 & 0 & \frac{2}{3}(2m_s + \hat{m}) & \frac{2\sqrt{2}}{3}(\hat{m} - m_s) \\ 0 & 0 & \frac{2\sqrt{2}}{3}(\hat{m} - m_s) & \frac{2}{3}(m_s + 2\hat{m}) \end{pmatrix} \tag{3.5}$$

in the basis (π, K, η_8, η_0). If this were diagonalized, one would find an isoscalar state degenerate with the pion. This is a manifestation of the $U(1)$ problem, which arises when there is no anomaly. However, at the next order in large N_c, the matrix picks up an extra contribution in the $SU(3)$-singlet channel due to the anomaly, yielding

$$\mathbf{m}^2 = B_0 \begin{pmatrix} 2\hat{m} & 0 & 0 & 0 \\ 0 & m_s + \hat{m} & 0 & 0 \\ 0 & 0 & \frac{2}{3}(2m_s + \hat{m}) & \frac{2\sqrt{2}}{3}(\hat{m} - m_s) \\ 0 & 0 & \frac{2\sqrt{2}}{3}(\hat{m} - m_s) & \frac{2}{3}(m_s + 2\hat{m}) + \frac{\epsilon}{N_c B_0} \end{pmatrix}, \tag{3.6}$$

---

[1] This result depends on the assumption that topologically nontrivial aspects of vacuum structure are smooth in the $N_c \to \infty$ limit.

where $\epsilon = \mathcal{O}(N_c^0)$. This mass matrix yields an interesting prediction. The quantities $B_0\hat{m}$ and $B_0 m_s$ are fixed as usual by using the π and K masses. Also the trace of the full matrix must yield $m_\pi^2 + m_K^2 + m_\eta^2 + m_{\eta'}^2$, which fixes $\epsilon = 2.16\,\text{GeV}^2$. The remaining diagonalization then predicts $m_{\eta'} = 0.98$ GeV, $m_\eta = 0.50$ GeV with a mixing angle of $18°$. This is a remarkably accurate representation of the situation in the real world. Although ϵ/N_c is suppressed in a technical sense, note how large it actually is. One is hard pressed to imagine any sense in which the physical η' mass can be taken as a small parameter.

X–4 The *OZI* rule

In the 1960s, an empirical property, called the Okubo–Zweig–Iizuka (*OZI*) rule [Ok 63, Zw 65, Ii 66], was developed for mesonic coupling constants. Its usual statement is that flavor-disconnected processes are suppressed compared to those in which quark lines are connected. In the language which we are using here, flavor disconnected processes are those with an extra quark loop. Unfortunately, the phenomenological and theoretical status of this so-called rule is ambiguous. We briefly describe it here because it is part of the common lore of particle physics.

The empirical motivation for the *OZI* rule is best formulated in the decays of mesons. Let us accept that $\varphi(1020)$ and $f_2'(1525)$ are primarily states with content $\bar{s}s$ whereas $\omega(783)$ and $f_2(1270)$ have content $(\bar{u}u + \bar{d}d)/\sqrt{2}$. Mixing between the $\bar{s}s$ and nonstrange components can take place with a small mixing angle, such that

$$\frac{\text{Amp}\,(\bar{s}s)}{\text{Amp}\left([\bar{u}u + \bar{d}d]/\sqrt{2}\right)} \equiv \tan\theta, \qquad (4.1)$$

with $\theta = \theta_V$ for the vector mesons and $\theta = \theta_T$ for the tensor mesons. In both cases, θ is small. Experimentally, the φ and f_2' decay dominantly into strange particles even though phase space (abbreviated as 'p.s.' below) considerations would strongly favor nonstrange modes,

$$\frac{\Gamma_{\varphi \to 3\pi + \rho\pi}}{\Gamma_{\varphi \to K\bar{K}}} \simeq 0.18, \qquad \frac{\Gamma_{\varphi \to 3\pi + \rho\pi}}{\Gamma_{\omega \to 3\pi}} \simeq 0.09,$$

$$\frac{\Gamma_{f_2' \to \pi\pi}}{\Gamma_{f_2' \to K\bar{K}}} = 0.012 \pm 0.002 \qquad \frac{\Gamma_{f_2' \to \pi\pi}}{\Gamma_{f_2 \to \pi\pi}} = 0.004 \pm 0.001 \qquad (4.2)$$

$$\simeq 0.003 \times \text{p.s.}, \qquad\qquad \simeq 0.002 \times \text{p.s.}$$

This suggests the hypothesis '$\bar{s}s$ states do not decay into final states not containing strange quarks'. Diagrammatically this leads to a pictorial representation of the

Fig. X–8 *OZI* (a) allowed, (b) forbidden amplitudes.

OZI rule, viz., the dominance of Fig. X–8(a) over Fig. X–8(b). Some scattering processes also show such a suppression. For example, we have

$$\frac{\sigma_{\pi^- p \to \varphi n}}{\sigma_{\pi^- p \to \omega n}} \simeq 0.03, \tag{4.3}$$

which can be interpreted as an *OZI* suppression. A stronger version of the *OZI* rule would have the φ/ω and f_2'/f_2 ratios equal to a *universal* factor of $\tan^2 \theta$ (cf. Eq. (4.1)) once kinematic phase space factors are extracted.

The narrow widths of the J/ψ and Υ states are also cited as evidence for the *OZI* rule, since these hadronic decays involve the annihilation of the $c\bar{c}$ or $b\bar{b}$ constituents. This can be correct almost as a matter of definition, but it is not very enlightening. Indeed, the small widths of heavy-quark states can be understood within the framework of perturbative *QCD* without invoking any extra dynamical assumptions. However, perturbative *QCD* certainly cannot explain the *OZI* rule in light mesons. It must have a different explanation for these states.

There actually exist several empirical indications counter to the *OZI* rule [Li 84, ElGK 89, RPP 12]. Among the more dramatic examples of *OZI*-forbidden reactions, expressed as ratios, are

$$\begin{aligned}
\frac{\Gamma_{J/\psi \to \varphi \pi^+ \pi^-}}{\Gamma_{J/\psi \to \varphi K^+ K^-}} &= 1.2 \pm 0.5, & \frac{\sigma_{\gamma p \to p \varphi \pi^+ \pi^-}}{\sigma_{\gamma p \to p \omega K^+ K^-}} &= 2.0 \pm 0.7, \\
\frac{\sigma_{\pi^- p \to f_2' n}}{\sigma_{\pi^- p \to f_2 n}} &= 0.23^{+0.14}_{-0.13}, & \frac{\sigma_{\gamma p \to p \varphi \pi^+ \pi^-}}{\sigma_{\gamma p \to p \varphi K^+ K^-}} &\geq 5 \; (90\% \text{ C.L.}).
\end{aligned} \tag{4.4}$$

The universal-mixing model is incorrect more often than not, with counterexamples being

$$\begin{aligned}
\frac{\Gamma_{J/\psi \to \varphi \pi^+ \pi^-}}{\Gamma_{J/\psi \to \omega \pi^+ \pi^-}} &= 0.11 \pm 0.02, & \frac{\sigma_{\gamma p \to p \varphi \pi^+ \pi^-}}{\sigma_{\gamma p \to p \omega \pi^+ \pi^-}} &= 0.10 \pm 0.02, \\
\frac{\sigma_{\bar{p} p \to f_2' \pi^+ \pi^-}}{\sigma_{\bar{p} p \to f_2 \pi^+ \pi^-}} &= 0.029^{+0.011}_{-0.007},
\end{aligned} \tag{4.5}$$

instead of the values 0.03, 0.03, and 0.006 expected from the previous ratios. The empirical η–η' mixing angle $\theta_{\eta-\eta'} \simeq -20°$ also violates the *OZI* rule, which would require a mixing angle of $-35°$.

There is also an intrinsic logical flaw with the simplest formulation of the *OZI* rule. This is because *OZI*-forbidden processes can take place as the product of two *OZI*-allowed processes. For example, each of the following transitions is *OZI*-allowed:

$$f_2' \to K\bar{K}, \quad K\bar{K} \to \pi\pi,$$
$$f_2' \to \eta\eta, \quad \eta\eta \to \pi\pi. \tag{4.6}$$

Hence the *OZI*-forbidden reaction $f_2' \to \pi\pi$ can take place by the chains

$$f_2' \to K\bar{K} \to \pi\pi, \qquad f_2' \to \eta\eta \to \pi\pi. \tag{4.7}$$

These two-step processes are in fact required by unitarity to the extent that the individual scattering amplitudes are nonzero.

The large N_c limit provides the only known dynamical explanation of the *OZI* rule at low energies. Although the gluonic coupling constant is not small at these scales and suppressed diagrams have ample energy to proceed, they are predicted to be of order $1/N_c^2$ in rate because of the extra quark loop. Yet large N_c arguments need not suggest a universal suppression factor of $\tan^2 \theta$, because there is no need for the $1/N_c$ corrections to be universal. Note that the large N_c framework also forbids the mixing of η and η' and, more generally, the scattering of mesons.

Thus, the *OZI* rule in light-meson systems remains somewhat heuristic. It has a partial justification in large N_c counting rules, but it also has known violations. It is not possible to predict with certainty whether it will work in any given new application.

X–5 Chiral lagrangians

The large N_c limit places restrictions on the structure of chiral lagrangians [GaL 85a]. To describe these, we must first allow for an enlarged number $N_f > 3$ of quark flavors. The three-flavor $\mathcal{O}(E^4)$ lagrangian is expanded as

$$\mathcal{L}_4 = \sum_{i=1}^{10} L_i O_i, \tag{5.1}$$

where the $\{O_i\}$ can be read off from Eq. (VII–2.7). Recall that in constructing $\mathcal{L}_4$, we removed the $\mathcal{O}(E^4)$ operator

$$O_0 \equiv \text{Tr} \left(D_\mu U D_\nu U^\dagger D^\mu U D^\nu U^\dagger \right), \tag{5.2}$$

because for $N_f = 3$ it is expressible (cf. Eq. (VII–2.3)) as a linear combination of $O_{1,2,3}$. However, if the number of flavors exceeds three, one must append O_0 to the lagrangian of Eq. (5.1),

$$\mathcal{L}_4 = \sum_{i=1}^{10} L_i O_i \xrightarrow[N_f > 3]{} \sum_{i=0}^{3} B_i O_i + \sum_{i=4}^{10} L_i O_i. \tag{5.3}$$

In view of the linear dependence of O_0 on $O_{1,2,3}$, note that we have needed to modify the coefficients $L_{1,2,3} \to B_{1,2,3}$. Upon returning to three flavors, we regain the original coefficients,

$$L_1 = \frac{B_0}{2} + B_1, \quad L_2 = B_0 + B_2, \quad L_3 = -2B_0 + B_3. \tag{5.4}$$

We can now study the large N_c behavior of the extended $\mathcal{O}(E^4)$ chiral lagrangian. The distinguishing feature is the number of traces in a given $\mathcal{O}(E^4)$ operator. Each such trace is taken over *flavor* indices and amounts to a sum over the quark flavors, which in turn can arise only in a *quark loop*. In particular, those operators with two flavor traces ($O_{1,2,4,6,7}$) will require at least two quark loops, while those with one flavor trace need only one quark loop. However, our study of the large N_c limit has taught us that every quark loop leads to a $1/N_c$ suppression. Thus, the $\mathcal{O}(E^4)$ chiral contributions having two traces will be suppressed relative to those with one trace by a power of $1/N_c$, and provided $B_3 \neq 0$ we can write[2]

$$\frac{B_1}{B_3} = \frac{B_2}{B_3} = \frac{L_4}{L_3} = \frac{L_6}{B_3} \xrightarrow[N_c \to \infty]{} 0. \tag{5.5a}$$

Alternatively, this N_c-counting rule implies (provided $B_0/B_3 \neq 1/2$) for the $\{B_i\}$ coefficients of flavor $SU(3)$,

$$\frac{2L_1 - L_2}{L_3} = \frac{L_4}{L_3} = \frac{L_6}{L_3} = \mathcal{O}(N_c^{-1}). \tag{5.5b}$$

The overall power of N_c for the remaining terms can be found by noting that the $\pi\pi$ scattering amplitude should be of order N_c^{-1}, implying $L_{1,2,3} = \mathcal{O}(N_c)$.

The only exception to the above counting behavior is the operator with coefficient L_7. This exception occurs because the operator can be generated by an η' pole, and the η' mass-squared is $\mathcal{O}(1/N_c)$. In particular, the coefficient of this term is absolutely predicted in the large N_c limit. This follows if we include the large N_c result for mixing between η and η' shown in Eq. (3.6) as a chiral lagrangian

$$\mathcal{L}_{\eta\eta'} = \frac{F_\pi}{2\sqrt{6}} \eta_0 \operatorname{Tr} \left(\chi^\dagger U - U^\dagger \chi \right), \tag{5.6}$$

which when expanded to order $\eta_0 \eta_8$ will yield the off-diagonal term in the mass mixing matrix of Eq. (3.6). Integrating out the $\eta_0 \sim \eta'$ leads to the effective lagrangian

---

[2] The operator O_7 presents a special case and is discussed below.

$$\mathcal{L}_{\text{eff}} = -\frac{1}{48}\frac{F_\pi^2}{m_{\eta'}^2}\left[\text{Tr}\left(\chi^\dagger U - U^\dagger \chi\right)\right]^2. \tag{5.7}$$

It is the factor of $m_{\eta'}^{-2}$ which overcomes the counting rules. Although the double trace suggests that this operator is suppressed in the large N_c limit, we have $m_{\eta'}^{-2} \propto N_c$. Thus, at least formally, an extra enhancement would be predicted.

The large N_c limit then predicts the following ordering of the chiral coefficients in $\mathcal{L}_4$:

$$L_7 = \mathcal{O}(N_c^2),$$
$$L_1,\ L_2,\ L_3,\ L_5,\ L_8,\ L_9,\ L_{10} = \mathcal{O}(N_c),$$
$$2L_1 - L_2,\ L_4,\ L_6 = \mathcal{O}(1). \tag{5.8}$$

An existing empirical test involves the occurrence of $2L_1 - L_2$ in $K \to \pi\pi e\bar{\nu}_e$ decays [Bi 90, RiGDH 91], and the prediction works quite well. The large N_c enhancement of L_7 is probably just a curiosity in that the physical value of the η' mass is not small compared to other masses in the theory, and hence the technical advantage of $m_{\eta'}^2 \propto N_c^{-1}$ is probably not useful phenomenologically.

Problems

(1) **The large N_c weak hamiltonian**

Retrace the calculation of the QCD renormalization of the weak nonleptonic hamiltonian described in Sect. VIII–3, but now in the limit $N_c \to \infty$ with $g_3^2 N_c$ fixed. Show that the penguin operators do not enter and that all short-distance effects are of order N_c^{-1}, with the operator-product coefficients $c_1 = 1$, $c_2 = 1/5$, $c_3 = 2/15$, $c_4 = 2/3$, $c_5 = c_6 = 0$.

(2) **The strong CP problem in the large N_c limit**

In the large N_c limit, the η_0 can be united with the Goldstone octet in the effective lagrangian. Generalizing the chiral matrix to nine fields we write $\mathcal{L} = \mathcal{L}_0 + \mathcal{L}_{N_c^{-1}}$, where

$$\mathcal{L}_0 = \frac{F^2}{4}\,\text{Tr}\left(\partial_\mu \tilde{U}\partial^\mu \tilde{U}^\dagger\right) + \frac{F^2}{4}B_0\,\text{Tr}\left(\mathbf{m}(\tilde{U} + \tilde{U}^\dagger)\right),$$

$$\mathcal{L}_{N_c^{-1}} = \frac{\epsilon}{N_c}\frac{F^2}{24}\left[\text{Tr}\left(\ln\tilde{U} - \ln\tilde{U}^\dagger\right)\right]^2,$$

$$\tilde{U} = \exp\left(i\boldsymbol{\lambda}\cdot\boldsymbol{\varphi}/F\right)\exp\left(i\sqrt{\frac{2}{3}}\frac{\varphi^0}{F}\right).$$

(a) Confirm that this reproduces the mixing matrix of Eq. (3.6).

(b) Another way to obtain this result is to employ an auxiliary pseudoscalar field $q(x)$ (with no kinetic energy term) to rewrite $\mathcal{L}_{N_c^{-1}}$ as

$$\mathcal{L}_{N_c^{-1}} = \frac{N_c}{4\epsilon} q^2(x) + i\frac{F}{2\sqrt{6}} q(x) \, \mathrm{Tr} \, (\ln \tilde{U} - \ln \tilde{U}^\dagger).$$

Identify the $SU(3)$-singlet axial current and calculate its divergence to show that $q(x)$ plays the same role as $F\tilde{F}$, i.e., $q(x) \sim \alpha F\tilde{F}/8\pi$. Integrate out $q(x)$ to show that this is equivalent to the form of part (a).

(c) Several authors [RoST 80, DiV 80] suggest adding the θ term through

$$\mathcal{L} = \mathcal{L}_0 + \mathcal{L}_{N_c^{-1}} - \theta q(x).$$

From this starting point, integrate out $q(x)$ and show that a chiral rotation can transfer θ to $\arg(\det \mathbf{m})$. However, in the sense described in Sect. IX–4, this theory is unstable about $\tilde{U} = 1$. The stable vacuum corresponds to $\tilde{U}_{jk} = \delta_{jk} \exp(i\beta_j)$. For small θ, solve for β_j in terms of θ.

(d) Using $\tilde{U} = e^{i\beta/2} U e^{i\beta/2}$, define the fields about the correct vacuum to find the *CP*-violating terms of the form

$$\mathcal{L}_{CP} = i\theta \left[a \, \mathrm{Tr} \, (U - U^\dagger) + b \, \mathrm{Tr} \, (\ln U - \ln U^\dagger) \right],$$

identifying a and b and showing they vanish if any quark mass vanishes. Calculate the *CP*-violating amplitude for $\eta \to \pi^+ \pi^-$.

XI

Phenomenological models

QCD has turned out to be a theory of such subtlety and difficulty that a concerted effort over an extended period has not yielded a practical procedure for obtaining analytic solutions. At the same time, vast amounts of hadronic data which require theoretical analysis and interpretation have been collected. This has spurred the development of accessible phenomenological methods. We devote this chapter to a discussion of three dynamical models (potential, bag, and Skyrme) along with a methodology based on sum rules. Although the dynamical models are constructed to mimic aspects of QCD, none of them *is QCD*. That is, none contains a rigorous program of successive approximations which, for arbitrary quark mass, can be carried out to arbitrary accuracy. Therefore, our treatment will emphasize issues of basic structure rather than details of numerical fits. By using all of these methods, one hopes to gain physical insight into the nature of hadron dynamics. Despite its inherent limitations the program of model building, fortified by the use of sum rules, has been generally successful, and there is now a reasonable understanding of many aspects of hadron spectroscopy.

XI–1 Quantum numbers of $Q\overline{Q}$ and Q^3 states

Among the states conjectured to lie in the spectrum of the QCD hamiltonian are mesons, baryons, glueballs, hybrids, dibaryons, etc. However, since practically all currently known hadrons can be classified as either $Q\overline{Q}$ states (*mesons*) or Q^3 states (*baryons*), it makes sense to focus on just these systems. We shall begin by determining the quark model construction of the light-hadron ground states. Much of the material will be valid for heavy-quark systems as well.

Hadronic flavor–spin state vectors

In many respects, the language of quantum field theory provides a simple and flexible format for implementing the quark model. Let us assume that for any given dynamical model, it is possible to solve the field equations of motion and obtain a complete set of spatial wavefunctions, $\{\psi_\alpha(x)\}$ for quarks and $\{\psi_{\overline{\alpha}}(x)\}$ for anti-quarks, where the labels α and $\overline{\alpha}$ refer to a complete set of observables. A quark field operator can then be expanded in terms of these wavefunctions,

$$\psi(x) = \sum_\alpha \left[\psi_\alpha(x) e^{-i\omega_\alpha t} b(\alpha) + \psi_{\overline{\alpha}}(x) e^{i\omega_{\overline{\alpha}} t} d^\dagger(\overline{\alpha}) \right], \qquad (1.1)$$

where ω_α, $\omega_{\overline{\alpha}}$ are the energy eigenvalues, $b(\alpha)$ destroys a quark and $d^\dagger(\overline{\alpha})$ creates the corresponding antiquark. The quark creation and annihilation operators obey

$$\{b(\alpha), b^\dagger(\alpha')\} = \delta_{\alpha\alpha'}, \qquad \{d(\overline{\alpha}), d^\dagger(\overline{\alpha}')\} = \delta_{\overline{\alpha}\overline{\alpha}'},$$
$$\{b(\alpha), b(\alpha')\} = 0, \qquad \{d(\overline{\alpha}), d(\overline{\alpha}')\} = 0,$$
$$\{b(\alpha), d^\dagger(\overline{\alpha}')\} = 0, \qquad (1.2)$$

which are the usual anticommutation relations for fermions.

In all practical quark models, an assumption is made which greatly simplifies subsequent steps in the analysis, that *the spatial, spin, and color degrees of freedom factorize*, at least in lowest-order approximation. This is true provided the zeroth-order hamiltonian is spin-independent and color-independent. Spin-dependent inter-actions are then taken into account as perturbations. This assumption allows us to write the sets $\{\alpha\}$ and $\{\overline{\alpha}\}$ in terms of the spatial (n), spin (s, m_s), flavor (q), and color (k) degrees of freedom respectively, i.e., $\alpha = (n, s, m_s, q, k)$. If we are concerned with just the ground state, we can suppress the quantum number n, and for simplicity replace the symbols b, $d^\dagger$, etc., for annihilation and creation operators with the flavor symbol q ($q = u, d, s$ for the light hadrons),

$$b^\dagger(n = 0, q, m_s, k) \rightarrow q^\dagger_{k,m_s},$$
$$d^\dagger(n = 0, \overline{q}, m_s, k) \rightarrow \overline{q}^\dagger_{k,m_s}. \qquad (1.3)$$

Hadrons are constructed in the Fock space defined by the creation operators for quarks and antiquarks. Light hadrons are labeled by the spin $(\mathbf{S}^2, S_3)$, isospin $(\mathbf{T}^2, T_3)$, and hypercharge (Y) operators as well as by the baryon number (B). Other observables like the electric charge Q_{el} and strangeness S are related to these,

$$Q_{\text{el}} = T_3 + Y/2, \qquad S = Y - B. \qquad (1.4)$$

Since quarks have spin one-half, the baryon (Q^3) and meson $(Q\overline{Q})$ configurations can carry the spin quantum numbers $S = 1/2, 3/2$ and $S = 0, 1$, respectively. If we neglect the mass difference between strange and nonstrange quarks, then flavor $SU(3)$ is a symmetry of the theory, and both quarks and hadrons occupy $SU(3)$

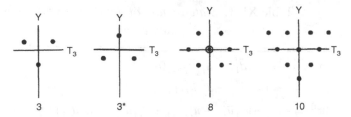

Fig. XI–1 Some $SU(3)$-flavor representations.

multiplets. The quarks are assigned to the triplet representation **3** and the antiquarks to **3***. The $Q\overline{Q}$ and Q^3 constructions then involve the group products

$$\mathbf{3} \times \mathbf{3^*} = \mathbf{8} \oplus \mathbf{1},$$

$$(\mathbf{3} \times \mathbf{3}) \times \mathbf{3} = (\mathbf{6} \oplus \mathbf{3^*}) \times \mathbf{3} = \mathbf{10} \oplus \mathbf{8} \oplus \mathbf{8} \oplus \mathbf{1}, \tag{1.5}$$

so that baryons appear as decuplets, octets, and singlets whereas mesons appear as octets and singlets. The $SU(3)$-flavor representations **3**, **3***, **8**, **10** are depicted in Y vs. T_3 plots in Fig. XI–1. The circle around the origin for the eight-dimensional representation denotes the presence of *two* states with identical Y, I_3 values. Finally, quarks and antiquarks transform as triplets and antitriplets of the *color* $SU(3)$ gauge group, and all baryons and mesons are color singlets.

Two simple states to construct are the ρ_1^+ meson and the $\Delta_{3/2}^{++}$ baryon,

$$|\rho_1^+\rangle = \frac{1}{\sqrt{3}} u_{i\uparrow}^\dagger \overline{d}_{i\uparrow}^\dagger |0\rangle, \qquad |\Delta_{3/2}^{++}\rangle = \frac{1}{6} \epsilon_{ijk} u_{i\uparrow}^\dagger u_{j\uparrow}^\dagger u_{k\uparrow}^\dagger |0\rangle, \tag{1.6}$$

where the superscript and subscript on the hadrons denote electric charge and spin component, and a summation over color indices for the creation operators is implied. The normalization constants are fixed by requiring that the hadrons $\{H_n\}$ form an orthonormal set, $\langle H_m | H_n \rangle = \delta_{mn}$. The other ground-state hadrons can be reached from those in Eq. (1.6) by means of ladder operations in the spin and fla-vor variables. In this manner, one can construct the flavor–spin–color state vec-tors of the 0^- octet and singlet mesons and the $\frac{1}{2}^+$ octet baryons displayed in Tables XI–1 and XI–2.

A convenient notation for fields which transform as $SU(3)$ octets involves the use of a cartesian basis rather than the 'spherical' basis of Tables XI–1,2. In fact, we have already encountered this description in Sect. VII–1 during our discussion of $SU(3)$ Goldstone bosons where the quantity $U = \exp(i\boldsymbol{\varphi} \cdot \boldsymbol{\lambda})$ played a central role. The eight cartesian fields $\{\varphi_a\}$ are related to the usual pseudoscalar fields by

Table XI–1. *State vectors of the pseudoscalar octet and singlet mesons.*

$$|\pi^+\rangle = \tfrac{1}{\sqrt{6}} [u_{i\uparrow}^\dagger \bar{d}_{i\downarrow}^\dagger - u_{i\downarrow}^\dagger \bar{d}_{i\uparrow}^\dagger] |0\rangle$$

$$|\pi^-\rangle = \tfrac{1}{\sqrt{6}} [d_{i\uparrow}^\dagger \bar{u}_{i\downarrow}^\dagger - d_{i\downarrow}^\dagger \bar{u}_{i\uparrow}^\dagger] |0\rangle$$

$$|\pi^0\rangle = \tfrac{1}{\sqrt{12}} [-u_{i\uparrow}^\dagger \bar{u}_{i\downarrow}^\dagger + u_{i\downarrow}^\dagger \bar{u}_{i\uparrow}^\dagger + d_{i\uparrow}^\dagger \bar{d}_{i\downarrow}^\dagger - d_{i\downarrow}^\dagger \bar{d}_{i\uparrow}^\dagger] |0\rangle$$

$$|K^+\rangle = \tfrac{1}{\sqrt{6}} [u_{i\uparrow}^\dagger \bar{s}_{i\downarrow}^\dagger - u_{i\downarrow}^\dagger \bar{s}_{i\uparrow}^\dagger] |0\rangle$$

$$|K^0\rangle = \tfrac{1}{\sqrt{6}} [s_{i\uparrow}^\dagger \bar{d}_{i\downarrow}^\dagger - s_{i\downarrow}^\dagger \bar{d}_{i\uparrow}^\dagger] |0\rangle$$

$$|\overline{K}^0\rangle = \tfrac{1}{\sqrt{6}} [s_{i\uparrow}^\dagger \bar{u}_{i\downarrow}^\dagger - s_{i\downarrow}^\dagger \bar{u}_{i\uparrow}^\dagger] |0\rangle$$

$$|K^-\rangle = \tfrac{1}{\sqrt{6}} [d_{i\uparrow}^\dagger \bar{s}_{i\downarrow}^\dagger - d_{i\downarrow}^\dagger \bar{s}_{i\uparrow}^\dagger] |0\rangle$$

$$|\eta_8\rangle = \tfrac{1}{\sqrt{36}} [u_{i\uparrow}^\dagger \bar{u}_{i\downarrow}^\dagger - u_{i\downarrow}^\dagger \bar{u}_{i\uparrow}^\dagger + d_{i\uparrow}^\dagger \bar{d}_{i\downarrow}^\dagger - d_{i\downarrow}^\dagger \bar{d}_{i\uparrow}^\dagger - 2s_{i\uparrow}^\dagger \bar{s}_{i\downarrow}^\dagger + 2s_{i\downarrow}^\dagger \bar{s}_{i\uparrow}^\dagger] |0\rangle$$

$$|\eta_1\rangle = \tfrac{1}{\sqrt{18}} [u_{i\uparrow}^\dagger \bar{u}_{i\downarrow}^\dagger - u_{i\downarrow}^\dagger \bar{u}_{i\uparrow}^\dagger + d_{i\uparrow}^\dagger \bar{d}_{i\downarrow}^\dagger - d_{i\downarrow}^\dagger \bar{d}_{i\uparrow}^\dagger + s_{i\uparrow}^\dagger \bar{s}_{i\downarrow}^\dagger - s_{i\downarrow}^\dagger \bar{s}_{i\uparrow}^\dagger] |0\rangle$$

$$\pi^\pm = \frac{1}{\sqrt{2}}(\varphi_1 \mp i\varphi_2), \qquad \pi^0 = \varphi_3, \qquad \eta_8 = \varphi_8,$$

$$K^\pm = \frac{1}{\sqrt{2}}(\varphi_4 \mp i\varphi_5),$$

$$K^0 = \frac{1}{\sqrt{2}}(\varphi_6 - i\varphi_7), \qquad \overline{K}^0 = \frac{1}{\sqrt{2}}(\varphi_6 + i\varphi_7), \qquad (1.7)$$

which is an alternative way of stating the content of Eq. (VIII–1.12). The physical spin one-half baryons $p, n, \ldots$ can likewise be expressed in terms of an octet of states $\{B_i\}$ ($i = 1, \ldots, 8$) in cartesian basis as

$$\Sigma^\pm = \frac{1}{\sqrt{2}}(B_1 \mp i B_2), \qquad \Sigma^0 = B_3, \qquad \Lambda = B_8,$$

$$p = \frac{1}{\sqrt{2}}(B_4 - i B_5), \qquad n = \frac{1}{\sqrt{2}}(B_6 - i B_7),$$

$$\Xi^0 = \frac{1}{\sqrt{2}}(B_6 + i B_7), \qquad \Xi^- = \frac{1}{\sqrt{2}}(B_4 + i B_5). \qquad (1.8)$$

In the quark model, hadron observables have simple interpretations, e.g., the baryon number is simply one-third the difference in the number of quarks and antiquarks, etc. Thus, writing quark and antiquark number operators as $N(q)$ and $N(\bar{q})$ for a quark flavor q, we have

$$B = [N(u) + N(d) + N(s) - N(\bar{u}) - N(\bar{d}) - N(\bar{s})]/3,$$

$$T_3 = [N(u) - N(d) - N(\bar{u}) + N(\bar{d})]/2,$$

$$Y = [N(u) + N(d) - 2N(s) - N(\bar{u}) - N(\bar{d}) + 2N(\bar{s})]/3,$$

$$Q_{el} = [2N(u) - N(d) - N(s) - 2N(\bar{u}) + N(\bar{d}) + N(\bar{s})]/3, \qquad (1.9)$$

Table XI–2. *State vectors of baryon
spin-one-half octet.*

$$|p_\uparrow\rangle = \tfrac{\epsilon_{ijk}}{\sqrt{18}} [u^\dagger_{i\downarrow} d^\dagger_{j\uparrow} - u^\dagger_{i\uparrow} d^\dagger_{j\downarrow}] u^\dagger_{k\uparrow} |0\rangle$$

$$|n_\uparrow\rangle = \tfrac{\epsilon_{ijk}}{\sqrt{18}} [d^\dagger_{i\uparrow} u^\dagger_{j\downarrow} - d^\dagger_{i\downarrow} u^\dagger_{j\uparrow}] d^\dagger_{k\uparrow} |0\rangle$$

$$|\Lambda_\uparrow\rangle = \tfrac{\epsilon_{ijk}}{\sqrt{12}} [u^\dagger_{i\uparrow} d^\dagger_{j\downarrow} - u^\dagger_{i\downarrow} d^\dagger_{j\uparrow}] s^\dagger_{k\uparrow} |0\rangle$$

$$|\Sigma^+_\uparrow\rangle = \tfrac{\epsilon_{ijk}}{\sqrt{18}} [s^\dagger_{i\downarrow} u^\dagger_{j\uparrow} - s^\dagger_{i\uparrow} u^\dagger_{j\downarrow}] u^\dagger_{k\uparrow} |0\rangle$$

$$|\Sigma^0_\uparrow\rangle = \tfrac{\epsilon_{ijk}}{6} [s^\dagger_{i\uparrow} d^\dagger_{j\downarrow} u^\dagger_{k\uparrow} + s^\dagger_{i\uparrow} d^\dagger_{j\uparrow} u^\dagger_{k\downarrow} - 2 s^\dagger_{i\downarrow} d^\dagger_{j\uparrow} u^\dagger_{k\uparrow}] |0\rangle$$

$$|\Sigma^-_\uparrow\rangle = \tfrac{\epsilon_{ijk}}{\sqrt{18}} [s^\dagger_{i\uparrow} d^\dagger_{j\downarrow} - s^\dagger_{i\downarrow} d^\dagger_{j\uparrow}] d^\dagger_{k\uparrow} |0\rangle$$

$$|\Xi^0_\uparrow\rangle = \tfrac{\epsilon_{ijk}}{\sqrt{18}} [s^\dagger_{i\downarrow} u^\dagger_{j\uparrow} - s^\dagger_{i\uparrow} u^\dagger_{j\downarrow}] s^\dagger_{k\uparrow} |0\rangle$$

$$|\Xi^-_\uparrow\rangle = \tfrac{\epsilon_{ijk}}{\sqrt{18}} [s^\dagger_{i\uparrow} d^\dagger_{j\downarrow} - s^\dagger_{i\downarrow} d^\dagger_{j\uparrow}] s^\dagger_{k\uparrow} |0\rangle$$

and the hadronic spin operator is

$$\mathbf{S} = \sum_q q^\dagger_{i,m'_s} \frac{(\sigma)_{m'_s m_s}}{2} q_{i,m_s}. \tag{1.10}$$

Quark spatial wavefunctions

Many applications of the quark model require the knowledge of the quark spatial wavefunctions within hadrons. It is here that the greatest variation in the different models can occur, but several general features still remain. Indeed, in many instances it is the general features that are primarily tested.

For example, the ground state in all models is a spatially symmetric S state in which the wavefunction peaks at $r = 0$. The normalization condition of the quark spatial wavefunction,

$$\int d^3x\, \psi^\dagger(x)\psi(x) = 1, \tag{1.11}$$

ensures that the magnitude of ψ will be similar in those models having wavefunctions of comparable spatial extent. This accounts for the agreement which can be found among diverse quark models in specific applications. How does one fix the spatial extent? One approach is to use an observable like the hadronic electromagnetic charge radius, e.g.,

$$\langle r^2\rangle^{1/2}_{\text{proton}} = 0.87 \pm 0.02 \text{ fm}, \qquad \langle r^2\rangle^{1/2}_{\text{pion}} = 0.66 \pm 0.02 \text{ fm}. \tag{1.12}$$

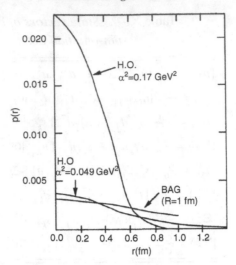

Fig. XI–2 Quark probability density in the bag and oscillator models.

Viewed this way, the bound states are seen to define a scale of order 1 fm. For example, we display two models in Fig. XI–2, the oscillator result with $\alpha^2 = 0.17$ GeV^2 and the bag profile, which are each obtained by fitting to ground-state baryon observables like the charge radius. Not surprisingly, their behaviors are quite similar. Also shown in Fig. XI–2 is an oscillator model wavefunction whose parameter $(\alpha^2 = 0.049$ GeV$^2)$ was determined by using data from decays of excited hadrons. The difference is rather striking, and serves to demonstrate that the most important general feature in setting the scale in quark model predictions of dimensional matrix elements is the spatial extent of the wavefunction.[1]

Another aspect of quark wavefunctions involves the issue of relativistic motion. A relativistic quark moving in a spin-independent central potential has a ground-state wavefunction of the form

$$\psi_{\text{gnd}}(x) = \begin{pmatrix} i\, u(r)\chi \\ \ell(r)\boldsymbol{\sigma} \cdot \hat{\mathbf{r}}\chi \end{pmatrix} e^{-iEt}, \qquad (1.13)$$

where u, ℓ signify 'upper' and 'lower' components. As we shall see, in the bag model these radial wavefunctions are just spherical Bessel functions. The above form also appears in some relativized harmonic-oscillator models, which use a central potential. If we allow for relativistic motion, then the major remaining difference in the quark wavefunctions concerns the lower two components of the Dirac wavefunction. Nonrelativistic models automatically set these equal to zero, while relativistic models can have sizeable lower components. Which description is the

---

[1] We could obtain a bag result which behaves similarly by employing a charge radius of 0.5 fm rather than the 1 fm value shown.

correct one? Quark motion in light hadrons must be at least somewhat relativistic since quarks confined to a region of radius R have a momentum given by the uncertainty principle,[2]

$$p \geq \sqrt{3}R^{-1} \simeq 342\,\text{MeV} \qquad (\text{for } R \simeq 1\,\text{fm}). \qquad (1.14)$$

Since this momentum is comparable to or larger than all the light-quark masses, relativistic effects are unavoidable. A more direct indication of the relativistic nature of quark motion comes from the hadron spectrum. Nonrelativistic systems are characterized by excitation energies which are small compared to the constituent masses. In the hadron spectrum, typical excitation energies lie in the range 300–500 MeV, again comparable to or larger than light-quark masses. Such considerations have motivated relativistic formulations of the quark model.

Interpolating fields

In the LSZ procedure (App. B–3) for analyzing scattering amplitudes the central role is played by interpolating fields. These are the quantities which experience the dynamics of the theory in the course of evolving between the asymptotic *in*-states and *out*-states. They turn out to be also useful as a kind of bookkeeping device. That is, one way to characterize the spectrum of observed states is to use operators made of appropriate combinations of quark fields $\psi(x)$. For example, corresponding to the meson sector of $\overline{Q}Q$ states, one could employ a sequence of quark bilinears, the simplest of which are

$$\overline{\psi}\psi, \ \overline{\psi}\gamma_5\psi, \ \overline{\psi}\gamma_\mu\psi, \ \overline{\psi}\gamma_\mu\gamma_5\psi, \ \overline{\psi}\sigma_{\mu\nu}\psi, \ldots. \qquad (1.15)$$

Any of these operators acting on the vacuum creates states with its own quantum numbers. The lightest states in the quark spectrum will be associated with those operators which remain nonzero for static quarks, i.e., with creation operators and Dirac spinors of the form

$$\psi \sim \begin{pmatrix} 0 \\ \chi_{\overline{m}} \end{pmatrix} d^\dagger_{\overline{m}}, \qquad \overline{\psi} \sim \begin{pmatrix} \chi_m \\ 0 \end{pmatrix} b^\dagger_m. \qquad (1.16)$$

Only the pseudoscalar operators $\overline{\psi}\gamma_5\psi$, $\overline{\psi}\gamma_0\gamma_5\psi$ and the vector operators $\overline{\psi}\gamma_i\psi$, $\overline{\psi}\sigma_{0i}\psi$ are nonvanishing in this limit. All the other operators have a nonrelativistic reduction proportional to spatial momentum, indicating the need for a unit of orbital angular momentum in forming a state.

The interpolating-field approach is particularly useful in situations where the imposition of gauge invariance determines whether a given field configuration can occur in the physical spectrum. We shall return to this point in Sect. XIII–4 in the

---

[2] The $\sqrt{3}$ factor is associated with the fact that there are *three* dimensions.

course of discussing glueball states. We now turn to a summary, carried throughout the next three sections, of various attempts to *model* the dynamics of light-hadronic states.

XI–2 Potential model

The potential model posits that there is a relatively simple effective theory in which the quarks move nonrelativistically within hadrons. In the light of our previous comments on relativistic motion, this would seem to be acceptable only for truly massive quarks like the b quark and certainly questionable for the light quarks u, d, s. However, in the potential model it is assumed that QCD interactions dress each quark with a cloud of virtual gluons and quark–antiquark pairs, and that the resulting dynamical mass contribution is so large that quarks move nonrelativistically. These 'dressed' degrees of freedom are called *constituent quarks*, and their masses are called 'constituent masses'. Constituent masses are not to be directly identified with the mass parameters occurring in the QCD lagrangian.[3] Energy levels and wavefunctions are then obtained by solving the nonrelativistic Schrödinger equation in terms of the constituent masses and some assumed potential energy function.

The potential model is not without flaws. For light-quark dynamics, it is far from clear that a static potential can adequately describe the QCD interaction. Even with the use of constituent masses, one finds from fits to the mass spectrum and/or the charge radius that quark velocity is nevertheless near the speed of light (cf. Prob. XI–1). Also, although it is possible [LeOPR 85] to make a connection between the lightest pseudoscalar mesons as Goldstone bosons on the one hand and $Q\bar{Q}$ composites on the other, this is not ordinarily done. Such criticisms notwithstanding, the nonrelativistic quark model does provide a framework for describing both ground and excited hadronic states, and brings a measure of order to a spectrum containing hundreds of observed levels. Besides, virtually all physicists are familiar with the Schrödinger equation, and find the potential model to be an understandable and intuitive language.

Basic ingredients

One begins by expressing the mass M_α of a hadronic state α as

$$M_\alpha = \sum_i M_i + E_\alpha, \qquad (2.1)$$

---

[3] We shall continue to denote the QCD mass parameter of quark q_i as m_i, and shall write the corresponding constituent mass as M_i.

where the sum is over the constituent quarks and antiquarks in α. The internal energy E_α is an eigenvalue of the Schrödinger equation

$$H\psi_\alpha = E_\alpha \psi_\alpha, \tag{2.2}$$

with hamiltonian

$$H = \sum_i \frac{1}{2M_i}\mathbf{p}_i^2 + \sum_{i<j} V_{\text{color}}(\mathbf{r}_{ij}), \tag{2.3}$$

where $\mathbf{r}_{ij} \equiv \mathbf{r}_i - \mathbf{r}_j$, and the subscript 'color' on the potential energy indicates that the dynamics of quarks necessarily involves the color degree of freedom in some manner. It is standard to assume that the potential energy is a sum of two-body interactions. Although there exists no unique specification of the interquark potential V_{color} from *QCD*, the following features are often adopted:

(1) a spin-and flavor-independent long-range confining potential,
(2) a spin-and flavor-dependent short-range potential,
(3) basis mixing in the baryon and meson sectors, and
(4) relativistic corrections.

We shall discuss specific models of the potential energy function in Sect. XIII–1. They all have in common the color dependence in which *the two-particle potential is twice as strong in mesons as it is in baryons*,

$$V_{\text{color}}(\mathbf{r}_{ij}) = \begin{cases} V(\mathbf{r}_{ij}) & \text{(mesons)}, \\ \frac{1}{2} V(\mathbf{r}_{ij}) & \text{(baryons)}. \end{cases} \tag{2.4}$$

We shall describe a simple empirical test for such behavior at the end of this section. To appreciate its theoretical basis, note that the quark–antiquark pair in a meson must occur in the **1** representation of color, whereas any two quarks in a baryon must be in a **3*** representation (in order that the three-quark composite be a color singlet),

$$\begin{aligned} V_{\text{color}} &\propto \begin{cases} F(\mathbf{3}) \cdot F(\mathbf{3}^*) & \text{(mesons)}, \\ F(\mathbf{3}) \cdot F(\mathbf{3}) & \text{(baryons)}, \end{cases} \\ &\propto \begin{cases} (F^2(\mathbf{1}) - F^2(\mathbf{3}) - F^2(\mathbf{3}^*))/2 = -4/3 & \text{(mesons)}, \\ (F^2(\mathbf{3}^*) - 2F^2(\mathbf{3}))/2 = -2/3 & \text{(baryons)}, \end{cases} \end{aligned} \tag{2.5}$$

where $F^a(\mathbf{R})$ is a color generator for $SU(3)$ representation $\mathbf{R}$. Thus, the color dependence in Eq. (2.4) is that which one would naturally associate with the interaction between two quarks or a quark–antiquark pair.

Table XI–3. *Quantum numbers of*
$Q\bar{Q}$ composites.

L	Singlet	Triplet
0	$^1S_0(0^{-+})$	$^3S_1(1^{--})$
1	$^1P_1(1^{+-})$	$^3P_{0,1,2}(0^{++}, 1^{++}, 2^{++})$
2	$^1D_2(2^{-+})$	$^3D_{1,2,3}(1^{--}, 2^{--}, 3^{--})$
3	$^1F_3(3^{+-})$	$^3F_{2,3,4}(2^{++}, 3^{++}, 4^{++})$

Mesons

For the two-particle $Q\bar{Q}$ system, it is straightforward to remove the center-of-mass dependence. In the center-of-mass frame the Schrödinger equation becomes

$$\left(\frac{\mathbf{p}^2}{2M} + V(\mathbf{r})\right)\psi_\alpha(\mathbf{r}) = E_\alpha\psi_\alpha(\mathbf{r}), \tag{2.6}$$

where $\mathbf{r} = \mathbf{r}_Q - \mathbf{r}_{\bar{Q}}$ and $M^{-1} = M_Q^{-1} + M_{\bar{Q}}^{-1}$ is the inverse reduced mass. The LS coupling scheme is typically employed to classify the eigenfunctions of this problem. One constructs the total $Q\bar{Q}$ spin, $\mathbf{S} = \mathbf{s}_Q + \mathbf{s}_{\bar{Q}}$, and adds the orbital angular momentum $\mathbf{L}$ to form the total angular momentum $\mathbf{J} = \mathbf{S} + \mathbf{L}$. There is an infinite tower of eigenstates, each labeled by the radial quantum number n and the angular momentum quantum numbers J, J_z, L, S.

The $Q\bar{Q}$ states are sometimes described in terms of spectroscopic notation $^{2S+1}L_J(J^{PC})$, where P is the parity and C is the charge conjugation,

$$P = (-)^{L+1}, \qquad C = (-)^{L+S}. \tag{2.7}$$

Strictly speaking, although only electrically neutral particles like π^0 can be eigenstates of the charge conjugation operation, C is often employed as a label for an entire isomultiplet, like $\pi = (\pi^+, \pi^0, \pi^-)$. The lowest $Q\bar{Q}$ orbital configurations, expressed in $^{2S+1}L_J(J^{PC})$ notation, are displayed in Table XI–3. The $0^+, 1^-, 2^+, \ldots$ series of J^P states is called *natural*, and has the same quantum numbers as would occur for two spinless mesons of a common intrinsic parity. The alternate sequence, $0^-, 1^+, 2^-, \ldots$ is referred to as *unnatural*. There are a number of J^{PC} configurations, called *exotic* states, which cannot be accommodated within the $Q\bar{Q}$ construction. For example, the 0^{--} state is exotic because any state with $J = 0$ must have $L = S$, and according to the $Q\bar{Q}$ constraint of Eq. (2.7) must therefore carry $C = +$. Likewise, the $CP = -1$ sequence $0^{+-}, 1^{-+}, 2^{+-}, \ldots$ is forbidden because the $Q\bar{Q}$ model requires $CP = (-)^{S+1}$, implying $S = 0$ and hence $J = L$. Thus, one would obtain $P = (-)^{J+1}$ in the $Q\bar{Q}$ model and not $P = (-)^J$.

Baryons

Most applications of the quark model for Q^3 baryons involve the light quarks. If, for simplicity, we assume degenerate constituent mass M, the Schrödinger equation is

$$H_0 = \frac{1}{2M} \sum_{i=1}^{3} \mathbf{p}_i^2 + \frac{1}{2} \sum_{i<j} V(\mathbf{r}_{ij}), \qquad (2.8)$$

where the prefactor of $1/2$ in the potential energy term follows from Eq. (2.4). It is convenient to define a center-of-mass coordinate $\mathbf{R}$ and internal coordinates λ and ρ by

$$\mathbf{R} = (\mathbf{r}_1 + \mathbf{r}_2 + \mathbf{r}_3)/3,$$
$$\rho = (\mathbf{r}_1 - \mathbf{r}_2)/\sqrt{2},$$
$$\lambda = (\mathbf{r}_1 + \mathbf{r}_2 - 2\mathbf{r}_3)/\sqrt{6}. \qquad (2.9)$$

Because it is not possible to remove the three-particle center-of-mass dependence for an arbitrary potential, the following approach is often followed [IsK 78]. The potential $V(\mathbf{r}_{ij})$ is rewritten as

$$V(\mathbf{r}_{ij}) = V_{\mathrm{osc}}(r_{ij}) + U(\mathbf{r}_{ij}), \qquad (2.10)$$

where

$$V_{\mathrm{osc}} = \frac{k}{2}\mathbf{r}_{ij}^2, \qquad U \equiv V - V_{\mathrm{osc}}. \qquad (2.11)$$

The Schrödinger equation is solved in terms of the oscillator potential and U is evaluated perturbatively in the oscillator basis. Having removed the center-of-mass coordinate, we are left with the following hamiltonian for the internal energy:

$$H_{\mathrm{int}} = \left(\frac{\mathbf{p}_\rho^2}{2m} + \frac{3k}{2}\rho^2\right) + \left(\frac{\mathbf{p}_\lambda^2}{2m} + \frac{3k}{2}\lambda^2\right), \qquad (2.12)$$

which is just that of two independent quantum oscillators each with spring constant $3k$. For later purposes, we write the number of excitation quanta for the two oscillators as N_ρ and N_λ ($N_{\rho,\lambda} = 0, 1, 2, \dots$) and let $N \equiv N_\rho + N_\lambda$. The angular momentum for the three-quark system is found in a similar manner as for the $Q\bar{Q}$ mesons, $\mathbf{J} = \mathbf{L} + \mathbf{S}$. The total quark spin is $\mathbf{S} = \sum \mathbf{s}_i$, the orbital angular momentum is given by $\mathbf{L} = \mathbf{L}_\rho + \mathbf{L}_\lambda$, and the parity is $P = (-)^{\ell_\rho + \ell_\lambda}$. The ground-state wavefunction has the form

$$\psi_{\mathrm{gnd}}(\mathbf{r}_1, \mathbf{r}_2, \mathbf{r}_3) = \left(\frac{\alpha^2}{\pi}\right)^{3/2} e^{i\mathbf{P}\cdot\mathbf{R}} e^{-\alpha^2(\rho^2+\lambda^2)/2}, \qquad (2.13)$$

where $\alpha^2 = (3km)^{1/2}$. A cautionary remark is in order. One should not misinterpret the use of an oscillator potential – it is *not* the intent to model the observed baryon spectrum as that of a system of quantum oscillators because such a picture would fail. For example, the oscillator spectrum has $E_N \sim N$, whereas the baryon spectrum obeys the law of linear Regge trajectories (cf. Sect. XIII–2), $E_N^2 \sim N$. The oscillator potential provides a convenient basis for structuring the calculation and nothing more.

Color dependence of the interquark potential

Short of doing a complete spectroscopic analysis, we can find experimental support in the following simple example for the assertion that the two-particle interquark potential is twice as strong in mesons as it is in baryons.

A potential model description for the meson and baryon mass splittings $\rho(770) - \pi(138)$ and $\Delta(1232) - N(939)$ is given by a *QCD* hyperfine interaction, H_{hyp}, akin to the delta function contribution in the *QED* hyperfine potential of Eq. (V–1.16),

$$H_{\text{hyp}} = k_\alpha \sum_{i<j} \bar{\mathcal{H}}_{ij} \, \mathbf{s}_i \cdot \mathbf{s}_j \, \delta^{(3)}(\mathbf{r}) \qquad (\alpha = M, B), \qquad (2.14)$$

where the $\{\bar{\mathcal{H}}_{ij}\}$ are constants and, assuming the color dependence is that given by Eqs. (2.4), (2.5), $k_M = 1$ for mesons and $k_B = 1/2$ for baryons. We shall discuss in Sect. XIII–2 how this effect could arise from gluon exchange. Although there is ordinarily dependence on quark mass in the $\{\bar{\mathcal{H}}_{ij}\}$, it suffices to treat the $\{\bar{\mathcal{H}}_{ij}\}$ as an overall constant since the hadrons in this example contain only light nonstrange quarks. The point is then to see whether the condition $k_M = 2k_B$ is in accord with phenomenology. Noting that for mesons the spin factors yield

$$\mathbf{s}_1 \cdot \mathbf{s}_2 = \frac{2S^2 - 3}{4} = \begin{cases} 1/4 & (S = 1), \\ -3/4 & (S = 0), \end{cases} \qquad (2.15a)$$

whereas for baryons one has

$$\sum_{i<j} \mathbf{s}_i \cdot \mathbf{s}_j = \frac{4S^2 - 9}{8} = \begin{cases} 3/4 & (S = 3/2), \\ -3/4 & (S = 1/2), \end{cases} \qquad (2.15b)$$

we find after taking expectation values that

$$\frac{m_\rho - m_\pi}{m_\Delta - m_N} = \frac{2k_M}{3k_B} \frac{|\psi_M(0)|^2}{|\psi_B(0)|^2} \simeq \frac{2k_M}{3k_B} \frac{(\text{Volume})_B}{(\text{Volume})_M} \simeq \frac{2k_M}{3k_B} \left[\frac{\langle r^2 \rangle_B}{\langle r^2 \rangle_M} \right]^{3/2}. \qquad (2.16)$$

The measured values (cf. Eq. (1.13)) of the proton and pion charge radii imply that $k_M/k_B \simeq 2$. This example, along with others, lends credence to the assumed color dependence of Eq. (2.4).

At this point we shall temporarily leave our discussion of the potential model to consider other descriptions of hadronic structure. We shall return to the potential model for the discussion of hadron spectroscopy in Chaps. XII–XIII.

XI–3 Bag model

A superconductor has an ordered quantum mechanical ground state which does not support a magnetic field (*Meissner effect*) and which is brought about by a condensation of dynamically paired electrons (*Cooper pairs*). An order parameter for this medium is provided by the Landau–Ginzburg wavefunction of a Cooper pair. Even at zero temperature, a sufficiently strong magnetic field, B_{cr}, can induce a transition from the superconducting phase to the normal phase. For example, in tin the critical field is $B_{cr}(\text{tin}) \simeq 3.06 \times 10^{-2}$ tesla, and the energy density of superconducting pairing (*condensation* energy) is $U_{\text{super}}/V \simeq 373 \text{ J/m}^3$.

Chromodynamics exhibits similar behavior, and this is the basis for the bag model [ChJJTW 74]. The QCD ground state evidently does not support a chromo-electric field, and is thus analogous to the superconducting state, although a compelling description of the QCD pairing mechanism has not yet been provided. In the bag model, the analog of the normal conducting ground state is called the *perturbative vacuum*. The vacuum expectation value of the quark bilinear $\bar{q}q$ ($q = u, d, s$) plays the role of an order parameter by distinguishing between the two vacua,

$$_{QCD}\langle 0|\bar{q}q|0\rangle_{QCD} < 0, \qquad _{\text{pert}}\langle 0|\bar{q}q|0\rangle_{\text{pert}} = 0. \tag{3.1}$$

Hadrons are represented as color-singlet 'bags' of perturbative vacuum occupied by quarks and gluons. The bag model employs as its starting point the lagrange density [Jo 78]

$$\mathcal{L}_{\text{bag}} = (\mathcal{L}_{QCD} - B)\,\theta(\bar{q}q), \tag{3.2}$$

where the θ function (which vanishes for negative argument) defines the spatial volume encompassed by the perturbative vacuum. B is called the *bag constant*, and is often expressed in units of $(\text{MeV})^4$. Physically, it represents the difference in energy density between the QCD and perturbative vacua. Phenomenological determinations of B yield $B^{1/4} \simeq 150$ MeV, which translates to a QCD condensation energy of $U_{QCD}/V \simeq 1.0 \times 10^{34} \text{ J m}^{-3}$. Although huge on the scale of the condensation energy for superconductivity, this value appears less remarkable in more natural units, $B \simeq 66 \text{ MeV fm}^{-3}$.

Static cavity

To obtain the equations of motion and boundary conditions for the bag model, we must minimize the action functional of the theory. We shall consider at first a simplified model consisting of a bag which contains only quarks of a given flavor q and mass m. The equations of motion that follow from the lagrangian of Eq. (3.2) are

$$(i\slashed{\partial} - m)q = 0, \tag{3.3}$$

within the bag volume V and

$$in^{\mu}\gamma_{\mu}q = q, \tag{3.4a}$$

$$n_{\mu}\partial^{\mu}(\bar{q}q) = 2B \tag{3.4b}$$

on the bag surface S, where n_{μ} is the covariant inward normal to S. Eq. (3.3) describes a Dirac particle of mass m moving freely within the cavity defined by volume V. Since the order parameter $\bar{q}q$ vanishes at the surface of the bag, the linear boundary condition in Eq. (3.4a) amounts to requiring that the normal component of the quark vector current also vanish at the surface. Thus, quarks are confined within the bag. The nonlinear boundary condition represents a balance between the outward pressure of the quark field and the inward pressure of B.

Spherical-cavity approximation

In principle, the bag surface should be determined dynamically. However, the only manageable approximation for light-quark dynamics is one in which the shape of the bag is taken as spherical with some radius R. For such a static configuration, the nonlinear boundary condition becomes equivalent to requiring that the energy be minimized as a function of R. The static-cavity hamiltonian is

$$H = \int_{V} d^3x \left[q^{\dagger}(-i\boldsymbol{\alpha} \cdot \boldsymbol{\nabla})q + q^{\dagger}\beta mq + B \right]. \tag{3.5}$$

Observe that B plays the role of a constant energy density at all points within the bag. As in Eq. (1.1), the normal modes of the cavity-confined quarks and antiquarks provide a basis for expanding quantum fields. They are determined by solving the Dirac equation Eq. (3.3) in a spherical cavity. We characterize each mode in terms of a radial quantum number n, an orbital angular momentum quantum number ℓ (as would appear in the nonrelativistic limit), and a total angular momentum, j. Only $j = 1/2$ modes are consistent with the nonlinear boundary condition since the rigid spherical cavity cannot accommodate the angular variation of $j > 1/2$ modes. Such nonspherical orbitals can be treated only approximately, by implementing the nonlinear boundary condition as an angular average or by minimizing the solution with respect to the energy. In addition, since neither $p_{1/2}$ modes nor radially excited

$s_{1/2}$ modes are orthogonal to a translation of the ground state, they must be admixed with some of the $j = 3/2$ modes to construct physically acceptable excitations. For these reasons, the bag model has been most widely applied in modeling properties of the ground-state hadrons rather than their excited states.

Let us consider the $s_{1/2}$ case in some detail. Even with the restriction to a single spin-parity state, there are still an infinity of eigenfrequencies ω_n. Each ω_n is fixed by the linear boundary condition, expressible as the transcendental equation

$$\tan p_n = -\frac{p_n}{\omega_n + mR - 1} \qquad (n = 1, 2, \ldots), \qquad (3.6)$$

where $p_n \equiv \sqrt{\omega_n^2 - m^2 R^2}$. For zero quark mass, the lowest eigenfrequencies are $\omega = 2.043,\ 4.611, \ldots$. For light-quark mass ($mR \leq 1$) the lowest mode frequency is approximated by $\omega_1 \simeq 2.043 + 0.493mR$, and in the limit of heavy-quark mass ($mR \gg 1$) becomes $\omega_1 \to \sqrt{m^2 R^2 + \pi^2}$. The spatial wavefunction which accompanies destruction of an $s_{1/2}$ quark with spin alignment λ and mode n is

$$\psi_n(\mathbf{x}) = \frac{1}{\sqrt{4\pi}} \begin{pmatrix} ij_0(p_n r/R)\chi_\lambda \\ -\epsilon j_1(p_n r/R)\sigma \cdot \hat{r}\chi_\lambda \end{pmatrix}, \qquad (3.7)$$

while for creation of an $s_{1/2}$ antiquark we have

$$\psi_{\bar{n}}(\mathbf{x}) = \frac{1}{\sqrt{4\pi}} \begin{pmatrix} -i\epsilon j_1(p_n r/R)\sigma \cdot \hat{r}\bar{\chi}_\lambda \\ j_0(p_n r/R)\bar{\chi}_\lambda \end{pmatrix}, \qquad (3.8)$$

where $\epsilon \equiv ((\omega_n - mR)/(\omega_n + mR))^{1/2}$, χ_λ is a two-component spinor, and $\bar{\chi}_\lambda \equiv i\sigma_2 \chi_\lambda$. The full quark field $q(x)$, expanded in terms of the $s_{1/2}$ modes, is given by

$$q(x) = \sum_n N(\omega_n) \left[\psi_n(\mathbf{x})e^{-i\omega_n t/R} b(n) + \psi_{\bar{n}}(\mathbf{x})e^{i\omega_n t/R} d^\dagger(\bar{n}) \right], \qquad (3.9)$$

where

$$N(\omega_n) = \left(\frac{p_n^4}{R^3(2\omega_n^2 - 2\omega_n + mR)\sin^2 p_n} \right)^{1/2} \qquad (3.10)$$

is a normalization factor which is fixed by demanding that the number operator $N_q = \int_{\text{bag}} d^3x\, q^\dagger(x)q(x)$ for quark flavor q have integer eigenvalues.

By computing the expectation value of the hamiltonian in a state of N quarks and/or antiquarks of a given flavor, one obtains

$$\langle H \rangle = N\omega/R + 4\pi B R^3/3 - Z_0/R. \qquad (3.11)$$

In the final term, Z_0 is a phenomenological constant that has been used in the literature to summarize effects having a $1/R$ dimension, most notably the effect of zero-point energies, which for an infinite-volume system would be unobservable. However, just as the Casimir effect is present for a finite-volume system

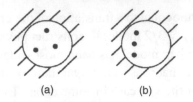

(a)   (b)

Fig. XI–3 Quarks in a bag.

with fixed boundaries, such a term must be present in the static cavity bag model [DeJJK 75]. Unfortunately, a precise calculation of this effect has proven to be rather formidable, and so one treats Z_0 as a phenomenological parameter.

Upon solving the condition $\partial \langle H \rangle / \partial R = 0$, we obtain expressions for the bag radius

$$R^4 = \frac{1}{4\pi B}(N\omega - Z_0), \tag{3.12}$$

and the bag energy

$$E = \frac{4}{3}(4\pi B)^{1/4}(N\omega - Z_0)^{3/4}. \tag{3.13}$$

The bag energy E is *not* precisely the hadron mass. Although the bag surface remains fixed in the cavity approximation, the quarks within move freely as independent particles. Thus, at one instant, the configuration of quarks might appear as in Fig. XI–3(a), whereas at another time, the quarks occupy the positions of Fig. XI–3(b). As a result, there are unavoidable fluctuations in the bag center-of-mass position. The bag energy is thus $E = \langle \sqrt{\mathbf{p}^2 + M^2} \rangle$, where M is the hadron mass and $\mathbf{p}$ represents the instantaneous hadron momentum. Although the average momentum vanishes ($\langle \mathbf{p} \rangle = 0$), the fluctuations do not, ($\langle \mathbf{p}^2 \rangle \neq 0$). For all hadrons but the pion, it is reasonable to expand the bag energy in inverse powers of the hadron mass,

$$E = M + \langle \mathbf{p}^2 \rangle / 2M + \cdots . \tag{3.14}$$

For the pion, one should instead expand as

$$E = \langle |\mathbf{p}| \rangle + M_\pi^2 \langle |\mathbf{p}|^{-1} \rangle / 2 + \cdots . \tag{3.15}$$

One can employ the method of wave packets, to be explained in Sect. XII–1, to estimate that $\langle |\mathbf{p}| \rangle \simeq 2.3R^{-1}$, $\langle |\mathbf{p}|^{-1} \rangle \simeq 0.7R$ for the pion bag, and $\langle \mathbf{p}^2 \rangle \simeq N\omega_1^2 R^{-2}$ for a bag containing N quarks and/or antiquarks in the $s_{1/2}$ mode.

Gluons in a bag

Any detailed phenomenological fit of the bag model to hadrons must include the spin–spin interaction between quarks. One way to incorporate this effect is to posit that gluons, as well as quarks, can exist within a bag. With only gluons present, the lagrangian is taken to be [Jo 78]

$$\mathcal{L}_{\text{bag}}^{\text{gluon}} = \left[-\frac{1}{4}F_{\mu\nu}^a F^{a\mu\nu} - B\right]\theta\left(-F_{\mu\nu}^a F^{a\mu\nu}/4 - B\right), \qquad (3.16)$$

and the Euler–Lagrange equations are

$$\partial^\mu F_{\mu\nu}^a = 0 \qquad (3.17)$$

in the bag volume V, and

$$n^\mu F_{\mu\nu}^a = 0 \qquad (3.18\text{a})$$

$$F_{\mu\nu}^a F^{a\mu\nu} = -4B \qquad (3.18\text{b})$$

on the bag surface S. In the limit of zero coupling, the gluon field strength becomes $F_{\mu\nu}^a = \partial_\mu A_\nu^a - \partial_\nu A_\mu^a$. The field equations in V are sourceless Maxwell equations with boundary conditions $\mathbf{x} \cdot \mathbf{E}^a = 0$ and $\mathbf{x} \times \mathbf{B}^a = 0$ on S, where $\mathbf{E}^a$ and $\mathbf{B}^a$ are the color electric and magnetic fields, respectively. It is convenient to work directly with the gluon field $\mathbf{A}^a(x)$, and with a gauge choice to restrict the dynamic degrees of freedom to the spatial components. In mode n, these obey

$$\left[\nabla^2 + (k_n/R)^2\right]\mathbf{A}_n^a = 0, \qquad (3.19)$$

and

$$\nabla \cdot \mathbf{A}_n^a = 0 \qquad (3.20)$$

within the bag. The gluon eigenfrequencies k_n are determined by the linear boundary condition

$$\mathbf{r} \times \left(\nabla \times \mathbf{A}_n^a\right) = 0. \qquad (3.21)$$

Restricting our attention to modes of positive parity, we have for the gluon field operator

$$\mathbf{A}^a(x) = \sum_{n,\sigma} N_G(k_n)\left(j_1(k_n r/R)\mathbf{X}_{1\sigma}(\Omega)a_{n,\sigma}^a + \text{h.c.}\right), \qquad (3.22)$$

where $\mathbf{X}_{1\sigma}$ is a vector spherical harmonic. The gluon normalization factor is obtained, analogously to $N(\omega_n)$ for quarks, by constraining the gluon number operator to be integer-valued and we find

$$[N_G(k_n)]^{-2} = \left[3(1 - \sin(2k_n)/2k_n) - 2(1 + k_n^2)\sin^2(k_n)\right]R^2. \qquad (3.23)$$

The quark–gluon interaction

In the following, we shall work with the lowest positive parity mode, for which $k_1 = 2.744$. The quark hyperfine interaction in hadron H can be computed from the second-order perturbation theory formula,

$$E_{\text{hyp}} = \langle H | H_{\text{q-g}} (E_0 - H_0 + i\epsilon)^{-1} H_{\text{q-g}} | H \rangle, \tag{3.24}$$

where the unperturbed hamiltonian H_0 is given in Eq. (3.5) and $H_{\text{q-g}}$ is the quark-gluon interaction

$$H_{\text{q-g}} = -g_3 \int_V d^3x \, \mathbf{J}^a(x) \cdot \mathbf{A}^a(x), \tag{3.25}$$

defined in terms of the quark color current density

$$\mathbf{J}^a(x) = \frac{1}{2} \bar{q}_i(x) \boldsymbol{\gamma} \lambda^a_{ij} q_j(x). \tag{3.26}$$

Implicit in Eq. (3.24) is an infinite sum over all intermediate states. In practice, the sum can be well approximated by the lowest-energy intermediate state, and we find for hadron H

$$E_{\text{hyp}} = \langle H | H_{\text{hyp}} | H \rangle = \alpha_s h_H R^{-1}, \tag{3.27}$$

where

$$h_H = -0.177 \langle H | \sum_{i<j} \boldsymbol{\sigma}_i \cdot \boldsymbol{\sigma}_j \, \mathbf{F}_i \cdot \mathbf{F}_j | H \rangle. \tag{3.28}$$

The numerical factor arises from an overlap integral of quark and gluon spatial wavefunctions, and $\mathbf{F}_i$, $\boldsymbol{\sigma}_i$ are, respectively, the color and spin operators for quark i. It is straightforward to demonstrate that $h_\pi = 0.708$, $h_N = -h_\Delta = h_\pi/2$, and $h_\rho = -h_\pi/3$.

We have described the primary ingredients of the bag model. Fits to the masses of the ground-state hadrons can be accomplished within this framework, for example in [DeJJK 75, DoJ 80]. These reproduce many of the features of these particles, and we return to baryon properties in the next chapter.

XI–4 Skyrme model

In Chap. X, we explored the $N_c \to \infty$ limit of *QCD*. In some respects the world thus defined is not unlike our own. Mesons and glueballs exist with masses which are $\mathcal{O}(1)$ as $N_c \to \infty$. To lowest order, these particles are noninteracting because their coupling strength is $\mathcal{O}(N_c^{-1})$. What becomes of baryons in this world? It takes N_c quarks to form a totally antisymmetric color-singlet composite, so baryon mass is expected to be $\mathcal{O}(N_c)$. Note the inverse correlation between interparticle coupling $\mathcal{O}(N_c^{-1})$ and baryon mass $\mathcal{O}(N_c)$. This is reminiscent of soliton behavior in theories with nonlinear dynamics.

Sine–Gordon soliton

An example is afforded by the Sine–Gordon model, defined in one space and one time dimension by the lagrangian,

$$\mathcal{L}_{SG} = \frac{1}{2}(\partial_\mu \varphi)^2 - \frac{\alpha}{\beta^2}(1 - \cos \beta\varphi), \tag{4.1}$$

where α and β are constants. For small-amplitude field excitations, an expansion in powers of φ,

$$\mathcal{L}_{SG} = \frac{1}{2}(\partial_\mu \varphi)^2 - \frac{\alpha}{2}\varphi^2 + \frac{\alpha\beta^2}{4!}\varphi^4 + \mathcal{O}\left(\beta^4\varphi^6\right), \tag{4.2}$$

identifies the parameter α as the boson squared mass and β as a coupling strength. For $\beta \to 0$ we recover the free field theory. The Sine–Gordon lagrangian has also a nonperturbative static solution,

$$\varphi_0(x) = \frac{4}{\beta} \tan^{-1}\left(\exp\left(\sqrt{\alpha}x\right)\right), \tag{4.3a}$$

with energy

$$E_0 = 8\sqrt{\alpha}/\beta^2. \tag{4.3b}$$

This solution is a Sine–Gordon soliton. The natural unit of length for the soliton is $\alpha^{-1/2}$, and the energy E_0 diverges as the coupling is turned off ($\beta \to 0$). The potential energy in this theory has an infinity of equally spaced minima, with $\varphi^{(n)} = 2\pi n/\beta$ ($n = 0, \pm 1, \pm 2, \dots$). As the coordinate x is varied continuously from $-\infty$ to $+\infty$, the soliton amplitude $\varphi_0(x)$, starting from the minimum $\varphi^{(0)} = 0$, moves to the adjoining minimum $\varphi^{(1)} = 2\pi/\beta$. An index ΔN, the *winding number*, counts the number of minima shifted. It can be expressed as the charge associated with a current density,

$$J^\mu = \frac{\beta}{2\pi} \epsilon^{\mu\nu} \partial_\nu \varphi, \tag{4.4}$$

such that

$$\Delta N = \int_{-\infty}^{\infty} dx \, J^0(x) = \frac{\beta}{2\pi} [\varphi(+\infty) - \varphi(-\infty)]. \tag{4.5}$$

For $\varphi = \varphi_0$ as in Eq. (4.3a) we see that $\Delta N = 1$. The current density is conserved, $\partial_\mu J^\mu = 0$. Thus its charge, the winding number ΔN, does not change with time.

This is an example of a *topological* conservation law, whose origin lies in the non-trivial boundary conditions (viz. Eq. (4.5)) which a given field configuration is constrained to obey.

Chiral SU(2) soliton

Let us now seek a soliton solution for an $SU(2)_L \times SU(2)_R$ invariant theory in a spacetime of dimension four. It is natural to consider first the lowest-order chiral lagrangian $\mathcal{L}_2$,

$$\mathcal{L}_2 = \frac{F_\pi^2}{4} \operatorname{Tr} \left(\partial_\mu U \partial^\mu U^\dagger \right), \qquad (4.6)$$

where U is an $SU(2)$ matrix which transforms as $U \rightarrow LUR^{-1}$ under a chiral transformation for $L \in SU(2)_L$ and $R \in SU(2)_R$. Unfortunately, $\mathcal{L}_2$ cannot support an acceptable soliton, as the soliton would have zero size and zero energy. To see why, recall that the Sine–Gordon soliton has a natural unit of length $\alpha^{-1/2}$. Suppose there is an analogous quantity, R, for the chiral soliton. Then we can write the radial variable as $r = \tilde{r}R$, where $\tilde{r}$ is dimensionless. For a static solution, the energy becomes

$$E = \int d^3x \, \mathcal{H} = -\int d^3x \, \mathcal{L} = \frac{F_\pi^2}{4} \int d^3x \, \operatorname{Tr} \left(\nabla U \cdot \nabla U^\dagger \right). \qquad (4.7)$$

Upon expressing the integral in terms of the dimensionless variable $\tilde{r}$, we find $E = aR$, where a is a nonnegative number. The energy is minimized at $R = 0$ to the value $E = 0$. This trivial solution is unacceptable, and thus the model must be extended.

The Skyrme model [Sk 61] employs, in addition to $\mathcal{L}_2$, a quartic interaction of a certain structure,

$$\mathcal{L} = \frac{F_\pi^2}{4} \operatorname{Tr} \left(\partial_\mu U \partial^\mu U^\dagger \right) + \frac{1}{32e^2} \operatorname{Tr} \left[\partial_\mu U \, U^\dagger, \partial_\nu U \, U^\dagger \right]^2, \qquad (4.8)$$

where e (not to be confused with the electric charge!) is a dimensionless real-valued parameter. The above chiral lagrangian should look familiar, since it is part of the general fourth-order chiral lagrangian used in Chap. VII. In particular, Eq. (4.8) is reproduced if $2L_1^r + 2L_2^r + L_3^r = 0$, in which case $(32e^2)^{-1} = (L_2^r - 2L_1^r - L_3^r)/4$. The comparison with the phenomenology of Chap. VII is not completely straightforward, as the pion physics was treated to one-loop order while the Skyrme lagrangian is used at tree level. We note, however, that the coefficients in Table VII–1 give

$$\frac{2L_1^r + 2L_2^r + L_3^r}{L_2^r - 2L_1^r - L_3^r} = 0.685, \qquad L_2^r - 2L_1^r - L_3^r = 0.0040. \qquad (4.9)$$

The latter combination, which is independent of renormalization scale, numerically gives $e \simeq 5.6$. In the following development, we shall follow standard practice by taking the parameter e as arbitrary.

We seek a static solution of the Skyrme model. Our strategy shall be to first determine the energy functional of the theory, and then minimize it. Following the procedure leading to Eq. (4.7), we can write the energy as

$$E = \int d^3x \ \text{Tr} \left[\frac{F_\pi^2}{4} X_i X_i^\dagger + \frac{1}{16e^2}(\epsilon_{ijk} X_i X_j)(\epsilon_{abk} X_a X_b)^\dagger \right], \qquad (4.10)$$

where $X_\mu \equiv U\partial_\mu U^\dagger$ and $X_\mu = -X_\mu^\dagger$. It is necessary that $X_i \to 0$ as $|\mathbf{x}| \to \infty$ in order that the energy be finite. Thus, U must approach a constant element of $SU(2)$, which we are free to choose as the identity I. For the mesonic sector of the theory, the vacuum state corresponds to $U(\mathbf{x}) = I$ for *all* $\mathbf{x}$. In this state, both the field variable X_i and the energy E vanish. The form $U \simeq I + i\boldsymbol{\pi} \cdot \boldsymbol{\tau}/F_\pi$, used extensively in earlier chapters, corresponds to small-amplitude pionic excitations of the vacuum.

To see that the Skyrme model does support a nontrivial soliton, we cast the energy integrals of Eq. (4.10) in terms of a natural length scale R and find

$$E = aR + bR^{-1}, \qquad (4.11)$$

where a, b are nonnegative. For $a, b \neq 0$, the energy is minimized at nonzero R and nonzero E. Thus, the quartic term of Eq. (4.8) is seen to have the desired effect of inducing soliton stability. Moreover, for arbitrary U a lower bound on the energy is provided by applying the Schwartz inequality to Eq. (4.10),

$$E \geq \frac{F_\pi}{4e} \int d^3x \,|\, \text{Tr} \ \epsilon_{ijk} X_i X_j X_k|. \qquad (4.12)$$

It is not hard to show that the integrand of Eq. (4.12) is proportional to the zeroth component of a four-vector current,

$$B^\mu = \frac{\epsilon^{\mu\nu\alpha\beta}}{24\pi^2} \ \text{Tr} \ X_\nu X_\alpha X_\beta, \qquad (4.13)$$

which is divergenceless, $\partial_\mu B^\mu = 0$, and thus has conserved charge

$$B = \int d^3x \ B^0(\mathbf{x}). \qquad (4.14)$$

It turns out that the current B^μ can be identified as the baryon current density and B is the baryon number of the theory. Note that this is consistent with our prescription $U(\mathbf{x}) = I$ for the meson vacuum, where we see from Eq. (4.13) that $B = 0$. Interestingly, B turns out to have an *additional* significance. It is the topological winding number for the Skyrme model, analogous to ΔN for the Sine–Gordon model. The point is, by having associated spatial infinity with a group element of $SU(2)$ to ensure that the field energy is finite, we have placed the elements of physical space into a correspondence with the elements of the compact group $SU(2)$. The parameter space of each set is S^3, the unit sphere in four dimensions, and it is precisely the field U which implements the mapping. The mappings from S^3 to S^3 are known to fall into classes, each labeled by an integer-valued winding number. In this context, B serves to measure the number of times that the set of space points covers the group parameters of $SU(2)$ for some solution U of the theory.

The Skyrme soliton

The Skyrme *ansatz* for a chiral soliton (*skyrmion*) has the functional form [BaNRS 83, AdNW 83]

$$U_0(\mathbf{x}) = \exp\left[i F(r)\boldsymbol{\tau} \cdot \hat{\mathbf{x}}\right]. \tag{4.15}$$

The unknown quantity is the skyrmion profile function $F(r)$. To specify it, we first determine the energy functional by substituting U_0 into Eq. (4.10),

$$E[F] = 4\pi \int_0^\infty dr\, r^2 \left[\frac{F_\pi^2}{2}\left(F'^2 + 2\frac{\sin^2 F}{r^2}\right) \right.$$
$$\left. +\frac{1}{2e^2}\frac{\sin^2 F}{r^2}\left(\frac{\sin^2 F}{r^2} + 2F'^2\right)\right], \tag{4.16}$$

where a prime signifies differentiation with respect to the argument. For a static solution, the minimization of the energy generates an extremum of the action, and is hence equivalent to the equations of motion. The variation $\delta E/\delta F = 0$ generates a differential equation for F,

$$\left(\frac{\tilde{r}^2}{4} + 2\sin^2 F\right) F'' + \frac{\tilde{r}}{2}F' + F'^2 \sin 2F - \frac{\sin 2F}{4} - \frac{\sin^2 F \sin 2F}{\tilde{r}^2} = 0, \tag{4.17}$$

as expressed in terms of a dimensionless variable $\tilde{r} = r/R$, with $R^{-1} \equiv 2eF_\pi$. This nonlinear equation must be solved numerically, subject to certain boundary conditions. The condition $U = I$ at spatial infinity implies $F(\infty) = 0$. The boundary condition at $r = 0$ is fixed by requiring that the soliton corresponds to baryon number 1. For the Skyrme ansatz, the baryon-number charge density is

$$B^0(r) = -\frac{1}{2\pi^2}\frac{F' \sin^2 F}{r^2}, \tag{4.18}$$

and corresponds to a baryon number

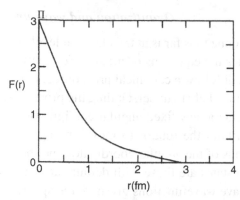

Fig. XI–4 Radial profile of the skyrmion.

$$B = \frac{1}{2\pi} \left[2F(0) - 2F(\infty) - \sin 2F(0) + \sin 2F(\infty) \right]. \qquad (4.19)$$

This leads to the choice $F(0) = \pi$. Although the profile $F(r)$ cannot be determined analytically over its entire range, it is straightforward to show that

$$F(r) \sim \begin{cases} \pi - \text{const.} \, r & (r \to 0), \\ \text{const.} \, r^{-2} & (r \to \infty). \end{cases} \qquad (4.20)$$

We display $F(r)$ in Fig. XI–4. Insertion of the solution to Eq. (4.17) into the energy functional $E[F]$ yields the mass M of the skyrmion, and from a numerical integration we obtain $M \simeq 73 \, F_\pi/e$. There is an important point to be realized about the skyrmion – it represents a use of chiral lagrangians outside the region of validity of the energy expansion. Recall that the full chiral lagrangian is written as a power series, $\mathcal{L} = \mathcal{L}_2 + \mathcal{L}_4 + \cdots$ in the number of derivatives. When matrix elements of pions are taken, terms with n derivatives produce n powers of the energy. Hence, at low energy, one may consistently ignore operators with large n, as their contributions to matrix elements are highly suppressed. However, in forming the skyrmion one employs only $\mathcal{L}_2$ and a subset of $\mathcal{L}_4$. The relative effects of the two are balanced in the minimization of the energy functional, and as a result both contribute equally. In an extended model containing $\mathcal{L}_6$, one would expect the import of $\mathcal{L}_6$ to be analogously comparable to $\mathcal{L}_4$, etc. Higher-derivative lagrangians thus will contribute to skyrmion matrix elements, and the result cannot be considered a controlled approximation. However, this is not sufficient cause for abandoning the skyrmion approach. It simply becomes a phenomenological *model* rather than a rigorous method, and thus has a status similar to potential or bag models.

Quantization and wavefunctions

The analysis done thus far is at the classical level, and merely shows that the chiral soliton satisfies the equations of motion. To determine the quantum version of the theory, we shall follow a canonical procedure. An analogy with quantization of the *rigid rotator* may help in understanding the process. A classical solution consists of the rotator being at any fixed angular configuration $\{\theta, \varphi\}$. To obtain the quantum theory, one allows the rotator to move among these solutions, and describes its motion in terms of the angular coordinates and their conjugate momenta $\{p_\theta, p_\varphi\}$. The quantum states are those with definite angular momentum quantum numbers $\{\ell, m\}$, and have wavefunctions given by the spherical harmonics,

$$\langle \theta, \varphi | \ell, m \rangle = Y_{\ell,m}(\theta, \varphi). \tag{4.21}$$

The classical skyrmion solutions consist not only of U_0 (cf. Eq. (4.15)), but also of any constant $SU(2)$ rotation thereof, $U_0' = AU_0A^{-1}$ with $A \epsilon SU(2)$. A particularly simple approach to quantization is then to allow the soliton to rotate rigidly in the space of these solutions,

$$U = A(t)U_0A^{-1}(t), \tag{4.22}$$

where now $A(t)$ is an arbitrary time-dependent $SU(2)$ matrix. One proceeds to define a set of coordinates $\{a_k\}$, their conjugate momenta $\{\pi_k \equiv \partial \mathcal{L}/\partial a_k\}$, and a hamiltonian constructed via Legendre transformation

$$H = \pi_k \dot{a}_k - L. \tag{4.23}$$

We shall presently describe how to choose quantum numbers and determine the associated wavefunctions. Note that this approach is approximate in that it neglects the possibility of spacetime-dependent excitations such as pion emission. As such, it would be most appropriate for a weakly coupled theory (as occurs for $N_c \to \infty$) where the soliton rotates slowly, but is only approximate in the real world.

In general, an $SU(2)$ matrix like A can be written in terms of three unconstrained parameters $\{\theta_k\}$ as

$$A(t) = \exp(i\tau \cdot \theta) = \mathbf{I} \cos\theta + i\tau \cdot \hat{\theta} \sin\theta. \tag{4.24}$$

However, we can equivalently employ the four constrained parameters,

$$a_0 = \cos\theta, \qquad \mathbf{a} = \hat{\theta} \sin\theta, \tag{4.25a}$$

where

$$\sum_{k=0}^{3} a_k^2 = 1. \tag{4.25b}$$

Substitution of the rotated quantity U into Eq. (4.7) and evaluation of the spatial integration yields

$$L = -M + \lambda \operatorname{Tr}(\partial_0 A^\dagger \partial_0 A) = -M + 2\lambda \sum_{k=0}^{3} \dot{a}_k^2, \tag{4.26}$$

where $\lambda = \pi \Lambda / 3 e^3 F_\pi$, with

$$\Lambda = \int d\tilde{r}\, \tilde{r}^2 \sin^2 F \left[1 + 4\left(F'^2 + \sin^2 F / \tilde{r}^2\right)\right] \simeq 50.9. \tag{4.27}$$

As written in terms of the conjugate momenta $\pi_k = 4\lambda \dot{a}_k$, the hamiltonian is

$$H = M + \frac{1}{8\lambda} \sum_{k=0}^{3} \pi_k^2. \tag{4.28}$$

Adopting the canonical quantization conditions

$$[a_k,\ \pi_l] = i\delta_{kl}, \tag{4.29}$$

we see that the canonical momenta can be expressed as differential operators, $\pi_k = -i\partial/\partial a_k$. Thus, the hamiltonian has the form

$$H = M - \frac{1}{8\lambda} \nabla_4^2, \tag{4.30}$$

where ∇_4^2 is the four-dimensional laplacian restricted to act on the three-sphere by the constraint of Eq. (4.25b).

We can determine the eigenvalues and eigenvectors of H by working in analogy with the more familiar three-dimensional laplacian,

$$\nabla_3^2 = \frac{\partial^2}{\partial r^2} + \frac{2}{r}\frac{\partial}{\partial r} - \frac{1}{r^2}\mathbf{L}^2. \tag{4.31}$$

If constrained to the unit two-sphere by the condition $\sum_{k=1}^{3} x_k^2 = r^2 = 1$, the three-dimensional laplacian ∇_3^2 reduces to $-\mathbf{L}^2$. As is well known, the three components of $\mathbf{L}$ are operators $L_1,\ L_2,\ L_3$ which satisfy

$$[L_j, L_k] = i\epsilon_{jkl} L_l, \tag{4.32}$$

and generate rotations in the 2–3, 3–1, 1–2 planes respectively. The underlying symmetry group is $SO(3)$, and the eigenfunctions are the spherical harmonics.

The four-dimensional problem is treated by analogy. Upon adding an extra dimension labeled by the index 0, we encounter the additional operators K_1, K_2, K_3, which generate rotations in the 0–1, 0–2, 0–3 planes. The full set of six rotational generators can be represented as

$$L_k = \epsilon_{ijk} a_i \pi_j, \qquad K_k = a_0 \pi_k - a_k \pi_0. \tag{4.33}$$

The extended symmetry group is $SO(4)$ and the commutator algebra of the rotation generators is

$$\left[L_j, \ L_k\right] = i\epsilon_{jkl}L_l, \quad \left[L_j, \ K_k\right] = i\epsilon_{jkl}K_l, \quad \left[K_j, \ K_k\right] = i\epsilon_{jkl}L_l. \quad (4.34)$$

The mathematics of this algebra is well known, underlying, for example the symmetry of the Coulomb hamiltonian in nonrelativistic quantum mechanics. By the substitutions

$$\mathbf{T} = (\mathbf{L} - \mathbf{K})/2, \qquad \mathbf{J} = (\mathbf{L} + \mathbf{K})/2, \qquad (4.35)$$

we arrive at operators $\mathbf{T}$ and $\mathbf{J}$, which generate commuting $SU(2)$ algebras. We associate $\mathbf{T}$ with the isospin and $\mathbf{J}$ with the angular momentum. The explicit operator representations,

$$T_k = i(-\epsilon_{ijk}a_i\partial_j + a_0\partial_k - a_k\partial_0),$$
$$J_k = i(-\epsilon_{ijk}a_i\partial_j - a_0\partial_k + a_k\partial_0), \qquad (4.36)$$

follow immediately from Eq. (4.33), and the Skyrme hamiltonian becomes

$$H = M + \left(\mathbf{T}^2 + \mathbf{J}^2\right)/4\lambda. \qquad (4.37)$$

It follows from the commutator algebra of Eq. (4.34) that $\mathbf{T}^2 = \mathbf{J}^2$. Thus, the quantum spectrum consists of states with equal isospin and angular momentum quantum numbers, $T = J$. This is no surprise. After all, in the Skyrme ansatz of Eq. (4.15), the isospin and spatial coordinates appear symmetrically, and we expect the quantum spectrum to respect this reciprocity. Our final form for the hamiltonian,

$$H = M + \mathbf{J}^2/2\lambda, \qquad (4.38)$$

has the eigenvalue spectrum

$$E = M + J(J+1)/2\lambda, \qquad (4.39)$$

where in general $J = 0, 1/2, 1, 3/2, \ldots$.

By analogy with the usual spherical harmonics, the eigenfunctions of H are seen to be traceless symmetric polynomials in the $\{a_k\}$. However, both $\{a_k\}$ and $\{-a_k\}$ describe the same solution U (cf. Eq. (4.22)). In the quantum theory, eigenfunctions thus fall into either of two classes, $\psi(\{-a_k\}) = \pm \psi(\{a_k\})$. Since fermions correspond to the antisymmetric choice, we select only the half-integer values in Eq. (4.39). In the Skyrme model, the N and Δ baryons will have wavefunctions which are respectively linear and cubic in the $\{a_k\}$. To construct such states, it is convenient to employ the differential representations of Eq. (4.36) to prove

$$\begin{aligned} L_3(a_1 \pm ia_2) &= \pm(a_1 \pm ia_2), & L_3a_{0,3} &= 0, \\ K_3(a_0 \pm ia_3) &= \pm(a_0 \pm ia_3), & K_3a_{1,2} &= 0. \end{aligned} \qquad (4.40)$$

From these and Eq. (4.36), the $T_3 = J_3 = 1/2$ eigenstate of a proton with spin up is found to be

$$\langle A | \, p_\uparrow \rangle = \frac{1}{\pi} \, (a_1 + ia_2). \tag{4.41}$$

The normalization of this state is obtained from the angular integral over the three-sphere

$$1 = \langle p_\uparrow | \, p_\uparrow \rangle = \int d\Omega_3 \, \langle p_\uparrow | A \rangle \langle A | \, p_\uparrow \rangle = \frac{1}{\pi^2} \int d\Omega_3 \, (a_1^2 + a_2^2), \tag{4.42}$$

where the angular measure is

$$\int d\Omega_3 = \int_0^{2\pi} d\varphi \int_0^{\pi} d\theta \, \sin \theta \int_0^{\pi} d\chi \, \sin^2 \chi, \tag{4.43}$$

and spherical coordinates in four dimensions are defined by

$$\begin{aligned} a_1 &= \sin \chi \, \sin \theta \, \cos \varphi, & a_2 &= \sin \chi \, \sin \theta \, \sin \varphi, \\ a_3 &= \sin \chi \, \cos \theta, & a_0 &= \cos \chi. \end{aligned} \tag{4.44}$$

The remaining nucleon states can be found by application of the spin and isospin lowering sperators

$$\begin{aligned} J_- &= [(a_1 - ia_2)\partial_3 - (a_3 + ia_0)\partial_1 + (-a_0 + ia_3)\partial_2 + (a_2 + ia_1)\partial_0]/2, \\ T_- &= [(a_1 - ia_2)\partial_3 + (-a_3 + ia_0)\partial_1 + (a_0 + ia_3)\partial_2 - (a_2 + ia_1)\partial_0]/2, \end{aligned} \tag{4.45}$$

where $\partial_k \equiv \partial/\partial a_k$. The $T - J = 3/2$ Δ states are formed by employing analogous ladder operations on

$$\langle A | \, \Delta_{3/2}^{++} \rangle = \frac{i\sqrt{2}}{\pi} \, (a_1 + ia_2)^3. \tag{4.46}$$

It is remarkable that fermions can be constructed from a chiral lagrangian which contains nominally bosonic degrees of freedom. However, the presence of a nonzero fermion quantum number can be easily verified by direct calculation.

The wavefunctions for the eigenstates (the equivalents of $Y_{\ell,m}(\theta, \varphi)$ for the rigid rotator) are given by $SU(2)$ rotation matrices with half-integer values. These are defined by the transformation properties of states under an $SU(2)$ rotation A,

$$|T, T_3'\rangle = \sum_{T_3} \mathcal{D}_{T_3' T_3}^{(T)}(A)|T, T_3\rangle. \tag{4.47}$$

The simplest case is then just the $T = 1/2$ representation, which we know is rotated by the matrix A,

$$\mathcal{D}_{ij}^{(1/2)}(A) = \mathbf{A}_{ij} = \begin{pmatrix} (a_0 + ia_3) & i(a_1 - ia_2) \\ i(a_1 + ia_2) & (a_0 - ia_3) \end{pmatrix}_{ij}. \tag{4.48}$$

Comparison with Eq. (4.41) and with the results of Eq. (4.47) shows that the properly normalized nucleon wavefunctions are

$$\langle A| \, N_{T_3, S_3} \rangle = \frac{1}{\pi} (-)^{T_3 + 1/2} \, \mathcal{D}^{(1/2)}_{-T_3, S_3}(A). \tag{4.49}$$

The general case for a nonstrange baryon B of isospin T and spin S ($S = T$) is given by

$$\langle A| \, B_{T, T_3, S_3} \rangle = \left[\frac{2T + 1}{2\pi^2} \right]^{1/2} (-)^{T + T_3} \, \mathcal{D}^{(T)}_{-T_3, S_3}(A), \tag{4.50}$$

of which the Δ states are specific examples.

Finally, the N and Δ masses are

$$M_N = M + 3/8\lambda = 73 F_\pi / e + e^3 F_\pi / 45.2\pi,$$
$$M_\Delta = M + 15/8\lambda = 73 F_\pi / e + 5 e^3 F_\pi / 45.2\pi. \tag{4.51}$$

If the measured N, Δ masses are used as input, one obtains $e = 5.44$ and $F_\pi = 65$ MeV. Alternatively, from the empirical value for F_π and the determination $e \simeq 5.6$ from pion–pion scattering data, the model implies $M_N \simeq 1.27$ GeV, $M_\Delta \simeq 1.80$ GeV. In either case, agreement between theory and experiment is at about the 30% level. The next state in the spectrum would have quantum numbers $T = J = 5/2$ and is predicted by the first of the above fitting procedures to have mass $M_{5/2} = M + 35/8\lambda \simeq 1.72$ GeV. There is no experimental evidence for such a baryon.

Although the development of the skyrmion and its quantization have been motivated by large-N_c ideas, we know of no proof that requires the skyrmion to come arbitrarily close to the baryons of QCD in the $N_c \to \infty$ limit. An oft-cited counter-example is the existence of a one-flavor version of QCD. Such a theory still contains baryons, such as the Δ^{++}. However, it makes no sense to speak of a one-flavor Skyrme model, as an $SU(2)$ group is required for the underlying soliton U_0. The Skyrme model remains an interesting picture for nucleon structure because it is in many ways orthogonal to the quark model, and thus offers opportunities for new insights.

XI–5 *QCD* sum rules

Low-energy QCD involves a regime where the degrees of freedom are hadrons, and where it is futile to attempt perturbative calculations of hadronic masses and decay widths. Contrasted with this is the short-distance asymptotically free limit in which quarks and gluons are the appropriate degrees of freedom, and in which perturbative calculations make sense. The method of QCD sum rules represents an attempt to bridge the gap between the perturbative and nonperturbative sectors by employing the language of dispersion relations [ShVZ 79a].

The existence of sum rules in *QCD* is quite general, and some might dispute the classification on these sum rules as a phenomenological method. However, in practice, to utilize the sum rules involves the introduction of various approximations and heuristic procedures. Like quark model methods, these are motivated by physical intuition but are not always rigorous consequences of *QCD*. As a result, there remains a certain degree of uncontrollable approximation in their use. Nonetheless, they have been employed in a large number of applications; some early reviews are [ReRY 85, Na 89] and for somewhat more recent entries see [Ra 98, CoK 00, Sh 10].

Correlators

It is convenient to approach the subject by considering the relatively simple two-point functions. Thus, we consider the quark bilinear,

$$J_\Gamma(x) = \bar{q}_1(x)\Gamma q_2(x), \tag{5.1}$$

where Γ is a Dirac matrix, and analyze the *correlator*,

$$i \int d^4x \, e^{iq\cdot x} \langle 0|T(J_\Gamma(x)J_\Gamma^\dagger(0))|0\rangle. \tag{5.2}$$

Such quantities can be expressed in terms of invariant functions $\Pi_\Gamma(q^2)$ and attendant kinematical factors, e.g., as for the correlators of pseudoscalar currents (J_P) and of conserved vector currents (J_V),

$$\Pi_P(q^2) = i \int d^4x \, e^{iq\cdot x} \langle 0|T\,(J_P(x)J_P(0))\,|0\rangle, \tag{5.3a}$$

$$(q^\mu q^\nu - q^2 g^{\mu\nu})\Pi_V(q^2) = i \int d^4x \, e^{iq\cdot x} \langle 0|T\,(J_V^\mu(x)J_V^\nu(0))\,|0\rangle. \tag{5.3b}$$

Analogous structures occur for other currents.

There are several means for analyzing a quantity like $\Pi_\Gamma(q^2)$. One is to write a dispersion relation based on its singularity structure in the complex q^2 plane. The singularities are just those imposed by unitarity. For example, by inserting a complete set of intermediate states into Eq. (5.3a) for the pseudoscalar function $\Pi_P(q^2)$ and invoking the constraints of Lorentz invariance and positivity of energy, we obtain

$$\Pi_P(q^2) = \int_{s_0}^\infty ds \, \frac{\rho_P(s)}{s - q^2 - i\epsilon},$$

$$\theta(q^0)\rho_P(q^2) = (2\pi)^3 \sum_n \delta^4(p_n - q)|\langle 0|J_P(0)|n\rangle|^2, \tag{5.4}$$

where s_0 is the threshold for the physical intermediate states. Such considerations, together with the application of Cauchy's theorem in the complex q^2 plane, imply a dispersion relation for $\Pi_\Gamma(q^2)$,

$$\Pi_\Gamma(q^2) = \frac{(q^2)^N}{\pi} \int_{s_0}^\infty ds \, \frac{\mathrm{Im}\,\Pi_\Gamma(s)}{s^N(s - q^2 - i\epsilon)} + \sum_{n=0}^{N-1} (q^2)^n a_n, \tag{5.5}$$

where the $\{a_n\}$ are N subtraction constants.[4] One attempts to introduce a phenomenological component to the dispersion relation by expressing $\mathrm{Im}\,\Pi_\Gamma(s)$ in terms of measureable quantities, e.g., with cross section-data as in the case of the charm contribution $\bar{c}\gamma^\mu c$ to the vector current,

$$\mathrm{Im}\,\Pi_V^{(\mathrm{chm})} = \frac{1}{12\pi e_c^2} \frac{\sigma_{e^+e^- \to \mathrm{charm}}}{\sigma_{e^+e^- \to \mu^+\mu^-}} = \frac{9s}{64\pi^2\alpha^2} \sigma_{e^+e^- \to \mathrm{charm}}, \tag{5.6}$$

where e_c is the c-quark electric charge and s is the squared center-of-mass energy. If such data are not available, another means must be found for expressing $\mathrm{Im}\,\Pi_\Gamma(s)$ in the range $s_0 \leq s < \infty$.

To approximate the low-s part of $\mathrm{Im}\,\Pi_\Gamma(s)$, one usually employs one or more single-particle states. As an illustration, let us determine the contribution to $\Pi_P(q^2)$ of a flavored pseudoscalar meson M, which is a bound state or a narrow-width resonance of the quark–antiquark pair $\bar{q}_1 q_2$. In this instance, we take the pseudoscalar current in the form of an axial-vector divergence, $J_P \to \partial_\mu A_-^\mu$, with

$$\partial_\mu A_-^\mu = i(m_1 + m_2)\bar{q}_1 \gamma_5 q_2,$$
$$\langle 0|\partial_\mu A_-^\mu(0)|M\rangle = \sqrt{2} F_M m_M^2, \tag{5.7}$$

where m_M and F_M are the meson's mass and decay constant. Then Eq. (5.4) implies

$$\theta(q_0)\rho_P(q^2) = (2\pi)^3 \int \frac{d^3 p}{(2\pi)^3 2\omega_p} 2F_M^2 m_M^4 \delta^4(p - q)$$
$$= 2F_M^2 m_M^4 \delta(q^2 - m_M^2)\theta(q_0), \tag{5.8}$$

which yields $\rho_P(q^2) = 2F_M^2 m_M^4 \delta(q^2 - m_M^2)$ for the spectral function or

$$\mathrm{Im}\,\Pi_P|_{\mathrm{meson}} = 2F_M^2 m_M^4 \pi \delta(s - m_M^2) \tag{5.9}$$

for the dispersion kernel. Thus, bound-state or narrow-resonance contributions give rise to delta-function contributions. It is not difficult to take resonant finite-width effects into account if desired. One or more of these single-particle contributions are then used to represent the low-s part of the dispersion integral.

---

[4] The number of subtraction constants needed depends on the behavior of $\mathrm{Im}\,\Pi_\Gamma(s)$ in the $s \to \infty$ limit, with $\Pi_\Gamma(q^2) \sim q^{2N} \ln q^2$ requiring N subtractions.

Table XI–4. *Local operators of low dimension.*

$d:$	0	4	4	6	6	6
$O_n:$	1	$m_q \bar{q} q$	$G^a_{\mu\nu} G^{\mu\nu}_a$	$\bar{q} \Gamma q \bar{q} \Gamma q$	$m \bar{q} \sigma_{\mu\nu} \frac{\lambda^a}{2} q G^{\mu\nu}_a$	$f_{abc} G^a_{\mu\nu} G^{\nu\lambda}_b G^{\mu c}_\lambda$

Proceeding to higher s values in the dispersion integral, one enters the continuum region, where multiparticle intermediate states become significant and the bound-state (or resonance) approximation breaks down. Although, as described below, one ordinarily attempts to suppress the large-s part of Im $\Pi_\Gamma(s)$ by taking moments or transforms of the dispersion integral, it has been common to add to the low-s contribution a 'QCD continuum' approximation,

$$\text{Im } \Pi_\Gamma(s) \xrightarrow[\text{large-}s]{} \theta(s - s_c) \text{Im } \Pi_{\text{cont}}(s), \qquad (5.10)$$

taken from discontinuities of QCD loop amplitudes and their $\mathcal{O}(\alpha_s)$ corrections. In Eq. (5.10), s_c parameterizes the point where the continuum description begins and the form of Im Π_{cont} depends on the specific correlator. Experience has shown that this 'parton' description can yield reasonable agreement of scattering data even down into the resonance region, provided the resonances are averaged over (*duality*).

Operator-product expansion

A representation for correlators which is distinct from the above phenomenological approach can be obtained by employing an operator-product expansion for the product of currents,

$$i \int d^4x \, e^{iq \cdot x} \, T \left(J_\Gamma(x) J_\Gamma^\dagger(0) \right) = \sum_n \mathcal{C}_n^\Gamma(q^2) O_n. \qquad (5.11)$$

The $\{O_n\}$ are local operators and the $\{\mathcal{C}_n^\Gamma(q^2)\}$ are the associated Wilson coefficients. As usual, the $\{O_n\}$ are organized according to their dimension and, aside from the unit operator I, are constructed from quark and gluon fields. Table XI–4 exhibits the operators up to dimension six which might contribute to the correlator of Eq. (5.2).

Although one may naively expect all the operators but the identity to have vanishing vacuum expectation values (as is the case for normal-ordered local operators in perturbation theory), nonperturbative long-distance effects like those discussed in Sect. III–5 generally lead to nonzero values. Most often, the operator-product approach contains vacuum expectation values like $\langle \frac{\alpha_s}{\pi} G^a_{\mu\nu} G^{\mu\nu}_a \rangle_0 \equiv \langle \frac{\alpha_s}{\pi} G^2 \rangle_0$ and

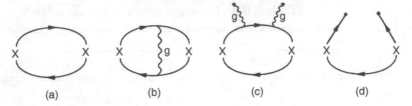

Fig. XI–5 Contributions to coefficient functions.

$\langle m_q \overline{q} q \rangle_0$ as universal parameters, 'universal' in the sense that the same few parameters appear repeatedly in applications. Calculation reveals that the quantity $\langle \frac{\alpha_s}{\pi} G^2 \rangle_0$ is divergent in perturbation theory, so the perturbative infinities must be subtracted off if one is working beyond tree-level. In principle, all the vacuum expectation values should be computable from lattice gauge theory once the renormalization prescriptions are specified. At present, the only theoretically determined combinations are the products

$$\langle \hat{m}(\overline{u}u + \overline{d}d) \rangle_0 \simeq -2F_\pi^2 m_\pi^2, \qquad \langle m_s \overline{s}s \rangle_0 \simeq -F_\pi^2 m_K^2, \qquad (5.12)$$

which follow from the lowest-order chiral analysis in Chap. VII. We caution that only the *product* $m\overline{\psi}\psi$ is renormalization-group invariant (the gluon condensate $\langle \frac{\alpha_s}{\pi} G^2 \rangle_0$ does not, however, depend on scale). It is difficult to separate out the quark masses uniquely, and values for input parameters like quark masses and condensates tend to vary throughout the literature.

Use of the short-distance expansion must be justified. We have seen in previous chapters how a given hadronic system is characterized in terms of the energy scales of confinement (Λ) and quark mass ($\{m_q\}$). Given these, it is indeed often possible to choose the momentum q such that short-distance, asymptotically free kinematics obtain. Two situations which have received the most attention are the heavy-quark limit ($m_q^2 \gg \Lambda^2, q^2$) and the light-quark limit ($q^2 \gg \Lambda^2 \gg m_q^2$). Once in the asymptotically free domain, it is legitimate to apply *QCD* perturbation theory to the $C_n^\Gamma(q^2)$, with the expansion being carried out to one or more powers of α_s,

$$C_n^\Gamma(q^2) = A_n^\Gamma(q^2) + B_n^\Gamma(q^2)\alpha_s + \cdots . \qquad (5.13)$$

Rather extensive lists of Wilson coefficients already appear in the literature. Fig. XI–5 depicts contributions to a few of the Wilson coefficients. Denoting there the action of a current by the symbol '$\times$', we display in (a)–(b) the lowest-order and an $\mathcal{O}(\alpha_s)$ correction to operator I and in (c)–(d), the lowest-order contributions to $\langle \frac{\alpha_s}{\pi} G^2 \rangle_0$ and to $\langle m\overline{q}q \rangle_0$, respectively.

Finally, as seen in Eq. (5.13), besides the vacuum expectation values, additional parameters which generally occur in the operator-product representation are the quark mass m_q and the strong coupling α_s. Since these quantities will depend on the momentum q, one must interpret them as running quantities whose renormalization is to be specified. Due to asymptotic freedom, they too can be treated perturbatively, e.g., as in the familiar expression Eq. (II–2.78) for the running coupling α_s or Eq. (XIV–1.9) for the running mass $\bar{m}$.

Master equation

The essence of the QCD sum rule approach is to equate the dispersion and the operator-product expressions to obtain a 'master equation',

$$\frac{(q^2)^N}{\pi} \int_{s_0}^{\infty} ds \, \frac{\text{Im } \Pi_\Gamma(s)}{s^N (s - q^2 - i\epsilon)} + \cdots = \sum_n C_n^\Gamma (q^2) \langle O_n \rangle_0. \qquad (5.14)$$

It is important to restrict use of this equation to a range of q^2 for which both the short-distance expansion and also any 'resonance + continuum' approximation to Im Π_Γ are jointly valid. To satisfy these twin constraints, it is common practice not to analyze Eq. (5.14) directly, but rather first to perform certain differential operations leading to either *moment* or *transform* representations. The nth *moment* $M_n^\Gamma (Q_0^2)$ is defined as

$$M_n^\Gamma (Q_0^2) \equiv \frac{1}{n!} \left(-\frac{d}{dQ^2} \right)^n \Pi_\Gamma (Q^2) \bigg|_{Q^2 = Q_0^2} = \frac{1}{\pi} \int_{s_0}^{\infty} ds \, \frac{\text{Im } \Pi_\Gamma(s)}{(s + Q_0^2)^{n+1}}, \qquad (5.15)$$

where, in the spacelike region $q^2 < 0$, one usually works with the variable $Q^2 = -q^2$. By taking sufficiently many derivatives, one can remove unknown subtraction constants from the analysis and at the same time, enhance the contribution of a single-particle state at low s in the dispersion integral.

Alternatively, one can express the dispersion integral as a kind of transform. The *Borel transform* is constructed from the moment $M_n^\Gamma (Q^2)$ as

$$n \, Q^{2n} M_n^\Gamma (Q^2) \xrightarrow[n, Q^2 \to \infty]{} \frac{1}{\pi \tau} \int_{s_0}^{\infty} ds \, e^{-s/\tau} \text{Im } \Pi_\Gamma(s), \qquad (5.16)$$

where $Q^2/n \equiv \tau$ remains fixed in the limiting process and defines the transform variable τ. To obtain the factor $e^{-s/\tau}$ in the above dispersion integral, we note

$$\frac{n \, Q^{2n}}{(s + Q^2)^{n+1}} = \frac{n}{s + Q^2} (1 + s/Q^2)^{-n} \xrightarrow[n, Q^2 \to \infty]{} \frac{e^{-s/\tau}}{\tau}. \qquad (5.17)$$

A slightly different version of exponential transform which has appeared in the literature is defined analogously,

$$Q^{2(n+1)} M_n^{\Gamma}\left(Q^2\right) \xrightarrow[n,Q^2\to\infty]{} \frac{1}{\pi} \int_{s_0}^{\infty} ds\, e^{-s\sigma} \operatorname{Im} \Pi_{\Gamma}(s), \qquad (5.18)$$

where the transform variable is now $\sigma = n/Q^2$. The transform method serves to remove the subtraction constants and to suppress the contributions from operators of higher dimension in the operator-product expansion.

Examples

Applications of the QCD sum rule approach generally proceed according to the following steps.

(1) Choose the currents and write a dispersion relation for the correlator.
(2) Model the dispersion integrals with phenomenological input, usually some combination of single-particle states and continuum.
(3) Employ the operator-product expansion, including all appropriate operators up to some dimension at which one truncates the series.
(4) Obtain the Wilson coefficients as an expansion in α_s.
(5) Use the moment or transform technique to extract information from the master equation.
(6) Vary the underlying parameters until stability of output is achieved.

Let us consider several examples, keeping the treatment on an elementary footing to better emphasize the kinds of relationships which QCD sum rules entail. In fact, modern calculations can be quite technical, involving issues such as optimizing the organization of input data, inclusion of ever higher orders of both perturbation theory and vacuum condensates.

(i) *Rho meson decay constant f_ρ*: This was among the first applications of the QCD sum rule approach [ShVZ 79a]. The ρ isovector current $J_\mu^{(\rho)}$ and decay constant f_ρ are

$$J_\mu^{(\rho)} = \frac{\bar{u}\gamma_\mu u - \bar{d}\gamma_\mu d}{2}, \qquad \langle \rho^0(\mathbf{p},\lambda)|J_\mu^{(\rho)}|0\rangle = \epsilon_\mu^{\dagger}(\mathbf{p},\lambda)\frac{f_\rho m_\rho}{\sqrt{2}}. \qquad (5.19a)$$

The sum rule which gives f_ρ is [CoK 00]

$$f_\rho^2 = \frac{e^{m_\rho^2 \tau}}{\tau}\left[\frac{1}{4\pi^2}\left(1 - e^{-s_0\tau}\right)\left(1 + \frac{\alpha_s(\tau^{-1})}{\pi}\right)\right.$$

$$\left. + (m_u + m_d)\tau^2\langle \bar{q}q\rangle + \cdots \right], \qquad (5.19b)$$

where τ is the Borel parameter, s_0 is threshold above which $\Pi_\rho(s)$ is to be approximated via perturbation theory and ellipses represent additional condensate contributions. The variation of f_ρ vs. $1/\tau$ turns out to display little variation in f_ρ for, say,

$0.6 \leq \tau^{-2}(\text{GeV}^2) \leq 1.3$; this is the Borel window of stability. The range of values $f_\rho \sim 208 \to 218$ MeV which occur within the stability window is in accord with the experimental determination cited in Eq. (V–3.14) for the equivalent quantity $g_\rho = f_\rho m_\rho / \sqrt{2}$.

(ii) *Mass of the charm quark:* We consider the correlator for the charm-quark vector current $J_\mu^{(\text{chm})} = \bar{c}\gamma_\mu c$,

$$\left(q_\mu q_\nu - q^2 g_{\mu\nu}\right) \Pi_V^{(\text{chm})}\left(q^2\right) = i \int d^4x \, e^{iq\cdot x} \langle 0| T \left(J_\mu^{(\text{chm})}(x) J_\nu^{(\text{chm})}(0)\right) |0\rangle,$$

$$(5.20\text{a})$$

and the corresponding dispersion relation,

$$\frac{\partial}{(-\partial Q^2)} \Pi_V^{(\text{chm})}\left(Q^2\right) = \frac{1}{\pi} \int_{s_0}^{\infty} ds \, \frac{\text{Im} \, \Pi_V^{(\text{chm})}(s)}{\left(s + Q^2\right)^2}. \qquad (5.20\text{b})$$

Following the original treatment of this system [ShVZ 79a], we work at $Q^2 = 0$ and employ a moment analysis of the short-distance expansion containing just the identity and gluon contributions. The experimental input is obtained from

$$M_n^{(\text{expt})} = \int \frac{ds}{s^{n+1}} \frac{\sigma_{e^+e^- \to c\bar{c}}(s)}{\sigma_{e^+e^- \to \mu^-\mu^+}(s)}, \qquad (5.21\text{a})$$

whereas the theory side involves

$$M_n^{(\text{thy})} = \frac{12\pi^2 e_c^2}{n!} \cdot \frac{d}{dq^{2n}} \Pi_V^{(\text{chm})}(q^2 = 0). \qquad (5.21\text{b})$$

The moments $M_n^{(\text{thy})}$ can be determined in terms of an operator-product expansion, which is dominated by the *QCD*-perturbative contribution provided the value of n is not too large, i.e., $m_c/n > \Lambda_{QCD}$. Perturbative contributions have long been studied and a library of exact results is now available (see [DeHMZ 11]):

(1) $\mathcal{O}(\alpha_s^0)$ and $\mathcal{O}(\alpha_s^1)$: known for all n.
(2) $\mathcal{O}(\alpha_s^2)$: known up to $n = 30$.
(3) $\mathcal{O}(\alpha_s^3)$: known up to $n = 3$.

Below, we cite two specific $\mathcal{O}(\alpha_s^3)$ determinations of the charm quark mass [KuSS 07], [DeHMZ 11]. Both adopt the gluon condensate value $\langle \frac{\alpha_s}{\pi} G^2 \rangle_0 = 0.006 \pm 0.012$ GeV^4 (each analysis obtains only a minor effect for this term). The results obtained in $\overline{\text{MS}}$ renormalization are

$$\overline{m}_c(\overline{m}_c) = (1.286 \pm 0.013) \text{ GeV} \qquad [\text{KuSS 07}],$$
$$\overline{m}_c(\overline{m}_c) = (1.277 \pm 0.026) \text{ GeV} \qquad [\text{DeHMZ 11}], \qquad (5.22)$$

whereas the value $\overline{m}_c(\overline{m}_c) = (1.275 \pm 0.025)$ GeV is cited in [RPP 12]. The issue of how to assign uncertainty in QCD-sum-rule-determinations of charm mass, especially involving the data side of the calculation, is currently a topic of some interest,

(iii) *Weak decay constant of the D^+ meson*: Consider first the axial-current divergence and correlator associated with a heavy quark Q and a light antiquark $\overline{q}$, respectively of mass m_Q and m_q, which comprise a heavy meson M_Q,

$$\partial_\mu A_-^\mu = i(m_Q + m_q)\overline{q}\gamma_5 Q,$$

$$\Pi_P(q^2) = i \int d^4x \, e^{iq\cdot x} \langle 0|T(\partial_\mu A_-^\mu(x)\partial_\nu A_-^{\nu\dagger}(0))|0\rangle. \tag{5.23}$$

A transformation with Borel variable τ yields

$$\Pi_P(\tau) = \int_{(m_Q+m_q)^2}^{\infty} ds \, e^{-s\tau} \rho_{\text{pert}}(s, \mu) + \Pi_{\text{pwr}}(\tau, m_Q, \mu) \tag{5.24}$$

where $\rho = \mathcal{I}m \, \Pi/\pi$ and $\rho_{\text{pert}}(s, \mu)$ and $\Pi_{\text{pwr}}(\tau, m_Q, \mu)$ represent respectively, the perturbative and nonperturbative contributions.

Let us now consider specifically the decay constant of the D^+, where symbolically $D^+ \sim (c\overline{u})$. The experimental value, $f_D \equiv \sqrt{2}F_D = 206.7 \pm 8.9$ MeV is found from $D^+ \rightarrow \mu^+\nu_\mu$ via decay formulas akin to the tree-level Eq. (VII–1.24) or radiatively corrected Eq. (VII–1.34) for pion leptonic decay. On the theory side, it is shown in [LuMS 11] that a straightforward QCD sum rule approach yields

$$f_{D^+}^2 M_{D^+}^4 e^{-M_{D^+}^2\tau} = \int_{(m_c+m_u)^2}^{s_{\text{eff}}(\tau)} ds \, e^{-s\tau} \rho_{\text{pert}}(s, \mu) + \Pi_{\text{pwr}}(\tau, m_Q, \mu), \tag{5.25}$$

where the condensate values adopted are

$$\langle \overline{q}q\rangle(\mu) = -(267 \pm 17 \text{ MeV})^3, \qquad \left\langle\frac{\alpha_s}{\pi}G^2\right\rangle = (0.024 \pm 0.012)\,\text{GeV}^4, \tag{5.26}$$

with $\mu = 2$ GeV being the $\overline{\text{MS}}$ renormalization scale. The most novel part of the expression in Eq. (5.25) is the presence of an 'effective continuum threshold' $s_{\text{eff}}(\tau)$. The τ-dependence of s_{eff} supplants the traditional form of Eq. (5.10) in which a constant cut-off s_c is used to describe the onset of continuum contributions. We leave a detailed discussion of the effective threshold to [LuMS 11] and simply state the final result,

$$f_D^{(\text{thy})} = \left(206.2 \pm 7.3_{(\text{OPE})} \pm 5.1_{(\text{sys})}\right) \text{MeV},$$

which is consistent with the experimental finding shown above.

(iv) *Nucleon mass*: It is not necessary to restrict oneself to mesonic currents as in Eq. (5.1). Here, we consider a current η_N (and its correlator), which carries the quantum numbers of the nucleon,

$$\eta_N = \epsilon_{ijk} u^i C\gamma^\mu u^j \gamma_5 \gamma_\mu d^k,$$

$$\Pi(q^2) = \Pi_1(q^2) + \slashed{q}\Pi_2(q^2) = i \int d^4x \, e^{iq \cdot x} \langle 0|T\left(\eta_N(x)\overline{\eta}_N(0)\right)|0\rangle, \qquad (5.27)$$

where C is the charge-conjugation matrix. The simplest approximation to the dispersion integral comes from the nucleon pole,

$$\Pi\left(q^2\right)\big|_{\text{pole}} = \lambda_N^2 \frac{M_N + \slashed{q}}{q^2 - M_N^2}, \qquad (5.28)$$

where the coupling λ_N^2 is proportional to the 'nucleon decay constant', i.e., the probability of finding all three quarks within the nucleon at one point. Upon making a simple approximation to the operator-product expansion,

$$\Pi_1\left(q^2\right) \simeq -\frac{q^2}{4\pi^2} \ln\left(-q^2\right) \langle \overline{q}q \rangle_0, \qquad \Pi_2\left(q^2\right) \simeq \frac{q^4}{64\pi^2} \ln\left(-q^2\right), \qquad (5.29)$$

and employing a Borel transform, one obtains an amusing relation between nucleon mass and quark condensate [Io 81],

$$M_N = \left(-8\pi^2\langle \overline{q}q \rangle_0\right)^{1/3} + \ldots \simeq 1 \text{ GeV}, \qquad (5.30)$$

and implies the vanishing of the former with the latter. However, it should be realized that this result is subject to important corrections in a more complete treatment.

Each of the above examples has involved two-point functions. It is possible to apply the method to three-point functions as well, where one can obtain coupling-constant relations. The underlying principles are the same, but some technical details are modified owing to the larger number of variables, e.g., one encounters double-moments or double-transforms.

QCD sum rules work best when there is a reliable way to estimate the dispersion integral, most often with ground-state single-particle contributions. However, the method has its limitations. It is not at its best in probing radial excitations since their dispersion effects are generally rather small. Even having a good approximation to the dispersion integral is not sufficient to guarantee success. For example, the method has trouble in dealing with high-spin ($J > 3$) mesons because, even with dispersion integrals which are dominated by ground-state contributions, power corrections in the operator-product expansion become unmanageable.

Problems

(1) **Velocity in potential models**

Truly nonrelativistic systems have excitation energies small compared to the masses of their constituents. However, fitting the observed spectrum of light hadrons requires excitation energies comparable to or larger than the constituent masses.

Assuming nonrelativistic kinematics, consider a particle of reduced mass m moving in a harmonic-oscillator potential of angular frequency ω. Expressing ω in terms of the energy splitting $E_1 - E_0$ between the first-excited state and the ground state, use the virial theorem to determine the 'rms' velocities of the ground state $(v_{rms}^{(0)})$ and of the first-excited state $(v_{rms}^{(1)})$ in terms of $E_1 - E_0$. Compute the magnitude of $v_{rms}^{(0)}/c$ and $v_{rms}^{(1)}/c$ using as inputs (i) $m_{f_2} - m_\rho \simeq 500$ MeV for light hadrons and (ii) $m_{\psi(2S)} - m_{J/\psi} \simeq 590$ MeV for charmed quarks.

Your results should demonstrate that the kinematics of quarks in light hadrons is not truly nonrelativistic. However, one tends to overlook this flaw given the potential model's overall utility.

(2) **Nucleon mass and the Skyrme model**

(a) Use the Skyrme ansatz of Eq. (4.15) to derive the expression Eq. (4.16) for the nucleon energy $E[F]$.

(b) Using the simple trial function $F(r) = \pi \exp(-r/R)$, scale out the range factor R to put $E[F]$ in the form of Eq. (4.11), where $a \simeq 30.8 F_\pi^2$ and $b \simeq 44.7/e^2$ are determined via numerical integration.

(c) Minimize $E[F]$ by varying R and compare your result with the value $73 F_\pi/e$ determined with a more complex variational function.

(d) Using the numerical value of the nucleon mass, determine e and compare with the value

$$\frac{1}{32e^2} \cdot \frac{4}{F_\pi^2} \sim \frac{1}{(4\pi F_\pi)^2}$$

expected from chiral-scaling arguments.

(3) **A 'QCD sum rule' for the isotropic harmonic oscillator**

Consider three-dimensional isotropic harmonic motion with angular frequency ω of a particle of mass m.

(a) Using ordinary quantum mechanics or more formal path-integral methods, determine the exact Green's function $G(\tau)$ for propagation from time $t = 0$ to imaginary time $t = -i\tau$ at fixed spatial point $\mathbf{x} = 0$. $G(\tau)$ is the analog of the 'correlator' for our quantum mechanical system.

(b) From the representation $G(\tau) = \langle \mathbf{0}, -i\tau | \mathbf{0}, 0 \rangle$, use completeness to express $G(\tau)$ in terms of the S-wave radial wavefunctions $\{R_n(0)\}$ evaluated at the origin and the energy eigenvalues $\{E_n\}$. What values of n contribute? This representation is the analog of the dispersion relation expression for a correlator in which one takes into account an infinity of resonances.

(c) Plot the negative logarithmic derivative $-d[\ln G(\tau)]/d\tau$ for the range $0 \le \omega\tau \le 5$ and interpret the large $\omega\tau$ behavior in terms of your result in part (b).

(d) Obtain the first three terms in a power series for $-d[\ln G(\tau)]/d\tau$, expanded about $\tau = 0$. This is the analog of the series of operator-product 'power corrections' to $-d[\ln G(\tau)]/d\tau$. Assume, as is the case in *QCD*, that you know only a limited number of terms in this series, first two terms and then four terms. Is there a common range of $\omega\tau$ for which (i) your truncated series reasonably approximates the exact behavior, and (ii) the approximation for keeping just the lowest bound state in part (b) is likewise reasonable? It is this compromise between competing demands of the resonance and operator-product approximations which must be satisfied in sucessfully applying the *QCD* sum rules to physical systems.

An important sector of hadron phenomenology is associated with the electroweak interactions. Baryons provide a particularly rich source of information, with data on vector and axial-vector couplings, magnetic moments, and charge radii. In Sect. XII–1, we describe the procedure for computing matrix elements in the constituent quark model, and then turn to a variety of applications in the succeeding sections.[1]

XII–1 Matrix-element computations

Much of the application of the quark model to physical systems involves the calculation of matrix elements. The subject divides naturally into two parts. On the one hand, many quantities of interest follow from just the flavor and spin content of the hadronic states. On the other, it is often necessary to have a detailed picture of the quark spatial wavefunction.

Flavor and spin matrix elements

For the first of these, the quark model is particularly appealing because of the intuitive physical picture which it provides. For example, consider the quark content of the proton state vector, which we reproduce here from Table XI–2,

$$|p_\uparrow\rangle = \frac{1}{\sqrt{18}}\epsilon_{ijk}[(u_{i\downarrow}^\dagger d_{j\uparrow}^\dagger - u_{i\uparrow}^\dagger d_{j\downarrow}^\dagger)u_{k\uparrow}^\dagger]\,|0\rangle. \tag{1.1}$$

The first two quarks form a spin-zero, isospin-zero pair with the net spin and isospin of the proton being given by the final quark. The prefactor of $1/\sqrt{18}$ ensures that the state vector has unit normalization. Calculation reveals that one-third of the magnitude of this normalization factor comes from the $u_\uparrow u_\downarrow d_\uparrow$ term

---

[1] The reader can also consult the $N_c \to \infty$ studies as described in [DaJM 94, Je 98].

Table XII–1. *Some baryon octet expectation values.*

	p	n	Λ	Σ^+	Σ^0	Σ^-	Ξ^0	Ξ^-
$\langle Q \rangle$	1	0	0	1	0	-1	0	-1
$\langle Q\sigma_z \rangle$	1	$-2/3$	$-1/3$	1	$1/3^a$	$-1/3$	$-2/3$	$-1/3$
$\langle \lambda_3\sigma_z \rangle$	$5/3$	$-5/3$	$2/\sqrt{6}$	$4/3$	0	$-4/3$	$-1/3$	$1/3$

^{a}The off-diagonal transition $\Sigma^0 \to \Lambda$ has $|\langle Q\sigma_z \rangle| = 1/\sqrt{3}$.

and two-thirds from the $u_\uparrow u_\uparrow d_\downarrow$ term, i.e. one concludes that 'the proton is twice as likely to be found in the configuration with the u-quark spins aligned than anti-aligned',

$$\text{Prob.} = \begin{cases} 2/3 & (u_\uparrow u_\uparrow d_\downarrow), \\ 1/3 & (u_\uparrow u_\downarrow d_\uparrow). \end{cases} \tag{1.2}$$

The 'six parts in eighteen' of the $u_\uparrow u_\downarrow d_\uparrow$ configuration arises entirely from the six ways that color can be distributed among three distinct entities. The configuration $u_\uparrow u_\uparrow d_\downarrow$ is twice as large due to the presence of two $u_\uparrow$ states. Similar kinds of inferences can be drawn for the remaining baryon state vectors in Table XI–2.

We can proceed analogously in deriving and interpreting various matrix-element relationships. It is instructive to work at first in the limit of $SU(3)$ invariance because more predictions become available. The effect of symmetry breaking is addressed in Sect. XII–2. Let us consider matrix elements, taken between members of the spin one-half baryon octet, of the operators

$$\text{squared charge-radius}: \int d^3x \, r^2 \psi^\dagger Q \psi \qquad \propto \langle Q \rangle,$$

$$\text{axial-vector current}: \int d^3x \, \psi^\dagger \gamma_3 \gamma_5 \lambda_3 \psi \qquad \propto \langle \lambda_3\sigma_z \rangle,$$

$$\text{magnetic moment}: \int d^3x \, \frac{1}{2}(\mathbf{r} \times \psi^\dagger \boldsymbol{\alpha} Q\psi)_3 \qquad \propto \langle Q\sigma_z \rangle. \tag{1.3}$$

Along with the definition of each operator is indicated the flavor–spin attribute of an individual quark which is being averaged over. For example, a magnetic moment is sensitive to the combination $Q\sigma_z$ of each quark within the baryon. Matrix elements will then be products of such averages times quark wavefunction overlap integrals. The flavor–spin averages for the baryon octet are displayed in Table XII–1.

To see how these values are arrived at, let us compute the value $5/3$ obtained for the proton axial-vector matrix element. For the configuration $u_\uparrow u_\uparrow d_\downarrow$, which occurs with a probability of $2/3$, the average value of $\lambda_3\sigma_z$ equals $(1 + 1 + 1) \times 2/3 = 2$, whereas for the configuration $u_\uparrow u_\downarrow d_\uparrow$ one finds $(1-1-1)\times 1/3 = -1/3$. Together they sum to the value $5/3$.

Overlaps of spatial wavefunctions

The spatial description of quark wavefunctions is less well understood than the spin/flavor aspect of the phenomenology.[2] The most extensive studies of the spatial wavefunctions are associated with matrix elements of *currents*. Because these are bilinear in quark fields and because of the wavefunction normalization condition, the magnitudes of these amplitudes are constrained to be nearly correct. Dimensional matrix elements are primarily governed by the *radius* of the bound state. As long as the proper value is fed into the calculation, the scale should come out right.

As noted in Sect. XI–1, a relativistic quark moving in a spin-independent central potential has a ground-state wavefunction of the form

$$\psi(\mathbf{x})\bigg|_{\text{gnd}} = \begin{pmatrix} i\, u(r)\,\chi \\ \ell(r)\boldsymbol{\sigma}\cdot\hat{\mathbf{x}}\,\chi \end{pmatrix} e^{-iEt}, \qquad (1.4)$$

where u, ℓ signify 'upper' and 'lower' components. For the bag model, these radial wavefunctions are just spherical Bessel functions. This form also appears in some relativistic harmonic oscillator models, which use a central potential. To characterize different types of relativistic behavior, it is worthwhile to express matrix elements in terms of u and ℓ without specifying them in detail. The normalization condition for the spatial wavefunction is then

$$\int d^3x\, \psi^\dagger(\mathbf{x})\psi(\mathbf{x}) = \int d^3x\, (u^2(r) + \ell^2(r)) = 1. \qquad (1.5)$$

In the nonrelativistic regime, the lower component vanishes ($\ell = 0$).

Let us consider the size of the lower components which occur in various approaches. In the bag model one obtains for massless quarks the integrated value

$$\int d^3x\, \ell^2(r) \simeq 0.26. \qquad (1.6)$$

Relativistic effects are often included in potential models by working in momentum space and employing the spinor appropriate for a quark q in momentum eigenstate $\mathbf{p}$,

---

[2] Even the *experimental* value of the proton charge radius r_E is in question. The historical approach, to measure the differential cross section in elastic electron–proton scattering at low Q^2, gives $r_E = 0.879(8)$ fm and $r_E = 0.875(11)$ fm [Zh *et al.* 11] in recent experiments. By contrast, measurement of the $2S_{1/2}^{F=0} - 2P_{3/2}^{F=1}$ energy difference in muonic hydrogen [An *et al.* 13] yields (using a consistent definition of charge radius) $r_E = 0.84087(39)$ fm, which is at 7σ variance relative to the scattering value.

$$u(\mathbf{p}) = \sqrt{E + m_q} \begin{pmatrix} \chi \\ \dfrac{\sigma \cdot \mathbf{p}}{E + m_q} \chi \end{pmatrix}. \tag{1.7}$$

In this case the relevant prescription is

$$\int d^3x \, \ell^2(r) \rightarrow \left\langle \frac{\mathbf{p}^2}{2E(E + m_q)} \right\rangle, \tag{1.8}$$

where the averaging is taken over the momentum-space wavefunction of the particular model. Using the uncertainty principle relation of Eq. (XI–1.14) to estimate $\langle \mathbf{p}^2 \rangle$, we find typical values

$$\left\langle \frac{\mathbf{p}^2}{2E(E + m_q)} \right\rangle \simeq 0.13 \rightarrow 0.20 \tag{1.9}$$

for a confinement scale of 1 fm. Larger effects are found in the harmonic-oscillator model if one uses the value $\alpha^2 = 0.17 \text{ GeV}^2$ (see Fig. XI–2). Generally, the lower component is found to be significant but not dominant in quark wavefunctions.

Connection to momentum eigenstates

In all cases except for the nonrelativistic version of the harmonic oscillator model, one cannot explicitly separate out the center-of-mass motion. The result of a quark model description of a bound state is a configuration localized in coordinate space, i.e., a position eigenstate. However, the analysis of scattering and decay deals with the plane waves of momentum eigenstates.

The basic assumption made in all quark models is that the bound state with a given set of quantum numbers is related to only those momentum eigenstates of the same type. If we denote $|H(\mathbf{x})\rangle$ as a unit-normalized hadron state centered about point $\mathbf{x}$ and $|H(\mathbf{p})\rangle$ as a plane-wave state, then we have

$$|H(\mathbf{x})\rangle = \int d^3p \, \varphi(\mathbf{p}) e^{i\mathbf{p} \cdot \mathbf{x}} |H(\mathbf{p})\rangle. \tag{1.10}$$

We shall give a prescription for obtaining a functional form for $\varphi(\mathbf{p})$ shortly. Let us normalize the plane-wave states for both mesons and baryons as

$$\langle H(\mathbf{p}')|H(\mathbf{p})\rangle = 2\omega_{\mathbf{p}}(2\pi)^3 \delta^{(3)}(\mathbf{p}' - \mathbf{p}). \tag{1.11}$$

The constraint of unit normalization then implies

$$\int d^3p \, 2\omega_{\mathbf{p}}(2\pi)^3 \, |\varphi(\mathbf{p})|^2 = 1. \tag{1.12}$$

We can employ the above wavepacket description to derive a general procedure within the quark model for calculating matrix elements [DoJ 80]. Many matrix elements of interest involve a local operator O evaluated between initial and final single-hadron states. Let us characterize the magnitude of the matrix element in terms of a constant g. Then, for baryons in the momentum basis, the spatial dependence is given by

$$\langle B'(\mathbf{p}')\,|O(x)|\,B(\mathbf{p})\rangle = g\,\bar{u}(\mathbf{p}')\Gamma_O u(\mathbf{p})\,e^{i(p'-p)\cdot x}, \qquad (1.13)$$

where Γ_O is a Dirac matrix appropriate for the operator O. By comparison, one obtains in any bound-state quark model (QM) calculation a spatial dependence whose specific form is model-dependent,

$$_{\text{QM}}\langle B'\,|O(x)|\,B\rangle_{\text{QM}} = f(x). \qquad (1.14)$$

Hereafter, let us center all quark model states at the origin. The method of wavepackets then implies

$$_{\text{QM}}\langle B'|\int d^3x\; O(x)|B\rangle_{\text{QM}} = g\int d^3x \int d^3p' d^3p\; \varphi^*(\mathbf{p}')\varphi(\mathbf{p})$$

$$\times \bar{u}(\mathbf{p}')\Gamma_O u(\mathbf{p})e^{i(p'-p)\cdot x}$$

$$= g\int d^3p\; (2\pi)^3\, |\varphi(\mathbf{p})|^2\, \bar{u}(\mathbf{p})\Gamma_O u(\mathbf{p}). \qquad (1.15)$$

For sufficiently heavy bound states the fluctuation in squared momentum $\langle \mathbf{p}^2\rangle$ is small, and one may expand about $|\mathbf{p}| = 0$,

$$\bar{u}(\mathbf{p})\Gamma_O u(\mathbf{p}) = \bar{u}(\mathbf{0})\Gamma_O u(\mathbf{0}) + \mathcal{O}\left(\langle \mathbf{p}^2\rangle/m_B^2\right). \qquad (1.16)$$

A common approach consists of keeping only the leading term to obtain

$$\frac{g}{2m_B}\,\bar{u}(\mathbf{0})\Gamma_O u(\mathbf{0}) = {}_{\text{QM}}\langle B'|\int d^3x\; O(x)\,|B\rangle_{\text{QM}}. \qquad (1.17)$$

It is interesting to note that this relation, often thought of as fundamental, is in fact only an approximation.

As an example, let us perform the complete quark model procedure for the neutron–proton axial-vector current matrix element. We begin by defining as usual

$$\langle p(\mathbf{p}_2, s_2)\,|A_\mu(x)|\,n(\mathbf{p}_1, s_1)\rangle = g_A\bar{u}(\mathbf{p}_2, s_2)\gamma_\mu\gamma_5 u(\mathbf{p}_1, s_1)e^{i(p_2-p_1)\cdot x} + \cdots. \qquad (1.18)$$

For spin-up nucleons the choice $\mu = 3$ gives

$$\bar{u}(\mathbf{0}, \uparrow)\gamma_3\gamma_5 u(\mathbf{0}, \uparrow) = 2m_N, \qquad (1.19)$$

yielding for Eq. (1.18) the basic formula,

$$g_A = {}_{QM}\langle p_\uparrow | \int d^3x \, \bar{u}(x)\gamma_3\gamma_5 d(x) \, |n_\uparrow\rangle_{QM}. \tag{1.20}$$

The field operator for any quark q is expanded as in Eq. (XI–1.1),

$$q_\alpha(x) = \sum_{n,s} \left[\psi_{n,s}(x)e^{-i\omega_n t} q_{n,\alpha}(s) + \psi_{\bar{n},\bar{s}}(x)e^{i\omega_{\bar{n}}t} \bar{q}^\dagger_{\bar{n},\alpha}(\bar{s}) \right]. \tag{1.21}$$

Substituting, we have

$$g_A = {}_{QM}\langle p_\uparrow | \int d^3x \, \bar{\psi}_{0,s'}(x) \, \gamma_3\gamma_5 \psi_{0,s}(x) \, u^\dagger_\alpha(s) d_\alpha(s')|n_\uparrow\rangle_{QM}, \tag{1.22}$$

where only the $n = 0$ ground-state mode contributes. At this stage, one can factorize the spin and space components by using the general ground-state wavefunction of Eq. (1.4). This leads to

$$\int d^3x \, \bar{\psi}_{0,s}\gamma_3\gamma_5\psi_{0,s'} = \int d^3x \, \chi^\dagger_s (u^2\sigma_3 - \ell^2\hat{r}_3\boldsymbol{\sigma} \cdot \hat{\mathbf{r}})\chi_{s'}$$

$$= \sigma_3^{ss'} \int d^3x \, (u^2 - \frac{1}{3}\ell^2), \tag{1.23}$$

and thus

$$g_A = \int d^3x \left(u^2 - \frac{1}{3}\ell^2 \right) {}_{QM}\langle p_\uparrow \left| u^\dagger(s, \alpha)\sigma_3^{ss'} d(s', \alpha) \right| n_\uparrow\rangle_{QM}. \tag{1.24}$$

Finally, upon dealing with the spin dependence in Eq. (1.24), we obtain

$$g_A = \frac{5}{3} \int d^3x \left(u^2 - \frac{1}{3}\ell^2 \right) = \frac{5}{3} \left(1 - \frac{4}{3} \int d^3x \, \ell^2 \right). \tag{1.25}$$

Any nonrelativistic quark model, having zero lower components, would simply yield $g_A = 5/3$. If one desires to make relativistic corrections to such a model, the result can be inferred from the above general formula with the appropriate substitution of Eq. (1.8). Clearly, the procedure just given can be extended to matrix elements of any physical observable.

The wavepacket formalism also allows for the estimation of the 'center-of-mass' correction. This arises from the $\langle \mathbf{p}^2 \rangle$ modifications to Eq. (1.16). For the axial current, the zero-momentum relation in Eq. (1.19) is extended for nonzero momentum to

$$\frac{\bar{u}_2(\mathbf{p}, \uparrow)\gamma_3\gamma_5 u_1(\mathbf{p}, \uparrow)}{2E} = 1 - \frac{\mathbf{p}^2}{3m_1m_2}\left(\frac{1}{4} + \frac{3}{8}\frac{m_2}{m_1} + \frac{3}{8}\frac{m_1}{m_2} \right) + \mathcal{O}(\mathbf{p}^4), \tag{1.26}$$

where an average over the direction of $\mathbf{p}$ has been performed. This expression generalizes Eq. (1.25) to

$$g_A \left[1 - \frac{\langle \mathbf{p}^2 \rangle_{np}}{3m_n m_p} \left(\frac{1}{4} + \frac{3}{8} \frac{m_p}{m_n} + \frac{3}{8} \frac{m_n}{m_p} \right) \right] = \frac{5}{3} \int d^3x \left(u^2 - \frac{1}{3} \ell^2 \right), \quad (1.27)$$

where $\langle \mathbf{p}^2 \rangle_{np} \simeq 0.5 \text{ GeV}^2$ is a typical bag model value.

It is possible to argue that in the transition from the current quarks of the *QCD* lagrangian to the constituent quarks of the quark model, the couplings to currents should be modified. For example, one might suspect that the coupling of a constituent quark to the axial current occurs not with strength unity, but with a strength $g_A^{(q)}$ such that the nonrelativistic expectation is not $g_1 = 5/3$ but rather $g_1 = 5g_A^{(q)}/3$. The choice $g_A^{(q)} \simeq 3/4$ would then yield the experimental value. This is not unreasonable but, if fully adopted, leads to a lack of predictivity. In such a picture, not only can the magnetic moments and weak couplings be renormalized, but also the spin and flavor structures. That is, in the 'dressing' process which a constituent quark undergoes, there could be 'sea' quarks, such that the constituent u quark could have gluonic, d-quark, or s-quark content. Likewise, some of the spin of the constituent quarks could be carried by gluons. One is then at a loss to know how to calculate matrix elements of currents. In practice, however, the naive quark model, with no rescaling of g_A or of the magnetic moment, does a reasonable job of describing current matrix elements. It is then of interest to study both the structure and limitations of this simple approach.

Calculations in the Skyrme model

There are several differences between taking matrix elements in the quark model and in the Skyrme model [Sk 62]. To begin, in the quark model a current is expressed in terms of a bilinear covariant in the quark fields (cf. Eq. (1.3)), whereas in the Skyrme model the representation of a current is rather different. As an example, application of either Noether's theorem or the external source method of Sect. IV–6 identifies the $SU(2)$ vector and axial-vector currents to be

$$\left(J_{\substack{v \\ a}} \right)^a_\mu = \frac{i F_\pi^2}{4} \text{Tr} \left(\tau^a (\partial_\mu U \, U^\dagger \pm \partial_\mu U^\dagger \, U) \right)$$

$$- \frac{i}{16e^2} \big[\text{Tr} \left([\tau^a, \partial_\nu U \, U^\dagger] [\partial_\mu U \, U^\dagger, \partial^\nu U \, U^\dagger] \right)$$

$$\pm \text{Tr} \left([\tau^a, \partial_\nu U^\dagger \, U] [\partial_\mu U^\dagger \, U, \partial^\nu U^\dagger \, U] \right) \big], \quad (1.28)$$

where $U = A(t) U_0 A^{-1}(t)$ is the quantized skyrmion form and $A(t)$ is an $SU(2)$ matrix. We shall neglect derivatives of $A(t)$, as the quantization hypothesis corresponds to slow rotations. This leads to a result similar in form to Eq. (1.28), but with $U \to U_0$ and $\tau^a \to A^{-1}(t)\tau^a A(t)$. The answer may be simplified by use of the explicit form of U_0 appearing in Eq. (XI–4.15).

Let us use Eq. (1.28) to compute the spatial integral of the axial current. After some algebra, we obtain a product of spatial and internal factors,

$$\int d^3x\,(J_A)^a_j = -G_5\,\mathrm{Tr}\,(\tau^a A\tau^j A^{-1}),$$

$$G_5 = -\frac{\pi}{3e^2}\int_0^\infty d\tilde{r}\,\tilde{r}^2\left[F' + \frac{\sin 2F}{\tilde{r}} + \frac{4\sin 2F}{\tilde{r}}(F')^2\right.$$

$$\left.+\frac{8\sin^2 F}{\tilde{r}^2}F' + \frac{4\sin^2 F}{\tilde{r}^3}\sin 2F\right], \tag{1.29}$$

where a is the isospin component and j is the Lorentz component. This is now suitable for taking matrix elements, such as

$$\langle p_\uparrow|\int d^3x\,(J_A)^a_j\,|p_\uparrow\rangle = \int d^3x\int d\Omega_3\,\langle p_\uparrow|A\rangle\,(J_A)^a_j\,\langle A|p_\uparrow\rangle$$

$$= G_5\int d\Omega_3\,D^{(\frac{1}{2})*}_{-\frac{1}{2},\frac{1}{2}}(A)\,\mathrm{Tr}\,(\tau^a A\tau^j A^{-1})D^{(\frac{1}{2})}_{-\frac{1}{2},\frac{1}{2}}(A), \tag{1.30}$$

where we have used the completeness relation of Eq. (XI–4.42). Upon expressing the trace in Eq. (1.30) as a rotation matrix, $\mathrm{Tr}\,(\tau^k A\tau^l A^{-1})/2 = D^{(1)}_{kl}$, we can determine the group integration in Eq. (1.30) in terms of $SU(2)$ Clebsch–Gordan coefficients,

$$\int d\Omega_3\,D^{(T'')*}_{mn}(A)D^{(T')}_{kl}(A)D^{(T)}_{ij}(A) = (-)^{2(T'-T+m)}\frac{2\pi^2}{2T''+1}C^{T'TT''}_{kim}C^{T'TT''}_{ljn}. \tag{1.31}$$

Alternatively, one can work directly with the collective coordinates, e.g., with the aid of Eqs. (XI–4.41–4.44) we obtain for $a = j = 3$

$$-\frac{2G_5}{\pi^2}\int d\Omega_3\,(a_1 - ia_2)(a_0^2 + a_3^2 - a_1^2 - a_2^2)(a_1 + ia_2) = \frac{2}{3}G_5. \tag{1.32}$$

Before one can infer a Skyrme model prediction for g_A from this calculation, there is a subtlety not present for the quark calculation, which must be addressed. Due to the original chirally invariant lagrangian, the Skyrme model is unique among phenomenological models in being completely compatible with the constraints of chiral symmetry. As a consequence, the near-static axial-vector matrix element is constrained to obey

$$q_j\langle p(\mathbf{p}')|(J_A)^3_j|p(\mathbf{p})\rangle = 0, \qquad (q = p - p') \tag{1.33a}$$

and hence must be of the form [AdNW 83],

$$\langle p(\mathbf{p}')|(J_A)_j^3|p(\mathbf{p})\rangle = 2m_p g_A \left(\delta_{jk} - \frac{q_j q_k}{|\mathbf{q}|^2}\right)\langle\sigma_k\rangle. \tag{1.33b}$$

The term containing $|\mathbf{q}|^{-2}$ arises from the pion pole, as will be discussed in Sect. XII–3 in connection with the Goldberger–Treiman relation. An angular average of Eq. (1.33b) then yields $2g_A/3$, which from comparison with Eq. (1.32) implies $g_A = G_5$. Thus in the Skyrme model, the axial-vector coupling constant equals the radial integral in Eq. (1.29) which defines G_5. Use of the profile given in Sect. XI–4 leads to the prediction $g_A = 0.61$, which is about only one-half the experimental value and constitutes a well-known deficiency of skyrmion phenomenology. Presumably, consideration of a more general chiral lagrangian could modify this result by including higher derivative components in the weak current.

Pions may be added to the Skyrme description through introduction of the matrix ξ described in App. B–4 [Sc 84],

$$U = \xi A(t) U_0 A^{-1}(t)\xi, \qquad \xi = \exp\left[i\boldsymbol{\tau}\cdot\boldsymbol{\pi}/(2F_\pi)\right]. \tag{1.34}$$

If currents are formed using this ansatz, some terms occur without derivatives on the pion field, while others contain one or more factors of $\partial^\mu\pi$. Since $\partial^\mu\pi$ gives rise to a momentum factor q_π^μ when matrix elements are taken and soft-pion theorems deal with the limit $q_\pi^\mu \to 0$, the lowest-order soft-pion contribution will consist of keeping only terms without derivatives. Thus in the process $\nu_\mu + N \to N + \pi + \mu$ the final-state pion is produced by a hadronic weak current and the soft-pion theorem relates the $N \to N\pi$ matrix element to the $N \to N$ current form factors. Expanding the currents to first order in the pion field yields

$$\left(J_{\underset{a}{v}}\right)_\mu^a = \frac{iF_\pi}{2}\left[\operatorname{Tr}\left(\tau^a A^{-1}\left(\partial_\mu U_0^\dagger U_0 \pm \partial_\mu U_0 U_0^\dagger\right)A\right)\right.$$
$$\left. - \frac{i\pi^b}{2F_\pi}\operatorname{Tr}\left([\tau^a,\tau^b]A^{-1}\left(\partial_\mu U_0^\dagger U_0 \mp \partial_\mu U_0 U_0^\dagger\right)A\right) + \cdots\right], \tag{1.35}$$

where for notational simplicity we have displayed only the first term in the current. Note the sign flip in the second line. This form is in accord with the soft-pion theorem (see App. B–3)

$$\lim_{q_\pi^\lambda \to 0}\langle N'(\mathbf{p}')\pi^b(\mathbf{q}_\pi)|\left(J_{\underset{a}{v}}\right)_\mu^a|N(\mathbf{p})\rangle = -\frac{i}{F_\pi}\langle N'(\mathbf{p}')|\left[Q_5^b,\left(J_{\underset{a}{v}}\right)_\mu^a\right]|N(\mathbf{p})\rangle$$
$$= -\frac{\epsilon^{abc}}{F_\pi}\langle N'(\mathbf{p}')|\left(J_{\underset{a}{v}}\right)_\mu^c|N(\mathbf{p})\rangle, \tag{1.36}$$

where the current commutation rules of App. B–3 have been used.

XII–2 Electroweak matrix elements

The static properties of baryons can be determined from their coupling to the weak and electromagnetic currents. In this section, we shall describe these features in terms of the quark model.

Magnetic moments

The generic quark model assumption for the magnetic moment is that the individual quarks couple independently to a photon probe. For ground-state baryons where all the quarks move in relative S waves, the magnetic moment is thus the vector sum of the quark magnetic moments,

$$\mu_{\text{baryon}} = \sum_{i=1}^{3} \mu_i \sigma_i, \qquad (2.1)$$

where σ_i is the Pauli matrix representing the spin state of the ith quark and μ_i is the magnitude of the quark magnetic moment.[3] Since the light hadrons contain three quark flavors, the most general fitting procedure to the moments of the baryon octet will involve the magnetic moments μ_u, μ_d, μ_s.

It is straightforward to infer baryon magnetic-moment predictions in the quark model directly from the state vectors of Table XI–2. For example, we have seen that the proton occurs in the two configurations $u_\uparrow u_\uparrow d_\downarrow$ and $u_\uparrow u_\downarrow d_\uparrow$ with probabilities $2/3$ and $1/3$, respectively. This can be used to carry out the construction defined by Eq. (2.1) as follows:

$$\mu_p = \frac{2}{3} \mu(u_\uparrow u_\uparrow d_\downarrow) + \frac{1}{3} \mu(u_\uparrow u_\downarrow d_\uparrow)$$

$$= \frac{2}{3}[2\mu(u_\uparrow) + \mu(d_\downarrow)] + \frac{1}{3}[\mu(u_\downarrow) + \mu(u_\uparrow) + \mu(d_\uparrow)] = \frac{4}{3}\mu_u - \frac{1}{3}\mu_d, \quad (2.2)$$

and similarly for the other baryons. Experimental and quark model values are displayed in Table XII–2.

It is of interest to see how well the assumption of $SU(3)$ symmetry fares. In the limit of degenerate quark mass (denoted by a superbar), the quark magnetic moments are proportional to the quark electric charges,

$$\bar{\mu}_d = \bar{\mu}_s = -\frac{1}{2}\bar{\mu}_u \qquad (SU(3) \text{ limit}), \qquad (2.3a)$$

---

[3] When referring to the 'magnetic moment' of a quantum system, one means the maximum component along a quantization axis (often chosen as the 3-axis). Thus, the magnetic moment is sensitive to the third component of quark spin as weighted by the quark magnetic moment.

Table XII–2. *Baryon magnetic moments.*

Mode	Experiment[a]	Quark model	Fit A[b]	Fit B[c]
μ_p	2.792847386(63)	$\frac{4\mu_u - \mu_d}{3}$	2.79	2.79
μ_n	−1.91304275(45)	$\frac{4\mu_d - \mu_u}{3}$	−1.86	−1.91
μ_Λ	−0.613(4)	μ_s	−0.93	−0.61
μ_{Σ^+}	2.458(10)	$\frac{4\mu_u - \mu_s}{3}$	2.79	2.67
$|\mu_{\Sigma^0\Lambda}|$	1.61(8)	$\frac{|\mu_u - \mu_d|}{\sqrt{3}}$	1.61	1.63
μ_{Σ^-}	−1.160(25)	$\frac{4\mu_d - \mu_s}{3}$	−0.93	−1.09
μ_{Ξ^0}	−1.250(14)	$\frac{4\mu_s - \mu_u}{3}$	−1.86	−1.44
μ_{Ξ^-}	−0.651(3)	$\frac{4\mu_s - \mu_d}{3}$	−0.93	−0.49

[a]Expressed in units of the nucleon magneton $\mu_N = e\hbar/2M_p$.
[b]$SU(3)$ symmetric fit.
[c]μ_u, μ_d, μ_s taken as independent parameters.

while isospin symmetry would imply

$$\bar{\mu}_d = -\frac{1}{2}\bar{\mu}_u \qquad (SU(2) \text{ limit}). \qquad (2.3b)$$

If we determine the one free parameter by fitting to the very precisely known proton moment, we obtain the $SU(3)$ symmetric Fit A shown in Table XII–2. More generally, allowing μ_u, μ_d, μ_s to differ and determining them from the proton, neutron, and lambda moments yields

$$\mu_u = 1.85\,\mu_N, \qquad \mu_d = -0.972\,\mu_N, \qquad \mu_s = -0.613\,\mu_N, \qquad (2.4)$$

and leads to the improved (but not perfect) agreement of Fit B in Table XII–2. We see from Eq. (2.4) that the main effect of $SU(3)$ breaking is to substantially reduce the magnetic moment of the strange quark relative to that of the down quark. The deviation of μ_d/μ_u from the isospin expectation of $\mu_d/\mu_u = -1/2$ is smaller and perhaps not significant. Observe that the famous prediction of the $SU(2)$ limit, $\mu_n/\mu_p = -2/3$, is very nearly satisfied.

The magnetic moment as derived from the multipole expansion of the electric current is defined by

$$\mu = \frac{1}{2} \int d^3x\, \mathbf{r} \times \mathbf{J}_{em}(x). \qquad (2.5)$$

It follows from this expression that the contribution of a nonrelativistic quark 'q' to the hadronic magnetic moment is just the Dirac result,

$$\mu_q = \frac{Q}{2M_q}, \qquad (2.6)$$

where M_q is the quark's constituent mass and Q is its charge. We can use this together with Eq. (2.4) to determine the constituent quark masses, with the result

$$M_u \simeq M_d \simeq 320 \text{ MeV}, \qquad M_s \simeq 510 \text{ MeV}. \qquad (2.7)$$

As we shall see in Sect. XIII–1, these masses are comparable to those extracted from mass spectra of the light hadrons.

One can also construct models involving relativistic quarks. For these, the magnetic-moment contribution of an individual quark becomes

$$\mu = \frac{2Q}{3}\sigma \int d^3x \, r \, u(r) \, \ell(r). \qquad (2.8)$$

Note the absence of an explicit dependence on quark mass. This is compensated by some appropriate dimensional quantity. The inverse radius R^{-1} plays this role in the bag model, and other determinations of R allow for a prediction of the hadronic magnetic moment. For example, the bag model defined by taking zero quark mass (corresponding to the ultrarelativistic limit) and $R = 1$ fm yields the value $\mu_p \simeq 2.5$ in a treatment which takes center-of-mass corrections into account [DoJ 80]. Although this specific value is somewhat too small, it is fair to say that quark models give a reasonable first approximation to baryon magnetic moments.

Semileptonic matrix elements

The most general form for the hadronic weak current in the transition $B_1 \to B_2 \ell \bar{\nu}_\ell$ is

$$\langle B_2(\mathbf{p}_2)|J_\mu^{(\text{wk})}|B_1(\mathbf{p}_1)\rangle$$
$$= \bar{u}(\mathbf{p}_2)\left[f_1(q^2)\gamma_\mu - \frac{i f_2(q^2)}{m_1 + m_2}\sigma_{\mu\nu}q^\nu + \frac{f_3(q^2)}{m_1 + m_2}q_\mu \right.$$
$$\left. + g_1(q^2)\gamma_\mu\gamma_5 - \frac{i g_2(q^2)}{m_1 + m_2}\sigma_{\mu\nu}q^\nu\gamma_5 + \frac{g_3(q^2)}{m_1 + m_2}q_\mu\gamma_5 \right]u(\mathbf{p}_1), \qquad (2.9)$$

where the $\{f_i\}$ and $\{g_i\}$ form factors correspond respectively to the vector and axial-vector current matrix elements, and $q = p_1 - p_2$ is the momentum transfer.[4] The form factors are all functions of q^2 and the phases are chosen so that each form factor is real-valued if time-reversal invariance is respected. In practice, the form factors accompanying the two terms with the kinematical factor q_μ are difficult to observe because each such contribution is multiplied by a (small) lepton mass upon being contracted with a leptonic weak current. Thus, we shall drop these until Sect. XII–4.

[4] Given the context of application, there should be no confusion between the QCD strong coupling constant g_3 and the axial-vector form factor $g_3(q^2)$.

As regards the remaining form factors, we have already presented the ingredients for performing a quark model analysis (see also [DoGH 86b]). Using the $n \to p$ transition as a prototype, we have

$$f_1^{np} = \langle p_\uparrow | \int d^3x \, \bar{u}\gamma^0 d \, |n_\uparrow\rangle = \int d^3x \, (u_u u_d + \ell_u \ell_d) = 1 \,, \quad (2.10a)$$

$$\frac{f_1^{np} + f_2^{np}}{m_p + m_n} = \langle p_\uparrow | \int d^3x \, \frac{1}{2} [\mathbf{r} \times (\bar{u}\boldsymbol{\gamma} d)]_3 \, |n_\uparrow\rangle$$

$$= \frac{1}{3} \int d^3x \, r(u_u \ell_d + u_d \ell_u) = \frac{1}{2} \left(\frac{1}{2M_u} + \frac{1}{2M_d} \right), \quad (2.10b)$$

$$g_1^{np} = \langle p_\uparrow | \int d^3x \, \bar{u}\gamma_3\gamma_5 d \, |n_\uparrow\rangle = \frac{5}{3} \int d^3x \, (u_u u_d - \frac{1}{3}\ell_u \ell_d), \quad (2.10c)$$

$$\frac{g_2^{np}}{m_n + m_p} + \left(\frac{1}{2m_n} - \frac{1}{2m_p} \right) \frac{g_1^{np} + g_3^{np}}{2} = \langle p_\uparrow | -i \int d^3x \, z\bar{u}\gamma^0\gamma_5 d \, |n_\uparrow\rangle$$

$$= \frac{1}{3} \int d^3x \, z(u_d \ell_u - u_u \ell_d) = \frac{1}{2} \left(\frac{1}{2M_d} - \frac{1}{2M_u} \right). \quad (2.10d)$$

In each case, we first give the defining relation, then the general Dirac wavefunction (cf. Eq. (1.4)) and, finally, the nonrelativistic quark model limit. The vanishing of g_2^{np} in the limit of exact isospin symmetry is a consequence of *G-parity* (cf. Sect. V–3).

Predictions for the other baryonic transitions are governed by $SU(3)$ invariance, amended by small departures from $SU(3)$ invariance as suggested by the quark model, i.e., $s \to u$ transitions are similar to those of $d \to u$ as given above, but with the down-quark mass and wavefunction replaced by those of the strange quark. $SU(3)$ breaking in the form factors arises from this difference in the wavefunction. As a quark gets heavier, its wavefunction is more concentrated near the origin and the lower component becomes less important. The form factors of the matrix element $\langle B_b | J_c^\mu | B_a \rangle$ evaluated in the $SU(3)$ limit at $q^2 = 0$ give for the vector current,

$$f_1(0) = i f_{abc}, \qquad\qquad f_2(0) = i f_{abc} f + d_{abc} d,$$

$$f = \frac{1}{2}(\mu_p + \mu_n - 1), \qquad d = -\frac{3}{2}\mu_n, \qquad (2.11a)$$

with $f/d = 0.29$, and for the axial-vector current,

$$g_1(0) = i f_{abc} F + d_{abc} D, \qquad (2.11b)$$

with $F + D = g_1^{np} = g_A = 1.27$. In the above, the indices $a, b, c = 1, \ldots, 8$ label the $SU(3)$ of flavor, with $c = (1 + i2)$ for $\Delta S = 0$ and $c = 4 + i5$ for $\Delta S = 1$. There is no $SU(3)$ parameterization for the g_2 form factor because it vanishes in the

$SU(3)$ limit. An important result specific to the quark model is $D/(D + F) = 3/5$ for the $SU(3)$ structure of the axial-current $\{g_1\}$ form factors.

$SU(3)$ breaking in the $\{f_1\}$ form factors is required by the Ademollo–Gatto theorem to occur only beginning at second order (see Sect. VIII–1). In practice, the quark model prediction for $SU(3)$ breaking yields an extremely small effect. This is not true for the $\{f_2\}$ form factors of weak magnetism, where inclusion of the strange-quark mass lowers all $s \to u$ transitions by 20% compared to the $d \to u$ transition. The wavefunction overlaps in g_1 lead to a slight increase in the strength of the $s \to u$ transition compared to $d \to u$ because of the reduced lower component of the s quark. For g_2, a nonzero but highly model-dependent value is generated, typically of order $|g_2/g_1| \simeq 0.3$.

XII–3 Symmetry properties and masses

In our discussion of baryon properties, we have first discussed quark models because they are generally simple and have predictive power. However, effective field theory methods are also useful when applied to the study of baryons.[5] We shall combine the two descriptions in this section.

Effective lagrangians for baryons

We begin by writing effective lagrangians which include *baryon* fields, using the procedure described in App. B–4. The lowest-order $SU(2)$-invariant lagrangian describing the nucleon and its pionic couplings has the form

$$\mathcal{L}_N = \bar{N}\left(i\not{\mathcal{D}} - g_A \not{\overline{A}}\gamma_5 - \mathbf{m}_0\right)N$$
$$- \frac{Z_0}{2}\bar{N}\left(\xi\hat{\mathbf{m}}\xi + \xi^\dagger\hat{\mathbf{m}}\xi^\dagger\right)N - \frac{Z_1}{2}\bar{N}N\;\mathrm{Tr}\left(\hat{\mathbf{m}}U + U^\dagger\hat{\mathbf{m}}\right),$$
$$\mathcal{D}_\mu \equiv \partial_\mu + i\overline{V}_\mu, \qquad \xi \equiv \exp\left[i\boldsymbol{\tau}\cdot\boldsymbol{\pi}/(2F_\pi)\right], \qquad U \equiv \xi\xi,$$
$$\overline{V}_\mu \equiv -\frac{i}{2}\left(\xi^\dagger\partial_\mu\xi + \xi\partial_\mu\xi^\dagger\right), \qquad \overline{A}_\mu \equiv -\frac{i}{2}\left(\xi^\dagger\partial_\mu\xi - \xi\partial_\mu\xi^\dagger\right), \qquad (3.1)$$

where $N = \binom{p}{n}$ is the nucleon field, $\hat{\mathbf{m}}$ is the mass matrix for current quarks (with $m_u = m_d \equiv \hat{m}$), Z_0 and Z_1 are arbitrary constants which parameterize terms proportional to the quark mass matrix, and the constant g_A is the nucleon axial-vector coupling constant $g_A \simeq 1.27$ (cf. Prob. XII–1). The mass parameter m_0 represents the nucleon mass in the $SU(2)$ chiral limit.

---

[5] There is also an effective field theory treatment of the few nucleon case [We 90, Va 08, EpM 12] which helps understand nuclei in a systematic manner.

For the full $SU(3)$ octet of baryons, the analog of 'N' is

$$B = \frac{1}{\sqrt{2}} \sum_{a=1}^{8} \lambda^a B^a = \begin{pmatrix} \frac{\Sigma^0}{\sqrt{2}} + \frac{\Lambda}{\sqrt{6}} & \Sigma^+ & p \\ \Sigma^- & -\frac{\Sigma^0}{\sqrt{2}} + \frac{\Lambda}{\sqrt{6}} & n \\ \Xi^- & \Xi^0 & -\frac{2\Lambda}{\sqrt{6}} \end{pmatrix}, \qquad (3.2)$$

where the phases have been adjusted to match our quark model phase convention of Eq. (XI–1.8). The $SU(3)$ version of Eq. (3.1) becomes

$$\mathcal{L}_B = \mathrm{Tr}\left(\bar{B}\left(i\slashed{\mathcal{D}} - \bar{m}_0\right)B - D\left(\bar{B}\gamma^\mu\gamma_5\{\overline{A}_\mu, B\}\right) - F\left(\bar{B}\gamma^\mu\gamma_5[\overline{A}_\mu, B]\right)\right)$$
$$- \frac{Z_0}{2}\,\mathrm{Tr}\left(d_m\left(\bar{B}\{\xi\mathbf{m}\xi + \xi^\dagger\mathbf{m}\xi^\dagger, B\}\right) + f_m\left(\bar{B}\left[\xi\mathbf{m}\xi + \xi^\dagger\mathbf{m}\xi^\dagger, B\right]\right)\right)$$
$$- \frac{Z_1}{2}\,\mathrm{Tr}\left(\bar{B}B\right)\mathrm{Tr}\left(\mathbf{m}U + U^\dagger\mathbf{m}\right), \qquad (3.3)$$

where the covariant derivative is now $\mathcal{D}_\mu B \equiv \partial_\mu B + i[\overline{V}_\mu, B]$, ξ is the $SU(3)$ generalization of the quantity in Eq. (3.1) with τ replaced by λ, $\mathbf{m}$ is the diagonal $SU(3)$ quark mass matrix,

$$\mathbf{m} = \left(\hat{m},\, \hat{m},\, m_s\right)_{\mathrm{diag}} = \frac{1}{3}(2\hat{m} + m_s)\mathbf{1} + \frac{1}{\sqrt{3}}(\hat{m} - m_s)\lambda_8, \qquad (3.4)$$

and $\bar{m}_0$ is the degenerate baryon mass in the $SU(3)$ chiral limit. Consistency of the $SU(2)$ and $SU(3)$ lagrangians requires

$$D + F = g_\mathrm{A}, \qquad d_m + f_m = 1,$$
$$m_0 = \bar{m}_0 + Z_1 m_s - Z_0 m_s(f_m - d_m). \qquad (3.5)$$

The description thus far is based on symmetry. It includes quark mass, but not higher powers of derivatives.

Baryon mass splittings and quark masses

The various parameters ($\hat{m}, m_s, Z_0$ etc.) appearing in the chiral lagrangians of Eqs. (3.1), (3.3) can be determined from baryon mass and scattering data. In the nonstrange sector, the nucleon mass is given in the notation of Eq. (3.1) as

$$m_N = m_0 + (Z_0 + 2Z_1)\hat{m}. \qquad (3.6)$$

To isolate the effect of the nonstrange quark mass $\hat{m}$ and of the constants Z_0, Z_1, it will prove useful to define a quantity σ,

$$\sigma = m_N - m_0 = \hat{m}\frac{\langle N|\bar{u}u + \bar{d}d|N\rangle}{2m_N} = \hat{m}\left(Z_0 + 2Z_1\right). \qquad (3.7)$$

Shortly, we shall see how this quantity can be determined from pion–nucleon scattering data.

However, let us first consider the baryonic mass splittings generated by the mass difference $m_s - \hat{m}$. Upon using Eq. (3.3) to obtain expressions for the baryon masses and working with isospin-averaged masses, it is possible by adopting the numerical values

$$Z_0(m_s - \hat{m}) = 132\,\text{MeV}, \qquad d_m/f_m = -0.31, \qquad (3.8)$$

to obtain the following good fit:

$$m_\Sigma - m_N = (f_m - d_m)Z_0(m_s - \hat{m}) = 251\,\text{MeV} \quad (\text{expt.}: 254.2\,\text{MeV}),$$

$$m_\Sigma - m_\Lambda = -\frac{4}{3}d_m Z_0(m_s - \hat{m}) = 79\,\text{MeV} \qquad (\text{expt.}: 77.5\,\text{Mev}), \qquad (3.9)$$

$$m_\Xi - m_N = 2f_m Z_0(m_s - \hat{m}) = 383\,\text{MeV} \qquad (\text{expt.}: 379.2\,\text{MeV}).$$

Observe that these mass splittings depend on Z_0 but not on Z_1. The three relations of Eq. (3.9) imply the Gell-Mann–Okubo formula [Ge 61, Ok 62],

$$m_\Sigma - m_N = \frac{1}{2}(m_\Xi - m_N) + \frac{3}{4}(m_\Sigma - m_\Lambda)$$

$$(\text{Expt.}: 254\,\text{MeV} = 248\,\text{MeV}), \qquad (3.10)$$

which displays an impressive level of agreement ($\simeq 3\%$) with experimental values.

The above analysis, based on a chiral lagrangian, can be enhanced by using ideas taken from the quark model. In the limit of *noninteracting* quarks, the quark model yields for a general spatial wavefunction,[6]

$$m_\Lambda - m_N = m_\Sigma - m_N = m_\Xi - m_\Sigma = (m_s - \hat{m})\int d^3x\,(u^2 - \ell^2). \qquad (3.11)$$

However, observe that $m_\Sigma = m_\Lambda$ (corresponding in the chiral lagrangian description to $d_m = 0$) for noninteracting quarks. Of course, the actual Λ and Σ baryons are not degenerate, so additional physics is required. A quark model source of the $\Lambda - \Sigma$ mass splitting lies in the hyperfine interaction of Eq. (XI–2.14),

$$H_{\text{hyp}}^{(\text{baryon})} = \frac{1}{2}\sum_{i<j}\bar{\mathcal{H}}_{ij}\mathbf{s}_i \cdot \mathbf{s}_j\,\delta^{(3)}(\mathbf{r}), \qquad (3.12)$$

where the prefactor of $1/2$ is associated with the color dependence of Eq. (XI–2.4). Matrix elements of this operator give rise to the additive mass contributions,

---

[6] One could equivalently use the language of the potential model, where these baryon mass splittings arise from the constituent quark mass difference $M_s - \hat{M}$.

$$m_N = \cdots - \frac{3}{8}\mathcal{H}_{nn}, \qquad\qquad m_\Lambda = \cdots - \frac{3}{8}\mathcal{H}_{nn},$$

$$m_\Sigma = \cdots + \frac{1}{8}\mathcal{H}_{nn} - \frac{1}{2}\mathcal{H}_{ns}, \quad m_\Xi = \cdots - \frac{1}{2}\mathcal{H}_{ns} + \frac{1}{8}\mathcal{H}_{ss}, \qquad (3.13)$$

where $\bar{\mathcal{H}}_{ij}$ and $\mathcal{H}_{ij}$ are related by $\mathcal{H}_{ij} \equiv \bar{\mathcal{H}}_{ij}|\Psi(0)|^2$ and the subscripts 'n', 's' denote an interaction involving a nonstrange quark and a strange quark respectively. For $\mathcal{H}_{nn} \neq \mathcal{H}_{ns}$, the Σ and Λ will not be degenerate. Treating both quark mass splittings and hyperfine effects as first-order perturbations (e.g. $\mathcal{H}_{ss} - \mathcal{H}_{ns} = \mathcal{H}_{ns} - \mathcal{H}_{nn}$), one obtains quark model mass relations

$$m_\Lambda - m_N = (m_s - \hat{m})\int d^3x\,(u^2 - \ell^2),$$

$$m_\Sigma - m_\Lambda = \frac{1}{2}(\mathcal{H}_{nn} - \mathcal{H}_{ns}),$$

$$m_\Xi - m_N = \frac{1}{4}(\mathcal{H}_{nn} - \mathcal{H}_{ns}) + 2(m_s - \hat{m})\int d^3x\,(u^2 - \ell^2) \qquad (3.14)$$

in accord with the sum rule of Eq. (3.10). These formulae can provide an estimate of quark mass. For the usual range of quark model wavefunctions (encompassing both bag and potential descriptions), the overlap integral has magnitude

$$\int d^3x\,(u^2 - \ell^2) \simeq \frac{1}{2} \to \frac{3}{4}. \qquad (3.15)$$

To the extent that this estimate is valid, it produces the values

$$m_s - \hat{m} \simeq 230 \to 350\,\text{MeV}, \qquad \hat{m} \simeq 11 \to 14\,\text{MeV}, \qquad (3.16)$$

where the chiral symmetry mass ratio of Eq. (VII–1.15a) has been used to obtain $\hat{m}$. In general, quoting absolute values of quark masses is dangerous as one must specify how the operator $\bar{q}q$, which occurs in the mass term $m_q\bar{q}q$, has been renormalized. It is all too common in the literature to ignore this point by using $m_s - \hat{m} = m_\Lambda - m_N$. The values quoted here are actually current-quark mass differences, renormalized at a hadronic scale using quark model matrix elements.

The parameter Z_1 which appears in the $SU(3)$ lagrangian of Eq. (3.3) is difficult to constrain in a quark model. For example, one might consider the matrix element

$$\frac{\langle N\,|m_s\bar{s}s|\,N\rangle}{2m_N} = m_s\,(Z_1 - Z_0(f_m - d_m)). \qquad (3.17)$$

The most naive assumption, that $\langle N\,|m_s\bar{s}s|\,N\rangle$ vanishes, would imply $Z_1 = Z_0$ $(f_m - d_m) \simeq 1.9\,Z_0$. However, one may legitimately question whether such an assumption is reasonable. We shall return to the issue of the 'strangeness content' of the nucleon later in this section.

Goldberger–Treiman relation

Moving from the study of baryon masses to the topic of interactions, let us consider the coupling of pions and nucleons. The $SU(2)$ lagrangian of Eq. (3.1), expanded to order π^2, becomes

$$\mathcal{L}_N = \bar{N}(i\not{\partial} - m_N)N + \frac{g_A}{F_\pi}\bar{N}\gamma^\mu\gamma_5\frac{\tau}{2}N \cdot \partial_\mu\pi$$

$$- \frac{1}{4F_\pi^2}\bar{N}\gamma^\mu\tau \cdot \pi \times \partial_\mu\pi \; N + \frac{1}{2F_\pi^2}\pi^2\bar{N}N\sigma + \cdots, \qquad (3.18)$$

where σ is defined in Eq. (3.7). The second term describes the $NN\pi$ vertex. Upon using Eq. (3.18) to compute the pion emission amplitude $N \to N\pi^i$ and comparing with the Lorentz invariant form

$$\mathcal{M}_{N \to N\pi^i} = -ig_{\pi NN}\bar{u}(\mathbf{p}')\gamma_5\tau^i u(\mathbf{p}), \qquad (3.19)$$

one immediately obtains the Goldberger–Treiman relation [GoT 58],

$$g_{\pi NN} = \frac{g_A m_N}{F_\pi}. \qquad (3.20)$$

Inserting the experimental value, $g_{\pi NN}^2/4\pi \simeq 13.8$, for the πNN coupling constant, one finds the Goldberger–Treiman relation to be satisfied to about 2.5%.

There also exist important implications for the g_3 term in the general expression given in Eq. (2.9) for the axial-current matrix element. In forming the $n \to p$ axial matrix element, one encounters a direct $\gamma_\mu\gamma_5$ contribution and also a pion-pole term which corresponds to pion propagation from the $n \to p\pi^-$ emission vertex to the axial current. Making use of Eq. (3.20) and Prob. XII–1, we have

$$\langle p(\mathbf{p}')|A_\mu^+|n(\mathbf{p})\rangle = \bar{u}(\mathbf{p}')\left[g_A\gamma_\mu\gamma_5 - \frac{g_A}{\sqrt{2}F_\pi}\not{q}\gamma_5\frac{\sqrt{2}F_\pi q_\mu}{q^2 - m_\pi^2}\right]u(\mathbf{p})$$

$$= \bar{u}(\mathbf{p}')\left[g_A\gamma_\mu\gamma_5 + \frac{2m_N g_A}{q^2 - m_\pi^2}q_\mu\gamma_5\right]u(\mathbf{p}). \qquad (3.21)$$

where $q = p - p'$. It is this induced pseudoscalar modification which allows the axial current to be conserved in the chiral limit $m_\pi^2 \to 0$,

$$-i\partial^\mu\langle p(\mathbf{p}')|A_\mu^+|n(\mathbf{p})\rangle = 2m_N g_A\left[1 - \frac{q^2}{q^2 - m_\pi^2}\right]\bar{u}(\mathbf{p}')\gamma_5 u(\mathbf{p})$$

$$= -\frac{2m_N g_A m_\pi^2}{q^2 - m_\pi^2}\bar{u}(\mathbf{p}')\gamma_5 u(\mathbf{p}). \qquad (3.22)$$

Note that for nonzero pion mass, the above is consistent with the PCAC relation of Eq. (B–3.7),

$$F_\pi m_\pi^2 \pi^k = \partial^\mu A_\mu^k, \qquad (3.23)$$

as both sides have the same matrix element,

$$-i\langle p(\mathbf{p}')|F_\pi m_\pi^2 \pi^+(0)|n(\mathbf{p})\rangle = i\sqrt{2}\,g_{\pi NN}\bar{u}(\mathbf{p}')\gamma_5 u(\mathbf{p})\frac{i}{q^2-m_\pi^2}\sqrt{2}\,F_\pi m_\pi^2$$

$$= -\frac{2m_N g_A m_\pi^2}{q^2-m_\pi^2}\bar{u}(\mathbf{p}')\gamma_5 u(\mathbf{p}). \tag{3.24}$$

The pion-pole contribution of the axial-vector current-matrix element has been probed in nuclear muon capture, as will be described in Sect. XII–4.

The nucleon sigma term

One of the features immediately apparent from the effective lagrangian of Eq. (3.1) is that all the couplings of pions to nucleons, with the exception of the quark mass terms, are derivative couplings. Before turning to the sigma term, which appears in the nonderivative sector, let us briefly consider the expansion in powers of the number of derivatives for pion-nucleon scattering. Recall for pion–pion scattering (cf. Sect. VI–4), there were no large masses and the chiral expansion was expressed in terms of m_π^2 or E_π^2. However, correction terms in the chiral expansion for nucleons will enter at relatively low energies since a term like $2p \cdot q \simeq 2m_p E_\pi$ can get large quickly (it is linear in the energy and has a large coefficient, e.g., $E_\pi = 250$ MeV gives $2m_p E_\pi = (700\,\mathrm{MeV})^2$). To combat this difficulty, additional (but still general) inputs such as analyticity and crossing symmetry are often invoked. Fortified with these theoretical constraints, one then matches intermediate-energy data to the low-energy chiral parameterizations. The low-energy chiral results thereby obtained appear to be well satisfied [Hö 83, GaSS 88].

The *nonderivative* pion–nucleon coupling coming from the quark mass terms in Eq. (3.1) is of particular interest. To determine this contribution from experiment, one must be able to suppress the various derivative couplings. Thus, if one extrapolated in the chiral limit to zero four-momentum, the derivative couplings would vanish. Not surprisingly then, a soft-pion analysis reveals that the nonderivative coupling can be isolated by extrapolating the isospin-even πN scattering amplitude with the Born term subtracted (called $\bar{D}^+$ in the literature) to the so-called 'Cheng–Dashen point' $t = m_\pi^2$, $s = m_N^2$ [ChD 71]. It is conventional to multiply the extrapolated amplitude by F_π^2 and thus define a quantity Σ,

$$\Sigma \equiv F_\pi^2\,\bar{D}_{CD}^+. \tag{3.25}$$

To lowest order in the chiral expansion, the measured quantity Σ is just the matrix element σ defined in Eq. (3.7),

$$\Sigma = \sigma = \hat{m}\frac{\langle N|\bar{u}u+\bar{d}d|N\rangle}{2m_N}. \tag{3.26}$$

It is this isospin-even scattering amplitude $\bar{D}^+$ which provides a unique window on the nonstrange quark mass $\hat{m}$. Because Σ is proportional to the small mass $\hat{m}$, it is difficult to determine this quantity precisely, and considerable effort has gone into its extraction. The Cheng–Dashen point lies outside the physical kinematic region, and extrapolation from the experimental region must be done carefully with dispersion relations. A recent estimate is [AlCO 13]

$$\Sigma = 59 \pm 7 \text{MeV}. \tag{3.27}$$

The result $\sigma = \Sigma - 15\,\text{MeV}$ has been obtained from studies of higher-order chiral corrections, implying

$$\sigma \simeq 44\,\text{MeV} \tag{3.28}$$

as the measure of light-quark mass [GaLS 91].

Strangeness in the nucleon

In light of the above discussion, it is tempting to interpret various contributions to the nucleon mass by making use of the energy-momentum trace. Recall the trace anomaly of Eq. (III–4.16),

$$\theta^\mu_\mu = \frac{\beta_{QCD}}{2g_3} F^a_{\mu\nu} F^{a\mu\nu} + m_u \bar{u}u + m_d \bar{d}d + m_s \bar{s}s. \tag{3.29}$$

Taking the nucleon matrix element gives

$$m_N = \frac{\langle N \left| \theta^\mu_\mu \right| N \rangle}{2m_N} = m_0 + \sigma,$$

$$m_0 = (2m_N)^{-1} \langle N \left| \frac{\beta_{QCD}}{2g_3} F^a_{\mu\nu} F^{a\mu\nu} + m_s \bar{s}s \right| N \rangle \simeq 894 \pm 8\,\text{MeV},$$

$$\sigma = \hat{m} \frac{\langle N | \bar{u}u + \bar{d}d | N \rangle}{2m_N} \simeq 44\,\text{MeV}. \tag{3.30}$$

This result is already quite interesting in that the largest contributions, the gluon and strange-quark terms in m_0, appear to be 'nonvalence'. At this stage, the separation is essentially model-independent.

One can explore the 'strangeness content of the nucleon' by using an $SU(3)$ analysis of hyperon masses. Thus, we introduce a mass-splitting operator, which transforms as the eighth component of an octet,

$$\mathcal{L}_{\text{m-s}} = \frac{1}{3} (\hat{m} - m_s)(\bar{u}u + \bar{d}d - 2\bar{s}s). \tag{3.31}$$

Since the hyperon mass splittings are governed by this octet operator, we find

$$\delta_s \equiv \frac{\langle p \left| (m_s - \hat{m})(\bar{u}u + \bar{d}d - 2\bar{s}s) \right| p \rangle}{2m_p} = \frac{3}{2}(m_\Xi - m_N) = 574 \, \text{MeV}. \quad (3.32a)$$

When scaled by the quark mass ratio $\hat{m}/m_s$, Eq. (3.32a) becomes

$$\delta \equiv \hat{m} \frac{\langle N \left| \bar{u}u + \bar{d}d - 2\bar{s}s \right| N \rangle}{2m_N}$$

$$= \frac{3}{2} \frac{m_\pi^2}{m_K^2 - m_\pi^2} (m_\Xi - m_\Lambda) \simeq 25 \, \text{MeV} \quad (35 \, \text{MeV}), \quad (3.32b)$$

where the figure in parentheses includes higher-order chiral corrections [Ga 87]. Comparison of δ and σ immediately indicates that they are compatible only if the strange-quark matrix element does *not* vanish. Indeed, one requires

$$\frac{\langle N \left| \bar{s}s \right| N \rangle}{\langle N \left| \bar{u}u + \bar{d}d + \bar{s}s \right| N \rangle} \simeq 0.18 \quad (0.09). \quad (3.33)$$

This gives for the constant Z_1 of Eq. (3.1) the value $Z_1 \simeq 3.9 Z_0$ $(2.9 Z_0)$ to be contrasted with the estimate which follows Eq. (3.17). At the same time, one can separate out the following matrix elements

$$(2m_N)^{-1} \langle N \left| \frac{\beta_{QCD}}{2g_3} F_{\mu\nu}^a F^{a\mu\nu} \right| N \rangle \simeq 634 \, \text{MeV} \quad (764 \, \text{MeV}),$$

$$(2m_N)^{-1} \langle N \left| m_s \bar{s}s \right| N \rangle \simeq 260 \, \text{MeV} \quad (130 \, \text{MeV}), \quad (3.34)$$

where figures in brackets use the corresponding bracketed quantity in Eq. (3.32b). Note the surprisingly large effect of the strange quarks. These results are controversial because they draw a counter-intuitive conclusion from the use of $SU(3)$ symmetry. However, even with $SU(3)$ breaking, the difference between σ and the $SU(3)$ value of δ is large enough that some $\bar{s}s$ contribution is likely to be required.

This analysis does not go well with the naive interpretation of the quark model as embodied, for example, in the proton-state vector formula which began this chapter. However, it is possibly compatible with a more sophisticated interpretation of the constituent quarks which enter into quark models. In the process of forming a constituent quark, the quark is 'dressed' by gluonic and even $\bar{s}s$ quark fields. It is no longer the naive object that occurs in the QCD lagrangian. It is this dressed object which may then easily generate gluonic and perhaps strange quark matrix elements. Recall that even the vacuum state has gluonic and quark matrix elements. Similar explanations exist in bag and Skyrme models [DoN 86]. This issue remains unresolved at present.

Based on the possible existence of a substantial nonzero value for the scalar density matrix element $\langle N | \bar{s}s | N \rangle$, a major program was launched to investigate the possibility for a similar nonzero value for the strange vector-current matrix element $\langle N|\bar{s}\gamma_\mu s|N\rangle$ which can be characterized in terms of charge and magnetic form factors $F_1^s(q^2)$, $F_2^s(q^2)$ via

$$\langle N|\bar{s}\gamma_\mu s|N\rangle = \bar{u}(p')\left[\gamma_\mu F_1^s(q^2) - \frac{i}{2m_N}\sigma_{\mu\nu}q^\nu F_2^s(q^2)\right]u(p). \qquad (3.35)$$

The form factor $F_1^s(q^2)$ obeys $F_1^s(0) = 0$, whereas $F_2^s(q^2)$ has no such constraint.[7] In order to determine the size of the $\bar{u}\gamma_\mu u$, $\bar{d}\gamma_\mu d$, $\bar{s}\gamma_\mu s$ contributions to the corresponding nucleon matrix elements, three experimental inputs are required. Two of these come from well-known electromagnetic form factors of the proton and neutron. The third can be found by performing parity-violating electron-scattering experiments from the proton, by measuring the difference in the cross sections for the scattering of electrons with left- and right-handed helicities. This is sensitive to the strange-quark current because the electromagnetic current and the neutral weak current involve strange quarks with different strengths. In this case there exists an interference between the electromagnetic (γ-exchange) and weak (Z^0-exchange) contributions and the resultant asymmetry will have the form

$$A_{LR} = \frac{d\sigma_R - d\sigma_L}{d\sigma_R + d\sigma_l} \sim \frac{Gq^2}{4\pi\sqrt{2}\alpha}(M_E + M_M + \cdots), \qquad (3.36)$$

where M_E, M_M involve the interference of the electromagnetic and electric, magnetic weak form factors and the ellipses indicate a small piece involving the vector electron coupling and the axial current. In this asymmetry, the electron side involves an axial current while the nucleon side involves a vector current. This asymmetry has been studied as a function of q^2 in a series of experiments at electron laboratories at MIT-Bates, at Jefferson Laboratory, and at the Mainz microtron. The result is that no signal for a strange vector-current matrix element has been seen and limits have been placed on the strange form factors. Numerically, strange quarks contribute less than 5% of the mean square charge radius and less than 10% of the magnetic moment of the proton. Reviews of this body of work can be found in [ArM 12] and [BeH 01].

Quarks and nucleon spin structure

The constituent quark model provides a simple picture of the contents of baryons as systems composed of three constituent quarks and nothing else. A rigorous description using the quark and gluon degrees of freedom which appear in the fundamental

---

[7] The condition on $F_1^s(0)$ is a consequence of current conservation. Equivalently, taking $\mu = 0$ in Eq. (3.35) and integrating over the proton volume, one encounters the strangeness 'charge' $S \equiv \int d^3x \, s^\dagger(\mathbf{x})s(\mathbf{x})$ and $S|N\rangle = 0$ since the nucleon carries no net strangeness.

lagrangian is in general more complex, but it is often nevertheless instructive to explore the constituent picture of a given observable. An interesting example is the spin structure of the nucleon.

For any Lorentz invariant theory, Noether's theorem requires that there exist an angular momentum tensor $M^{\mu\alpha\beta}$ which is conserved ($\partial_\mu M^{\mu\alpha\beta} = 0$) and which gives rise to three angular momentum charges associated with rotational invariance,

$$J^{\alpha\beta} \equiv \int d^3x \, M^{0\alpha\beta}(x). \tag{3.37}$$

In the rest frame of a particle, the $\{J^{\alpha\beta}\}$ are related to the three components of angular momenta via

$$J^i = \frac{1}{2}\epsilon^{ijk} J^{jk}. \tag{3.38}$$

For the example of a free fermion, the above quantities take the form

$$M^{\mu\alpha\beta} = i\bar{\psi}\gamma^\mu \left(x^\alpha \partial^\beta - x^\beta \partial^\alpha\right)\psi + \frac{1}{2}\bar{\psi}\gamma^\mu\sigma^{\alpha\beta}\psi, \tag{3.39}$$

up to total derivatives which do not contribute to the charges, and

$$\mathbf{J} = \int d^3x \left[-i\psi^\dagger (\mathbf{x} \times \partial)\psi + \frac{1}{2}\bar{\psi}\boldsymbol{\gamma}\gamma_5\psi\right] \equiv \mathbf{L} + \mathbf{S}. \tag{3.40}$$

The two contributions in Eq. (3.40) may be labeled the orbital and spin components of the angular momentum.

The quarks in the Noether current are lagrangian (current) quarks, not constituent quarks. Nevertheless, in the spirit of the quark model let us apply Eq. (3.40) to the quarks in a spin-up proton. As expressed in terms of upper (u) and lower (ℓ) components (cf. Eq. (XI–1.13)), the orbital and spin contributions are found to be

$$\langle \mathbf{L} \rangle = \frac{2}{3}\int d^3x \, \ell^2\langle\sigma\rangle, \qquad \langle \mathbf{S} \rangle = \int d^3x \left(u^2 - \frac{1}{3}\ell^2\right)\frac{\langle\sigma\rangle}{2}. \tag{3.41}$$

Aside from the factor $1/2$ occurring in $\sigma/2$, the quark spin contribution to $\mathbf{S}$ is just the axial-vector matrix element of Eq. (1.24), whereas the orbital angular momentum contains just the lower component ℓ because the $\mathbf{x} \times \partial$ operator has a nonzero effect only when acting on the $\sigma \cdot \hat{\mathbf{x}}$ factor in the lower component of Eq. (XI–1.13). Observe that the orbital angular momentum is nonvanishing and proportional to the quark spin. The spin and orbital portions for the individual u, d flavors are easily computed to yield

$$\langle S_z^{(u)} \rangle = \frac{2}{3}\int d^3x \left(u^2 - \frac{1}{3}\ell^2\right), \qquad \langle S_z^{(d)} \rangle = -\frac{1}{6}\int d^3x \left(u^2 - \frac{1}{3}\ell^2\right),$$

$$\langle L_z^{(u)} \rangle = \frac{8}{9}\int d^3x \, \ell^2, \qquad \langle L_z^{(d)} \rangle = -\frac{2}{9}\int d^3x \, \ell^2. \tag{3.42}$$

A first lesson is that, despite the spin wavefunction of the protons being written entirely in terms of quarks as in Table XI–2, the quark spin averages of Eq. (3.42) do *not* add up to yield the proton spin. The sum is reduced from the anticipated value of $1/2$ by the lower component ℓ in the Dirac spinor. It is the total angular momentum **J** which has the expected result,

$$\langle \mathbf{J} \rangle = \frac{1}{2} \langle \sigma \rangle, \tag{3.43}$$

but the total is split up between the orbital and spin components. The bag model, for example, yields

$$\langle \mathbf{S} \rangle \simeq 0.65 \, \langle \mathbf{J} \rangle, \tag{3.44}$$

so about 35% of the nucleon spin arises from orbital angular momentum.

Of course, *QCD* is a full interacting theory and the discussion of the angular momenta of the quarks and the gluons cannot be fully separated because these fields interact with each other. The total angular momentum can be decomposed into several terms, including the interactions between the fields [JaM 90]. These can be grouped in various ways. In the current quark–gluon description, it is common to write

$$\frac{1}{2} = \frac{1}{2} S_q + L_q + J_g, \tag{3.45}$$

where S_q, L_q are the spin and angular momentum components carried by the quarks and J_g is that carried by the gluons. Thus, we have

$$\mathbf{J_q} = \int d^3x \left[\psi^\dagger \frac{\mathbf{\Sigma}}{2} \psi + \psi^\dagger \, \mathbf{x} \times (-\mathbf{i}\mathbf{D}) \psi \right],$$

$$\mathbf{J_g} = \int d^3x \, \mathbf{x} \times (\mathbf{E} \times \mathbf{B}), \tag{3.46}$$

where $\mathbf{\Sigma}$ is the usual Dirac spin matrix and $D_\mu \psi = [\partial_\mu + ig A_\mu]\psi$ is the covariant derivative of ψ and therefore, in this definition, the quark angular momentum has a gluonic component [JiTH 96].

Polarized deep-inelastic electron scattering from the nucleon can measure spin effects of the quarks. The study of spin dependent deep inelastic scattering involves the antisymmetric component of the nucleon tensor,[8] which can be written in the form

$$W^{[\mu\nu]} = \frac{1}{4\pi} \int d^4x \, e^{-q \cdot x} \langle p, s | [J_\mu^{\text{em}}(x), J_\nu^{\text{em}}(0)] | p, s \rangle$$

$$= -i\epsilon_{\mu\nu\alpha\beta} q^\alpha \left[G_1(\nu, Q^2) \cdot \frac{s^\beta}{m_N^2} + G_2(\nu, Q^2) \cdot \frac{m_N \nu s^\beta - s \cdot q p^\beta}{m_N^4} \right], \tag{3.47}$$

---

[8] More details can be found in the review [Ba 05].

where $\nu = p \cdot q / m_N$ and $Q^2 = -q^2$. The scaling behavior of the two structure functions is

$$g_1(x, Q^2) = \frac{\nu}{m_N} G_1(\nu, Q^2), \qquad g_2(x, Q^2) = \left(\frac{\nu}{m_N}\right)^2 G_2(\nu, Q^2), \qquad (3.48)$$

where $x = Q^2 / 2m_N \nu$ is the Bjorken scaling variable. In the parton model, neglecting QCD renormalization, one determines

$$\int_0^1 dx \, g_1^p(x, Q^2) = \frac{1}{2} \sum_q e_q^2 \Delta q = \frac{1}{12} g_A^{(3)} + \frac{1}{36} g_A^{(8)} + \frac{1}{9} g_A^{(0)}, \qquad (3.49)$$

where $g_A^{(3)}$, $g_A^{(8)}$, $g_A^{(0)}$ are the isovector, $SU(3)$ octet, and flavor-singlet axial charges respectively. The axial charges are written in terms of their quark spin content as

$$2m_N s_\mu \Delta q = \langle p, s | \bar{q} \gamma_\mu \gamma_5 q | p, s \rangle, \qquad (3.50)$$

with

$$\Delta q = \int_0^1 dx \, (q_\uparrow(x) - q_\downarrow(x)), \qquad (3.51)$$

where $q_s(x)$ is the parton distribution function carrying spin s. In terms of the light quarks we have then

$$g_A^{(3)} = \Delta u - \Delta d, \quad g_A^{(8)} = \Delta u + \Delta d - 2\Delta s, \quad g_A^{(0)} = \Delta u + \Delta d + \Delta s. \quad (3.52)$$

The first two of these are well defined from the study of hyperon beta decay,

$$g_A^{(3)} = F + D = 1.27 \pm 0.003 \qquad \text{(from neutron beta decay)},$$

$$g_A^{(8)} = 3F - D = 0.58 \pm 0.03 \qquad \text{(from semileptonic hyperon decay)}. \quad (3.53)$$

The first of these is directly measured and the second comes from an $SU(3)$ rotation from the values that are obtained in an $SU(3)$ fit to $\Delta S = 1$ hyperon decay. Such a partonic analysis leads to a decomposition, $\Delta u = 0.84 \pm 0.01 \pm 0.02$, $\Delta d = -0.43 \pm 0.01 \pm 0.02$ and $\Delta s = -0.08 \pm 0.01 \pm 0.02$, where these numbers are from recent COMPASS data [Qu 12]. The sum of these, $\Delta u + \Delta d + \Delta s \sim 0.33$, is about half of what would be expected for the nucleon spin in the naive quark model, Eq. (3.44), and of course the quark model predicts that the Δs should be zero.

However, there is reason for caution in this interpretation. The singlet axial current,

$$J_\mu^0 = \bar{u} \gamma_\mu \gamma_5 u + \bar{d} \gamma_\mu \gamma_5 d + \bar{s} \gamma_\mu \gamma_5 s, \qquad (3.54)$$

whose matrix element is said to be represented by $\Delta u + \Delta d + \Delta s$, is anomalous, as seen in Sect. III–3. This has important consequences [AlR 88, Sh 08]. While the axial currents which transform as $SU(3)$ octets have only finite multiplicative renormalization, the singlet current mixes with gluonic fields under radiative corrections. Different renormalization schemes yield different mixtures of the quark and gluon components [Sh 08]. Moreover, the quark component is not scale-independent; there is renormalization group running as a function of Q^2. Note that the other currents do not suffer from these problems. In particular, the Bjorken sum rule [Bj 66] involves the difference of the proton and neutron matrix elements, which then cancels out the isosinglet contributions, such that the first moment is independent of Q^2,

$$\int_0^1 dx \, g_1^{p-n}(x, Q^2) = \frac{1}{6} g_A^{(3)}. \tag{3.55}$$

This sum rule yields a value $g_A^{(3)} = 1.28 \pm 0.07 \pm 0.01$, which agrees well with the number $g_A^{(3)} = 1.270 \pm 0.003$ measured in neutron beta decay. The anomaly in the singlet current complicates the discussion of the quark contribution to the proton spin.

The partonic analysis of the quark spins has led to further studies. Attempts at the experimental study of the gluonic contributions has revealed only a small contribution to the nucleon spin [AiBHM 13]. There may be the possibility of studying the angular-momentum components through the concept of generalized parton distributions [Ji 94]. However, the experimental determination of these generalized parton distributions is yet to be achieved.

XII–4 Nuclear weak processes

One area in which the structure of the weak hadronic current has received a great deal of attention is that of nuclear beta decay and muon capture. Although in some sense this represents simply a nuclear modification of the basic weak transitions $n \to p + e^- + \bar{\nu}_e$, $p \to n + e^+ + \nu_e$, the use of nuclei allows specific features to be accented by the choice of levels possessing particular spins and/or parities [Ho 89]. Here, we shall confine our attention to *allowed* decays ($\Delta J = 0, \pm 1$, no parity change) and will emphasize those aspects which stress the structure of the weak current rather than that of the nucleus itself. In particular, nuclear beta decay provides the best determination of V_{ud}, while muon capture provides the only measurement of the pseudoscalar axial weak form factor predicted by chiral symmetry.

Measurement of V_{ud}

There are many occurrences in nuclei of an isotriplet of $J^P = 0^+$ states. Examples are found with $A = 10, 14, 26, 34, 42, \ldots$. Because Coulombic effects raise the mass of the proton-rich $I_3=1$ state with respect to that with $I_3=0$, the positron emission process $N_1(I_z = 1) \rightarrow N_2(I_z = 0) + e^+ + \nu_e$ can occur. These transitions are particularly clean theoretically, and this is the reason why they are important. Since the transition is $0^+ \rightarrow 0^+$, only the vector current is involved, and because of the lack of spin there can be no weak magnetic form factor. The vector-current matrix element involves but a single form factor $a(q^2)$,

$$\langle N_2(p_2)|V_\mu|N_1(p_1)\rangle = a(q^2)(p_1 + p_2)_\mu. \tag{4.1}$$

This form factor is known at $q^2 = 0$ because the charged vector weak current V_μ is just the isospin rotation of the electromagnetic current,

$$[I_-, J_{em}^\mu] = \bar{d}\gamma^\mu u. \tag{4.2}$$

This relation is often called the *conserved vector current hypothesis* or CVC, and requires for each of the $0^+ \rightarrow 0^+$ transitions,

$$a(0) = \sqrt{2}. \tag{4.3}$$

What is generally quoted for such decays is the $\mathcal{F}t_{1/2}$ *value*, essentially the half-life $t_{1/2}$ multiplied by the (kinematic) phase space factor f plus various radiative and Coulomb corrections [WiM 72]. Theoretically, one expects a universal form

$$\mathcal{F}t_{1/2} = \frac{2\pi^3 \ln 2}{G_\mu^2 m_e^5 |V_{ud}|^2 a^2(0)} \left(1 - \frac{\alpha}{2\pi}(4\ln(M_Z/m_N) + \cdots)\right), \tag{4.4}$$

which should be identical for each isotriplet transition. G_μ is the weak decay constant measured in muon decay while the logarithmic correction arises from 'hard'-photon corrections, as discussed in Chap. VII. The 'soft'-photon piece as well as finite-size and Coulombic corrections are contained in the phase space factor $\mathcal{F}$. Much careful experimental and theoretical study has been given to this problem, and the current situation is summarized in Table XII–3 where the experimental $\mathcal{F}t_{1/2}$ values are tabulated. A fit to these and additional Fermi decays produces the value $\mathcal{F}t_{1/2} = 3072.08 \pm 0.79$ s with chi-squared per degree of freedom $\chi^2/\nu = 0.28$. This excellent agreement over a wide range of Z values is evidence that soft-photon corrections are under control.

Comparison of the experimental $\mathcal{F}t_{1/2}$ value with the theoretical expression given in Eq. (4.4) yields the determination

$$V_{ud} = 0.97425(22), \tag{4.5}$$

which makes V_{ud} the most precisely measured component of the CKM matrix.

Table XII–3. *Energy release and* $\mathcal{F}t_{1/2}$ *values for*
$0^+ \rightarrow 0^+$ *Fermi decays [HaT 09].*

Nucleus	E_0(KeV)	$\mathcal{F}t_{1/2}$ (s)
^{10}C	885.87(11)	3076.7(4.6)
^{14}O	1809.24(23)	3071.5(3.3)
^{26m}Al	3210.66(06)	3072.4(1.4)
^{34}Cl	4469.64(23)	3070.2(2.1)
^{38m}K	5022.40(11)	3072.5(2.4)
^{42}Sc	5404.28(30)	3072.4(2.7)
^{46}V	6030.49(16)	3073.3(2.7)
^{50}Mn	6612.45(07)	3070.9(2.8)
^{54}Co	7222.37(28)	3069.9(3.2)

The pseudoscalar axial form factor

Chiral symmetry predicts a rather striking result for the form factor $g_3(q^2)$ of Eq. (2.9), namely that it is determined by the pion pole with a coupling fixed by the PCAC condition. One cannot detect this term in either neutron or nuclear beta decay because when the full matrix element is taken, one obtains

$$\frac{g_3}{2m_N}\overline{v}(\mathbf{p}_\nu)q_\mu\gamma^\mu(1+\gamma_5)u(\mathbf{p}_e) = \frac{g_3 m_e}{2m_N}\overline{v}(\mathbf{p}_\nu)(1-\gamma_5)u(\mathbf{p}_e), \qquad (4.6)$$

which is proportional to the electron mass and is thus too small to be seen (effects in the spectra are $\mathcal{O}(m_e^2/m_N E_e) \ll 1$). However, in the muon capture process $\mu^- p \rightarrow \nu_\mu n$, the corresponding effect is $\mathcal{O}(m_\mu/m_N) \sim 10\%$. Thus, muon capture is a feasible arena in which to study the chiral symmetry prediction [CzM 07]. The drawback in this case is that typically one has available from experiment only a *single* number, the capture rate. In order to interpret such experiments, one needs to know the value of each nuclear form factor at $q^2 \simeq -0.9 \, m_\mu^2$, which introduces some uncertainty since these quantities are determined in beta decay only at $q^2 \simeq 0$. Nevertheless, predicted and experimental capture rates are generally in good agreement provided one assumes (i) the $q^2 \simeq 0$ value of form factors from the analogous beta decay, (ii) q^2 dependence of form factors from CVC and electron scattering results, (iii) the CVC value for the weak magnetic term f_2, and (iv) the PCAC value of Eq. (3.21) for g_3. The results are summarized in Table XII–4. Obviously, agreement is good except for ^{6}Li, for which the origin of the discrepancy is unknown, although it has been speculated that perhaps the spin mixture is not statistical. Also, in the case of ^{3}He there remains a small disagreement between the elementary particle model (EPM) and impulse approximation (IA) predictions for the capture rate.

Table XII–4. *Muon capture rates.*

Reaction	Theory (10^3 s^{-1})	Experiment (10^3 s^{-1})
$\mu^- + p \to \nu_\mu + n$	0.712 ± 0.005^a	$0.715 \pm 0.005 \pm 0.005^a$
$\mu^- + {}^3\text{He} \to \nu_\mu + {}^3\text{H}$	$1.537 \pm 0.022^{\text{EPM}}$ $1.506 \pm 0.015^{\text{IA}}$	1.496 ± 0.004
$\mu^- + {}^6\text{Li} \to \nu_\mu + {}^6\text{He}$	0.98 ± 0.15	$1.60^{+0.33}_{-0.12}$
$\mu^- + {}^{12}\text{C} \to \nu_\mu + {}^{12}\text{B}$	7.01 ± 0.16	$6.75^{+0.3}_{-0.75}$

$^a S_{\mu^- p} = 0.$

Before proceeding, we should emphasize one relevant point. When PCAC is applied, it is for the nucleon

$$2m_N g_1(q^2) - \frac{q^2}{2m_N} g_3(q^2) = 2F_\pi g_{\pi NN}(q^2) \left(1 - \frac{q^2}{m_\pi^2}\right)^{-1}. \qquad (4.7)$$

Then, at $q^2 = 0$, we have

$$1.27 = g_1(0) \simeq \frac{F_\pi g_{\pi NN}(m_\pi^2)}{m_N} = 1.30, \qquad (4.8)$$

which is the Goldberger–Treiman relation. On the other hand, taking similar q^2 dependence for $g_1(q^2)$ and $g_{\pi NN}(q^2)$, we find

$$\frac{m_\mu}{2m_N} \frac{g_3(-0.9m_\mu^2)}{g_1(-0.9m_\mu^2)} = \frac{2m_N m_\mu}{m_\pi^2 + 0.9m_\mu^2} - \frac{1}{3} r_A^2 m_\mu m_N \simeq 6.45. \qquad (4.9)$$

PCAC is generally applied in nuclei in the context of a simple impulse approximation, and it is this version of PCAC which is tested by the muon capture rates listed in Table XII–4. The direct application of PCAC in nuclei cannot generally be utilized since the pion couplings are unknown.

In the case of muon capture on ^{12}C, additional experimental data are available. One class of experiment involves measurement of the polarization of the recoiling ^{12}B nucleus. Combining this measurement with that of the total capture rate yields a separate test of CVC as well as of PCAC. The results,

$$\frac{f_2^{\text{expt}}}{f_2^{CVC}} = 1.00 \pm 0.05, \qquad \frac{m_\mu}{2m_N} \frac{g_3(-0.9m_\mu^2)}{g_1(-0.9m_\mu^2)} = 8.0 \pm 3.0, \qquad (4.10)$$

are in good agreement with both symmetry assumptions.

In addition, one can measure the average and longitudinal recoil polarizations in the ^{12}C muon capture, yielding a value for the induced pseudoscalar coupling,

$$\frac{m_\mu}{2m_N} \frac{g_3(-0.9m_\mu^2)}{g_1(-0.9m_\mu^2)} = 9.0 \pm 1.7, \tag{4.11}$$

which is again in good agreement with PCAC.

The most precise value comes from the recent measurement of the singlet-muon capture rate in hydrogen, which yields

$$\frac{m_\mu}{2m_N} \frac{g_3(-0.9m_\mu^2)}{g_1(-0.9m_\mu^2)} = 5.75 \pm 0.95, \tag{4.12}$$

which is excellent agreement with PCAC.

XII–5 Hyperon semileptonic decay

The goals in studying hyperon semileptonic processes are to confirm the value of V_{us} obtained in kaon decay and to use the form factors to better understand hadronic structure. These two goals are interconnected. In earlier days when data were not very precise, fits to hyperon decays were made under the assumption of perfect $SU(3)$ invariance in order to extract V_{us}. Presently, the experiments are precise enough that exact $SU(3)$ no longer provides an acceptable fit. The desire to learn about V_{us} is thus impacted by the need to understand the $SU(3)$ breaking.

We have already described in Sect. XII–1 the physics ingredients which lead to $SU(3)$ breaking within a simple quark description. These include recoil or center-of-mass corrections, wavefunction mismatch (in which a normalization condition realized in the symmetry limit no longer holds), and generation of the axial form factor g_2. For hyperons, because of the presence of the axial current, $SU(3)$ breaking can occur in first order. This means that hyperon decays are more difficult to use for determining V_{us} than are kaon decays, where the Ademollo–Gatto theorem reduces the amount of symmetry breaking. Thus, at the moment it is probably best to use the value of V_{us} determined from kaon decay, and require that hyperon decays yield a consistent value.

The clearest evidence on $SU(3)$ breaking comes from the $\Sigma^- \rightarrow \Lambda + e^- + \bar{\nu}_e$ rate. Since this is a $\Delta S = 0$ process, V_{us} does not enter and, in addition, the vector current matrix element must vanish. Thus, the rate is determined by the axial-current contribution alone, for which the theoretical prediction is

$$g_1^{\Sigma^- \Lambda} = \rho \sqrt{\frac{2}{3}} \frac{D}{D+F} g_1^{np}, \tag{5.1}$$

where ρ is a $SU(3)$ breaking factor due to the center-of-mass effect. A bag-model estimate yields $\rho = 0.939$. Taking $\rho = 1$, the best $SU(3)$ symmetric fit to all the data [CaSW 03] would require $D/(D+F) = 0.635 \pm 0.006$, and hence $g_1^{\Sigma^- \Lambda} = 0.658$ if $SU(3)$ were exact. On the other hand, the data on $\Sigma^- \to \Lambda e \bar{v}_e$ requires $g_1^{\Sigma^- \Lambda} = 0.591 \pm 0.014$, which implies the correction $\rho = 0.931 \pm 0.022$. There seems to be no way to avoid this need for $SU(3)$ breaking.

The full pattern of $SU(3)$ breaking is more difficult to uncover. One problem is experimental. When the g_1 values are extracted from the data, they have generally been analyzed under the assumptions that the f_1 and f_2 form factors have exactly their $SU(3)$ values and that $g_2 = 0$. If these assumptions are not correct, then the values cited in [RPP 12] do not reflect the true g_1 but rather some combination of g_1, f_1, f_2, and g_2. The correlation with g_2 is particularly strong. Thus, quoted values of g_1 must be treated with caution.

The present status of these decays is reviewed in [CaSW 03]. The data can be fit well either by the center-of-mass correction described above, with $g_2 = 0$, or by the full corrections including wavefunction mismatch, with $g_2/g_1 = 0.20 \pm 0.07$ in $\Lambda \to p + e + \bar{v}_e$. Without an independent measurement of g_2 one cannot decide between these. We note, however, that either option yields a value of V_{us} consistent with that found in kaon decays,

$$V_{us} = 0.2250 \pm 0.0027. \tag{5.2}$$

XII–6 Nonleptonic decay

The dominant decays of hyperons are the nonleptonic $B \to B'\pi$ modes. Because of the spin of the baryons and the many decay modes available, the nonleptonic hyperon decays present a richer opportunity for study than do the nonleptonic kaon decays.

Phenomenology

The $B \to B'\pi$ matrix elements can be written in the form

$$\mathcal{M}_{B \to B'\pi} = \bar{u}(\mathbf{p}') \left[A + B\gamma_5 \right] u(\mathbf{p}), \tag{6.1}$$

with parity-violating (A) and parity-conserving (B) amplitudes. Watson's theorem implies that if CP is conserved, the phase of these amplitudes is given by the strong $B'\pi$ scattering phase shifts in the final-state S wave (for A) or P wave (for B), i.e.,

$$A = A_0 \exp\left(i\delta^S_{B'\pi}\right), \qquad B = B_0 \exp\left(i\delta^P_{B'\pi}\right), \tag{6.2}$$

with A_0, B_0 real. Aside from the πN system, these phase shifts are not known precisely, but are estimated to be $\simeq 10°$ in magnitude. The decay rate is expressed in terms of the partial wave amplitudes by

$$\Gamma_{B \to B'\pi} = \frac{|\mathbf{q}|(E' + m_{B'})}{4\pi m_B} \left(|A|^2 + |\bar{B}|^2\right), \tag{6.3}$$

where $\mathbf{q}$ is the pion momentum in the parent rest frame and we define $\bar{B} \equiv (E' - m_{B'}/E' + m_{B'})^{1/2} B$. Additional observables are the decay distribution $W(\theta)$,

$$W(\theta) = 1 + \alpha \mathbf{P}_B \cdot \hat{\mathbf{p}}_{B'}, \qquad \alpha = \frac{2\mathrm{Re}\,(A^* \bar{B})}{|A|^2 + |\bar{B}|^2}, \tag{6.4}$$

and the polarization $\langle \mathbf{P}_{B'} \rangle$ of the final-state baryon,

$$\langle \mathbf{P}_{B'} \rangle = \frac{\left(\alpha + \mathbf{P}_B \cdot \hat{\mathbf{p}}_{B'}\right) \hat{\mathbf{p}}_{B'} + \beta \left(\mathbf{P}_B \times \hat{\mathbf{p}}_{B'}\right) + \gamma \left[\hat{\mathbf{p}}_{B'} \times \left(\mathbf{P}_B \times \hat{\mathbf{p}}_{B'}\right)\right]}{W(\theta)},$$

$$\beta = \frac{2\mathrm{Im}(A^* \bar{B})}{|A|^2 + |\bar{B}|^2}, \qquad \gamma = \frac{|A|^2 - |\bar{B}|^2}{|A|^2 + |\bar{B}|^2} = \pm\sqrt{1 - \alpha^2 - \beta^2}, \tag{6.5}$$

where $\mathbf{P}_B$ is the polarization of B and $\hat{\mathbf{p}}_{B'}$ is a unit vector in the direction of motion of B'. Experimental studies of these distributions lead to the amplitudes listed in Table XII–5.

The nonleptonic amplitudes may be decomposed into isospin components in a notation where superscripts refer to $\Delta I = 1/2, 3/2$,

$$
\begin{aligned}
A_{\Lambda \to p\pi^-} &= \sqrt{2} A_\Lambda^{(1)} - A_\Lambda^{(3)}, & A_{\Sigma^- \to n\pi^-} &= A_\Sigma^{(1)} + A_\Sigma^{(3)}, \\
A_{\Lambda \to n\pi^0} &= -A_\Lambda^{(1)} - \sqrt{2} A_\Lambda^{(3)}, & A_{\Sigma^+ \to n\pi^+} &= \frac{1}{3} A_\Sigma^{(1)} - \frac{2}{3} A_\Sigma^{(3)} + X_\Sigma, \\
A_{\Xi^0 \to \Lambda\pi^0} &= -A_\Xi^{(1)} - \sqrt{2} A_\Xi^{(3)}, & & \\
A_{\Xi^- \to \Lambda\pi^-} &= \sqrt{2} A_\Xi^{(1)} - A_\Xi^{(3)}, & \sqrt{2} A_{\Sigma^+ \to p\pi^0} &= -\frac{2}{3} A_\Sigma^{(1)} + \frac{4}{3} A_\Sigma^{(3)} + X_\Sigma,
\end{aligned}
\tag{6.6}
$$

and X_Σ is of mixed symmetry. Similar relations hold for the B amplitudes. From the entries in Table XII–5 it is not hard to see that the $\Delta I = 1/2$ rule, described previously for kaon decays, is also present here. Table XII–6 illustrates that the dominance of $\Delta I = 1/2$ amplitudes compared to those with $\Delta I = 3/2$ holds in the six possible tests in S-wave and P-wave hyperon decay, at about the same level (several per cent) as occurs in kaon decay.[9] Thus, the $\Delta I = 1/2$ rule is not an accident of kaon physics, but is rather a universal feature of nonleptonic decays. This makes the failure to clearly understand it all the more frustrating.

---

[9] For P waves, the observed smallness of $B_{\Sigma^- \to n\pi^-}$ indicates that $B_\Sigma^{(1)}$ is small, presumably accidentally so. In this case the measure of $\Delta I = 3/2$ to $\Delta I = 1/2$ effects is given by $B_\Sigma^{(3)}/X_\Sigma$.

Table XII–5. *Hyperon decay amplitudes.[a]*

Mode	A amplitudes		B amplitudes	
	Expt.	Thy.[b]	Expt.	Thy.
$\Lambda \to p\pi^-$	3.25	3.38	22.1	23.0
$\Lambda \to n\pi^0$	−2.37	−2.39	−15.8	−16.0
$\Sigma^+ \to n\pi^+$	0.13	0.00	42.2	4.3
$\Sigma^+ \to p\pi^0$	−3.27	−3.18	26.6	10.0
$\Sigma^- \to n\pi^-$	4.27	4.50	−1.44	−10.0
$\Xi^0 \to \Lambda\pi^0$	3.43	3.14	−12.3	3.3
$\Xi^- \to \Lambda\pi^-$	−4.51	−4.45	16.6	−4.7

[a]In units of 10^{-7}.
[b]Lowest-order chiral fit.

The assumption that the dominant $\Delta I = 1/2$ hamiltonian is a member of an $SU(3)$ octet leads to an additional formula, called the Lee-Sugawara relation,

$$\sqrt{3}\, A_{\Sigma^+ \to p\pi^0} = 2A_{\Xi^- \to \Lambda\pi^-} + A_{\Lambda \to p\pi^-}, \tag{6.7}$$

which also is well satisfied by the data. In this case, the corresponding formula for the B amplitudes is *not* a symmetry prediction [MaRR 69], although for unknown reasons it is in qualitative accord there also.

Lowest-order chiral analysis

Chiral symmetry provides a description of hyperon nonleptonic decay, which is of mixed success when truncated at lowest order in the energy expansion. Given our comments on the convergence of the energy expansion for baryons made in Sect. XII–3, the need for corrections to the lowest-order results is not surprising. We shall present the lowest-order analysis here, as it forms the starting point for most theoretical analyses.

Recalling from Sect. IV–7 the procedure for adding baryons to the chiral analysis, one finds that the two following nonderivative lagrangians have the chiral $(8_L, 1_R)$ transformation property:

Table XII–6. *Ratio of $\Delta I = 3/2, 1/2$ amplitudes.*

	S wave	P wave
Λ	0.014	0.006
Σ	−0.017	−0.047
Ξ	0.034	0.023

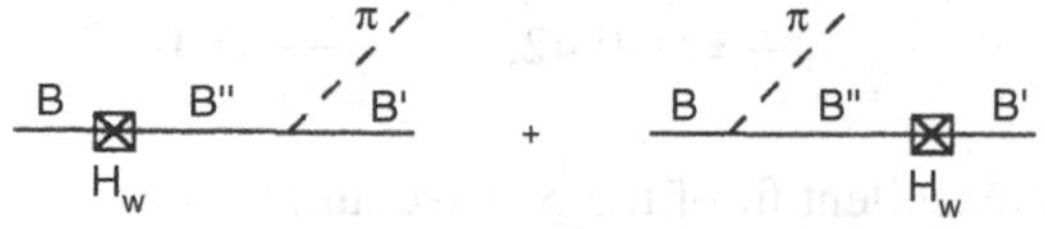

Fig. XII–1 P-wave hyperon decay amplitudes.

$$\mathcal{L}_W^{(S)} = D \operatorname{Tr} \left(\bar{B} \{ \xi^\dagger \lambda_6 \xi, B \} \right) + F \operatorname{Tr} \left(\bar{B} \left[\xi^\dagger \lambda_6 \xi, B \right] \right),$$
$$\mathcal{L}_W^{(P)} = D_5 \operatorname{Tr} \left(\bar{B} \gamma_5 \{ \xi^\dagger \lambda_6 \xi, B \} \right) + F_5 \operatorname{Tr} \left(\bar{B} \gamma_5 \left[\xi^\dagger \lambda_6 \xi, B \right] \right), \qquad (6.8)$$

where ξ, B are defined in Eqs. (3.1), (3.2) respectively. However, the operator $\mathcal{L}_W^{(P)}$ must vanish, as it has the wrong transformation property under CP [LeS 64]. That is, a CP transformation implies

$$B \to \left(i \gamma_2 \bar{B} \right)^T, \qquad \xi \to \left(\xi^\dagger \right)^T, \qquad (6.9)$$

and including the anticommutation of B and $\bar{B}$, $\mathcal{L}_W^{(S)}$ is seen to return to itself, but $\mathcal{L}_W^{(P)}$ changes sign and hence must vanish. This leaves $\mathcal{L}_W^{(S)}$ as the only allowed chiral lagrangian at lowest order. Observe that $\mathcal{L}_W^{(S)}$ lacks a γ_5 factor. Thus, its $B \to B'\pi$ matrix elements will be parity-violating, leading to only A amplitudes. The parity-conserving B amplitudes in $B \to B'\pi$ are produced through pole diagrams as in Fig. XII–1, and are proportional to the parity-conserving $B \to B'$ matrix elements of $\mathcal{L}_W^{(S)}$.

The counting of powers of energy (momentum transfer) in the energy expansion goes as follows. Both the $B \to B'$ transition and the A amplitudes in $B \to B'\pi$ are obtained as matrix elements of $\mathcal{L}_W^{(S)}$, which is zeroth order in the energy. The pole diagrams are likewise of zeroth order in the energy, being the product of the $\mathcal{L}_W^{(S)}$ vertex ($\mathcal{O}(1)$), a baryon propagator ($\mathcal{O}(q^{-1})$) and an $NN\pi$ vertex ($\mathcal{O}(q)$). Since the kinematic part of the pole diagrams, $\bar{u}'\gamma_5 u \sim \sigma \cdot \mathbf{q}$, is of first order in q, the B amplitudes themselves are of order $B \sim q^{-1} \sim 1/\Delta m$ for the baryon pole. Kaon poles and higher-order chiral lagrangians enter at next order, i.e., having one power of the momentum transfer.

The lowest-order chiral $SU(3)$ analysis provides a fit to the data in terms of two parameters, called F and D,

$$iA_{\Lambda \to n\pi^0} = -\frac{1}{2F_\pi}(3F + D), \qquad iA_{\Sigma^+ \to p\pi^0} = \frac{\sqrt{6}}{2F_\pi}(D - F),$$
$$iA_{\Sigma^+ \to n\pi^+} = 0, \qquad\qquad iA_{\Xi^0 \to \Lambda\pi^0} = \frac{1}{2F_\pi}(3F - D), \qquad (6.10)$$

with other amplitudes being predicted by the $\Delta I = 1/2$ rule. Use of the numerical values

$$\frac{D}{F} = -0.42, \qquad \frac{F}{2F_\pi} = 0.92 \times 10^{-7}, \tag{6.11}$$

leads to the excellent fit of the S-wave amplitudes seen in Table XII–5. Note that this form has one less free parameter than the general $SU(3)$ structure [MaRR 69]. Thus, the prediction of chiral symmetry that $A_{\Sigma^+ \to n\pi^+} \simeq 0$ is independent of the D/F ratio (up to $\Delta I = 3/2$ effects), and represents a successful explanation of the smallness of this amplitude.

In principle, the A amplitudes, together with the strong $BB'\pi$ vertices, determine the baryon pole contributions to the B amplitudes. These are then parametrized by the same d/f ratio as in the axial current,[10] e.g.,

$$\mathcal{M}_{\Sigma^+ \to \Sigma^+ \pi^0} = \frac{2f}{F_\pi} \bar{u} \gamma_\mu \gamma_5 u \, q^\mu = \frac{g_A^{\Sigma^+ \Sigma^+}}{F_\pi} \bar{u} \gamma_\mu \gamma_5 u \, q^\mu$$

$$= \frac{2g_{\pi NN}}{2m_N} \frac{f}{f+d} \bar{u} \gamma_\mu \gamma_5 u \, q^\mu . \tag{6.12}$$

Using this parameterization for the pole diagrams, one finds contributions such as

$$B_{\Sigma^+ \to p\pi^0} = -\frac{m_N + m_\Sigma}{2m_N F_\pi} \cdot \frac{(d-f)\mathcal{M}_{\Sigma^+ p}}{m_\Sigma - m_N}. \tag{6.13}$$

Taking $d + f = 1.27$, $d/f = 1.8$, one obtains from relations like this the disappointing P-wave predictions quoted in Table XII–5. This failure to simultaneously fit the S waves and P waves is a deficiency of the lowest-order chiral analysis. Perhaps this is not too surprising, as the chiral expansion converges slower in baryons than in mesons [BoH 99]. At the next order in the energy expansion, the chiral analysis contains enough free parameters to accommodate the data, but is not predictive. Lattice studies are just beginning to explore this topic [BeBPS 05].

Problems

(1) **The axial-vector coupling**

Consider the effective lagrangian in Eq. (3.1) for nucleons and pions. For combined left-handed and right-handed transformations of the fields, we have

$$U \to LUR^\dagger, \quad \xi \to L\xi V^\dagger = V\xi R^\dagger, \quad N \to VN,$$

where $L[R]$ are the spacetime independent $SU(2)$ matrices corresponding to global transformations in $SU(2)_L[SU(2)_R]$ and $V = V(\pi(x))$ is an $SU(2)$

---

[10] This statement is the $SU(3)$ generalization of the Goldberger–Treiman relation, Eqs. (3.20), (3.24).

matrix describing a vectorial transformation of the nucleons. For the lagrangian of Eq. (3.1), use Noether's theorem to generate the $SU(2)$ axial-vector current,

$$A_\mu^j = \frac{g_A}{4} \bar{\psi} \gamma_\mu \gamma_5 \left(\xi \lambda^j \xi^\dagger + \xi^\dagger \lambda^j \xi \right) \psi,$$

where ξ is the 'square root' of U (cf. Eq. (3.1)), and thereby show that the axial-vector coupling constant for beta decay is given by $g_1 = g_A$.

(2) **CP violation and nonleptonic hyperon decay**

Although the $\Delta S = 1$ hamiltonian of the Standard Model contains a *CP*-violating component, there is no *practical* way to see this in any single hyperon decay mode. Rather, one must compare the decays of hyperons with those of antihyperons [DoHP 86]. In the presence of *CP* violation, there are two sources of phases in the weak matrix elements, e.g., for the Λ decay modes,

$$A_{\Lambda \to p\pi^-} = A_1 \, e^{i\varphi_1^S} \, e^{i\delta_1^S} + A_3 \, e^{i\varphi_3^S} \, e^{i\delta_3^S},$$

$$B_{\Lambda \to p\pi^-} = B_1 \, e^{i\varphi_1^P} \, e^{i\delta_1^P} + B_3 \, e^{i\varphi_3^P} \, e^{i\delta_3^P},$$

where the isospin (I) subscripts '1, 3' stand for $\Delta I = 1/2, 3/2$, the angular momentum (J) superscripts 'S, P' stand for S waves or P waves, A_I are real amplitudes, δ_I^J are strong final-state phases, and φ_I^J are the weak *CP*-violating phases. Observe that there are three small parameters in these amplitudes – the weak phases φ_I^J, the strong phases $\delta_I^J \simeq 10°$, and the ratio of $\Delta I = 3/2$ to $\Delta I = 1/2$ effects. To leading order in these quantities, show that one has the *CP*-odd observables,

$$\frac{\beta + \bar{\beta}}{\alpha - \bar{\alpha}} = \sin\left(\varphi_1^S - \varphi_1^P\right), \quad \frac{\alpha + \bar{\alpha}}{\alpha - \bar{\alpha}} = -\sin\left(\varphi_1^S - \varphi_1^P\right)\sin\left(\delta_1^S - \delta_1^P\right),$$

$$\frac{\Gamma_{p\pi^-} - \overline{\Gamma}_{\bar{p}\pi^+}}{\Gamma_{p\pi^-} + \overline{\Gamma}_{\bar{p}\pi^+}} = -2\frac{A_1 A_3 \sin(\delta_1^S - \delta_3^S)\sin(\varphi_1^S - \varphi_3^S)}{|A_1|^2 + |\bar{B}_1|^2}$$

$$- 2\frac{\bar{B}_1 \bar{B}_3 \sin(\delta_1^P - \delta_3^P)\sin(\varphi_1^P - \varphi_3^P)}{|A_1|^2 + |\bar{B}_1|^2}.$$

A hierarchy is apparent in these three signals. The $\beta + \bar{\beta}$ asymmetry requires only the weak phase, the $\alpha + \bar{\alpha}$ asymmetry requires both the weak and final-state phases, while $\Gamma - \overline{\Gamma}$ has both phases plus a $\Delta I = 3/2$ suppression. Present experiments are not sufficiently sensitive to test for *CP* violation in these observables at the required accuracy.

XIII

Hadron spectroscopy

Studies of hadron masses, and of both strong and electromagnetic decays of hadrons, provide insights regarding QCD dynamics over a variety of distance scales. Among various possible theoretical approaches, the potential model has most heavily been employed in this area. We shall start our discussion by considering heavy-quark bound states, which begin to approximate truly nonrelativistic systems and for which the potential model is expected to provide a suitable basis for discussion.

XIII–1 The charmonium and bottomonium systems

Quarkonium is the bound state of a heavy quark Q with its antiparticle. Two such systems, charmonium ($c\bar{c}$) and bottomonium ($b\bar{b}$) have been the subject of much experimental and theoretical study; a comprehensive overview is provided by [Br *et al.* 11]. Due to weak decay of the top quark, the $t\bar{t}$ system has rather different properties from these, and thus constitutes a special case (cf. Sect. XIV–2).

Since the quarkonium systems are quark–antiquark composites, we shall employ the set of quantum numbers n, L, S, J introduced in Sect. XI–2. One generally refers to the individual quarkonium levels with the nomenclature of Table XIII–1,

Table XIII–1. *Nomenclature for S-wave and P-wave states in the $c\bar{c}$ and $b\bar{b}$ systems.*

L	S	Charmonium	Bottomonium
0	1	$\psi(nS)^a$	$\Upsilon(nS)$
	0	$\eta_c(nS)$	$\eta_b(nS)$
1	1	$\chi_{cJ}(nP)$	$\chi_{bJ}(nP)$
	0	$h_c(nP)$	$h_b(nP)$

[a] For historical reasons, the spin-one charmonium ground state is called J/ψ.

366

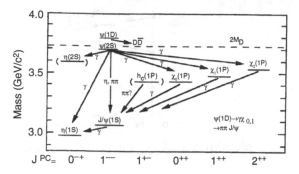

Fig. XIII–1 The low-lying spectrum of charmonium.

although the nL identification is sometimes replaced by either the degree of excitation or the mass, e.g., $\psi(2S)$ is called ψ' or $\psi(3686)$. The $n^{2S+1}L_J$ spectroscopic notation is also invoked on occasion.

Figs. XIII–1,2 give a summary of the lightest observed $c\bar{c}$ and $b\bar{b}$ states. Most of these states, as well as their transitions have been detected both in the charmonium and bottomonium systems [Br *et al.* 11]. The largest set of observed excitations comes from the $\psi(nS)$ and $\Upsilon(nS)$ radial towers, reaching up to $n = 6$ for the Υ system. Excitation energies are relatively small on the scale of the bottomonium reduced mass $\mu_b \simeq 2.5$ GeV, but not that of charmonium $\mu_c \simeq 0.8$ GeV.

Phenomenological potentials: Historically, the success of potential models in charmonium was of major importance in convincing the community that quarks were simple dynamical objects and that QCD provides a manageable theory of the strong interactions. Because of this success, we describe the states by the spectroscopic classification of nonrelativistic quantum mechanics. Thus, quarkonium mass values are often expressed as

$$m_{[nLSJ]} = 2M_Q + E_{[nLSJ]}, \tag{1.1}$$

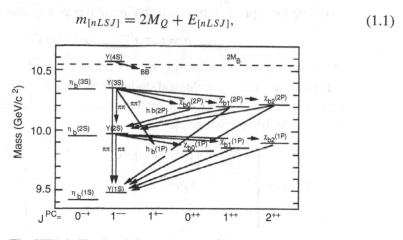

Fig. XIII–2 The low-lying spectrum of bottomonium.

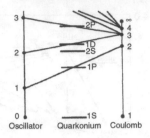

Fig. XIII–3 Energy levels of various potential functions.

where $E_{[nLSJ]}$ is obtained by solving the Schrodinger equation for a particle of reduced mass $\mu_Q = M_Q/2$ moving in the field of an assumed potential energy function. The shape of the potential is chosen via a combination of theoretical and phenomenological considerations.

The spectra of quarkonium states already hints at the radial dependence of the $Q\bar{Q}$ potential, with the progression in nL levels suggesting an interaction which lies 'between' Coulomb and harmonic oscillator potentials, as depicted in Fig. XIII–3. Conceptually, the simplest potential that matches QCD to this behavior is

$$V(r) = br - \frac{a}{r} + V_0, \tag{1.2}$$

where a, b, V_0 are constants and the color dependence between quark and antiquark is that in Eq. (XI–2.4). The Coulomb-like $1/r$ component is designed to reproduce one-gluon exchange at short distance. The confining linear 'br' term models a color-flux tube of constant energy density, as noted in Sect. XI–2. The coefficient b is commonly described in the literature as the *string tension*, in reference to the string model of hadrons, and its value is estimated from a string model relation involving the typical slope α' of a hadronic Regge trajectory (cf. Table XIII–2),

$$b = (2\pi\alpha')^{-1} \simeq 0.18 \text{ GeV}^2. \tag{1.3}$$

This is equivalent to a restoring force of about 16 tons!

In practice, phenomenological studies of quarkonium can be carried out by adopting the potential of Eq. (1.2) or another assumed potential energy functions. Examples include the following, e.g.,[1]

$$V(r) = \begin{cases} -\frac{64\pi^2}{27}\mathcal{F}\left\{\left[q^2 \ln\left(1 + (q^2/\Lambda^2)\right)\right]^{-1}\right\} & \{\Lambda \simeq 0.4 \text{ GeV}\}, \\ br - a/r & \left\{\begin{matrix} b \simeq 0.18 \text{ GeV}^2 \\ a \simeq 0.52 \end{matrix}\right\}, \\ cr^d & \left\{\begin{matrix} c \simeq 6.87 \text{ GeV} \\ d \simeq 0.1 \end{matrix}\right\}, \end{cases} \tag{1.4}$$

---

[1] The second and third potentials provide fits only up to an additive constant.

Table XIII–2. *Regge trajectories.*

Trajectory	N	Slope[a]	J-intercept
N	3	0.99	−0.34
Δ	3	0.92	0.07
Λ	3	0.94	−0.64
Σ	3	1.1	−1.2
Σ^*	2	0.91	−0.24
π	3	0.72	−0.05
ρ	4	0.84	0.54
K	4	0.69	−0.22
K^*	4	0.86	0.29

[a]In units of GeV^{-2}

where $\mathcal{F}\{\dots\}$ denotes a Fourier transform. The first two of the potentials in Eq. (1.4) are commonly called the 'Richardson' [Ri 79] and 'Cornell' [EiGKLY 80] potentials, respectively. They are constructed to mimic QCD by exhibiting a linear confining potential at long distances and single gluon exchange at short distances. The Richardson potential even incorporates the asymptotic freedom property for the strong interaction coupling. The third is a power-law potential [Ma 81] which, although not motivated by QCD, can be of use in analytical work or in obtaining simple scaling laws. The power-law potential also serves as a reminder of how alternative forms can achieve a reasonable success in fitting $b\bar{b}$ and $c\bar{c}$ spectra, which, after all, are primarily sensitive to the limited length scale $0.25 \le r(\text{fm}) \le 1$.

From the viewpoint of phenomenology, it is ultimately more useful to appreciate the general features of the $Q\bar{Q}$ static potential than to dwell on the relative virtues and shortcomings of individual models.

Effective field theories: The full theory of QCD is richer than can be captured in a single potential function. Gluon degrees of freedom can be dynamically active, and field-theoretic corrections introduce subtle modification to masses and couplings. Effective field theory techniques provide a modern way of understanding both the perturbative and nonperturbative properties of heavy-quark systems.

There are various scales associated with quarkonium systems. The heavy-quark mass sets a hard scale. Degrees of freedom associated with this scale may be treated perturbatively. Scales connected to the momentum transfer in the bound state, $p \sim mv$, are related to the typical spatial extent, $\langle r \rangle$, of the bound state. The time scales involved for quarkonium dynamics are related to the nonrelativistic kinetic energy $E \sim mv^2/2$. For large quark mass, the velocity, typically of order

$v^2 \sim 0.1 \rightarrow 0.3$, can be treated as a small parameter such that each of these scales is technically distinct, with

$$m_Q \gg m_Q v \gg m_Q v^2. \tag{1.5}$$

Different versions of effective field theories can be invoked to treat the different scales [BrPSV 05].

In Non-Relativistic *QCD*, abbreviated as *NRQCD*, degrees of freedom of order m_Q are integrated out from the theory [CaL 86]. This leaves the light degrees of freedom being the full set of particles of *QCD*. The gluons (and light quarks) are included dynamically, but are treated with an ultraviolet cut-off of order m_Q because their high-momentum components have been integrated out.[2] The heavy quark itself is treated nonrelativistically. Because the hard modes have been integrated out, there appear higher-order gauge-invariant interactions with Wilson coefficients that parameterize the strength of the new terms. The effective lagrangian then starts out as

$$\mathcal{L} = \mathcal{L}_{0G} + \mathcal{L}_{0Q} + \mathcal{L}_{(\text{h.o.})Q} \tag{1.6}$$

where $\mathcal{L}_{0G}$ is the usual lagrangian for gluons and $\mathcal{L}_{0Q}$ is the lowest-order lagrangian for nonrelativistic quarks

$$\mathcal{L}_{0Q} = \psi^\dagger \left[i D_0 + \frac{C_k}{2m_Q} \mathbf{D}^2 \right] \psi \tag{1.7}$$

where D_0, $\mathbf{D}$ give the coupling of the heavy quarks to gluons. To lowest order in the both the *QCD* coupling constant and in the heavy-quark expansion one can set the Wilson coefficient $C_k = 1$, but perturbative corrections lead to different definitions of the heavy-quark mass (see Sect. XIV–1) and C_k can account for matching onto these definitions. Operators that are higher order in the $1/m_Q$ expansion also emerge. Examples are

$$\mathcal{L}_{(\text{h.o.})Q} = \psi^\dagger \left[\frac{C_4}{8m_Q^3} \mathbf{D}^4 + \frac{g_3 C_G}{2m_Q} \boldsymbol{\sigma} \cdot \mathbf{B} \right] \psi + \frac{C_0}{m_Q^2} \psi^\dagger \psi \psi^\dagger \psi + \cdots \tag{1.8}$$

The first two terms here describe higher-order interactions with gluons, while the last term is a contact interaction which mimics the effect of a potential. In effective field theory, the contact interaction is appropriate because it comes from the higher-momentum modes above the scale $p \sim mv$. The gluonic Coulomb interaction is still treated perturbatively. There will be further contact interactions for different spin and color combinations.

---

[2] See the discussion of Sec. IV–7.

Because this effective theory includes gluons, there are perturbative corrections also to the heavy-quark mass. These are discussed more fully in Sect. XIV–1. For the purposes here, we will note that the definition of the quark mass is tied up with an overall energy shift in the potential, previously denoted by V_0 in Eq. (1.2). Definitions of the mass which are perturbatively well-behaved are those that are tied to physical thresholds [HoSSW 98]. Effectively, this absorbs V_0 into the definition of the quark mass within some specific prescription. Because this prescription may vary, the appropriate kinetic energy mass in Eq. (1.7) may have a different value, leading to $C_k \neq 1$.

One can go further and integrate out degrees of freedom between $p \sim mv$ and $E \sim mv^2$. Since these modes are below the spatial scale of the bound state, contact interactions are no longer appropriate, but they must be replaced by a spatially dependent potential [PiS 98]. Such an effective field theory is labeled *pNRQCD*, with the '*p*' referring to the potential. This starts to make closer contact with the phenomenological potential models. However, it remains a field theory and there are controlled perturbative modifications from the so-called '*ultra-soft*' modes which remain dynamical at this scale [HoS 03].

The effective field theory treatments put many of the early successes of phenomenological potential models onto a firmer footing. Moreover, they have also been successful at helping to connect lattice calculations to the phenomenology of quarkonium.

Lattice studies: Lattice-gauge theory is well suited to the exploration of the heavy-quark potential [DeD 10, GaL 10, Ro 12]. In the heavy-quark limit, the quarks become static and their interaction energy can be measured by numerical methods. Such studies confirm the general picture of a 'Coulomb plus linear' interaction. However, the lattice calculations can also provide the connection between the physical values of the parameters to the underlying scale of *QCD*, Λ_{QCD}.

In general, the static interaction can be described by a function

$$V(r) = -\int \frac{d^3q}{(2\pi)^3} e^{i\mathbf{q}\cdot\mathbf{r}} \frac{a\left(q^2\right)}{q^2} \qquad (1.9)$$

At large q, the coefficient $a(q)$ is determined by the perturbative expansion of *QCD*, which has now been accomplished to three-loop order [AnKS 10]. Numerical studies must then match on to the perturbative results at short distance, and this can be accomplished.[3] In doing so, the residual interactions can be mapped onto the operators of *NRQCD* and/or the potential of *pNRQCD*. While the state of the art continues to advance, the present connection between theory and experiment in the quarkonium spectrum is impressive [Br *et al.* 11].

---

[3] See, e.g., [Le 98]

Transitions in quarkonium

All quarkonium states are unstable. Among the decay mechanisms are annihilation processes, hadronic transitions, and radiative transitions. Roughly speaking, the lightest quarkonium states are relatively narrow, but those lying above the *heavy-flavor threshold*, defined as twice the mass of the lightest heavy-flavored meson and depicted by dashed lines in Figs. XIII–1 and XIII–2, are broader. This pattern is particularly apparent for the $^3S_1$ states – below the heavy-flavor threshold, widths are typically tens of keV, whereas above, they are tens of MeV. The primary reason for this difference is that above the heavy-flavor threshold, quarkonium can rapidly 'fall apart' into a pair of heavy-flavored mesons, e.g., $\Upsilon[4S] \to B\bar{B}$, whereas below, this mode is kinematically forbidden.

In the following, we shall describe only decays which occur beneath the heavy-flavor threshold, and shall limit our discussion to annihilation processes and hadronic decays. Radiative electric and magnetic dipole transitions are adequately described in quantum mechanics textbooks.

Annihilation transitions: To motivate a procedure for computing annihilation rates in quarkonium, let us consider the simple case of a charged lepton of mass m moving nonrelativistically with its antiparticle in a $^1S_0$ state, and undergoing a transition to a two-photon final state.[4] First, we write down the invariant amplitude for the pair annihilation process,

$$\mathcal{M} = -ie^2\bar{v}(\mathbf{p}_+, \lambda_+)\left[\not{\epsilon}_2^*\frac{i}{\not{p}_- - \not{q}_1 - m}\not{\epsilon}_1^* + \not{\epsilon}_1^*\frac{i}{\not{p}_- - \not{q}_2 - m}\not{\epsilon}_2^*\right]u(\mathbf{p}_-, \lambda_-),$$

(1.10)

for momentum eigenstates. In the lepton rest frame, we are free to choose the *transverse gauge* $\epsilon_1^* \cdot p_- = \epsilon_2^* \cdot p_- = 0$, i.e. $\epsilon_{1,2}^0 = 0$. Since $^3S_1$ states can make no contribution to the two-photon mode, we can compute the squared-amplitude for a $^1S_0$ transition by summing over initial state spins,

$$\sum_{\lambda_\pm} |\mathcal{M}|^2 = \frac{e^4}{2m^2}\left[2 + \frac{\omega_1}{\omega_2} + \frac{\omega_2}{\omega_1} - 4\left(\epsilon_1^* \cdot \epsilon_2^*\right)^2\right],$$

(1.11)

where $\omega_{1,2}$ are the photon energies in the lepton rest frame. Near threshold, the photons emerge back to back, and the differential cross section is found to be

$$\frac{d\sigma}{d\Omega} = \frac{\alpha^2}{2m^2 v_+}\left(1 - \left(\epsilon_1^* \cdot \epsilon_2^*\right)^2\right).$$

(1.12)

Likewise, near threshold, a sum on photon polarizations gives

---

[4] The $^1S_0$ ($^3S_1$) states have even (odd) charge conjugation, and can therefore give rise to even (odd) numbers of photons in an annihilation process.

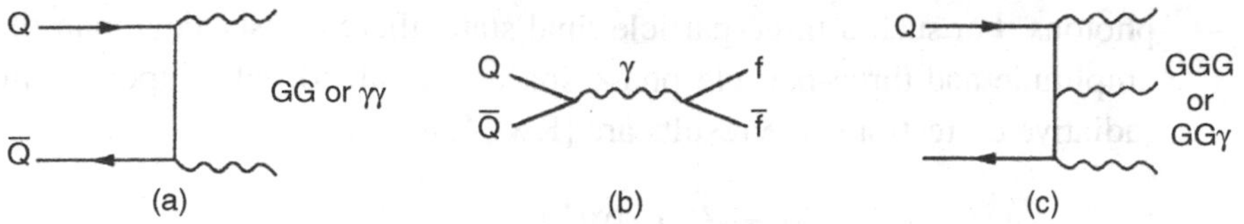

(a) (b) (c)

Fig. XIII–4 Decay of quarkonium through annihilation.

$$\sum_{\sigma_{1,2}} \left(1 - \left(\boldsymbol{\epsilon}_1^* \cdot \boldsymbol{\epsilon}_2^*\right)^2\right)_{\text{thr}} = 2, \qquad (1.13)$$

and upon integrating over half the solid angle (due to photon indistinguishability) we obtain the cross section,

$$\sigma = \frac{4\alpha^2 \pi}{m^2 v_+}. \qquad (1.14)$$

This is the transition rate per incident flux of antileptons. Since the flux is just the antilepton velocity v_+ times a unit lepton density, we interpret $v_+ \bar{\sigma}$ as the transition rate for a density of *one* lepton per volume. For a bound state with radial quantum number n and wavefunction $\Psi_n(\mathbf{x})$, the density is $|\Psi_n(0)|^2$ and the lowest-order expression for the electromagnetic decay rate $\Gamma_{\gamma\gamma}^{(\text{em})}[^1S_0]$ becomes

$$\Gamma_{\gamma\gamma}^{(\text{em})}\left[^1S_0\right] = v_+\bar{\sigma}|\Psi_n(0)|^2 = \frac{4\pi\alpha^2}{m^2}|\Psi_n(0)|^2. \qquad (1.15)$$

The corresponding rate for $\gamma\gamma$ emission from $^1S_0$ states of the $b\bar{b}$ (Υ) system is obtained from Eq. (1.15) by including a factor $e_b^4 = 1/81$, which accounts for the b-quark charge, and a color factor of three. Determination of the two-gluon emission is found similarly (cf. Fig. XIII–4(a)) provided the gluons are taken to be massless free particles, and is left for a problem at the end of the chapter. Including also the effects of *QCD* radiative corrections, referred to a common renormalization point $\mu_R = m_b$, we have [KwQR 87]

$$\Gamma_{\Upsilon\rightarrow\gamma\gamma}\left[n^1S_0\right] = \frac{48\pi\alpha^2|\Psi_n(0)|^2}{81(2m_b)^2}\left[1 - 3.4\frac{\alpha_s(m_b)}{\pi}\right],$$

$$\Gamma_{\Upsilon\rightarrow gg}\left[n^1S_0\right] = \frac{32\pi\alpha_s^2(m_b)|\Psi_n(0)|^2}{3(2m_b)^2}\left[1 + 4.4\frac{\alpha_s(m_b)}{\pi}\right]. \qquad (1.16)$$

Decays can also occur from the n^3S_1 states.[5] The single-photon intermediate state of Fig. XIII–4(b) leads to emission of a lepton pair, whereas Fig. XIII–4(c) describes final states consisting of three gluons, two gluons and a photon, or three

---

[5] There are annihilations from higher partial waves as well. These involve derivatives of the wavefunction at the origin.

photons. For such a three-particle final state, there are six Feynman diagrams per amplitude and three-particle phase space to contend with. Upon including *QCD* radiative corrections, the results are [KwQR 87]

$$\Gamma_{\Upsilon \to \ell\bar{\ell}}[n^3S_1] = \frac{16\pi\alpha^2|\Psi_n(0)|^2}{9(2m_b)^2}\left[1 - \frac{16}{3}\frac{\alpha_s(m_b)}{\pi}\right],$$

$$\Gamma_{\Upsilon \to 3g}[n^3S_1] = \frac{160\left(\pi^2 - 9\right)\alpha_s^3(m_b)|\Psi_n(0)|^2}{81(2m_b)^2}\left[1 - 4.9\frac{\alpha_s(m_b)}{\pi}\right],$$

$$\Gamma_{\Upsilon \to 3\gamma}[n^3S_1] = \frac{64\left(\pi^2 - 9\right)\alpha^3|\Psi_n(0)|^2}{2187(2m_b)^2}\left[1 - 12.6\frac{\alpha_s(m_b)}{\pi}\right],$$

$$\Gamma_{\Upsilon \to gg\gamma}[n^3S_1] = \frac{128\left(\pi^2 - 9\right)\alpha\alpha_s^2(m_b)|\Psi_n(0)|^2}{81(2m_b)^2}\left[1 - 1.7\frac{\alpha_s(m_b)}{\pi}\right]. \quad (1.17)$$

The *QCD* contributions in Eq. (1.17) are of interest in several respects. They contribute, on the whole, with rather sizeable coefficients and can substantially affect the annihilation rates. Also, they have come to be used as one of several standard inputs for phenomenological determinations of α_s. To eliminate the model-dependent factors $|\Psi_n(0)|^2$, one works with ratios of annihilation rates,

$$\frac{\Gamma_{\Upsilon \to gg\gamma}\left[n^3S_1\right]}{\Gamma_{\Upsilon \to 3g}\left[n^3S_1\right]} = \frac{4}{5}\frac{\alpha}{\alpha_s(m_b)}\left(1 - 2.6\frac{\alpha_s(m_b)}{\pi}\right),$$

$$\frac{\Gamma_{\Upsilon \to 3g}\left[n^3S_1\right]}{\Gamma_{\Upsilon \to \mu\bar{\mu}}\left[n^3S_1\right]} = \frac{10\left(\pi^2 - 9\right)\alpha_s^3(m_b)}{9}\left(\frac{M_\Upsilon}{2m_b}\right)^2\left(1 + 0.43\frac{\alpha_s(m_b)}{\pi}\right). \quad (1.18)$$

In reality, there are a number of theoretical and experimental concerns which make the extraction of $\alpha_s(m_b)$ a rather more subtle process than it might at first appear: (i) the contribution of $|\Psi_n(0)|^2$ in Eqs. (1.16), (1.17) as a strictly multiplicative factor is a consequence of the nonrelativistic approximation and may be affected by relativistic corrections; (ii) there is no assurance that $\mathcal{O}(\alpha_s)^2$ terms are negligible; particularly in the light of the large first-order corrections; (iii) experiments see not gluons but rather gluon *jets*, and at the mass scale of the upsilon system, jets are not particularly well defined; and (iv) the γ spectrum observed in the γgg mode is softer than that predicted by perturbative *QCD*, implying the presence of important nonperturbative effects. Nevertheless, determinations of this type lead to the central value (and its uncertainties) $\Lambda_{\overline{MS}}^{(4)} = 296 \pm 10$ MeV as extracted from upsilon data and cited earlier in Table II–2. This example indicates how demanding a task it is to obtain a precise experimental determination of $\alpha_s(q^2)$.

Hadron transitions: The transitions $V' \to V + \pi^0$ and $V' \to V + \eta$ involving the decay of an excited $^3S_1$ quarkonium level (V') down to the $^3S_1$ ground state (V) are interesting because they are forbidden in the limits of flavor-$SU(2)$ and flavor-$SU(3)$ symmetry, respectively. Their rates are therefore governed by quark mass differences, and a ratio of such rates provides a determination of quark mass ratios. There is a modest theoretical subtlety in extracting the rates, as degenerate perturbation theory must be used [IoS 80]. The leading-order effective lagrangian for these P-wave transitions must be linear in the quark mass matrix $\mathbf{m}$,

$$
\begin{aligned}
\mathcal{L}_{\text{VVM}} &= -i \frac{c}{2\sqrt{2}} F_\pi \, \text{Tr} \left(\mathbf{m} \left(U - U^\dagger \right) \right) \epsilon^{\mu\nu\alpha\beta} \partial_\mu V_\nu \partial_\alpha V'_\beta \\
&= c \left[(m_d - m_u) \frac{\pi_3}{\sqrt{2}} + (2m_s - m_d - m_u) \frac{\eta_8}{\sqrt{6}} + \cdots \right] \epsilon^{\mu\nu\alpha\beta} \partial_\mu V_\nu \partial_\alpha V'_\beta,
\end{aligned}
\tag{1.19}
$$

where c is a constant. Here, π_3 and η_8 are the pure $SU(3)$ states which appear prior to mixing

$$
\pi^0 = \cos\theta \, \pi_3 + \sin\theta \, \eta_8, \qquad \eta = -\sin\theta \, \pi_3 + \cos\theta \, \eta_8,
\tag{1.20}
$$

where $\tan\theta \simeq \theta = \sqrt{3}(m_d - m_u)/[2(2m_s - m_d - m_u)]$ describes the quark mixing. Upon calculating the transition amplitudes and then substituting for the small mixing angle θ, we obtain

$$
\begin{aligned}
\mathcal{M}_{V' \to V\pi^0} &= \frac{\mathcal{M}_0}{\sqrt{2}} \left[m_d - m_u + \frac{2m_s - m_d - m_u}{\sqrt{3}} \theta \right] = \frac{3\mathcal{M}_0}{2\sqrt{2}} (m_d - m_u), \\
\mathcal{M}_{V' \to V\eta^0} &= \frac{\mathcal{M}_0}{\sqrt{2}} \left[(m_d - m_u)\theta + \frac{2m_s - m_d - m_u}{\sqrt{3}} \right] \\
&= \frac{2\mathcal{M}_0}{\sqrt{6}} (m_s - \hat{m}) + \mathcal{O}\left(\frac{(m_d - m_u)^2}{m_s} \right),
\end{aligned}
\tag{1.21}
$$

where $\mathcal{M}_0 \equiv ic\,\epsilon^{\mu\nu\alpha\beta} k_\mu \epsilon_\nu^* k'_\alpha \epsilon_\beta$. The ratio of decay rates is found to be

$$
\Omega \equiv \frac{\Gamma_{V' \to V\pi^0}}{\Gamma_{V' \to V\eta}} = \frac{27}{16} \left| \frac{m_d - m_u}{m_s - \hat{m}} \right|^2 \left| \frac{\mathbf{p}_\pi}{\mathbf{p}_\eta} \right|^3.
\tag{1.22}
$$

We can extract a quark mass ratio from charmonium data involving $\psi(2S) \to J/\psi$ transitions. From the measured value $\Omega = 0.0396 \pm 0.0033$ [RPP 12], we find

$$
\frac{m_d - m_u}{m_s - \hat{m}} = 0.0354 \pm 0.0015,
\tag{1.23}
$$

which is rather larger than the value in Eq. (VII–1.19) extracted from pion and kaon masses.

XIII–2 Light mesons and baryons

In the quark model, the light baryons and mesons are Q^3 and $Q\bar{Q}$ combinations of the u, d, s quarks. The resulting spectrum is very rich, containing both orbital and radial excitations of the $L = 0$ ground-state hadrons. For mesons, the Q and $\bar{Q}$ spins couple to the total spins $S = 0, 1$, and each $(\mathbf{L}, \mathbf{S})$ combination occurs in the nine flavor configurations of the flavor-$SU(3)$ multiplets $\mathbf{8}, \mathbf{1}$. Analogous statements can be made for baryon states.

In the face of such complex spectra, we are mainly interested in the regularities that allow us to extract the essential physics. A tour through the database in [RPP 12] reveals some general patterns.[6] Both radial and orbital excitations of the light hadrons appear $0.5 \rightarrow 0.7$ GeV above the ground states. As pointed out in Sect. XI–1, this indicates that the light quarks move relativistically. Other striking regularities are (i) the existence of quasi-degenerate *supermultiplets* of particles with differing flavors and equal (or adjoining) spins, and (ii) excitations of a given flavor having increasingly large mass (M) and angular momentum (J) values, which obey $J = \alpha' M^2 + J_0$.

SU(6) classification of the light hadrons

To the extent that the potential is spin-independent and we work in the limit of equal u, d, s mass, the quark hamiltonian is invariant under flavor-$SU(3)$ and spin-$SU(2)$ transformations. To lowest order, hadrons are thus placed in irreducible representations of $SU(6)$, and quarks are assigned to the fundamental representation $\mathbf{6}$,

$$\mathbf{6} = (u\uparrow \quad d\uparrow \quad s\uparrow \quad u\downarrow \quad d\downarrow \quad s\downarrow). \qquad (2.1)$$

We can also write the $SU(6)$ quark multiplet in terms of the $SU(3)$ flavor representation and the spin multiplicity as $\mathbf{6} = (\mathbf{3}, 2)$. Although the $SU(6)$, invariant limit forms a convenient basis for a classification of the meson and baryon states, it cannot be a full symmetry of Nature since the spin is a spacetime property of particles whereas $SU(3)$ flavor symmetry is not. Thus, it is impossible to unite the flavor and spin symmetries in a relativistically invariant manner [CoM 67]. Although we shall avoid making detailed predictions based on $SU(6)$, it is nonetheless useful in organizing the multitude of observed hadronic levels.

Meson supermultiplets: The $L=0$ $Q\bar{Q}$ composites are contained in the $SU(6)$ group product $\mathbf{6} \times \mathbf{6}^* = \mathbf{35} \oplus \mathbf{1}$, where the representations $\mathbf{35}, \mathbf{1}$ have flavor–spin content

---

[6] Our discussion will focus on hadron masses. Strong and electromagnetic transitions are described in [LeOPR 88].

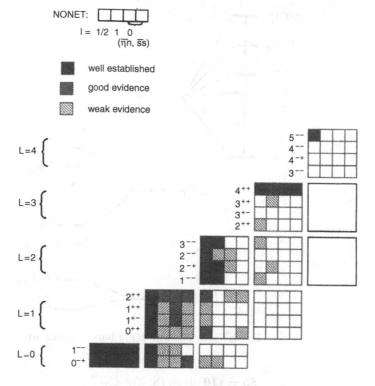

Fig. XIII–5 Spectrum of the light mesons.

$$\mathbf{35} = (\mathbf{8}, \mathbf{3}) \oplus (\mathbf{8}, \mathbf{1}) \oplus (\mathbf{1}, \mathbf{3}), \qquad \mathbf{1} = (\mathbf{1}, \mathbf{1}). \tag{2.2}$$

The $L = 0$ ground state consists of a vector octet, a pseudoscalar octet, a vector singlet, and a pseudoscalar singlet. For excited states, the meson supermultiplets constitute an $SU(6) \times O(3)$ spectrum of particles. The $O(3)$ label refers to how the total angular momentum is obtained from $\mathbf{J} = \mathbf{L} + \mathbf{S}$, giving rise to the pattern of rotational excitations displayed previously in Table XI–3. Roughly speaking, mesons occur in mass bands having a common degree of radial and/or orbital excitation.

Fig. XIII–5 provides a view of the mass spectrum for the lightest mesons. The $SU(6) \times O(3)$ structure of the ground state and a sequence of orbitally excited states are observed to the extent that sufficient data are available for particle assignments to be made. Note that the S-wave $Q\bar{Q}$ states are all accounted for, but gaps appear in all higher partial waves. Even after many years of study, meson phenomenology below 2 GeV is far from complete!

Baryon supermultiplets: The $SU(6)$ baryon multiplet structure arises from the Q^3 group product $(\mathbf{6} \times \mathbf{6}) \times \mathbf{6} = (\mathbf{21} \oplus \mathbf{15}) \times \mathbf{6} = \mathbf{56} \oplus \mathbf{70} \oplus \mathbf{70} \oplus \mathbf{20}$, and has flavor–spin content

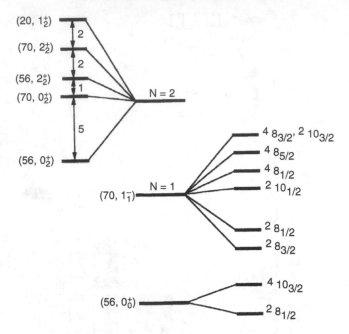

Fig. XIII–6 The low-lying baryon spectrum.

$$\mathbf{56} = (\mathbf{10}, 4) \oplus (\mathbf{8}, 2),$$
$$\mathbf{70} = (\mathbf{8}, 4) \oplus (\mathbf{10}, 2) \oplus (\mathbf{8}, 2) \oplus (\mathbf{1}, 2),$$
$$\mathbf{20} = (\mathbf{8}, 2) \oplus (\mathbf{1}, 4). \tag{2.3}$$

A three-quark system must adhere to the constraint of Fermi statistics. Each baryon-state vector is thus antisymmetric under the interchange of any two quarks. A Young-tableaux analysis of the above group product reveals that the spin–flavor parts of the $\mathbf{56}$, $\mathbf{70}$ and $\mathbf{20}$ multiplets are, respectively, symmetric, mixed, and antisymmetric under interchange of pairs of quarks. Since the color part of any Q^3 color-singlet-state vector is antisymmetric under interchange of any two quarks, the $\mathbf{56}$-plet has a totally symmetric space wavefunction, with zero orbital angular momentum between each quark pair. The $\mathbf{70}$ and $\mathbf{20}$ multiplets require either radial excitations and/or orbital excitations. Recall the characterization of the baryon spectrum in terms of the basis defined by an independent pair of oscillators (cf. Eq. (XI–2.12)). In this context, a standard notation for a baryon supermultiplet is $(\mathbf{R}, L_N^{\mathcal{P}})$, where $\mathbf{R}$ labels the $SU(6)$ representation, $\mathcal{P}$ is the parity, N labels the number of oscillator quanta and L is the orbital angular momentum quantum number (cf. Sect. XI–2).

Like meson masses, baryon masses tend to cluster in bands having a common value of N. The first three bands are shown in Fig. XIII–6, and effects of $SU(6)$ breaking are displayed for the first two. The lowest-lying $SU(6) \times O(3)$

supermultiplet is the positive-parity $(\mathbf{56}, 0_0^+)$, having content as in Eq. (2.3). Next comes the negative-parity $(\mathbf{70}, 1_1^-)$ supermultiplet. This contains more states than the **70**-plet shown in Eq. (2.3) because the extension from $L = 0$ to $L = 1$ requires addition of angular momenta,

$$
\begin{aligned}
(\mathbf{10}, 2) &\to (\mathbf{10}, 4) \oplus (\mathbf{10}, 2), & (\mathbf{8}, 4) &\to (\mathbf{8}, 6) \oplus (\mathbf{8}, 4) \oplus (\mathbf{8}, 2), \\
(\mathbf{1}, 2) &\to (\mathbf{1}, 4) \oplus (\mathbf{1}, 2), & (\mathbf{8}, 2) &\to (\mathbf{8}, 4) \oplus (\mathbf{8}, 2).
\end{aligned}
\tag{2.4}
$$

The number of supermultiplets grows per unit of excitation thereafter. There are five $SU(6)$ multiplets in the $N = 2$ band, $(\mathbf{56}, 2_2^+)$, $(\mathbf{56}, 0_2^+)$, $(\mathbf{70}, 2_2^+)$, $(\mathbf{70}, 0_2^+)$, and $(\mathbf{20}, 1_2^+)$. Recall that the baryonic inter-quark potential was expressed in Eq. (XI–2.10) as $V = V_{\rm osc} + U$, where $V_{\rm osc}$ is the potential energy of a harmonic oscillator and $U \equiv V - V_{\rm osc}$. If the potential energy were purely $V_{\rm osc}$, the supermultiplets within the $N = 2$ band would all be degenerate. In the potential model, assuming that the largest part of U is purely radial, this degeneracy is removed by the first-order perturbative effect of U, and the splittings in the $N = 2$ band are shown at the top of Fig. XIII–6. Aside from choosing the $(\mathbf{56}, 0_2^+)$ supermultiplet to have the lowest mass, one finds the pattern of splitting to be as shown in Fig. XIII–6, *independent* of the particular form of U.

Regge trajectories

It is natural to classify together a ground-state hadron and its rotational excitations, e.g., the isospin one-half positive-parity baryons $N(939)_{J=1/2}$ (the nucleon), $N(1680)_{J=5/2}$, $N(2220)_{J=9/2}$ and $N(2700)_{J=13/2}$. Although no higher-spin entries have been detected in this particular set of nucleonic states (presumably due to experimental limitations), there is no theoretical reason to expect any such sequence to end. The database in [RPP 12] contains a number of similar structures, each characteristically containing three or four members.

Each such collection of states is said to belong to a given *Regge trajectory*. To see how this concept arises, let us consider the simplest case of two spinless particles with scattering amplitude $f(E, z)$ (i.e. $d\sigma/d\Omega = |f(E, z)|^2$), where E is the energy and $z = \cos \theta$ is the scattering angle. It turns out that analytic properties of the scattering amplitude in the complex angular momentum (J) plane are of interest. One may obtain a representation of $f(E, z)$ in the complex J-plane by converting the partial wave expansion into a so-called *Watson–Sommerfeld transform*,

$$
\begin{aligned}
f(E, z) &= \sum_{\ell=0}^{\infty} (-)^{\ell} (2\ell + 1) a(E, \ell) P_{\ell}(-z) \\
&\to \frac{1}{2\pi i} \oint_C dJ \, \frac{\pi}{\sin \pi J} (2J + 1) a(E, J) P_J(-z),
\end{aligned}
\tag{2.5}
$$

where P_ℓ is a Legendre polynomial and $\mathcal{C}$ is a contour enclosing the nonnegative integers. Suppose that as $\mathcal{C}$ is deformed away from the Re J-axis to, say, a line of constant Re J, a pole in the partial wave amplitude $a(E, J)$ is encountered. Such a singularity is referred to as a *Regge pole* and contributes (cf. Eq. (2.5)) to the full scattering amplitude as

$$f(E, z) = \frac{\beta[E]P_{\alpha[E]}(-z)}{\sin(\pi\alpha[E])} + \cdots, \tag{2.6}$$

where $\alpha[E]$ is the energy-dependent pole position in the complex J-plane and $\beta[E]$ is the pole residue.

The Regge-pole contribution of Eq. (2.6) can manifest itself physically in both the direct channel as a resonance and a crossed channel as an exchanged particle. Here, we discuss just the former case by demonstrating how a given Regge pole can be related to a *sequence* of rotational excitations. Suppose that at some energy E_R, the real part of the pole position equals a nonnegative integer ℓ, i.e., Re $\alpha[E_R] = \ell$. Then, with the aid of the identity,

$$\frac{1}{2}\int_{-1}^{1} dz\, P_\ell(z)P_\alpha(-z) = \frac{1}{\pi}\frac{\sin(\pi\alpha)}{(\ell - \alpha)(\ell + \alpha + 1)}, \tag{2.7}$$

we can infer from Eq. (2.6) the Breit–Wigner resonance form,

$$a_\ell^{\text{(Rg.-ple.)}} = \frac{\beta}{\pi}\frac{1}{(\alpha[E] - \ell)(\alpha[E] + \ell + 1)} \simeq \frac{\Gamma/2}{E - E_R + i\Gamma/2}, \tag{2.8}$$

provided Re $\alpha[E_R] \gg$ Im $\alpha[E_R]$. A physical resonance thus appears if $\alpha[E]$ passes near a nonnegative integer and, if the Regge pole moves to ever-increasing J values in the complex J-plane as the energy E is increased, it generates a tower of high-spin states. Except in instances of so-called exchange degeneracy, parity dictates that there be two units of angular momentum between members of a given trajectory. In this manner, a single Regge pole in the angular-momentum plane gives rise to the collection of physical states called a *Regge trajectory*.

A plot of the angular momentum vs. squared-mass for the states on any meson or baryon trajectory reveals the linear behavior,

$$J \simeq \alpha' M^2 + J_0. \tag{2.9}$$

A compilation of slopes (α') and intercepts (J_0) appears in Table XIII–2, with each trajectory labeled by its ground-state hadron. Such linearly rising trajectories have been interpreted as a consequence of QCD [JoT 76]. In this picture, hadrons undergoing highly excited rotational motion come to approach color-flux tubes, whereupon it becomes possible to relate the angular momentum of rotation to the energy contained in the color field. This line of reasoning leads to the behavior of Eq. (2.9), and accounts for the universality seen in the slope values displayed in Table XIII–2.

SU(6) breaking effects

Although an $SU(6)$-invariant hamiltonian provides a convenient basis for describing light hadrons, the physical spectrum exhibits substantial departures from the mass degeneracies which occur in this overly symmetric picture. In the following, we shall consider some simple models for explaining the many $SU(6)$-breaking effects observed in the real world.

The QCD Breit–Fermi model: If one ascribes the nonconfining part of the quark interaction to single-gluon exchange, the nonrelativistic limit yields the 'QCD Breit–Fermi potential' [DeGG 75]

$$
\begin{aligned}
V_{\text{one-gluon}} = & -\frac{4k\alpha_s}{3r} \\
& + \frac{4k\alpha_s}{3} \sum_{i<j} \left[\frac{8\pi}{3M_iM_j} \mathbf{s}_i \cdot \mathbf{s}_j \delta^3(\mathbf{r}) + \frac{\pi}{2}\delta^3(\mathbf{r}) \left(\frac{1}{M_i^2} + \frac{1}{M_j^2} \right) \right. \\
& + \frac{1}{M_iM_jr^3} \left[3(\mathbf{s}_i \cdot \hat{\mathbf{r}})(\mathbf{s}_j \cdot \hat{\mathbf{r}}) - \mathbf{s}_i \cdot \mathbf{s}_j \right] \\
& + \frac{1}{r^3} \left(\frac{\mathbf{s}_i \cdot \mathbf{r} \times \mathbf{p}_i}{2M_i^2} - \frac{\mathbf{s}_j \cdot \mathbf{r} \times \mathbf{p}_j}{2M_j^2} - \frac{\mathbf{s}_j \cdot \mathbf{r} \times \mathbf{p}_i - \mathbf{s}_i \cdot \mathbf{r} \times \mathbf{p}_j}{M_iM_j} \right) \\
& \left. + \frac{1}{2M_iM_jr} \left(\mathbf{p}_i \cdot \mathbf{p}_j + \hat{\mathbf{r}}(\hat{\mathbf{r}} \cdot \mathbf{p}_i) \cdot \mathbf{p}_j \right) \right],
\end{aligned}
\tag{2.10}
$$

where α_s is the strong fine structure constant, $\mathbf{r} \equiv \mathbf{r}_{ij}$, and k denotes the color dependence of the potential (cf. Sect. XI–2) with $k = 1$ $(1/2)$ for mesons (baryons). In keeping with the potential model, the mass parameters $\{M_i\}$ are interpreted as constituent quark masses. Although the QCD Breit–Fermi model incorporates $SU(6)$ breaking by means of both quark mass splittings and spin-dependent interactions, it lacks a rigorous theoretical foundation. One might argue on the grounds of asymptotic freedom that Eq. (2.10) does justice to physics at very short distances (in the approximation that α_s is constant), but there is no reason to believe that it suffices at intermediate-length scales. It also does not account for mixing between isoscalar mesons, so such states must be considered separately.

Meson masses: The gluon-exchange model can be used to obtain information on constituent quark mass. In the following, we shall temporarily ignore the minor effect of isospin breaking by working with $\hat{M} \equiv (M_u + M_d)/2$. To compute meson masses, we take the expectation value of the full hamiltonian between $SU(6)$ eigenstates, specifically the $L = 0$ $Q\bar{Q}$ states.[7] Although the form of Eq. (2.10) implies the presence of spin–spin, spin–orbit, and tensor interactions, the spin–orbit and

---

[7] An analysis of spin dependence in the $L = 1$ states is the subject of a problem at the end of the chapter (cf. Prob. XIII–3)).

tensor terms do not contribute here because each quark pair moves in an S wave, and it is the spin–spin (hyperfine) interaction which lifts the vector meson states relative to the pseudosclar mesons. We can parameterize the nonisoscalar $L = 0$ meson masses as

$$m_{Q\bar{Q}}^{(L=0)} = \hat{n}\hat{M} + n_s M_s + \frac{\langle \mathbf{p}_Q^2 \rangle}{2M_Q} + \frac{\langle \mathbf{p}_{\bar{Q}}^2 \rangle}{2M_{\bar{Q}}} + \mathcal{H}_{Q\bar{Q}}\langle s_Q \cdot s_{\bar{Q}} \rangle, \quad (2.11)$$

where $\hat{n}$ and n_s are the number of nonstrange (n) and strange consituents (s) respectively, and $\mathcal{H}_{Q\bar{Q}}$ refers to the hyperfine interaction in the second line of Eq. (2.10).

One consequence of Eq. (2.11) is a relation involving the mass ratio $\hat{M}/M_s$. Fitting the four masses $\pi(138)$, $K(496)$, $\rho(770)$, $K^*(892)$ to the parameters in Eq. (2.11) yields

$$\frac{m_{K^*} - m_K}{m_\rho - m_\pi} = \frac{\mathcal{H}_{ns}}{\mathcal{H}_{nn}} = \frac{\hat{M}}{M_s} \simeq 0.63. \quad (2.12)$$

The origin of this result lies in the inverse dependence of the hyperfine interaction upon constituent quark mass, which affects the mass splitting between $S = 1$ and $S = 0$ mesons differently for strange and nonstrange mesons. The numerical value of $\hat{M}/M_s$ in Eq. (2.12) graphically demonstrates the difference between constituent quark masses and current quark masses, the latter having a mass ratio of about 0.04. In earlier sections of this book, which stressed the role of chiral symmetry, the pion was given a special status as a quasi-Goldstone particle. In the $Q\bar{Q}$ model, the small pion mass is seen to be a consequence of severe cancelation between the spin-independent and spin-dependent contributions. However, the parameterization of Eq. (2.11) cannot explain the large $\eta'(960)$ mass.

In addition to the $SU(6)$ symmetry-breaking effects of mass and spin, there is an additive contribution present in the isoscalar channel, which is induced by quark–antiquark annihilation into gluons. In the basis of u, d, s quark flavor states, this annihilation process produces a 3×3 mass matrix of the form

$$\begin{pmatrix} 2M_u + X & X & X \\ X & 2M_d + X & X \\ X & X & 2M_s + X \end{pmatrix}, \quad (2.13)$$

where for $C = +1(-1)$ mesons, X is the two-gluon (three-gluon) annihilation amplitude, and for simplicity we display just the quark mass contribution ($2M_i$) as the nonmixing mass contribution. The annihilation process is a short-range phenomenon, so the magnitude of X depends sharply on the orbital angular momentum L of the $Q\bar{Q}$ system. For $L \neq 0$ waves (where the wavefunction vanishes at zero relative separation), and $C = -1$ channels (where the annihilation amplitude is suppressed by the three powers of gluon coupling), we expect $M_s - \hat{M} \gg X$. In this limit, diagonalization of Eq. (2.13) yields to leading order the set of basis states $(\bar{u}u \pm \bar{d}d)/\sqrt{2}$ and $\bar{s}s$. Only the $L = 0$ pseudoscalar channel experiences opposite limit $X \gg M_s - \hat{M}$, wherein to leading order the basis vectors are the $SU(3)$ singlet state $(\bar{u}u + \bar{d}d + \bar{s}s)/\sqrt{3}$ and octet states $(\bar{u}u - \bar{d}d)/\sqrt{2}$, $(\bar{u}u + \bar{d}d - 2\bar{s}s)/\sqrt{6}$. The overall picture that emerges is one of relatively unmixed light pseudoscalar states, and heavily mixed vector, tensor, etc., states.

Baryon masses: Applying the one-gluon exchange potential to the ground-state baryons of $(\mathbf{56}, 0_0^+)$ yields a mass formula analogous to Eq. (2.11),

$$m_{Q^3}^{(L=0)} = \hat{n}\hat{M} + n_s M_s + \sum_{i=1}^{3} \frac{\langle \mathbf{p}_i^2 \rangle}{2M_i} + \frac{1}{2} \sum_{i<j} \mathcal{H}_{ij} \langle \mathbf{s}_i \cdot \mathbf{s}_j \rangle. \tag{2.14}$$

For the system of $1/2^+$ and $3/2^+$ (iospin-averaged) baryons, there are eight mass values and since the above mass formula contains five parameters, one should obtain three relations. The additional perturbative assumption $\mathcal{H}_{ss} - \mathcal{H}_{ns} = \mathcal{H}_{ns} - \mathcal{H}_{nn}$ for the hyperfine mass parameters yields the Gell-Mann–Okubo relation of Eq. (XII–3.10) for the $1/2^+$ baryons and the *equal spacing rule* for $3/2^+$ states,

$$m_{\Sigma^*} - m_\Delta = m_{\Xi^*} - m_{\Sigma^*} = m_\Omega - m_{\Xi^*}.$$
$$\text{(Expt. 153 MeV} = 149 \text{ MeV} = 139 \text{ MeV)} \tag{2.15}$$

A third relation which relates the $3/2^+$ and $1/2^+$ masses and is independent of further perturbative assumptions has the form

$$3m_\Lambda - m_\Sigma - 2m_N = 2(m_{\Sigma^*} - m_\Delta)$$
$$\text{(Expt. : } 276 \text{ MeV} = 305 \text{ MeV)} \tag{2.16}$$

In addition, one can obtain estimates for $\hat{M}/M_s$, among them

$$\frac{\hat{M}}{M_s} = \frac{2(m_{\Sigma^*} - m_\Sigma)}{2m_{\Sigma^*} + m_\Sigma - 3m_\Lambda} \simeq 0.62,$$
$$\frac{\hat{M}}{M_s} = \frac{m_{\Sigma^*} - m_\Sigma}{m_\Delta - m_N} \simeq 0.65, \tag{2.17}$$

both in accord with Eq. (2.12).

Isospin-breaking effects: The above description of $SU(6)$ breaking assumes isospin conservation. In fact, hadrons exhibit small mass splittings within isospin multiplets, arising from electromagnetism and the $u - d$ mass difference. In the pion and kaon systems, we were able to use chiral $SU(3)$ symmetry to isolate each of these separately. Unfortunately, this is not possible in general, and models are required to address this issue.

There are a few consequences which follow purely from symmetry considerations. Since the mass difference $m_u - m_d$ is $\Delta I = 1$, the $\Delta I = 2$ combinations

$$m_{\Sigma^+} + m_{\Sigma^-} - 2m_{\Sigma^0} = 1.7 \pm 0.1 \text{ MeV}, \quad m_{\rho^+} - m_{\rho^0} = -0.3 \pm 2.2 \text{ MeV},$$
$$\text{(2.18)}$$

arise only from the electromagnetic interaction. In addition, both electromagnetic and quark mass contributions satisfy the *Coleman–Glashow relation* [CoG 64],

$$m_{\Sigma^+} - m_{\Sigma^-} + m_n - m_p + m_{\Xi^-} - m_{\Xi^0} = 0$$
$$\text{[Expt. } 0.4 \pm 0.6 \text{ MeV} = 0]. \tag{2.19}$$

For electromagnetism, this is a consequence of the U-spin-singlet character of the current, whereas for quark masses it follows from the $\Delta I = 1$ and $SU(3)$-octet character of the current.

We proceed further by using a simple model, based on the *QED* Coulomb and hyperfine effects, to describe the electromagnetic interaction of quarks,

$$\Delta m_{\text{coul}} = \mathcal{A}_{\text{coul}} \sum_{i<j} Q_i Q_j,$$

$$\Delta m_{\text{hyp}} = -\mathcal{A}_{\text{hyp}} \sum_{i<j} \frac{Q_i Q_j}{M_i M_j} \mathbf{s}_i \cdot \mathbf{s}_j, \tag{2.20}$$

where $\mathcal{A}_{\text{coul}}, \mathcal{A}_{\text{hyp}}$ are constants, $\{Q_i\}$ are quark electric charges, and the sums are taken over constituent quarks. In Δm_{hyp}, we shall neglect further isospin breaking in the masses and use $M_u = M_d = \hat{M}$, and assume electromagnetic self-interactions of a quark to be already accounted for in the mass parameter of that quark. For any values of $\mathcal{A}_{\text{coul}}$ and $\mathcal{A}_{\text{hyp}}$, this model contains the sum rule

$$(m_n - m_p)_{\text{em}} = -\frac{1}{3}(m_{\Sigma^+} + m_{\Sigma^-} - 2m_{\Sigma^0}) = -0.57 \pm 0.03 \text{ MeV}, \tag{2.21}$$

leaving the excess due to the quark mass difference,

$$(m_n - m_p)_{\text{qm}} = \frac{m_u - m_d}{2} \cdot \langle n|\bar{u}u - \bar{d}d|n\rangle - \frac{m_u - m_d}{2} \cdot \langle p|\bar{u}u - \bar{d}d|p\rangle$$
$$\equiv (m_d - m_u)(d_m + f_m)Z_0$$
$$= (m_n - m_p) - (m_n - m_p)_{\text{em}} = 1.86 \pm 0.03 \text{ MeV}, \tag{2.22}$$

where the second line in the above uses the parameterization of hyperon mass splittings given in Eq. (XII–3.9). To the extent that this estimate of quark mass differences is meaningful, one obtains the mass ratio,

$$\frac{m_d - m_u}{m_s - \hat{m}} = \frac{(m_n - m_p)_{\text{qm}}}{m_{\Xi} - m_{\Sigma}} \simeq 0.015, \tag{2.23}$$

to be compared to the chiral-symmetry extraction from meson masses, which yielded 0.023. With further neglect of terms $\mathcal{O}(\alpha(M_s - \hat{M}))$ in the hyperfine interaction, this exercise can be repeated for vector mesons to yield

$$(m_{K^{*0}} - m_{K^{*+}})_{\text{em}} = -\frac{2}{3}(m_{\rho^+} - m_{\rho^0}) = 0.2 \pm 1.5 \text{ MeV},$$

$$(m_{K^{*0}} - m_{K^{*+}})_{\text{qm}} = (m_{K^{*0}} - m_{K^{*+}}) - (m_{K^{*0}} - m_{K^{*+}})_{\text{em}}$$
$$= 6.5 \pm 1.9 \text{ MeV},$$

$$\frac{m_d - m_u}{m_s - \hat{m}} = \frac{m_{K^{*0}} - m_{K^{*+}}}{m_{K^*} - m_\rho} = 0.053 \pm 0.016. \tag{2.24}$$

The additional assumption that the constants $\mathcal{A}_{\text{coul}}$ and $\mathcal{A}_{\text{hyp}}$ are the *same* in the decuplet baryons and the octet baryons, as is true in the $SU(6)$ limit, leads to

$$(m_{\Delta^{++}} - m_{\Delta^0})_{\text{em}} = \frac{5}{3}(m_{\Sigma^+} + m_{\Sigma^-} - 2m_{\Sigma^0}) = 2.8 \pm 0.2 \text{ MeV},$$

$$(m_{\Delta^{++}} - m_{\Delta^0})_{\text{qm}} = (m_{\Delta^{++}} - m_{\Delta^0}) - (m_{\Delta^{++}} - m_{\Delta^0})_{\text{em}}$$
$$= -5.5 \pm 0.4 \text{ MeV},$$

$$\frac{m_d - m_u}{m_s - \hat{m}} = \frac{1}{2}\frac{m_{\Delta^0} - m_{\Delta^{++}}}{m_{\Sigma^*} - m_\Delta} = 0.018 \pm 0.002. \tag{2.25}$$

Of course, the spread of values for the mass ratios raises a concern about the validity of this simple model. However, all methods of calculation agree on the smallness of the ratio $(m_d - m_u)/(m_s - \hat{m})$.

XIII–3 The heavy-quark limit

In the quark description, a heavy-flavored hadron contains at least one of the heavy quarks c, b, t. An effective field theory, Heavy Quark Effective Theory (HQET), has been developed which provides a powerful tool for heavy quark physics. This involves a study of the limit $(m_Q \to \infty)$ in which the theory is expanded in powers of m_Q^{-1}. We describe a simple introduction to the topic and much more detail can be found in [MaW 07].

Heavy-flavored hadrons in the quark model

The spectroscopy of heavy-flavored hadrons should qualitatively follow that of the light hadronic spectrum, with states containing a single heavy-quark Q occurring as either mesons $(Q\bar{q})$ or baryons (Qq_1q_2). The lowest-energy state for a given hadronic flavor will have zero orbital angular momentum between the quarks, leading to ground-state spin values $S = 0, 1$ for mesons and $S = 1/2, 3/2$ for

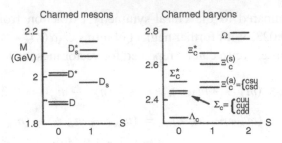

Fig. XIII–7 Spectrum of charmed (a) mesons, (b) baryons.

baryons. The hyperfine interaction will lower the $S = 0$ meson and $S = 1/2$ baryon masses, and both orbital and radial hadronic excitations of the ground state will be present.

Although it is possible to contemplate extended flavor transformations which involve interchange of the light and heavy quarks, e.g., as in the $SU(4)$ of the light and charmed hadrons, such symmetries are so badly broken by the difference in energy scales $M_Q \gg M_q$ and $M_Q \gg \Lambda_{QCD}$ as to be rendered useless. The $SU(3)$- and $SU(2)$-flavor symmetries associated with the light hadrons are still viable, but multiplet patterns become modified. The mesons $Q\bar{q}$ will exist in the $SU(3)$ multiplet $\mathbf{3^*}$, whereas in the baryonic Qq_1q_2 configurations the two light quarks q_1, q_2 will form the flavor-$SU(3)$ multiplets $\mathbf{6}$ and $\mathbf{3^*}$. For example, the charmed system has the meson ground state

$$\mathbf{3^*}: \quad D^+ \left[c\bar{d} \right], \ D^0 \left[c\bar{u} \right], \ D^s \left[c\bar{s} \right],$$

which displays the mass pattern of an $SU(2)$ doublet (D^+_{1869}, D^0_{1865}) and an $SU(2)$ singlet (D^s_{1969}). The charmed-baryon multiplets are

$$\mathbf{6}: \quad \Sigma_c^{++}[uuc], \ \Sigma_c^+[udc], \ \Sigma_c^0[ddc], \ \Xi_c^{+(s)}[usc], \ \Xi_c^{0(s)}[dsc], \ \Omega_c^0[ssc]$$
$$\mathbf{3^*}: \quad \Lambda_c^+[udc], \ \Xi_c^{+a}[usc], \ \Xi_c^{0a}[dsc].$$

Fig. XIII–7 displays the anticipated charmed-meson and charmed-baryon levels, including the effect of $SU(3)$ breaking.

Heavy-quark constituent mass values can be inferred from the D^*-D and B^*-B hyperfine splittings. That the former splitting is about three times the latter is a consequence of $M_b \simeq 3M_c$ and of the inverse dependence of the hyperfine effect upon quark mass. Analogously to Eq. (2.17), we find

$$\frac{\hat{M}}{M_c} = \frac{m_{D^*} - m_D}{m_\rho - m_\pi} \simeq 0.22, \qquad \frac{\hat{M}}{M_b} = \frac{m_{B^*} - m_B}{m_\rho - m_\pi} \simeq 0.08, \qquad (3.1)$$

where $\hat{M} \equiv (M_u + M_d)/2$. These findings depend to some extent on how the fit is done, e.g., with mesons or with baryons, and we leave further study for Prob. XIII–4.

Spectroscopy in the $m_Q \to \infty$ limit

In a hadron which contains a single heavy quark Q along with light degrees of freedom, the heavy quark is essentially static. The best analogy is with atoms, where the nucleus can in the first approximation be treated as a static, electrically charged source. Likewise, for heavy hadrons the heavy quark is a static source with color charge, and the light degrees of freedom provide a nonstatic hadronic environment around Q. This scenario can be formalized by partitioning the heavy-quark lagrangian as [CaL 86, Ei 88, LeT 88]

$$\mathcal{L}_Q = \bar{\psi} \left(i \not{D} - m_Q \right) \psi \equiv \mathcal{L}_0 + \mathcal{L}_{\text{space}}$$
$$\mathcal{L}_0 = \bar{\psi} \left(i\gamma_0 D_0 - m_Q \right) \psi, \qquad \mathcal{L}_{\text{space}} = -i\bar{\psi}\boldsymbol{\gamma} \cdot \mathbf{D}\psi, \qquad (3.2)$$

where $D_\mu\psi$ is the covariant derivative of $SU(3)_c$. Since the spatial γ matrices connect upper and lower components, we see that the effect of $\mathcal{L}_{\text{space}}$ is $\mathcal{O}(m_Q^{-1})$.

Observe that the static lagrangian $\mathcal{L}_0$ of Eq. (3.2) is invariant under spin rotations of the heavy quark Q. In the world defined by $\mathcal{L}_0$, with both $\mathcal{O}(\Lambda_{QCD}/M_Q)$ effects and $\mathcal{O}(\alpha_s(M_Q))$ effects (associated with hard-gluon exchange) ignored, heavy-hadronic energy levels and couplings are constrained by the $SU(2)$ spin symmetry. It is helpful to visualize the situation. A heavy-flavored hadron of spin $\mathbf{S}$ will contain a static quark Q having a constant spin vector $\mathbf{S}_Q$ (with $S_Q = 1/2$) and light degrees of freedom having a constant angular momentum vector $\mathbf{J}_\ell \equiv \mathbf{S} - \mathbf{S}_Q$.[8] For a meson of this type, we assume that J_ℓ behaves as it does in the quark model, with $J_\ell = 1/2$ in the ground state and $J_\ell = L \pm 1/2$ for $L > 0$ rotational excitations. From the decoupling of the heavy-quark spin, it follows that *there will be a two-fold degeneracy between mesons having spin values $S = J_\ell \pm 1/2$*. The meson $L = 0$ ground state will have $J_\ell = 1/2$ and thus degenerate states with $S = 0, 1$. The $L = 1$ first rotational excitation with $J_\ell = 1/2$ will give rise to degenerate $S = 0, 1$ levels, whereas for $J_\ell = 3/2$ one obtains degenerate levels having $S = 1, 2$. Moreover, the energy differences between different levels should be independent of heavy-quark flavor. Analogous conditions hold for heavy flavored baryons, and hadronic transitions between levels of differing L can be similarly analyzed.

---

[8] Although the light degree(s) of freedom in the simple quark model is an antiquark $\bar{q}$ for mesons and two quarks $q_1 q_2$ for baryons, the physical (i.e. actual) light degrees of freedom could entail unlimited numbers of gluons and/or quark–antiquark pairs.

Let us explicitly demonstrate that the splitting between the $J^P = 1^-$ and $J^P = 0^-$ states of a $Q\bar{q}$ meson must vanish in the limit of infinite quark mass. We note that the mathematical condition for spin-independence is

$$\left[H_0, S_3^Q \right] = 0, \tag{3.3}$$

where S_3^Q is the generator of spin rotations about the 3-axis for quark Q and H_0 is the hamiltonian obtained from $\mathcal{L}_0$. Since the action of S_3^Q on a 0^- state produces a 1^- state, i.e., $|M_{1-}\rangle = 2S_3^Q |M_{0-}\rangle$, we then have

$$H_0 |M_{1-}\rangle = m_{1-} |M_{1-}\rangle = 2S_3^Q H_0 |M_{0-}\rangle = m_{0-} |M_{1-}\rangle , \tag{3.4}$$

implying that $m_{1-} - m_{0-} \to 0$ as $m_Q \to \infty$.

Another consequence of working in the static limit of $\mathcal{L}_0$ is that the propagator, $S_\infty(x, y)$, of the heavy quark in an external field can be determined exactly. From the defining equations,

$$(i\gamma_0 D_0 - m_Q) \, S_\infty(x, y) = \delta^{(4)}(x - y) \qquad (D_0 \equiv \partial_0 + ig_3 \mathbf{A}_0 \cdot \boldsymbol{\lambda}/2), \tag{3.5}$$

one has the solution

$$S_\infty(x, y) = -i \, P(x_0, y_0) \delta^{(3)}(\mathbf{x} - \mathbf{y}) \left[\theta(x^0 - y^0) e^{-im_Q(x^0 - y^0)} \left(\frac{1 + \gamma_0}{2} \right) \right.$$
$$\left. + \theta(y^0 - x^0) e^{im_Q(x^0 - y^0)} \left(\frac{1 - \gamma_0}{2} \right) \right], \tag{3.6}$$

where $P(x_0, y_0)$ is the path-ordered exponential along the time direction,

$$P(x_0, y_0) \equiv P \exp \left[i \frac{g_3}{2} \int_{y^0}^{x_0} dt \, \boldsymbol{\lambda} \cdot \mathbf{A}_0(\mathbf{x}, t) \right]. \tag{3.7}$$

In this approximation, the heavy quark is static at point $\mathbf{x}$ and the only time-dependence is that of a phase.

This discussion can be generalized to a frame where the heavy quark is moving at a fixed velocity $\mathbf{v}$, described by a velocity-four vector $v^\mu = p^\mu / m_Q$, with $v^\mu v_\mu = 1$. One can define projection operators

$$\Gamma_{v\pm} = \frac{1}{2} \, (1 \pm \not{p}), \tag{3.8}$$

where $\Gamma_{v\pm}^2 = \Gamma_{v\pm}$, $\Gamma_{v\pm} \Gamma_{v\mp} = 0$, and $\Gamma_{v+} + \Gamma_{v-} = 1$. The $\Gamma_{v\pm}$ generalize the usual projection of 'upper' and 'lower' components into the moving frame. A quark moving with velocity $\mathbf{v}$ will have the leading description of its wavefunction contained in the 'upper' component described by a field h_v [Ge 90, Wi 91],

$$\Gamma_{v+} \psi \equiv e^{-im_Q v \cdot x} h_v(x), \tag{3.9}$$

where the main dependence on the quark mass has been factored out, and h_v obviously satisfies $\Gamma_{v+}h_v = h_v$. Substituting into the Dirac lagrangian, neglecting lower components, and using $\Gamma_{v+}\!\!\not{D}\Gamma_{v+} = v \cdot D$ yields

$$\mathcal{L}_Q = \bar{\psi}\left(i\!\!\not{D} - m_Q\right)\psi \simeq \bar{\psi}\Gamma_{v+}\left(i\!\!\not{D} - m_Q\right)\Gamma_{v+}\psi = \bar{h}_v iv \cdot Dh_v, \qquad (3.10)$$

which generates the lowest-order equation of motion $v \cdot Dh_v = 0$. This approximation can be systematically improved by inclusion of a 'lower' component for the heavy-quark field [EiH 90, Lu 90, GeGW 90],

$$\Gamma_{v-}\psi \equiv e^{-im_Q v \cdot x}\ell_v(x), \qquad (3.11)$$

with $\Gamma_{v-}\ell_v = \ell_v$. The equations of motion allow us to solve for ℓ_v by following the sequence of steps,

$$\begin{aligned}
0 = \left(i\!\!\not{D} - m_Q\right)\psi &= \left(i\!\!\not{D} - m_Q\right)e^{-im_Q v \cdot x}\left[h_v + \ell_v\right] \\
&= e^{-im_Q v \cdot x}\left(m_Q\left(\not{v} - 1\right) + ie^{-im_Q v \cdot x}\!\!\not{D}\right)\left[h_v + \ell_v\right] \\
&= e^{-im_Q v \cdot x}\left[(-2m_Q + i\!\!\not{D})\ell_v + i\!\!\not{D}h_v\right], \qquad (3.12)
\end{aligned}$$

which yields ℓ_v and ψ as

$$\ell_v = \frac{i}{2m_Q}\!\!\not{D}h_v + \mathcal{O}\left(m_Q^{-2}\right)$$

$$\psi = e^{-im_Q v \cdot x}\left[1 + \frac{i}{2m_Q}\!\!\not{D}\right]h_v + \mathcal{O}\left(m_Q^{-2}\right). \qquad (3.13)$$

Inserting these forms into Eq. (3.10) and using $\Gamma_{v+}h_v = h_v$ and Eq. (III–3.50) for $\not{D}\!\!\not{D}$ yields

$$\begin{aligned}
\mathcal{L}_v^Q &= \bar{h}_v\left[i\!\!\not{D} - \frac{\not{D}\!\!\not{D}}{m_Q} - \frac{\not{D}(\not{v} - 1)\!\!\not{D}}{4m_Q}\right]h_v \\
&= \bar{h}_v\left[iv \cdot D - \frac{1}{2m_Q}\left(D_\mu D^\mu + \frac{1}{4}g_3\lambda^a\sigma^{\mu\nu}F^a_{\mu\nu}\right) - \frac{(v \cdot D)^2}{2m_Q}\right]h_v, \qquad (3.14)
\end{aligned}$$

which is the desired expansion in terms of the heavy-quark mass. Because the last term in this expression is second order in $v \cdot D$ and noting that $v \cdot Dh_v = 0$ to lowest order, it will not contribute to matrix elements at order $1/m_Q$ and can be dropped. The lagrangian of Eq. (3.14) corresponds to a quark moving at fixed velocity. Antiquark solutions can be constructed with the mass dependence $e^{+im_Q v \cdot x}$, with the result

$$\mathcal{L}_v^{\bar{Q}} = \bar{k}_v\left[-iv \cdot D - \frac{1}{2m_Q}\left(D_\mu D^\mu + \frac{1}{4}g_3\lambda^a\sigma^{\mu\nu}F^a_{\mu\nu}\right) - \frac{(v \cdot D)^2}{2m_Q}\right]k_v, \qquad (3.15)$$

where the field k_v satisfies $\Gamma_v{-}k_v = k_v$. It is legitimate to neglect the production of heavy $Q\bar{Q}$ pairs. However, one should superpose the lagrangians for different velocities in a Lorentz-invariant fashion,

$$\mathcal{L} = \int d^4v \, \delta \left(v_\mu v^\mu - 1\right) \theta(v_0) \left[\mathcal{L}_v^Q + \mathcal{L}_v^{\bar{Q}}\right] = \int \frac{d^3v}{2v_0} \left[\mathcal{L}_v^Q + \mathcal{L}_v^{\bar{Q}}\right]. \quad (3.16)$$

The nature of the approximation at this stage is more of a classical limit rather than a nonrelativistic limit. To be sure, for any given quark one can work in the quark's rest frame, in which case the quark will be nonrelativistic. However, when external currents act on the fields, transitions from one frame to another occur for which $\Delta \mathbf{v}$ is *not* small. On the other hand, the result can be said to be classical because quantum corrections have not yet been included and these can renormalize the coefficients in $L_v^{Q\bar{Q}}$. Also, diagrams involving the exchange of hard gluons can produce nonstatic intermediate states. Such corrections can be accounted for in perturbation theory [Wi 91].

XIII–4 Nonconventional hadron states

Many suggestions have been made regarding the possibility of hadronic states beyond those predicted by the simple quark model of $Q\bar{Q}$ and Q^3 configurations. The study of such states is hampered by the fact that we still have very little idea why the quark model works. *QCD* at low energy is a strongly interacting field theory, and we would expect a very rich and complicated description of hadronic structure. That the result should be describable in terms of a simple $Q\bar{Q}$ and Q^3 picture as even a first approximation remains a mystery. Quark models have been popular because they seem to work phenomenologically, not because they are a controlled approximation to *QCD*. This weakness becomes all the more evident when one tries to generalize quark model ideas to new areas.

Much of the theoretical work on nonconventional states has involved the concept of a *constituent gluon G*, analogous to a constituent quark Q, and we shall cast our discussion with respect to this degree of freedom.[9] It is clear that there should be a cost in energy to excite a constituent gluon. The energy should not be extremely large, else it would be difficult to understand the early onset of scaling in deep-inelastic scattering. However, it cannot be less than the uncertainty principle bound on a massless particle confined to a radius $R \sim 1$ fm of $E = p \gtrsim \sqrt{3}/R \simeq 342$ MeV (cf. Sect. XI–1). Model calculations have tended to use a somewhat larger effective gluon 'mass'.

---

[9] However, it should be understood that such a concept has not been shown to follow rigorously from *QCD*, nor indeed is a configuration of definite numbers of consitituent gluons a gauge-invariant entity (cf. Sect. X–2).

Table XIII–3. *Gauge-invariant color-singlet*
interpolating fields.

Operator	Dimension	J^{PC}
$\bar{q}\Gamma q$	3	$0^{-+}, 1^{--}, 0^{++}, 1^{+-}, 1^{++}$
$\bar{q}\Gamma \mathcal{D}q$	4	$2^{++}, 2^{-\pm}$
FF	4	$0^{++}, 2^{++}, 0^{-+}, 2^{-+}$
$\bar{q}\Gamma q F$	5	$0^{\pm+}, 0^{+-}, 1^{\pm+}, 1^{\pm-}, 2^{\pm\,\mathsf{I}}, 2^{\pm}$
$F\mathcal{D}F$	5	$1^{++}, 3^{++}$

The basic idea of confinement is that only color-singlet states exist as physical hadrons. If we identify those states which are color singlets and which contain few quark or gluon quanta, we can easily find other possible configurations besides $Q\bar{Q}$ and Q^3. Some of the more well-known examples are

(1) Gluonia (or glueballs) – quarkless G^2 or G^3 states, which we shall discuss in more detail below,
(2) Hybrids – color-singlet mixtures of constituent quarks and gluons like $Q\bar{Q}G$ mesons or Q^3G baryons,
(3) Dibaryons – six-quark configurations in which the quarks have *similar* spatial wavefunctions rather than two separate three-quark clusters,
(4) Meson molecules – loosely bound deuteron-like composites of mesons.
(5) Tetraquark states – strongly bound states with quark structures $qq\bar{q}\bar{q}$.

A convenient framework for describing the quantum numbers of possible hadronic states is obtained by considering gauge-invariant, color-singlet operators of low dimension [JaJR 86], as was discussed in Sect. XI–1. Table XIII–3 lists all such operators up to dimension five which can be constructed from quark fields, QCD covariant derivatives, and the gluon field strength, denoted respectively by $q, \mathcal{D}q, \mathcal{D}F$, and F. Also appearing in Table XIII–3 is the collection of J^{PC} quantum numbers associated with each such operator. Particular spin-parity values are obtained from these operators by choosing indices in appropriate combinations.

The first resonance – $\sigma(440)$

The lightest resonance encountered in the meson spectrum has long been one of the most controversial states. This state is officially known as $f_0(500)$, but it is almost universally referred to as σ. The existence of this resonance has finally been established unambiguously. However, the interpretation remains remarkably subtle.

The scattering of two pions in the $I = 0$ and $J = 0$ channel becomes strong at low energies. The amplitude is described by chiral perturbation theory, as described in Sect. VII–3. At first order in the energy expansion, the scattering amplitude is[10]

$$T_{00}^{(0)} \equiv t_0 = \frac{s}{16\pi F_\pi^2}.$$ (4.1)

This amplitude is purely real, while under the general principle of unitarity of the S matrix the elastic amplitude must have the form

$$T_{00} = e^{i\delta_{00}} \sin \delta_{00},$$ (4.2)

and has to satisfy

$$\mathrm{Im}\, T_{00} = |T_{00}|^2.$$ (4.3)

The lowest-order amplitude of Eq. (4.1) has no imaginary part. However, in chiral perturbation theory, the imaginary part starts at order E^4, and the first contribution to this appears through one-loop diagrams. Chiral perturbation theory satisfies unitarity order by order in the energy expansion.

The σ appears as a resonance when exact unitarity is applied to the scattering amplitude. The pole can be seen in an exceptionally simple approximation. If one simply iterates the lowest-order amplitude one can produce a fully unitary result

$$T_{00} = \frac{t_0}{1 - i t_0},$$ (4.4)

which satisfies Eq. (4.3) exactly and also reproduces the chiral result to first order. The use of Eq. (4.1) with a complex value for s as the input for Eq. (4.4) produces a pole on the second sheet at

$$\sqrt{s} = (1 - i)\sqrt{8\pi}\, F_\pi = (460 - i460)\ \mathrm{MeV}.$$ (4.5)

This is the first approximation to the σ.

The complete analysis is much more subtle, but carries a similar result. By including not only unitarity, but also crossing symmetry and analyticity, one can obtain a dispersive representation of the scattering amplitude [Ro 71]. When evaluated using chiral constraints at low energy and data at high energy, the $\pi\pi$ data can be fully described [CoGL 01]. When extended into the complex plane, the real σ pole appears at [CaCL 06]

$$\sqrt{s} = m_\sigma - i\frac{\Gamma_\sigma}{2} = (441 - i272)\ \mathrm{MeV}.$$ (4.6)

---

[10] In order to keep the formulas simple and physically transparent in this introductory section, we present them with the pion mass set equal to zero.

However, this does not appear as a typical resonance. In contrast to others, the σ width is larger than its mass, indicating that the pole is far from the physical values of s. Moreover, in the scattering amplitude itself, there is no sign of a resonant bump. The phase shift rises almost linearly from $\delta_{00} = 0$ at threshold to $\delta_{00} = 100°$ around 900 MeV. The phase shift does go through $90°$, traditionally a sign of a resonance in elastic scattering, but at an energy $\sqrt{s} \sim 850$ MeV which is far removed from the pole position. These unusual features had long created confusion about the existence of the σ, which has been cleared up only through the rigorous combination of chiral and dispersive techniques.

The σ is a dynamical strong-coupling resonance. The resonance does not fit naturally into the quark model and it does not seem profitable to try to force the σ into that framework. While we do expect to see quark model bound states as resonances, there is no requirement that all resonant behavior must be associated with quark model states. Indeed, there is a strong theoretical argument that the σ is different from the bound states of *QCD* [Pe 04]. Recall two features of the large N_c limit discussed in Chap. X – that the meson bound states stay constant in mass when the large N_c limit is taken, but scattering amplitudes fall like $1/N_c$. This latter requirement is satisfied for the $\pi\pi$ amplitudes; in the lowest-order amplitude of Eq. (4.1), the amplitude falls with N_c because $F_\pi \sim \sqrt{N_c}$ appears squared in the denominator. Because the $\pi\pi$ amplitude is smaller at larger N_c, the ampltitude becomes of order unity at a higher energy. If the σ is indeed connected with the strong coupling of $\pi\pi$ scattering, its mass will shift to higher energy as N_c increases. While we cannot change N_c in the scattering data themselves, there are straightforward analytic methods, such as the inverse-amplitude method [DoP 97], which is a variant of Padé techniques,[11] to closely describe the data including chiral perturbation theory and exact unitarity. Use of such techniques is able to reproduce the σ found in the data, and then when N_c is varied one finds [Pe 04],

$$m_\sigma \sim \sqrt{N_c}, \qquad (4.7)$$

as expected by the general argument. Indeed, even our simplified approximation of Eq. (4.5) has this behavior, again due to $F_\pi \sim \sqrt{N_c}$. Because the bound states of *QCD* should behave as a constant, $m \sim N_c^0$, the σ appears distinct from these. It appears to be a resonance associated with the unitarity of elastic scattering.[12]

Some caveats and cautions about this result are appropriate. This experimental resonance does not appear to be the σ of the linear σ model. As described in Chaps. IV and VII, the coefficients of the chiral lagrangian are sensitive to the underlying fundamental theory, and the coefficients found for *QCD* do not resem-

[11] Our approximation of Eq. (4.5) above is equivalent to the lowest order of the inverse amplitude method.

[12] Other states that may have a related origin include the $\kappa(800)$ seen in $K\pi$ scattering and the $N(1405)$ in πN scattering.

ble those of the linear σ model. Nor is the existence of this state a justification to use a fundamental σ field in field-theoretic calculations. While the use of σ exchange with a particular coupling may be a proxy for $\pi\pi$ effects in a given reaction, this use is not necessarily valid in general. The use of a fundamental σ is much more restrictive than the variety of pionic effects. Moreover, it is neither an accurate nor controlled approximation, and may double-count the pionic contributions, which must also be included.

In addition, the above discussion provides a cautionary counterexample to a widely used argument. It is common to use the violation of tree unitarity of an effective theory as an indication of the energy at which New Physics should be seen [LeQT 77], with the expectation that the New Physics would restore unitarity. In the situation discussed above, the usual measure of tree-unitarity violation, $\mathrm{Re}\, T_{00} \le 1/2$, occurs at 460 MeV, which is well below the production threshold of the quarks and gluons of QCD. Also, the energy of tree-unitarity violation varies as $\sqrt{N_c}$ in units where the scale of QCD is held fixed [AyAD 12]. Thus, any 'New Physics' does not have the same N_c scaling. The strongly coupled effective theory manages to respect unitarity without new degrees of freedom. The situation above indicates that, while the violation of tree unitarity does indicate the existence of a strongly coupled region, its use as an indicator of New Physics must be treated with caution.

Gluonia

The existence of a gluon degree of freedom in hadrons is beyond dispute, with evidence from deep-inelastic lepton scattering and jet structure in hadron–hadron collisions. However, trying to predict the properties of a new class of hadrons whose primary ingredient is gluonic is nontrivial. Hypothetically, if quarks could be removed from QCD the resulting hadron spectrum would consist only of *gluonia* (or 'glueballs').

Gluonic configurations should be signaled by the existence of extra states beyond the expected nonets of $Q\bar{Q}$ hadrons. However, mixing with $Q\bar{Q}$ hadrons is generally possible (cf. Sect. X–2). Although predicted by the $1/N_c$ expansion to be suppressed, such mixing effects serve to cloud the interpretation of data vis-à-vis gluonium states. Referring to the interpolating fields mentioned above, we see that for gluons the gauge-invariant combinations

$$F^a_{\mu\nu} F^{a\mu\nu}, \qquad F^a_{\mu\lambda} F^{\lambda}_{a\nu}, \qquad F^a_{\mu\nu} \tilde{F}^{a\mu\nu}, \qquad F^a_{\mu\lambda} \tilde{F}^{\lambda}_{a\nu} \qquad (4.8)$$

can be formed out of *two* factors of a gluon field-strength tensor $F^a_{\mu\nu}$ or its dual $\tilde{F}^{a\mu\nu}$. The spin, parity, and charge conjugation carried by these these operators are respectively $J^{PC} = 0^{++}, 2^{++}, 0^{-+}, 2^{-+}$, and are thus the quantum numbers expected for the lightest glueballs,[13] i.e., such operators acting on the vacuum state produce states with these quantum numbers. Although there is no *a priori* guarantee that one obtains a single particle state (e.g., a 2^{++} operator could in principle create two 0^{++} glueballs in a D wave), the simplicity of the operators leads one to suspect that this will be the case. There is one, somewhat controversial, construct missing from the above list. Two massive spin-one particles in an S wave can have $J^{PC} = 1^{-+}$ as well as $J^{PC} = 0^{++}, 2^{++}$, and some models predict such a gluonium state. However, a 1^{-+} combination of two massless on-shell vector particles is forbidden by a combination of gauge invariance plus rotational symmetry [Ya 50]. The lack of a 1^{-+} gauge-invariant, two-field operator is an indication of this.

Aside from a list of quantum numbers and some guidance as to relative mass values, theory does not provide a very clear profile of gluonium phenomenology. Lattice-gauge methods offer the best hope for future progress. Present quenched lattice studies predict that in a quarkless version of QCD the lightest glueball is a 0^{++} state of mass 1.7 ± 0.1 GeV and while the 2^{++} and 0^{-+} glueballs are about 1.4 ± 0.1 times heavier [Ba *et al.* 93], [MoP 99], [Ch *et al.* 06].

The challenge arises when couplings to quark degrees of freedom are introduced, in which case substantial mixing between quark and gluonium states must occur. Lattice studies of the mixing with the 0^{++} state have yielded mixed results, some indicating a lowering of the mass by as much as several hundred MeV [Ha *et al.* 06], while others show little effect [Ri *et al.* 10]. It is generally agreed that inclusion of quarks has little effect on the mass of the 2^{++} and 0^{-+} states [Ri *et al.* 10], [HaT 02]. The problem has also been studied via QCD sum rules with inclusion of instanton effects, but again there exists considerable uncertainty [Fo 05], [Ha *et al.* 11].

Gluonium states would be classified as flavor-$SU(3)$ singlets and if mixing with quark states exist there should exist 'extra' such states. An example of this phenomenon exists in the 1.5 GeV region where the states

$$f_0(1370), \ f_0(1500), \ f_0(1710), \ K_0^*(1430), \ a_0(1450)$$

can be interpreted as a nonet of $q\bar{q}$ states plus a glueball [AmC 96]. In this picture the three f_0 states are mixtures of the 0^{++} glueball and the two $q\bar{q}$ states from the nonet. The validity of this description relies on the existence of these three f_0 resonances. While the $f_0(1500)$ and $f_0(1710)$ are reasonably well established and have significant two-meson decay channels, the same is not true of the $f_0(1370)$, which, if it does exist, has a large ($>80\%$) decay fraction into 4π. For this reason the interpretation in terms of three-channel mixing of these states is still

---

[13] Gluonic operators with *three* field-strength tensors produce states with $J^{PC} = 0^{\pm+}, 1^{\pm+}, 2^{\pm+}, 1^{\pm+}, 2^{\pm-}, 3^{\pm-}$. Because of the extra gluon field, one expects these states to be somewhat heavier.

Table XIII–4. *Spectroscopy of six-quark configurations.*

$SU(6)$ of color-spin	$SU(3)$ of flavor	Spin
490	1	0
896	8	1,2
280	10	1
175	10*	1,3
189	27	0,2
35	35	1
1	28	0

controversial. Thus, despite 30 years of work on the problem of glueballs, the situation remains confused. A recent review of the subject can be found in [Oc 13].

Additional nonconventional states

There is a widespread belief that gluonium states *must* appear in the spectrum of the QCD hamiltonian, though as discussed above it has proved challenging to identify them. For other kinds of nonconventional configurations, it is also difficult to reach a meaningful consensus, although experimental efforts to detect such states are ongoing. We briefly review several such possibilities.

(i) *Hybrids*: From Table XIII–3, we see that among the $\bar{Q}QG$ meson hybrids is one with the quantum numbers $J^{PC} = 1^{-+}$. This would-be hadron is of particular interest because comparison with Table XI–3 reveals that it cannot be a $\bar{Q}Q$ configuration. Model calculations suggest that the lightest such state should be isovector, with mass in the range 1.5–2.0 GeV, and that such states may largely decouple from $L = 0$ $\bar{Q}Q$ meson final states. A study of $Q^3 G$ baryon hybrids reveals that *none* of the states is exotic in the sense of lying outside the usual Q^3 spectrum [GoHK 83].

(ii) *Dibaryons*: The most remarkable aspect learned yet about the dibaryon states is how much six-quark configurations are restricted by Fermi–Dirac statistics. Table XIII–4 lists the possible six-quark $SU(3)$ multiplets along with their spin values [Ja 77]. Of this collection of states, the most attention has been given to the spinless $SU(3)$-singlet state, called the *H-dibaryon*. This particle, which has strangeness $S = -2$ and isospin $I = 0$, is predicted to be the lightest dibaryon, and if bound would to be unstable to weak decay. A series of experiments has failed to find the H, so at this time there is no evidence for the existence of dibaryons.

(iii) *Hadronic molecules and tetraquarks*: Particles with the quark content $qq\bar{q}\bar{q}$ also form color singlets. The literature distinguishes two types of such states:

hadronic molecules and tetraquarks. Roughly speaking, the molecular states refer to two separate $q\bar{q}$ color-singlet states that are lightly bound. Since the binding energy is small, such states could be expected to be found right near the threshold for the two mesons. Tetraquarks refer to configurations where the $qq\bar{q}\bar{q}$ constituents are more compactly intertwined, with the details of the configuration varying in different models. Clearly, there can be a continuum interpolating between these extremes. We will not enter into the debate about the signals for the two classes of four-quark states.

There appears to be clear evidence for the existence of a state in this category. The $Z_c(3900)$ [Li *et al.* 13] [Ab *et al.* 13] has mass and production properties that indicate that it contains a $c\bar{c}$ pair. However, it also carries a charge which proves that it also contains light quarks with the $u\bar{d}$ combination producing the positive charge. The internal configuration has not been sorted out yet.

Among the particles that have been discussed as molecules are the isovector $a_0(980)$ and isoscalar $f_0(975)$ mesons. Nominally, these particles have the quantum numbers of the $L = 1$ sector of the $Q\bar{Q}$ model, and their near equality in mass suggests an internal composition similar to that of the $\rho(770)$ and $\omega(783)$, i.e., orthogonal configurations of nonstrange quark–antiquark pairs. However, among properties which argue against this are their relatively strong coupling to modes which contain strange quarks, their narrower-than-expected widths, and their $\gamma\gamma$ couplings. The proximity of the $K\bar{K}$ threshold and the importance of the $K\bar{K}$ modes has motivated their interpretation as $K\bar{K}$ molecules [Wei 83]. However, interpretation of scattering data near the 1 GeV region is not clear, and indeed a strong case has been made for the alternative $qq\bar{q}\bar{q}$ picture ['tHoIMPR 08] and for heavier states as well.

A clearer situation is provided by the $X(3872)$, which has been interpreted in terms of a D_0–$\bar{D}_0^*$ hadronic molecule, which is bound by π^0 exchange at long distance and quark/color exchange at short distances. That $X(3872)$ is not a simple charmonium state is indicated by large isospin violation seen in the data. This occurs in the molecule interpretation because the mass of the resonance is essentially identical to $m_{D^0} + m_{D^{0*}}$ and considerably lighter than $m_{D^+} + m_{D^{-*}}$. Thus, the molecular state would predominantly involve D_0–$\bar{D}_0^*$ containing $c\bar{c}u\bar{u}$ quarks, so that this structure is a mixture of isospin states

$$c\bar{c}u\bar{u} = c\bar{c}\sqrt{\frac{1}{2}}\left[\sqrt{\frac{1}{2}}(u\bar{u} + d\bar{d}) + \sqrt{\frac{1}{2}}(u\bar{u} - d\bar{d})\right] \qquad (4.9)$$

In this picture there should be nearly comparable decays to final states with $I = 0$ and $I = 1$, and this is indeed indicated by significant branching ratios of the $X(3872)$ to both $J/\psi\rho$ and $J\psi\omega$ modes.

Other examples of four-quark states may occur in the $b\bar{b}$ system and the resonances $X_b^+(10610)$ and $X_b^+(10650)$, which appear to be a bound states of $B^+–\bar{B}^{0*}$ and $B^{0*}–\bar{B}^{+*}$, respectively. In this case the states are charged, with quark content $b\bar{b}u\bar{d}$, so that both states are clearly exotic–they cannot be excited bottomonium.

The overall interpretation of these states is complicated by the fact that molecules and tetraquarks have the same quark content and are distinguished only by details of their internal configuration. In some cases, both interpretations have advocates [AlHW 12, Du *et al.* 10].

Problems

(1) **Power-law potential in quarkonium**

Consider an interquark potential of the form $V(r) = cr^d$.

(a) Use the virial theorem to determine $\langle T \rangle / \langle V \rangle$ for the ground state.

(b) Given the form $E_{2S} - E_{1S} = f(d)M^{-d/(2+d)}$, where M is the reduced mass, determine d from the observed mass differences in the $c\bar{c}$ and $b\bar{b}$ systems, using Eq. (3.1) to supply heavy-quark mass values.

(c) Assuming this model is used to fit the spin-averaged ground-state $c\bar{c}$ and $b\bar{b}$ mass values, determine $\mathbf{v}^2/c^2$ for each system.

(2) **Quarkonium annihilation from the $^1S_0$ state**

Modify Eq. (1.15) to obtain the leading-order contributions appearing in Eq. (1.16).

(3) **Mass relations involving heavy quarks**

(1) Repeat the analysis of Eq. (3.1) but using the masses of the charmed/strange mesons D_s, D_s^* instead. Infer a value for $\hat{M}/M_c$ by referring to the result obtained in Eq. (2.17). Compare with the determination of Eq. (3.1).

(2) Extend the procedure of Eqs. (2.20–2.25) to isospin-violating mass differences of c-flavored and b-flavored hadrons.

XIV

Weak interactions of heavy quarks

Heavy quarks provide a valuable guide to the study of weak interactions. Measurements of decay lifetimes and of semileptonic decay spectra of heavy, flavored mesons[1] yield information on individual elements of the CKM matrix, as does the observation of heavy-meson particle–antiparticle transitions such as B_d–$\bar{B}_d$ mixing. Long anticipated data involving detection of *CP*-violating signals have been found to be in accord with expectations of the Standard Model and have played a crucial role in constraining the sole complex phase in the CKM matrix.

XIV–1 Heavy-quark mass

At the level of the Standard Model lagrangian, the six quark masses are equivalent; they are all just input parameters that must each be determined experimentally. In the real world of particle phenomenology, quark mass divides into two sectors, 'light' (u, d, s) and 'heavy' (c, b, t). It is a hallmark of light-quark spectroscopy that hadron mass is not a direct reflection of quark mass. However, for hadrons which contain a heavy quark, the energy scale is set by the mass of the heavy quark. In the following, we discuss topics of special relevance to heavy-quark mass.

Running quark mass

Heretofore we have described the renormalization of quark mass in terms of the mass shift $\delta m = m - m_0$, where m_0 is the bare mass. We can also, for convenience, employ a multiplicative mass renormalization constant Z_m with $m_0 = Z_m m$. In minimal subtraction, Z_m will have an ϵ-expansion,

---

[1] Note that in the conventions of the Particle Data Group the quantum numbers of the neutral mesons are
$K^0 = (d\bar{s})$, $D^0 = (c\bar{u})$, $B^0 = (d\bar{b})$ and $B_s^0 = (s\bar{b})$.

$$Z_m\left(\alpha_s, \epsilon^{-1}\right) = 1 + \sum_{n=1}^{\infty} \frac{Z_{m,n}(\alpha_s)}{\epsilon^n} = 1 - 3C_2(3)\frac{\alpha_s}{4\pi}\frac{1}{\epsilon} + \cdots. \tag{1.1}$$

Both $m = m(\mu)$ and $Z_m(\mu)$ will depend implicitly on a scale μ, but not the bare mass m_0. A sequence of steps follows from this simple observation,

$$m_0 = Z_m m \quad \text{with} \quad \frac{dm_0}{d\ln\mu} = 0,$$

$$\frac{dm}{d\ln\mu} = -\frac{m(\mu)}{Z_m}\frac{dZ_m}{d\ln\mu} \equiv -\gamma_m(g(\mu))m(\mu),$$

$$\gamma_m = \frac{1}{Z_m}\frac{dZ_m}{d\ln\mu} = \gamma_m^{(0)}\frac{\alpha_s}{4\pi} + \gamma_m^{(1)}\left(\frac{\alpha_s}{4\pi}\right)^2 + \cdots, \tag{1.2}$$

where γ_m is called the *anomalous dimension* of the quark mass operator. Since there is no explicit dependence in Z_m on either quark mass m or a renormalization scale μ, the anomalous dimension γ_m is the same in any minimally subtracted regularization scheme, such as $\overline{\text{MS}}$.

Let us determine the leading coefficient $\gamma_m^{(0)}$. From Eq. (1.2) we have[2]

$$Z_m\gamma_m(g) = \frac{dZ_m}{d\ln\mu} = 2g\frac{dg}{d\ln\mu}\frac{dZ_m}{dg^2}. \tag{1.3}$$

To proceed, we shall require an extension to $\epsilon \neq 0$ of Eq. (II–2.57b),

$$\frac{dg}{d\ln\mu} \equiv \beta(g(\mu), \epsilon) = -\epsilon g - \beta_0\frac{g^3}{16\pi^2} + \cdots = -\epsilon g + \cdots, \tag{1.4}$$

where we recall that $\beta_0 = 11 - 2n_f/3 > 0$. We then obtain from Eq. (1.3),

$$(1 + \cdots)\left(\gamma_m^{(0)} + \cdots\right) = 2g(-\epsilon g + \cdots)\left(\frac{dZ_{m,1}}{dg^2}\frac{1}{\epsilon} + \cdots\right) \tag{1.5}$$

or, finally, the desired result

$$\gamma_m^{(0)} = 6C_2(3) = 8. \tag{1.6}$$

At this point, we have a differential equation whose integration gives the scale dependence of the quark mass,

$$\frac{dm(\mu)}{m(\mu)} = -\gamma_m(g(\mu))d\ln\mu,$$

$$d\ln\mu = \frac{d\ln\mu}{dg}dg = \frac{dg}{\beta(g)},$$

$$m(\mu) = m(\mu_0)\exp\left[-\int_{g(\mu_0)}^{g(\mu)} dg'\frac{\gamma_m(g')}{\beta(g')}\right], \tag{1.7}$$

---

[2] For notational simplicity, we suppress the subscript in g_3 and use instead g.

where $\beta(g)$ is the beta function of Eq. (II–2.57b). This equation is ordinarily used for situations for which QCD perturbation theory is applicable (i.e. short distances). Here, we consider the leading-order expressions, with

$$\beta = -\beta_0 \frac{\alpha_s}{4\pi} g, \qquad \gamma_m = \gamma_m^{(0)} \frac{\alpha_s}{4\pi}, \tag{1.8}$$

the insertion of which into Eq. (1.7) yields

$$m(\mu) = m(\mu_0) \left[\frac{\alpha_s(\mu)}{\alpha_s(\mu_0)} \right]^{\gamma_m^{(0)}/2\beta_0}. \tag{1.9}$$

We hasten to note that the concept of a running quark mass is valid for all six flavors, not just heavy quarks. For heavy quarks, it has become standard to express the $\overline{\text{MS}}$ mass in the form $\overline{m}(\overline{m})$, i.e., to refer to the scale $\mu = \overline{m}$ which equals the $\overline{\text{MS}}$ mass itself. This is convenient because any experimental determination $\overline{m}(\mu_{\text{expt}})$ can always be 'run' to the scale $\mu = \overline{m}$. A compilation of various phenomenological inputs yields [RPP 12]

$$\overline{m}_c(\overline{m}_c) = 1.275 \pm 0.025 \text{ GeV}, \qquad \overline{m}_b(\overline{m}_b) = 4.18 \pm 0.03 \text{ GeV}. \tag{1.10}$$

Equation (1.9) represents the leading-order expression for the running mass. Extensive work on higher-order corrections has been carried out, to the level of four loops [Ch 97, VeLR 97]. An accessible recipe for a running mass at four loops is given by

$$m(\mu) = m(\mu_0) \cdot \frac{c(a_s(\mu))}{c(a_s(\mu_0))}, \tag{1.11}$$

where[3] $a_s(\mu) \equiv \alpha_s(\mu)/\pi$. In the above, the argument of the function $c(a_s(\mu))$ requires a running strong-coupling $\alpha_s(\mu)$ also evaluated at four-loop order, but this has been addressed earlier in Eqs. (II–2.77), (II–2.78). Useful numerical forms of $c(x)$ are given in [Ch 97] for each of the s, c, b, and t quarks. For example, we shall refer in Chap. XV to the b-quark version,

$$c_b(x) = x^{12/23} \left(1 + 1.17549\, x + 1.50071\, x^2 + 0.172478\, x^3 \right). \tag{1.12}$$

This can be applied to run the b-quark mass from the scale $\mu_0 = \overline{m}_b(\overline{m}_b)$ to $\mu = M_H$, where M_H is the mass of the Higgs boson. We find $\overline{m}_b(M_H) \simeq 0.665\, \overline{m}_b(\overline{m}_b)$.

The pole mass of a quark

Since quarks do not exist as free particles, it should perhaps not be surprising that different theoretical definitions of quark mass appear in the literature. In the above,

---

[3] Note this is *not* the same as the quantity a_s appearing in Eq. (II–2.76).

we have discussed quark mass as it is defined in the $\overline{\text{MS}}$ renormalization scheme. Another definition, the *pole mass*, is simply the renormalized quark mass in on-shell renormalization. As an example where use of pole mass seems natural, consider the top quark. Top-quark mass is measured 'directly' in collider experiments, primarily via the production of $t\bar{t}$ pairs. The t quarks will each decay as $t \to W^+ b$, which ultimately gives rise to lepton + jet, dilepton, and all-jet final states. The top mass obtained by fitting invariant mass distributions of final-state particles has been interpreted as a pole mass, with recent Tevatron and LHC evaluations [Mu 12]

$$M_t = \begin{cases} 173.18 \pm 0.94 \text{ GeV} & \text{[Tevatron]} \\ 173.3 \pm 1.4 \text{ GeV} & \text{[LHC]}. \end{cases} \tag{1.13}$$

There is also an 'indirect' way of determining top mass by performing a global fit of Standard Model observables in which the top quark contributes as a virtual particle.

An interesting theoretical issue is the relation between pole mass and $\overline{\text{MS}}$ mass. This has been carried out in *QCD* perturbation theory as far as three-loop order [MeR 00]. In the following we shall review this process to leading order in α_s. We begin with the inverse renormalized quark propagator, expressed as

$$S_{F,\text{ren}}^{-1}(p) = \not{p} B_{\text{ren}}(p^2, \overline{m}^2) - \overline{m} A_{\text{ren}}(p^2, \overline{m}^2), \tag{1.14}$$

where the functions A_{ren} and B_{ren} are calculated in *QCD* perturbation theory, with $\overline{m}$ being the $\overline{\text{MS}}$ mass. We must seek a zero in $S_{F,\text{ren}}^{-1}(p)$ for the on-shell conditions of $\not{p} = M$ and $p^2 = M^2$ with M being the pole mass. Following [FlJTV 99], we have for the $\mathcal{O}(\alpha_s)$ renormalized propagator amplitudes in the on-shell limit,

$$A_{\text{o-s}} \equiv A_{\text{ren}}\Big|_{\not{p}=M,\ p^2=M^2} = 1 + \frac{\alpha_s}{4\pi} 2C_2(3)(2+\xi) + \cdots$$

$$B_{\text{o-s}} \equiv B_{\text{ren}}\Big|_{\not{p}=M,\ p^2=M^2} = 1 + \frac{\alpha_s}{4\pi} 2C_2(3)\xi + \cdots, \tag{1.15}$$

where ξ is the gauge parameter and $C_2(3)$ is given below Eq. (II–2.12). We can now obtain the desired relation between pole mass M and $\overline{\text{MS}}$ mass $\overline{m}$ in terms of the $\overline{\text{MS}}$ coupling $\hat{\alpha}_s$. The condition for a zero, $0 = mA_{\text{o-s}} - MB_{\text{o-s}}$, implies the relation,

$$M = \overline{m}\frac{A_{\text{o-s}}}{B_{\text{o-s}}} = \overline{m}(M)\left[1 + C_2(3)\frac{\hat{\alpha}_s(M)}{\pi} + \cdots\right], \tag{1.16}$$

where we exhibit scale dependence in $\overline{m}$ or α_s as it would appear in a more general treatment. Notice that the explicit gauge dependence has canceled, as it must.

For the top quark, a comparison between $\overline{\text{MS}}$ mass and pole mass at NNLO level in the *QCD* perturbation theory gives

$$\overline{m}_t(\overline{m}_t) = 163.3 \pm 2.7 \, \text{GeV}, \qquad M_t = 173.3 \pm 2.8 \, \text{GeV}, \qquad (1.17)$$

as inferred from Tevatron data [AlDM 12].

Actually, it would appear that the very concept of pole mass for a quark is paradoxical because, after all, quarks are *not* free particles, and it is, in fact, the case that due to confinement the exact nonperturbative quark propagator will not have a pole. The pole mass exists as a creature of perturbation theory and phenomenology. There is, however, a price to pay for this convenience. Calculation has shown that there will be higher orders which grow factorially in the perturbation expansion [BeB 95, BiSUV 94]. Because of this, the pole mass itself cannot be determined to an accuracy better than the confinement scale Λ_{QCD}. Other definitions of the mass parameter include the $1S$ *mass*, defined as one-half the energy of the $1S \, Q\bar{Q}$ state [HoLM 99], and the *kinetic mass*, defined via a threshold in weak decay [BiSUV 94]. Because these include the effects of confinement, they turn out to be better behaved in many perturbative calculations [ElL 02]. Indeed, even for the top quark the $1S$ mass is prefered for a proper theoretical description of the $t\bar{t}$ production cross section near threshold [HoT 99].

Our lack of understanding of the large magnitude of the top mass illustrates how little we actually know about the mechanism of mass generation. If all fermion masses arise from the Yukawa interaction of a single Higgs doublet, then the Yukawa coupling constants must vary by the factor $g_t/g_e = m_t/m_e \sim 3 \times 10^5$. There is nothing inconsistent about such a variation, but it is so striking as to beg for a logical explanation, one which is presently lacking.

XIV–2 Inclusive decays

Heavy quarks decay to a large number of final states, often containing many particles. As the mass of the heavy quark gets larger, it makes increasing sense to treat the final states inclusively. We discuss this approach in this section.

The spectator model

Consider the weak beta decay, $Q \to q\bar{e}\nu_e$, of an isolated heavy quark Q into a lighter quark q. By analogy with muon decay, this proceeds with decay rate (if radiative corrections are ignored)

$$\Gamma_{Q \to q\bar{e}\nu_e} = \frac{G_F^2 m_Q^5}{192\pi^3} \left| V_{qQ} \right|^2 f(m_q/m_Q),$$

$$f(x) = 1 - 8x^2 + 8x^6 - x^8 - 24x^4 \ln x \qquad (2.1)$$

where $f(x)$ is the phase-space factor already encountered in our discussion of muon decay in Sect. V–2 and of tau decay in Sect. V–3. Under what circumstances would this be a good representation for the beta decay of a heavy-*meson*-containing quark Q? For it to be accurate, the final state must develop independently of the other (so-called *spectator*) quark in the heavy meson. Experience with deep-inelastic scattering suggests that this occurs when the recoiling quark q carries energy and momentum larger than typical hadronic scales, i.e., in the range $E_q > 1$–1.5 GeV. For D decays, the average light-quark energy is $\langle E_q \rangle \sim m_D/3 \simeq 0.5$ GeV, in which case this approximation is suspect. It should be considerably better in B decays, but still not perfect.

Let us explore the consequences of adopting the spectator model for D and B decays. If we neglect CKM-suppressed modes, the main decay channels for b quarks are $b \to c\bar{u}d, c\bar{c}s, c\ell\bar{\nu}_\ell$ ($\ell = e, \mu, \tau$), while for c quarks they are restricted to $c \to s\bar{d}u, s\bar{\mu}\nu_\mu, s\bar{e}\nu_e$. Relative to the lepton modes, each hadronic decay channel picks up an additional factor of 3 upon summing over the final-state colors. Two of the B-meson final states ($c\bar{c}s$ and $c\tau\bar{\nu}_\tau$) have significant phase-space suppressions (reducing them to about 20% of the $c\bar{u}d$ mode) due to the heavy masses involved. The simplest spectator model then predicts branching ratios

$$\text{Br}_{D \to \bar{e}\nu_e X} \simeq \frac{1}{3+2} = 0.2,$$

$$\text{Br}_{B \to e\bar{\nu}_e X} \simeq \frac{1}{3 \times (1 + 0.2) + 2 + 0.2} = 0.17 \tag{2.2}$$

where X denotes a sum over the remaining final-state particles. Also, this picture predicts the absolute rates of the D and B decays to be

$$\tau_D = \left[5\frac{G_F^2 m_c^5}{192\pi} |V_{cs}|^2 f(x_c) \right]^{-1} \simeq 1.1 \times 10^{-12} \text{ s}, \tag{2.3a}$$

$$\tau_B = \left[5.8\frac{G_F^2 m_b^5}{192\pi^2} |V_{cb}|^2 f(x_b) \right]^{-1} \simeq 1.8 \times 10^{-12} \text{ s} \left| \frac{0.041}{V_{cb}} \right|^2, \tag{2.3b}$$

where $f(x_c) \simeq 0.7$ and $f(x_b) \simeq 0.5$ are phase-space factors. For definiteness, we have taken $m_c = 1.5$ MeV and $m_b = 4.9$ GeV in the above. However, note the quintic dependence on quark mass; the B-lifetime prediction would be 10% lower if $m_b = 5.0$ GeV were used!

For D decays, the D^+ and D^0 lifetimes differ by a factor of about 2.5,

$$\tau_{D^+} = (10.40 \pm 0.07) \times 10^{-13} \text{ s}, \qquad \tau_{D^+}/\tau_{D^0} = 2.54 \pm 0.02, \tag{2.4}$$

whereas the spectator model requires them to be equal. This failure is not surprising, as the D-meson mass lies in the region of strong hadronic resonances; final-state interactions can seriously disturb the spectator picture. Thus, we expect the spectator model to reveal only gross features of the D system. It is remarkable, given its simplicity, that the spectator model predicts (roughly) the correct magnitudes of the lifetime and of the inclusive branching ratios,

$$\mathrm{Br}_{D^0 \to e\bar{\nu}_e X} = (6.49 \pm 0.11)\%, \qquad \mathrm{Br}_{D^+ \to e\bar{\nu}_e X} = (16.07 \pm 0.30)\%. \qquad (2.5)$$

We see that the decays of the D^+ correspond more closely to the spectator predictions than do those of the D^0. The D^0-hadronic decay modes are notably greater that the expectation of the spectator model.

Even for B mesons, the spectator model provides only a rough guide. The lifetimes of the different-flavored B mesons are reasonably similar

$$\tau_{B^0} = (1.519 \pm 0.0.007) \times 10^{-12}\,\mathrm{s},$$

$$\frac{\tau_{B^+}}{\tau_{B^0}} = 1.079 \pm 0.007,$$

$$\frac{\tau_{B_s^0}}{\tau_{B^0}} = 0.986 \pm 0.011, \qquad (2.6)$$

and the spectator estimate differs from these by less than 20%. However, the spectator prediction for the leptonic branching ratio is about 60% larger than the experimental value

$$\mathrm{Br}_{B \to e\bar{\nu}_e X} = (10.72 \pm 0.13)\%, \qquad (2.7)$$

where the number quoted corresponds to roughly an equal mixture of B^+ and B^0. The shorter lifetime and lower leptonic branching ratio point to a modest enhancement of the hadronic modes.

The heavy-quark expansion

The spectator model can be transformed into a solid QCD calculation through the use of the operator-product expansion (OPE) [ChGG 90, Ne 05]. This allows the inclusion of perturbative and nonperturbative corrections.

Using the B meson as our example, the treatment starts by considering the current matrix element, squared and summed over all final states,

$$W_{\alpha\beta} = (2\pi)^4 \sum_X \delta^4(P_B - q - P_X)\langle B(v)|J_\alpha^\dagger|X\rangle\langle X|J_\beta|B(v)\rangle, \qquad (2.8)$$

where q is the momentum carried by the current. The total decay rate is obtained by combining $W_{\alpha\beta}$ with the squared lepton current matrix element $L^{\alpha\beta}$,

$$L^{\alpha\beta} = 4\left(p_\ell^\alpha p_\nu^\beta + p_\ell^\beta p_\nu^\alpha - g^{\alpha\beta} p_\ell \cdot p_\nu + i\epsilon^{\alpha\beta\gamma\delta} p_{\ell\gamma} p_{\nu\delta}\right), \qquad (2.9)$$

and integrating over phase space.

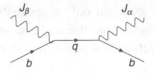

Fig. XIV–1 The leading contribution to the heavy-quark expansion.

The on-shell tensor $W_{\alpha\beta}$ is given by the discontinuity in the full tensor

$$T_{\alpha\beta} = -i \int d^4x \, e^{-iq \cdot x} \langle B(v) | T(J_\alpha^\dagger(x) J_\beta(0)) | B(v) \rangle, \qquad (2.10)$$

related by $W_{\alpha\beta} = -\pi \, \mathrm{Im} \, T_{\alpha\beta}$. The discontinuity is evaluated at the physical cut, which extends over the region

$$m_B \sqrt{q^2} \leq m_B v \cdot q \leq \frac{1}{2}(m_B^2 + q^2 - m_j^2), \qquad (2.11)$$

where m^j is the lightest hadron for the final-state quark q_j, i.e., m_π for $q_j = u$ or m_D for $q^j = c$. In this formalism, the spectator calculation arises from the evaluation of the diagram in Fig. XIV–1 using the free intermediate-state propagator. For a current $\bar{q}\Gamma_\alpha b = \bar{q}\gamma_\alpha(1 + \gamma_5)b$, the tensor $T_{\alpha\beta}$ becomes

$$T_{\alpha\beta} = -i \int d^4x \, e^{-iq \cdot x} \langle B(v) | b(\bar{x}) \Gamma_\alpha S_F(x) \Gamma_\beta b(0) | B(v) \rangle$$

$$= \langle B(v) | \frac{2}{p^2 - m_q^2 + i\epsilon} \cdot \mathcal{M}_{\alpha\mu\beta} \cdot \bar{b}\gamma^\mu(1 + \gamma_5)b | B(v) \rangle, \qquad (2.12)$$

where $\mathcal{M}_{\alpha\mu\beta} \equiv g_{\alpha\mu}p_\beta + g_{\beta\mu}p_\alpha - g_{\alpha\beta}p_\mu - i\epsilon_{\alpha\beta\mu\nu}p^\nu$ with $p^\mu = mv^\mu - q^\mu$ being the momentum carried by the intermediate propagator. The only nonzero matrix element for a B hadron at rest is $\langle B | \bar{b}\gamma^0 b | B \rangle = 1$. In this case, the amplitude is equivalent to the free decay of a b quark.

However, one can do better because the short-distance behavior of the full tensor can be described by an OPE. Because the heavy b quark carries a high energy and transfers that energy to the intermediate states, the region of validity of the OPE is somewhat different than our previous discussion for the weak hamiltonian [ChGG 90]. As $v \cdot q$ approaches the upper range given in Eq. (2.11), the overall hadronic mass becomes smaller and enters the region where binding becomes important and perturbation theory fails.

There are two key improvements that can be accomplished by this method. One is the addition of perturbative corrections. Included in this process is the ability to connect the b-quark mass to a perturbatively well-defined definition of that mass. This tames the strong m_b^5 dependence found in the spectator model by relating the b mass to a well-defined observable. In practice, mass definitions which are tied to measurements that already include confinement effects, such as the $1S$ mass or the kinetic mass mentioned in Sect. XIV–1, provide the most stable perturbative definition [BiSUV 94, HoLM 99]. The other path of improvement is to include new operators that describe nonperturbative hadronic matrix elements [BiSUV 93, Ma 94, MaW 94]. These new operators enter in an expansion in the inverse of the heavy-quark mass. The leading operators are those discussed for the heavy-quark expansion in the preceding chapter. We can see how these arise by expanding the tensor $T_{\alpha\beta}$ around the heavy-quark limit including interactions. The interactions can be seen in the full propagator

$$S_q(x) = \langle x|\frac{1}{\not{D} - m_q + i\epsilon}|0\rangle = \langle x|(\not{D} + m_q)\frac{1}{D^2 + \frac{g_3\lambda^a}{4}\sigma^{\mu\nu}F^a_{\mu\nu} - m_q + i\epsilon}|0\rangle,$$

(2.13)

where $\not{D}$ contains the full covariant derivative including the gauge potential and we have used Eq. (III–3.50) in obtaining the second form. When the matrix element is taken, the derivative turns into $D^\mu = (mv^\mu - q^\mu) + d^\mu$ where d^μ contains the residual momenta and the gauge field. The result is an OPE of the form

$$T(J^{\dagger\alpha}(x)J^\beta(0)) = c_1^{\alpha\beta}\bar{b}b + \frac{c_2^{\alpha\beta}}{m_b^2}\bar{b}(iD)^2b + \frac{c_3^{\alpha\beta}}{m_b^2}\bar{b}\frac{it^a}{2}\sigma_{ij}F^{aij}b.$$

(2.14)

To leading order in $1/m_b$, the result can then be expressed in terms of the two matrix elements

$$\mu_\pi^2 = \langle B(v)|\bar{b}(iD)^2b|B(v)\rangle, \qquad \mu_G^2 = \langle B(v)|\bar{b}\frac{it^a}{2}\sigma_{\mu\nu}F^{a\mu\nu}b|B(v)\rangle.$$

(2.15)

The overall inclusive result has the form [BeBMU 03]

$$\Gamma(B \to X_c e\nu) = \frac{G_F^2 m_b^5(\mu)}{192\pi^3}|V_{bc}|^2\left[f\left(\frac{m_q}{m_Q}\right)(1 + O(\alpha_s))\left(1 - \frac{\mu_\pi^2 - \mu_G^2}{2m_b^2}\right)\right.$$
$$\left.-2\left(1 - \frac{m_c^2}{m_b^2}\right)^4\frac{\mu_G^2}{m_b^2} + \cdots\right].$$

(2.16)

The gluonic operator also appears in the description of the spectroscopy of heavy quarks, as described in Chap. XIII, and its value can be estimated from the mass splittings in heavy hadrons. The kinetic operator can be fit as part of the energy distribution of semileptonic B decay in a combined fit with the total decay rate. The perturbative corrections depend on which definition of the renormalized mass is employed. Further refinements include the perturbative scaling of the coefficients of μ_i and the addition of $1/m_b^3$ effects.

Inclusive measurements can be used to extract the CKM elements [RPP 12],

$$V_{cb} = (41.88 \pm 0.44 \pm 0.59) \times 10^{-3}, \qquad V_{ub} = (4.41 \pm 0.15 \pm 0.16) \times 10^{-3}.$$

$$(2.17)$$

The top quark

The top quark is the real heavyweight of the quarks and presents a rather novel decay pattern. Because $m_t > M_W + m_b$ and the CKM element $|V_{tb}|$ is near unity, the dominant decay is the *semiweak* transition $t \to b + W^+$. The amplitude and transition rate for this process are

$$\mathcal{M}_{t \to bW^+} = -i \frac{g_2}{\sqrt{8}} V_{tb}^* \epsilon_\mu^*(\mathbf{p}_W) \overline{u}(\mathbf{p}_b) \gamma^\mu (1 + \gamma_5) u(\mathbf{p}_t),$$

$$\Gamma_{t \to bW^+} = \frac{G_F m_t^3}{8\pi\sqrt{2}} |V_{tb}|^2 \left[1 - 3\frac{M_W^2}{m_t^2} + 2\frac{M_W^4}{m_t^4} \right], \qquad (2.18)$$

where we have neglected the b quark mass in the decay rate. The question of which definition of m_t to use can be answered only when including *QCD* radiative corrections, and the convergence of the perturbative series is best when using a short-distance definition of the mass, such as the $\overline{\text{MS}}$ mass, rather than the pole mass [BeB 95]. *QCD* corrections including gluon radiation have now been carried out to second order in α_s [CzM 99, ChHSS 99]. Including these, the top width is [BeE *et al.* 00]

$$\Gamma_t = 1.42 \text{ GeV}, \qquad (2.19)$$

corresponding to a lifetime of $\tau = 4.6 \times 10^{-25}$ s. For such a large t-quark mass, the emitted W^+ bosons are predominantly longitudinally polarized, exceeding production of transversely polarized W^+ bosons by a factor $\sim m_t^2/M_W^2$. This is a reflection of the large Yukawa coupling of the t quark to the (unphysical) charged Higgs scalar, which becomes the longitudinal component of the W^+. Other decay modes of the t quark will be highly suppressed by weak mixing factors, e.g., for the mode $t \to s + W^+$ the suppression amounts to $|V_{ts}/V_{tb}|^2 \simeq 1.6 \times 10^{-3}$.

An interesting consequence of the large $t \to b + W^+$ quark decay rate is that there will not be sufficient time for the top quark to form bound-state hadrons. In view of the large top-quark mass, the $t\overline{t}$ system (*toponium*) is nonrelativistic and sits in an effectively Coulombic potential, $V = -4\alpha_s/3r$. In the ground state, one finds the quark velocity $v_{rms} = 4\alpha_s/3$ and atomic radius $r_0 = 3/(2\alpha_s m_t)$. A characteristic orbital period is then $T = 2\pi r_0/v_{rms} = 9\pi/(4\alpha_s^2 m_t)$. Using $\alpha_s(r_0) = 0.12$, we estimate $T = 19 \times 10^{-25}$ s. In contrast, the toponium lifetime would be one-half the t lifetime given above, since either t or $\overline{t}$ could decay first. These comparisons imply that the top quark has an appreciable probability of decaying before completion of even a single bound-state orbit. An equivalent indication of the same effect is the observation that the toponium weak decay width (twice that of a single top quark) is larger than the spacing between energy levels, such as $E_{2S} - E_{1S} = \alpha_s^2 m_t/3 \sim 0.9$ GeV. The production cross section does not then occur through sharp resonances. Instead, there exists a rather broad and weak threshold enhancement, due to the attractive nature of the Coulombic potential. This permits the production and decay of top quarks to be analyzed perturbatively, with Γ_t serving as the infrared cut-off. A heavy top quark can then provide a new laboratory for perturbative QCD studies.

XIV–3 Exclusive decays in the heavy-quark limit

The spectator model calculates the decay rates as if the final-state quarks were free. However, the actual decays take place to physical hadronic final states. For the total rate, there is absolutely no hope of reliably calculating and summing all the individual nonleptonic decays. For semileptonic decays, the situation is somewhat better. The data show that the quasi-one-hadron states, i.e., $D \to K\bar{e}\nu_e$, $K^*\bar{e}\nu_e$ and $B \to De\bar{\nu}_e$, $D^*e\bar{\nu}_e$, form the largest component of the semileptonic rates,

$$\frac{\Gamma_{D^+ \to K\bar{e}\nu_e + K^*\bar{e}\nu_e}}{\Gamma_{D^+ \to X\bar{e}\nu_e}} = 0.89 \pm 0.03, \qquad \frac{\Gamma_{B^+ \to De\bar{\nu}_e + D^*e\bar{\nu}_e}}{\Gamma_{B^+ \to X_c e\bar{\nu}_e}} = 0.74 \pm 0.05. \quad (3.1)$$

These transitions can be addressed by quark model calculations, so that we have an independent handle on such decays. The hadronic-current matrix elements are described by form factors such as

$$\langle K^-(\mathbf{p}') |\bar{s}\gamma_\mu c| D^0(\mathbf{p}) \rangle = f_+ (p + p')_\mu + f_- (p - p')_\mu \ ,$$

$$\langle K^{*-}(\mathbf{p}') |\bar{s}\gamma_\mu c| D^0(\mathbf{p}) \rangle = ig\epsilon_{\mu\nu\alpha\beta}\epsilon^{*\nu} (p + p')^\alpha (p - p')^\beta \ ,$$

$$\langle K^{*-}(\mathbf{p}') |\bar{s}\gamma_\mu\gamma_5 c| D^0(\mathbf{p}) \rangle = f_1\epsilon_\mu^* + \epsilon^* \cdot q \left[f_2 (p + p')_\mu + f_3 q_\mu \right], \quad (3.2)$$

with analogous definitions for the B decays. All form factors are functions of the four-momentum transfer $q^2 = (p - p')^2$. The physics underlying these form factors is two-fold:

(1) If the final-state meson does not recoil, the amplitude is determined by an overlap of the quark wavefunctions, as described in Sect. XII–2.
(2) As the final-state meson recoils, the wavefunction overlap becomes smaller, so that the form factors fall off with increasing recoil momentum.

For D decays, the CKM element is known to a high degree of accuracy from the unitarity of the CKM matrix. In this case, lattice or quark model calculations serve to check whether the experimental rate can be reproduced. For B decays involving the $b \to c$ transition, the exclusive rates are treated using Heavy Quark Effective Theory, which we will describe below.

In the case of *nonleptonic B, D* decays, we have considerably less confidence in our ability to predict the decay amplitudes. This is especially true in D nonleptonic decay because the rescattering corrections required by unitarity can play a major role. Unitarity predicts (cf. Eq. (C–3.14)) for the $D \to f$ matrix element of the transition operator,

$$i(\mathcal{T} - \mathcal{T}^\dagger)_{D \to f} = \sum_n \mathcal{T}^*_{n \to f} \mathcal{T}_{n \to D}, \qquad (3.3)$$

where n are the physically allowed intermediate states. The scattering matrix elements are evaluated at the mass of the D, which happens to lie in an energy range where many strong resonances lie. The scattering elements $\mathcal{T}_{n \to f}$ are therefore expected to be of order unity, implying that rescattering can mask the underlying pattern of weak matrix elements. This makes calculation of nonleptonic D decays particularly suspect.

Inclusive vs. exclusive models for $b \to c e \bar{\nu}_e$

Inclusive and exclusive techniques appear conceptually quite different, even if we know that the total inclusive rate is made from a sum of exclusive individual modes. However, the following observation [ShV 88] is instructive for connecting the two methods.

Consider the semileptonic decay of a heavy quark into another heavy quark, $Q_a \to Q_b e \bar{\nu}_e$, such that their mass difference Δm is small compared to the average of their masses $((m_a + m_b)/2 \gg \Delta m)$, yet large compared to the *QCD* scale ($\Delta m \gg \Lambda_{QCD}$). Because of the second condition, one might use the spectator model result,

$$\Gamma_{Q_a \to Q_b e \bar{\nu}_e} \simeq \frac{G_F^2 (m_a - m_b)^5}{15 \pi^3} |V_{ab}|^2, \qquad (3.4)$$

where V_{ab} is the appropriate weak-mixing matrix element. However, if the first condition is satisfied, the quark recoil will be nonrelativistic. This leads to a nonrelativistic calculation of the transitions from a pseudoscalar $Q_a \bar{q}$ state to pseudoscalar and to vector $Q_b \bar{q}$ states. In this limit, $\bar{\psi}_b \gamma_0 \psi_a \to \psi_b^\dagger \psi_a$ is proportional to the normalization operator, while the axial current $\bar{\psi}_b \gamma_i \gamma_5 \psi_a \to \psi_b^\dagger \sigma_i \psi_a$ is proportional to the spin operator. For states normalized as

$$\langle (Q_a \bar{q})^{0^-}_{\mathbf{p}'} | (Q_a \bar{q})^{0^-}_{\mathbf{p}} \rangle = 2m \, \delta^3 (\mathbf{p} - \mathbf{p}'), \qquad (3.5)$$

one then has

$$\langle (Q_b\bar{q})^{0^-}_{\mathbf{p'}} | \bar{\psi}_b\gamma_0\psi_a | (Q_a\bar{q})^{0^-}_{\mathbf{p}} \rangle = 2m ,$$

$$\langle (Q_b\bar{q})^{1^-}_{\mathbf{p'}} | \bar{\psi}_b\gamma_i\gamma_5\psi_a | (Q_a\bar{q})^{0^-}_{\mathbf{p}} \rangle = 2m\,\epsilon^\dagger_i(\mathbf{p'}),$$

(3.6)

where m is either m_a or m_b. This translates into invariant form factors

$$\langle (Q_b\bar{q})^{0^-}_{\mathbf{p'}} | \bar{\psi}_b\gamma_\mu\psi_a | (Q_a\bar{q})^{0^-}_{\mathbf{p}} \rangle = (p + p')_\mu ,$$

$$\langle (Q_b\bar{q})^{1^-}_{\mathbf{p'}} | \bar{\psi}_b\gamma_\mu\gamma_5\psi_a | (Q_a\bar{q})^{0^-}_{\mathbf{p}} \rangle = 2m\,\epsilon^\dagger_\mu(\mathbf{p'}),$$

(3.7)

which are the correct relativistic results. Using these to calculate the semileptonic decays, one finds

$$\Gamma_{(Q_a\bar{q})_{0^-}\to(Q_b\bar{q})_{0^-}e\bar{\nu}_e} = \frac{G_F^2}{60\pi^3}\,(m_a - m_b)^5\,|V_{ab}|^2 ,$$

$$\Gamma_{(Q_a\bar{q})_{0^-}\to(Q_b\bar{q})_{1^-}e\bar{\nu}_e} = \frac{G_F^2}{20\pi^3}\,(m_a - m_b)^5\,|V_{ab}|^2 .$$

(3.8)

Comparing these, one sees that the sum of the pseudoscalar and vector widths exactly saturates the spectator result of Eq. (3.4). In this combined set of limits, it seems that both types of calculations can be valid simultaneously. Direct application of this insight to $b \to ce\bar{\nu}_e$ decays is somewhat marginal, as the nonrelativistic condition is not well satisfied. A velocity as large as $v = 0.8c$ is reached in portions of the decay region, although on the average a lower value is obtained. However, it is likely that the near equality of spectator versus quark model results is a remnant of the situation described above.

Heavy Quark Effective Theory and exclusive decays

The discussion of the previous section leaned heavily on the use of models to describe quark weak decay. However, many aspects of weak transitions can be obtained in a model-independent fashion through the use of the $m_Q \to \infty$ limit, which was introduced in Sect. XIII–3. This effective theory provides a variety of qualitative and quantitative insights of considerable value.

The heavy-quark approximation manages to justify many results which have become part of the standard lore of quark models. For example, consider the decay constant of a $Q\bar{q}$ pseudoscalar meson M,

$$\langle 0 | \bar{q}(x)\gamma^\mu\gamma_5 Q(x) | M(\mathbf{p}) \rangle = i\sqrt{2}\,F_M p^\mu\,e^{-ip\cdot x}.$$

(3.9)

In the quark model one finds that $F_M \propto (m_M)^{-1/2}$. This follows from the normalization of momentum eigenstates,

$$\langle M(\mathbf{p'}) | M(\mathbf{p}) \rangle = 2E_{\mathbf{p}}\delta^{(3)}(\mathbf{p} - \mathbf{p'}),$$

(3.10)

such that

$$\langle 0|\bar{q}\gamma_0\gamma_5 Q|M(0)\rangle = \begin{cases} i\sqrt{2}\, F_M m_M & \text{(decay const defn.)}, \\ i\sqrt{2m_M}\, \psi(0)\sqrt{2N_c} & \text{(quark model reln.)}, \end{cases} \qquad (3.11)$$

where $\psi(0)$ is the $Q\bar{q}$ wavefunction at the origin and N_c is the number of colors. Since, as $m_Q \to \infty$, the $Q\bar{q}$ reduced mass approaches the constant value $\mu \to m_q$, we expect that $\psi(0)$ itself approaches a constant in this limit,[4] and the scaling behavior $F_M \propto (m_M)^{-1/2}$ then follows immediately from Eq. (3.3). Alternatively, the dependence of F_M on m_M can be derived using the wavepacket formalism introduced in Chap. XII.

This quark model result can be validated in the heavy-quark limit [Ei 88]. Consider the contribution of meson M to the correlation function

$$C(t) = \int d^3x \,\langle 0 \left| A_0(t, \mathbf{x}) A_0^\dagger(0) \right| 0 \rangle, \qquad (3.12)$$

where $A_0 \equiv \bar{q}\gamma_0\gamma_5 Q$. Inserting a complete set of intermediate states and isolating the contribution of meson M, we have

$$C(t) = \int d^3x \int \frac{d^3\mathbf{p}}{(2\pi)^3 2E_\mathbf{p}} \langle 0 |A_0(t, \mathbf{x})| M(\mathbf{p})\rangle\langle M(\mathbf{p}) |A_0(0)|0\rangle + \cdots, \qquad (3.13)$$

where the ellipses denote other intermediate states. From the definition of F_M, one finds

$$C(t) = \frac{F_M^2 m_M^2}{2m_M} e^{-im_M t} + \cdots. \qquad (3.14)$$

Alternatively, the heavy quark develops in time in this correlation function according to the static propagator of Eq. (XIII–3.6),

$$C(t) = -\frac{i}{2} e^{-im_Q t} \langle 0|\bar{q}(t, 0)\gamma_0\gamma_5 P(t, 0)(1 + \gamma_5)\gamma_0\gamma_5 q(0)|0\rangle, \qquad (3.15)$$

with all the dynamics being contained in the light degrees of freedom. The matrix element is independent of m_M, and the scaling behavior,

$$F_M \propto (m_M)^{-1/2}, \qquad (3.16)$$

follows immediately. This technique is applicable to lattice theoretic calculations of F_M. There, one considers euclidean ($t \to -i\tau$) correlation functions, and identifies the M contribution by the $e^{-m_M \tau}$ behavior. At present, lattice calculations attempting to obtain physical results from the $m_Q \to \infty$ limit and from the light-quark limit do not agree in regions of overlap. We thus feel it is premature to quote

---

[4] For example, in the nonrelativistic potential model, the S-wave wavefunction at the origin is related to the reduced mass by $|\psi(0)|^2 = \mu\langle dV/dr\rangle/2\pi\hbar^2$.

theoretical values of F_D, F_B. Another piece of quark model lore which can be justified by this correlation function is that the mass difference $m_M - m_Q$ approaches a constant value in the $m_Q \to \infty$ limit. This can be inferred by comparing the exponential time dependences in Eq. (3.14) and Eq. (3.15), and noting that the difference must be be independent of the heavy quark.

The heavy-quark limit also makes predictions [IsW 89] for transition form factors between two heavy quarks (which for definiteness we shall call b and c). Recall the lagrangian developed in Eq. (XIII–3.15), the leading term of which is

$$\mathcal{L}_v = \bar{h}_v^{(c)} iv \cdot D \, h_v^{(c)} + \bar{h}_v^{(b)} iv \cdot D \, h_v^{(b)}. \tag{3.17}$$

This lagrangian exhibits an $SU(2)$-flavor symmetry involving rotation of $h_v^{(c)}$ and $h_v^{(b)}$. It is also spin-independent, and thus contains an additional $SU(2)$-spin symmetry. The two $SU(2)$s may be combined to form an $SU(4)$ flavor–spin invariance. Physically, the internal structure of hadrons containing a heavy quark and moving at a common velocity is seen to become independent of the quark flavor and spin. This property leads to many relations between transition amplitudes.

An example of a process appropriate for the heavy-quark technique is the weak semileptonic transition $B \to D$ induced by a vector current. For a static matrix element (i.e., both B and D at rest), the weak current transforms quark flavor $b \to c$, but leaves the remaining contents unchanged, resulting in unit wavefunction overlap. This can be seen calculationally by noting that the time component of the spatially integrated current is the conserved charge of the $SU(2)$-flavor group mentioned above,

$$\int d^3x \, \langle D(\mathbf{p}') | \bar{c}(x)\gamma_0 b(x) | \bar{B}(\mathbf{p}) \rangle = \delta(\mathbf{p} - \mathbf{p}') \sqrt{4 m_D m_B}$$
$$= \delta(\mathbf{p} - \mathbf{p}') \left[f_+(t_m)(m_D + m_B) + f_-(t_m)(m_B - m_D) \right], \tag{3.18}$$

where $t_m = (m_B - m_D)^2$ is the value of $t \equiv (p - p')^2$ at the point of zero recoil, and the general decomposition of a vector-current matrix element,

$$\langle D(\mathbf{p}') | \bar{c}\gamma_\mu b | \bar{B}(\mathbf{p}) \rangle = f_+(t)(p + p')_\mu + f_-(t)(p - p')_\mu, \tag{3.19}$$

has been used in the second line of Eq. (3.18). We have seen results similar to Eq. (3.18) in the discussion of the Shifman–Voloshin limit in the previous section. However, there the restriction $m_B - m_D \ll m_B + m_D$ was required, whereas here no restriction is implied as long as both quarks are sufficiently heavy.

This framework may be extended to nonstatic transitions [IsW 90] with the observation that the heavy-quark symmetry can be applied in any frame moving at fixed velocity. First, in addition to Eq. (3.19) for the $B \to D$ transition, we require also the $D \to D$ and $B \to B$ vector form factors,

$$\langle D(\mathbf{p}'_D)\, |\bar{c}\gamma_\mu c|\, D(\mathbf{p}_D)\rangle = f_D(t_D)\left(p_D + p'_D\right)_\mu ,$$

$$\langle \bar{B}(\mathbf{p}'_B)\, |\bar{b}\gamma_\mu b|\, \bar{B}(\mathbf{p}_B)\rangle = f_B(t_B)\left(p_B + p'_B\right)_\mu , \qquad (3.20)$$

where $f_B(0) = f_D(0) = 1$. Considering the momentum transfers t_D, t_B, and $t_{BD} \equiv (p_B - p'_D)^2$ in terms of the velocities, using $p_j^\mu = m_j v^\mu$, $\mathbf{p}_B = m_B \mathbf{v}$, and $\mathbf{p}_D = m_D \mathbf{v}$, we have

$$t_B = \left(p_B - p'_B\right)^2 = 2m_B^2\left(1 - v\cdot v'\right),$$

$$t_D = \left(p_D - p'_D\right)^2 = 2m_D^2\left(1 - v\cdot v'\right),$$

$$t_{BD} = \left(p_B - p_D\right)^2 = (m_B - m_D)^2 + 2m_B m_D\left(1 - v\cdot v'\right). \qquad (3.21)$$

If each transition has common velocity factors, the various momentum transfers are related by

$$t_D = \frac{m_D^2}{m_B^2} t_B = \frac{m_D}{m_B}\left(t_{BD} - t_m\right). \qquad (3.22)$$

In view of the normalization convention of Eq. (3.2), one must divide the state vector of particle i by $\sqrt{2m_i}$ (assuming $m_i \gg |\mathbf{p}|$) before applying the $b \leftrightarrow c$ symmetry. Upon doing so and requiring the resulting expressions to be identical functions of the velocities $\mathbf{v}$ and $\mathbf{v}'$ leads to the relations

$$\frac{\langle D(\mathbf{p}'_D)\,|\bar{c}\gamma_i c|\, D(\mathbf{p}_D)\rangle}{2m_D} = \frac{\langle \bar{B}(\mathbf{p}'_B)\,|\bar{b}\gamma_i b|\, \bar{B}(\mathbf{p}_B)\rangle}{2m_B} = \frac{\langle D(\mathbf{p}_D)'\,|\bar{c}\gamma_i b|\, \bar{B}(\mathbf{p}_B)\rangle}{\sqrt{4m_D m_B}},$$

$$f_D(t_D)\frac{(\mathbf{v}+\mathbf{v}')_i}{2} = f_B(t_B)\frac{(\mathbf{v}+\mathbf{v}')_i}{2} = f_+(t_{BD})\frac{(m_B\mathbf{v}+m_D\mathbf{v}')_i}{\sqrt{4m_D m_B}}$$

$$+ f_-(t_{BD})\frac{(m_B\mathbf{v}-m_D\mathbf{v}')_i}{\sqrt{4m_D m_B}}. \qquad (3.23)$$

After simple algebra, this results in the form-factor relations

$$f_B(t) = f_D\left[\frac{m_D^2}{m_B^2}t\right],$$

$$f_+(t) = \frac{m_B+m_D}{\sqrt{4m_B m_D}} f_D\left[\frac{m_D}{m_B}(t - t_m)\right],$$

$$f_-(t) = -\frac{m_B-m_D}{\sqrt{4m_B m_D}} f_D\left[\frac{m_D}{m_B}(t - t_m)\right]. \qquad (3.24)$$

Although consistent with Eq. (3.21), this manages to separate out $f_\pm$. The results are expressible in terms of a single function of velocity. It is notationally simpler to express the kinematic dependence using $v \cdot v'$ instead of t, i.e., $f_i(t) \to f_i(v \cdot v')$. Thus, we have

$$f_B(v \cdot v') = f_D(v \cdot v') = \sqrt{\frac{4m_B m_D}{m_B + m_D}} f_+(v \cdot v'),$$

$$= -\sqrt{\frac{4m_B m_D}{m_B - m_D}} f_-(v \cdot v') \equiv \xi(v \cdot v'), \qquad (3.25)$$

where, aside from the constraint $\xi(1) = 1$, the function $\xi(v \cdot v')$ is unknown and must thus be determined phenomenologically. If we exploit the full $SU(4)$-flavor–spin symmetry, then all the weak-current form factors involving B, B^*, D, and D^* can be expressed in terms of the quantity $\xi(v \cdot v')$, e.g.,

$$\langle D^*(\mathbf{p}'_D) | \bar{c} \gamma_\mu b | \bar{B}(\mathbf{p}_B) \rangle = i \sqrt{m_{D^*} m_B} \xi(v \cdot v') \epsilon_{\mu\nu\alpha\beta} \epsilon_\nu^*(\mathbf{p}'_D) v'_\alpha v_\beta,$$

$$\langle D^*(\mathbf{p}'_D) | \bar{c} \gamma_\mu \gamma_5 b | \bar{B}(\mathbf{p}_B) \rangle = \sqrt{m_{D^*} m_B} \xi(v \cdot v') \left[(1 + v \cdot v') \epsilon_\mu^* - \epsilon^* \cdot v \, v'_\mu \right]$$

$$(3.26)$$

The symmetry language is appropriate here because, similar to the symmetry relations detailed in the first part of this book, we have *related* different processes even though there remains an uncalculated ingredient to be determined from experiment. However, effective field theory techniques allow a more detailed study of the same matrix elements beyond just the leading symmetry relation. Hard perturbative effects can also be included [Wi 91, CzM 97]. Suppressed corrections due to deviations from the heavy-quark limit can be calculated in the effective theory. The shape of the form factors [CaLN 98] can be determined experimentally, but what is most important phenomenologically is the normalization of these form factors at the zero-recoil point $v \cdot v' = 1$. This deviation is second order in the inverse masses [Lu 90] which, since $m_c \ll m_b$, means that it is of order $1/m_c^2$. While analytic estimates of this deviation can be achieved [ShUV 95, GaMU 12], lattice methods now can provide well-controlled calculations of this effect [Be *et al.* 09].

For the $b \to u$ semileptonic transition, there is no corresponding heavy-quark theory that provides a solid starting point for analysis of the $B \to \pi e \nu$ decay. Quark model calculations are particularly unreliable for this transition. Fortunately, improved lattice calculations now appear capable of calculating the transition matrix element in the region of small recoil [DaGWDLS 06, Ba *et al.* 09]. Supplemented by theoretical constraints [BeH 06], experimental work can measure the q^2 variation and use the lattice matrix element to provide the normalization when using this process to measure V_{ub}.

Phenomenologically, exclusive decays are key ingredients to the extraction of the CKM elements. The present best values from exclusive decays are [RPP 12]:

$$V_{\text{cb}} = (39.6 \pm 0.9) \times 10^{-3}, \qquad V_{\text{ub}} = (3.23 \pm 0.31) \times 10^{-3}. \qquad (3.27)$$

The reader will note there is a modest disagreement between the values of these elements between the inclusive determination of Eq. (2.17) and the exclusive values of Eq. (3.27). For V_{ub}, the effect is sizeable and may be indicative of a gap in our theoretical methods. The smaller disagreement seen in V_{cb} may also be an indication that more theoretical work is needed at understanding the duality between inclusive and exclusive methods.

XIV–4 B^0–$\bar{B}^0$ and D^0–$\bar{D}^0$ mixing

Just as $K^0 - \bar{K}^0$ mixing occurs due to the weak interactions, so does mixing exist in the $B_d - \bar{B}_d$, $B_s - \bar{B}_s$ and $D^0 - \bar{D}^0$ systems. We shall discuss first the $B_d - \bar{B}_d$ and $B_s - \bar{B}_s$ mixings, then conclude with the D^0 case. The formalism is the same in all situations and can be taken directly from the discussion of $K^0 - \bar{K}^0$ mixing in Sect. IX–1.

B^0–$\bar{B}^0$ mixing

The mixing occurring in B_d and B_s mesons is short-distance dominated. This is because (i) the dominant weak coupling of the b quark is to the t quark, and (ii) the short-distance box diagram (Fig. XIV–2) grows roughly with the squared-mass of the intermediate-state quarks. Since the very heavy mass of the top quark greatly enhances its contribution, the top intermediate state dominates B-meson mixing.

The effective hamiltonians for B_d, and B_s mixing are[5]

$$\mathcal{H}_{\text{W}}^{\Delta B_d = 2} = \frac{G_F^2}{16\pi^2} \left(V_{\text{tb}} V_{\text{td}}^* \right)^2 m_t^2 H(x_t) \eta_B O^{B_d} + \text{h.c.,}$$

$$\mathcal{H}_{\text{W}}^{\Delta B_s = 2} = \frac{G_F^2}{16\pi^2} \left(V_{\text{tb}} V_{\text{ts}}^* \right)^2 m_t^2 H(x_t) \eta_B O^{B_s} + \text{h.c.,}$$

$$O^{B_d} = \bar{d}\gamma_\mu(1 + \gamma_5)b \, \bar{d}\gamma^\mu(1 + \gamma_5)b,$$

$$O^{B_s} = \bar{s}\gamma_\mu(1 + \gamma_5)b \, \bar{s}\gamma^\mu(1 + \gamma_5)b, \qquad (4.1)$$

where $\eta_B \simeq 0.9$ is the QCD correction and $H(x_t)$ is given in Eq. (IX–1.20). The matrix elements of O^{B_d} and O^{B_s} can be parameterized analogously to that used in kaon mixing,

$$\langle B_d | O^{B_d} | \bar{B}_d \rangle = \frac{16}{3} F_{B_d}^2 m_{B_d}^2 B_{B_d}, \qquad \langle B_s | O^{B_s} | \bar{B}_s \rangle = \frac{16}{3} F_{B_s}^2 m_{B_s}^2 B_{B_s}, \qquad (4.2)$$

[5] A more advanced treatment of $B_s - \bar{B}_s$ mixing than given here appears in [LeN 07].

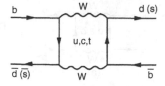

Fig. XIV–2 Box diagram contribution to B-meson mixing.

where the pseudoscalar decay constants are normalized as

$$\langle 0 | \bar{d} \gamma^\mu \gamma_5 b | \bar{B}_d(\mathbf{p}) \rangle = i \sqrt{2} F_{B_d} p^\mu, \qquad \langle 0 | \bar{s} \gamma^\mu \gamma_5 b | \bar{B}_s(\mathbf{p}) \rangle = i \sqrt{2} F_{B_s} p^\mu. \quad (4.3)$$

These correspond to the normalization $F_\pi \simeq 92$ MeV.

Both B_d and B_s mixing have been observed, with the results,[6]

$$x_d \equiv \frac{\Delta m_{B_d}}{\Gamma_{B_d}} = 0.775 \pm 0.006, \qquad x_s \equiv \frac{\Delta m_{B_s}}{\Gamma_{B_s}} = 26.82 \pm 0.23. \quad (4.4)$$

The width difference of B_d is consistent with zero, while that of B_s is nonzero but small,

$$\frac{\Delta \Gamma_d}{\Gamma_d} = 0.015 \pm 0.018, \qquad \frac{\Delta \Gamma_s}{\Gamma_s} = 0.123 \pm 0.017. \quad (4.5)$$

In Eqs. (4.4)–(4.5) above, we have denoted $\Delta m \equiv m_H - m_L$ and $\Delta \Gamma \equiv \Gamma_H - \Gamma_L$, where H (L) refers to the heavier (lighter) of the neutral B CP eigenstates,

The large magnitude of x_s/x_d is readily understood in the Standard Model to be mainly due to the CKM elements, as the ratio is predicted to be

$$\frac{\Delta m_{B_s}}{\Delta m_{B_d}} = \left[\frac{F_{B_s}^2 B_{B_s}}{F_{B_d}^2 B_{B_d}} \right] \left| \frac{V_{ts}}{V_{td}} \right|^2. \quad (4.6)$$

The $SU(3)$ breaking in the matrix elements is well under control in lattice calculations [LaLV 10],

$$\frac{F_{B_s} \sqrt{B_{B_s}}}{F_{B_d} \sqrt{B_{B_d}}} = 1.237 \pm 0.032. \quad (4.7)$$

The remaining dependence in the ratio of the mass splittings comes from the CKM elements and in fact this ratio is the most precise measurement of the relative sizes of these CKM elements

$$\left| \frac{V_{ts}}{V_{td}} \right| = 4.739 \pm 0.126, \quad (4.8)$$

consistent with other determinations. This is an important test of the Standard

---

[6] We use the updated version of [Am *et al.* (Heavy Flavor Averaging Group collab.) 12] found in www.slac.stanford.edu/xorg/hfag.

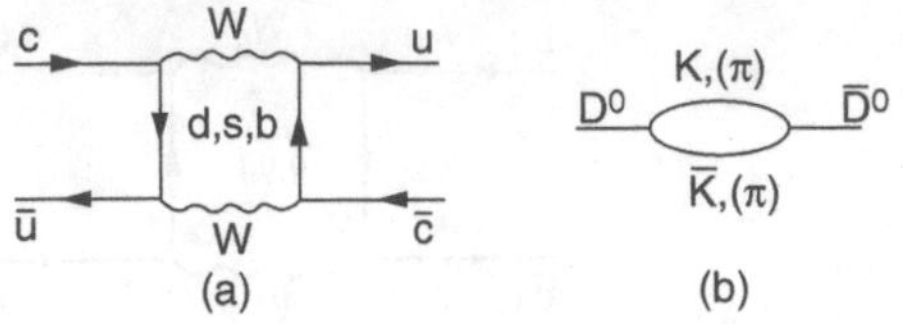

Fig. XIV–3 Short-distance (a) and long-distance (b) contributions to D-meson mixing.

Model as New Physics could readily contribute to Δm_{B_d} and/or Δm_{B_s}. The absolute magnitudes of these mixings are also compatible with the Standard Model. Using the mixing formula developed in Chap. IX and the lattice magnitude [LaLV 10] $F_{B_d}\sqrt{B_{B_d}} = (149 \pm 9)$ MeV, the experimental number for Δm_d is reproduced with $|V_{td}| = (8.4 \pm 0.6) \times 10^{-3}$, which becomes a tight constraint on fits of the unitarity triangle, to be discussed shortly.

The width differences are smaller than the mass differences because real on-shell intermediate states are required; thus, top-quark intermediate states do not contribute to $\Delta\Gamma_{d,s}$. For this reason, the widths $\Delta\Gamma_{d,s}$ are suppressed compared to $\Delta m_{d,s}$ by a factor of roughly m_b^2/m_t^2. The width difference for B_d is smaller than that for B_s because the CKM favored decay mechanism $b \to c\bar{c}s$ when active for a $b\bar{d}$ meson leads to an intermediate state $(c\bar{c}s\bar{d})$ that cannot convert to a $d\bar{b}$ meson, while when occurring in the decay of $b\bar{s}$ leads to intermediate states $(c\bar{c}s\bar{s}$ or $c\bar{c})$ that can transition back to $s\bar{b}$. Thus, $\Delta\Gamma_d$ is CKM-suppressed compared to $\Delta\Gamma_s$. The measurements of $\Delta\Gamma_{d,s}$ are also compatible with theoretical expectations [LeN 07].

D^0–$\bar{D}^0$ *mixing*

The analysis of $D^0 - \bar{D}^0$ transitions is considerably more complex than that involving $B_{d,s}$ mesons because the mixing is *not* short-distance-dominated [Wo 85, DoGH 86a]. To see this, we display the corresponding box diagram in Fig. XIV–3(a), and some possible long-distance contributions in Fig. XIV–3(b). The GIM cancelation in the intermediate state is between the two light quarks d, s (the b-quark contribution is suppressed by CKM angles). However, there is no compensating large mass factor here; long-distance and short-distance effects contribute at the same order of magnitude. As a result, reliable *quantitative* predictions of Δm_D have eluded theorists thus far, despite the attempts of many to solve the problem. Even such basic issues as correctly predicting the sign of Δm_D or determining to what extent a component from New Physics could be present [GoHPP 07] remain unresolved.

For example, consider the application of the OPE (which has worked so well for $B_{d,s}$ mixings) to D^0 mixing [Ge 92, OhRS 93, BiU 01, BoLRR 10],

$$\langle \bar{D}^0 | \mathcal{H}_{|\Delta C|=2} | D^0 \rangle = G \sum_i \mathcal{C}_i(\mu) \langle \bar{D}^0 | \mathcal{Q}_i | D^0 \rangle, \qquad (4.9)$$

where the prefactor G has the unit of inverse squared mass, the sum is over operator dimension, and both Standard Model and New Physics operators are included. The expansion begins at dimension six, with two operators for just the Standard Model and eight upon including New Physics. However, even within just the Standard Model, the number of operators increases sharply as the dimension grows, e.g., there are about a dozen at dimension nine and more than twenty at dimension twelve. This introduces a multitude of unknown parameters. It is also the case that the sum in Eq. (4.9) is not expected to converge rapidly because the ratio $\Lambda_{QCD}/m_c \simeq 0.25$ is not sufficiently small.

Some aspects of D^0–$\bar{D}^0$ mixing can, however, be understood. For example, the Standard Model clearly requires that $\Delta m_D / \Gamma_D \ll 1$ because Δm_D is twice Cabibbo-suppressed (i.e. $\Delta m_D = \mathcal{O}(\lambda^2)$) while Γ_D suffers no such suppression. Hence, upon counting CKM factors and noting that the *GIM* cancelation is a measure of the breaking of $SU(3)$ symmetry, one is led to estimate that[7]

$$\frac{\Delta m_D}{\Gamma_D} \sim \lambda^2 \times [SU(3) \text{ breaking}] = \mathcal{O}(10^{-2}). \qquad (4.10)$$

Of the various meson-mixing systems, the D^0–$\bar{D}^0$ transitions were the last to be detected experimentally. However, by studying the decay time dependence of $D^0 \to K^+\pi^- / D^0 \to K^-\pi^+$, a recent experiment [Aa *et al.* (LHCb collab.) 13*a*] excludes the no-mixing hypothesis with a probability of over nine standard deviations. The current-mixing values in [RPP 12] are

$$x_D \equiv \frac{\Delta m_D}{\Gamma_D} = \left(0.63^{+0.19}_{-0.20}\right) \times 10^{-2}, \qquad \frac{\Delta \Gamma_D}{\Gamma_D} = (1.50 \pm 0.24) \times 10^{-2}. \quad (4.11)$$

The suppression in D^0–$\bar{D}^0$ mixing is evident upon comparing the above value for x_D with those for x_d and x_s in Eq. (4.4).

Observation of D^0–$\bar{D}^0$ mixing motivates the search for *CP* violation in the D-meson system. Here, we cite two recent results. In one, the *CP*-violating asymmetry A_D in the time dependent transition $D^0 \to K^+\pi^-$ is measured to be $A_D = (-0.7 \pm 1.9)\,\%$, which is consistent with zero [Aa *et al.* (LHCb collab.) 13*c*]. In the other, the *CP*-violating asymmetry $A_\Gamma(f)$, between the D^0 and $\overline{D}^0$ decay rates to a given final state f, yields results also consistent with zero [Aa *et al.* (LHCb collab.) 13*d*],

---

[7] Actually, it can be proved that if $SU(3)$ violation in D^0 mixing enters perturbatively, then a group theoretic analysis of $\langle 0 | D \mathcal{H}_w \mathcal{H}_w D | 0 \rangle$ shows that $SU(3)$ breaking occurs only at second order [FaGLP 02].

$$A_\Gamma(\pi^+\pi^-) = (0.33 \pm 1.06 \pm 0.14) \times 10^{-3},$$
$$A_\Gamma(K^+K^-) = (-0.35 \pm 0.62 \pm 0.12) \times 10^{-3}. \qquad (4.12)$$

The uncertainties in the above determinations are dominated by statistical, rather than by systematic, effects. Thus, although the current status of *CP* violation in charm is inconclusive, there is reason to be optimistic that additional statistics as obtained in forthcoming studies will yield nonzero results.

XIV–5 The unitarity triangle

The *B*-meson transitions form a nontrivial system and provide much of our information on the pattern of weak mixing. The overall *B* lifetime and $b \rightarrow c$ semileptonic decays are governed by V_{cb}, the suppressed $b \rightarrow u$ modes by V_{ub}, $B_d - \bar{B}_d$ mixing by V_{td}, and $B_s - \bar{B}_s$ mixing by V_{ts}. Together with the V_{us} element, these form all of the 'interesting' sectors of weak mixing.

There is a useful pictorial representation of the constraints of unitarity on these elements. Consider the effect of the unitarity relation

$$V_{ub}V_{ud}^* + V_{cb}V_{cd}^* + V_{tb}V_{td}^* = 0. \qquad (5.1)$$

Of the components to this equation, V_{ud}, V_{td} and V_{cd} are known up to corrections of second order in $\lambda = |V_{us}|$, yielding

$$V_{ub} - \lambda V_{cb} + V_{td}^* = 0. \qquad (5.2)$$

If we treat these elements as complex vectors, this relation is equivalent to a triangle in the complex plane. In the Wolfenstein parameterization the various elements are

$$V_{cb} = -V_{ts} = A\lambda^2, \quad V_{ub} = \lambda^2 A(\rho - i\eta), \quad V_{td} = \lambda^3 A(1 - \rho - i\eta). \qquad (5.3)$$

The unitarity triangle is shown in Fig. XIV–4. Note that the unitarity triangle can be constructed knowing only the *magnitude* of the elements $|V_{cb}|$, $|V_{ub}|$, and $|V_{td}|$. The existence of such a closed triangle is independent of the parameterization. Other unitarity triangles, corresponding to the other unitarity constraints, also exist but are either less useful than this one or are equivalent to it [Ja 89].

The unitarity triangle has an important connection with *CP* violation. If the *CP*-violating parameter η vanishes, the triangle is reduced down to a line since all the angles go to either $0°$ or $180°$. In fact, the area $\lambda^6 A^2 \eta$ of this triangle is exactly the unique rephasing invariant measure of *CP* violation. The angles α, β, γ are themselves indicators of nonconservation of *CP* and play a role in the *B* studies to be described in the next section.[8] Note that the magnitudes of the sides of the

---

[8] In the literature there is an alternate naming of angles $\varphi_1 = \beta$, $\varphi_2 = \alpha$, $\varphi_3 = \gamma$. We are following the conventions of the Particle Data Group.

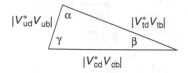

Fig. XIV–4 The unitarity triangle.

triangle and the interior angles of the triangle are all independently measurable and the fact that the separate measurements are consistent is a powerful test of the Standard Model. Our Fig. XIV–4 is drawn using the present fits of the sides and angles, and illustrates the relative magnitudes of these elements.

XIV–6 *CP* violation in *B*-meson decays

The decays of B mesons exhibit a rich variety of *CP*-violating signals, some of which are rather large [BiS 81]. These reactions have provided dramatic confirmation of the validity of the CKM mixing scheme as the dominant origin of *CP* violation. Recall that the value of ϵ cannot be regarded as a prediction of the Standard Model because there is an unknown parameter, the CKM phase δ, which must be adjusted to fit experiment. The value of ϵ'/ϵ is consistent with the Standard Model and is an important verification of the existence of direct *CP* violation, but theoretical uncertainties are presently too large for this to be a precision test. However, the Standard Model, with its single *CP*-odd parameter, makes clear predictions for the patterns of *CP* violations in B decays, and observation has confirmed many of these.

There is an important division in the study of *CP* violations for B mesons: (i) processes which proceed via $B^0 - \bar{B}^0$ mixing, and (ii) those which do not. We shall discuss those involving mixing first, and then return to those not related to mixing.

CP-odd signals induced by mixing

General formalism: The analysis of time evolution for a B^0 or $\bar{B}^0$ meson parallels that of a neutral kaon. Given the conventions for Δm and $\Delta\Gamma$ following Eq. (4.5), one obtains for states that start out at $t = 0$ being either B^0 or $\bar{B}^0$,

$$|B^0(t)\rangle = g_+(t)|B^0\rangle + \frac{q}{p}g_-(t)|\bar{B}^0\rangle,$$

$$|\bar{B}^0(t)\rangle = \frac{p}{q}g_-(t)|B^0\rangle + g_+(t)|\bar{B}^0\rangle,$$

$$\frac{p}{q} \equiv \sqrt{\frac{M_{12} - i\Gamma_{12}}{M_{12}^* - i\Gamma_{12}^*}},$$

$$g_{\pm}(t) \equiv \frac{1}{2}e^{-\Gamma_L t/2}e^{im_L t}\left[1 \pm e^{-\Delta\Gamma t/2}e^{i\Delta m t}\right]. \tag{6.1}$$

The strategy for observing *CP*-violating asymmetries is to compare the decay $B^0(t) \to f$, where f is some given final state, to that of $\bar{B}^0(t) \to \bar{f}$, where $\bar{f}$ is the *CP*-conjugate of f,

$$|\bar{f}\rangle = \mathcal{CP}|f\rangle. \tag{6.2}$$

Let us define the matrix elements

$$A(f) = \langle f|\mathcal{H}_W|B^0\rangle, \qquad \bar{A}(\bar{f}) = \langle \bar{f}|\mathcal{H}_W|\bar{B}^0\rangle,$$
$$\bar{A}(f) = \langle f|\mathcal{H}_W|\bar{B}^0\rangle, \qquad A(\bar{f}) = \langle \bar{f}|\mathcal{H}_W|B^0\rangle, \tag{6.3}$$

and their ratios,[9]

$$\bar{\rho}(f) = \frac{\bar{A}(f)}{A(f)}, \qquad \rho(\bar{f}) = \frac{A(\bar{f})}{\bar{A}(\bar{f})}. \tag{6.4}$$

The decay rates for the two processes are easily found to be [BiKUS 89]

$$\Gamma_{B^0(t) \to f} \propto \left[a + be^{-\Delta\Gamma t} + c\,e^{-\frac{1}{2}\Delta\Gamma t}\cos\Delta m\,t + d\,e^{-\frac{1}{2}\Delta\Gamma t}\sin\Delta m\,t\right]e^{-\Gamma_L t},$$

$$a = |A(f)|^2\left(\frac{1}{2}\left[1 + \left|\frac{q}{p}\bar{\rho}(f)\right|^2\right] + \mathrm{Re}\left[\frac{q}{p}\bar{\rho}(f)\right]\right),$$

$$b = |A(f)|^2\left(\frac{1}{2}\left[1 + \left|\frac{q}{p}\bar{\rho}(f)\right|^2\right] - \mathrm{Re}\left[\frac{q}{p}\bar{\rho}(f)\right]\right),$$

$$c = |A(f)|^2\left(1 - \left|\frac{q}{p}\bar{\rho}(f)\right|^2\right),$$

$$d = 2|A(f)|^2\,\mathrm{Im}\left[\frac{q}{p}\bar{\rho}(f)\right], \tag{6.5a}$$

and

$$\Gamma_{\bar{B}^0(t) \to \bar{f}} \propto \left[\bar{a} + \bar{b}\,e^{-\Delta\Gamma t} + \bar{c}\,e^{-\frac{1}{2}\Delta\Gamma t}\cos\Delta m\,t + \bar{d}\,e^{-\frac{1}{2}\Delta\Gamma t}\sin\Delta m\,t\right]e^{-\Gamma_L t},$$

$$\bar{a} = |\bar{A}(\bar{f})|^2\left(\frac{1}{2}\left[1 + \left|\frac{p}{q}\rho(\bar{f})\right|^2\right] + \mathrm{Re}\left[\frac{p}{q}\rho(\bar{f})\right]\right),$$

---

[9] We caution the reader not to confuse the notation for these ratios with the CKM element ρ in the Wolfenstein parameterization of Eq. (II–4.19).

$$\bar{b} = |\bar{A}(\bar{f})|^2 \left(\frac{1}{2} \left[1 + \left| \frac{p}{q} \rho(\bar{f}) \right|^2 \right] - \mathrm{Re} \left[\frac{p}{q} \rho(\bar{f}) \right] \right),$$

$$\bar{c} = |\bar{A}(\bar{f})|^2 \left(1 - \left| \frac{p}{q} \right|^2 \rho(\bar{f})|^2 \right),$$

$$\bar{d} = 2 |\bar{A}(\bar{f})|^2 \, \mathrm{Im} \left[\frac{p}{q} \rho(\bar{f}) \right]. \tag{6.5b}$$

Any observed difference between these two quantities would indicate the presence of *CP* violation.

Before considering some examples, there is a simplifying approximation which it is useful to make. As seen in the previous section $M_{12} \gg \Gamma_{12}$ for B and B_s, so it is a good approximation to neglect Γ_{12} (and hence $\Delta\Gamma$) in almost all cases.[10] In this approximation q/p becomes a pure phase, $q/p = e^{i\varphi}$, so that $|q/p| = 1$.

Decays to CP eigenstates

The most striking processes are those where the final state f is a *CP* eigenstate, $|\bar{f}\rangle = \pm |f\rangle$, such as $f = \psi K_S, \psi K_L, D^+ D^-, \pi^+ \pi^-$. In this case one has $\bar{\rho}(f) = 1/\rho(\bar{f})$. Time-dependent *CP* asymmetries have two components

$$A_f(t) = \frac{\Gamma(\bar{B}^0(t) \to f) - \Gamma(B^0(t) \to f)}{\Gamma(\bar{B}^0(t) \to f) + \Gamma(B^0(t) \to f)} = S_f \sin(\Delta m t) - C_f \cos(\Delta m t), \tag{6.6}$$

where

$$S_f = \frac{2\mathrm{Im}\left[\frac{q}{p}\bar{\rho}(f)\right]}{1 + \left|\frac{q}{p}\bar{\rho}(f)\right|^2}, \qquad C_f = \frac{1 - \left|\frac{q}{p}\bar{\rho}(f)\right|^2}{1 + \left|\frac{q}{p}\bar{\rho}(f)\right|^2}. \tag{6.7}$$

We see that there are two possible ways that the asymmetry can be nonvanishing, corresponding to the S_f and C_f amplitudes.

The cleanest analysis occurs when $|\bar{\rho}(f)| = 1$, i.e., $|\bar{A}(f)| = |A(f)|$. An example is $B_d \to \psi K_s^0$, which proceeds dominantly through $b \to c\bar{c}s$, so that both factors are pure phases

$$\bar{\rho}(f) = \frac{V_{cs}^* V_{cb}}{V_{cs} V_{cb}^*}, \qquad \frac{q}{p} = \frac{V_{td} V_{tb}^*}{V_{td}^* V_{tb}}. \tag{6.8}$$

[10] The one exception is the semileptonic asymmetry to be discussed below.

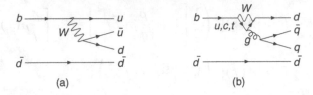

Fig. XIV–5 Tree (a) and penguin (b) diagrams for $B \to \pi\pi$.

In this case, it is clear that $C_{\psi K} = 0$. The asymmetry involves the relative phases of $V_{cs}^* V_{cb}$ and $V_{td}^* V_{tb}$, which we see from Fig. XIV–4 is the angle β, such that the result becomes

$$S_{\psi K} = \sin 2\beta. \tag{6.9}$$

This prediction is independent of hadronic uncertainties and depends only on the phases in the CKM matrix. The result is large, with the resulting measurement [RPP 12] of the angle β of $\sin 2\beta = 0.679 \pm 0.020$, consistent with other constraints on the unitarity triangle. *CP* violation in this mode is one of the cleanest and most direct confirmations of the Standard Model.

One might at first expect that $|\bar{\rho}(f)|^2 = 1$ is automatic if f is a *CP* eigenstate. However, it is possible to obtain $|\bar{\rho}(f)| \neq 1$ if there are two different ways to reach the same final state. For example, one could have the decay $\bar{B}^0 \to \pi^+\pi^-$ either directly through $b \to u\bar{u}d$ or through the penguin diagram, which includes the CKM elements for c or t intermediate states, cf. Fig. XIV–5.[11] By CKM unitarity, we have $V_{cb}^* V_{cd} = -(V_{ub}^* V_{ud} + V_{tb}^* V_{td})$. Therefore, if we absorb the portion of the penguin diagram proportional to $V_{ub}^* V_{ud}$ into the tree-amplitude reduced matrix element, which carries the same CKM factor, we have the amplitude expressed in terms of two CKM elements,

$$\bar{A}(\pi^+\pi^-) = V_{ud}^* V_{ub} \, |T| \, e^{i\delta_T} + V_{td}^* V_{tb} \, |P| \, e^{i\delta_P} \,,$$
$$A(\pi^+\pi^-) = V_{ud} V_{ub}^* \, |T| \, e^{i\delta_T} + V_{td} V_{tb}^* \, |P| \, e^{i\delta_P}, \tag{6.10}$$

where T and P are tree and penguin amplitudes and δ_T, δ_P are strong-interaction phase shifts. Because the weak phases change sign under *CP* and the strong phases do not, we have the ratio of amplitudes $|\bar{\rho}(f)| \neq 1$. Indeed, experimentally one finds

$$S_{\pi^+\pi^-} = -0.65 \pm 0.07, \qquad C_{\pi^+\pi^-} = -0.38 \pm 0.06, \tag{6.11}$$

[11] In discussions such as this, it is understood that the weak hamiltonian receives *QCD* radiative corrections, which can mix operators with identical quantum numbers. However, since we are using only the CKM factors and symmetry properties of the amplitudes, these corrections do not influence the analysis and are absorbed into the reduced matrix elements.

Table XIV–1. *Standard Model pattern for CP violation in B decays.*

Transitions	Examples	Im $(q/p)\bar{\rho}(f)^a$
$b \to c\bar{c}s$	$B_d \to \psi K_S$	$\sin 2\beta$
	$B_s \to \psi\varphi$	~ 0
$b \to c\bar{c}d$	$B_d \to D\bar{D}$	$\sin 2\beta$
	$B_s \to \psi K_S$	~ 0
$b \to u\bar{u}d$	$B_d \to \pi^+\pi^-$	$\sin 2\alpha$
	$B_s \to \pi^0 K_S$	$\sin 2\alpha$
$b \to u\bar{u}s$	$B_d \to \pi^0 K_S$	$\sin 2\alpha$
	$B_s \to \pi^0\varphi^0$	$\sin 2\gamma$

^{a}The angles α, β, γ are defined by the unitarity triangle of Fig. XIV–4 and we take $|\bar{\rho}(f)| = 1$.

indicating the presence of both *CP*-violating phases and sizeable strong rescattering phases. The solution to this 'penguin pollution' involves looking at other $\pi\pi$ modes. There is an isospin relation among the three-pion channels (cf. Eq. (VIII–4.1)

$$A(\pi^+\pi^-) - A(\pi^0\pi^0) = \sqrt{2}A(\pi^+\pi^0), \qquad (6.12)$$

similar to the kaon decay analysis of Chap. VIII. The penguin amplitude is purely $\Delta I = 1/2$ and hence only the tree amplitude can contribute to the $I = 2$ final state $\pi^{\perp}\pi^0$. Measurement of branching ratios and *CP* asymmetries $S_{\pi\pi}$, $C_{\pi\pi}$ allows one to disentangle the *CP* violation due to tree and penguin amplitudes [GrL 90]. For the tree amplitude, involving $V_{ub}^* V_{ud}$, the interference is with the B_d-mixing amplitude, dominated by the top quark, so that the measurement is of the CKM phase α.

At this stage we can categorize the decays of neutral B mesons to *CP* eigenstates. For this purpose it is most convenient to use the Wolfenstein form of the CKM matrix. In this parameterization, the elements V_{tb}, V_{cb}, V_{ts}, V_{cs} are all almost purely real. The B_d and B_s decays can proceed either through the CKM-favored transition $b \to c\bar{c}s$ or the CKM-suppressed transitions $b \to u\bar{u}d$, $b \to c\bar{c}d$, $b \to u\bar{u}s$. In the former category are included $B_d \to \psi K_S$ and also $B_s \to \psi\varphi$, $\psi\eta$, $D_s^+ D_s^-$. The B_s decays pick up no phase since

$$\frac{q}{p} = \frac{V_{ts}V_{tb}^*}{V_{ts}^* V_{tb}} = 1 \text{ and } \bar{\rho}(f) = \frac{V_{cb}}{V_{cb}^*} = 1 \Rightarrow \text{Im}\left[\frac{q}{p}\bar{\rho}(f)\right] = 0. \qquad (6.13)$$

However, the B_d decay does pick up a phase, leading to a distinctive signature of the Standard Model. The CKM-suppressed decays can also be analyzed in terms of the angles which appear in the unitarity triangle, and are given in Table XIV–1 for the case $|\bar{\rho}(f)| = 1$. However, in some cases we know that $|\bar{\rho}(f)| \neq 1$, such that further efforts are required to extract the given angle, as described for the $\pi\pi$ system above. It should also be pointed out that under all circumstances, asymmetries for B_s are more difficult to observe because x_s is large due to the rapid oscillations in the $B_s \leftrightarrow \bar{B}_s$ system. Thus, regardless of whether one starts out at $t = 0$ with B_s or $\bar{B}_s$, after a few oscillation lengths one will have roughly equal amounts of B_s and $\bar{B}_s$.

Decays to non-CP eigenstates

There may also exist *CP* violation in final states which are *not CP* eigenstates. Consider, for example, the final state $B_d \to \pi^- K^+$. This transition can occur both through tree amplitudes, with the CKM factor $V_{ub}^* V_{us}$ and through penguin decays of the form $b \to s\bar{q}q$. Because the CKM elements satisfy $V_{tb}^* V_{ts} = -(V_{cb}^* V_{cs} + V_{ub}^* V_{us})$, we can write the amplitude in terms of two reduced matrix elements such that the corresponding decays of the B^0 and $\bar{B}^0$ will have the form

$$A(\pi^- K^+) = V_{ub}^* V_{us} |U| e^{i\delta_U} + V_{cb}^* V_{cs} |C| e^{i\delta_C} \,,$$
$$\bar{A}(\pi^+ K^-) = V_{ub} V_{us}^* |U| e^{i\delta_U} + V_{cb} V_{cs}^* |C| e^{i\delta_C}, \tag{6.14}$$

where the reduced matrix element C comes from the penguin diagram alone and U comes from a mixture of tree and penguin effects. The decay rates for these two processes will then be different by a factor

$$|A(\pi^- K^+)|^2 - |\bar{A}(\pi^+ K^-)|^2 = -4|U||C| \sin(\delta_U - \delta_C)\lambda^6 A^2 \eta, \tag{6.15}$$

where we have used $\mathrm{Im}\, V_{ub}^* V_{us} V_{cb} V_{cs}^* = \lambda^6 A^2 \eta$ in the Wolfenstein parameterization. This effect has required two paths to the given final state, with differing strong phases and differing weak phases. Because the hadronic matrix elements are difficult to calculate reliably, this rate difference cannot by itself be a precision test of the Standard Model.

However, there is a way to make an approximate test of the Standard Model using corresponding decays of the B_s meson. The key point [He 99, Gr 00] is that the tree process $b \to u\bar{u}d$ and the penguin amplitude for $b \to d\bar{q}q$ proceed identically to the corresponding processes used above for $b \to u\bar{u}s$ and $b \to s\bar{q}q$ aside from CKM factors. In the U-spin subgroup of $SU(3)$ the d and s quarks form a doublet, and all other quarks are singlets. The two sets of interactions then form two components of a U-spin doublet, and their matrix elements are related. B_d and B_s are also related by U-spin, so that the matrix elements for $B_d \to \pi^- K^+$ and $B_s \to K^- \pi^+$ are U-spin reflections of each other. The corresponding rates for B_s decay are given in the U-spin limit by

$$A_s(K^-\pi^+) = V_{ub}^* V_{ud} |U| e^{i\delta_U} + V_{cb}^* V_{cd} |C| e^{i\delta_C},$$

$$\bar{A}_s(K^+\pi^-) = V_{ub} V_{ud}^* |U| e^{i\delta_U} + V_{cb} V_{cd}^* |C| e^{i\delta_C}. \tag{6.16}$$

The weak CKM elements are different, but the hadronic matrix elements are the same. However, the Standard Model has only a single *CP*-violating phase, so the the imaginary parts of the products of CKM elements are always related. In this case, they are identical up to a sign Im $V_{ub}^* V_{ud} V_{cb} V_{cd}^* = -\lambda^6 A^2 \eta$, such that the decay rate differences are the same

$$|A(\pi^-K^+)|^2 - |\bar{A}(\pi^+K^-)|^2 = -(|A_s(K^-\pi^+)|^2 - |\bar{A}_s(K^+\pi^-)|^2). \tag{6.17}$$

However, asymmetries are defined by dividing by the the sum of the decay rates, and the overall decay rates are different in these two cases. Correcting for the overall rates yields a sum-rule [Li 05]

$$Q = A_{CP}(B_s \to K^-\pi^+) + A_{CP}(B_d \to \pi^-K^+)\frac{\text{Br}(B_d \to \pi^-K^+)\tau_s}{\text{Br}(B_s \to K^-\pi^+)\tau_d} = 0, \tag{6.18}$$

where Br is the *CP*-averaged branching ratio. Despite the individual rates and asymmetries being different, the sum-rule appears valid within error bars [Aa *et al.* (LHCb collab.) 13c]

$$A_{CP}(B_s \to K^-\pi^+) = 0.27 \pm 0.04 \pm 0.01,$$

$$A_{CP}(B_d \to \pi^-K^+) = -0.080 \pm 0.007 \pm 0.003,$$

$$Q = -0.02 \pm 0.05 \pm 0.04. \tag{6.19}$$

While the use of U-spin symmetry is only approximately accurate, this sum-rule nevertheless is a strong test of the overall pattern of direct *CP* violation within the Standard Model, including loop diagrams.

Semileptonic asymmetries

For a final example involving mixing, let us consider *CP* violation in semileptonic decays. In much of our previous analysis, we have neglected the quantity Γ_{12}. However, for semileptonic decays, the whole effect vanishes if we neglect Γ_{12}, so we must include it. For this case, only the transitions $B^0 \to \ell^+\nu_\ell X$, $\bar{B}^0 \to \ell^-\bar{\nu}_\ell X$ ($\ell = e, \mu, \tau$) can occur. The 'wrong sign' transitions in the time developments, $B^0(t) \to \ell^-\bar{\nu}_\ell X$, $\bar{B}^0(t) \to \ell^+\nu_\ell X$, are then uniquely due to mixing. The appropriate formulas can be obtained from our general result Eqs. (6.5a), (6.5b) by the substitutions

$$A(e^-) \to 0, \quad A(e^-)\bar{\rho}(e^-) \to \bar{A}(e^-),$$
$$\bar{A}(e^+) \to 0, \quad \bar{A}(e^+)\rho(e^+) \to A(e^+) = \bar{A}(e^-). \tag{6.20}$$

The integrated rate is

$$A_{\mathrm{SL}} = \frac{\int_0^\infty dt \left[\Gamma_{B^0(t) \to \ell^- \bar{\nu}_\ell X} - \Gamma_{\bar{B}^0(t) \to \ell^+ \nu_\ell X} \right]}{\int_0^\infty dt \left[\Gamma_{B^0(t) \to \ell^- \bar{\nu}_\ell X} + \Gamma_{\bar{B}^0(t) \to \ell^+ \nu_\ell X} \right]} = \frac{\left| \frac{q}{p} \right|^2 - \left| \frac{p}{q} \right|^2}{\left| \frac{q}{p} \right|^2 + \left| \frac{p}{q} \right|^2}. \tag{6.21}$$

This sort of *CP* violation is thus solely sensitive to mixing in the mass matrix, as was the semileptonic K_L^0 asymmetry. Unfortunately, in the Standard Model it is small for reasons connected to the CKM elements. Expanding in powers of Γ_{12} and defining $\varphi_\Gamma \equiv \arg \left(\Gamma_{12}/M_{12} \right)$, one has

$$A_{\mathrm{SL}} \simeq -\mathrm{Im} \frac{\Gamma_{12}}{M_{12}} = - \left| \frac{\Delta\Gamma}{\Delta m} \right| \sin \varphi_\Gamma. \tag{6.22}$$

We have seen that $\Delta\Gamma/\Delta m$ is suppressed by factors of m_b^2/m_t^2 since the top quark cannot contribute to the real intermediate states required for $\Delta\Gamma$. For B_s, there is a further suppression in the Standard Model because the dominant contributions to Γ_{12} ($c\bar{c}$ intermediate states coming with CKM elements $(V_{cb}^* V_{cs})^2$) and M_{12} ($t\bar{t}$ intermediate states with $(V_{tb}^* V_{ts})^2$) have almost the same phase because $V_{tb}^* V_{ts} = -V_{cb}^* V_{cs}[1 + \mathcal{O}(\lambda^2)]$. Thus, φ_{Γ_s} is also suppressed to a fraction of a percent. These features are seen in the theoretical predictions [LeN 11]

$$A_d^{\mathrm{SL}}[\mathrm{Thy}] = (4.1 \pm 0.6) \times 10^{-4}, \qquad A_s^{\mathrm{SL}}[\mathrm{Thy}] = (1.9 \pm 0.3) \times 10^{-5}. \tag{6.23}$$

The present experimental results [RPP 12, Ve (LHCb collab.) 13],

$$A_d^{\mathrm{SL}}[\mathrm{Expt}] = 0.0007 \pm 0.0027, \qquad A_s^{\mathrm{SL}}[\mathrm{Expt}] = -0.0024 \pm 0.0054 \pm 0.0033, \tag{6.24}$$

are not yet precise enough to confirm the Standard Model predictions.

CP-odd signals not induced by mixing

Situations where *CP* violation occurs *without* the presence of mixing can occur in $B^\pm$ decays through the interference of different decay mechanisms. The requirements are the same as we saw previously in a different context, i.e., there must be two different paths to the same final state, these paths must have different strong-interaction final-state phases, and the two paths must also have different weak phases. Consider, for example, the decays $B^+ \to D^0 K^+$ and $B^+ \to \bar{D}^0 K^+$. While initially one might think that these two reactions are distinct, if the D^0 and $\bar{D}^0$ decay to a common final state, such as $K_S^0 \pi^+ \pi^-$, the overall amplitudes to that final state will in fact interfere. The decay with a D^0 in the final state involves the $\bar{b} \to \bar{u}c\bar{s}$ reaction, with CKM elements $V_{ub}^* V_{cs}$, while the $\bar{D}^0$ reaction proceeds through $\bar{b} \to \bar{c}u\bar{s}$ and $V_{cb}^* V_{us}$. The relative phase between these amplitudes is the angle γ.

Interestingly, despite the need for final-state phases in this reaction, the *CP* violation can be extracted without hadronic uncertainties [GrW 91, GiGSZ 03]. The key to this is that the subreaction $D^0 \to K_S^0 \pi^+ \pi^-$ can be separately measured in tagged D reactions as a function of the kinematic variables, and then can be treated as a known quantity. In addition the D^0 and $\bar{D}^0$ decay amplitudes are related to each other[12] at mirror kinematic values. In particular, if the decay $D^0 \to K_S^0 \pi^+ \pi^-$ is given the name $A(m_+^2, m_-^2)$ with $m_\pm^2 = (p_K + p_\pm)^2$ then the corresponding $\bar{D}^0$ amplitude is $\bar{A}(m_+^2, m_-^2) = A(m_-^2, m_+^2)$. The amplitudes, including the possibility of final-state interaction phases, have the form

$$|A_{B^+ \to (K_S \pi^+ \pi^-)K^+}|^2 = |A_0|^2 |A(m_+^2, m_-^2) + r\bar{A}(m_+^2, m_-^2)e^{\delta+\gamma}|^2,$$
$$|A_{B^- \to (K_S \pi^+ \pi^-)K^-}|^2 = |A_0|^2 |\bar{A}(m_+^2, m_-^2) + rA(m_+^2, m_-^2)e^{\delta-\gamma}|^2, \qquad (6.25)$$

where an overall amplitude for $A_0 \equiv A_{B^+ \to D^0 K^+}$ has been factored out and where r is the ratio of the magnitudes of the amplitudes $r = |A_{B^+ \to \bar{D}^0 K^+}|/|A_{B^+ \to D^0 K^+}|$. Here, the possible strong-phase difference δ has been made explicit. Knowledge of the D decay amplitudes plus the observation of both $B^\pm$ decays then lets one separate the strong phase from the weak phase and also divide out the underlying weak matrix elements. This has become a favored way to measure the angle γ with the present result [Aa *et al.* (LHCb collab.) 12],

$$\gamma = (71.1^{+16.6}_{-15.7})^o, \qquad (6.26)$$

when all related channels are included.

To summarize, we have discussed thus far a variety of tests for *CP*-violating signals in the system of B mesons. The partial rate differences can be quite large. At first, this seems to go against the general dictum that all *CP* violations in the Standard Model must be proportional to a single, numerically small product of CKM angles. However, B decays satisfy this stricture in the sense that the mixing and decay of B mesons are in themselves proportional to small CKM angles. Overall, the product of mixing, decay, and *CP* violation does turn out to be proportional to all of these CKM angles. However, in forming the asymmetry by dividing out the rates themselves, one is canceling the small CKM angles, thus leaving a rather large effect. This argument also explains why there is little *CP* violation in D decays in the Standard Model. The *CP* observables must be small due to the usual product of CKM angles. However, the overall decay rate itself has no small

---

[12] Here we neglect *CP* violation in the D-meson system, which is a good approximation for CKM-favored decays.

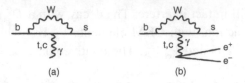

(a)   (b)

Fig. XIV–6 Some one-loop diagrams for rare B decays.

angles, so that the signal remains small. B-meson decays have proven to be optimal
for the exploration of the rich CP-violating structure of the Standard Model.

XIV–7 Rare decays of B mesons

The number of B-decay modes is so large that *any* single mode will be 'rare' in
the sense of having a small branching ratio. Nonetheless, considerable attention
has been given to modes that proceed *only* at one loop, as in Fig. XIV–6, and these
are the ones that are normally labeled as rare decays. The expectation is that, by
measuring the transition rates of such processes, one can test the Standard Model
at loop level, and hopefully observe deviations due to New Physics. Moreover,
since prediction of rare decays involves many of the techniques we have developed
for calculating weak transitions, these decays can provide a nontrivial test of our
ability to apply the Standard Model.

The quark transition $b \to s\gamma$

The process $b \to s\gamma$ is described by the magnetic-dipole transition

$$\mathcal{M}_{b \to s\gamma} = \frac{G_F}{\sqrt{2}} \frac{e}{8\pi^2} F_2 V_{cb} V_{cs}^* \epsilon^*(\mathbf{q})^\mu q^\nu$$

$$\times \bar{u}(\mathbf{p}_s) \sigma_{\mu\nu} \left[m_b (1 - \gamma_5) + m_s (1 + \gamma_5) \right] u(\mathbf{p}_b), \tag{7.1}$$

where the quark mass factors occur in the combination shown because the $\sigma_{\mu\nu}$
Dirac matrix connects left-handed fields to right-handed fields, and a factor of mass
must appear whenever a chirality change $L \to R$ occurs.

The quantity F_2, which represents the quark-level loop amplitude with numerical
factors containing G_F and e extracted, is given by

$$F_2 \simeq \bar{F}_2(x_t) - \bar{F}_2(x_c) \simeq \bar{F}_2(x_t), \tag{7.2}$$

with $x_i = m_i^2 / M_W^2$ and

$$\bar{F}_2(x) = \frac{x}{(x-1)^3} \left[\frac{2x^2}{3} + \frac{5x}{12} - \frac{7}{13} - \left(\frac{3x^2}{2} - x \right) \ln x \right]. \tag{7.3}$$

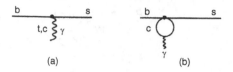

Fig. XIV–7 Standard Model diagrams for $b \to s\gamma$.

The flavor content of F_2 and the overall factor of $V_{cb}V_{cs}^*$ in Eq. (7.1) can be easily understood. The overall loop amplitude, which involves a sum over the intermediate quark flavors t, c, u, must vanish in the limit of equal quark mass from a GIM cancelation since it involves a neutral flavor-changing process. In reality, however, the contribution from the very light u quark is negligible, and the top-quark contribution to F_2 clearly dominates. The CKM unitarity relation $V_{tb}V_{ts}^* = -V_{cb}V_{cs}^* - V_{ub}V_{us}^*$ can be used to substitute for $V_{tb}V_{ts}^*$ upon neglecting the small factor $V_{ub}V_{us}^*$. The $b \to s\gamma$ decay rate, relative to the $b \to ce\bar{\nu}_e$ semileptonic rate can be expressed in the simple form

$$\frac{\Gamma_{b \to s\gamma}}{\Gamma_{b \to ce\bar{\nu}_e}} = \frac{3\alpha |F_2|^2}{f(m_c/m_b)}, \tag{7.4}$$

where $f(x)$ is the phase-space factor given in Eq. (2.1), and factors of m_s^2/m_b^2 arising from phase space and from the amplitude of Eq. (7.1) have been dropped.

Short-distance *QCD* corrections can be used to improve this free-quark calculation. These produce a surprisingly large modification to the analysis of $b \to s\gamma$, and the reason is instructive. The t quark is so heavy that, at all scales relevant to the weak decay, its effect may be treated as a point $bs\gamma$ vertex, with renormalizations as in Fig. XIV–7(a). However, the c quark is light on all scales from M_W to m_b so that in its renormalization one must also include the diagrams of Fig. XIV–7(b), where the dot represents the $b \to c\bar{c}s$ weak hamiltonian. That is, there is mixing between the $b \to s\gamma$ vertex and the $b \to c\bar{c}s$ transition. The theoretical prediction is [Mi *et al.* 07],

$$\mathcal{B}_{b \to s\gamma}[\text{Thy}] = (3.15 \pm 0.23) \times 10^{-4}, \tag{7.5}$$

for photon energies above 1.6 GeV. The corresponding measurement (highly nontrivial) is [Am *et al.* (Heavy Flavor Averaging Group collab.) 12]

$$\mathcal{B}_{b \to X_s\gamma}[\text{Expt}] = (3.55 \pm 0.24 \pm 0.09) \times 10^{-4}, \tag{7.6}$$

where the last error bar is due to uncertainties in the treatment of the photon energy distribution.

At the hadronic level, the quark transition $b \to s\gamma$ is observed in channels such as $B \to K\pi\gamma$, $K\pi\pi\gamma$, etc. The simplest final state occurs when the $K\pi$ system

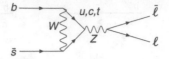

Fig. XIV–8 The penguin diagram for $B_s \to \ell^+\ell^-$.

forms a resonant $J^P = 1^-$ state, $K^*(890)$.[13] As the inclusive rate appears to be in agreement with the Standard Model, this effort is a test of the calculation of exclusive transitions. Within the same class of decays is the transition $B \to K^*\ell^+\ell^-$. Theoretical interest in this transition comes from the hope that New Physics not present in $B \to X_s\gamma$ could show up here [DeHMV 13]. The amplitude includes Z^0 as well as photon exchange, and the loops could be sensitive to new interactions. Experimentally, the decay is rich and challenging because a full angular distribution can be probed, with the possibility of sensitivity to different physics in different kinematic regions.

The decay $B_s \to \ell^+\ell^-$

The leptonic transition $B_s \to \ell^+\ell^-$ is also particularly promising as a sensitive test of the Standard Model. The rate is suppressed even more by a factor of m_ℓ^2 due to a helicity argument which relies on the current–current structure of the theory, and this allows New Physics to be present.

The decay proceeds through the Z^0 penguin diagram of Fig. XIV–8 with the dominant contribution from the top quark due to its large mass. The photon penguin does not contribute because the photon as a vector has $C = -1$, while the lepton–antilepton pair with zero angular momentum has $C = +1$. The transition then occurs through the axial-vector Z^0 current, with an effective hamiltonian,

$$H = \frac{G_F \alpha}{2\sqrt{2}\pi \sin^2 \theta_W} V_{tb}^* V_{ts} C_A \bar{b}\gamma^\mu \gamma_5 s \; \bar{\ell}\gamma_\mu \gamma_5 \ell, \qquad (7.7)$$

where, as usual, C_A is a coefficient which includes the QCD short-distance corrections. When computing the decay amplitude, we encounter the matrix element

$$\langle 0|\bar{b}\gamma^\mu \gamma_5 s|B_s(q)\rangle = i F_{B_s} q^\mu, \qquad (7.8)$$

and the q^μ contracted with the lepton current produces a factor of m_ℓ in direct analogy to the pion decay discussed in Chap. VII. Note that scalar or pseudoscalar interactions would not have such suppression and so these New Physics possibilities

---

[13] The $B \to K\gamma$ transition is forbidden because it is a spin-zero to spin-zero transition.

could potentially have a large enhancement over the Standard Model prediction. The theoretical prediction [BuGGI 12, DeFKKMPT 12]

$$\mathcal{B}^{(\text{theo})}_{B_s \to \mu^+ \mu^-} = (3.54 \pm 0.30) \times 10^{-9} \qquad (7.9)$$

is quite robust, with the major uncertainty being the lattice calculation of F_{B_s}. This mode has recently been measured [Aa *et al.* (LHCb collab.) 13*b*] with the result,

$$\mathcal{B}^{(\text{expt})}_{B_s \to \mu^+ \mu^-} = \left(3.2^{+1.5}_{-1.2}\right) \times 10^{-9}. \qquad (7.10)$$

An even more recent result, although preliminary, shows that combined LHCb and CMS data agree with the Standard Model prediction by more than 5σ. This clearly indicates that there is no large effect from New Physics.

Problems

(1) **Patterns of *CP* violation**

All signals of *CP* violation involve the interference of two or more amplitudes. Identify the origin of the interference in partial rate asymmetries for the decays (a) $B_s \to \varphi\varphi$, (b) $B_s \to \rho^\pm \pi^\mp$, (c) $B_d \to \bar{K}^{*0}\varphi$, (d) $B^\pm \to \rho^\pm \pi^0$, (e) $B^\pm \to K^\pm \pi^0$.

(2) **Amplitude relations in the heavy-quark limit**

In the heavy-quark limit, a static b quark in a B meson can be described in terms of just the two upper components of its four-component Dirac field. This can simplify various matrix elements or be used to relate them. Use this feature to show that the $\bar{B} \to K^* \gamma$ matrix element of the $\sigma^{\mu\nu}$ operator,

$$\langle K^*(\epsilon, \mathbf{k}) | \bar{s} \sigma^{\mu\nu} b | B(\mathbf{p}) \rangle = \epsilon^{\mu\nu\alpha\beta} \left[A\, \epsilon^\dagger_\alpha p_\beta + B\, \epsilon^\dagger_\alpha k_\beta + \epsilon^\dagger \cdot p\, C\, p_\alpha k_\beta \right],$$

can be related to the vector and axial-vector form factors of $\bar{B} \to \rho \ell \bar{\nu}_\ell$,

$$\langle \rho^+(\epsilon, \mathbf{k}) | \bar{u} \gamma^\mu b | \bar{B}^0(\mathbf{p}) \rangle = i D\, \epsilon^{\mu\nu\alpha\beta} p_\nu \epsilon^\dagger_\alpha k_\beta,$$

$$\langle \rho^+(\epsilon, \mathbf{k}) | \bar{u} \gamma^\mu \gamma_5 b | \bar{B}^0(\mathbf{p}) \rangle = E\, \epsilon^{\dagger \mu} + \epsilon^\dagger \cdot p\, [F p^\mu + G k^\mu],$$

through

$$A = -(E - k_0 m_B D)/m_B, \quad B = -m_B D, \quad C = (D + G)/m_B,$$

under the assumptions of a static b quark and of $SU(3)$ symmetry. In this relation, all form factors must be evaluated at the same momentum transfer, $q^2 = (p - k)^2$.

XV

The Higgs boson

On July 4, 2012, the LHC collaborations ATLAS and CMS announced the discovery of a resonance which, despite limited statistics, seemed to have characteristics expected of a Standard Model Higgs boson. Mass determinations presented at the 2013 Lepton-Photon Conference are

$$M_H(\text{GeV}) = \begin{cases} 125.5 \pm 0.2 \text{ (stat) } ^{+0.5}_{-0.6}(\text{syst}) & [\text{Ja (ATLAS collab.) 13}] \\ 125.7 \pm 0.4 & [\text{De (CMS collab.) 13}]. \end{cases} \quad (1.1)$$

Since this resonance has a nonzero branching fraction for decay into two photons, it must be a boson, one not having spin-one. In fact, current spin/parity analyses are compatible with $J^P = 0^+$ but not with $J^P = 0^-, 1^+, 1^-, 2^+$ [Aa *et al.* (ATLAS collab.) 13*b*], [Ch *et al.* (CMS Collab.) 13]. Its couplings to bosons and fermions appear to be consistent with Standard Model expectations, in particular that the Higgs should couple to mass. At present, the overall precision is limited to about 25%, so an extended period of careful study will be necessary to reveal the anomalous properties, if any, of this particle. In this chapter, we will consider the basics of the Standard Model Higgs, with the intent of describing its phenomenology and also addressing certain theoretical issues.

XV–1 Introduction

A central feature of the Standard Model is the spontaneous symmetry breaking in the electroweak sector which gives mass to fermions and to the $W^\pm$ and Z^0 gauge bosons. Although a complex doublet of Higgs fields is initially introduced in the Weinberg–Salam model, there remains following spontaneous symmetry breaking precisely one *physical* Higgs state, a neutral scalar particle H^0. That is, if we define the number of degrees of freedom for Higgs and gauge-boson states, respectively, as N_H and N_G, then before the symmetry breaking we have $N_H = 4$, $N_G = 8$

whereas afterwards we find $N_H = 1$, $N_G = 11$. To obtain these values, recall that massive vector particles have three spin components whereas massless vector particles have just two. Although the total of Higgs and gauge-boson degrees of freedom remains fixed ($N_H + N_G = 12$), there is a transfer of three states from the Higgs sector to the gauge-boson sector. These Higgs states become the longitudinal spin modes of the $W^\pm$, Z^0 particles.

This transfer can be displayed analytically by first performing a contact transformation to cast the two complex Higgs states φ^0, φ^+ in terms of four real fields H^0 and $\chi = \{\chi_i\}$ ($i = 1, 2, 3$)

$$\Phi = U^{-1}(\chi) \begin{pmatrix} 0 \\ (v + H^0)/\sqrt{2} \end{pmatrix}, \tag{1.2}$$

where

$$U(\chi) = \exp(i\chi \cdot \tau/v), \tag{1.3}$$

and we recall that $v = 1/\sqrt{2^{1/2}G_F} \simeq 246$ GeV. One completes the procedure with the gauge transformation,

$$\Phi' = U(\chi)\Phi = \begin{pmatrix} 0 \\ (v + H^0)/\sqrt{2} \end{pmatrix},$$

$$\psi'_L = U(\chi)\psi_L, \quad \psi'_R = \psi_R, \quad B'_\mu = B_\mu,$$

$$\frac{\tau}{2} \cdot \mathbf{W}'_\mu = U(\chi)\frac{\tau}{2} \cdot \mathbf{W}_\mu U^{-1}(\chi) + ig_2^{-1}\partial_\mu U(\chi) \cdot U^{-1}(\chi), \tag{1.4}$$

for all fermion weak isodoublets ψ_L and weak isosinglets ψ_R. Within this unitary gauge, the physical content of the theory is manifest, and the quantity Φ' is seen to contain *a single Higgs field* H^0.[1] In the following, we shall employ this gauge but with the primes in Eq. (1.4) suppressed.

XV–2 Mass and couplings of the Higgs boson

We have already specified in Chap. II how the Higgs boson H fits into the Standard Model. The various lagrangians written down there provide the basis for a complete phenomenological portrait to be drawn for the H boson. In this section, and the ones to follow, we present the theory for this program.

---

[1] For notational simplicity, we shall hereafter omit the superscript '0' and denote the Higgs field simply as H.

Higgs mass term

Consider first the Higgs potential of Eq. (II–3.19) which, when expressed in terms of the field H, becomes

$$V = -\frac{\mu^2 v^2}{4} + \mu^2 H_0^2 + \lambda v H_0^3 + \frac{\lambda}{4} H_0^4, \tag{2.1}$$

where the parameters μ, λ are *a priori* unknown. The term quadratic in the Higgs field determines the Higgs mass to be

$$M_H = \sqrt{2}\,\mu = v\sqrt{2\lambda}. \tag{2.2}$$

This does not provide a numerical value for the Higgs mass M_H because only the quantity v, but not λ, is phenomenologically determined.

This fact places the burden of determining the Higgs mass on experiment. We will interpret the LHC finding of an unstable boson as indeed the Standard Model Higgs boson and for definiteness adopt the value

$$M_H = (126.0 \pm 0.5)\ \text{GeV} \tag{2.3}$$

for subsequent discussion. If so, the remaining parameters in Eq. (2.1) become

$$\mu = 89.1 \pm 0.3\,\text{GeV} \qquad \text{and} \qquad \lambda = 0.131 \pm 0.001. \tag{2.4}$$

The naturalness problem

Radiative corrections to the Higgs mass raise a question of the 'naturalness' of the Standard Model. To motivate the discussion, let us first consider one-loop electromagnetic corrections to the electron mass. If we impose a cut-off Λ_e on the momentum flowing through the loop, the mass shift,

$$m_e = m_{e,0}\left[1 + \frac{3}{2}\frac{\alpha}{\pi}\ln\frac{\Lambda_e}{m_{e,0}} + \cdots \right], \tag{2.5}$$

is obtained. The magnitude of this first-order correction, although cut-off dependent, is generally tiny. Taking for Λ_e the entire mass of the observable universe, $\Lambda_e \simeq 10^{79}$ GeV, results in only the modest mass shift $m_e \simeq 1.7 m_{e,0}$. This teaches us that, with logarithmic behavior, the renormalization program of absorbing divergences into renormalized parameters is not implausible.

However, radiative corrections to the Higgs mass are not as tame. We display in Fig. XV–1 one-loop self-energy processes which shift the Higgs boson mass.

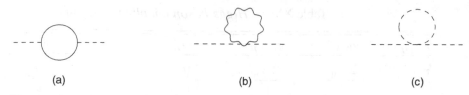

(a)   (b)   (c)

Fig. XV–1 Some quadratically divergent Higgs self-energy diagrams.

Considering for definiteness diagram (c), which involves a Higgs loop with quartic self-coupling, we have

$$-i\,\Sigma_H(p) = -3i\lambda \int \frac{d^4k}{(2\pi)^4} \frac{i}{k^2 - M_{H,0}^2 + i\epsilon}. \tag{2.6}$$

This expression is quadratically divergent, $\Sigma_H \sim \Lambda_H^2$, where Λ_H is the cut-off parameter for the above integral, and leads to a shift of the Higgs mass,

$$M_H^2 = M_{H,0}^2 + \frac{3}{16}\frac{\lambda}{\pi^2}\Lambda_H^2. \tag{2.7}$$

If Λ_H is as large as, say, the Planck mass $E_{\text{Planck}} \simeq 10^{19}$ GeV, then in order to obtain a renormalized mass as given by Eq. (2.3), the parameter $M_{H,0}^2$ must be negative and have a magnitude which equals the correction up to 31 decimal places! This is referred to as *fine tuning*. While technically possible, it is surely unnatural. Including the other contributions of Fig. XV–1 we obtain the Higgs mass shift

$$M_H^2 = M_{H,0}^2 + \frac{3}{16}\frac{\lambda}{\pi^2}\Lambda_H^2 \left[M_H^2 + 2M_W^2 + M_Z^2 - 4m_t^2\right]. \tag{2.8}$$

It is possible to cancel this mass shift by arranging the value of M_H contained within the brackets in Eq. (2.8). This strategy gives $M_H \simeq 314$ GeV, which is ruled out by experiment.

The inability to make sense of Higgs mass corrections is perhaps the most serious flaw in the fabric of the Standard Model. At present, there are no known compelling mechanisms for curing this ailment. Accordingly, many physicists have been motivated by this 'unnaturalness problem' to search for alternatives to the Standard Model description, and to suggest that New Physics must exist not very far above the weak scale $v \sim 250$ GeV.

Higgs coupling constants

There are a variety of ways that the Higgs can interact, including vacuum energy, Higgs self-couplings, Higgs couplings to massive particles, and finally Higgs couplings to massless particles.

Table XV–1. *Higgs-boson coupling constants.*

$g_{\bar{f}fH}$	g_{WWH}	g_{WWH^2}	g_{ZZH}	g_{ZZH^2}	g_{H^3}	g_{H^4}
$\dfrac{m_f}{v}$	$\dfrac{2M_W^2}{v}$	$\dfrac{2M_W^2}{v^2}$	$\dfrac{2M_Z^2}{v}$	$\dfrac{2M_Z^2}{v^2}$	$\dfrac{3M_H^2}{v}$	$\dfrac{3M_H^2}{v^2}$

Vacuum Higgs energy: The first term in V, the Higgs potential of Eq. (2.1), is a constant energy density, which can be interpreted as a contribution $\Lambda^{(\text{Higgs})}$ to the full cosmological constant Λ. Inserting known values for μ and v, we have

$$|U_{\text{Higgs}}^{(\text{vac})}| = \Lambda^{(\text{Higgs})} = \frac{\mu^2 v^2}{4} \simeq 1.2 \times 10^8 \text{ GeV}^4, \qquad (2.9a)$$

which is huge compared to the observed value [RPP 12],

$$|\Lambda^{(\text{obs})}| \simeq (2.3 \times 10^{-3} \text{ eV})^4 = 2.8 \times 10^{-47} \text{ GeV}^4. \qquad (2.9b)$$

This should not, however, be viewed as a defect of the Higgs mechanism, as there are many such contributions to the vacuum energy. Presumably, there is some overriding issue of physics which forces the suppression or cancelation of the vacuum energy by so many orders of magnitude.

Higgs coupling to massive particles: Next, we express couplings of the Higgs boson to particles which have nonzero mass. In cases where n identical fields appear, a numerical factor $1/n!$ is introduced to account for the number of identical fields. The set of all such coupling constants is collected in Table XV–1.

The Higgs potential of Eq. (2.1) contains cubic and quartic Higgs interactions, which we express as

$$\mathcal{L}_{\text{self}} = -\frac{g_{H^3}}{3!} H^3 - \frac{g_{H^4}}{4!} H^4. \qquad (2.10a)$$

There are also couplings of the Higgs to massive fermions. From Eq. (2.3) and Eq. (II–3.20), we find for the interaction to fermion f,

$$\mathcal{L}_{f\bar{f}H} = -g_{\bar{f}fH} H \bar{\psi}_f \psi_f. \qquad (2.10b)$$

The catalog of Higgs particle interactions is extended by presenting its couplings to the $W^\pm$ and Z^0 bosons, including both trilinear and quadrilinear terms for each,

$$\mathcal{L}_{WWH} = W_\mu^- W_+^\mu \left[\frac{g_{WWH^2}}{2!} H^2 + g_{WWH} H \right],$$

$$\mathcal{L}_{ZZH} = Z_\mu Z^\mu \left[\frac{g_{ZZH^2}}{(2!)^2} H^2 + \frac{g_{ZZH}}{2!} H \right], \qquad (2.10c)$$

where we have employed Eqs. (II–3.18), (II–3.29), (II–3.32). Observe that each of the couplings $g_{H^4}, g_{\bar{f}fH}, g_{WWH^2}, g_{ZZH^2}$ are pure numbers whereas $g_{H^3}, g_{WWH}, g_{ZZH}$ have the unit of energy.

Higgs coupling to massless particles: The coupling between the Higgs boson and a particle depends on the particle's mass. This means that at the basic level of the Higgs lagrangian, there is no coupling to photons and gluons because these particles are massless. However, such couplings are induced through quantum effects. This is a phenomenon we have seen already in Chap. IV, in which the photon-photon interaction, $\gamma\gamma \rightarrow \gamma\gamma$, although zero at a fundamental level, is described to one-loop order by the Euler–Heisenberg effective lagrangian of Eq. (IV–8.5).

Higgs–photon–photon vertex: A Higgs boson will couple to a two-photon final state through $W^{\pm}$-boson and charged-fermion loops. The decay rate

$$\Gamma_{H\rightarrow\gamma\gamma} = \frac{M_H^3}{4\pi} \cdot \left| \frac{\alpha}{8\pi v} \left[\mathcal{A}_1(x_W) + \sum_{f=q,\ell} N_c q_f^2 \mathcal{A}_{1/2}(x_f) \right] \right|^2 , \tag{2.11}$$

contains the loop functions $\mathcal{A}_1(x)$ and $\mathcal{A}_{1/2}(x)$,

$$\mathcal{A}_1(x) = -\frac{1}{x^2}\left[2x^2 + 3x + 3(2x-1)f(x)\right],$$

$$\mathcal{A}_{1/2}(x) = \frac{2}{x^2}\left[x + (x-1)f(x)\right] ,$$

$$f(x) = \begin{cases} \arcsin^2(\sqrt{x}) & (x \leq 1) \\ -\frac{1}{4}\left(\ln\left[\frac{1+(1-1/x)^{1/2}}{1-(1-1/x)^{1/2}}\right] - i\pi\right)^2, & (x > 1) \end{cases} \tag{2.12}$$

where x is the dimensionless variable $x \equiv M_H^2/(4m^2)$ and the subscripts on $\mathcal{A}_1(x)$ and $\mathcal{A}_{1/2}(x)$ denote the respective spins of the loop particles. The sum over fermions f in Eq. (2.11) is taken over both quarks q and leptons ℓ.

The above procedure is based on calculating the decay amplitude from Feynman diagrams as in Fig. XV–2. It is worthwhile to consider the possibility of an alternative approach. Throughout this book, we have emphasized the use of effective field theories. Can we employ this method here, via a local effective lagrangian, to describe the Higgs–photon–photon vertex? Note that the function $f(x)$ defined in Eq. (2.12) develops an imaginary part for $m < M_H/2$, which is the case for all the loop fermions except the t quark. The imaginary part signals that H would be able to physically decay into any of the light fermion–antifermion loop pairs. If so, the conversion of a Higgs into two photons is nonlocal and cannot possibly be described with a local lagrangian defined at scale $\mu = M_H$. Although the $W^{\pm}$ and t quark evade such a prohibition, the issue remains whether it would be a good

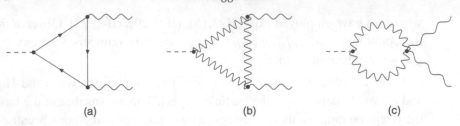

(a)   (b)   (c)

Fig. XV–2 $H \to \gamma\gamma$ via (a) charged fermion, (b)–(c) W boson.

numerical approximation to use a local lagrangian for either. Let us compare the loop functions $\mathcal{A}_1$ and $\mathcal{A}_{1/2}$ evaluated both in the heavy mass limit $x \to 0$ and also using the physical values $x_W \simeq 0.60$ and $x_t \simeq 0.13$,[2]

$$\left| \frac{\mathcal{A}_1(0) - \mathcal{A}_1(x_W)}{\mathcal{A}_1(0)} \right| \simeq 0.16 \qquad \text{vs.} \qquad \left| \frac{\mathcal{A}_{1/2}(0) - \mathcal{A}_{1/2}(x_t)}{\mathcal{A}_{1/2}(0)} \right| \simeq 0.03.$$

Since the difference between the infinite-mass and physical t-quark amplitudes is only 3%, most would agree that an effective lagrangian description for the t-quark contribution is appropriate, and we write

$$\mathcal{L}_{\text{eff}} = g_{\gamma\gamma}^{(t)} H F^{\mu\nu} F_{\mu\nu} \qquad \text{with} \qquad g_{\gamma\gamma}^{(t)} = \frac{2\alpha}{9\pi v}, \tag{2.13}$$

where α is the fine-structure constant and $F_{\mu\nu}$ is the electromagnetic field strength tensor (cf. Eq. (I–5.9)). Note that the heavy top quark evades the decoupling theorem of Sect. IV–2 because the $t\bar{t}H$ vertex is proportional to the large mass parameter m_t.

An alternate derivation of Eq. (2.13) begins by considering the contribution of a $t\bar{t}$ loop to the photon vacuum polarization [ShVVZ 79],

$$\Pi^{\mu\nu}(q) \Big|_{\text{t-quark}} = (q^\mu q^\nu - q^2 g^{\mu\nu}) \left[\frac{N_c q_t^2 \alpha}{3\pi} \ln \frac{\Lambda^2}{m_t^2} + \cdots \right], \tag{2.14}$$

where $q_t = 2/3$ is the top-quark electric charge in units of e and we have chosen regularization with cut-off Λ here (instead of the dimensional approach used elsewhere in this book) to keep the notation compact. The photon vacuum polarization of Eq. (2.14) can equivalently be expressed via the effective lagrangian,

$$\mathcal{L}_{\text{ph. vac. pol.}}^{(\text{t-loop})} = -\frac{1}{4} F_{\mu\nu} F^{\mu\nu} \cdot \frac{q_t^2 \alpha}{\pi} \ln \frac{\Lambda^2}{m_t^2}, \tag{2.15}$$

---

[2] For reference we note the expansions about $x = 0$: $\mathcal{A}_{1/2}(x) \simeq 4/3(1 + 7x/30 + 2x^2/21 + \cdots)$ and $\mathcal{A}_1(x) \simeq -7 - 22x/15 - 76x^2/105 + \cdots$.

as can be shown by taking its photon-to-photon matrix element. Now, writing the top-quark mass term together with its Higgs interaction,

$$\mathcal{L}_{Ht\bar{t}} = -\left(m_t + \frac{m_t}{v}H\right)\bar{t}t = -m_t\left(1 + \frac{H}{v}\right)\bar{t}t, \qquad (2.16)$$

suggests treating the Higgs field as a constant and thus formally extending the top-quark mass as $m_t \to m_t(1 + H/v)$. Inserting this into Eq. (2.15) and considering only the term linear in H yields precisely the effective lagrangian of Eq. (2.13). This 'background field' derivation is valid if the momenta involved are small compared to the top-quark mass, which is not perfect but a good first approximation.[3]

Higgs–Z^0–photon vertex: This process, too, occurs first as a loop amplitude ([CaCF 79], who assume $M_H < M_Z$ and study $Z^0 \to H\gamma$; see also [BeH 85]) via triangle diagrams dominated by $W^\pm$-boson and t-quark contributions. We refer the reader to the literature for the explicit, somewhat cumbersome, analytic form of the vertex.

Higgs–gluon–gluon interaction: The Higgs two-gluon amplitude has similarities with the Higgs two-photon interaction. One calculates Feynman amplitudes for triangle diagrams, although now summed over only quarks $\{q\}$ since gluons couple neither to leptons nor to the electroweak gauge bosons, leading to

$$\Gamma_{H\to gg} = \frac{2M_H^3}{\pi} \cdot \left|\frac{\alpha_s}{16\pi v}\left[\sum_q \mathcal{A}_{1/2}(x_q)\right]\right|^2. \qquad (2.17)$$

The top-quark amplitude is by far the largest in the above sum, and so we can again turn to the effective lagrangian description. The contribution of a $t\bar{t}$ loop to the gluon vacuum polarization in cut-off regularization is

$$\Pi^{\mu\nu}(q)_{ab}\bigg|_{\text{t-quark}} = \delta_{ab}(q^\mu q^\nu - q^2 g^{\mu\nu})\left[\frac{\alpha_s}{6\pi}\ln\frac{\Lambda^2}{m_t^2} + \cdots\right], \qquad (2.18)$$

which leads, as explained earlier, to

$$\mathcal{L}_{\text{gl. vac. pol.}}^{(\text{t-loop})} = -\frac{1}{4}F_{\mu\nu}^a F^{a\mu\nu} \cdot \frac{\alpha_s}{6\pi}\ln\frac{\Lambda^2}{m_t^2}, \qquad (2.19)$$

and finally, from Eq. (2.16), to

$$\mathcal{L}_{\text{eff}} = g_{gg}^{(t)} H F^{a\mu\nu} F_{\mu\nu}^a \qquad \text{with} \qquad g_{gg}^{(t)} = -\frac{\alpha_s}{12\pi v}, \qquad (2.20)$$

---

[3] For completeness, we take note of yet another derivation [ElGN 76] of Eq. (2.13) which uses the *QED* trace anomaly (see Eq. (III–4.16) for the *QCD* version),

$$\theta_\mu^\mu = \frac{\alpha_s}{12\pi}F_{\mu\nu}F^{\mu\nu} + m_t\bar{t}t,$$

taking into account only the t-quark part of the fermion contribution.

where $F^{a\mu\nu}$ is the chromodynamic field strength tensor of Eq. (II–2.2a). This represents the linear term in an expansion in powers of the Higgs field H. Higher powers provide the two-gluon coupling to an arbitrary number of Higgs bosons. The quadratic term in this expansion would be a prediction for $gg \rightarrow HH$. There, in addition to the direct coupling of Eq. (2.20), one encounters a pole diagram (i.e. $gg \rightarrow H \rightarrow HH$) which contains the triple Higgs coupling. The direct and pole contributions cancel exactly at threshold and, more generally, the residual effect remains small.

XV–3 Production and decay of the Higgs boson

Following the discovery of the top quark, finding the Standard Model Higgs boson became a primary goal of experimental particle physics. The search strategy was based on Standard Model predictions of both production and decay amplitudes. We discuss each of these in turn, beginning with the topic of Higgs decay.

Decay

One begins calculation of a Higgs decay mode with the lowest-order amplitude, and then incorporates higher-order QCD and electroweak (EW) corrections. These higher-order effects are described, with many references, in [Dj 08]. Here, we display branching fraction predictions in Table XV–2 [He *et al.* 13], but restrict our presentation here to only the lowest-order analysis (except for two decays $H \rightarrow b\bar{b}$ and $H \rightarrow gg$, which have especially large corrections). The major two-body Standard Model decay branching fractions in Table XV–2 correspond to a total width,

$$\Gamma_H^{\text{(tot)}} \simeq 4.21 \ (\pm 3.9\%) \ \text{MeV}. \tag{3.1}$$

The individual branching fractions in Table XV–2 are purely theoretical quantities. An experimental reality at LHC is that detection of the modes $b\bar{b}$, gg, $c\bar{c}$ is greatly inhibited by huge hadronic backgrounds. As a consequence, other modes (e.g. $\gamma\gamma$) can play a central role in Higgs phenomenology at the LHC, despite their smaller branching fractions.

Table XV–2. *Two-body Higgs branching fractions.*[a]

$b\bar{b}$	WW^{*b}	gg	$\tau^+\tau^-$	$\bar{c}c$	ZZ^{*b}	$\gamma\gamma$	γZ	$\mu^+\mu^-$
56.1	23.1	8.48	6.15	2.83	2.89	0.23	0.16	0.02

[a] All branching fractions are in % and the value $M_H = 126.$ GeV is assumed.
[b] The asterisk denotes a virtual vector boson.

Decay into fermion–antifermion pairs: For transitions of the type $H \to f\bar{f}$, the leading-order (LO) decay rate is

$$\Gamma^{(LO)}_{H \to f\bar{f}} = \frac{N_c}{8\pi} \frac{m_f^2}{v^2} M_H \left(1 - 4x_f^2\right)^{3/2}, \tag{3.2a}$$

where m_f is the fermion mass (which arises from the Yukawa coupling), $x_f \equiv m_f/M_H$ and $N_c = 1$ for leptons and $N_c = 3$ for quarks. We already know that in the Standard Model the Higgs coupling to a fermion–antifermion pair is linear in the fermion mass m_f. The factor of m_f^2 in Eq. (3.2a) reflects this and ensures that the $b\bar{b}$ mode is largest amongst all fermions with $2m_f < M_H$ (the mode $H \to t\bar{t}$ is kinematically forbidden).

Let us consider the $H \to b\bar{b}$ mode in a bit more detail. If Eq. (3.2a) is used to determine the $b\bar{b}$ decay rate and Eq. (3.1) is used for $\Gamma_H^{(tot)}$, then a branching fraction $\simeq 104\%$ is predicted. This unphysical result is disconcerting to say the least! The flaw in our numerical exercise is that we have ignored corrections to the tree-level prediction of Eq. (3.2a). Ordinarily, one expects a 'correction' to be no more than a few tens of percent and usually much smaller. This case is not like that; it turns out that the most important correction is to replace the m_f^2 factor by the squared running mass $\overline{m}_f^2(\mu)$ with $\mu = M_H$,

$$\Gamma_{H \to f\bar{f}} = \frac{N_c}{8\pi} \frac{\overline{m}_b^2(M_H)}{v^2} M_H \left(1 - 4x_f^2\right)^{3/2} \left[1 + 5.67 \frac{\alpha_s(M_H)}{\pi} + \cdots\right], \tag{3.2b}$$

where the $\mathcal{O}(\alpha_s)$ correction is also displayed. For the b quark, we have already found below Eq. (XIV–1.12) that $\overline{m}_b(M_H) \simeq 0.665\, \overline{m}_b(\overline{m}_b)$, implying a corrected $H \to b\bar{b}$ branching fraction of 56%. This means that all the remaining corrections for this mode amount to a rather more modest effect. The moral of this lesson is to not place unwarranted trust in tree-level estimates.

Decay into three-body states: Although the Higgs boson couples to the electroweak gauge bosons, a Higgs with mass $M_H \simeq 126$ GeV is too light to decay into WW and ZZ final states. However, a transition like $H \to WW^* \to Wf\bar{f}'$ (or $H \to Zf\bar{f}$) can occur, e.g., $H \to W^+d\bar{u}$ or $H \to W^-c\bar{s}$ and so on. We shall consider this possibility here. If dependence on fermion mass (such as m_f/M_H or m_f/M_W) is ignored, the energy distribution of the final state W is [KeM 84]

$$\frac{d\Gamma^{(LO)}_{H \to Wf\bar{f}'}}{dx} = \frac{1}{192\pi^3} \left(\frac{M_W}{v}\right)^4 M_H \frac{(x^2 - 4\epsilon^2)^{1/2}}{(1-x)^2} \left(x^2 - 4\epsilon^2 x + 8\epsilon^2 + 12\epsilon^4\right), \tag{3.3}$$

where $x = 2E_W/M_H$ and $\epsilon = M_W/M_H$. Integration over the W-boson energy yields

$$\Gamma^{(LO)}_{H \to W f \bar{f}'} = \frac{1}{192\pi^3} \left(\frac{M_W}{v} \right)^4 M_H F(\epsilon)$$

$$F(\epsilon) = \frac{3(1 - 8\epsilon^2 + 20\epsilon^4)}{(4\epsilon^2 - 1)^{1/2}} \arccos \left[\frac{3\epsilon^2 - 1}{2\epsilon^3} \right]$$

$$- (1 - \epsilon^2) \left[\frac{47}{2}\epsilon^2 - \frac{13}{2} + \frac{1}{\epsilon^2} \right] - 3 \left(1 - 6\epsilon^2 + 4\epsilon^4 \right) \ln \epsilon. \quad (3.4)$$

Thus far, we have kept the final state fixed as $W f \bar{f}'$. To obtain the inclusive rate $\Gamma_{H \to W^\pm X}$, we sum over all distinct final states (like the ones displayed above Eq. (3.3)) to find

$$\Gamma^{(LO)}_{H \to W^\pm X} = \frac{3}{32\pi^3} \left(\frac{M_W}{v} \right)^4 M_H F(\epsilon). \quad (3.5)$$

The case of $H \to Z f \bar{f}$ is obtained from the above relations via insertion of a factor $\eta_Z = \frac{7}{12} - \frac{10}{9} \sin^2 \theta_w + \frac{40}{9} \sin^4 \theta_w$.

Decay into four-body states: The degrees of freedom appearing in Table XV–2 are those occurring at the primary vertex, at which the Higgs decay process begins. However, these are often not the final states which are actually detected. For example, the quark–antiquark states will hadronize into jets whereas the vector gauge bosons will quickly decay and be observed as four-fermion final states, e.g., as in final states containing leptons and antileptons. We do not display analytic formulae here for such modes, but numerical examples are displayed in Table XV–3. The leptons and neutrinos there are summed respectively over $\ell = e, \mu, \tau$ and $\nu = \nu_e, \nu_\mu, \nu_\tau$.

Decay into massless final-state particles: The general leading-order $H \to \gamma\gamma$ decay rate is given in Eq. (2.11). Approximating this with the W-boson and top-quark contributions gives

$$\Gamma^{(LO)}_{H \to \gamma\gamma} \simeq \frac{\alpha^2}{256\pi^3} \cdot \frac{M_H^3}{v^2} \left| A_1(x_W) + N_c q_t^2 A_{1/2}(x_t) \right|^2, \quad (3.6)$$

where the quantities $A_1(x_W)$ and $A_{1/2}(x_t)$ are the loop functions defined in Eq. (2.11), with arguments $x_W = M_H^2/(4M_W^2)$ and $x_t = M_H^2/(4m_t^2)$. Although the

Table XV–3. *Four-body Higgs branching fractions.*[a]

$(qqqq)$	$(qq\ell\nu_\ell)$[b]	$(qq\nu\nu)$	$(qq\ell^+\ell^-)$	$(\ell^+\ell^-\ell^+\ell^-)$
11.8	3.38	0.81	0.27	0.03

[a] All branching fractions are in % and the value $M_H = 126.$ GeV is assumed.
[b] Here, $\ell = e, \mu$.

top-quark contribution dominates that of the other fermions (due to its much larger Yukawa coupling to the Higgs), that of the W-boson is even larger, $|N_c q_t^2 \mathcal{A}_{1/2}(x_t)/\mathcal{A}_1(x_W)| \simeq 0.22$.

For the transition $H \to gg$, decay products would appear as jets consisting of light hadrons. The $H \to gg$ decay rate has already been given in Eq. (2.17). Approximating this with the dominant top-quark contribution in the heavy m_t limit yields the tree-level expression,

$$\Gamma_{H \to gg}^{(\mathrm{LO})} = \frac{\alpha_s^2 M_H^3}{72\pi^3 v^2}. \tag{3.7}$$

Virtual gluon exchanges will modify the above. Unlike the case for $H \to \gamma\gamma$ the next-to-leading-order $H \to gg$ amplitude will experience gluon self-interactions such as triple-gluon vertices and turns out to have a large numerical effect [SpDGZ 95],

$$\Gamma_{H \to gg} \simeq \Gamma_{H \to gg}^{(\mathrm{LO})} \left[1 + \left(\frac{95}{4} - \frac{7}{6} n_f \right) \alpha_s(M_H) + \dots \right] \simeq 1.64\, \Gamma_{H \to gg}^{(\mathrm{LO})}, \tag{3.8}$$

with $n_f = 5$ and $\alpha_s(M_H)$ given previously in Eq. (II–2.79).

Production

Next, we consider the most important of the mechanisms at LHC energies for producing the Higgs boson in the inclusive process $p + p \to H + X$, where X represents a sum over all the other final-state particles. The scattering which yields the Higgs production will involve the basic degrees of freedom (*partons*) occurring within a proton, the quarks and gluons. Because the partons are not physical entities, the cross section must be expressed as

$$\sigma = \sum_{i,j} \int_0^1 dx_1\, dx_2\, f_i(x_i, Q) f_j(x_2, Q) \hat{\sigma}_{ij}, \tag{3.9}$$

where the indices i, j refer to the two initial-state protons and the quantities f_i and f_j are parton distribution functions ('PDFs'). A hadron's PDF $f(x, Q)$ gives the probability density for finding a parton carrying a fraction x of the hadronic longitudinal momentum at momentum reference scale Q. Given the difficulty presented by nonperturbative QCD, a PDF is commonly inferred from experimental data, e.g., as with

$$f_i(x, Q) = N x^{\alpha_i} (1 - x)^{\beta_i} g_i(x). \tag{3.10}$$

where α_i, β_i are fit parameters. The function $g_i(x)$ is defined to approach constants at $x = 0, 1$ e.g., $g_i(x) = 1 + \epsilon_i \sqrt{x} + D_i x + E_i x^2$ and itself contains the fit parameters

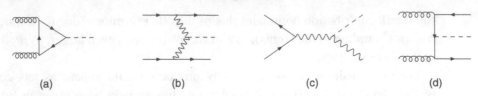

(a)   (b)   (c)   (d)

Fig. XV–3 Higgs production via: (a) gg fusion, (b) VBF, (c) HV, (d) $t\bar{t}H$.

ϵ_i, D_i, E_i. The parton cross section $\hat{\sigma}_{ij}$ is calculated at leading order from various Standard Method processes and corrected by both *QCD* and EW perturbations.

Within this phenomenological framework, one has at the *Higgs mass scale and LHC energies* the following Standard Model mechanisms, depicted to leading order in Figure XV–3 and listed here according to cross-section magnitude:

(1) Gluon–gluon fusion (*gg* fusion): $gg \to t\bar{t} \to H$

(2) Vector–boson fusion (VBF): $qq \to qq + V^*V^* \to qq + H$

(3) Vector–boson-associated production (HV): $q\bar{q} \to V^* \to H + V$

(4) $t\bar{t}$ associated-production ($t\bar{t}H$): $gg \to t\bar{t} + H$,

Numerical values [He *et al.* 13] for each of these contributions at the energies $\sqrt{s} = 8, 14$ TeV appear in Table XV–4. Table XV–4 contains not only cross-section values but also uncertainties for each, given numerically in per cent. These arise mainly from aspects of *QCD*, such as uncertainties in *QCD* parameters (e.g. α_s, m_c, etc.), parton PDFs and a significant uncertainty from the uncalculated higher-order *QCD* corrections.

The gluon–gluon fusion reaction proceeding via top-quark loops is the dominant component of the $p + p \to H + X$ cross section.[4] It also has the interesting property of being sensitive to certain types of virtual heavy particles. We saw in the derivation of Eq. (2.20) that the top-quark contribution to the triangle graph for $H \to gg$ does not decouple, despite having $4m_t^2 \gg M_H^2$, because the coupling

Table XV–4. *Standard Model Higgs production cross sections.*[a]

$\sqrt{s}$ (TeV)	gg Fusion	VBF	HW	HZ	$t\bar{t}H$
8	18.97 $\left(^{+7.2\%}_{-7.8\%}\right)$	1.57 $\left(^{+0.3\%}_{-0.1\%}\right)$	0.69 $(\pm 1.0\%)$	0.41 $(\pm 3.2\%)$	0.13 $\left(^{+3.8\%}_{-9.3\%}\right)$
14	49.85 $\left(^{+19.6\%}_{-14.6\%}\right)$	4.18 $\left(^{+2.8\%}_{-3.0\%}\right)$	1.50 $\left(^{+4.1\%}_{-4.4\%}\right)$	0.88 $\left(^{+6.4\%}_{-5.5\%}\right)$	0.61 $\left(^{+14.8\%}_{-18.2\%}\right)$

[a] All cross sections are in *pb* units; the value $M_H = 126$ GeV is used for $\sqrt{s} = 8$ (TeV) and $M_H = 125$ GeV for $\sqrt{s} = 14$ (TeV).

[4] The next most important contribution, that of the *b*-quark loop, is estimated at leading order to be at most a 10% effect.

between $t\bar{t}H$ is proportional to m_t. Thus, what if there were a very heavy fourth generation of Standard Model fermions (a situation often denoted as SM$_4$) with all else the same (i.e. same Standard Model couplings, only one physical Higgs boson) as with the known fermions? The new generation would contain two new, very heavy quarks, say u_4, d_4, which likewise would not decouple in the $H \to gg$ vertex. The $H \to gg$ amplitude would then be about a factor three larger than in the Standard Model case, and the gluon–fusion production cross section about nine times as large. Moreover, using LHC and Tevatron data as input, it has been concluded from an analysis of Higgs decay modes that SM$_4$ is excluded at more than 5σ [EbHLLMNW 12].

Earlier, in the discussion following Eq. (3.1), we pointed out that detection of final states like $b\bar{b}$, gg, $c\bar{c}$ at the LHC, where the $gg \to H$ is the dominant production mechanism, is greatly hindered by hadronic backgrounds. However, a $b\bar{b}$ final state can be relatively more accessible if the Higgs particle is predominantly produced in association with a vector boson ($V = W, Z$) or a $t\bar{t}$ pair, a strategy which has been pursued by the detectors CDF and D0 (Tevatron) and ATLAS and CMS (LHC). This can lead to detection of $H \to b\bar{b}$ via more easily identifiable configurations like

$$HW \to b\bar{b}\ell\nu_\ell, \qquad HZ \to b\bar{b}\ell\bar{\ell}, \qquad HW, HZ \to \not{E}_T b\bar{b}$$

where $\ell = e, \mu$ and $\not{E}_T$ represents missing transverse energy. Some promising results have been obtained thus far, e.g., a reported excess of events at 3.1σ with $M_H = 125.$ GeV [Aa *et al.* (CDF and D0 Collabs.) 13] and a $> 3\sigma$ significance in the combined $\tau\bar{\tau} + b\bar{b}$ channels reported by the CMS collaboration at the 2013 Lepton–Photon Conference.

Comparison of Standard Model expectations with LHC data

Statistical data analyses have been performed to test the extent to which collected data agree with the Standard Model Higgs boson scenario. Such testing can be done directly by experimental collaboration or as a theoretically motivated exercise:

(1) *Experimental*: One can define a *global signal strength factor* μ_i for a given final state '*i*' by folding together the production cross section and branching fraction for the observed signal relative to the Standard Model prediction,

$$\mu_i = \frac{\left[\sum_j \sigma_{j \to H} \, \mathrm{Br}_{H \to i}\right]_{\mathrm{obs}}}{\left[\sum_j \sigma_{j \to H} \, \mathrm{Br}_{H \to i}\right]_{\mathrm{SM}}}. \tag{3.11}$$

There is a label '*j*' because a given final state '*i*' might be summed over a subset of Higgs production processes '*i*'. The value $\mu = 0$ corresponds to the background-only hypothesis whereas $\mu = 1$ corresponds to the Standard Model Higgs boson signal in addition to the background. Announced results from the LHC detectors have been found, thus far, to be statistically consistent with the Standard Model hypothesis.

(2) *Theoretical*: There are a number of ways to parameterize couplings to include non-Standard Model behavior. Suppose Standard Model Higgs couplings to fermion f and to vector boson V are generalized to have the forms [ElY 12],

$$g_f = \sqrt{2}\frac{m_f}{v} \to \sqrt{2}\left(\frac{m_f}{M}\right)^{1+\epsilon}, \qquad g_V = 2\frac{M_V^2}{v} \to 2\frac{M_V^{2(1+\epsilon)}}{M^{(1+2\epsilon)}}, \qquad (3.12)$$

where ϵ and M are purely phenomenological parameters. In the Standard Model, they become $\epsilon = 0$ and $M = v \simeq 246$ GeV. A global fit to LHC data yields results consistent with these values, $\epsilon = 0.05 \pm 0.08$ and $M = 241 \pm 18$ GeV.

Another procedure is to consider an effective lagrangian for the electroweak symmetry-breaking sector, which modifies couplings to vector mesons and fermions in terms of universal parameters 'a' and 'c'.

$$\mathcal{L}_{\text{eff}} = \sum_{V=W,Z} \eta_V M_V^2 V_\mu^\dagger V^\mu \left[1 + 2a\frac{H}{v}\right] - \sum_i m_i \bar{f}_i f_i \left[1 + c\frac{H}{v}\right] + \cdots$$

$$(3.13)$$

where $\eta_W = 1$, $\eta_Z = 1/2$, and the ellipses represent a sum over all remaining Standard Model contributions as well as possible higher-order terms in the field variable H. In the Standard Model, we have $a = c = 1$. Fits to the current dataset again yield results consistent with Standard Model expectations [ElY 12, EsGMT 12].

The above parameterizations are just two examples of Higgs-related phenomenology. These tests, and others, will continue into the future as the Higgs database expands.

XV–4 Higgs contributions to electroweak corrections

Prior to the discovery of a new boson at the LHC, direct Higgs searches yielded only upper bounds, e.g., as with $M_H < 114.4$ GeV obtained at LEP2. However, the calculation of quantum corrections to Standard Model predictions came to play a central role in particle phenomenology and Higgs physics in particular. The procedure is straightforward; a collection of observables ($M_W, \ldots$) is measured and then compared to predictions expressed in terms of a set of input parameters ($G_\mu, \alpha, \ldots$) including the Higgs mass M_H (cf. Sect. XVI–6). Although the dependence on Higgs mass in such analyses is somewhat weak, being logarithmic $\sim \ln M_H^2$ at leading order, it has continued to show for quite some time that the Higgs boson is 'light'. A recent $\Delta\chi^2$ fit gives [Ba *et al.* (Gfitter group) 12] $M_H = 94^{+25}_{-22}$ GeV. That this value is consistent with the LHC determinations of M_H is generally regarded as a noteworthy success of the Standard Model. To observe the role of the Higgs boson in this procedure, let us next consider a few specific examples of such corrections.

The corrections $\Delta\rho$ and Δr

Higgs contributions to $\Delta\rho$: We begin with the so-called effective weak mixing angle

$$\bar{s}_w^2 = 1 - \frac{M_W^2}{M_Z^2} + c_w^2 \Delta\rho, \tag{4.1}$$

which is discussed at length in Sect. XVI–1. The corrections to $\bar{s}_w^2$ are contained within the quantity $\Delta\rho$. For arbitrary M_H, the one-loop Higgs contribution to $\Delta\rho$ is

$$\Delta\rho_H^{\text{1-loop}} = -\frac{3}{4}\left(\frac{M_W^2}{4\pi^2 v^2}\right) f(M_H^2/M_Z^2), \tag{4.2a}$$

where

$$f(x) = x\left[\frac{\ln c_w^2 - \ln x}{c_w^2 - x} + \frac{\ln x}{c_w^2(1-x)}\right]. \tag{4.2b}$$

The leading dependence on M_H for $M_H \gg M_W$ is logarithmic,

$$\Delta\rho_H^{\text{1-loop}} \sim -\frac{3}{4}\left(\frac{M_W^2}{4\pi^2 v^2}\right)\frac{s_w^2}{c_w^2}\ln\frac{M_H^2}{M_W^2}, \tag{4.3}$$

as are all the other leading one-loop Higgs contributions.[5] A term like $\ln M_H^2/M_W^2$ does not respond sensitively to changes in M_H^2, so the shift $\Delta\rho_H^{\text{1-loop}}$ by itself does not lead to a precise estimate for M_H.

There are also multi-loop Higgs contributions. In contrast to the $\ln M_H^2/M_W^2$ logarithmic dependence of the one-loop amplitude, these also contain *power-law* dependence on M_H,

$$\Delta\rho_H^{\text{2-loop}} \sim 0.1499\left(\frac{M_W^2}{4\pi^2 v^2}\right)^2 \frac{s_w^2 M_H^2}{c_w^2 M_W^2}, \tag{4.4a}$$

---

[5] It is, however, not the case that one-loop corrections for all the remaining Standard Model particles are logarithmic, e.g., $\Delta\rho$ has a $\mathcal{O}(G_\mu m_t^2)$ dependence on the *t*-quark mass (viz. Sect. XVI–6).

Table XV–5. *Higgs contribution to* $\Delta\rho$.

Order:	One-loop	Two-loop	Three-loop
	-1.8×10^{-3}	8.1×10^{-7}	-6.2×10^{-8}

and the three-loop amplitude gives

$$\Delta\rho_H^{3\text{-loop}} \sim -1.728 \left(\frac{M_W^2}{4\pi^2 v^2}\right)^3 \frac{s_w^2 M_H^4}{c_w^2 M_W^4}. \tag{4.4b}$$

Observe that common to all terms is the coefficient,

$$\frac{M_W^2}{4\pi^2 v^2} \simeq 0.0027. \tag{4.5}$$

An extra power of this small quantity will accompany each additional loop and thus suppress the multi-loop contributions, at least for moderate values of M_H. Note also that the two-loop and three-loop amplitudes have opposite sign. The values of the one-loop, two-loop, and three-loop amplitudes are summarized in Table XV–5 using $M_H = 126$. GeV. The one-loop amplitude is dominant and gives an accurate estimate of the Higgs contribution to $\Delta\rho$.

Higgs contributions to Δr: A second class of Standard Model corrections affects the relation between the Fermi constant and M_W, given to leading order by Eq. (II–3.43). Upon using Eq. (II–3.42) and Eq. (II–3.33), we can express this as

$$M_W^2 \left(1 - \frac{M_W^2}{M_Z^2}\right) = \frac{\pi\alpha}{\sqrt{2}G_\mu}. \tag{4.6}$$

The one-loop Higgs correction to this relation,

$$M_W^2 \left(1 - \frac{M_W^2}{M_Z^2}\right) = \frac{\pi\alpha}{\sqrt{2}G_\mu} \left(1 + \Delta r_H^{1\text{-loop}}\right), \tag{4.7}$$

is given by

$$\Delta r_H^{1\text{-loop}} = \frac{11}{48\pi^2} \cdot \frac{M_W^2}{v^2} \left(\ln\frac{M_H^2}{M_W^2} - \frac{5}{6}\right). \tag{4.8}$$

Custodial symmetry

As part of our discussion of chiral symmetry in Chap. IV, we obtained a representation of the linear sigma model by expressing an $SU(2)_L \times SU(2)_R$ invariant lagrangian (cf. Eq. (IV–1.4)) in terms of two chiral fermions ψ_L, ψ_R and a 2×2 matrix $\Sigma = \sigma + i\tau \cdot \pi$ of four scalar fields. The $SU(2)_L \times SU(2)_R$ transformation properties were $\psi_L \to L\psi_L$, $\psi_R \to R\psi_R$ and $\Sigma \to L\Sigma R^\dagger$ with L, R in $SU(2)$.

Somewhat analogously, we can express the Higgs doublet as a matrix $\mathbf{H}$ via the construction[6]

$$\mathbf{H} = \frac{1}{\sqrt{2}} \begin{pmatrix} \widetilde{\Phi} & \Phi \end{pmatrix} \equiv \frac{1}{\sqrt{2}} \begin{pmatrix} \varphi^{0*} & \varphi^+ \\ -\varphi^- & \varphi^0 \end{pmatrix}, \tag{4.9}$$

where Φ is the Higgs field of Eq. (II–3.16) and $\widetilde{\Phi}$ is its conjugate.[7] This will be convenient for considering transformations of both $SU(2)_L$ and $SU(2)_R$.

Even though this chapter is, for the most part, a discussion/celebration of the $M \simeq 125$ GeV particle, which could well be the Standard Model Higgs boson, we shall, for the remainder of this section, instead emphasize the symmetry aspect of the Higgs sector. In the notation introduced above, a Higgs lagrangian invariant under $SU(2)_L \times U(1)_Y$ gauge symmetry is

$$\mathcal{L}_{\text{Higgs}} = \text{Tr}\left[(D^\mu \mathbf{H})^* D_\mu \mathbf{H}\right] - V(\mathbf{H}^\dagger \mathbf{H}), \tag{4.10a}$$

where the covariant derivative is

$$D_\mu \mathbf{H} = (\partial_\mu + i\frac{g_1}{2}B_\mu \tau_3 + ig_2 \frac{\vec{\tau}}{2} \cdot \vec{W}_\mu)\mathbf{H}, \tag{4.10b}$$

and the potential has the form

$$V(\mathbf{H}^\dagger \mathbf{H}) = -\mu^2 \text{Tr}\left[\mathbf{H}^\dagger \mathbf{H}\right] + \lambda \left(\text{Tr}\left[\mathbf{H}^\dagger \mathbf{H}\right]\right)^2. \tag{4.10c}$$

The matrix τ_3 in Eq. (4.10b) accounts for the opposite relative weak hypercharge of Φ and its conjugate $\widetilde{\Phi}$. That the lagrangian $\mathcal{L}_{\text{Higgs}}$ of Eq. (4.10a) is indeed gauge-invariant can be verified by noting

$$SU(2)_L : \mathbf{H} \to L\mathbf{H} \text{ and } D_\mu \mathbf{H} \to L(D_\mu \mathbf{H}) \qquad U(1)_Y : \mathbf{H} \to \mathbf{H}e^{-i\tau_3 \theta_Y}. \tag{4.11}$$

Actually, the potential energy $V(\mathbf{H}^\dagger \mathbf{H})$ of Eq. (4.10c) (but not the kinetic part in Eq. (4.10a)) is invariant under the larger set of $SU(2)_L \times SU(2)_R$ transformations.

Thus far, we have simply used a new notation to reproduce what we already know. In order to learn something new, however, consider the limit $g_1 \to 0$. There is now present the symmetry, $SU(2)_R$, under which

$$SU(2)_R : \mathbf{H} \to \mathbf{H}R^\dagger \text{ and } D_\mu \mathbf{H} \to (D_\mu \mathbf{H})R^\dagger. \tag{4.12}$$

Thus, for the combined $SU(2)_L \times SU(2)_R$ transformations, we have $\mathbf{H} \to L\mathbf{H}R^\dagger$, like the sigma model matrix Σ mentioned at the beginning of this section. Although true, the above analysis is incomplete; we must address the Higgs spontaneous

---

[6] In the following, we adopt the general approach of [SiSVZ 80] and [Wi 04].
[7] In the language of group theory, the conjugate spinor $\widetilde{\Phi} = i\tau_2 \Phi^*$ is *equivalent* to Φ.

symmetry breaking of Eq. (II–3.25), for which the ground-state configuration of **H** becomes

$$\langle \mathbf{H} \rangle = \frac{1}{2} \begin{pmatrix} v & 0 \\ 0 & v \end{pmatrix}, \tag{4.13}$$

with $v \equiv (\mu^2/\lambda)^{1/2}$ as in Eq. (II–3.24). Although this ground state does not respect the full $SU(2)_L \times SU(2)_R$ symmetry,

$$L\langle \mathbf{H} \rangle \neq \langle \mathbf{H} \rangle, \qquad \langle \mathbf{H} \rangle R^\dagger \neq \langle \mathbf{H} \rangle, \tag{4.14a}$$

it *does* remain invariant under $SU(2)_{L+R}$ transformations, i.e., those having $R = L$,

$$L\langle \mathbf{H} \rangle L^\dagger = \langle \mathbf{H} \rangle. \tag{4.14b}$$

This $SU(2)_{L+R}$ invariance is often referred to as *custodial* symmetry [SiSVZ 80].

In Chap. II, the basis of our discussion of the electroweak sector was the Higgs effect, i.e., the spontaneous breaking of the gauge symmetry $SU(2)_L \times U(1)_Y$. Here, let us instead use elementary group theory to see what the $g_1 = 0$ world, with its exact $SU(2)_{L+R}$ global symmetry, would be like.[8] Eq. (II–3.31) shows that setting $g_1 = 0$ would cause the weak mixing angle to vanish, $\theta_w \to 0$, and so from Eq. (II–3.30) for $Z^0 \to W_3$.

It follows from Eq. (I–5.17) that the three **W**-boson fields would transform as an isotriplet under the (global!) $SU(2)_L$ transformations, and as an isosinglet under $SU(2)_R$ (since $g_1 = 0$). They would thus transform as an isotriplet under $SU(2)_{L+R}$ and, since the $SU(2)_{L+R}$ symmetry is exact, the **W** triplet would be degenerate. The above remarks imply the equality

$$\rho = (M_W/(M_Z \cos \theta_w))^2 = 1 \qquad \text{(in the } g_1 \to 0 \text{ limit).} \tag{4.15a}$$

When viewed as a statement of invariance, this equality is a consequence of the $SU(2)_{L+R}$ symmetry, which is called 'custodial' for this reason. As we then return to the real world of $g_1 \neq 0$ and allow for higher-order Standard Model corrections, we would expect corrections to $\rho = 1$ to be modest [SiSVZ 80],

$$\rho = 1 + \mathcal{O}(\alpha) + \mathcal{O}(\alpha(m_u^2 - m_d^2)/M_W^2). \tag{4.15b}$$

XV–5 The quantum Higgs potential and vacuum stability

Our treatment of the Higgs potential has thus far been at the classical level. We have simply taken the quadratic and quartic terms that appear in the bare lagrangian,

---

[8] For example, the electric charge would vanish (cf. Eq. (II–3.42)), so modest mass shifts would occur, e.g., the leading-order contribution to pion mass splitting would vanish, etc.

minimized the energy, and found the vacuum expectation value and the Higgs mass. However, quantum effects modify this form significantly, most importantly through a top-quark loop. Even more remarkably, the presently indicated value of the Higgs and top masses indicate that we are very close to the border where the Higgs potential is actually unstable. In this section, we explore the nature of the quantum effects. Our focus is on the role of the top quark, which is the major contributor to the potential instability.

The Higgs potential describes the vacuum energy as a function of a constant Higgs field. Since the top-quark mass and the Higgs Yukawa coupling to the top quark are related, it is convenient to define a *background field* $h(x) = v + H(x)$. In the following we take H (and hence h) as constant and thus omit any spacetime dependence,

$$-\mathcal{L}_t = \frac{\Gamma_t}{\sqrt{2}}(v + H)\bar{t}t \equiv \frac{\Gamma_t}{\sqrt{2}}h\bar{t}t \equiv m_t(h)\bar{t}t, \tag{5.1}$$

where $m_t(h) = \Gamma_t h/\sqrt{2}$ is the field-dependent mass. We then calculate the vacuum energy as a function of $m_t(h)$. This can be done relatively simply by studying the $t\bar{t}$ contribution to the vacuum matrix element of the energy-momentum tensor $\mathcal{T}_{\mu\nu}$,

$$\begin{aligned}
\langle 0|\mathcal{T}_{\mu\nu}|0\rangle_{\text{top}} &= -N_c \int \frac{d^d p}{(2\pi)^d} \frac{1}{2}\text{Tr}\left[(\gamma_\mu p_\nu + \gamma_\nu p_\mu)\frac{i}{\not{p} - m_t(h) + i\epsilon}\right] \\
&= -12 \int \frac{d^d p}{(2\pi)^d} p_\mu p_\nu \frac{i}{p^2 - m_t^2(h) + i\epsilon} \\
&= \delta V(h)g_{\mu\nu},
\end{aligned} \tag{5.2}$$

where the important minus sign comes from the Feynman rule for a closed fermion loop. This leads to a result

$$\delta V(h) = \frac{3m_t^4(h)}{16\pi^2}\left[\frac{2}{4-d} - \gamma + \ln 4\pi - \ln\frac{m_t^2(h)}{\mu_d^2} + \frac{3}{2}\right], \tag{5.3}$$

with μ_d being the scale that enters in dimensionally regularized integrals.[9] The divergence is proportional to $m_t^4(h) \sim h^4$ and thus goes into the renormalization of the $\lambda\varphi^4$ term in the Higgs potential. In the $\overline{\text{MS}}$ scheme, one then arrives at the potential,

$$V(h) = -\frac{1}{2}\mu^2 h^2 + \frac{1}{4}\lambda(\mu_d)h^4 - \frac{3m_t^4(h)}{16\pi^2}\left[\ln\frac{m_t^2(h)}{\mu_d^2} - \frac{3}{2}\right]. \tag{5.4}$$

The $-m_t^4(h)\ln m_t^2(h) \sim -h^4 \ln h^2$ term from the loop diagram is the key new feature.

---

[9] In this context we add the subscript to μ_d to avoid confusion with the $-\mu^2\varphi^2$ term in Higgs potential.

We note that the logarithmic term produces an instability for large enough values of the field h. No matter what the coefficient $\lambda(\mu_d)$ of the h^4 term is, the $-h^4 \ln h^2$ term eventually will overpower it and lead to a potential that is unbounded below at large enough values of h. However, New Physics (NP) beyond the Standard Model could modify this result, for example by generating an effective operator

$$-\mathcal{L}_{NP} = \frac{1}{\Lambda^2}(\varphi^\dagger\varphi)^3 = \frac{1}{8\Lambda^2}h^6. \tag{5.5}$$

At the very least, such effects should be generated at the Planck scale $\Lambda \sim M_P$, so that we should not be concerned if the apparent instability occurs beyond the Planck scale. However, if the instability occurs at a lower scale, it implies either that the vacuum is at best meta-stable – a very dramatic conclusion – or that other New Physics must come in before the Planck scale – also important.

To use the quantum effective potential, one minimizes the energy with the vacuum expectation value constrained to equal 246 GeV and the top-quark mass equal to its physical value, and determines the Higgs mass parameter from the quadratic term in the expansion. However, unlike at tree level, the curvature of the potential near the minimum does not give the physical Higgs mass. In order to get the Higgs pole mass one must include finite momentum effects from the vacuum polarization diagrams.

Given the physical values of these parameters, indications are that the potential is close to being unstable below the Planck scale. A more detailed treatment must include the effects of the Higgs itself and of the other particles. The state of the art includes the inclusion of more loops and the use of running couplings [De *et al.* 12]. Moreover, if the seesaw mechanism is at play for neutrino masses, the neutrino Yukawa couplings provide an extra unknown destabilizing influence [CaDIQ 00]. It remains very interesting that the parameters of the Standard Model place us so close to this prediction of an unstable Higgs potential, implying yet another suggestion of New Physics below the Planck scale.

XV–6 Two Higgs doublets

Earlier in this chapter we briefly discussed the issue of a very heavy fourth quark generation, assumed to otherwise resemble the observed three generations. On the one hand, it would introduce new particles and thus lie beyond the Standard Model; on the other, it would respect the twin pillars of gauge symmetry and spontaneous symmetry breaking of a scalar doublet on which the Standard Model is based. Here, we proceed analogously by briefly considering the replacement of a single Higgs doublet Φ by *two* Higgs doublets (Φ_1, Φ_2) having the same $SU(2) \otimes U(1)$

quantum numbers.[10] A two-Higgs-doublet theory would enlarge the spectrum of Higgs bosons and also considerably enrich the content of the Higgs potential.

Spectrum: Since each Higgs doublet corresponds to four real fields as in Eq. (1.2), then two Higgs doublets will amount to eight real fields. Of these, three will become the longitudinal degrees of freedom of the Z^0 and $W^\pm$ gauge bosons. There will also be five spinless Higgs particles: a charged pair ($H^\pm$), two $CP = +1$ neutrals (H, h), and one $CP = -1$ neutral (A). If we associate H with the Higgs boson of the one-doublet theory, then the two-doublet model predicts the four new particles h, A, $H^\pm$. At present, there is no experimental evidence for any of these four. Current lower-mass bounds are in the range of roughly 100 GeV for each [RPP 12].

Consider, for example, charged Higgs particles [Le 73] whose rich phenomenology was realized early on [DoL 79, GoY 79]. The $H^\pm$ particles can be sought directly or indirectly:

(1) *Direct*: Charged Higgs-pair production, $e^+ e^- \to H^+ H^-$ would arise via $H^\pm$ coupling to photons and Z^0 bosons. A charged Higgs could also couple semi-weakly to the known fermions with strength proportional to the fermion mass. Thus, at the LHC, a study [Aa *et al.* (ATLAS collab.) 13a] of $gg \to t\bar{t}$ followed by a decay chain such as

$$t \to H^+ b \to c\bar{s}\, b \qquad \text{and} \qquad \bar{t} \to H^- \bar{b} \to \bar{c}s\, \bar{b}$$

has yielded sharp upper limits on $Br_{t \to H^\pm b}$ for the mass range $90 < M_H$ (GeV) < 150.

(2) *Indirect*: A charged Higgs could contribute as a virtual particle, as with the leptonic decay of a B meson,

$$Br_{B^+ \to \ell^+ \nu_\ell} = Br^{(\text{SM})}_{B^+ \to \ell^+ \nu_\ell} \left[1 - \tan^2 \beta \, \frac{m_B^2}{M_{H^\pm}^2} \right]^2,$$

where $\tan \beta \equiv \langle \varphi_2^0 \rangle / \langle \varphi_1^0 \rangle$.

Higgs potential: The Standard Model Higgs potential energy of Eq. (II–3.19) is based on one quadratic mass term and one quartic Higgs self-coupling. The most general renormalizable $SU(2) \otimes U(1)$ two-Higgs-doublet version has three quadratic mass terms and seven quartic Higgs self-couplings,

---

[10] The possibility of two Higgs-doublets is usually associated with supersymmetry, but this is not necessary.

$$V_{\text{2-Higgs}} = m_{11}^2 \Phi_1^\dagger \Phi_1 + m_{22}^2 \Phi_2^\dagger \Phi_2 - \left[m_{12}^2 \Phi_1^\dagger \Phi_2 + \text{h.c.} \right]$$

$$+ \frac{\lambda_1}{2} \left(\Phi_1^\dagger \Phi_1 \right)^2 + \frac{\lambda_2}{2} \left(\Phi_2^\dagger \Phi_2 \right)^2$$

$$+ \lambda_3 \left(\Phi_1^\dagger \Phi_1 \right) \left(\Phi_2^\dagger \Phi_2 \right) + \lambda_4 \left(\Phi_1^\dagger \Phi_2 \right) \left(\Phi_2^\dagger \Phi_1 \right) + \left[\frac{\lambda_5}{2} \left(\Phi_1^\dagger \Phi_2 \right)^2 + \text{h.c.} \right]$$

$$+ \left[\lambda_6 \Phi_1^\dagger \Phi_2 + \text{h.c.} \right] \Phi_1^\dagger \Phi_1 + \left[\lambda_7 \Phi_1^\dagger \Phi_2 + \text{h.c.} \right] \Phi_2^\dagger \Phi_2. \tag{6.1}$$

Since this most general structure has the potential to produce overly large flavor-changing neutral currents (FCNCs) or gross violations of custodial symmetry, it cannot be realized in Nature without restrictions on the ten free parameters. A great deal of research on $V_{\text{2-Higgs}}$ has been reported in the literature; two recent works citing many earlier contributions are [MaM 10] and [HaO 11]. Two additional items of interest deserve mention. One is that the above potential energy allows for *CP* violation. A careful discussion appears in Chapter 22 of [BrLS 99]. Another involves the vexing strong *CP* problem of *QCD*. It has been shown that introduction of a 'Peccei–Quinn' global $U(1)_{\text{PQ}}$ symmetry [PeQ 77], which becomes spontaneously broken, can lead to a solution of the problem. The two-Higgs framework provides a natural platform for the $U(1)_{PQ}$ symmetry.

Problems

(1) **The rho parameter**

   (a) Show that for an arbitrary number of Higgs multiplets $(\langle \varphi_i \rangle_0 \neq 0,$ $(i = 1, \ldots))$, the rho parameter becomes

$$\rho_0 = \frac{\sum_i \left[(I_{\text{w}})_i^2 + (I_{\text{w}})_i - (I_{\text{w}3}^2)_i \right] \langle \varphi_i \rangle_0^2}{2 \sum_i (I_{\text{w}3}^2)_i \langle \varphi_i \rangle_0^2}.$$

   (b) Given two Higgs fields, with quantum numbers $I_{\text{w}} = -I_{\text{w}3} = 1/2$ and $I_{\text{w}} = 1$, $I_{\text{w}3} = 0$ respectively, and with nonvanishing vacuum expectation values $\langle \varphi_{1/2} \rangle$ and $\langle \varphi_1 \rangle$, obtain a bound for $|\langle \varphi_1 \rangle / \langle \varphi_{1/2} \rangle|$ assuming an experimental value $\rho_0 = 1.0004 \pm 0.0003$.

(2) **Higgs–gluon coupling**

In the text we used the background field method to show that, at lowest order in the momenta, the effective Higgs coupling to gluons is

$$\mathcal{L}_{\text{eff}} = \frac{\alpha_s}{24\pi} \ln \left(\frac{h^2}{v^2} \right) F_{\mu\nu}^a F^{a\mu\nu},$$

with $h = v + H$. As mentioned briefly in the text, this coupling implies a cancelation in the Standard Model prediction for the reaction in which two gluons produce two Higgs bosons, which makes the residual effect small.

In addition to the direct coupling from the above effective lagrangian, there is a pole diagram of $GG \to H \to HH$, which utilizes the triple Higgs coupling. Show that these two contributions cancel *exactly* at threshold.

(3) **Higgs sector and the cosmological constant**

The Higgs sector makes several contributions to the cosmological constant, Λ, which is defined as the energy density of the vacuum. The observed value of the cosmological constant is $\Lambda = U_{vac} = 2.8 \times 10^{-47}$ GeV^4. In Eq. (2.9) we displayed one contribution that is 51 orders of magnitude larger than the observed value. Other calculable contributions also come from the Higgs sector. For example, show that if one changes the up-quark Yukawa coupling by a few parts in 10^{-43}, one changes the cosmological constant by 100%. The leading change is linear in the Yukawa coupling, and to uncover this you may use the effective lagrangians of Chap. VII. Specifically, compare the Yukawa coupling's effect on the vacuum expectation value of the lagrangian to the contribution of the Yukawa coupling to the mass of the pion, expressing the result in terms of F_π, m_π and ratios of the quark masses.

XVI

The electroweak sector

Early studies of the weak interactions were confined to processes, like nuclear beta decay and muon decay, which concern just the charged weak current. Starting from the mid-1970s, the field of weak interaction phenomenology was broadened by experiments involving neutral weak currents. The advent of collider experiments made possible direct studies of the $W^{\pm}$ and Z^0 gauge bosons themselves. This chapter will first address the topic of low-energy neutral-current phenomenology and then consider physical processes at the higher mass scales M_W and M_Z. To conclude, we turn to the more theoretical topic of electroweak radiative corrections and renormalization. Throughout, we shall keep our treatment at a relatively simple introductory level.

XVI–1 Neutral weak phenomena at low energy

The words '*low energy*' in the title of this section denote processes with $Q^2 \ll M_Z^2$. We shall focus on three of these:

(1) deep-inelastic neutrino scattering (DIνS),
(2) atomic parity violation (APV),
(3) parity-violating (PV) Møller scattering.

In each case, the main finding is a determination of the weak mixing angle at the kinematical scale $\mu = Q$ appropriate to that experiment. In this context, it is convenient to use a scale-dependent version of the weak mixing angle, such as the $\overline{\text{MS}}$ quantity $\hat{s}_{\text{w}}^2(\mu)$.[1] Then, we display in Fig. XVI–1 the dependence of $\hat{s}_{\text{w}}^2$ on Q^2 found from both low-energy and high-energy studies. Fig. XVI–1, although not yet reaching the iconic status of Fig. II–6 (which displays the asymptotic freedom

---

[1] We employ the common abbreviations $s_{\text{w}} \equiv \sin\theta_{\text{w}}$, $c_{\text{w}} \equiv \cos\theta_{\text{w}}$ and also employ $\hat{s}_{\text{w}}$ for $\overline{\text{MS}}$ renormalization. For convenience, we shall refer (admittedly loosely) to the quantity s_{w}^2 as the 'weak mixing angle'.

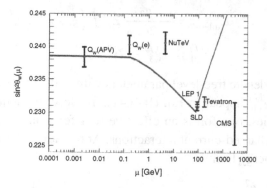

Fig. XVI–1 Scale dependence of $\hat{s}_w^2$, from [RPP 12] (used with permission).

property of *QCD*), has become an apt representation of this field. It has, indeed, been a major achievement of low-energy neutral-current studies to verify (within experimental uncertainties) the variation of $\hat{s}_w^2(\mu)$ with scale μ expected from the Standard Model.

One can use the renormalization group to 'run' each of the low-energy determinations up to a standard reference scale, say $\mu = M_Z$, to provide the values for $\hat{s}_w^2(M_Z)$ shown in Table XVI–1 [KuMMS 13]. For comparison's sake is also included the quantity $\hat{s}_w^2(M_Z)$ obtained by using an average of data from experiments carried out directly at the Z^0 scale, e.g., Z^0 decays and cross-section asymmetries, cf. Sect. XVI–2. At present, the high-energy determination is far more accurate than the low-energy determination due to its dominance in statistics.

Neutral-current effective lagrangians

To provide a theoretical language for such low-energy experiments, let us identify effective lagrangians for some neutral-current processes. Recall from Eq. (II–3.40) that the neutral weak interaction between the gauge boson Z^0 and a fermion f is given at tree level by

Table XVI–1. *Weak mixing angle from neutral-current experiments*

Experiment	$\langle Q^2 \rangle (\text{GeV}^2)$	$\hat{s}_w^2(M_Z)$
DIνS	20	0.2356(16)
APV (in Cs)	5.8×10^{-6}	0.2383(20)
PV Møller	2.6×10^{-2}	0.2329(13)
Average at Z mass scale	$M_Z^2 \simeq 8.3 \times 10^3$	0.23125(016)

$$\mathcal{L}_{\text{ntl-wk}}^{(f)} = -\frac{g_{2,0}}{2c_{\text{w},0}} Z_\mu \, \bar{f} \left(g_{\text{v},0}^{(f)} \gamma^\mu + g_{\text{a},0}^{(f)} \gamma^\mu \gamma_5 \right) f$$

$$g_{\text{v},0}^{(f)} = T_{\text{w}3}^{(f)} - 2s_{\text{w},0}^2 \, \mathcal{Q}_{\text{el}}^{(f)}, \qquad g_{\text{a},0}^{(f)} = T_{\text{w}3}^{(f)}, \qquad (1.1)$$

where we denote tree-level parameters with a '0' subscript. Examples of individual $g_{\text{v},0}^{(f)}$ and $g_{\text{a},0}^{(f)}$ appear in Eq. (II–3.41). To describe neutral-current interactions at low energies, one forms an effective four-fermion lagrangian, akin to the Fermi model of charged-current interactions. At tree level, the Z^0-mediated interaction in the low-energy limit is

$$\mathcal{L} = -\frac{1}{2} \frac{g_{2,0}^2}{4c_{\text{w},0}^2} \sum_{f,f'} \bar{f} \left(g_{\text{v},0}^{(f)} \gamma^\mu + g_{\text{a},0}^{(f)} \gamma^\mu \gamma_5 \right) f \, \frac{1}{M_{Z,0}^2} \, \bar{f}' \left(g_{\text{v},0}^{(f')} \gamma_\mu + g_{\text{a},0}^{(f')} \gamma_\mu \gamma_5 \right) f'$$

$$= -\rho_0 \frac{G_\mu}{\sqrt{2}} \sum_{f,f'} \bar{f} \left(g_{\text{v},0}^{(f)} \gamma^\mu + g_{\text{a},0}^{(f)} \gamma^\mu \gamma_5 \right) f \, \bar{f}' \left(g_{\text{v},0}^{(f')} \gamma_\mu + g_{\text{a},0}^{(f')} \gamma_\mu \gamma_5 \right) f', \quad (1.2)$$

where ρ_0 is the tree-level *rho parameter*,

$$\rho_0 \equiv \frac{1}{c_{\text{w},0}^2} \frac{M_{W,0}^2}{M_{Z,0}^2}. \qquad (1.3)$$

Comparison of the second of the relations in Eq. (1.2) with Eq. (V–2.1) shows that ρ_0 governs the relative strengths of the neutral and charged weak-current effective lagrangians. In the Standard Model, it has the tree-level value unity, $\rho_0^{\text{(SM)}} = 1$. The reader might wonder – why include a quantity, ρ_0, whose Standard Model value is unity? There are actually two reasons: (i) although $\rho_0^{\text{(SM)}} = 1$, ρ_0 is *not* unity in general, e.g., alternative choices for Higgs structure can lead to different values for ρ_0 (cf. Prob. XV–1), and (ii) even in the Standard Model, electroweak corrections will change its value away from unity (cf. Sect. XVI–6).

The set of low-energy neutral-current processes includes neutrino–electron, neutrino–quark, and parity-violating electron–quark interactions. There is an effective lagrangian for each of these, two examples being

$$\mathcal{L}_{\nu q} = -\frac{G_\mu}{\sqrt{2}} \, \bar{\nu}_\ell \gamma^\mu (1 + \gamma_5) \nu_\ell \left[\epsilon_L^{(\alpha)} \bar{q}_\alpha \gamma_\mu (1 + \gamma_5) q_\alpha + \epsilon_R^{(\alpha)} \bar{q}_\alpha \gamma_\mu (1 - \gamma_5) q_\alpha \right]$$

$$\mathcal{L}_{\text{eq}}^{\text{(p.v.)}} = -\frac{G_\mu}{\sqrt{2}} \left[C_1^\alpha \bar{e} \gamma^\mu \gamma_5 e \, \bar{q}_\alpha \gamma_\mu q_\alpha + C_2^\alpha \bar{e} \gamma^\mu e \, \bar{q}_\alpha \gamma_\mu \gamma_5 q_\alpha \right], \qquad (1.4)$$

where the index $\alpha = u, d, \ldots$ denotes quark flavor. Of course, contributions other than neutral weak effects also enter, e.g., parity-conserving eq scattering experiences the electromagnetic interaction.

In Eq. (1.4), we have implicitly included the effect of radiative corrections and thus omit the subscript '0'. Table XVI–2 gives a compilation of the radiatively

Table XVI–2. *Radiatively corrected coefficients.*

Coefficient	General form[a]
$\epsilon_L^{(u)}$	$\rho_{\nu N}\left(\frac{1}{2} - \frac{2}{3}\kappa_{\nu N}s_{\mathrm{w}}^2\right)$
$\epsilon_L^{(d)}$	$\rho_{\nu N}\left(-\frac{1}{2} + \frac{1}{3}\kappa_{\nu N}s_{\mathrm{w}}^2\right)$
$\epsilon_R^{(u)}$	$\rho_{\nu N}\left(-\frac{2}{3}\kappa_{\nu N}s_{\mathrm{w}}^2\right)$
$\epsilon_R^{(d)}$	$\rho_{\nu N}\left(\frac{1}{3}\kappa_{\nu N}s_{\mathrm{w}}^2\right)$
C_1^u	$\rho_{\mathrm{eq}}\left(-\frac{1}{2} + \frac{4}{3}\kappa_{\mathrm{eq}}s_{\mathrm{w}}^2\right)$
C_1^d	$\rho_{\mathrm{eq}}\left(\frac{1}{2} - \frac{2}{3}\kappa_{\mathrm{eq}}s_{\mathrm{w}}^2\right)$

[a]Small additive terms are omitted.

corrected coefficients (with renormalization scheme left unspecified). The quantities ρ_i and κ_i in Table XVI–2 reduce at tree level to unity, $\rho_{i,0} = \kappa_{i,0} = 1$. The ρ_i are overall multiplicative factors and the κ_i multiply the weak mixing angle, which itself has become renormalized, $s_{\mathrm{w},0}^2 \to s_{\mathrm{w}}^2$. The presence of such quantities in the effective lagrangians can be traced back to the underlying neutral-current couplings,

$$g_{\mathrm{v},0}^{(f)} \to g_{\mathrm{v}}^{(f)} = \sqrt{\rho_f}\left(T_{\mathrm{w}3}^{(f)} - 2\kappa_f s_{\mathrm{w}}^2 Q_{\mathrm{el}}^{(f)}\right), \qquad g_{\mathrm{a},0}^{(f)} \to g_{\mathrm{a}}^{(f)} = \sqrt{\rho_f}\, T_{\mathrm{w}3}^{(f)}, \quad (1.5)$$

where, again, we leave the renormalization scheme unspecified. However, see [MaS 80] for the introduction of $\overline{\mathrm{MS}}$ renormalization to electroweak corrections. The quantities ρ_i and κ_i will be discussed in more detail later in Sect. XVI–6.

Deep-inelastic neutrino scattering from isoscalar targets

In deep-inelastic scattering, one measures the ratios of neutral to charged-current neutrino/antineutrino cross sections,

$$R_\nu \equiv \sigma_{\nu N}^{\mathrm{NC}}/\sigma_{\nu N}^{\mathrm{CC}}, \qquad R_{\bar\nu} \equiv \sigma_{\bar\nu N}^{\mathrm{NC}}/\sigma_{\bar\nu N}^{\mathrm{CC}}. \quad (1.6)$$

Under the conditions of 'deep-inelastic' kinematics ([BaP 87]), theoretical calculations of R_ν and $R_{\bar\nu}$ are carried out in terms of quark, rather than hadronic, degrees of freedom. It is plausible that by working with *ratios* like those in Eq. (1.6), theoretical uncertainites associated with hadron structure tend to cancel. At tree level, R_ν and $R_{\bar\nu}$ are straightforwardly computed if scattering from an isoscalar target is assumed and antiquark contributions are ignored. It is useful to express the $\epsilon_{L,R}^{(\alpha)}$ coefficients of Eq. (1.4) as

$$g_L^2 \equiv \left(\epsilon_L^{(u)}\right)^2 + \left(\epsilon_L^{(d)}\right)^2 \simeq \frac{1}{2} - s_{w,0}^2 + \frac{5}{9}\left(1 + r_0\right) s_{w,0}^4,$$

$$g_R^2 \equiv \left(\epsilon_R^{(u)}\right)^2 + \left(\epsilon_R^{(d)}\right)^2 \simeq \frac{5}{9} s_{w,0}^4. \tag{1.7}$$

These quantities can be determined from the combination of neutrino and anti-neutrino cross sections,

$$R_\pm \equiv \frac{R_\nu \pm r R_{\bar{\nu}}}{1 \pm r} = g_L^2 \pm g_R^2, \tag{1.8}$$

where $r = 1/\bar{r} \equiv \sigma_{\bar{\nu}N}^{CC}/\sigma_{\nu N}^{CC}$ are measurable quantities with tree-level values $r_0 = \bar{r}_0^{-1} = 3$. The NuTeV experiment [Ze *et al.* 01] at Fermilab, carried out at an average momentum-squared transfer $\langle Q^2 \rangle = \langle -q^2 \rangle \simeq 20$ GeV^2, has yielded the most precise determination to date,

$$g_L^2 = 0.3005 \pm 0.0014, \quad g_R^2 = 0.0310 \pm 0.0011. \tag{1.9}$$

This translates into a determination of the weak mixing angle, which lies nearly 3σ above the stated Standard Model prediction, a finding which has spurred much discussion since then.

Atomic parity violation in cesium

The Z^0-mediated electron–nucleus interaction, expressed here in the electron spin space, contains a component which is parity-violating,

$$\mathcal{H}_{PNC}(r) = \frac{G_\mu}{2\sqrt{2}} Q_w \gamma_5 \rho_{nucl}(r), \tag{1.10}$$

where γ_5 signals the presence of parity violation and $\rho_{nucl}(r)$ reminds us that the electron feels the effect only where the nuclear density is nonvanishing.[2] The quantity Q_w is the 'weak nuclear charge' to which the electron couples, and is given to lowest order by

$$Q_{w,0}(N, Z) = -2\left(N_u C_{1,0}^u + N_d C_{1,0}^d\right) = Z\left(1 - 4s_{w,0}^2\right) + N, \tag{1.11}$$

where Z and N are, respectively, the nuclear proton and neutron number. The fact that $s_{w,0}^2 \simeq 0.25$ suppresses the proton contribution, leaving the coupling of the atomic electron to neutrons as dominant.

Consider the effect in atomic cesium, ^{138}Ce. Because of the neutral weak-current interaction, the single valence electron in cesium contains small admixtures of P wave in its $6S$ (ground) and $7S$ (excited) states. We write these mixed states as $|\overline{6S}\rangle$

---

[2] The abbreviation 'PNC' stands for *parity nonconservation*.

and $|\overline{7S}\rangle$. As a consequence, there occurs a measurable parity-violating $7S \rightarrow 6S$ electric-dipole (E1) transition matrix element [NoMW 88],

$$\mathcal{I}m\ E_{\text{PNC}} = \langle \overline{7S}\,|\mathbf{D}|\overline{6S}\rangle \equiv \frac{Q_{\text{w}}}{N}k_{\text{PNC}}, \qquad (1.12a)$$

where **D** is the electric-dipole operator and

$$k_{\text{PNC}} \equiv \frac{N}{Q_{\text{w}}} \sum_n \left[\frac{\langle 7S|\mathbf{D}|nP\rangle \langle nP|H_{\text{PNC}}|6S\rangle}{E_{6S} - E_{nP}} + \frac{\langle 7S|H_{\text{PNC}}|nP\rangle \langle nP|\mathbf{D}|6S\rangle}{E_{7S} - E_{nP}} \right].$$
$$(1.12b)$$

The experiments involve finding the ratio of the PNC amplitude E_{PNC} to the vector transition probability β. The most accurate results to date on the $6S \rightarrow 7S$ transition are $E_{\text{PNC}}/\beta = 1.5935(56)$ mV cm^{-1} [BcCMRTWW 97] and $\beta = 26.957(51)$ a_{B}^3 [BeW 99]. However, interpretation of the PNC measurements requires evaluating Eq. (1.12b) and this contains intractable aspects of the atomic many-body problem. There has, however, been recent progress [PoBD 09] and the latest calculation gives [DzBFR 12] $E_{\text{PNC}} = 0.08977(40)i(-Q_{\text{w}}/N)$, implying the weak-charge value

$$Q_{\text{w}}(^{138}\text{Cc}) = -72.58(29)_{\text{expt}}(32)_{\text{thy}}, \qquad (1.13)$$

where the uncertainties refer respectively to statistical and theoretical contributions. This result lies about 1.5σ beneath the Standard Model prediction $Q_{\text{w}}^{(\text{SM})}$ $(^{138}\text{Ce}) = -73.23(2)$.

Polarized Møller scattering

Another experiment which has probed the weak mixing angle at a low-energy scale is polarized Møller scattering [An *et al.* (SLAC E158 collab.) 05], where we remind the reader that Møller scattering is the elastic scattering of electrons on electrons. In SLAC E158, a 50 GeV beam of longitudinally polarized electrons was scattered from an unpolarized fixed target. The parity-violating observable is the asymmetry

$$A_{\text{pv}} = \frac{\sigma_R - \sigma_L}{\sigma_R + \sigma_L}, \qquad (1.14)$$

where $\sigma_{R(L)}$ is the cross section for incident right (left) polarized electrons. Relevant kinematic variables are the center-of-mass-squared energy $s = (p + p')^2$, the momentum transfer $Q^2 = -q^2 = -(p - p')^2$, and the ratio of the two, $y \equiv Q^2/s = (1 - \cos\theta)/2$, where θ is the scattering angle in the center of mass. The experiment was carried out with average values $\langle Q^2 \rangle = 0.026$ GeV^2 and $\langle y \rangle \simeq 0.6$; the tiny asymmetry $A_{\text{pv}} = -131(14)_{\text{stat}}(10)_{\text{sys}} \times 10^{-9}$ was found.

There are three parts to the theory analysis. First is the tree-level amplitude, where the Møller scattering amplitude arises from t-channel and u-channel γ and Z^0 exchange diagrams. Parity violation is due to the interference of the electromagnetic and weak neutral-current amplitudes. An approximate tree-level expression for A_{pv} which is valid for the conditions of the E158 experiment is

$$A_{\mathrm{pv}}^{(\mathrm{tree})} \simeq \frac{G_\mu Q^2}{\sqrt{2}\pi\alpha} \cdot \frac{1-y}{1+y^4+(1-y)^4} \left(1 - 4\sin^2\theta_{\mathrm{w},0}\right). \tag{1.15}$$

The dependence on the weak mixing angle suppresses $A_{\mathrm{pv}}^{(\mathrm{tree})}$ due to the proximity of $\sin^2\theta_{\mathrm{w},0}$ to $1/4$.

Next are the one-loop corrections, due mainly to the γ–Z^0 propagator-mixing terms induced by fermion and W-boson loop amplitudes [CzM 96]. These are absorbed by the $\overline{\mathrm{MS}}$ running weak mixing angle,

$$\hat{s}_{\mathrm{w}}^2(\mu) = (1 + \Delta\kappa(\mu))\,\hat{s}_{\mathrm{w}}^2(M_Z), \tag{1.16a}$$

as parameterized by $\Delta\kappa(\mu)$. In particular, one finds $\Delta\kappa(0) \simeq 0.03$, so that

$$1 - 4\sin\theta_{\mathrm{w}}^2 \simeq 0.075 \implies 1 - \kappa(0)\hat{s}_{\mathrm{w}}^2(M_Z) \simeq 0.046. \tag{1.16b}$$

The rather small ($\sim3\%$) effect of $\Delta\kappa(0)$ translates into a major ($\sim40\%$) change in $1 - 4\sin\theta_{\mathrm{w}}^2$! Finally comes the renormalization-group-improved analysis [ErR-M 05]. This serves to ameliorate the dependence on large logarithms ($\ln(m_f^2/Q^2)$ and $\ln(M_W^2/Q^2)$ for fermion and W-boson loops respectively), which appear in the one-loop amplitudes. This results in the improved determinations,

$$\Delta\kappa(0) = 0.03232 \pm 0.00029, \qquad \hat{s}_{\mathrm{w}}^2(0) = 0.23867 \pm 0.00016. \tag{1.17}$$

XVI–2 Measurements at the Z^0 mass scale

The collection of resonances observed in $e\bar{e}$ collisions as a function of the total center-of-mass energy is displayed in Fig. XVI–2. At the Z^0 mass scale, it is the *weak* interaction which dominates the physics in this reaction, with strong and electromagnetic effects merely supplying modest corrections. An enormous database has been established at the Z^0 factories with the LEP1 experiments at CERN and the SLD collaboration at SLAC. There also exists data from $p\bar{p} \to f^+f^-$ measured at the Tevatron as well as that from $pp \to \ell^+\ell^- + X$ taken by the LHC detectors. These experiments have come to be analyzed in terms of a so-called 'effective description' wherein the renormalized vector and axial-vector couplings of Eq. (1.5) are written as

$$\bar{g}_{\mathrm{v}}^{(f)} = \sqrt{\rho_f}\left(T_{\mathrm{w}3}^{(f)} - 2\bar{s}_f^2 Q_{\mathrm{el}}^{(f)}\right), \qquad \bar{g}_{\mathrm{a}}^{(f)} = \sqrt{\rho_f}\,T_{\mathrm{w}3}^{(f)}, \tag{2.1}$$

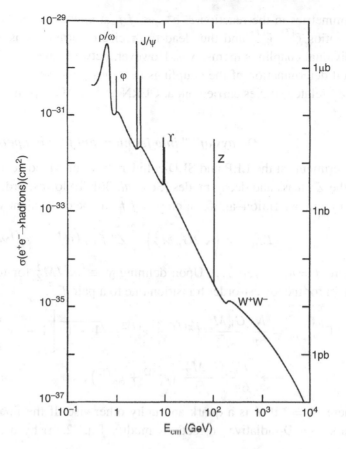

Fig. XVI–2 Resonances in $e\bar{e}$ collisions.

where the superbars denote evaluation in the effective renormalization scheme associated with the scale $\mu = M_Z$. In this approach, the effective weak mixing angle $\bar{s}_f$ for fermion f is defined so as to absorb the κ_f factor in Eq. (1.5) [Sc *et al.* 06],

$$\bar{s}_f^2 \equiv \kappa_f s_{\rm w}^2, \tag{2.2a}$$

and can be measured experimentally by

$$\bar{s}_f^2 = \frac{1}{4}\left(1 - \bar{g}_{\rm v}^{(f)}/\bar{g}_{\rm a}^{(f)}\right), \tag{2.2b}$$

independent of the quantity κ_f.[3]

We shall discuss two kinds of measurements in the following: Z^0 decay into fermion–antifermion pairs, which is sensitive to $(\bar{g}_{\rm v}^{(f)})^2 + (\bar{g}_{\rm a}^{(f)})^2$, and cross-section

---

[3] Although both ρ_f and κ_f are both generally *complex-valued*, we shall tacitly use just the real part without further comment.

asymmetries in the reaction $e^- e^+ \to f \bar{f}$ at energy $\sqrt{s} = M_Z$, which determine the ratios $\bar{g}_v^{(f)} / \bar{g}_a^{(f)}$ and thus lead to precise determinations of $\bar{s}_f^2$ but not of the individual couplings themselves. However, between the two kinds of experiments a full determination of the couplings becomes possible. A comprehensive review of Z^0-related studies carried out at CERN and SLAC appears in [Sc *et al.* 06].

Decays of Z^0 into fermion–antifermion pairs

Experiments at the LEP and SLD colliders have provided accurate determinations of the Z^0 mass and decay modes [Sc *et al.* 06]. To lowest order, the decay of a Z^0 boson into a fermion–antifermion pair $f \bar{f}$ can be conveniently expressed as

$$\mathcal{L}_{\text{ntl}}^{(f\bar{f})} = \left(\sqrt{2} G_\mu M_Z^2 \right)^{1/2} Z^\mu \bar{f} \gamma_\mu \left(g_v^{(f)} + g_a^{(f)} \gamma_5 \right) f, \tag{2.3}$$

where $f = u, d, \nu_e, e, \dots$. Upon defining $y \equiv m^2 / M_Z^2$ for fermion mass m, we obtain for the lowest-order transition rate to a pair $f \bar{f}$,

$$\Gamma_{Z^0 \to f \bar{f}}^{(0)} = \frac{N_c}{6\pi} \frac{G_\mu M_Z^3}{\sqrt{2}} \left(g_v^{(f)2} + g_a^{(f)2} \right) \sqrt{1 - 4y} \left[1 + 2y \frac{g_v^{(f)2} - 2g_a^{(f)2}}{g_v^{(f)2} + g_a^{(f)2}} \right]$$

$$\xrightarrow[y \to 0]{} \frac{N_c}{6\pi} \frac{G_\mu M_Z^3}{\sqrt{2}} \left(g_v^{(f)2} + g_a^{(f)2} \right), \tag{2.4a}$$

where $N_c = 3$ if f is a quark and unity otherwise. If the final-state fermions are quarks, *QCD*-radiative corrections modify Eq. (2.4a) by a multiplicative factor δ_{QCD},

$$\delta_{QCD} = 1 + \frac{\alpha_s (M_W^2)}{\pi} + 1.41 \left(\frac{\alpha_s (M_W^2)}{\pi} \right)^2 + \cdots \simeq 1.04, \tag{2.4b}$$

where $\alpha_s (M_W^2) \simeq 0.12$ has been used in the above.

There exist also electroweak radiative effects, which we can take into account by employing the effective weak coupling constants $\bar{g}_v^{(f)}$ and $\bar{g}_a^{(f)}$ of Eq. (2.1). Upon including both strong and electroweak corrections, the tree-level relation of Eq. (2.4a) is replaced (shown here in the limit of massless final-state fermions) by

$$\Gamma_{Z^0 \to f \bar{f}} = \eta_f \frac{N_c}{6\pi} \frac{G_\mu M_Z^3}{\sqrt{2}} \left(\bar{g}_v^{(f)2} + \bar{g}_a^{(f)2} \right), \tag{2.5}$$

where $\eta_f = \delta_{QCD}$ if f is a quark and $\eta_f = 1$ otherwise.

Some Z^0 decay-related quantities are listed in Table XVI–3. These results are taken from [Sc *et al.* 06], but many others are provided in this source. Although there will be some adjustments from more recent studies (e.g. see [RPP 12]), the overall picture provided by [Sc *et al.* 06] bears testimony to an impressive advance

Table XVI–3. Z^0 decay [Sc et al. 06].

Measurable[a]	Experiment	Standard Model prediction
$\Gamma_{e\bar{e}}$	83.9 ± 0.1	84.00 ± 0.01
Γ_{inv}	496.2 ± 8.8	501.66 ± 0.03
$\Gamma_{b\bar{b}}$	377.3 ± 0.3	375.98 ± 0.03
Γ_{tot}	2495.2 ± 2.3	2496.0 ± 0.2
$\bar{g}_{\text{v}}^{(\ell)2}$	0.0012 ± 0.0003	$0.0011 \rightarrow 0.0013$
$\bar{g}_{\text{a}}^{(\ell)2}$	0.2492 ± 0.0012	$0.2513 \rightarrow 0.2518$

[a]Decay widths are expressed in units of MeV.

in particle physics. One application, among many, of the Z^0 decays is to use leptonic modes to test the concept of lepton universality, and one finds

$$\frac{\Gamma_{\mu\bar{\mu}}}{\Gamma_{e\bar{e}}} = 1.0009 \pm 0.0027, \qquad \frac{\Gamma_{\tau\bar{\tau}}}{\Gamma_{e\bar{e}}} = 1.0021 \pm 0.0030, \qquad (2.6)$$

which is seen to be consistent with universality.

Asymmetries at the Z^0 peak

For the reaction $e^- e^+ \rightarrow f\bar{f}$ carried out at the Z^0 peak a natural variable is the *asymmetry parameter* for fermion f,

$$\mathcal{A}_f \equiv 2\frac{\bar{g}_{\text{v}}^{(f)} \bar{g}_{\text{a}}^{(f)}}{\bar{g}_{\text{v}}^{(f)2} + \bar{g}_{\text{a}}^{(f)2}} = 2\frac{g_{\text{v}}^{(f)}/g_{\text{a}}^{(f)}}{1 + \left(\bar{g}_{\text{v}}^{(f)2}/\bar{g}_{\text{a}}^{(f)2}\right)}, \qquad (2.7)$$

which can be determined experimentally from angular distribution and/or polarization data, as discussed below. In the case that the final-state fermion f is a quark q, then it is hadrons which are detected and the final-state hadronic charge asymmetry which is measured. It is to be understood that the measured data have been corrected for contributions such as initial-state *QED* corrections, γ exchange, γ–Z^0 interference, etc., leaving asymmetries which are purely electroweak in origin. Finally, let the incident electron beam carry a polarization $\mathcal{P}_e$ but the positron beam be unpolarized. For LEP1, the incident electron beam is unpolarized ($\mathcal{P}_e = 0$), whereas for SLC one has partial polarization ($\mathcal{P}_e \simeq 0.75$).

In the following, the symbols σ_F and σ_B refer to

$$\sigma_F = 2\pi \int_0^1 d\cos\theta \, \frac{d\sigma}{d\Omega}, \qquad \sigma_B = 2\pi \int_{-1}^0 d\cos\theta \, \frac{d\sigma}{d\Omega}, \qquad (2.8)$$

and σ_L, σ_R denote the cross section for an incident left-handed and right-handed polarized electron. Then three types of asymmetry are:

$$A_{FB} \equiv \frac{\sigma_F - \sigma_B}{\sigma_F + \sigma_B} \qquad \text{[forward–backward]}, \tag{2.9a}$$

$$A_{LR} \equiv \frac{\sigma_L - \sigma_R}{\sigma_L + \sigma_R} \qquad \text{[left–right]}, \tag{2.9b}$$

$$A_{LRFB} \equiv \frac{(\sigma_F - \sigma_B)_L - (\sigma_F - \sigma_B)_R}{(\sigma_F - \sigma_B)_L + (\sigma_F - \sigma_B)_R} \qquad \text{[left–right forward-backward]}, \tag{2.9c}$$

and the relation of these to the asymmetry parameter of Eq. (2.7) is

$$A_{FB}^{(f)} = \frac{3}{4} \mathcal{A}_f \frac{\mathcal{A}_e + \mathcal{P}_e}{1 + \mathcal{A}_e \mathcal{P}_e} \qquad A_{LR} = \mathcal{A}_e \mathcal{P}_e \qquad A_{FBLR}^{(f)} = \frac{3}{4} \mathcal{A}_f. \tag{2.10}$$

Yet another approach is to exploit the fact that final-state tau leptons themselves carry a polarization $\mathcal{P}_\tau$, which affects the tau angular distribution as well as its FB asymmetry $\mathcal{P}_{FB}^{(\tau)}$. As such, the LEP1 experiments were able to extract the asymmetry parameters $\mathcal{A}_\tau$ and $\mathcal{A}_e$ via the polarization measurements

$$\mathcal{A}_\tau = -\mathcal{P}_\tau, \qquad \mathcal{A}_e = -\frac{4}{3} \mathcal{P}_{FB}^{(\tau)}. \tag{2.11}$$

The above set of asymmetries were the subject of much study for a number of years. One interesting example is the high-precision measurement of $\mathcal{A}_{LR}$ carried out by the SLD collaboration [Ab *et al.* (SLD collab.) 00]. The left–right asymmetry was measured from the e^+e^- production cross section by counting (mainly) hadronic final states for each of the two longitudinal polarizations of the incident electron beam at energies near the Z^0 mass. Despite the emphasis on detecting final-state hadrons, this measurement actually probes the asymmetry parameter of the incident-state electrons,

$$A_{LR} = \frac{1 - 4\bar{s}_\ell^2}{1 - 4\bar{s}_\ell^2 + 8\bar{s}_\ell^2}, \tag{2.12}$$

where we have assumed lepton universality in writing the weak mixing angle as $\bar{s}_\ell^2$. The precision measurement of A_{LR} then leads to the following determination of $\bar{s}_\ell^2$,

$$A_{LR}^{(e)} = 0.15138 \pm 0.00216, \qquad \bar{s}_\ell^2 = 0.23097 \pm 0.00027. \tag{2.13}$$

In summary, the collection of measurements taken at scale $\mu = M_Z$ has, on the whole, been in agreement with Standard Model expectations.

Let us conclude by commenting on just a few topics:

(1) Effective weak mixing angle $\bar{s}_\ell$: Adopting a Higgs mass value $M_H = 125$ GeV, the Standard Model prediction [ErS 13] $\bar{s}_\ell^2 = 0.23158$ is consistent with the experimental determination $\bar{s}_\ell^2 = 0.23153 \pm 0.00016$.

(2) Unresolved issue: A long-standing item is the roughly 3σ difference between the two most precise individual measurements $\bar{s}_\ell = 0.23097 \pm 0.00027$ (via $\mathcal{A}_{LR}$ from the SLD production cross sections discussed earlier) and $\bar{s}_\ell^2 = 0.23221 \pm 0.00029$ (via $\mathcal{A}_{FB}$ from the $Z^0 \to b\bar{b}$ transition found at LEP). Despite much discussion, the issue remains unresolved.

(3) Quantum corrections: The large collection of high-quality Z^0 data has provided determinations which are sensitive to quantum corrections. For example, the result $g_a^{(\ell)} = -0.50125 \pm 0.00026$ (found in part by assuming lepton universality) implies via Eq. (2.1) that $\rho_\ell = 1.005 \pm 0.001$. This differs from the bare value $\rho_\ell^{(\text{tree})} = 1.000$ by 5σ and attests that quantum corrections have indeed been probed. Some even more impressive examples appear in Sect. I of [FeS 12].

Definitions of the weak mixing angle

Thus far in this chapter, we have made reference to three different versions of (and notations for) the weak mixing angle,

$$\text{Effective}: \bar{s}_f^2 \qquad \text{On-shell}: s_w^2 \qquad \overline{\text{MS}}: \hat{s}_w^2(\mu). \qquad (2.14)$$

Since there is, in principle, an unlimited number of renormalization prescriptions for a given quantity in quantum field theory, it is no surprise to come across the three above usages in the literature (several others, not covered here, also exist). Let us briefly consider their relation to each other, starting with the effective weak angle for a lepton ℓ.

Given the definition for $\bar{s}_\ell^2$ in Eq. (2.2b), it's clear that this quantity is tied to the ratio $\bar{g}_v^{(\ell)}/\bar{g}_a^{(\ell)}$ as measured at the scale $\mu = M_Z$. The motivation for doing things this way is a matter of convenience for the massive experimental effort by the Z^0 factories – one reads off a basic quantity of the Standard Model directly in terms of Z^0-related data. The current precise determination, given earlier and repeated here, of $\bar{s}_\ell^2 = 0.23153 \pm 0.00016$ attests to the success achieved by the Z^0-factory experimentalists in doing precision physics.

We have already seen (cf. Sect. II–1) how modified minimal subtraction ($\overline{\text{MS}}$) can be implemented in dimensional regularization for the electric charge $e(q^2)$, and one proceeds accordingly for the weak mixing angle $\hat{s}_w^2(q^2)$ (or $s_w(q^2)_{\overline{\text{MS}}}$) by adopting the scale-dependent definition [Ma 79, MaS 81],

$$\hat{s}_w^2\left(q^2\right) \equiv \frac{e^2\left(q^2\right)}{g_2^2\left(q^2\right)}. \qquad (2.15)$$

A fit to the current database yields the value appearing already in Table XVI–1, viz., $\hat{s}_w^2(M_Z) = 0.23125 \pm 0.00016$.

The effective and $\overline{\text{MS}}$ descriptions of the weak mixing angle can be related [GaS 94]. As pointed out in [GaS 94] there was, at the time of the LEP1 operation, 'considerable confusion among theorists and experimentalists alike as to the precise conceptual and numerical relation between the two'. The analysis in [GaS 94] established that

$$\bar{s}_\ell^2 = Re\hat{\kappa}_\ell(M_Z)\, \hat{s}_{\text{w}}^2(M_Z) \simeq 1.0012\, \hat{s}_{\text{w}}^2(M_Z) \simeq \hat{s}_{\text{w}}^2(M_Z) + 0.0003. \qquad (2.16)$$

This is in accord with the individual values for $\bar{s}_\ell^2$ and $\hat{s}_{\text{w}}^2(M_Z)$ given above.

Finally, the on-shell weak mixing angle is *defined* in terms of the physical gauge-boson masses,

$$s_{\text{w}}^2 \equiv 1 - M_W^2/M_Z^2. \qquad (2.17a)$$

Thus, the on-shell weak mixing angle can be experimentally determined directly from M_W and M_Z. Inserting the gauge-boson mass values from Table I–1 into Eq. (2.17a), one has

$$s_{\text{w}}^2\big|_{M_W,M_Z} = 0.2229 \pm 0.0003. \qquad (2.17b)$$

The current uncertainty in $s_{\text{w}}^2\big|_{M_W,M_Z}$, about twice that in $\bar{s}_\ell^2$ and $\hat{s}_{\text{w}}^2(M_Z)$, is due largely to the $W^\pm$ mass uncertainty, $\delta M_W = 15\,\text{MeV}$, compared to the much smaller $\delta M_Z = 2.1\,\text{MeV}$. With the completion of the Tevatron data analysis, along with the resumption of LHC operations, the precision gap between the direct on-shell determination and the alternative $\bar{s}_\ell^2$ and $\hat{s}_{\text{w}}^2(M_Z)$ schemes is expected to be narrowed. Even so, the fact that the on-shell scheme contains some relatively large $\mathcal{O}(G_\mu m_t^2)$ corrections (see Sect. XVI–6 for a discussion) not present in $\overline{\text{MS}}$ renormalization lessens its appeal for use in electroweak perturbation theory.

Returning to the idea of scale-dependent (or running) quantities, consider the possibility of relating gauge coupling constants $\hat{g}_k$ ($k = 1, 2, 3$) in the $\overline{\text{MS}}$ scheme at the Z^0 scale with those of a 'grand unified' theory defined at an energy $E_{\text{GUT}} \gg M_Z$. The so-called GUT scale signals the existence of a gauge group undergoing spontaneous symmetry breaking to $SU(3)_c \times SU(2)_L \times U(1)_Y$. The condition

$$\hat{g}_1 = \hat{g}_2 = \hat{g}_3 \qquad (E = E_{\text{GUT}}) \qquad (2.18)$$

leads to a prediction [GeQW 74] for the weak mixing angle at the scale E_{GUT}. In the grand unified theory of $SU(5)$ [GeG 74, La 81] and its supersymmetric extension ($SUSY$–$SU(5)$), the $\overline{\text{MS}}$ weak mixing angle obeys

$$\hat{s}_{\text{w}}^2(E_{\text{GUT}}) = 3/8. \qquad (2.19)$$

At the much lower energy scale $\mu = M_Z$, this value is reduced by a calculable amount,[4]

$$\hat{s}_w^2(M_Z) \equiv \bar{s}_w^2 = \frac{3}{8}\left[1 - C\frac{\bar{\alpha}}{2\pi}\ln\frac{M_X}{M_Z} + \cdots\right], \tag{2.20}$$

where $\bar{\alpha} \equiv \hat{\alpha}(M_Z)$, M_X is the mass scale of the superheavy gauge bosons, and C is a constant which depends upon the number n_H of Higgs doublets,

$$C = \begin{cases} \dfrac{110 - n_H}{9} & (SU(5)) \\ \dfrac{30 - n_H}{3} & (SUSY-SU(5)). \end{cases} \tag{2.21}$$

The $SU(5)$ extension of the Standard Model has $n_H = 1$, whereas the minimal supersymmetric model takes $n_H = 2$.

The 'bare-bones' $SU(5)$ model turns out to be unacceptable. It is well known to give rise to an unacceptably short proton lifetime, and precision data indicate that the three coupling constants of the Standard Model *disagree* with a single unification point if evolved according to $SU(5)$ [AmBF 91]. Interestingly, the $SUSY$ extension improves matters in both respects. The rate at which $\hat{s}_w^2(\mu)$ 'runs' is decreased due to contributions from supersymmetric partners ('sparticles') of the known particles, and the unification scale is raised to a level ($M_X \simeq 10^{16}$ GeV) consistent with the observed proton stability. The unification condition of Eq. (2.18) is better satisfied. Studies continue on whether supersymmetry breaking yields insights regarding masses of the long-sought $SUSY$ 'sparticles'.

XVI–3 Some W± properties

We shall return to issues regarding the weak mixing angle and its several definitions in Sect. XVI–4. Before that, however, we consider some aspects of $W^\pm$ physics. The LEP2 (e^+e^-), the Tevatron ($\bar{p}p$) and the LHC (pp), colliders have provided copious $W^\pm$-related data.

Decays of W± into fermions

The decay of a W-boson into a lepton and neutrino pair $\ell\nu_\ell$ is governed by the lagrangian,[5]

$$\mathcal{L}_{ch}^{(lept)} = -\frac{g_2}{\sqrt{8}} W_\mu^+ \bar{\nu}_\ell \gamma^\mu (1 + \gamma_5)\ell + \text{h.c.} \tag{3.1}$$

---

[4] Actually, Eq. (2.20) represents a simplification in that (i) lowest-order estimates for the renormalization-group coefficients are employed, (ii) supersymmetry-breaking effects are ignored, and (iii) the fact that $m_t > M_Z$ is also ignored.

[5] Although we shall denote tree-level decay widths, cross sections, etc. with a zero superscript in this section, for the sake of notational simplicity, we shall suppress the zero subscript for bare parameters.

It is a straightforward exercise to compute the tree-level decay width,

$$\Gamma^{(0)}_{W \to \bar{\nu}_\ell \ell} = \frac{g_2^2}{8} \frac{M_W}{6\pi} (1-x) \left(1 - \frac{x}{2} - \frac{x^2}{2}\right) \xrightarrow[x \to 0]{} \frac{1}{6\pi} \frac{G_\mu}{\sqrt{2}} M_W^3, \qquad (3.2)$$

where $x \equiv m_\ell^2 / M_W^2$ and we have employed Eq. (II–3.43). Including a small electroweak correction, we have $\Gamma_{W \to \bar{e} \nu_e} \simeq 0.226$ GeV.

There exist also decays $W \to \bar{q}^{(i)} q^{(j)}$ into quark modes (the superscripts i, $j = 1, 2, 3$ are generation labels), induced by the lagrangian

$$\mathcal{L}^{(qk)}_{ch} = -\frac{g_2}{\sqrt{8}} W^+_\mu V_{ij} \, \bar{q}^{(i)}_k \gamma^\mu (1 + \gamma_5) q^{(j)}_k + \text{h.c.}, \qquad (3.3)$$

where V_{ij} is a CKM matrix element, and the index k labels color. The lowest-order decay width for quark emission is

$$\sum_{\text{color}} \Gamma^{(0)}_{W \to \bar{q}^{(i)} q^{(j)}} = \frac{1}{2\pi} \frac{G_\mu}{\sqrt{2}} M_W^3 \left|V_{ij}\right|^2 \left[1 - 2(x + \bar{x}) + (x - \bar{x})^2\right]^{1/2}$$

$$\times \left[1 - \frac{x + \bar{x}}{2} - \frac{(x - \bar{x})^2}{2}\right] \xrightarrow[x, \bar{x} \to 0]{} \frac{1}{2\pi} \frac{G_\mu}{\sqrt{2}} M_W^3 \left|V_{ij}\right|^2, \quad (3.4)$$

where $x, \bar{x}$ are mass ratios defined as above, and we assume that all emitted quarks eventually convert to hadrons. Since the t quark is too massive to be a product of W decay, a sum over accessible quark flavors yields $\sum_{i,j} \left|V_{ij}\right|^2 = 2$. For decay into quarks, these lowest-order partial decay widths are modified by δ_{QCD}, the QCD factor of Eq. (2.4b) introduced in our earlier discussion of Z^0 hadronic decays.

If all final-state masses are ignored, the predicted total width for $W^\pm$ decay into fermions is

$$\Gamma^{(\text{tot})}_{W^\pm} = \Gamma^{(\text{had})}_{W^\pm} + \Gamma^{(\text{lept})}_{W^\pm} \simeq 2.093 \text{ GeV} \left(\frac{M_W(\text{GeV})}{80.385}\right)^3. \qquad (3.5)$$

An average of data [RPP 12] yields the value $\Gamma^{(\text{tot})}_{W^\pm} = 2.085 \pm 0.042$ GeV, which is consistent with the prediction of Eq. (3.5). The current experimental uncertainty far exceeds that from theory. In the limit of massless final-state particles, the branching ratio for decay into a lepton pair $\ell \bar{\nu}_\ell$ is $(\text{Br})_\ell \simeq 1/9$ ($\ell = e, \mu, \tau$), while inclusive decay to a mode containing a positively charged quark q ($q = u, c$) gives $(\text{Br})_q \simeq 1/3$.

Triple-gauge couplings

The $SU(2) \times U(1)$ lagrangian of Eq. (II–3.10) and the $SU(2)$ field strength tensor of Eq. (II–3.11) alert us that there will be trilinear and quadrilinear couplings of the gauge bosons. We shall limit our discussion here to the so-called *charged* triple-gauge couplings (TGCs). Upon using Eq. (II–3.30) to replace the neutral gauge bosons B_μ, W_μ^3 with the physical fields A_μ, Z_μ^0, we can write an effective WWV ($V = Z^0, \gamma$) lagrangian as[6]

$$\mathcal{L}_{WWV} = i g_{WWV} \left[g_1^V \left(W_{\mu\nu}^\dagger W^\mu - W_{\mu\nu} W^{\mu\dagger} \right) V^\nu + \kappa_V W_\mu^\dagger W_\nu V^{\mu\nu} \right.$$

$$\left. + i \frac{\lambda_V}{M_W^2} W_{\rho\mu}^\dagger W_\nu^\mu V^{\nu\rho} \right], \tag{3.6a}$$

where $W_{\mu\nu} \equiv \partial_\mu W_\nu - \partial_\nu W_\mu$, $V_{\mu\nu} \equiv \partial_\mu V_\nu - \partial_\nu V_\mu$ and g_{WWV} represents the coupling strengths

$$g_{WW\gamma} = -e, \qquad g_{WWZ^0} = -e \cot\theta_{\mathrm{w}}. \tag{3.6b}$$

The above lagrangian is constrained to contain only terms which are invariant under charge-conjugation (C), parity (P), and $SU(2) \times U(1)$ gauge transformations. In the Standard Model, the individual couplings in Eq. (3.6a) become

$$g_1^V = 1, \qquad \kappa_V = 1, \qquad \lambda_V = 0 \qquad \left(V = Z^0, \gamma \right), \tag{3.6c}$$

and are consistent with the following constraint of gauge invariance,

$$\kappa_Z = g_1^Z - (\kappa_\gamma - 1) \tan^2\theta_{\mathrm{w}}, \qquad \lambda_Z = \lambda_\gamma. \tag{3.6d}$$

A recent review of LEP experiments gives [Sc *et al.* 13]

$$g_1^Z = 0.984_{-0.020}^{+0.018}, \qquad \kappa_\gamma = 0.982 \pm 0.042, \qquad \lambda_\gamma = -0.022 \pm 0.019, \tag{3.7}$$

consistent with Standard Model expectations.

We can read off static electromagnetic properties of the W boson upon taking $V = \gamma$. The decomposition in Eq. (3.6a) allows for the existence of a magnetic dipole moment μ_W and an electric quadrupole moment q_W,

$$\mu_W = \frac{e}{2M_W}(1 + \kappa_\gamma + \lambda_\gamma), \qquad q_W = -\frac{e}{M_W^2}(\kappa_\gamma - \lambda_\gamma), \tag{3.8a}$$

or to lowest order in the Standard Model (SM),

$$\mu_W^{\mathrm{SM}} = e/M_W, \qquad q_W^{\mathrm{SM}} = e/M_W^2. \tag{3.8b}$$

A number of experimental studies of the TGCs, especially data from the LEP2 e^+e^-, the Tevatron $\bar{p}p$, and the LHC pp colliders, has emphasized searching for *anomalous* TGCs, often expressed in terms of the five quantities,

$$\Delta g_1^Z \equiv g_1^Z - 1, \qquad \Delta\kappa_Z \equiv \kappa_Z - 1, \qquad \Delta\kappa_\gamma \equiv \kappa_\gamma - 1, \qquad \lambda_Z, \qquad \lambda_\gamma. \tag{3.9}$$

[6] Unlike TGCs with two $W^\pm$ bosons, purely neutral gauge-boson vertices are not present at tree level in the Standard Model.

These constitute anomalous behavior in that they vanish for the Standard Model values in Eq. (3.6c). Currently, no experimental evidence exists for any of the anomalous TGCs, for example

$$-0.038 < \lambda_Z < +0.031, \qquad -0.111 < \Delta\kappa_\gamma < 0.142 \qquad \text{[CMS]},$$
$$-0.074 < \lambda_Z < +0.073, \qquad -0.135 < \Delta\kappa_\gamma < 0.190 \qquad \text{[ATLAS]}. \qquad (3.10)$$

The status of recent bounds is indicated by the results the LHC detectors. ATLAS and CMS will be performing further studies at higher energies and, even lacking discovery of such effects, will supply ever more stringent bounds on anomalous behavior.

We can expand the preceding discussion to incorporate possible violations of parity and charge-conjugation invariance, for which an appropriate effective lagrangian $\widetilde{\mathcal{L}}_{WWV}$ which does just this is

$$\widetilde{\mathcal{L}}_{WWV} = g_{WWV}\left[i\widetilde{\kappa}_V\, W_\mu^\dagger W_\nu \widetilde{V}^{\mu\nu} + i\frac{\widetilde{\lambda}_V}{M_W^2} W_{\alpha\mu}^\dagger\, W_\nu^\mu \widetilde{V}^{\nu\alpha} \right.$$
$$\left. + g_4^V\, W_\mu^\dagger W_\nu\, (\partial^\mu V^\nu + \partial^\nu V^\mu) + g_5^V\, \epsilon^{\mu\nu\alpha\beta}\left(W_\mu^\dagger \partial_\alpha W_\nu - \partial_\alpha W_\mu^\dagger \cdot W_\nu \right) V_\beta \right].$$
$$(3.11)$$

Here, $\widetilde{\kappa}_\gamma$ and $\widetilde{\lambda}_\gamma$ are P-violating but C-invariant, whereas g_4^V respects P but not C and g_5^V respects neither P nor C. In particular, the W boson could itself have static properties which violate at least some of the discrete symmetries. For example, an electric dipole moment d_W or magnetic quadrupole moment $\widetilde{q}_W$ would be parameterized as

$$d_W = \frac{e}{2M_W}\left(\widetilde{\kappa}_\gamma + \widetilde{\lambda}_\gamma\right), \qquad \widetilde{q}_W = -\frac{e}{M_W^2}\left(\widetilde{\kappa}_\gamma - \widetilde{\lambda}_\gamma\right). \qquad (3.12)$$

Limits on the neutron electric dipole moment can be used to place a bound on the W electric dipole moment [MaQ 86], and an updated evaluation gives $|d_W| \leq 5 \times 10^{-21}$ e-cm.

XVI–4 The quantum electroweak lagrangian

In the following three sections, we shall give a simple description of how electroweak radiative corrections are calculated. We begin by quantizing the classical electroweak lagrangian to obtain certain of its Feynman rules. We also expand on earlier comments made in Sect. XVI–1 regarding on-shell renormalization.

Classical electroweak theory of three fermion generations is defined by an $SU(2)_L \times U(1)_Y$ gauge-invariant lagrangian,[7]

$$\mathcal{L}_{\text{ew}}^{(\text{cl})} = \mathcal{L}_{\text{ew}}^{(\text{cl})} \left(\left\{ \psi_{L,R}^{(f)} \right\}, \mathbf{W}_\mu, B_\mu, \Phi ; \{g_f\}, g_1, g_2, \lambda, v^2 \right), \tag{4.1}$$

where Φ is the Higgs doublet and the collection $\{g_f\}$ of Higgs–fermion coupling constants is flavor-nondiagonal. With spontaneous symmetry breaking, all particles but the photon become massive and diagonalization of the neutral gauge-boson mass matrix occurs in the basis of the photon A_μ and massive gauge-boson Z_μ^0 fields, as given at tree level by Eq. (II–3.30). In addition, diagonalization of the charged-fermion and neutrino mass matrices for the three-generation system involves additional mixing angles and phases. The physical degrees of freedom of the gauge and Higgs sectors become manifest in unitary gauge (cf. Sect. XV–1),

$$\mathcal{L}_{\text{ew}}^{(\text{cl})} = \mathcal{L}_{\text{ew}}^{(\text{cl})} \left(\{\psi^{(f)}\}, W_\mu^\pm, Z_\mu^0, A_\mu, H_0 ; \{m_f\}, M_W, M_Z, M_H, e \right), \tag{4.2}$$

where the fermion mixing parameters are included in the $\{m_f\}$.

Gauge fixing and ghost fields in the electroweak sector

The quantum electroweak lagrangian $\mathcal{L}_{\text{ew}}^{(\text{qm})}$ will contain, in addition to the classical lagrangian of Eq. (4.1), both gauge-fixing and ghost-field contributions,

$$\mathcal{L}_{\text{ew}}^{(\text{qm})} = \mathcal{L}_{\text{ew}}^{(\text{cl})} + \mathcal{L}_{\text{ew}}^{(\text{g-f})} + \mathcal{L}_{\text{ew}}^{(\text{gh})}. \tag{4.3}$$

Mixing between gauge fields and unphysical Higgs fields occurs in the covariant derivative of the Higgs doublet (cf. Eq. (II–3.18)),[8]

$$\mathcal{L}_{HG} = \left| \left(\mathbf{1} \left(\partial_\mu + \frac{i}{2} g_1 B_\mu \right) + \frac{i}{2} g_2 \vec{\tau} \cdot \vec{W}_\mu \right) \Phi \right|^2 + \cdots$$

$$= i \frac{g_1}{2} (\partial^\mu \Phi)^\dagger B_\mu \Phi + i \frac{g_2}{2} (\partial^\mu \Phi)^\dagger \vec{\tau} \cdot \vec{W}_\mu \, \Phi + \text{h.c.} + \cdots . \tag{4.4}$$

One can arrange the gauge-fixing term to cancel such mixing contributions. Expressing the complex Higgs doublet in terms of the physical field H_0, unphysical fields χ_+, χ_3, and the vacuum expectation value v as

$$\Phi = \frac{1}{\sqrt{2}} \begin{pmatrix} \sqrt{2}\chi_+ \\ H_0 + i\chi_3 + v \end{pmatrix}, \tag{4.5}$$

we write the gauge-fixing contribution in the form,

---

[7] We have replaced the Higgs parameter μ^2 by the equivalent quantity v^2.
[8] Mixing also occurs, of course, between the neutral gauge fields B_μ, W_μ^3.

$$\mathcal{L}_{\text{ew}}^{(\text{g-f})} = -\frac{1}{2\xi_+}\left|\partial_\mu W_+^\mu - \frac{\xi_+ g_2 v}{2}\chi_+\right|^2$$

$$- \frac{1}{2\xi_3}\left(\partial_\mu W_3^\mu - \frac{\xi_3 g_2 v}{2}\chi_3\right)^2 - \frac{1}{2\xi_0}\left(\partial_\mu B^\mu + \frac{\xi_0 g_1 v}{2}\chi_3\right)^2. \quad (4.6)$$

It is not hard to see that cancelation of the unwanted Higgs–gauge mixing terms occurs for arbitrary values of the gauge-fixing parameters $\xi_{+,3,0}$. Even with this cancelation, there remain in $\mathcal{L}_{\text{ew}}^{(\text{g-f})}$ quadratic terms containing the unphysical Higgs fields, and such terms will contribute to the propagators of these fields.

As explained in App. A–6, once the gauge fixing is specified as in Eq. (4.6), the structure of the Faddeev–Popov lagrangian $\mathcal{L}_{\text{ew}}^{(\text{gh})}$ of ghost fields is determined. For the electroweak sector, it turns out that there are four ghost fields,

$$\mathcal{L}_{\text{ew}}^{(\text{gh})} = \mathcal{L}_{\text{ew}}^{(\text{gh})}(\mathbf{c}_W, c_B). \quad (4.7)$$

These are associated with the four gauge fields $\mathbf{W}_\mu$, B_μ which appear in the original $SU(2)_L \times U(1)_Y$ symmetric lagrangian.

A subset of electroweak Feynman rules

The full set of electroweak Feynman rules is rather lengthy and we refer the reader to the detailed discussions in [BöHS 86, AoHKKM 82] or to the summary in [Ho 90]. A few of the more useful rules, expressed in terms of bare parameters are[9]

fermion W-boson vertex:

$$-i\frac{e}{2\sqrt{2}s_{\text{w}}}V_{ij}\left[\gamma_\mu(1+\gamma_5)\right]_{\alpha\beta} \qquad \beta\ j \longrightarrow \quad\alpha\ i \qquad (4.8a)$$

fermion Z-boson vertex:

$$-i\frac{e}{2s_{\text{w}}c_{\text{w}}}\left[\gamma_\mu(g_v^{(f)} + g_a^{(f)}\gamma_5)\right]_{\alpha\beta} \qquad \beta \longrightarrow \quad\alpha \qquad (4.8b)$$

---

[9] For notational simplicity, we suppress the zero subscript in the following discussion.

W-boson propagator $i D^{(W)}_{\mu\nu}(q)$:

$$\frac{i}{q^2 - M_W^2 + i\epsilon}\left[-g_{\mu\nu} + \frac{q_\mu q_\nu(1 - \xi_+)}{q^2 - \xi_+ M_W^2 + i\epsilon}\right] \qquad \text{(4.8c)}$$

Z-boson propagator $i D^{(Z)}_{\mu\nu}(q)$:

$$\frac{i}{q^2 - M_Z^2 + i\epsilon}\left[-g_{\mu\nu} + \frac{q_\mu q_\nu(1 - \xi_Z)}{q^2 - \xi_Z M_Z^2 + i\epsilon}\right] \qquad \text{(4.8d)}$$

unphysical charged Higgs propagator $i \Delta^{(\chi_+)}(q)$:

$$\frac{i}{q^2 - \xi_+ M_W^2 + i\epsilon} \qquad \text{(4.8e)}$$

In the above, (V_{ij}) is a matrix element for quark-mixing, $g^{(f)}_{(v,a)}$ are given in Eq. (II–3.41), and ξ_Z is defined by expressing the gauge fixing in the form of Eq. (4.6) but using the physical neutral fields.

As seen in Eqs. (4.8c), (4.8e), each boson propagator is explicitly gauge-dependent and, in particular, the propagator of the unphysical χ_+ vanishes in the $\xi_1 \to \infty$ limit of the unitary gauge. This is as expected, because only physical degrees of freedom appear in unitary gauge. In fact, the absence of unphysical degrees of freedom in unitary gauge would appear to be an appealing reason for carrying out the computation of radiative corrections in this gauge. However, there is a 'hidden cost'. In unitary gauge, the $W^\pm$ propagator of Eq. (4.8c) becomes

$$i D^{(W)}_{\mu\nu}(q)\Big|_{\text{unitary}} = i\frac{-g_{\mu\nu} + q_\mu q_\nu/M_W^2}{q^2 - M_W^2 + i\epsilon}, \qquad \text{(4.9)}$$

and the high-energy behavior produced by the $q_\mu q_\nu/M_W^2$ term makes this a questionable choice for doing higher-order calculations. Instead, as the price for acceptable high-energy behavior, many opt to accept the presence of unphysical fields. One popular choice of gauge fixing is the *'t Hooft–Feynman gauge*, defined by setting all the gauge-fixing parameters equal to unity, $\xi_i = 1$. In this gauge, the lowest-order propagators for the physical gauge bosons and unphysical Higgs and ghost fields have poles at either M_W^2 or M_Z^2. This condition can be maintained in higher orders by a suitable renormalization of the gauge-fixing parameters.

On-shell determination of electroweak parameters

The topic of electroweak radiative corrections to Standard Model quantities has been well developed over many years of research and by now there exists an impressively large literature. To focus our attention, there is one aspect that we will particularly address in the following. Given that the largest mass parameter in the Standard Model is that of the top quark, a natural question regards the effect m_t has on the set of electroweak corrections. The answer turns out to depend on the renormalization prescription followed, its effect being largest in the so-called on-shell scheme.

Two sets of electroweak parameters appear in the classical lagrangians of Eqs. (4.1), (4.2),

$$\text{classical parameter sets} = \begin{cases} \{g_f\}, g_1, g_2, \lambda, v^2 & \text{(Eq. (4.1))}, \\ \{m_f\}, M_W, M_Z, M_H, e & \text{(Eq. (4.2))}. \end{cases}$$

Considered as bare (input) parameters to the quantum theory, these obey the simple tree-level relations

$$M_{W,0} = v_0 \frac{g_{2,0}}{2}, \qquad M_{Z,0} = v_0 \frac{g_{1,0}^2 + g_{2,0}^2}{2}, \qquad e_0^{-2} = g_{1,0}^{-2} + g_{2,0}^{-2},$$

$$M_{H,0} = v_0 \sqrt{2\lambda_0}, \qquad m_{f,0} = v_0 \frac{g_{f,0}}{\sqrt{2}}. \tag{4.10}$$

At this stage, there are several (equivalent) expressions for the bare weak mixing angle, e.g.,

$$s_{w,0}^2 = 1 - \frac{M_{W,0}^2}{M_{Z,0}^2} \qquad \text{or} \qquad s_{w,0}^2 = \frac{g_{1,0}^2}{g_{1,0}^2 + g_{2,0}^2}. \tag{4.11}$$

The second relation becomes Eq. (2.15) in the $\overline{\text{MS}}$ renormalization.

Radiative corrections will generally modify tree-level relations and, as a result, necessitate a precise definition of the weak mixing angle. Following the analysis in [Si 80], let us compare the parameter subsets $(g_{1,0}, g_{2,0}, v_0^2)$ and $(e_0, M_{W,0}, M_{Z,0})$. Each of these bare quantities will experience a shift,

$$g_{1,0} = g_1 - \delta g_1, \qquad g_{2,0} = g_2 - \delta g_2, \qquad v_0^2 = v^2 - \delta v^2,$$

$$e_0 = e - \delta e, \qquad M_{W,0}^2 = M_W^2 + \delta M_W^2, \qquad M_{Z,0}^2 = M_Z^2 + \delta M_Z^2. \tag{4.12}$$

In on-shell renormalization, the theory is specified in terms of e, M_W, and M_Z. Moreover, the following relations are arranged to hold order by order,

$$e^{-2} = g_1^{-2} + g_2^{-2}, \qquad M_W^2 = v^2 \frac{g_2^2}{4}, \qquad M_Z^2 = v^2 \frac{(g_1^2 + g_2^2)}{4}. \tag{4.13}$$

These equations constrain the effects of radiative corrections upon the parameters.

By differentiating the three relations in Eq. (4.13), one finds after a modest amount of algebra the conditions,

$$
\begin{pmatrix} \frac{\delta g_1^2}{g_1^2} \\ \frac{\delta g_2^2}{g_2^2} \\ \frac{\delta v^2}{v^2} \end{pmatrix} = \begin{pmatrix} -1 & 1 & 1 \\ \frac{c_w^2}{s_w^2} & -\frac{c_w^2}{s_w^2} & 1 \\ \frac{s_w^2 - c_w^2}{s_w^2} & \frac{c_w^2}{s_w^2} & -1 \end{pmatrix} \begin{pmatrix} \frac{\delta M_W^2}{M_W^2} \\ \frac{\delta M_Z^2}{M_Z^2} \\ \frac{\delta e^2}{e^2} \end{pmatrix}.
\tag{4.14}
$$

Also in on-shell renormalization, one *defines* the weak mixing angle in terms of the masses M_W, M_Z as in Eq. (2.17a). Since this relation is to be maintained to all orders, the bare value $s_{w,0}^2$ will be modified by shifts in the W and Z masses,

$$
s_{w,0}^2 = 1 - \frac{M_{W,0}^2}{M_{Z,0}^2} = 1 - \frac{M_W^2 + \delta M_W^2}{M_Z^2 + \delta M_Z^2}
$$

$$
\simeq s_w^2 \left[1 - \cot^2 \theta_w \left(\frac{\delta M_W^2}{M_W^2} - \frac{\delta M_Z^2}{M_Z^2} \right) \right].
\tag{4.15}
$$

For any renormalizable field theory, it makes sense to express results in terms of the most accurately measured quantities available. Thus, it is preferable in the electroweak sector to replace M_W by G_μ and work with a modified parameter set,

$$
\text{Physical parameter set} = \begin{cases} \alpha^{-1} = 137.035999173(35), \\ G_\mu = 1.1663787(6) \times 10^{-5} \text{ GeV}^{-2}, \\ M_Z = 91.1876(21) \text{ GeV}. \end{cases}
\tag{4.16}
$$

To accomplish this, the relationship $G_\mu = G_\mu(\alpha, M_W, M_Z, \dots)$ can be used to replace M_W by G_μ.

XVI–5 Self-energies of the massive gauge bosons

It is is evident from Eq. (4.14) that the parameter shifts δe^2, δM_W^2 and δM_Z^2 play an important role in the study of electroweak radiative corrections. We have already determined from our analysis of *QED* (cf. Eq. (II–1.30)) that

$$
\frac{\delta e^2}{e^2} = -\Pi(0),
\tag{5.1}
$$

where the photon vacuum polarization $\Pi(q^2)$ appears in Eq. (II–1.26). In this section, we shall compute the portion of δM_W^2 and δM_Z^2 arising from the fermionic vacuum polarization contributions to the $W^\pm$ and Z^0 propagators. As a consequence, we shall be able to reveal the presence of propagator contributions which scale as $\mathcal{O}(G_\mu m_t^2)$.

The charged gauge bosons $W^\pm$

The radiative correction experienced by a $W^\pm$ gauge boson propagating at momentum q is expressed in terms of a self-energy function, $\Pi_{ww}^{\mu\nu}(q^2)$,

$$\Pi_{ww}^{\mu\nu}(q^2) \equiv A_{ww}(q^2) g^{\mu\nu} - B_{ww}(q^2) q^\mu q^\nu. \tag{5.2}$$

(For notational simplicity in this subsection we denote W and Z boson subscripts for the quantities $\Pi^{\mu\nu}$, A, and B in terms of lower-case Roman indices.) Although a vector-boson propagator $i D_{\mu\nu}(q)$ generally contains terms proportional to $g_{\mu\nu}$ and to $q_\mu q_\nu$, it will suffice to study just the $g_{\mu\nu}$ part. As indicated at the end of Sect. II–3, the $q_\mu q_\nu$ dependence is absent if the gauge boson couples to a conserved current or will give rise to suppressed contributions if the external particles have small mass. Thus, we have for the W propagator in 't Hooft–Feynman gauge,

$$\frac{-i g_{\mu\nu}}{q^2 - M_{W,0}^2} \rightarrow \frac{-i g_{\mu\nu}}{q^2 - M_{W,0}^2} + \frac{-i g_{\mu\alpha}}{q^2 - M_{W,0}^2} \left(-i A_{ww}(q^2) g^{\alpha\beta}\right) \frac{-i g_{\beta\nu}}{q^2 - M_{W,0}^2}$$

$$\rightarrow \frac{-i g_{\mu\nu}}{q^2 - M_{W,0}^2 + A_{ww}(q^2)}$$

$$= \frac{-i g_{\mu\nu}}{q^2 - M_W^2 + A_{ww}(q^2) - \delta M_W^2}, \tag{5.3}$$

where we have substituted for the bare W mass using Eq. (4.12).

Let us now calculate the loop contribution of a fermion–antifermion pair $f_1 \bar{f}_2$ to the self-energy $A_{ww}(q^2)$. We begin with

$$-i \Pi_{ww}^{\alpha\beta}(q^2)\Big|_{f_1 \bar{f}_2} = -\frac{(-i g_2)^2 \eta_{f_1 \bar{f}_2}}{8}$$

$$\times \int \frac{d^4 p}{(2\pi)^4} \operatorname{Tr}\left[\gamma^\alpha (1 + \gamma_5) \frac{i}{\not{p} - m_1} \gamma^\beta (1 + \gamma_5) \frac{i}{\not{p} - \not{q} - m_2}\right], \tag{5.4}$$

where $\eta_{f_1 \bar{f}_2} = N_c |V_{f_1 f_2}|^2$ for the case when the fermions are quarks. Aside from the occurrence of the $1 + \gamma_5$ chiral factor and the nondegeneracy in fermion masses m_1, m_2, the above Feynman integral is identical to the photon vacuum polarization function of Eq. (II–1.20). It is thus straightforward to evaluate this quantity in dimensional regularization, and we find for the $g^{\alpha\beta}$ component,

$$A_{\text{ww}}^{(f_1\bar{f}_2)}(q^2) = \frac{\eta_{f_1\bar{f}_2}g_2^2}{24\pi^2}\left[q^2\left\{\frac{2}{\epsilon} - \frac{\gamma}{2} + \ln\sqrt{4\pi}\right.\right.$$

$$\left. -3\int_0^1 dx\, x(1-x)\ln\frac{M^2 - q^2x(1-x)}{\mu^2}\right\}$$

$$\left. -\frac{3}{2}\left\{(m_1^2 + m_2^2)\left[\frac{2}{\epsilon} - \frac{\gamma}{2} + \ln\sqrt{4\pi}\right] - \int_0^1 dx\, M^2\ln\frac{M^2 - q^2x(1-x)}{\mu^2}\right\}\right],$$

$$(5.5)$$

where $M^2 \equiv m_1^2 x + m_2^2(1-x)$. Since the $W^\pm$ boson is an unstable particle with decay rate Γ_W, the function $A_{\text{ww}}(q^2)$ is complex-valued, and we consider its real and imaginary parts separately.

From Eq. (5.5), we see that Re $A_{\text{ww}}(q^2)$ is divergent. One can construct a finite quantity $\hat{A}_{\text{ww}}(q^2)$ by defining the field renormalization, $\mathbf{W}_{\mu,0} = (Z_2^W)^{1/2}\mathbf{W}_\mu$, and constraining δM_W^2 and δZ_2^W to cancel the ultraviolet divergence in Re $A_{\text{ww}}(q^2)$,

$$\hat{A}_{\text{ww}}(q^2) \equiv A_{\text{ww}}(q^2) - \delta M_W^2 + \delta Z_2^W\left(q^2 - M_W^2\right). \tag{5.6}$$

It follows from Eq. (5.6) that the mass shift δM_W^2 is fixed by

$$\delta M_W^2 = \text{Re } A_{\text{ww}}\left(M_W^2\right), \tag{5.7}$$

and the $f_1\bar{f}_2$ contribution to the field renormalization, which ensures that $\hat{A}_{\text{ww}}(M_W^2) = 0$ is

$$\delta Z_2^W[f_1\bar{f}_2] = \frac{\eta_{f_1\bar{f}_2}g_2^2}{8\pi^2}\left[\frac{2}{\epsilon} - \frac{\gamma}{2} + \ln\sqrt{4\pi}\right]. \tag{5.8}$$

To obtain a relation for the imaginary part of the self-energy, we recall that instability in a propagating state of mass M is described by the replacement $M \to M - i\Gamma/2$. This produces the following modification of a propagator denominator,

$$\frac{1}{q^2 - M^2} \to \frac{1}{q^2 - M^2 + iM\Gamma}, \tag{5.9}$$

where we ignore the $\mathcal{O}(\Gamma^2)$ term. Comparison with Eq. (5.5) then immediately yields

$$\text{Im } A_{\text{ww}}(M_W^2) = M_W\Gamma_W. \tag{5.10}$$

We can use Eq. (5.6) to check this relation by setting $q^2 = M_W^2$. If, for simplicity, we neglect the masses of the fermion–antifermion pair $f_1\bar{f}_2$, then the imaginary part comes from the logarithm contained in the first of the integrals in Eq. (5.5),

$$\text{Im}\int_0^1 dx\, x(1-x)\ln\frac{-q^2x(1-x)}{\mu^2 - i\epsilon} \xrightarrow[q^2=M_W^2]{} -\frac{\pi}{6}, \tag{5.11}$$

and we obtain

$$\Gamma_{W \to f_1 \bar{f}_2} = \frac{\mathrm{Im}\left[A_{\mathrm{ww}}^{(f_1 \bar{f}_2)}\left(M_W^2\right)\right]}{M_W} = \eta_{f_1 \bar{f}_2} \frac{G_\mu M_W^3}{6\sqrt{2}\pi}, \tag{5.12}$$

where we have substituted $g_2^2 = 4\sqrt{2}M_W^2 G_\mu$. This agrees with the results of our earlier decay width calculations for W decay in Sect. XVI–3.

The neutral gauge bosons Z^0, γ

The system of neutral gauge bosons is treated analogously to the charged case except that we must deal with a 2×2 propagator matrix, and the issue of particle mixing arises. Although the neutral channel was already diagonalized at tree level (cf. Eq. (II–3.30)), interactions reintroduce nondiagonal propagator contributions at higher orders. The $g_{\mu\nu}$ part of the neutral channel inverse propagator $\mathbf{D}_{[\mathrm{ntl}]\mu\nu}^{-1}\left(q^2\right)$, diagonal at tree level,

$$\mathbf{D}_{[\mathrm{ntl}]\mu\nu}^{(0)-1}\left(q^2\right) = i g_{\mu\nu} \begin{pmatrix} q^2 & 0 \\ 0 & q^2 - M_{Z,0}^2 \end{pmatrix}, \tag{5.13}$$

has the renormalized form,

$$\mathbf{D}_{[\mathrm{ntl}]\mu\nu}^{(0)-1}\left(q^2\right) \to \mathbf{D}_{[\mathrm{ntl}]\mu\nu}^{-1}\left(q^2\right) = g_{\mu\nu} \begin{pmatrix} q^2 + \hat{A}_{\gamma\gamma}\left(q^2\right) & \hat{A}_{\gamma z}\left(q^2\right) \\ \hat{A}_{\gamma z}\left(q^2\right) & q^2 - M_Z^2 + \hat{A}_{zz}\left(q^2\right) \end{pmatrix}. \tag{5.14}$$

Upon taking the inverse, we obtain for the individual neutral boson renormalized propagators,

$$D_{\gamma\gamma}^{\mu\nu}\left(q^2\right) = \frac{-i g^{\mu\nu}}{q^2 + \hat{A}_{\gamma\gamma}\left(q^2\right) - \hat{A}_{\gamma z}^2\left(q^2\right) / \left(q^2 - M_Z^2 + \hat{A}_{zz}^2\left(q^2\right)\right)},$$

$$D_{zz}^{\mu\nu}\left(q^2\right) = \frac{-i g^{\mu\nu}}{q^2 - M_Z^2 + \hat{A}_{zz}\left(q^2\right) - \hat{A}_{\gamma z}^2\left(q^2\right) / \left(q^2 + \hat{A}_{\gamma\gamma}^2\left(q^2\right)\right)},$$

$$D_{\gamma z}^{\mu\nu}\left(q^2\right) = \frac{i g^{\mu\nu} \hat{A}_{\gamma z}\left(q^2\right)}{\left[q^2 + \hat{A}_{\gamma\gamma}\left(q^2\right)\right]\left[q^2 - M_Z^2 + \hat{A}_{\gamma\gamma}^2\left(q^2\right)\right] - \hat{A}_{\gamma z}^2\left(q^2\right)}. \tag{5.15}$$

Observe that there is indeed a particle-mixing propagator, $D_{\gamma z}^{\mu\nu}$, proportional to the the reduced self-energy $\hat{A}_{\gamma z}(q^2)$. It might appear from Eq. (5.15) that Z^0-photon mixing gives rise to a photon mass contribution. However, one arranges as a renormalization condition that $\hat{A}_{\gamma z}(0) = 0$, and the photon remains massless under electroweak radiative corrections.

If we consider only the vacuum-polarization loop contribution due to a fermion of mass m, we obtain for the Z^0 self-energy,

$$-i\Pi_{zz}^{\alpha\beta}(q^2) = N_c \left(\frac{ig_2}{2c_w}\right)^2 \int \frac{d^4 p}{(2\pi)^4} \frac{N^{\alpha\beta}}{(p^2 - m^2)\left((p-q)^2 - m^2\right)}, \quad (5.16)$$

where m is the fermion mass, N_c is a quark color factor, and

$$N^{\alpha\beta} = g_v^{(f)2} \text{Tr} \left[\gamma^\alpha \not{p} \gamma^\beta (\not{p} - \not{q}) + m^2 \gamma^\alpha \gamma^\beta\right] + g_a^{(f)2} \text{Tr} \left[\gamma^\alpha \not{p} \gamma^\beta (\not{p} - \not{q}) - m^2 \gamma^\alpha \gamma^\beta\right]. \quad (5.17)$$

The quantities in $N^{\alpha\beta}$ are just those expected from the coupling of fermion f to the neutral weak current. We then obtain, using dimensional regularization,

$$A_{zz}^{(f\bar{f})}(q^2) = \frac{g_2^2 N_c}{16\pi^2 c_w^2} \left[\frac{2q^2(g_v^{(f)2} + g_a^{(f)2})}{3} \left\{\frac{2}{\epsilon} - \frac{\gamma}{2} + \ln\sqrt{4\pi}\right.\right.$$
$$\left. -3 \int_0^1 dx\, x(1-x) \ln\frac{m^2 - q^2 x(1-x)}{\mu^2}\right\}$$
$$\left. +4m^2 g_a^{(f)2} \left\{\frac{2}{\epsilon} - \frac{\gamma}{2} + \ln\sqrt{4\pi} - \frac{1}{2}\int_0^1 dx\, \ln\frac{m^2 - q^2 x(1-x)}{\mu^2}\right\}\right]. \quad (5.18)$$

It is also easy to demonstrate that the photon–Z^0 self-energy $A_{\gamma z}^{(f\bar{f})}$ is proportional to $A_{\gamma\gamma}^{(f\bar{f})}$ for the case of a charged-fermion loop contribution,

$$A_{\gamma z}^{(f\bar{f})}(q^2) = \frac{g_v^{(f)}}{2c_w s_w Q_f} A_{\gamma\gamma}^{(f\bar{f})}(q^2), \quad (5.19)$$

where Q_f is the electric charge of the fermion.

XVI–6 Examples of electroweak radiative corrections

All electroweak amplitudes will be affected by radiative corrections. We have already pointed out our interest in potentially large contributions arising from the heavy masses m_t and M_H. We shall find leading corrections which are quadratic in the top-quark mass ($\mathcal{O}(G_\mu m_t^2)$).[10] To begin this section, we consider corrections to the coefficients ρ_f and κ_f of Eq. (1.5), followed by an analysis of the quantum correction known as Δr, and finally the $Z \to b\bar{b}$ vertex correction. A historical overview of electroweak corrections appears in [FeS 12], and a thorough state-of-the-art presentation is given by Erler and Langacker in [RPP 12].

[10] Recall from Chap. XV that corrections at leading order are only logarithmic in the Higgs mass ($\mathcal{O}(\ln[M_H^2/M_Z^2])$).

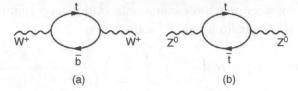

Fig. XVI–3 Top-quark corrections to the (a) $W^\pm$, (b) Z^0 propagators.

The $\mathcal{O}(G_\mu m_t^2)$ contribution to Δ_ρ

Contributions to ρ_f and κ_f can be classified as either independent of the external fermions (*universal*) or explicitly dependent on the fermion flavor f (*nonuniversal*). Recalling that at tree level these quantities reduce to unity, we have

$$\rho_f = 1 + \Delta\rho + (\Delta\rho)^{(f)}_{\text{nonuniv}}, \qquad \kappa_f = 1 + \Delta\kappa + (\Delta\kappa)^{(f)}_{\text{nonuniv}}, \qquad (6.1)$$

where $\Delta\rho$ and $\Delta\kappa$ denote universal pieces. It should be apparent that $W^\pm$- and Z^0-propagator corrections, like those in Fig. XVI–3, occur independent of the external fermions and are thus 'universal'. Nonuniversal effects have been found to be small (i.e. subdominant) except for the $Z^0 \to \bar{b}b$ vertex. The universal effects are of special interest because they turn out to be the primary source of $\mathcal{O}(G_\mu m_t^2)$ radiative corrections [Ve 77a, ChFH 78]. As such, in the following we shall approximate

$$\Delta\rho = (\Delta\rho)_t + \cdots, \qquad \Delta\kappa = \frac{c_{\text{w}}^2}{s_{\text{w}}^2}(\Delta\rho)_t + \cdots, \qquad (6.2)$$

where

$$(\Delta\rho)_t = \frac{3G_\mu m_t^2}{8\pi^2\sqrt{2}} \simeq 0.00942 \times \left(\frac{m_t}{173.4 \text{ GeV}}\right)^2. \qquad (6.3)$$

Observe in Eq. (6.2) that $\Delta\kappa$ is proportional to $\Delta\rho$. This is a result of the Standard Model; in general, these quantities are independent.

The quantity $\Delta\rho$ can be defined as a correction to the rho parameter of Eq. (1.3),

$$\rho_0 = \frac{1}{c_{\text{w},0}^2} \cdot \frac{D_Z(q^2 = 0)}{D_W(q^2 = 0)}$$

$$\rho_0 + \Delta\rho = \frac{M_Z^2 + \delta M_Z^2}{M_W^2 + \delta M_W^2} \cdot \frac{\left(-M_Z^2 - \delta M_Z^2 + A_{zz}(0)\right)^{-1}}{\left(-M_W^2 - \delta M_W^2 + A_{ww}(0)\right)^{-1}} \qquad (6.4)$$

or

$$\Delta\rho = \frac{A_{zz}(0)}{M_Z^2} - \frac{A_{ww}(0)}{M_W^2}. \qquad (6.5)$$

Observe that $\Delta\rho$ is finite since the singular terms in Eqs. (5.5), (5.18) cancel. If we set $m_1 = m_t$ and $m_2 = m_b$ in Eq. (5.5) and include both t-quark and b-quark loops in Eq. (5.18), a simple calculation reveals that $\Delta\rho = 0$ in the limit that $m_t = m_b$. However, the leading term in the small m_b limit gives

$$
\begin{aligned}
(\Delta\rho)_t &= \frac{g_2^2 N_c}{16\pi^2 M_W^2} \int_0^1 dx \left[\frac{m_t^2}{2} \ln \frac{m_t^2}{\mu^2} - x m_t^2 \ln \frac{x m_t^2}{\mu^2} \right] + \cdots \\
&= \frac{g_2^2 N_c}{64\pi^2} \frac{m_t^2}{M_W^2} + \cdots .
\end{aligned}
\tag{6.6}
$$

Substitution of $N_c = 3$ and $G_\mu/\sqrt{2} = g_2^2/8M_W^2$ yields the result shown in Eqs. (6.2), (6.3).

This quadratic dependence on the heavy-top-quark mass is in striking contrast with the behavior observed for the photon self-energy (cf. Eq. (II–1.26)). In the heavy-fermion limit, the photon vacuum polarization exhibits instead the decoupling result $\mathcal{O}(m_t^{-2})$. The reason for this difference is that QED is a vector theory, whereas the charged and neutral weak interactions are chiral. Indeed, one can show (cf. Prob. XVI–2) that the decoupling expected of a vector interaction results when left-handed and right-handed self-energies are averaged. However, equally important is the fact that as m_t grows while m_b is kept fixed, the weak doublet is being split in mass. Thus, decoupling of the top quark in the large m_t limit should not be expected because if we were to integrate out the top quark, we would no longer have a renormalizable theory – the remaining low-energy theory would have an incomplete weak doublet. Early contributions to this subject appear in [Ve 77b] and [ChFH 78]. As noted above, if both members of the doublet are taken to be *equally* heavy ($m_t = m_b \to \infty$), there would exist no quadratic dependence on the heavy-quark mass, and the decoupling theorem (cf. Sect. IV–2) would be satisfied. It is the large splitting in the weak doublet which leads to the observable violation of decoupling.

Even though two different renormalization schemes must give the same final set of results, intermediate details will generally differ. For example, the leading m_t behaviors for the coefficients ρ_f and κ_f of Eq. (1.5) are [RPP 12],

$$
\text{on-shell}: \quad \rho_f \sim 1 + (\Delta\rho)_t + \cdots \qquad \kappa_f \sim 1 + \frac{c_w^2}{s_w^2}(\Delta\rho)_t + \cdots
$$

$$
\overline{(\text{MS})}: \quad \hat{\rho}_f \sim 1 + \cdots \qquad \hat{\kappa}_f \sim 1 + \cdots ,
\tag{6.7}
$$

where $(\Delta\rho)_t$ is defined in Eq. (6.3).[11]

---

[11] The case $f = b$ is special; the leading behaviors are $\hat{\rho}_f \sim 1 - 4(\Delta\rho)_t/3$ and $\hat{\kappa}_f \sim 1 + 2(\Delta\rho)_t/3$.

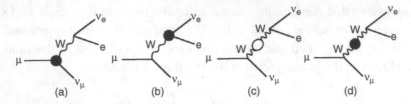

Fig. XVI–4 (a)–(b) Vertex, (c) propagator, and (d) mass-shift counterterm corrections to muon decay.

The $\mathcal{O}(G_\mu m_t^2)$ contribution to Δr

The quantity Δr describes the effect of electroweak corrections on the leading order relation which defines the muon decay constant. In particular, the tree-level relation of Eq. (II–3.43) becomes modified by the radiative corrections of Fig. XVI–4 [Si 80, BuJ 89],

$$\frac{G_{\mu,0}}{\sqrt{2}} = \frac{g_{2,0}^2}{8M_{W,0}^2} \quad \to \quad \frac{G_\mu}{\sqrt{2}} = \frac{g_2^2}{8M_W^2}[1 + \Delta r]. \tag{6.8a}$$

It is to be understood in Eq. (6.8a) that 'G_μ' is determined from the muon lifetime with the photonic corrections described in Sect. V–2 already taken into account. Thus, Δr contains only the remaining electroweak effects.

To trace the origin of the quantum correction, we observe first the effect of the $W^\pm$ self-energy on the bare relation in Eq. (6.8a),

$$\frac{G_\mu}{\sqrt{2}} = -\frac{g_{2,0}^2}{8}\frac{1}{q^2 - M_{W,0}^2 + A_{\mathrm{ww}}(0)} \simeq \frac{g_{2,0}^2}{8M_{W,0}^2}\left[1 + \frac{A_{\mathrm{ww}}(0)}{M_W^2} + \cdots\right], \tag{6.8b}$$

where we have taken $q^2 \simeq 0$. Next, we replace the bare parameters $g_{2,0}^2$ and $M_{W,0}^2$ by their physical forms as in Eq. (4.12). Comparison with Eq. (6.8a) directly yields

$$\Delta r = \frac{\delta M_W^2 - A_{\mathrm{ww}}(0)}{M_W^2} - \frac{\delta g_2^2}{g_2^2}. \tag{6.9}$$

Upon using Eq. (4.14) for δg_2^2, we can rewrite Eq. (6.9) as

$$\Delta r = -\frac{\delta e^2}{e^2} - \frac{c_{\mathrm{w}}^2}{s_{\mathrm{w}}^2}\left[\frac{\delta M_Z^2}{M_Z^2} - \frac{\delta M_W^2}{M_W^2}\right] + \frac{A_{\mathrm{ww}}(0) - \delta M_W^2}{M_W^2}. \tag{6.10}$$

Recalling that the $W^\pm$ and Z^0 mass shifts can be related to the self-energy functions $A_{\mathrm{ww}}(M_W^2)$ and $A_{\mathrm{zz}}(M_Z^2)$, it should be clear that Eq. (6.9) expresses Δr entirely in terms of calculable quantities.[12] Although each of the terms in Eq. (6.10) is

---

[12] There are additional radiative corrections, such as the 'box' diagrams, which we shall not discuss.

divergent, the overall combination is finite. A number of rearrangements and algebraic steps can be used to isolate the leading contributions, and one finds

$$\Delta r = \Delta\alpha + \Delta r_{\mathrm{w}} + (\Delta r)_{\mathrm{rem}}, \tag{6.11}$$

where

$$\Delta\alpha \equiv \frac{\alpha\left(M_Z^2\right) - \alpha}{\alpha} \simeq \hat{\Pi}\left(M_Z^2\right), \quad \text{and} \quad \Delta r_{\mathrm{w}} = -\frac{c_{\mathrm{w}}^2}{s_{\mathrm{w}}^2}\Delta\rho. \tag{6.12}$$

$\Delta\rho$ is given by Eqs. (6.5)–(6.6), and $(\Delta r)_{\mathrm{rem}}$ contains smaller finite contributions.

The largest contribution to Δr is $\Delta\alpha$, the shift in the fine-structure constant. Although we have previously expressed the variation in $\alpha(q^2)$ in terms of fermion masses (cf. Eq. (II–1.38)), the difficulty in precisely determining quark masses would appear to undermine an accurate evaluation of $\Delta\alpha$. However, one can use dispersion relations to relate the hadronic contribution to the vacuum polarization, $\hat{\Pi}_{\mathrm{had}}(q^2)$, directly to cross-section data. Recalling Prob. V–2, we have

$$\Pi_{\mathrm{had}}^{\mu\nu}(q^2) = ie^2 \int d^4x \, e^{iq\cdot x} \langle 0|T(J_{\mathrm{em}}^\mu(x)J_{\mathrm{em}}^\nu(0))|0\rangle$$
$$= \left(q^\mu q^\nu - q^2 g^{\mu\nu}\right)\Pi_{\mathrm{had}}\left(q^2\right). \tag{6.13}$$

The imaginary part of $\Pi_{\mathrm{had}}(q^2)$ is expressible in terms of cross-section data evaluated at invariant energy q^2,

$$\mathrm{Im}\,\Pi_{\mathrm{had}}\left(q^2\right) = \frac{\alpha}{3}\mathrm{R}\left(q^2\right) \quad \text{with} \quad \mathrm{R}\left(q^2\right) = \frac{\sigma(e\bar{e} \to \mathrm{hadrons})}{\sigma(e\bar{e} \to \mu\bar{\mu})}. \tag{6.14}$$

Thus, we obtain a dispersion relation for the subtracted quantity $\hat{\Pi}_{\mathrm{had}}(q^2)$,

$$\hat{\Pi}_{\mathrm{had}}\left(q^2\right) \equiv \Pi_{\mathrm{had}}\left(q^2\right) - \Pi_{\mathrm{had}}(0)$$
$$= \frac{\alpha q^2}{3\pi}\left[\int_{4m_\pi^2}^{s_0} + \int_{s_0}^\infty\right]ds\,\frac{\mathrm{R}(s)}{s(s - q^2 - i\epsilon)}, \tag{6.15}$$

where s_0 denotes the point at which data become unavailable. For energies above s_0, a perturbative representation is used to approximate $\mathrm{R}(s)$. The result of Eq. (6.15), when added to the lepton contributions, implies a value for $\alpha^{-1}(M_Z^2)$ [DaHMZ 11],[13]

$$\alpha^{-1}\left(M_Z^2\right) = 128.952 \pm 0.014. \tag{6.16}$$

Some feeling for the magnitudes of 'Δr' corrections is given in the following (the numerical values have been taken from [KuMMS 13]):

[13] There are minor differences in various evaluations cited in the literature, depending on how the perturbative estimate is performed or on the particular renormalization scheme.

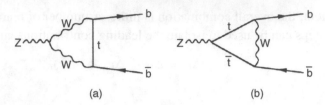

Fig. XVI–5 Top-quark corrections to the $Z^0 b \bar{b}$ vertex.

$$\Delta r = 1 - \frac{\pi \alpha}{\sqrt{2} G_\mu M_W^2 \left(1 - M_W^2/M_Z^2\right)} = 0.0350(9),$$

$$\Delta \hat{r}_{\rm w} = 1 - \frac{\pi \alpha}{\sqrt{2} G_\mu M_W^2 \hat{S}_{\rm w}^2(M_Z)} = 0.0699(7)(4),$$

$$\Delta \hat{r} = 1 - \frac{\pi \alpha}{\sqrt{2} G_\mu M_Z^2 \hat{C}_{\rm w}^2(M_Z) \hat{S}_{\rm w}^2(M_Z)} = 0.0598(4). \qquad (6.17)$$

The above relations, although exact at tree level (the '$\Delta r = 0$' limit), lead to the different values shown away from this limit. As before in this chapter, the quantities $\hat{s}_{\rm w}^2(M_Z)$ and $\hat{c}_{\rm w}^2(M_Z)$ are defined in $\overline{\rm MS}$ renormalization and evaluated at scale M_Z. In order to obtain the above form for Δr, we have replaced $[1 + \Delta r]$ in Eq. (6.8a) by $1/[1 - \Delta r]$, which is valid in our first-order analysis.

The $Z \to b \bar{b}$ vertex correction

The preceding analyses of $\Delta \rho$ and Δr could very well be carried out for any other electroweak observable. In most cases, we would again find important $\mathcal{O}(G_\mu m_t^2)$ radiative corrections. Thus, for example, the Z^0 width for decay into lepton ℓ ($\ell = e, \mu, \tau$) has the form

$$\Gamma_{Z^0 \to \ell \bar{\ell}} = \Gamma_{Z^0 \to \ell \bar{\ell}}^{(0)} [1 + (\Delta \rho)_t + \cdots], \qquad (6.18)$$

and grows quadratically with increasing m_t [AkBYR 86]. The origin of this effect, the one-loop $t \bar{t}$ contribution to the Z^0 propagator, is identical to that discussed earlier.

Interestingly, however, a more complete calculation reveals a slight *decrease* to occur in the decay rate $\Gamma_{Z^0 \to b \bar{b}}$ as m_t grows. This is because, although the decay amplitude contains a (universal) propagator contribution proportional to $(\Delta \rho)_t$, an even larger effect, the vertex correction of Fig. XVI–5, contributes with opposite sign [AkBYR 86, DjKZ 90],[14]

---

[14] Due to cancellations, the vertex correction turns out not to affect asymmetry phenomena, such as the b-quark forward–backward asymmetry $A_{FB}^{(b)}$.

$$\Gamma_{Z^0 \to b\bar{b}} = \Gamma^{(0)}_{Z^0 \to b\bar{b}} \left[1 + \frac{19}{13} \left((\Delta\rho)_t + \left(\Delta v^{(b)} \right)_t \right) + \cdots \right], \qquad (6.19)$$

where the $Z^0 b\bar{b}$ vertex correction is given by

$$\left(\Delta v^{(b)} \right)_t = -\frac{20}{19} (\Delta\rho)_t - \frac{130}{57} \frac{\alpha}{\pi} \ln \frac{m_t^2}{M_Z^2}. \qquad (6.20)$$

The $d\bar{d}, s\bar{s}$ modes also contain virtual t-quark vertex corrections, but they are greatly suppressed by the tiny accompanying CKM factors $|V_{ti}|^2$ $(i = d, s)$. Recalling the characterization given in Sect. XVI–1 of radiative corrections as either 'universal' or 'nonuniversal', one may interpet the $Z^0 b\bar{b}$ effect as a nonuniversal term which contributes as

$$(\Delta\rho)^{(b)}_{\text{nonuniv}} = -2(\Delta\kappa)^{(b)}_{\text{nonuniv}} = -\frac{4}{3}\Delta\rho - \frac{\alpha}{4\pi s_w^2} \left(\frac{8}{3} + \frac{1}{6c_w^2} \right) \ln \frac{m_t^2}{M_W^2}. \qquad (6.21)$$

Although $\mathcal{O}(m_t^2)$ corrections are the most important, $\mathcal{O}(\ln(m_t^2/M_Z^2))$ logarithmic dependence has been included in Eq. (6.20) because it has a nonnegligible numerical impact.

Precision tests and New Physics

In precision electroweak tests, about 20 (mainly $W^\pm$ or Z^0) observables are fit to Standard Model predictions (e.g. see [Ba *et al.* (Gfitter group) 12, RPP 12]). Such tests are based on the availability of high-quality data (with precision at the 1% level or better), multi-loop theoretical Standard Model predictions, and sophisticated software packages.[15] In view of the LHC discovery regarding the Higgs boson, the list of detected Standard Model particles is now complete. Consequently, there will be, more than ever, an emphasis on using precision tests to probe contributions from beyond the Standard Model.

As was noted ever since the first electroweak corrections were calculated (e.g. [Ve 77a]), physics associated with a large-energy scale Λ should affect the gauge-boson self-energies $-i\Pi^i_{\mu\nu}(q)$ $(i = \gamma\gamma, \gamma z, WW, ZZ)$. In particular, the $-i\Pi^i_{\mu\nu}(q)$ could contain loop corrections (sometimes referred to as *oblique* corrections) from

---

[15] Let us describe just a few of these. The Zfitter collaboration, begun in 1985 ([AkARR 13]), established a FORTRAN library of Standard Model predictions for $e^+e^- \to \bar{f}f(+\gamma's)$ at energies $\sqrt{s} = 20 \to 150$ GeV using the on-shell renormalization scheme. The LEP electroweak working group LEPEWWG was founded in 1993 to perform fits of LEP and Tevatron data, particularly of Z-pole observables such as the effective weak mixing angle $\bar{s}_f^2$ of Eqs. (2.2b),(2.14), using Zfitter in part as input. A more recent effort using on-shell renormalization to perform electroweak global fits is the Gfitter group. The Global Analysis of Particle Properties (GAPP) software is employed by the Particle Data Group [Er 00]. This is a special purpose FORTRAN package, which performs calculations and fitting procedures and utilizes $\overline{\text{MS}}$ renormalization, Finally, the Heavy Flavor Averaging Group provides updates to world averages of heavy-flavor quantities.

'new' particles. For $\Lambda^2 \gg q^2$, one would expect rapid convergence of an expansion for $-i\Pi^i_{\mu\nu}(q)$ in powers of q^2/Λ^2, yielding the following effective low-energy description,

$$-i\Pi^i_{\mu\nu}(q) = g_{\mu\nu}\left(A_i + q^2 A'_i\right) + \cdots . \tag{6.22}$$

This description involves eight free parameters, $A_{\gamma\gamma}, \ldots, A'_{zz}$. However, the conditions $\Pi_{\gamma\gamma}(0) = \Pi_{\gamma z}(0) = 0$ reduce this number to six. An additional three parameters can be absorbed into the renormalization of α, G_μ, M_Z, which experience the shifts [BaFGH 90],

$$\frac{\delta\alpha}{\alpha} = -A'_{\gamma\gamma}, \qquad \frac{\delta G_\mu}{G_\mu} = A_{ww}, \qquad \frac{\delta M_Z^2}{M_Z^2} = -\frac{A_{zz}}{M_Z^2} - A'_{zz}. \tag{6.23}$$

The three remaining parameters may be chosen to be quantities known as S, T, U and defined as [PeT 90] (we employ $\overline{\text{MS}}$ renormalization here [RPP 12])

$$\frac{\hat{\alpha}(M_Z)}{4\hat{s}_Z^2\hat{c}_Z^2} S \equiv \left[\frac{A_{zz}^{(NP)}\left(M_Z^2\right) - A_{zz}^{(NP)}(0)}{M_Z^2} - \frac{\hat{c}_Z^2 - \hat{s}_Z^2}{\hat{s}_Z^2\hat{c}_Z^2}\frac{A_{z\gamma}^{(NP)}(0)}{M_Z^2} - \frac{A_{\gamma\gamma}^{(NP)}(0)}{M_Z^2}\right],$$

$$\hat{\alpha}(M_Z)T \equiv \left[\frac{A_{ww}^{(NP)}(0)}{M_W^2} - \frac{A_{zz}^{(NP)}(0)}{M_Z^2}\right],$$

$$\frac{\hat{\alpha}(M_Z)}{4\hat{s}_Z^2}(S+U) \equiv \left[\frac{A_{ww}^{(NP)}\left(M_W^2\right) - A_{ww}^{(NP)}(0)}{M_W^2} - \frac{\hat{c}_Z}{\hat{s}_Z}\frac{A_{z\gamma}^{(NP)}(0)}{M_Z^2} - \frac{A_{\gamma\gamma}^{(NP)}(0)}{M_Z^2}\right],$$

$$\tag{6.24}$$

where the superscript (NP) refers to contributions only from New Physics. Clearly, these S, T, U parameters are defined so as to vanish in the limit of only Standard Model physics. If nonzero, they would appear as new contributions to various observables, e.g.,

$$\left[\frac{M_Z^{(\text{expt})}}{M_Z^{(\text{SM})}}\right]^2 = \frac{1 - \hat{\alpha}(M_Z)T}{1 - G_\mu M_Z^{(\text{SM})2}S/(2\sqrt{2}\pi)},$$

$$\left[\frac{M_W^{(\text{expt})}}{M_W^{(\text{SM})}}\right]^2 = \frac{1}{1 - G_\mu M_W^{(\text{SM})2}(S+U)/(2\sqrt{2}\pi)},$$

$$\left[\frac{M_Z^{(\text{SM})}}{M_Z^{(\text{expt})}}\right]^3 \cdot \frac{\Gamma_Z^{(\text{expt})}}{\Gamma_Z^{(\text{SM})}} = \frac{1}{1 - \hat{\alpha}(M_Z)T}, \tag{6.25}$$

all of which compare the experimental value with the Standard Model (SM) prediction. In this way, bounds are placed on the New Physics parameters and the results found in [RPP 12] are[16]

[16] The range of Higgs-boson masses $115.5 < M_H(\text{GeV}) < 127$ was used as input.

$$S = 0.00^{+0.11}_{-0.10}, \qquad T = 0.02^{+0.11}_{-0,12}, \qquad U = 0.08 \pm 0.11, \tag{6.26}$$

and are consistent with Standard Model expectations. At present, the precision electroweak fits do not yet display evidence for effects beyond those predicted by the Standard Model.

The literature contains several other possible New Physics parameterizations. For example, if the scale of New Physics is not much larger than the Standard Model weak scale, then parameters X, Y, V, W will, in principle, contribute to the fitting procedure [BuGKLM 94, BaPRS 04]. However, their determination requires data at energies higher than the scale set by the Z-boson mass and so, e.g., in the work of [Ba *et al.* (Gfitter group) 12], the quantities X, Y, V, W are set equal to zero.

As we have emphasized throughout this book (e.g. Sect. IV–9), the effects of heavy particles can be analyzed theoretically by using effective lagrangians and the preceding analysis can be expressed naturally in this language (e.g. see [Sk 10]). These must respect the $SU(2)_L \times U(1)_Y$ gauge symmetry, but may or may not include the extra custodial $SU(2)_L \times SU(2)_R$ invariance of the Higgs sector with doublet Higgs fields. There will be a tower of such operators, beginning with those of dimension-six. However, not all dimension-six operators are relevant to electroweak phenomenology. Examples of these are $\left(H^\dagger H\right)^3$ and $H^\dagger H \mathcal{D}_\mu H^\dagger \mathcal{D}^\mu H$, the point being that processes having Higgs bosons as external states are presently experimentally inaccessible. Instead, we consider the two operators.[17]

$$\mathcal{O}_S \equiv H^\dagger \sigma_i H F^i_{\mu\nu} B^{\mu\nu}, \qquad \mathcal{O}_T \equiv \left|H^\dagger \mathcal{D}_\mu H\right|^2, \tag{6.27}$$

where $B^{\mu\nu}$, $F^i_{\mu\nu}$ are, respectively, the field strength tensors defined in Eqs. (II–3.11), (II–3.12). These operators, together with the usual Standard Model lagrangian $\mathcal{L}_{SM}$, can be added together to form

$$\mathcal{L} = \mathcal{L}_{SM} + a_S \mathcal{O}_S + a_T \mathcal{O}_T. \tag{6.28}$$

The New Physics coefficients a_S and a_T will each carry units of inverse squared-energy and the Higgs fields in Eq. (6.27) will each contribute a factor of the symmetry-breaking energy v, so that in this approach the S, T parameters will obey

$$S \propto a_S \, v^2, \qquad T \propto a_T \, v^2, \tag{6.29}$$

and we leave evaluation of the proportionality factors to an exercise at the end of the chapter. Although we do not survey New Physics models in this book, a large variety is discussed in [Sk 10, RPP 12, Ba *et al.* 12], among others.

---

[17] It turns out that associated with the parameter U will be a dimension-eight operator.

Problems

(1) **Tree-level coefficients in effective lagrangians**

    (a) Using the simplest quark–parton description of protons and neutrons as uud and ddu composites, reproduce the content of the tree-level expressions of Eq. (1.7) by determining the quantities $R_{\nu,0}$ and $R_{\bar{\nu},0}$ for scattering from an isoscalar target. It might be helpful to first refer to a summary of parton phenomenology, e.g., as in [RPP 12], for guidance. Suppose a neutrino deep-inelastic experiment reports $R_{\nu} = 0.3072 \pm 0.0032$. Infer from this a central value and an error estimate for the tree-level quantity $s_{w,0}^2$.

    (b) Likewise, reproduce the tree-level expressions for the coefficients $C_{1,0}^{(u,d)}$ of Eq. (1.11), and infer a value for $s_{w,0}^2$ assuming the value $Q_w = -69.4 \pm 1.55 \pm 3.8$ for the weak nuclear charge.

(2) **Power-law radiative corrections**

    (a) Verify the statement that if $m_t = m_b \to \infty$, there is no quadratic mass dependence in the calculation of $\Delta\rho$.

    (b) From the combination of Dirac matrices appearing in Eq. (5.4), it is evident that the self-energy amplitude has a 'left–left' (LL) chiral structure. To see how this affects the result, repeat the analysis of Eqs. (5.4), (5.5) except now employing a 'left–right' (LR) chiral structure.

    (c) By averaging the LL and LR self-energies and passing to the limit $m_1 = m_2 \to m$, reproduce the $g^{\alpha\beta}$ part of the photon self-energy of Eq. (II–1.26).

    (d) If we had a purely left-handed $U(1)$ theory, the vacuum polarization would grow with m_t^2 as $m_t \to \infty$. How can this be consistent with the decoupling theorem?

(3) **Effective field theory and the S,T parameters**

    Determine the proportionality factors which were not provided in Eq. (6.29).

Appendix A
Functional integration

In this appendix we outline the basis of functional methods which are employed in the text. Path-integral techniques appear at first sight to be rather formal and abstract. However, it is remarkable how easy it is to obtain practical information from them. Very often they add insight or new results, which are difficult to obtain from canonical quantization.

A–1 Quantum-mechanical formalism

Before attempting to address the full field-theoretic formalism we first review the application of such techniques within the more familiar setting of nonrelativistic quantum mechanics in one spatial dimension. Unless otherwise specified we here-after set $\hbar = 1$.

Path-integral propagator

Simply stated, the functional integral is an alternative way of evaluating the quantity

$$D(x_f, t_f; x_i, t_i) = \langle x_f | e^{-iH(t_f - t_i)} | x_i \rangle \equiv \langle x_f, t_f | x_i, t_i \rangle. \qquad (1.1)$$

This matrix element, usually called the *propagator*, is the amplitude for a particle located at position x_i and time t_i to be found at position x_f and subsequent time t_f. The propagator can also be written as a functional integral

$$D(x_f, t_f; x_i, t_i) = \int \mathcal{D}[x(t)] e^{iS[x(t)]}, \qquad (1.2)$$

where the integration is over all histories (i.e. paths) of the system which begin at spacetime point x_i, t_i and end at x_f, t_f. The paths are identified by specifying the coordinate x at each intermediate time t, so that the symbol $\int \mathcal{D}[x(t)]$ represents a sum over all such trajectories. The contribution of each path to the integral is weighted by the exponential involving the classical action

$$S[x(t)] = \int_{t_i}^{t_f} dt \left(\frac{m}{2}\dot{x}^2(t) - V(x(t))\right), \tag{1.3}$$

which, since it depends on the detailed shape of $x(t)$, is a functional of the trajectory.[1] Although the validity of the path-integral representation, Eq. (1.2), may not be obvious, its correctness can be verified by beginning with Eq. (1.1) and breaking the time interval $t_f - t_i$ into N discrete steps of size $\epsilon = (t_f - t_i)/N$. Using the completeness relation

$$1 = \int_{-\infty}^{\infty} dx_n \, |x_n\rangle\langle x_n|,$$

one can write Eq. (1.1) as

$$D(x_f, t_f; x_i, t_i) = \int_{-\infty}^{\infty} dx_{N-1} \cdots \int_{-\infty}^{\infty} dx_1$$

$$\langle x_N|e^{-i\epsilon H}|x_{N-1}\rangle\langle x_{N-1}|e^{-i\epsilon H}|x_{N-2}\rangle \cdots \langle x_1|e^{-i\epsilon H}|x_0\rangle, \tag{1.4}$$

where $x_0 \equiv x_i$, $x_N \equiv x_f$. In the limit of large N the time slices become infinitesimal, implying

$$\langle x_\ell|e^{-iH\epsilon}|x_{\ell-1}\rangle = \langle x_\ell|e^{-i\epsilon\left(\frac{p^2}{2m}+V(x)\right)}|x_{\ell-1}\rangle$$

$$= e^{-i\epsilon V(x_\ell)}\langle x_\ell|e^{-i\epsilon\frac{p^2}{2m}}|x_{\ell-1}\rangle + \mathcal{O}(\epsilon^2). \tag{1.5}$$

Inserting a complete set of momentum states and introducing a convergence factor $e^{-\kappa p^2}$ for the resulting integral over momentum, we have

$$\langle x_\ell|e^{-i\epsilon\frac{p^2}{2m}}|x_{\ell-1}\rangle = \lim_{\kappa\to 0} \int_{-\infty}^{\infty} \frac{dp}{2\pi} \, e^{ip(x_\ell-x_{\ell-1})-i\epsilon p^2/2m-\kappa p^2}$$

$$= \sqrt{\frac{m}{2\pi i\epsilon}} \, e^{i\frac{m}{2\epsilon}(x_\ell-x_{\ell-1})^2}. \tag{1.6}$$

---

[1] It is important to understand the difference between the concept of a function and that of a functional. A real-valued function involves the mapping from the space of real numbers onto themselves

$$\text{reals} \longleftarrow [f : \text{reals}].$$

On the other hand, a real-valued functional such as $S[x(t)]$ is a mapping from the space of functions $x(t)$ onto real numbers

$$\text{reals} \longleftarrow [S : x(t)].$$

Upon taking the continuum limit we obtain

$$D\left(x_f, t_f; x_i, t_i\right)$$

$$= \lim_{N \to \infty} \left(\frac{m}{2\pi i \epsilon}\right)^{\frac{N}{2}} \left[\prod_{n=1}^{N-1} \int_{-\infty}^{\infty} dx_n\right] e^{i\sum_{\ell=1}^{N} \left(\frac{m}{2} \frac{(x_\ell - x_{\ell-1})^2}{\epsilon} - \epsilon V(x_\ell)\right)}. \qquad (1.7)$$

It is clear then that we can make connection with Eq. (1.2) by identifying each path with the sequence of locations $(x_1, \ldots, x_{N-1})$ at times $\epsilon, 2\epsilon, \ldots, (N-1)\epsilon$. Integration over these intermediate positions is what is meant by the symbol $\int \mathcal{D}[x(t)]$, viz.

$$\int \mathcal{D}[x(t)] \equiv \lim_{N \to \infty} \left(\frac{m}{2\pi i \epsilon}\right)^{N/2} \prod_{n=1}^{N-1} \int_{-\infty}^{\infty} dx_n. \qquad (1.8)$$

Each trajectory has an associated exponential factor $e^{iS[x(t)]}$, where the quantity

$$S[x(t)] = \sum_{\ell=1}^{N} \epsilon \left(\frac{m}{2} \frac{(x_\ell - x_{\ell-1})^2}{\epsilon^2} - V(x_\ell)\right) \qquad (1.9)$$

becomes the classical action in the limit $N \to \infty$. We have thus demonstrated the equivalence of the operator (Eq. (1.1)) and path-integral (Eq. (1.2)) representations of the propagator.[2] It is important to realize that in the latter all quantities are *classical* – no operators are involved.

The path-integral propagator contains a great deal of information, and there are a variety of techniques for extracting it. For example, the spatial wavefunctions and energies are all present, as can be seen by inserting a complete set of energy eigenstates $\{|\,n\rangle\}$ into the definition of the propagator given in Eq. (1.1),

$$D(x_f, t_f; x_i, t_i) = \sum_{n=0}^{\infty} \psi_n(x_f)\psi_n^*(x_i)e^{-iE_n(t_f-t_i)}. \qquad (1.10)$$

---

[2] For completeness, we note that by combining Eqs. (1.5)–(1.8), one can also write the propagator in a corresponding hamiltonian path-integral representation

$$D(x_f, t_f; x_i, t_i) = \lim_{N \to \infty} \int \frac{dp_0}{2\pi} dx_1 \frac{dp_1}{2\pi} dx_2 \cdots dx_{N-1} \frac{dp_{N-1}}{2\pi}$$

$$\times e^{i\sum_{\ell=1}^{N} \left(p_\ell(x_\ell - x_{\ell-1}) - \left(\frac{p_\ell^2}{2m} + V(x_\ell)\right)\epsilon\right)}$$

$$\equiv \int \mathcal{D}[x(t)]\mathcal{D}[p(t)] e^{i \int dt\, (p\dot{x} - H(p,x))}.$$

This form is useful when one is dealing with non-cartesian variables or with constrained systems.

In addition, other quantum-mechanical amplitudes can be found by use of the identity[3]

$$\langle x_f, t_f | T (x(t_1) \cdots x(t_n)) | x_i, t_i \rangle$$

$$= \int \mathcal{D}[x(t)] \; x(t_1) \cdots x(t_n) \; e^{i \int_{t_i}^{t_f} dt \left(\frac{m}{2} \dot{x}^2(t) - V(x(t)) \right)}, \tag{1.11}$$

where 'T' is the time-ordered product.

External sources

An important technique involves the addition of an external source. In the quantum-mechanical case this is added like an arbitrary external 'force' $j(t)$,

$$\langle x_f, t_f | x_i, t_i \rangle_{j(t)} = \int \mathcal{D}[x(t)] \, e^{i \int_{t_i}^{t_f} dt \left[\frac{m}{2} \dot{x}^2(t) - V(x(t)) + j(t)x(t) \right]}. \tag{1.12}$$

The amplitude is now a functional of the source $j(t)$. From this quantity one can obtain all matrix elements using *functional differentiation*, which can be defined by means of the relation

$$j(t) = \int dt' \delta(t - t') j(t') \quad \Rightarrow \quad \frac{\delta j(t)}{\delta j(t')} = \delta(t - t') \tag{1.13}$$

and yields the result we seek,

$$\langle x_f, t_f | T (x(t_1) \cdots x(t_n)) | x_i, t_i \rangle$$

$$= (-i)^n \frac{\delta^n}{\delta j(t_1) \dots \delta j(t_n)} \langle x_f, t_f | x_i, t_i \rangle_{j(t)} \Big|_{j=0}. \tag{1.14}$$

For many applications it is necessary only to consider matrix elements between the lowest energy states (*vacuum*) of the quantum system. This can be

---

[3] One can prove this relation by choosing a particular ordering, say

$$t_i < t_1 < t_2 < \cdots < t_f,$$

and noting that

$$\langle x_f, t_f | T (x(t_1) \cdots x(t_n)) | x_i, t_i \rangle = \langle x_f, t_f | x(t_n) x(t_{n-1}) \cdots x(t_1) | x_i, t_i \rangle$$

$$= \prod_{k=1}^{n} \int_{-\infty}^{\infty} dx_k \langle x_f, t_f | x_n, t_n \rangle \, x_n \langle x_n, t_n | x_{n-1}, t_{n-1} \rangle x_{n-1} \cdots x_1 \langle x_1, t_1 |, x_i, t_i \rangle,$$

where we have used completeness and have defined $x_k = x(t_k)$ ($k = 1, 2, \ldots, n$). The amplitudes $\langle x_k, t_k | x_{k-1}, t_{k-1} \rangle$ are simply free propagators as in Eq. (1.1), and can be evaluated by means of the time-slice methods outlined above. Thus, the above expression is identical to the right-hand side of Eq. (1.11). In the case of a different time ordering the same result goes through provided one always places the times such that the later time always appears to the left of an earlier counterpart. However, this is simply the definition of the time-ordered product and hence the proof holds in general.

accomplished in either of two ways. First, it is possible to explicitly project out this amplitude using the ground-state wavefunction

$$\langle x, t | 0 \rangle = \psi_0(x) e^{-i E_0 t}, \tag{1.15}$$

which implies

$$\langle 0 | T (x(t_1) \cdots x(t_n)) | 0 \rangle \equiv \int_{-\infty}^{\infty} dx_f \int_{-\infty}^{\infty} dx_i \, \psi_0^*(x_f) \, e^{i E_0 t_f}$$
$$\langle x_f, t_f | T (x(t_1) \cdots x(t_n)) | x_i, t_i \rangle \psi_0(x_i) \, e^{-i E_0 t_i}. \tag{1.16}$$

However, this amplitude can be isolated in a simpler fashion. If we consider the amplitude $\langle x_f, t_f | x_i, t_i \rangle$ in the unphysical limit $t_f \to -i\tau_f$, $t_i \to +i\tau_i$ we find for large $\tau_f + \tau_i$,

$$\langle x_f, t_f | x_i, t_i \rangle \to \sum_n \psi_n(x_f) \psi_n^*(x_i) \, e^{-E_n(\tau_f + \tau_i)}$$
$$\xrightarrow[\tau_f + \tau_i \to \infty]{} \psi_0(x_f) \psi_0^*(x_i) \, e^{-E_0(\tau_f + \tau_i)}. \tag{1.17}$$

Generalizing, we have

$$\lim_{\substack{t_f \to -i\infty \\ t_i \to i\infty}} \frac{e^{i E_0(t_f - t_i)}}{\psi_0(x_f) \psi_0^*(x_i)} \langle x_f, t_f | T (x(t_1) \cdots x(t_n)) | x_i, t_i \rangle$$

$$= \langle 0 | T (x(t_1) \cdots x(t_n)) | 0 \rangle \tag{1.18}$$

which is operationally a much simpler procedure than Eq. (1.16).

The generating functional

We may combine all these techniques in the so-called *generating functional*, defined by

$$Z[j] = \lim_{\substack{t_f \to -i\infty \\ t_i \to +i\infty}} \langle x_f, t_f | x_i, t_i \rangle_{j(t)}. \tag{1.19}$$

This has the path-integral representation

$$Z[j] = \lim_{\substack{t_f \to -i\infty \\ t_i \to i\infty}} \int \mathcal{D}[x(t)] \, e^{i \int_{t_i}^{t_f} dt \, \left(\frac{m}{2} \dot{x}^2(t) - V(x(t)) + x(t) j(t) \right)}. \tag{1.20}$$

Noting that for $t_i = i\tau_i$ and $t_f = -i\tau_f$,

$$\langle x_f, t_f | x_i, t_i \rangle \to \psi_0(x_f) \psi_0^*(x_i) \, e^{-E_0(\tau_f + \tau_i)} \xrightarrow[\tau_i, \tau_f \to \infty]{} Z[0], \tag{1.21}$$

we find that ground-state matrix elements as in Eq. (1.16) can be given in terms of the generating functional $Z[j]$,

$$\langle 0| T \left(x(t_1) \cdots x(t_n) \right) |0\rangle = (-i)^n \frac{1}{Z[0]} \frac{\delta^n}{\delta j(t_1) \cdots \delta j(t_n)} Z[j]\Big|_{j=0}. \qquad (1.22)$$

It often happens with path integrals that formal procedures are best defined, as above, by using the imaginary-time limits $t \to \pm i\infty$. However, in practice it is common instead to express the theory in terms of Minkowski spacetime. Thus, the generating functional will involve the real-time limits $t \to \pm\infty$. Does the dominance of the ground-state contribution, as in Eq. (1.21), continue to hold? The answer is 'yes'. At an intuitive level, one understands this as a consequence of the rapid variation of the phase $e^{-iE_n t}$ in the limit $t \to \infty$. The more rapid phase variation accompanying the increased energy E_n of any excited state washes out its contribution relative to that of the ground state. In a more formal sense, the real-time limit is defined by an analytic continuation from imaginary time. To properly define the continuation, one must introduce appropriate '$i\epsilon$' factors into the Green's functions in order to deal with various singularities. Beginning with the next section, we shall often employ the Minkowski formulation and thus explicitly display the '$i\epsilon$' terms in our formulae.

The prescription given in Eq. (1.22) represents a powerful but formal procedure for the generation of matrix elements in the presence of an arbitrary potential $V(x)$. Unfortunately, an explicit evaluation is no more generally accessible via this route than is an exact solution of the Schrödinger equation. In practice, aside from an occasional special case, the only path integrals which can be performed exactly are those in quadratic form. However, approximation procedures are generally available.

One of the most common of these is perturbation theory. Suppose that the full potential $V(x)$ is the sum of two parts $V_1(x)$ and $V_2(x)$, where $V_1(x)$ is such that the generating functional can be evaluated exactly while $V_2(x)$ is in some sense small. Then we can write

$$\begin{aligned}
Z[j] &= \lim_{\substack{t_f \to -i\infty \\ t_i \to i\infty}} \int \mathcal{D}[x(t)] e^{i \int_{t_i}^{t_f} dt \left[\frac{m}{2} \dot{x}^2(t) - V_1(x(t)) - V_2(x(t)) + x(t) j(t) \right]} \\
&= \lim_{\substack{t_f \to -i\infty \\ t_i \to i\infty}} e^{-i \int_{t_i}^{t_f} dt \, V_2\left(-i \frac{\delta}{\delta j(t)}\right)} Z^{(0)}[j] \\
&= \lim_{\substack{t_f \to -i\infty \\ t_i \to i\infty}} \sum_{n=0}^{\infty} \frac{(-i)^n}{n!} \left[\int_{t_i}^{t_f} dt \, V_2\left(-i \frac{\delta}{\delta j(t)}\right) \right]^n Z^{(0)}[j], \qquad (1.23)
\end{aligned}$$

where

$$Z^{(0)}[j] = \lim_{\substack{t_f \to -i\infty \\ t_i \to i\infty}} \int \mathcal{D}[x(t)] \, e^{i \int_{t_i}^{t_f} dt [\frac{m}{2}\dot{x}^2(t) - V_1(x(t)) + x(t)j(t)]} \tag{1.24}$$

is the generating functional for $V_1(x)$ alone. Obviously, Eq. (1.23) defines an expansion for $Z[j]$ in powers of the perturbing potential $V_2(x)$.

A–2 The harmonic oscillator

It is useful to interrupt our formal development by considering the harmonic oscillator as an example of these methods. This treatment turns out to reproduce known oscillator properties with the use of functional methods, which are very similar to corresponding field-theory techniques.

It is most convenient to address the problem by employing Fourier transforms,

$$x(t) = \int_{-\infty}^{\infty} \frac{dE}{2\pi} e^{-iEt} \tilde{x}(E), \tag{2.1}$$

whereby for $t_i = -\infty$ and $t_f = +\infty$,

$$S^j[x(t)] = \int_{-\infty}^{\infty} dt \left(\frac{m}{2}\dot{x}^2(t) - \frac{m\omega^2}{2}x^2(t) + x(t)j(t) \right)$$

$$= \int_{-\infty}^{\infty} \frac{dE}{2\pi} \left[\frac{m}{2}(E^2 - \omega^2)\tilde{x}(E)\tilde{x}(-E) + \frac{1}{2}\tilde{j}(E)\tilde{x}(-E) + \frac{1}{2}\tilde{j}(-E)\tilde{x}(E) \right]$$

$$= \int_{-\infty}^{\infty} \frac{dE}{2\pi} \left\{ \frac{m}{2}(E^2 - \omega^2)\tilde{x}'(E)\tilde{x}'(-E) - \frac{1}{2m}\tilde{j}(E)\frac{1}{E^2 - \omega^2 + i\epsilon}\tilde{j}(-E) \right\}, \tag{2.2}$$

with the definition $\tilde{x}'(E) \equiv \tilde{x}(E) + \tilde{j}(E)/(mE^2 - m\omega^2 + i\epsilon)$. An infintesimal imaginary part $i\epsilon$ has been introduced to make the integration precise. Upon taking the inverse Fourier transform

$$x'(t) = \int_{-\infty}^{\infty} \frac{dE}{2\pi} e^{-iEt} \tilde{x}'(E) = x(t) + \frac{1}{m} \int_{-\infty}^{\infty} dt' \, D(t - t')j(t'), \tag{2.3}$$

where

$$D(t - t') = \int_{-\infty}^{\infty} \frac{dE}{2\pi} e^{-iE(t-t')} \frac{1}{E^2 - \omega^2 + i\epsilon} = -\frac{i}{2\omega} e^{-i\omega|t'-t|}, \tag{2.4}$$

we have

$$S^j[x(t)] = \int_{-\infty}^{\infty} dt \left(\frac{m}{2} \dot{x}'^2(t) - \frac{m\omega^2}{2} x'^2(t) \right)$$
$$- \frac{1}{2m} \int_{-\infty}^{\infty} dt \int_{-\infty}^{\infty} dt' \, j(t) D(t - t') j(t'). \tag{2.5}$$

Finally, changing variables from $x(t)$ to $x'(t)$ we obtain the generating functional

$$Z[j] = \int \mathcal{D}[x'(t)] e^{i \int_{-\infty}^{\infty} dt \left(\frac{m}{2} \dot{x}'^2(t) - \frac{m\omega^2}{2} x'^2(t) \right)}$$
$$\times e^{-\frac{i}{2m} \int_{-\infty}^{\infty} dt \int_{-\infty}^{\infty} dt' \, j(t) D(t-t') j(t')}$$
$$= Z[0] e^{-\frac{i}{2m} \int_{-\infty}^{\infty} dt \int_{-\infty}^{\infty} dt' \, j(t) D(t-t') j(t')}. \tag{2.6}$$

Note that the above change of variables has left the measure invariant ($\int \mathcal{D}[x(t)] = \int \mathcal{D}[x'(t)]$).

We can use this result to calculate arbitrary oscillator matrix elements. Thus for $t_2 > t_1$, we have for the ground state

$$\langle 0|T\,(x(t_2)x(t_1))\,|0\rangle = (-i)^2 \frac{1}{Z[0]} \frac{\delta^2 Z[j]}{\delta j(t_2)\delta j(t_1)} \Bigg|_{j=0}$$
$$= \frac{i}{m} D(t_2 - t_1) = \frac{e^{-i\omega(t_2-t_1)}}{2m\omega}, \tag{2.7}$$

which, in the limit $t_2 \to t_1$, reproduces the familiar result

$$\langle 0|x^2|0\rangle = \frac{1}{2m\omega}. \tag{2.8}$$

Although only ground-state expectation values have been treated thus far, it is also possible to deal with arbitrary oscillator matrix elements with this formalism by generalizing the operator relation

$$|n\rangle = \frac{1}{\sqrt{n!}} \left(a^\dagger \right)^n |0\rangle, \tag{2.9}$$

where

$$a^\dagger = \sqrt{\frac{m\omega}{2}} \left(x - \frac{i}{m\omega} p \right) \tag{2.10}$$

is the usual creation operator. First, however, it is convenient to use the classical relation $p = m\dot{x}$ to rewrite the operator $a^\dagger$ as

$$a^\dagger = \sqrt{\frac{m\omega}{2}} \left(1 - \frac{i}{\omega} \frac{d}{dt} \right) x(t). \tag{2.11}$$

In a simple application, we calculate that

$$\langle 0|x|1\rangle = \lim_{t_2 \to t_1^+} \sqrt{\frac{m\omega}{2}} \left(1 - \frac{i}{\omega}\frac{\partial}{\partial t_1}\right) \langle 0|x(t_2)x(t_1)|0\rangle$$

$$= \lim_{t_2 \to t_1^+} (-i)^2 \sqrt{\frac{m\omega}{2}} \left(1 - \frac{i}{\omega}\frac{\partial}{\partial t_1}\right) \frac{1}{Z[0]} \frac{\delta^2}{\delta j(t_2)\delta j(t_1)} Z[j]$$

$$= \lim_{t_2 \to t_1^+} \sqrt{\frac{m\omega}{2}} \left(1 - \frac{i}{\omega}\frac{\partial}{\partial t_1}\right) \frac{i}{m} D(t_2 - t_1)$$

$$= \frac{1}{\sqrt{2m\omega}}, \tag{2.12}$$

which agrees with the result obtained by more conventional means,

$$\langle 0|x|1\rangle = \sqrt{\frac{1}{2m\omega}} \langle 0| \left(a + a^\dagger\right) |1\rangle = \frac{1}{\sqrt{2m\omega}}. \tag{2.13}$$

More complicated matrix elements can also be found, as with

$$\langle 1|x^2|1\rangle = \frac{m\omega}{2} \lim_{\substack{t_2 \to t^- \\ t_1 \to t^+}} \left(1 + \frac{i}{\omega}\frac{\partial}{\partial t_1}\right) \left(1 - \frac{i}{\omega}\frac{\partial}{\partial t_2}\right) \langle 0|x(t_1)x^2(t)x(t_2)|0\rangle$$

$$= \frac{(-i)^4}{Z[0]} \lim_{\substack{t_2 \to t'^-, t_1 \to t^+ \\ t' \to t^-}} \left(\frac{m\omega}{2}\right) \left(1 + \frac{i}{\omega}\frac{\partial}{\partial t_1}\right) \left(1 - \frac{i}{\omega}\frac{\partial}{\partial t_2}\right)$$

$$\frac{\delta^4}{\delta j(t_1)\delta j(t_2)\delta j(t)\delta j(t')} Z[j]\bigg|_{j=0}$$

$$= \frac{m\omega}{2} \left(\frac{i}{m}\right)^2 \lim_{\substack{t_2 \to t^- \\ t_1 \to t^+}} \left(1 + \frac{i}{\omega}\frac{\partial}{\partial t_1}\right) \left(1 - \frac{i}{\omega}\frac{\partial}{\partial t_2}\right)$$

$$\times \left[D(t_1 - t_2)D(0) + 2D(t_1 - t)D(t - t_2)\right] = \frac{3}{2m\omega}, \tag{2.14}$$

which agrees with

$$\langle 1|x^2|1\rangle = \frac{1}{2m\omega} \langle 1| \left(a + a^\dagger\right)\left(a + a^\dagger\right) |1\rangle = \frac{3}{2m\omega}. \tag{2.15}$$

In this manner, arbitrary oscillator matrix elements can be reduced to ground-state expectation values, which in turn can be determined from the generating functional $Z[j]$. The ground-state amplitude in the presence of an arbitrary source $j(t)$ contains *all* the information about the harmonic oscillator.

One should note the analogy of the above methods to those of quantum field theory. The 'one-particle' matrix elements involving $|1\rangle$ have been reduced to vacuum matrix elements by use of Eq. (2.9). This is similar to the LSZ reduction of fields (see App. B–3). As a result, all that one needs to deal with are the vacuum Green's functions. The generating functional is ideal for this purpose, as we shall see in our development of functional techniques in field theory.

A–3 Field-theoretic formalism

One of the advantages of the functional approach to quantum mechanics is that it can be taken over with little difficulty to quantum field theory. An important difference is that instead of trajectories $x(t)$, which pick out a particular point in space at a given time, one must deal with fields $\varphi(x, t)$ which are defined at *all* points in space at a given time t. Also, instead of a sum $\int \mathcal{D}[x(t)]$ over trajectories one has instead a sum $\int [d\varphi(x)]$ over all possible field configurations. Nevertheless, the analogy is rather direct.

Path integrals with fields

The formal transition from quantum mechanics to field theory can be accomplished by dividing spacetime, both time and space, into a set of tiny four-dimensional cubes of volume $\delta t \, \delta x \, \delta y \, \delta z$. Within each cube one takes the field

$$\varphi\left(x_i, y_j, z_k, t_\ell\right) \tag{3.1}$$

as a constant. Derivatives are defined in terms of differences between fields in neighboring blocks, e.g.,

$$\partial_t \varphi\big|_{x_i, y_j, x_k, t_l} \simeq \frac{1}{\delta t} \left(\varphi\left(x_i, y_j, z_k, t_l + \delta t\right) - \varphi\left(x_i, y_j, z_k, t_l\right)\right). \tag{3.2}$$

The lagrangian is easily found,

$$\mathcal{L}(\varphi, \partial_\mu \varphi)\big|_{x_i, y_j, z_k, t_l} \simeq \mathcal{L}\left(\varphi\left(x_i, y_j, z_k, t_l\right), \partial_\mu \varphi\left(x_i, y_j, z_k, t_l\right)\right), \tag{3.3}$$

and the action is written as

$$S \simeq \sum_{i,j,k,l} \delta x \delta y \delta z \delta t \; \mathcal{L}\left(\varphi\left(x_i, y_j, z_k, t_l\right), \partial_\mu \varphi\left(x_i, y_j, z_k, t_l\right)\right). \tag{3.4}$$

The field-theory analog of the path integral can then be constructed by summing over all possible field values in each cell

$$D \sim \prod_{i,j,k,l} \int_{-\infty}^{\infty} d\varphi\left(x_i, y_j, z_k, t_l\right) e^{iS[\varphi(x_i, y_k, z_k, t_l), \; \partial_\mu \varphi(x_i, y_j, z_k, t_l)]}. \tag{3.5}$$

Formally, in the limit in which the cell size is taken to zero, this is written as

$$\int [d\varphi(x)] \, e^{iS[\varphi(x), \; \partial_\mu \varphi(x)]}. \tag{3.6}$$

By analogy with the quantum mechanical case (cf. Eq. (1.18)), it is clear that, since the time integration for S in Eq. (3.4) is from $-\infty$ to $+\infty$, this amplitude is to be identified with the vacuum-to-vacuum amplitude of the field theory,

$$\langle 0|0 \rangle = N \int [d\varphi(x)] \, e^{iS[\varphi(x), \partial_\mu \varphi(x)]}. \tag{3.7}$$

Generally, quantum field theory is formulated in terms of vacuum expectation values of time-ordered products of the fields

$$G^{(n)}(x_1, \ldots, x_n) = \langle 0|T \, (\varphi(x_1) \cdots \varphi(x_n)) \, |0 \rangle \tag{3.8}$$

i.e., the *Green's functions* of the theory. By analogy with the quantum-mechanical case, one is naturally led to the path-integral definition

$$G^{(n)}(x_1, \ldots, x_n) = N \int [d\varphi(x)] \varphi(x_1) \cdots \varphi(x_n) e^{iS[\varphi(x), \partial_\mu \varphi(x)]}, \tag{3.9}$$

where N is a normalization factor. Again we emphasize that all quantities here are c numbers and no operators are involved. In terms of a functional representation, we then have from Eqs. (3.7), (3.9),

$$G^{(n)}(x_1, \ldots, x_n) = \frac{\int [d\varphi(x)] \, \varphi(x_1) \cdots \varphi(x_n) e^{iS[\varphi(x), \partial_\mu \varphi(x)]}}{\int [d\varphi(x)] \, e^{iS[\varphi(x), \partial_\mu \varphi(x)]}}. \tag{3.10}$$

Generating functional with fields

These Green's functions can most easily be evaluated by use of the generating functional

$$Z[j] = N \int [d\varphi(x)] \, e^{\left(iS[\varphi(x), \partial_\mu \varphi(x)] + i \int d^4x \, j(x)\varphi(x) \right)} \tag{3.11}$$

Functional differentiation for fields is defined by

$$\frac{\delta \varphi(y)}{\delta \varphi(x)} = \delta^{(4)}(x - y), \tag{3.12}$$

which lets us obtain (cf. Eq. (3.9))

$$G^{(n)}(x_1, \ldots, x_n) = (-i)^n \frac{1}{Z[0]} \frac{\delta^n}{\delta j(x_1) \cdots \delta j(x_n)} Z[j] \Big|_{j=0}. \tag{3.13}$$

As an example of this formalism consider the free scalar field theory

$$\mathcal{L}^{(0)}(x) = \frac{1}{2} \partial_\mu \varphi \partial^\mu \varphi - \frac{m^2}{2} \varphi^2. \tag{3.14}$$

In general, we have

$$Z^{(0)}[j] = Z^{(0)}[0] \sum_{n=0}^{\infty} \frac{i^n}{n!} \left[\prod_{k=1}^{n} \int_{-\infty}^{\infty} dx_k \, j(x_k) \right] G^{(n)}(x_1, x_2, \ldots, x_n), \qquad (3.15)$$

where the generating functional $Z^{(0)}[j]$ is given by

$$Z^{(0)}[j] = N \int [d\varphi(x)] \, e^{i \int d^4 x \left(\frac{1}{2} \partial_\mu \varphi \partial^\mu \varphi - \frac{m^2}{2} \varphi^2 + j\varphi \right)}. \qquad (3.16)$$

There exist two common ways in which to handle the issue of convergence for such functional integrals, i.e., to ensure acceptable behavior for large φ^2. One is to give the mass an infinitesimal negative imaginary part, $m^2 \to m^2 - i\epsilon$. This is the approach we shall employ in the discussion to follow. The second involves a continuation to euclidean space by means of $t \to -i\tau$ wherein the functional integral becomes

$$\langle 0|0 \rangle = N \int [d\varphi(x)] e^{-\int d^4 x_E \left(\frac{1}{2} \partial_\mu \varphi \partial_\mu \varphi + \frac{m^2}{2} \varphi^2 \right)}, \qquad (3.17)$$

and is now convergent due to the negative argument of the exponential. Continuation back to Minkowski space then yields the desired result.

Integrating by parts, we have from Eq. (3.16)

$$Z^{(0)}[j] = N \int [d\varphi] \, e^{-i \int d^4 x \left[\frac{1}{2} \varphi(x) O_x \varphi(x) - \varphi(x) j(x) \right]}$$

$$= N \int [d\varphi'] \, e^{-\frac{i}{2} \left[\int d^4 x \, \varphi'(x) O_x \varphi'(x) + \int d^4 x \int d^4 y \, j(x) \Delta_F(x-y) j(y) \right]} \qquad (3.18)$$

where $O_x = \Box_x + m^2 - i\epsilon$ and

$$\varphi'(x) = \varphi(x) + \int d^4 y \, \Delta_F(x - y) j(y),$$

$$i \Delta_F(x - y) = \int \frac{d^4 k}{(2\pi)^4} \, e^{-ik \cdot (x-y)} \frac{i}{k^2 - m^2 + i\epsilon},$$

$$(\Box_x + m^2) \Delta_F(x - y) = -\delta^{(4)}(x - y). \qquad (3.19)$$

Note that we have used invariance of the measure $\left(\int [d\varphi] = \int [d\varphi'] \right)$. Finally, we recognize a factor of $Z^{(0)}[0]$ in Eq. (3.18), thus leading to the expression

$$Z^{(0)}[j] = Z^{(0)}[0] e^{-\frac{i}{2} \int d^4 x \int d^4 y \, j(x) \Delta_F(x-y) j(y)}. \qquad (3.20)$$

We can now determine the Green's functions for the free field theory, e.g.,

$$G^{(2)}(x_1, x_2) = \frac{(-i)^2}{Z^{(0)}[0]} \frac{\delta^2}{\delta j(x_1)\delta j(x_2)} Z^{(0)}[j]\Big|_{j=0} = i\Delta_F(x_1 - x_2),$$

$$G^{(4)}(x_1, x_2, x_3, x_4) = \frac{(-i)^4}{Z^{(0)}[0]} \frac{\delta^4}{\delta j(x_1)\delta j(x_2)\delta j(x_3)\delta j(x_4)} Z^{(0)}[j]\Big|_{j=0}$$

$$= G^{(2)}(x_1, x_2)G^{(2)}(x_3, x_4) + G^{(2)}(x_1, x_3)G^{(2)}(x_2, x_4)$$

$$+ G^{(2)}(x_1, x_4)G^{(2)}(x_2, x_3). \tag{3.21}$$

More interesting is the case of a self-interacting field theory for which the lagrangian becomes

$$\mathcal{L}(x) = \frac{1}{2}\partial_\mu\varphi\partial^\mu\varphi - \frac{1}{2}m^2\varphi^2 + \mathcal{L}_{\text{int}}(\varphi) \equiv \mathcal{L}^{(0)}(\varphi) + \mathcal{L}_{\text{int}}(\varphi). \tag{3.22}$$

The theory is no longer exactly soluble, but one can find a perturbative solution by use of the generating functional

$$Z[j] = N \int [d\varphi(x)] \, e^{i\int d^4x \,(\mathcal{L}^{(0)}(\varphi)+\mathcal{L}_{\text{int}}(\varphi)+j(x)\varphi(x))}$$

$$= N e^{i\int d^4x \,\mathcal{L}_{\text{int}}\left(-i\frac{\delta}{\delta j(x)}\right)} Z^{(0)}[j]. \tag{3.23}$$

As before, the Green's functions of the theory are given by

$$G^{(n)}(x_1, \ldots, x_n) = \frac{1}{Z[0]} \left[\prod_{k=1}^{n} \frac{-i\delta}{\delta j(x_k)}\right] e^{i\int d^4x \,\mathcal{L}_{\text{int}}\left(-i\frac{\delta}{\delta j(x)}\right)} Z^{(0)}[j]\Big|_{j=0}. \tag{3.24}$$

For most purposes one requires only the *connected* portions of the Green's function, i.e., those diagrams which cannot be broken into two or more disjoint pieces. This is illustrated in Fig. A–1 which can be found by dividing the full Green's function

$$G^{(n)}(x_1, \ldots, x_n) = \langle 0|T(\varphi(x_1)\cdots\varphi(x_n))|0\rangle \tag{3.25}$$

into products of connected particle sectors and dividing by the vacuum-to-vacuum amplitude $\langle 0|0\rangle$ in each sector.

Mathematically, one eliminates the disconnected diagrams by defining

$$Z[j] = e^{iW[j]}. \tag{3.26}$$

Then one can show that $W[j]$ is the generating functional for connected Green's functions,

$$iW[j] = \sum_{n=0}^{\infty} \frac{i^n}{n!} \int_{-\infty}^{\infty} dx_1 \cdots \int_{-\infty}^{\infty} dx_n \, j(x_1)\cdots j(x_n) G^{(n)}_{\text{conn}}(x_1, \ldots, x_n), \tag{3.27}$$

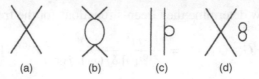

(a) (b) (c) (d)

Fig. A–1 Contributions to the four-point Green's function in φ^4 theory: (a)–(b) connected, (c)–(d) disconnected.

where

$$G_{\text{conn}}^{(n)}(x_1, \ldots, x_n) = (-i)^{n-1} \frac{\delta^n}{\delta j(x_1) \cdots \delta j(x_n)} W[j]\Big|_{j=0}. \tag{3.28}$$

A–4 Quadratic forms

The most important example of a soluble path integral is one that is quadratic in the fields because, at least formally, it can be solved exactly.

Let us consider an action quadratic in the fields,

$$S = -\int d^4x \, \varphi(x) O \varphi(x), \tag{4.1}$$

where O is some differential operator which may contain fields distinct from φ within it. The general result for the quadratic path integral is given by

$$I_{\text{quad}} = \int [d\varphi(x)] e^{-i \int d^4x \, \varphi(x) O \varphi(x)} = N[\det O]^{-1/2}, \tag{4.2}$$

where $\det O$ is the determinant of the operator O. In order to prove this, one can expand $\varphi(x)$ in terms of eigenfunctions of O,

$$\varphi(x) = \sum_n a_n \varphi_n(x), \tag{4.3}$$

where $\varphi_n(x)$ satisfies

$$O\varphi_n(x) = \lambda_n \varphi_n(x) \quad \text{and} \quad \int d^4x \, \varphi_n(x)\varphi_m(x) = \delta_{nm}. \tag{4.4}$$

The sum over all field values can then be performed by summing over all values of the expansion coefficients a_n,

$$I_{\text{quad}} = N \left[\prod_n \int_{-\infty}^{\infty} da_n \right] e^{-i \int d^4x \sum_{k=1}^{\infty} a_k \varphi_k(x) \sum_{l=1}^{\infty} a_l \varphi_\ell(x) \lambda_l}$$

$$= N \prod_n \int_{-\infty}^{\infty} da_n \, e^{-i\lambda_n a_n^2} = N' (\det O)^{-1/2}, \tag{4.5}$$

where N, N' are normalization constants and

$$\det O = \prod_{n=1}^{\infty} \lambda_n \tag{4.6}$$

denotes, as usual, the product of operator eigenvalues.

In general, some effort is required to evaluate the determinant of an operator. One valuable relation, easily proven for finite dimensional matrices and generalizable to infinite dimensional ones is[4]

$$\det O = \exp(\operatorname{tr} \ln O). \tag{4.7}$$

This trace now denotes a summation over spacetime points, i.e.,

$$\operatorname{tr} \ln O = \int d^4x \, \langle x \mid \ln O \mid x \rangle, \tag{4.8}$$

which is the most commonly used form in practice.

Background field method to one loop

We can illustrate one use of this result by constructing an expansion about a background field configuration (which satisfies the classical equation of motion) and retaining the quantum fluctuations up to quadratic order. Consider a scalar field theory with interaction $\mathcal{L}_{\text{int}}(\varphi(x))$. We define $\bar{\varphi}$ as a solution to

$$\left(\Box + m^2\right)\bar{\varphi}(x) - \mathcal{L}'_{\text{int}}(\bar{\varphi}(x)) = j(x). \tag{4.9}$$

Writing

$$\varphi(x) = \bar{\varphi}(x) + \delta\varphi(x), \tag{4.10}$$

leads to the generating functional

$$Z[j] = e^{\left(iS[\bar{\varphi}(x)] + i\int d^4x\, j(x)\bar{\varphi}(x)\right)}$$
$$\int [d\delta\varphi]\, e^{i\int d^4x \left(\frac{1}{2}\partial_\mu\delta\varphi\partial^\mu\delta\varphi - \frac{1}{2}\left(m^2 - \mathcal{L}''_{\text{int}}(\bar{\varphi}(x))\right)\delta\varphi^2\right)} + \cdots, \tag{4.11}$$

where

$$S[\bar{\varphi}(x)] = \int d^4x \left(\frac{1}{2}\partial_\mu\bar{\varphi}(x)\partial^\mu\bar{\varphi}(x) - \frac{m^2}{2}\bar{\varphi}^2(x) + \mathcal{L}_{\text{int}}(\bar{\varphi}(x))\right). \tag{4.12}$$

---

[4] For a discrete basis, this follows from the result

$$\exp(\operatorname{tr} \ln O) = \exp\sum_n \ln \lambda_n = \prod_n \exp(\ln \lambda_n) = \prod_n \lambda_n = \det O,$$

where λ_n are the eigenvalues of the operator O.

Integration by parts gives

$$Z[j] = e^{\left(iS[\bar{\varphi}(x)] + i \int d^4x \, \bar{\varphi}(x)j(x)\right)} \int [d\delta\varphi] \, e^{-\frac{i}{2} \int d^4x \, \delta\varphi(x) O_x \delta\varphi(x)}, \qquad (4.13)$$

where

$$O_x \equiv \Box_x + m^2 - \mathcal{L}''_{\text{int}}(\bar{\varphi}(x)). \qquad (4.14)$$

The functional integration can then be performed (cf. Eq. (4.5)) and we obtain

$$Z[j] = \text{const.} \, (\det O_x)^{-1/2} \, e^{\left(iS[\bar{\varphi}(x)] + i \int d^4x \, j(x)\bar{\varphi}(x)\right)}. \qquad (4.15)$$

It is convenient to normalize the determinant somewhat differently by defining

$$O_{0x} \equiv \Box_x + m^2. \qquad (4.16)$$

Then, suppressing the x subscript, we write

$$(\det O)^{-1/2} = \text{const.} \, \left(\det O_0^{-1} O\right)^{-1/2}, \qquad (4.17)$$

where

$$\text{const.} = (\det O_0)^{-1/2}, \qquad (4.18)$$

and

$$O_0^{-1} O = 1 + \Delta_F \mathcal{L}''_{\text{int}}(\varphi). \qquad (4.19)$$

Using Eq. (4.2) we have

$$Z[j] = N e^{\left[iS[\bar{\varphi}(x)] + i \int d^4x \, j(x)\bar{\varphi}(x) - \frac{1}{2} \text{Tr} \ln\left(1 + \Delta_F \mathcal{L}''(\bar{\varphi})\right)\right]}. \qquad (4.20)$$

The generating functional for connected diagrams can now be identified immediately as

$$\begin{aligned}
W[j] &= S[\bar{\varphi}] + \int d^4x \, j(x)\bar{\varphi}(x) + \frac{i}{2} \text{Tr} \ln\left(1 + \Delta_F \mathcal{L}''_{\text{int}}(\bar{\varphi})\right) \\
&= \int d^4x \left[\frac{1}{2} j(x)\bar{\varphi}(x) + \mathcal{L}_{\text{int}}(\bar{\varphi}) - \frac{1}{2}\bar{\varphi}(x)\mathcal{L}'_{\text{int}}(\bar{\varphi})\right] \\
&\quad + \frac{i}{2} \text{Tr} \ln\left(1 + \Delta_F \mathcal{L}''_{\text{int}}(\bar{\varphi})\right).
\end{aligned} \qquad (4.21)$$

The trace 'Tr' includes the integration over spacetime variables and can be interpreted as follows,

$$\mathrm{Tr}\ln\left[1+\Delta_F\mathcal{L}''_{\mathrm{int}}\left(\bar\varphi\right)\right]=\mathrm{Tr}\sum_{n=1}^{\infty}\frac{(-)^{n+1}}{n}\left(\Delta_F\mathcal{L}''_{\mathrm{int}}\left(\bar\varphi\right)\right)^n,$$

$$\mathrm{Tr}\left[\Delta_F\mathcal{L}''_{\mathrm{int}}\left(\bar\varphi\right)\right]=\int d^4x\,\Delta_F(x-x)\mathcal{L}''_{\mathrm{int}}\left(\bar\varphi\right),$$

$$\mathrm{Tr}\left[\Delta_F\mathcal{L}''_{\mathrm{int}}\left(\bar\varphi\right)\Delta_F\mathcal{L}''_{\mathrm{int}}\left(\bar\varphi\right)\right]=\int d^4x\int d^4y\,\Delta_F(x-y)\mathcal{L}''_{\mathrm{int}}\left(\bar\varphi(y)\right)$$

$$\times\Delta_F(y-x)\mathcal{L}''_{\mathrm{int}}\left(\bar\varphi(x)\right). \tag{4.22}$$

In this manner, one-loop diagrams containing arbitrary numbers of $\mathcal{L}''_{\mathrm{int}}\left(\bar\varphi\right)$ factors are generated. The physics associated with this approximation can be gleaned from counting arguments. The overall power of $\hbar$ attached to a particular diagram can be found by noting that associated with a propagator and a vertex are the powers $\hbar$ and $\hbar^{-1}$, respectively. There is also an overall factor of $\hbar$ for each diagram. Then with the relation

no. internal lines $-$ no. internal vertices $=$ no. loops -1

we see that this approximation corresponds to an expansion to one loop. The classical phase generates the tree diagram $\left(\mathcal{O}(\hbar^0)\right)$ contribution and the determinant yields the one-loop $\left(\mathcal{O}(\hbar^1)\right)$ correction to a given amplitude.

A–5 Fermion field theory

Thus far, our development has been performed within the simple context of scalar fields. It is important also to consider the case of fermion fields where the requirements of antisymmetry impose interesting modifications on functional integration techniques. The key to the treatment of anticommuting fields is the use of Grassmann variables. Thus, while ordinary c-number quantities (hereafter denoted by roman letters $a,b,\ldots$) commute with one another,

$$[a,a]=[a,b]=[a,c]=\cdots=0, \tag{5.1}$$

the Grassmann numbers (hereafter denoted by Greek letters $\alpha,\beta,\ldots$) anticommute,

$$\{\alpha,\alpha\}=\{\alpha,\beta\}=\{\alpha,\gamma\}=\cdots=0. \tag{5.2}$$

It follows that the square of a Grassmann quantity must vanish,

$$\alpha^2=\beta^2=\gamma^2=\cdots=0, \tag{5.3}$$

and that any function must have the general expansion

$$f(\alpha)=f_0+f_1\alpha\,,\qquad g(\alpha,\beta)=g_0+g_1\alpha+g_2\beta+g_3\alpha\beta. \tag{5.4}$$

Differentiation is defined correspondingly via

$$\frac{d\alpha}{d\alpha} = \frac{d\beta}{d\beta} = \cdots = 1, \qquad \frac{d\beta}{d\alpha} = \frac{d\alpha}{d\beta} = \cdots = 0, \tag{5.5}$$

so that in the notation of Eq. (5.4) we have

$$\frac{df}{d\alpha}(\alpha) = f_1, \qquad \frac{dg}{d\beta}(\alpha, \beta) = g_2 - g_3\alpha. \tag{5.6}$$

Second derivatives then have the property

$$\frac{d^2}{d\alpha d\alpha} = 0. \tag{5.7}$$

We must also define the concept of Grassmann integration. If we demand that integration have the property of translation invariance

$$\int d\alpha \, f(\alpha) = \int d\alpha \, f(\alpha + \beta), \tag{5.8}$$

it follows that

$$\int d\alpha \, f_1\beta = 0 \quad \text{or} \quad \int d\alpha = 0. \tag{5.9}$$

The normalization in the diagonal integral can be chosen for convenience,

$$\int d\alpha \, \alpha = 1, \qquad \int d\alpha \, f(\alpha) = f_1. \tag{5.10}$$

Let us extend this formalism to a matrix notation by considering the discrete sets $\alpha = \{\alpha_1, \ldots, \alpha_n\}$ and $\bar{\alpha} = \{\bar{\alpha}_1, \ldots, \bar{\alpha}_n\}$ of Grassmann variables. A class of integrals which commonly arises in a functional framework is

$$Z[M] = \int d\bar{\alpha}_n \cdots d\bar{\alpha}_1 \, d\alpha_n \cdots d\alpha_1 e^{i\bar{\alpha}M\alpha}. \tag{5.11}$$

As an example, the simple 2×2 case is calculated to be

$$Z[M] = \int d\bar{\alpha}_2 \, d\bar{\alpha}_1 \, d\alpha_2 \, d\alpha_1 \left[1 + i\bar{\alpha}_i M_{ij}\alpha_j \right. $$
$$\left. + \bar{\alpha}_2\bar{\alpha}_1\alpha_2\alpha_1 \left(M_{11}M_{22} - M_{12}M_{21} \right) \right]. \tag{5.12}$$

Only the final term survives the integration, and we obtain

$$Z[M] = \det M. \tag{5.13}$$

This result generalizes to the $n \times n$ system [Le 82] yielding essentially the inverse of the result found for Bose fields,

$$Z[M]_{\text{Fermi}} = \int d\bar{\alpha}_n \cdots d\bar{\alpha}_1 \, d\alpha_n \cdots d\alpha_1 e^{i\bar{\alpha}M\alpha} = \det M,$$

$$Z[M]_{\text{Bose}} = \int da_n^* \cdots da_1^* \, da_n \cdots da_1 e^{-a^*Ma} \propto (\det M)^{-1}. \tag{5.14}$$

We can now extend this formalism to the case of fermion *fields* $\psi(x)$ and $\bar{\psi}(x)$. Since such quantities always enter the lagrangian quadratically, the functional integral can be performed exactly to yield

$$Z[O] = \int [d\psi][d\bar{\psi}] e^{i \int d^4x \, \bar{\psi}(x) O \psi(x)} = N \det O. \tag{5.15}$$

The remaining development proceeds parallel to that given for scalar fields. Given the free field lagrangian

$$\mathcal{L}_0(\bar{\psi}, \psi) = \bar{\psi}(x)(i\not{\partial} - m)\psi(x), \tag{5.16}$$

the generating functional for the noninteracting spin one-half field becomes

$$Z[\eta, \bar{\eta}] = \int [d\psi][d\bar{\psi}] e^{i \int d^4x [\bar{\psi}(x) O_x \psi(x) + \bar{\eta}(x)\psi(x) + \bar{\psi}(x)\eta(x)]}, \tag{5.17}$$

where $O_x \equiv i\not{\partial}_x - m + i\epsilon$ and $\bar{\eta}(x), \eta(x)$ are Grassmann fields. Introducing the change of variables

$$\psi'(x) = \psi(x) - \int d^4y \, S_F(x, y)\eta(y),$$

$$\bar{\psi}'(x) = \bar{\psi}(x) - \int d^4y \, \bar{\eta}(y)S_F(y, x),$$

$$iS_F(x - y) = \int \frac{d^4k}{(2\pi)^4} e^{-ik\cdot(x-y)} \frac{i}{\not{k} - m + i\epsilon}$$

$$(i\not{\partial}_x - m)S_F(x - y) = \delta^{(4)}(x - y), \tag{5.18}$$

we find that an alternative form for the generating functional is

$$Z[\eta, \bar{\eta}] = \int [d\psi'][d\bar{\psi}'] e^{i \int d^4x \, \bar{\psi}'(x) O_x \psi'(x) - i \int d^4x \int d^4y \, \bar{\eta}(x) S_F(x,y)\eta(x)},$$

$$= Z[0, 0] e^{-i \int d^4x \int d^4y \, \bar{\eta}(x) S_F(x,y)\eta(x)}. \tag{5.19}$$

Thus, the generating functional for connected diagrams is

$$W[\eta, \bar{\eta}] = -\int d^4x \int d^4y \, \bar{\eta}(x) S_F(x, y)\eta(y), \tag{5.20}$$

and the only nonvanishing connected Green's function is

$$G^{(2)}_{\text{conn}}(x_1, x_2) = (-i)^2 \frac{\delta^2 W}{\delta \eta(x_2) \delta \bar{\eta}(x_1)}$$

$$= S_F(x_1, x_2) = \int \frac{d^4k}{(2\pi)^4} e^{-ik \cdot (x_1 - x_2)} \frac{i}{\not{k} - m + i\epsilon}, \qquad (5.21)$$

which is the usual Feynman propagator.

A–6 Gauge theories

For our final topic, we examine gauge theories within a functional framework. We shall employ *QED* as the archetypical example, for which the action is

$$S[A_\mu] = -\frac{1}{4} \int d^4x \, F_{\mu\nu} F^{\mu\nu} = \frac{1}{2} \int d^4x \, (A_\mu \Box_x A^\mu - A_\mu \partial_x^\mu \partial_x^\nu A_\nu)$$

$$\equiv \frac{1}{2} \int d^4x \, A_\mu O_x^{\mu\nu} A_\nu, \qquad (6.1)$$

where the second line follows from the first by an integration by parts and

$$O_x^{\mu\nu} \equiv g^{\mu\nu} \Box_x - \partial_x^\mu \partial_x^\nu. \qquad (6.2)$$

In the presence of a source j_μ, the generating functional is then

$$Z[j_\mu] = N \int [dA_\mu] e^{iS[A_\mu] + i \int d^4x \, j_\mu A^\mu}. \qquad (6.3)$$

Due to the bilinear form of Eq. (6.1), it would appear that one could perform the functional integration as usual, resulting in

$$Z[j_\mu] = Z[0] e^{-\frac{i}{2} \int d^4x \int d^4y \, j^\mu(x) D_{F\mu\nu}(x,y) j^\nu(y)}, \qquad (6.4)$$

where the inverse operator $D_{F\mu\nu}(x, y)$ is defined as

$$O_x^{\lambda\mu} D_{F\mu\nu}(x, y) \equiv \delta_\nu^\lambda \delta^{(4)}(x - y). \qquad (6.5)$$

However, this is illusory since the inverse does not exist. That is, acting on Eq. (6.5) from the left with the derivative ∂_λ^x yields

$$0 \times D_{F\mu\nu}(x, y) = \partial_\nu^x \delta^{(4)}(x - y), \qquad (6.6)$$

implying that $D_{F\mu\nu}$ must be infinite. An alternative way to demonstrate that $O_x^{\mu\nu}$ is a singular operator is to observe that

$$O_x^{\mu\nu} \partial_\nu^x \alpha = 0. \qquad (6.7)$$

Thus, any four-gradient $\partial_\nu^x \alpha$ is an eigenfunction of $O_x^{\mu\nu}$ having eigenvalue zero, and an operator having zero eigenvalues does not possess an inverse.

Gauge fixing

The occurrence of such a divergence in the generating functional of a gauge theory can be traced to gauge invariance. For *QED*, any gauge transformation of vector potentials (cf. Eq. (II–1.3)),

$$A_\mu(x) \to A'_\mu(x) = A_\mu(x) + \frac{1}{e}\partial_\mu \alpha(x), \tag{6.8}$$

leaves the action invariant,

$$S[A'_\mu(x)] = S[A_\mu(x)]. \tag{6.9}$$

If we partition the full field integration $[dA_\mu]$ into a component $[d\bar{A}_\mu]$ that includes only those configurations which are *not* related by a gauge transformation and a component $[d\alpha]$ that denotes all possible gauge transformations, then we have

$$\int [dA_\mu]\, e^{iS[A_\mu]} = \int [d\bar{A}_\mu]\, e^{iS[\bar{A}_\mu]} \times \int [d\alpha]. \tag{6.10}$$

But $\int [d\alpha]$ is clearly infinite and this is the origin of the problem. The solution, first given by Faddeev and Popov [FaP 67], involves finding a procedure which somehow isolates the integration over the distinctly different vector potentials $\bar{A}_\mu(x)$. In order to understand this technique, we shall first examine a finite-dimensional analog [Ra 89].

Consider the functional

$$Z[\mathbf{A}] = \left[\prod_{i=1}^{N} \int_{-\infty}^{\infty} dx_i \right] e^{-\sum_{k,l} x_k A_{kl} x_l}, \tag{6.11}$$

where $\mathbf{A}$ is an $N \times N$ matrix. Suppose that $\mathbf{A}$ is brought into diagonal form $\mathbf{A}^D$ by linear transformation $\mathbf{R}$,

$$\mathbf{A}^D \equiv \mathbf{R}\mathbf{A}\mathbf{R}^{-1}. \tag{6.12}$$

Letting $\vec{y} = \mathbf{R}\vec{x}$ denote the coordinates in the diagonal basis, we have

$$Z[\mathbf{A}] = \left[\prod_{i=1}^{N} \int dy_i \right] e^{-\sum_{k,l} y_k A_{kl}^D y_l} = \left[\prod_{i=1}^{N} \int dy_i \right] e^{-\sum_k y_k^2 A_{kk}^D}$$

$$= \prod_{i=1}^{N} \left(\frac{\pi}{A_{ii}^D} \right)^{1/2} = \pi^{N/2} [\det \mathbf{A}]^{-1/2}. \tag{6.13}$$

Suppose that the last n of the N eigenvalues belonging to $\mathbf{A}$ vanish. The exponential factor in Eq. (6.13) is then *independent* of the coordinates $y_{N-n+1}, \ldots, y_N$ and the corresponding integrations $\int dy_{N-n+1} \ldots \int dy_N$ diverge. This is reflected in the vanishing of $\det \mathbf{A}$, and causes the quantities in Eq. (6.13) to diverge. The infinity is removed if the integration is restricted to only variables associated with nonzero eigenvalues, in which case we obtain the finite result

$$Z^f[\mathbf{A}] = \left[\prod_{i=1}^{N-n} \int dy_i \right] e^{-\sum_{k,l} y_k A_{kl}^D y_l}. \tag{6.14}$$

It is possible to express $Z^f[\mathbf{A}]$ as an integral over the *full* range of indices $1 \le i \le N$ by defining variables

$$z_i = \begin{cases} y_i & (1 \le i \le N - n), \\ \text{arbitrary} & (N - n + 1 \le i \le N), \end{cases} \tag{6.15}$$

and writing for the generating functional

$$Z^f[\mathbf{A}] = \left[\prod_{i=1}^{N} \int dz_i \right] \delta(z_{N-n+1}) \cdots \delta(z_N) e^{-\sum_{k,l} z_k(x) A_{kl} z_l(x)}. \tag{6.16}$$

Upon tranforming back to an arbitrary set of coordinates $\{x_i\}$, we obtain the useful expression

$$Z^f[\mathbf{A}] = \left[\prod_{i=1}^{N} \int dx_i \right] \det \left| \frac{\partial \vec{z}}{\partial \vec{x}} \right| \prod_{j=N-n+1}^{N} \delta\left(z_j(\vec{x}) \right) e^{-\sum_{k,l} x_k A_{kl} x_l}. \tag{6.17}$$

Let us now return to the subject of gauge fields, broadening the scope of our discussion to include even nonabelian gauge theories. By analogy, corresponding to the variables $z_{N-n+1}, \ldots, z_N$ will be the gauge degrees of freedom and the prescription of Faddeev and Popov becomes for generic gauge fields $A_\mu^a(x)$,

$$Z^f = \prod_a \int [dA_\mu^a] \prod_{b=1}^{n} \delta(G_b(A_\mu^a)) \det |\delta G_b / \delta \alpha_a| \, e^{iS[A_\mu^a]}, \tag{6.18}$$

where the $\{\alpha_a\}$ are gauge-transformation parameters (cf. Sect. I–4) and the $\{G_b(A_\mu^a)\}$ are functions which vanish for some value of $A_\mu^a(x)$. Since the $\{G_b\}$ serve to define the gauge, such contributions to the generating functional are referred to as *gauge-fixing* terms. The variation $\delta G_b / \delta \alpha_a$ signifies the response of the gauge-fixing function G_b to a gauge-transformation parameter α_a.

For any gauge theory, there are a variety of choices possible for the gauge-fixing function G. In *QED*, one defines the *axial gauge* by

$$G(A_\mu) = n_\mu A^\mu, \tag{6.19}$$

where n_μ is an arbitrary spacelike four-vector. Due to the presence of the four-vector n^μ, one must forgo manifestly covariant Feynman rules in this approach. Thus, one often employs a *covariant* gauge-fixing condition such as

$$G(A_\mu) = \partial^\mu A_\mu - F, \tag{6.20}$$

where F is an arbitrary constant. Under the gauge transformation of Eq. (6.8), we find

$$G(A_\mu) \to G(A_\mu) + \Box\alpha, \tag{6.21}$$

so that

$$\delta G/\delta\alpha = \Box. \tag{6.22}$$

Referring back to the general formula of Eq. (6.18), we see in this case that $\det|\delta G/\delta\alpha|$ is independent of the gauge field and thus may be dropped from the functional integral. The *QED* generating functional then becomes

$$
\begin{aligned}
Z[j_\mu] &= N \int [dA_\mu]\, \delta(\partial^\mu A_\mu - F)\, e^{i \int d^4x \, \left(\frac{1}{2} A_\mu \Box_x A^\mu - \frac{1}{2} A_\mu \partial^\mu_x \partial^\nu_x A_\nu + j_\mu A^\mu\right)} \\
&= N \int [dA_\mu] e^{i \int d^4x \, \left(\frac{1}{2} A_\mu \Box_x A^\mu + j_\mu A^\mu\right)} \\
&= Z[0] e^{-\frac{i}{2} \int d^4x \int d^4y \, j_\lambda(x) D_F^{\lambda\nu}(x,y) j_\nu(y)}.
\end{aligned}
\tag{6.23}
$$

Note that, as promised, this result is finite and leads to a photon propagator in Feynman gauge

$$D_F^{\nu\lambda}(x, y) = \frac{1}{Z[0]} \frac{\delta^2 Z[j_\mu]}{\delta j_\nu(x)\delta j_\lambda(y)}\Big|_{j_\mu=0} = -i \int \frac{d^4q}{(2\pi)^4} e^{-iq\cdot(x-y)} \frac{g^{\nu\lambda}}{q^2 + i\epsilon}. \tag{6.24}$$

The result is independent of the choice of F. Consequently, even if the constant F is evaluated to the status of a field $F(x)$, one can functionally integrate over $F(x)$ with an arbitrary weighting factor since this will only affect the overall normalization of the generating functional. A common choice is

$$\int [dF]\, \delta(\partial^\mu A_\mu - F(x)) e^{-\frac{i}{2\xi} \int d^4x \, F^2(x)} = e^{-\frac{i}{2\xi} \int d^4x \, (\partial^\mu A_\mu)^2}, \tag{6.25}$$

where ξ is a real-valued parameter. In this case, the generating functional becomes

$$Z[j_\mu] = N \int [dA_\mu]\, e^{i \int d^4x \left(\frac{1}{2} A_\mu (\Box g^{\mu\nu} - \partial^\mu \partial^\nu) A_\nu - \frac{1}{2\xi} (\partial^\mu A_\mu)^2 + j_\mu A^\mu\right)}. \tag{6.26}$$

The integrand of the above spacetime integral can be regarded as the effective lagrangian of the theory, and the gauge-fixing term appears as one of its contributions. At this point, the functional integration can be carried out with impunity to obtain

$$Z[j_\mu] = Z[0] e^{-\frac{i}{2} \int d^4x \int d^4y \, j_\mu(x) D_F^{\mu\nu}(x,y) j_\nu(y)}, \tag{6.27}$$

where $D_F^{\mu\nu}$ is defined as

$$\left(\Box_x g^{\mu\nu} - (1 - \xi^{-1})\partial_x^\mu \partial_x^\nu\right) D_{F\nu\lambda}(x - y) = \delta_\lambda^\mu \delta^{(4)}(x - y). \tag{6.28}$$

We find in this way the form of the photon propagator in an arbitrary gauge, as appearing in Eq. (II–1.17).

Ghost fields

In the path-integral formalism, if the generating functional can be written in purely exponential form, then one can read off the lagrangian of the theory from the exponent. However, the general formula in Eq. (6.18) for a gauge-fixed generating functional contains a seemingly nonexponential factor, the determinant factor $\det|\delta G_b/\delta\alpha_a|$. A fruitful procedure, due to Faddeev and Popov, for expressing the determinant as an exponential factor is motivated by the identity (cf. Eq. (5.15)),

$$\det M = N \int [dc][d\bar{c}]\, e^{i\bar{c}Mc}, \tag{6.29}$$

where $c, \bar{c}$ are Grassmann fields. This identity suggests that we replace the determinant factor with an appropriate functional integration over Grassmann variables. For QED, the generating functional can then be written in the concise form

$$Z[j_\mu] = N \int [dA_\mu][dc][d\bar{c}]\, e^{i\int d^4x\,\left(A_\mu(\Box g^{\mu\nu} - \partial^\mu \partial^\nu)A_\nu - \frac{1}{2\xi}(\partial^\mu A_\mu)^2 + \bar{c}\Box_x c + j_\mu A^\mu\right)}. \tag{6.30}$$

As pointed out earlier, for this case the integration over $c, \bar{c}$ yields only an unimportant constant and may be discarded. However, for nonabelian gauge theory Eq. (6.30) generalizes to

$$Z[j_\mu^a] = \int \prod_{a,b,d} [dA_\mu^a][dc^b][d\bar{c}^d]\, e^{i\int d^4x\,\left[\mathcal{L}[A_\mu^a] + j_\mu^a A_a^\mu + \bar{c}^b M_{be} c^e - \frac{1}{2\xi}\sum_b F_b^2(A_\mu^a)\right]}, \tag{6.31}$$

where repeated indices are summed over. The quantities

$$M_{be} \equiv \frac{\delta F_b(A_\mu^a)}{\delta\alpha^e} \tag{6.32}$$

will generally depend upon the fields A_μ^a themselves. Thus, the fields $\{c_a\}$, $\{\bar{c}_a\}$ will appear as degrees of freedom in the defining lagrangian of the theory. However, although coupled to the gauge fields A_μ^a through $\bar{c}\mathbf{M}c$, they do not interact with any source terms and therefore can only appear in closed loops inside more complex diagrams.[5] Since these Grassmann quantities are unphysical, they are

---

[5] Such loops must include a multiplicative factor of -1 to account for the anticommuting nature of these variables.

often called *Faddeev–Popov ghost fields*. They are scalar, anticommuting variables, which transform as members of the regular representation of the gauge group, e.g., for the gauge group $SU(n)$, there are $n^2 - 1$ of the $\{c_a\}$ and $\{\bar{c}_a\}$ fields.

To complete the discussion, let us determine the ghost-field contribution to the QCD lagrangian. We choose $F_b = \partial_\mu A_b^\mu$ and note the form of a gauge transformation (cf. Eqs. (I–5.12), (I–5.17) with α_a infinitesimal),

$$A_b^\mu \to A_b^{\prime\mu} = A_b^\mu + \frac{1}{g_3}\partial^\mu\alpha_b - f_{bae}A_a^\mu\alpha_e. \tag{6.33}$$

Then we find from a direct evaluation of $\partial F_b/\partial\alpha_e$ followed by the rescaling $-g_3^{-1}\bar{c}c \to \bar{c}c$,

$$\mathcal{L}_{\text{gh}} = -\bar{c}_b\partial_\nu[\delta_{be}\partial^\nu - g_3 f_{bae}A_a^\nu]c_e. \tag{6.34}$$

Upon performing an integration by parts in the first term and relabeling the indices in the second, we obtain the ghost contribution to the QCD lagrangian of Eq. (II–2.25).

Problems

(1) The van Vleck determinant

The semiclassical approximation to the propagator (valid as $\hbar \to 0$) can be derived by expanding about the classical path. Writing

$$x(t) = x_{\text{cl}}(t) + \delta x(t),$$

we have

$$D(x_f, t_f; x_i, t_i) = e^{iS[x_{\text{cl}}(t)]}\int \mathcal{D}[\delta x(t)]e^{\frac{i}{2}\int dt dt'\,\delta x(t)\frac{\delta^2 S}{\delta x(t)\delta x(t')}\delta x(t')},$$

where

$$\frac{\delta^2 S}{\delta x(t)\delta x(t')} = -\left(\frac{\partial^2}{\partial t^2} + \frac{\partial^2 V[x_{\text{cl}}(t)]}{\partial x_{\text{cl}}^2(t)}\right)\delta(t - t')$$

and we have dropped the term linear in $\delta x(t)$ by Hamilton's condition. Performing the path integration we have then

$$D(x_f, t_f; x_i, t_i) = N\left[\det\frac{\delta^2 S}{\delta x(t)\delta x(t')}\right]^{-1/2}e^{iS[x_{\text{cl}}(t)]},$$

where N is a normalization constant and the quantity inside the square root is called the *van Vleck determinant*.

(a) Show that this can be written in the form

$$N\left[\frac{\delta^2 S}{\delta x(t)\delta x(t')}\right]^{-\frac{1}{2}} = \left[\frac{1}{2\pi i}\frac{\partial^2 S[x_{\mathrm{cl}}(t)]}{\partial x_f \partial x_i}\right]^{1/2}.$$

Hint: The following argument is hardly rigorous but leads to the correct answer. Write

$$D(x_f, t_f; x_i, t_i) \equiv A(x_f, x_i; t_f - t_i)e^{i S_{\mathrm{cl}}(x_f, x_i; t_f - t_i)}$$

and use completeness to show that at equal times

$$\delta(x_f - x_i)D(x_f, t_i; x_i, t_i)$$
$$= \int dx\, A(x_f, x; T)A^*(x_i, x; T)e^{i(S_{\mathrm{cl}}(x_f, x; T) - S_{\mathrm{cl}}(x_i, x; T))},$$

where T is an arbitrary positive time. Now define $\rho(x_i, x; T) \equiv \partial S_{\mathrm{cl}}(x_i, x; T)/\partial x_i$ so that

$$S_{\mathrm{cl}}(x_f, x; T) - S_{\mathrm{cl}}(x_i, x; T) \simeq (x_f - x_i)\rho(x_i, x; T).$$

Finally, change variables from x to ρ and compare with the free particle result to obtain

$$A(x_f, x_i; T) = \left[\frac{1}{2\pi i}\frac{\partial^2 S_{\mathrm{cl}}}{\partial x_f \partial x_i}\right]^{1/2}.$$

(b) Show that

$$S_{\mathrm{cl}}(x_f, x_i; T) = -ET + \int_{x_i}^{x_f} dx\sqrt{2m(E - V(x))}$$

and verify that

$$\left[\frac{1}{2\pi i}\frac{\partial^2 S_{\mathrm{cl}}}{\partial x_f \partial x_i}\right]^{\frac{1}{2}} = \left[\frac{m}{2\pi i\dot{x}_{\mathrm{cl}}(t_f)\dot{x}_{\mathrm{cl}}(t_i)\int_{x_i}^{x_f} dx\, \dot{x}_{\mathrm{cl}}^{-3}(x)}\right]^{1/2}.$$

Hint: Recall that t is an independent variable, so that

$$0 = \frac{\partial t}{\partial x_f} = \frac{\partial t}{\partial t_i}.$$

We thus have the result for the semiclassical propagator

$$D(x_f, t_f; x_i, t_i) = \left[\frac{m}{2\pi i\dot{x}_{\mathrm{cl}}(t_i)\dot{x}_{\mathrm{cl}}(t_f)\int_{x_i}^{x_f} dx\, \dot{x}_{\mathrm{cl}}^{-3}(x)}\right]^{1/2} e^{i S_{\mathrm{cl}}},$$

which is identical to that found from WKB methods.

(2) **Propagator for the charged scalar field**

The lagrangian for a charged scalar field φ of mass m and charge e in the presence of an external (c-number) potential A_μ is

$$\mathcal{L} = D^\mu \varphi^* D_\mu \varphi - m^2 \varphi^* \varphi,$$

where $D_\mu = \partial_\mu + ieA_\mu$ is the covariant derivative.

(a) Show that the full Feynman propagator,

$$D_F(x'; x) = \frac{\int [d\varphi][d\varphi^*] \varphi(x') \varphi^*(x) e^{i \int d^4 x \, \mathcal{L}(x)}}{\int [d\varphi][d\varphi^*] e^{i \int \mathcal{L}(x)}},$$

can be written as

$$D_F(x'; x) = -i \langle x' | (D^\mu D_\mu + m^2 - i\epsilon)^{-1} | x \rangle.$$

Suggestion: This is a quadratic form. Use the generating functional to integrate it.

(b) By expanding $D_F(x'; x)$ as a power series in $A_\mu(x)$, show that an alternative representation for the propagator is

$$D_F(x'; x) = \langle x' | \int_0^\infty ds \, e^{-is(D^\mu D_\mu + m^2 - i\epsilon)} | x \rangle.$$

(3) **Functional methods and φ^4 theory**

Consider a scalar field theory with the self-interaction

$$\mathcal{L}_{\text{int}} = -\frac{\lambda}{4!} \varphi^4(x).$$

(a) Show that the generating functional can be written as

$$Z[j] = N e^{-i \frac{\lambda}{4!} \int d^4 z \left(\frac{\delta^4}{\delta j^4(z)} \right)} e^{-\frac{1}{2} \int d^4 x \int d^4 y \, j(x) i \Delta_F(x,y) j(y)},$$

where the free field Feynman propagator $i\Delta_F(x, y)$ is as in Eq. (C–2.12).

(b) Evaluate the two-point function to $\mathcal{O}(\lambda^2)$. Associate a Feynman diagram with each term of this expansion and separate the connected and disconnected diagrams.

(c) Calculate the connected generating functional via

$$W[j] = W_0[j] - i \ln \left[1 + e^{-iW^{(0)}[j]} \left(e^{-i \frac{\lambda}{4!} \int d^4 z \frac{\delta^4}{\delta j^4(z)}} - 1 \right) e^{iW^{(0)}[j]} \right]$$

where

$$W^{(0)}[j] = \frac{1}{2} \int d^4 x \int d^4 y \, j(x) i \Delta_F(x, y) j(y).$$

(d) Compare the connected diagrams found in parts (b) and (c).

Appendix B
Advanced field-theoretic methods

B–1 The heat kernel

When using path-integral techniques one must often evaluate quantities of the form

$$H(x, \tau) \equiv \langle x \left| e^{-\tau \mathcal{D}} \right| x \rangle, \tag{1.1}$$

where $\mathcal{D}$ is a differential operator and τ is a parameter. In this section, we shall describe the *heat kernel* method by which $H(x, \tau)$ is expressed as a power series in τ. For example, if in d dimensions the differential operator $\mathcal{D}$ is of the form

$$\mathcal{D} = \Box + m^2 + V, \tag{1.2}$$

where V is some interaction, then the heat kernel expansion for $H(x, \tau)$ is

$$H(x, \tau) = \frac{i}{(4\pi)^{d/2}} \frac{e^{-\tau m^2}}{\tau^{d/2}} \left[a_0(x) + a_1(x)\tau + a_2(x)\tau^2 + \cdots \right], \tag{1.3}$$

where $a_i(x)$ are coefficients which will be determined below.

Let us begin by citing the two most common occurrences of $H(x, \tau)$. One is in the evaluation of the functional determinant

$$\det \mathcal{D} = e^{\text{tr} \ln \mathcal{D}} = e^{\int d^4x \ \text{Tr} \langle x | \ln \mathcal{D} | x \rangle}, \tag{1.4}$$

where 'Tr' is a trace over internal variables like isospin, Dirac matrices, etc., and 'tr' is a trace over these plus spacetime. The (generally singular) matrix element $\langle x | \ln \mathcal{D} | x \rangle$ appearing in Eq. (1.4) can be expressed in a variety of ways. For example, in dimensional regularization one can use the identity

$$\ln \frac{b}{a} = \int_0^\infty \frac{dx}{x} \left(e^{-ax} - e^{-bx} \right) \tag{1.5}$$

to write

$$\langle x | \ln \mathcal{D} | x \rangle = - \int_0^\infty \frac{d\tau}{\tau} \langle x \left| e^{-\tau \mathcal{D}} \right| x \rangle + C, \tag{1.6}$$

where C is a divergent constant having no physical consequences. Substituting Eq. (1.3) into the above yields

$$\langle x| \ln \mathcal{D}|x\rangle - C = -\frac{i}{(4\pi)^{d/2}} \sum_{n=0}^{\infty} m^{d-2n} \, \Gamma\left(n - \frac{d}{2}\right) a_n(x). \tag{1.7}$$

The divergences in the series representation arise from the Γ function and are restricted in four dimensions to the terms $a_0(x)$, $a_1(x)$, $a_2(x)$.

The heat kernel can likewise be used to analyze the functional determinant in alternative regularization procedures, such as zeta-function regularization. Here, one expresses the matrix element $\langle x| \ln \mathcal{D}|x\rangle$ as

$$\langle x| \ln \mathcal{D}|x\rangle = -\left\langle x \left| \left[\frac{d}{ds} e^{-s \ln \mathcal{D}} \right]_{s=0} \right| x \right\rangle$$

$$= -\left[\frac{d}{ds} \left\langle x \left| \frac{1}{\mathcal{D}^s} \right| x \right\rangle \right]_{s=0} = -\frac{d}{ds} \xi_{\mathcal{D}}(x, s) \bigg|_{s=0},$$

$$\xi_{\mathcal{D}}(x, s) \equiv \frac{1}{\Gamma(s)} \int_0^{\infty} d\tau \, \tau^{s-1} H(x, \tau). \tag{1.8}$$

The penultimate equality in Eq. (1.8) is obtained from repeated formal differentiation of Eq. (1.6) with respect to $\mathcal{D}$. Upon expanding the $H(x, \tau)$ term in $\xi_{\mathcal{D}}(x, s)$, one arrives at the desired power-series expansion of $\langle x| \ln \mathcal{D}|x\rangle$. This usage is applied in the next section.

The other main use of the heat kernel is in the regularization of anomalies. Often one is faced with making sense of $\mathrm{Tr} \, \langle x \, |O(x)| \, x\rangle$, where O is a local operator. Although such quantities are generally singular, they can be defined in a gauge-invariant manner by damping out the contributions of large eigenvalues,

$$\mathrm{Tr} \, \langle x \, |O(x)| \, x\rangle = \lim_{\epsilon \to 0} \mathrm{Tr} \, \langle x \, |O(x)e^{-\epsilon \mathcal{D}} | \, x\rangle, \tag{1.9}$$

where $\mathcal{D}$ is a gauge-invariant differential operator. Again, it is only the low-order coefficients, generally those up to $a_2(x)$, which contribute in the $\epsilon \to 0$ limit. We employ this technique in Sects. III–3,4.

As an example of heat-kernel techniques, let us consider the following operator defined in d dimensions:

$$\mathcal{D} = d_\mu d^\mu + m^2 + \sigma(x) \qquad \left(d_\mu \equiv \frac{\partial}{\partial x^\mu} + \Gamma_\mu(x) \right), \tag{1.10}$$

where $\Gamma_\mu(x)$ and $\sigma(x)$ are functions and/or matrices defined in some internal symmetry space. In particular, neither Γ_μ nor σ contains derivative operators. Employing a complete set of momentum eigenstates $\{|p\rangle\}$ allows us to express the heat kernel as

$$H(x, \tau) = \int \frac{d^d p}{(2\pi)^d} e^{-ip \cdot x} e^{-\tau \mathcal{D}} e^{ip \cdot x}, \tag{1.11}$$

where in d dimensions use is made of the relations

$$\langle p | x \rangle = \frac{1}{(2\pi)^{d/2}} e^{ip \cdot x},$$

$$\langle x | x' \rangle = \int \frac{d^d p}{(2\pi)^d} e^{ip \cdot (x'-x)} = \delta^{(d)}(x - x'),$$

$$\langle p' | p \rangle = \int \frac{d^d x}{(2\pi)^d} e^{i(p'-p) \cdot x} = \delta^{(d)}(p' - p). \tag{1.12}$$

From the identities

$$d_\mu e^{ip \cdot x} = e^{ip \cdot x} (ip_\mu + d_\mu),$$
$$d_\mu d^\mu e^{ip \cdot x} = e^{ip \cdot x} (ip_\mu + d_\mu)(ip^\mu + d^\mu), \tag{1.13}$$

we can then write

$$H(x, \tau) = \int \frac{d^d p}{(2\pi)^d} e^{-\tau \left[(ip_\mu + d_\mu)^2 + m^2 + \sigma \right]}$$

$$= \int \frac{d^d p}{(2\pi)^d} e^{\tau [p^2 - m^2]} e^{-\tau [d \cdot d + \sigma + 2ip \cdot d]}. \tag{1.14}$$

The first exponential factor is simply the free field result, while all the interesting physics is in the second exponential. The latter can be Taylor expanded in powers of τ, keeping those terms which contribute up to order τ^2 after the integration over momentum is performed. Note that each power of p^2 contributes a factor of $1/\tau$. Thus, we obtain the expansion

$$H(x, \tau) = \int \frac{d^d p}{(2\pi)^d} e^{\tau (p^2 - m^2)} \Big[1 - \tau [d \cdot d + \sigma]$$

$$+ \frac{\tau^2}{2} [(d \cdot d + \sigma)(d \cdot d + \sigma) - 4 p \cdot d \, p \cdot d]$$

$$+ \frac{4\tau^3}{3!} [p \cdot d \, p \cdot d (d \cdot d + \sigma) + p \cdot d (d \cdot d + \sigma) p \cdot d$$

$$+ (d \cdot d + \sigma) p \cdot d \, p \cdot d]$$

$$+ \frac{16\tau^4}{4!} p \cdot d \, p \cdot d \, p \cdot d \, p \cdot d + \cdots \Big], \tag{1.15}$$

where terms odd in p have been dropped and we have displayed only those $\mathcal{O}(\tau^3)$ and $\mathcal{O}(\tau^4)$ terms which contribute to H at order τ^2 after p is integrated over. To perform the integral, it is convenient to continue to euclidean momentum

$p_E = \{p_1, p_2, p_3, p_4 = -ip_0\}$. Then, with the replacement $p_\mu p^\mu \to -|p_E^\mu p_E^\mu| = -p_E^2$, we obtain

$$
\int \frac{d^d p_E}{(2\pi)^d} e^{-(p_E^2 + m^2)\tau} = \int \frac{d\Omega_d}{(2\pi)^d} \int dp_E \, p_E^{d-1} e^{-(p_E^2 + m^2)\tau}
$$

$$
= \frac{2\pi^{d/2}}{\Gamma(d/2)} \frac{1}{(2\pi)^d} \frac{e^{-m^2\tau} \Gamma(d/2)}{2\tau^{d/2}}
$$

$$
= \frac{1}{(4\pi)^{d/2}} \frac{e^{-m^2\tau}}{\tau^{d/2}},
$$

$$
\int \frac{d^d p_E}{(2\pi)^d} e^{-(p_E^2 + m^2)\tau} \, p_E^\mu p_E^\nu = \frac{\delta^{\mu\nu}}{d} \frac{1}{(4\pi)^{d/2}} \frac{e^{-m^2\tau}}{\tau^{d/2+1}} \frac{\Gamma(d/2+1)}{\Gamma(d/2)}
$$

$$
= \frac{\delta^{\mu\nu}}{2} \frac{e^{-m^2\tau}}{(4\pi)^{d/2} \tau^{d/2+1}},
$$

$$
\int \frac{d^d p_E}{(2\pi)^d} e^{-(p_E^2 + m^2)\tau} \, p_E^\mu p_E^\nu p_E^\lambda p_E^\sigma = \frac{e^{-m^2\tau}}{(4\pi)^{d/2} \tau^{d/2+2}}
$$

$$
\times \frac{\left(\delta^{\mu\nu}\delta^{\lambda\sigma} + \delta^{\mu\lambda}\delta^{\nu\sigma} + \delta^{\mu\sigma}\delta^{\lambda\nu} \right)}{4}. \tag{1.16}
$$

Employing these relations to evaluate Eq. (1.14) gives (to second order in τ),

$$
H(x, \tau) = \frac{ie^{-m^2\tau}}{(4\pi)^{d/2}\tau^{d/2}}
$$

$$
\times \left[1 - \tau\sigma + \tau^2 \left(\frac{1}{2}\sigma^2 + \frac{1}{12}[d_\mu, d_\nu][d^\mu, d^\nu] + \frac{1}{6}[d_\mu, [d^\mu, \sigma]] \right) \right],
\tag{1.17}
$$

or in the notation of Eq. (1.3),

$$
a_0(x) = 1, \qquad\qquad a_1(x) = -\sigma,
$$

$$
a_2(x) = \frac{1}{2}\sigma^2 + \frac{1}{12}[d_\mu, d_\nu][d^\mu, d^\nu] + \frac{1}{6}[d_\mu, [d^\mu, \sigma]]. \tag{1.18}
$$

Fermions are treated in a similar manner. For example, the identity

$$
\ln\!D = \frac{1}{2}\ln(D \, D) \tag{1.19}
$$

allows the same technique to be used for the operator $D \, D$. In particular let us consider the case where

$$
D = \partial\!\!\!/ + iV\!\!\!\!/ + iA\!\!\!\!/\gamma_5. \tag{1.20}
$$

With some work, one can cast this into the form of Eq. (1.10) with the identifications

$$\mathcal{D} \mathcal{D} \equiv \mathcal{D} = d_\mu d^\mu + \sigma,$$

$$d_\mu = \partial_\mu + i V_\mu + \sigma_{\mu\nu} A^\nu \gamma_5 \equiv \partial_\mu + \Gamma_\mu,$$

$$\sigma = \frac{1}{2} \sigma_{\mu\nu} V^{\mu\nu} - 2 A_\mu A^\mu + \left(i \partial_\mu A^\mu - [V_\mu, A^\mu] \right) \gamma_5,$$

$$V_{\mu\nu} = \partial_\mu V_\nu - \partial_\nu V_\mu + i [V_\mu, V_\nu] + i [A_\mu, A_\nu]. \tag{1.21}$$

The values of $a_i(x)$ appearing in Eq. (1.18) can also be used in this case. The heat-kernel coefficients have been worked out for more general situations [Gi 75].

B–2 Chiral renormalization and background fields

In this section, we illustrate the method described above while also proving an important result for the theory of chiral symmetry. The goal is to demonstrate that all the divergences encountered at one loop can be absorbed into a renormalization of the coefficients of the $\mathcal{O}(E^4)$ chiral lagrangian and to identify the renormalization constants. The technique used here, the *background field method*, is of considerable interest in its own right [Sc 51, De 67, Ab 82] and is applicable to areas such as general relativity [BiD 82].

The basic idea of the background field method is to calculate quantum corrections about some nonvanishing field configuration $\bar{\varphi}$,

$$\varphi(x) = \bar{\varphi}(x) + \delta\varphi(x), \tag{2.1}$$

rather than about the zero field,[1] and to then compute the path integral over the fluctuation $\delta\varphi(x)$. The result is an effective action for $\bar{\varphi}$. This effective action can be expanded in powers of $\bar{\varphi}$ and applied to matrix elements at tree level, resulting in a description of scattering processes at one-loop order. In the case of the chiral lagrangian, one expands the full chiral matrix

$$U = \bar{U} + \delta U, \tag{2.2}$$

where $\bar{U}$ satisfies the classical equation of motion. Upon integration over δU, one obtains the one-loop effective action for $\bar{U}$. This contains a great deal of information. In particular, $\bar{U}$ can be expanded in the usual way in terms of a set of external meson fields

$$\bar{U} = \exp(i \lambda^a \bar{\varphi}^a / F) \quad (a = 1, \ldots, 8). \tag{2.3}$$

---

[1] See the discussion in Appendix A–4.

Contained in $S_{\text{eff}}(\bar{U})$ is the effective one-loop action for arbitrary numbers of meson fields. Upon identification of renormalization constants, all processes become renormalized at the same time.

Our starting point is, in the notation of Sect. IV–6, the $\mathcal{O}(E^2)$ lagrangian

$$\mathcal{L}_2 = \frac{F_0^2}{4} \text{Tr} \left(D_\mu U D^\mu U^\dagger \right) + \frac{F_0^2}{4} \text{Tr} \left(\chi^\dagger U + U^\dagger \chi \right). \tag{2.4}$$

The procedure to follow is rather technical, so let us first quote the end result of the calculation. Upon performing the one-loop quantum corrections, the effective action will have the form

$$S_{\text{eff}} = S_2^{\text{ren}} + S_4^{\text{ren}} + S_4^{\text{finite}} + \cdots.$$

Here the lagrangians in S_2^{ren}, S_4^{ren} are the ones quoted in Sect. VII–2, but now with renormalized coefficients. In particular S_4^{ren} is the sum $S_4^{\text{ren}} = S_4^{\text{bare}} + S_4^{\text{div}}$ where, in chiral $SU(3)$ and employing dimensional regularization, S_4^{div} is given by

$$\begin{aligned}
S_4^{\text{div}} = -\lambda \int d^4x \Bigg[&\frac{3}{32} \left[\text{Tr} \left(D_\mu U D^\mu U^\dagger \right) \right]^2 \\
&+ \frac{3}{16} \text{Tr} \left(D_\mu U D_\nu U^\dagger \right) \text{Tr} \left(D^\mu U D^\nu U^\dagger \right) \\
&+ \frac{1}{8} \text{Tr} \left(D_\mu U D^\mu U^\dagger \right) \text{Tr} \left(\chi^\dagger U + U^\dagger \chi \right) \\
&+ \frac{3}{8} \text{Tr} \left[D_\mu U D^\mu U^\dagger \left(\chi U^\dagger + U \chi^\dagger \right) \right] \\
&+ \frac{11}{144} \left[\text{Tr} \left(\chi U^\dagger + U \chi^\dagger \right) \right]^2 + \frac{5}{48} \text{Tr} \left(\chi U^\dagger \chi U^\dagger + U \chi^\dagger U \chi^\dagger \right) \\
&+ \frac{i}{4} \text{Tr} \left(L_{\mu\nu} D^\mu U D^\nu U^\dagger + R_{\mu\nu} D^\mu U^\dagger D^\nu U \right) - \frac{1}{4} \text{Tr} \left(L_{\mu\nu} U R^{\mu\nu} U^\dagger \right) \Bigg]
\end{aligned} \tag{2.5}$$

with

$$\lambda \equiv \frac{1}{32\pi^2} \left\{ \frac{2}{d-4} - \ln 4\pi - 1 + \gamma \right\}. \tag{2.6}$$

The terms in S_4^{div} are all of the same form as the terms in the bare lagrangian at order E^4. Therefore, all the divergences can be absorbed into renormalized values of these constants. The finite remainder, S_4^{finite}, cannot be simply expressed as a local lagrangian, but can be worked out for any given transition. When S_4^{div} is added to the $\mathcal{O}(E^4)$ tree-level lagrangian of Eq. (VII–2.7), the result has the same form but with coefficients

$$L_i^r = L_i - \gamma_i \lambda, \tag{2.7}$$

Table B–1. *Renormalization coefficients.*

i	1	2	3	4	5	6	7	8	9	10
$SU(2)$ γ_i	$\frac{1}{12}$	$\frac{1}{6}$	0	$\frac{1}{8}$	$\frac{1}{4}$	$\frac{3}{32}$	0	0	$\frac{1}{6}$	$-\frac{1}{6}$
$SU(3)$ γ_i	$\frac{3}{32}$	$\frac{3}{16}$	0	$\frac{1}{8}$	$\frac{3}{8}$	$\frac{11}{144}$	0	$\frac{5}{48}$	$\frac{1}{4}$	$-\frac{1}{4}$

where the $\{\gamma_i\}$ are numbers which are given in Table B–1 for both the case of chiral $SU(2)$ and $SU(3)$. Thus, the divergences can all be absorbed into the redefined parameters and these in turn can be determined from experiment. Let us now turn to the task of obtaining this result.

In applying the background field method, there are a variety of ways to para-meterize δU, and several different ones are used in the literature. The prime consideration is to maintain the unitarity property $U^\dagger U = 1 = \left(\bar{U}^\dagger + \delta U^\dagger\right)\left(\bar{U} + \delta U\right)$ along with $\bar{U}^\dagger \bar{U} = 1$. We shall take

$$U = \bar{U}e^{i\Delta}, \tag{2.8}$$

with $\Delta \equiv \lambda^a \Delta^a$ representing the quantum fluctuations. This choice is made to simplify the algebra in the heat-kernel renormalization approach, which we shall describe shortly. Another possible choice is

$$U = \xi e^{i\eta}\xi \tag{2.9}$$

with $\eta = \lambda^a \eta^a$ and $\xi\xi \equiv \bar{U}$. These two forms are related by $\eta = \xi \Delta \xi^\dagger$. Since in the path integral, we integrate over all values of Δ (or η) at each point of spacetime, these two choices are equivalent.

The expansion of the lagrangian in terms of $\bar{U}$ and Δ is straightforward, and we find

$$\begin{aligned}
\mathrm{Tr}\left(D_\mu U D^\mu U^\dagger\right) &= \mathrm{Tr}\left(D_\mu \bar{U} D^\mu \bar{U}^\dagger\right) - 2i\,\mathrm{Tr}\left(\bar{U}^\dagger D_\mu \bar{U} \tilde{D}^\mu \Delta\right) \\
&\quad + \mathrm{Tr}\left[\tilde{D}_\mu \Delta \tilde{D}^\mu \Delta + \bar{U}^\dagger D_\mu \bar{U}\left(\Delta \tilde{D}^\mu \Delta - \tilde{D}^\mu \Delta \Delta\right)\right], \\
\mathrm{Tr}\left(\chi^\dagger U + U^\dagger \chi\right) &= \mathrm{Tr}\left(\chi^\dagger \bar{U} + \bar{U}^\dagger \chi\right) + i\,\mathrm{Tr}\left(\Delta\left(\chi^\dagger \bar{U} - \bar{U}^\dagger \chi\right)\right) \\
&\quad - \frac{1}{2}\mathrm{Tr}\left[\Delta^2\left(\chi^\dagger \bar{U} + \bar{U}^\dagger \chi\right)\right],
\end{aligned} \tag{2.10}$$

where

$$\tilde{D}_\mu \Delta \equiv \partial_\mu \Delta + i\left[r_\mu, \Delta\right], \tag{2.11}$$

where r_μ is the matrix source function of Eqs. (IV–6.1,6.2). Since $\bar{U}$ satisfies the equation of motion, there is no term linear in Δ. One may integrate various terms in the action by parts to obtain

$$S_2^{(0)} = \int d^4x \left\{ \mathcal{L}_2(\bar{U}) - \frac{F_0^2}{2} \Delta^a \left(d_\mu d^\mu + \sigma \right)^{ab} \Delta^b + \cdots \right\}, \qquad (2.12)$$

where

$$d_\mu^{ab} = \delta^{ab} \partial_\mu + \Gamma_\mu^{ab},$$

$$\Gamma_\mu^{ab} = -\frac{1}{4} \operatorname{Tr}\left([\lambda^a, \lambda^b] \left(\bar{U}^\dagger \partial_\mu \bar{U} + i\bar{U}^\dagger \ell_\mu \bar{U} + ir_\mu\right)\right),$$

$$\sigma^{ab} = \frac{1}{8} \operatorname{Tr}\left(\{\lambda^a, \lambda^b\} \left(\chi^\dagger \bar{U} + \bar{U}^\dagger \chi\right) + [\lambda^a, \bar{U}^\dagger D_\mu \bar{U}][\lambda^b, \bar{U}^\dagger D^\mu \bar{U}]\right). \quad (2.13)$$

The action is now a simple quadratic form, and the path integral may be performed. The only potential complication is the question of interpreting the integration variables. This is referred to as the 'question of the path-integral measure'. The integration over all the unitary matrices U can be accomplished by an integration over the parameters in the exponential

$$\int [dU] = N \int [d\Delta^a], \qquad (2.14)$$

where N is a constant which plays no dynamical role. With this identification one obtains

$$e^{i W_{\text{loop}}} = \int [d\Delta^a] e^{i \int d^4x \frac{F_0^2}{2} \Delta^a (d_\mu d^\mu + \sigma)^{ab} \Delta^b}$$

$$= \left(\det\left[d_\mu d^\mu + \sigma\right]\right)^{-1/2} = e^{-\frac{1}{2} \operatorname{tr} \ln(d_\mu d^\mu + \sigma)}. \qquad (2.15)$$

Here 'tr' indicates a trace over the spacetime indices as well as over the $SU(N)$ indices a, b.

The identification of divergences is most conveniently done by using the heat-kernel expansion derived earlier in App. B–1, where it is shown that all the ultra-violet divergences are contained in the first few expansion coefficients. The relevant terms are

$$W_{\text{loop}} = \frac{i}{2} \operatorname{tr} \ln \left(d_\mu d^\mu + \sigma\right)$$

$$= \frac{1}{2(4\pi)^{d/2}} \int d^4x \lim_{m \to 0} \left\{ \Gamma\left(1 - \frac{d}{2}\right) m^{d-2} \operatorname{Tr}\sigma \right.$$

$$\left. + m^{d-4} \Gamma\left(2 - \frac{d}{2}\right) \operatorname{Tr}\left(\frac{1}{12}\Gamma_{\mu\nu}\Gamma^{\mu\nu} + \frac{1}{2}\sigma^2\right) + \cdots \right\}, \qquad (2.16)$$

where

$$\Gamma_{\mu\nu}^{ab} = \partial_\mu \Gamma_\nu^{ab} - \partial_\nu \Gamma_\mu^{ab} + \Gamma_\mu^{ac}\Gamma_\nu^{cb} - \Gamma_\nu^{ac}\Gamma_\mu^{cb} = [d_\mu, d_\nu]^{ab}. \qquad (2.17)$$

For N_f flavors, the operator part of the first term in Eq. (2.16) is

$$\text{Tr}\,\sigma = \frac{N_f}{2}\,\text{Tr}\,\left(D_\mu \bar{U} D^\mu \bar{U}^\dagger\right) + \frac{N_f^2 - 1}{2N_f}\,\text{Tr}\,\left(\chi^\dagger \bar{U} + \bar{U}^\dagger \chi\right). \qquad (2.18)$$

The above two traces are just those which appear in $\mathcal{L}_2$; as such, they can only modify the quantities F_π and m_π^2. The remaining terms can be worked out with a bit more algebra. Using the identity

$$\partial_\mu \left(\bar{U}^\dagger \partial_\nu \bar{U}\right) - \partial_\nu \left(\bar{U}^\dagger \partial_\mu \bar{U}\right) = -\left[\bar{U}^\dagger \partial_\mu U, \bar{U}^\dagger \partial_\nu U\right], \qquad (2.19)$$

we find for the field strength,

$$\Gamma_{\mu\nu}^{ab} = \frac{1}{8}\,\text{Tr}\,\left\{\left[\lambda^a, \lambda^b\right]\left(\left[\bar{U}^\dagger D_\mu \bar{U}, \bar{U}^\dagger D_\nu \bar{U}\right] + i\bar{U}^\dagger L_{\mu\nu}\bar{U} + iR_{\mu\nu}\right)\right\}. \qquad (2.20)$$

This produces, for N_f flavors in chiral $SU(N_f)$,

$$\begin{aligned}
\text{Tr}\,\left(\Gamma_{\mu\nu}\Gamma^{\mu\nu}\right) &= \frac{N_f}{8}\,\text{Tr}\,\left(\left[\bar{U}^\dagger D_\mu \bar{U}, \bar{U}^\dagger D_\nu \bar{U}\right]\left[\bar{U}^\dagger D^\mu \bar{U}, \bar{U}^\dagger D^\nu \bar{U}\right]\right) \\
&\quad + iN_f\,\text{Tr}\,\left(R_{\mu\nu}\partial^\mu \bar{U}^\dagger \partial^\nu \bar{U} + L_{\mu\nu}\partial^\mu \bar{U}\partial^\nu \bar{U}^\dagger\right) \\
&\quad - N_f\,\text{Tr}\,\left(L_{\mu\nu}\bar{U}R^{\mu\nu}\bar{U}^\dagger\right) - \frac{N_f}{2}\,\text{Tr}\,\left(L_{\mu\nu}L^{\mu\nu} + R_{\mu\nu}R^{\mu\nu}\right), \\
\text{Tr}\,\sigma^2 &= \frac{1}{8}\left[\text{Tr}\,\left(D_\mu \bar{U}D^\mu \bar{U}^\dagger\right)\right]^2 + \frac{1}{4}\,\text{Tr}\,\left(D_\mu \bar{U}D_\nu \bar{U}^\dagger\right)\,\text{Tr}\,\left(D^\mu \bar{U}D^\nu \bar{U}^\dagger\right) \\
&\quad + \frac{N_f}{8}\,\text{Tr}\,\left(D_\mu \bar{U}D^\mu \bar{U}^\dagger D_\nu \bar{U}D^\nu \bar{U}^\dagger\right) + \frac{2 + N_f^2}{8N_f^2}\left[\text{Tr}\,\left(\chi\bar{U}^\dagger + \bar{U}^\dagger \chi\right)\right]^2 \\
&\quad + \frac{1}{4}\,\text{Tr}\,\left(D_\mu \bar{U}D^\mu \bar{U}^\dagger\right)\,\text{Tr}\,\left(\chi\bar{U}^\dagger + \bar{U}\chi^\dagger\right) \\
&\quad + \frac{N_f}{4}\,\text{Tr}\,\left(D_\mu \bar{U}D^\mu \bar{U}^\dagger\left(\chi\bar{U}^\dagger + \bar{U}\chi^\dagger\right)\right) \\
&\quad + \frac{N_f^2 - 4}{8N_f}\,\text{Tr}\,\left(\left(\chi\bar{U}^\dagger + \bar{U}\chi^\dagger\right)\left(\chi\bar{U}^\dagger + \bar{U}\chi^\dagger\right)\right). \qquad (2.21)
\end{aligned}$$

The only operator which is not of the same form as the basic $\mathcal{O}(E^4)$ lagrangian occurs in the first term of $\text{Tr}\,\Gamma^2$. However, by use of Eq. (VII–2.3) for $SU(3)$, it can be written as a linear combination of our standard forms. For $N_f = 3$, these add up to the result previously quoted in Eq. (2.5). Here, the divergence is in the parameter λ. For convenience in applications, we have added some finite terms to the definitions of λ in Eq. (2.6). The results for $N_f = 2$ are also quoted in Table B–1, although some of the operators are redundant for that case.

The reader who has understood the above development as well as the standard perturbative methods presented in the main text will be prepared for the use of the background field method in the full calculation of transition amplitudes. This procedure consists of writing

$$d_\mu d^\mu + \sigma = \mathcal{D}_0 + V$$
$$\mathcal{D}_0 = \Box + m^2$$
$$V = \{\partial_\mu, \Gamma^\mu\} + \Gamma_\mu \Gamma^\mu + \sigma - m^2, \tag{2.22}$$

where m^2 is the meson mass-squared matrix. The one-loop action is then expanded in powers of the interaction V

$$W_{\text{loop}} = \frac{i}{2} \text{tr} \, \ln(d_\mu d^\mu + \sigma) = \frac{i}{2} \text{tr} \left[\ln \mathcal{D}_0 + \ln(1 + \mathcal{D}_0^{-1} V) \right]$$
$$= \frac{i}{2} \text{tr} \left[\ln \mathcal{D}_0 + \mathcal{D}_0^{-1} V - \frac{1}{2} \mathcal{D}_0^{-1} V \mathcal{D}_0^{-1} V + \cdots \right]. \tag{2.23}$$

The first term is an uninteresting constant which may be dropped, and the remainder has the coordinate space form

$$W_{\text{loop}} = -\frac{i}{2} \int d^4x \, \text{Tr} \left[\Delta_F(x - x) V(x) \right]$$
$$- \frac{i}{4} \int d^4x d^4y \, \text{Tr} \left[\Delta_F(x - y) V(y) \Delta_F(y - x) V(x) \right] + \cdots . \tag{2.24}$$

When the matrix elements of this action are taken, the result contains not only the divergent terms calculated above, but also the finite components of the one-loop amplitudes. The resulting expressions are presented fully in [GaL 84, GaL 85a]. This method allows one to calculate the one-loop corrections to many processes at the same time and, in practice, is a much simpler procedure for some of the more difficult calculations.

B–3 PCAC and the soft-pion theorem

We have emphasized the use of effective lagrangians to elucidate the symmetry predictions of a theory. For a dynamically broken chiral symmetry such as *QCD*, these predictions will relate processes which have different numbers of Goldstone bosons. The machinery of effective lagrangians will correctly yield such predictions, but it is often useful to have an alternative technique for understanding or calculating these results. In the case of chiral symmetry, this is provided by the *soft-pion theorem*, which explicitly relates a process with a pion to one with that pion removed from the amplitude. Calculations performed this way uses current algebra methods which go by the name of *partial conservation of the axial current* or PCAC [AdD 68]. While these techniques are often more cumbersome, they often are useful. This section describes these methods.

We can again turn to the sigma model to introduce this subject. We return to the effective lagrangian treatment of Chap. IV, with a pion mass included and the S field integrated out. The lagrangian of Eq. (IV–6.12) gives rise to the vector and axial-vector currents

$$V_\mu^k = -i\frac{v^2}{4}\,\mathrm{Tr}\left[\tau^k\left(U^\dagger\partial_\mu U + U\partial_\mu U^\dagger\right)\right],$$

$$A_\mu^k = i\frac{v^2}{4}\,\mathrm{Tr}\left[\tau^k\left(U^\dagger\partial_\mu U - U\partial_\mu U^\dagger\right)\right], \tag{3.1}$$

with $k = 1, 2, 3$. The equation of motion is found to be

$$\partial^\mu\left(U^\dagger\partial_\mu U\right) + \frac{m_\pi^2}{2}\left(U - U^\dagger\right) = 0, \tag{3.2}$$

and two important matrix elements are

$$\left\langle 0\left|A_\mu^k\right|\pi^j(\mathbf{p})\right\rangle = ivp_\mu\delta^{kj}, \quad \left\langle 0\left|\partial^\mu A_\mu^k\right|\pi^j(\mathbf{p})\right\rangle = vm_\pi^2\delta^{kj}. \tag{3.3}$$

The former allows the identification $v = F_\pi$, where F_π is the pion decay constant $F_\pi \simeq 92$ MeV, while the latter follows either from Eq. (3.1) directly, or by use of the equation of motion for A_μ^k,

$$\partial^\mu A_\mu^k = -i\frac{v^2 m_\pi^2}{4}\,\mathrm{Tr}\left[\tau^k\left(U - U^\dagger\right)\right] = F_\pi m_\pi^2\pi^k + \cdots. \tag{3.4}$$

This last equation forms the heart of the PCAC method. It describes a situation covered by Haag's theorem (recall Sect. IV–1), and says that we may use either π^k or $\partial^\mu A_\mu^k$ (properly normalized) as the pion field. It is more general than the sigma model, which we used to motivate it. This, plus certain smoothness assumptions, gives rise to a *soft-pion theorem* for the following matrix element of a local operator O,

$$\lim_{q^\mu\to 0}\left\langle\pi^k(\mathbf{q})\beta|O|\alpha\right\rangle = -\frac{i}{F_\pi}\left\langle\beta|\left[Q_5^k, O\right]|\alpha\right\rangle, \tag{3.5}$$

where β, α are arbitrary states and $Q_5^k = \int d^3x\, A_0^k(x)$ is an axial charge.

The proof of Eq. (3.5) starts with the LSZ reduction formula. We consider the matrix element for the process $\alpha \to \beta + \pi^k(q)$ as the pion four-momentum q is taken off the mass-shell,

$$\left\langle\pi^k(q)\beta|O(0)|\alpha\right\rangle = i\int d^4x\, e^{iq\cdot x}\left(\Box + m_\pi^2\right)\left\langle\beta|T\left(\pi^k(x)O(0)\right)|\alpha\right\rangle$$

$$= i\int d^4x\, e^{iq\cdot x}(-q^2 + m_\pi^2)\left\langle\beta|T\left(\pi^k(x)O(0)\right)|\alpha\right\rangle, \tag{3.6}$$

The pion field can be replaced by using the PCAC relation (valid in the sense of Haag's theorem),

$$\pi^k = \frac{1}{F_\pi m_\pi^2} \partial^\mu A_\mu^k, \tag{3.7}$$

leading to

$$\langle \pi^k(q)\beta | O(0) | \alpha \rangle = i \frac{(m_\pi^2 - q^2)}{F_\pi m_\pi^2} \int d^4x\, e^{iq\cdot x} \langle \beta | T \left(\partial^\mu A_\mu^k(x) O(0) \right) | \alpha \rangle. \tag{3.8}$$

The derivative can be extracted from the time-ordered product by using

$$\partial^\mu \langle \beta | T \left(A_\mu^k(x) O(0) \right) | \alpha \rangle$$
$$= \langle \beta | T \left(\partial^\mu A_\mu^k(x) O(0) \right) | \alpha \rangle + \delta(x_0) \langle \beta | \left[A_0^k(x), O(0) \right] | \alpha \rangle, \tag{3.9}$$

where the last term arises from differentiating the functions $\theta(\pm x_0)$, which occur in the time-ordering prescription. Upon integrating by parts, we find

$$\langle \pi^k(q)\beta | O(0) | \alpha \rangle = i \frac{(m_\pi^2 - q^2)}{F_\pi m_\pi^2} \int d^4x\, e^{iq\cdot x} \left[-\langle \beta | \left[A_0^k(x), O(0) \right] | \alpha \rangle \delta(x_0) \right.$$
$$\left. - iq^\mu \langle \beta | T \left(A_\mu^k(x) O(0) \right) | \alpha \rangle \right]. \tag{3.10}$$

Up to this stage all the formulae are exact for physical processes, even if appearing rather senseless, since $\partial^\mu A_\mu^k$ has the same singularity for $q^2 \to m_\pi^2$ as does the field π^k. However, to obtain the soft-pion theorem one assumes that the matrix element does not vary much between its on-shell value and the point where the pion's four-momentum vanishes. In that circumstance, we have [NaL 62, AdD 68]

$$\lim_{q^\mu \to 0} \langle \pi^k(q)\beta | O | \alpha \rangle = -\frac{i}{F_\pi} \langle \beta | \left[Q_5^k, O(0) \right] | \alpha \rangle + \lim_{q^\mu \to 0} iq^\mu R_\mu^k, \tag{3.11}$$

where

$$R_\mu^k = -\frac{i}{F_\pi} \int d^4x\, e^{iq\cdot x} \langle \beta | T \left(A_\mu^k(x) O(0) \right) | \alpha \rangle. \tag{3.12}$$

The remainder term of Eq. (3.11) vanishes unless R_μ^k has a singularity as $q^\mu \to 0$. Such a singularity can occur if there are intermediate states in R_μ^k which are degenerate in mass with either α or β. This last statement can be proven by inserting a complete set of intermediate states in the time-ordered product in R_μ^k, and taking the $q^\mu \to 0$ limit. This caveat should be kept in mind as it is sometimes relevant.

The soft-pion theorem relates to the intuitive picture for dynamically broken symmetries mentioned in Sect. I–6. Since a chiral transformation corresponds in the symmetry limit to the addition of a zero-energy Goldstone boson, we expect the states $\langle \beta |$ and $\langle \pi_{q_\mu=0}^k \beta |$ to be related by the symmetry and, indeed, the soft-pion theorem expresses this. Although the soft-pion theorem is exact in the symmetry limit, a smoothness assumption is needed in the real world to pass from $q_\mu = 0$ to $q^2 = m_\pi^2$, implying that corrections of order q_μ or of order m_π^2 can be expected.

In the Standard Model, the charge commutation rules are commonly abstracted from those of the quark model. Upon expressing charge operators in terms of quark fields,

$$Q^k = \int d^3x \; \bar{\psi}\gamma_0\frac{\lambda^k}{2}\psi, \quad Q^k_5 = \int d^3x \; \bar{\psi}\gamma_0\gamma_5\frac{\lambda^k}{2}\psi, \tag{3.13}$$

one obtains the algebra

$$[Q^i, V_\mu^j] = if^{ijk}V_\mu^k, \qquad [Q^i_5, V_\mu^j] = if^{ijk}A_\mu^k,$$
$$[Q^i, A_\mu^j] = if^{ijk}A_\mu^k, \qquad [Q^i_5, A_\mu^j] = if^{ijk}V_\mu^k. \tag{3.14}$$

These commutation rules can be extended to equal-time commutators, which contain a charge density, e.g.,

$$\left[V_0^i(x), A_\mu^j(y)\right]_{x^0=y^0} = if^{ijk}A_\mu^k\delta^{(3)}(\mathbf{x} - \mathbf{y}). \tag{3.15}$$

However, commutators which involve two spatial components can be more problematic [AdD 68].

Sometimes, in the PCAC approach, if the matrix element is assumed to be strictly constant, the various soft-pion limits turn out to be contradictory. If so, the amplitude must be extended to include momentum dependence, as happens in nonleptonic kaon decay. By contrast, the effective lagrangian approach automatically gives the appropriate momentum dependence, and its predictions follow in a straightforward manner. Moreover, effective lagrangians are especially useful in identifying and parameterizing corrections to the lowest-order results. They allow a systematic expansion in terms of energy and mass.

B–4 Matching fields with different symmetry-transformation properties

In Chapter IV, we described the construction of an effective lagrangian for pion fields with chiral transformation properties. However, most particles do not transform in the same way as the pions of that chapter. In a broader context, we require a procedure for combining fields with different symmetry properties. For example, in the case of hadronic physics one often needs to consider particles such as nucleons, $\rho(770)$, etc., interacting with pions. A general approach for this was presented in a set of classic papers on the subject [We 68, CoWZ 69, CaCWZ 69]. We shall introduce this framework by again referring to the sigma model, and then we shall extend the results.

Heavy particles do not themselves exist in chiral multiplets. For example, the chiral partner of $\rho(770)$ would be the $J^{PC} = 1^{++}$ state $a(1260)$. The $a(1260)$–$\rho(770)$ mass difference is considerable, and attempts to pair these particles in a chiral multiplet would clearly be a matter of speculation. However, since each falls into vectorial flavor ($SU(2)$ or $SU(3)$) multiplets, it makes sense to build in only vectorial flavor invariance without invoking assumptions about chiral properties.

We shall proceed by first working out an example, the fermionic sector of the linear sigma model,

$$\mathcal{L}_f = \bar{\psi}\left[i\not{\partial} - g\left(\sigma - i\boldsymbol{\tau}\cdot\boldsymbol{\pi}\gamma_5\right)\right]\psi$$

$$= \bar{\psi}_L i\not{\partial}\psi_L + \bar{\psi}_R i\not{\partial}\psi_R - gv\left(1 + \frac{S}{v}\right)\left(\bar{\psi}_L U\psi_R + \bar{\psi}_R U^\dagger \psi_L\right). \tag{4.1}$$

We shall drop reference to the scalar field S in the following. The above lagrangian is invariant under the chiral transformations

$$\psi_L \to L\psi_L, \qquad \psi_R \to R\psi_R, \qquad U \to LUR^\dagger, \tag{4.2}$$

with L in $SU(2)_L$ and R in $SU(2)_R$. As always, we are free to change variables via contact transformations. In this instance, a useful choice of field redefinitions turns out to be

$$N_L \equiv \xi^\dagger \psi_L, \qquad N_R \equiv \xi\psi_R, \qquad U = \xi\xi, \tag{4.3}$$

where $\xi = \exp(i\boldsymbol{\tau}\cdot\boldsymbol{\pi}/2F_\pi)$. This is seen, after some algebra, to convert the fermion lagrangian to

$$\mathcal{L}'_f = \overline{N}\left(i\not{\mathcal{D}} - \overline{\not{A}}\gamma_5 - \mathbf{M}\right)N, \quad \mathcal{D}_\mu = \partial_\mu + i\overline{V}_\mu,$$

$$\overline{V}_\mu = -\frac{i}{2}\left(\xi^\dagger \partial_\mu\xi + \xi\partial_\mu\xi^\dagger\right), \quad \overline{A}_\mu = -\frac{i}{2}\left(\xi^\dagger \partial_\mu\xi - \xi\partial_\mu\xi^\dagger\right), \tag{4.4}$$

which is a theory of fermions of mass $M = gv$ having pseudovector coupling. The new fields transform as

$$\xi \to L\xi V^\dagger \equiv V\xi R^\dagger, \qquad N_L \to V N_L, \qquad N_R \to V N_R,$$

$$\overline{V}_\mu \to V\left(\overline{V}_\mu - i\partial_\mu V^\dagger \cdot V\right)V^\dagger, \quad \overline{A}_\mu \to V\overline{A}_\mu V^\dagger, \quad \mathcal{D}_\mu N \to V\mathcal{D}_\mu N. \tag{4.5}$$

For purely vector transformations we have $L = R = V$. For $L \neq R$, the property of V is more complicated, and Eq. (4.5) implies that it cannot be a simple global transformation, but must be a function of $\boldsymbol{\pi}(x)$ and hence a function of x. At first sight, the need to express an $SU(2)$-transformation matrix like V as a function of $\boldsymbol{\pi}(x)$ appears unnatural. However, it is in fact consistent with physical expectations. Recall from the general discussion of dynamical symmetry breaking in Sect. I–6 that, in the symmetry limit, axial transformations mix the proton not with the neutron (as in isospin transformations) but rather with states consisting of nucleons plus zero-momentum pions. Mathematically, the important point is that N_L and N_R transform in an identical fashion. This corresponds to the fact that heavy fields do not transform chirally, but have a common vectorial $SU(2)$ transformation. It can be directly verified that Eq. (3.5) is a symmetry of the lagrangian. Thus, we have obtained the expected result that the baryons can have a vectorial $SU(2)$ invariance, while maintaining a chiral invariance for pion couplings.

We see in the above example the ingredients of a general procedure for adding heavy fields to effective chiral lagrangians. The heavy fields are assumed to have an $SU(2)$ (or $SU(n)$, if desired) transformation described by the matrix V. A derivative ∂_μ acting on a heavy field must be incorporated as part of a covariant derivative $\mathcal{D}_\mu$ in order to maintain this invariance. Couplings to pions are described by the matrices ξ and U, with ξ having the same transformation as in Eq. (4.5). It is usually straightforward to combine factors of ξ and U in such a way that the overall lagrangian is invariant. In the general case, each invariant term will have an unknown coefficient which must be determined phenomenologically. For example, the $\bar{N}\,\slashed{A}\gamma_5\,N$ term in Eq. (4.4) would be expected to have a coefficient different from unity; the unit coefficient is a prediction specific to the linear sigma model. Effects which break the symmetry in an explicit fashion, like mass terms or electroweak interactions, can be added by using appropriate external sources. To date, heavy-field lagrangians have been used in applications primarily at tree level. The feature which is essential for their application is that the *pion* momenta are small, and hence the heavy fields are essentially static.

Appendix C

Useful formulae

C–1 Numerics

Conversion factors ($\hbar = c = k_B = 1$):

$$1 \text{ GeV}^{-1} = 6.582122 \times 10^{-25} \text{ s} \quad 1 \text{ GeV} = 1.16 \times 10^{13} \text{ K}$$
$$= 0.197327 \text{ fm} \quad = 1.78 \times 10^{-24} \text{ g}.$$

Physical constants ($\hbar = c = 1$):

$$G_\mu = 1.1663787(6) \times 10^{-5} \text{ GeV}^{-2} \qquad G_N^{-1/2} = M_{\text{Pl}} = 1.2 \times 10^{19} \text{ GeV}$$
$$\alpha^{-1} = 137.035999074(44) \qquad \sin^2 \theta_{\text{w}}^{\overline{\text{MS}}}(M_Z) = 0.23125(16)$$
$$m_W = 80.385(15) \text{ GeV} \qquad m_Z = 91.1876(21) \text{ GeV}$$
$$m_e = 0.510998928(11) \text{ MeV} \qquad m_p = 938.272046(21) \text{ MeV}$$
$$F_\pi = 92.2(2)\text{MeV} \qquad F_K = 110.4(8) \text{ MeV}$$
$$|\eta_{+-}| = 2.232(11) \times 10^{-3} \qquad |\eta_{00}| = 2.220(11) \times 10^{-3}.$$

CKM matrix elements:

$$|V_{\text{ud}}| = 0.97427(15) \qquad |V_{\text{us}}| = 0.22534(65) \qquad |V_{\text{ub}}| = 0.00351^{+0.00015}_{-0.00014}$$
$$|V_{\text{cd}}| = 0.22520(65) \qquad |V_{\text{cs}}| = 0.97344(16) \qquad |V_{\text{cb}}| = 0.0412^{+0.0011}_{-0.0005}$$
$$|V_{\text{td}}| = 0.00867^{+0.00029}_{-0.00031} \qquad |V_{\text{ts}}| = 0.0404^{+0.0011}_{-0.0005} \qquad |V_{\text{tb}}| = 0.999146^{+0.000021}_{-0.000046}.$$

C–2 Notations and identities

Metric tensor:

$$g^{\mu\nu} = \begin{pmatrix} 1 & 0 & 0 & 0 \\ 0 & -1 & 0 & 0 \\ 0 & 0 & -1 & 0 \\ 0 & 0 & 0 & -1 \end{pmatrix} \qquad g^\mu_{\ \mu} = 4. \tag{2.1}$$

535

Totally antisymmetric four-tensor:

$$\epsilon^{\mu\nu\alpha\beta} = \begin{cases} +1 & \{\mu, \nu, \alpha, \beta\} \text{ even permutation of } \{0, 1, 2, 3\} \\ -1 & \text{odd permutation} \\ 0 & \text{otherwise} \end{cases}$$

$$\epsilon^{\mu\nu\alpha\beta}\epsilon_{\mu}^{\ \nu'\alpha'\beta'} = g^{\nu\alpha'}g^{\alpha\nu'}g^{\beta\beta'} + g^{\nu\nu'}g^{\alpha\beta'}g^{\beta\alpha'} + g^{\nu\beta'}g^{\alpha\alpha'}g^{\beta\nu'}$$
$$- g^{\nu\nu'}g^{\alpha\alpha'}g^{\beta\beta'} - g^{\nu\beta'}g^{\alpha\nu'}g^{\beta\alpha'} - g^{\nu\alpha'}g^{\alpha\beta'}g^{\beta\nu'}. \tag{2.2}$$

Totally antisymmetric three-tensor:

$$\epsilon_{ijk} = \begin{cases} +1 & \{i, j, k\} \text{ even permutation of } \{1, 2, 3\} \\ -1 & \text{odd permutation} \\ 0 & \text{otherwise} \end{cases}$$

$$\epsilon^{0ijk} = -\epsilon_{0ijk} = \epsilon^{ijk} = \epsilon_{ijk}$$

$$\epsilon_{ijk}\epsilon_{ilm} = \delta_{jl}\delta_{km} - \delta_{jm}\delta_{kl}. \tag{2.3}$$

Pauli matrices:

$$\sigma^j\sigma^k = \delta^{jk}I + i\epsilon^{jkl}\sigma^l \qquad (j, k, l = 1, 2, 3)$$
$$\sigma^j_{ab}\sigma^j_{cd} = 2\delta_{ad}\delta_{bc} - \delta_{ab}\delta_{cd} \qquad (a, b, c, d = 1, 2). \tag{2.4}$$

Dirac matrices:

$$\gamma_5 = -i\gamma^0\gamma^1\gamma^2\gamma^3$$
$$\sigma^{\mu\nu} = \frac{i}{2}\left[\gamma^\mu, \gamma^\nu\right]$$
$$\gamma^\mu\gamma^\nu\gamma^\alpha = g^{\mu\nu}\gamma^\alpha + g^{\nu\alpha}\gamma^\mu - g^{\alpha\mu}\gamma^\nu - i\epsilon^{\mu\nu\alpha\beta}\gamma_\beta\gamma_5$$
$$\gamma^0\Gamma_i^\dagger\gamma^0 = \Gamma_i \qquad (\Gamma_i = 1, \gamma^\mu, \gamma^\mu\gamma_5, \sigma^{\mu\nu})$$
$$\gamma^0\Gamma_i^\dagger\gamma^0 = -\Gamma_i \qquad (\Gamma_i = \gamma_5). \tag{2.5}$$

Trace relations:

$$\text{Tr}\,(\gamma^\mu) = 0$$
$$\text{Tr}\,(\gamma_5) = 0$$
$$\text{Tr}\,(\gamma^\mu\gamma^\nu) = 4g^{\mu\nu}$$
$$\text{Tr}\,(\gamma^\mu\gamma^\nu\gamma_5) = 0$$
$$\text{Tr}\,(\gamma^\mu\gamma^\nu\gamma^\alpha\gamma^\beta) = 4\left(g^{\mu\nu}g^{\alpha\beta} - g^{\mu\alpha}g^{\nu\beta} + g^{\mu\beta}g^{\nu\alpha}\right)$$
$$\text{Tr}\,(\gamma_5\gamma^\mu\gamma^\nu\gamma^\alpha\gamma^\beta) = 4i\epsilon^{\mu\nu\alpha\beta}$$
$$\text{Tr}\,(\displaystyle{\not}a_1 \ldots \displaystyle{\not}a_{2n+1}) = 0$$
$$\text{Tr}\,(\displaystyle{\not}a_1 \ldots \displaystyle{\not}a_{2n}) = \text{Tr}\,(\displaystyle{\not}a_{2n} \ldots \displaystyle{\not}a_1). \tag{2.6}$$

Plane wave solutions:

The Dirac spinor $u(\mathbf{p}, s)$ is a positive-energy eigenstate of the momentum $\mathbf{p}$ and energy $E = \sqrt{\mathbf{p}^2 + m^2}$. Antifermions are described in terms of the Dirac spinor $v(\mathbf{p}, s)$. The adjoint solutions are denoted by $\bar{u} \equiv u^\dagger \gamma^0$ and $\bar{v} \equiv v^\dagger \gamma^0$. Note that our normalization of Dirac spinors behaves smoothly in the massless limit.

$$(\not{p} - m)u(\mathbf{p}, s) = 0$$
$$\bar{u}(\mathbf{p}, s)(\not{p} - m) = 0$$
$$(\not{p} + m)v(\mathbf{p}, s) = 0$$
$$\bar{v}(\mathbf{p}, s)(\not{p} + m) = 0$$
$$\bar{u}(\mathbf{p}, r)u(\mathbf{p}, s) = 2m\delta_{rs}$$
$$\bar{v}(\mathbf{p}, r)v(\mathbf{p}, s) = -2m\delta_{rs}$$
$$u^\dagger(\mathbf{p}, r)u(\mathbf{p}, s) = 2E\delta_{rs}$$
$$v^\dagger(\mathbf{p}, r)v(\mathbf{p}, s) = 2E\delta_{rs}$$
$$\sum_s u(\mathbf{p}, s)\bar{u}(\mathbf{p}, s) = \not{p} + m$$
$$\sum_s v(\mathbf{p}, s)\bar{v}(\mathbf{p}, s) = \not{p} - m. \tag{2.7}$$

Gordon decomposition for a fermion of mass m:

$$\bar{u}(\mathbf{p}', r) \gamma^\mu u(\mathbf{p}, s) = \bar{u}(\mathbf{p}', r) \left(\frac{(p' + p)^\mu}{2m} + \frac{i\sigma^{\mu\nu}(p' - p)_\nu}{2m} \right) u(\mathbf{p}, s). \tag{2.8}$$

Dirac representation:

$$\gamma^0 = \begin{pmatrix} 1 & 0 \\ 0 & -1 \end{pmatrix} \quad \boldsymbol{\gamma} = \begin{pmatrix} 0 & \boldsymbol{\sigma} \\ -\boldsymbol{\sigma} & 0 \end{pmatrix} \quad \gamma_5 = \begin{pmatrix} 0 & -1 \\ -1 & 0 \end{pmatrix} \tag{2.9}$$

$$u(\mathbf{p}, s) = \sqrt{E + m} \begin{pmatrix} \chi_s \\ \dfrac{\boldsymbol{\sigma} \cdot \mathbf{p}}{E + m} \chi_s \end{pmatrix} \quad v(\mathbf{p}, s) = \sqrt{E + m} \begin{pmatrix} \dfrac{\boldsymbol{\sigma} \cdot \mathbf{p}}{E + m} \chi_s \\ \chi_s \end{pmatrix}. \tag{2.10}$$

Fierz relations:

The anticommutativity of fermion fields and the algebra of Dirac matrices imply the (particularly useful) Fierz relations,

$$\bar{\psi}_1 \gamma^\mu (1 + \gamma_5) \psi_2 \bar{\psi}_3 \gamma_\mu (1 + \gamma_5) \psi_4 = \bar{\psi}_1 \gamma^\mu (1 + \gamma_5) \psi_4 \bar{\psi}_3 \gamma_\mu (1 + \gamma_5) \psi_2$$
$$\bar{\psi}_1 \gamma^\mu (1 + \gamma_5) \psi_2 \bar{\psi}_3 \gamma_\mu (1 - \gamma_5) \psi_4 = -2 \bar{\psi}_1 (1 - \gamma_5) \psi_4 \bar{\psi}_3 (1 + \gamma_5) \psi_2. \tag{2.11}$$

Propagators:

The propagators associated with fields $\varphi(x)$, $\psi(x)$, $W_\lambda(x)$ having spins 0, 1/2, 1 and masses μ, m, M are, respectively,

$$i\Delta_F(x) = \langle 0|T\left(\varphi(x)\varphi^\dagger(0)\right)|0\rangle = \int \frac{d^4p}{(2\pi)^4} e^{-ip\cdot x} \frac{i}{p^2 - \mu^2 + i\epsilon}$$

$$iS_{F\beta\alpha}(x) = \langle 0|T\left(\psi_\beta(x)\bar{\psi}_\alpha(0)\right)|0\rangle = \int \frac{d^4p}{(2\pi)^4} e^{-ip\cdot x} \frac{i\,(\not{p}+m)_{\beta\alpha}}{p^2 - m^2 + i\epsilon}$$

$$iD_{F\lambda\nu}(x) = \langle 0|T\left(W^\lambda(x)W^{\dagger\nu}(0)\right)|0\rangle$$

$$= \int \frac{d^4p}{(2\pi)^4} e^{-ip\cdot x} \frac{i\left(-g_{\lambda\nu} + (1-\xi)p_\lambda p_\nu/\left(p^2 - \xi M^2 + i\epsilon\right)\right)}{p^2 - M^2 + i\epsilon}, \quad (2.12)$$

where ξ is a gauge-dependent parameter.

Feynman parameterization:

$$\frac{1}{a^n b^m} = \frac{\Gamma(n+m)}{\Gamma(n)\Gamma(m)} \int_0^1 dx \frac{x^{n-1}(1-x)^{m-1}}{[ax + b(1-x)]^{n+m}} \quad (n, m > 0)$$

$$\frac{1}{abc} = 2\int_0^1 x\,dx \int_0^1 dy \frac{1}{[a(1-x) + bxy + cx(1-y)]^3}. \quad (2.13)$$

C–3 Decay lifetimes and cross sections

Parameters of choice for quantum fields:

The literature reveals a variety of conventions employed in quantum field theory. We can characterize all of these with certain parameters of choice, J_i, K_i, L_i ($i = B, F$ distinguishes bosons from fermions), occurring in the normalization of spin zero and spin one-half fields,

$$\varphi(x) = \int \frac{d^3k}{J_B} \left(a(\mathbf{k})e^{-ik\cdot x} + a^\dagger(\mathbf{k})e^{ik\cdot x}\right)$$

$$\psi(x) = \sum_s \int \frac{d^3p}{J_F} \left(b(\mathbf{p}, s)u(\mathbf{p}, s)e^{-ip\cdot x} + d^\dagger(\mathbf{p}, s)v(\mathbf{p}, s)e^{ip\cdot x}\right), \quad (3.1)$$

in momentum space algebraic relations, e.g.,

$$\left[a(\mathbf{k}), a^\dagger(\mathbf{k}')\right] = K_B\delta^3(\mathbf{k} - \mathbf{k}'),$$

$$\left\{b(\mathbf{p}, r), b^\dagger(\mathbf{p}', s)\right\} = K_F\delta_{rs}\delta^3(\mathbf{p} - \mathbf{p}'), \quad (3.2)$$

and in the normalization of single-particle states

$$|\mathbf{k}\rangle_B = L_B a^\dagger(\mathbf{k})|0\rangle, \qquad |\mathbf{p}, s\rangle_F = L_F b^\dagger(\mathbf{p}, s)|0\rangle. \quad (3.3)$$

It is convenient to introduce an additional parameter N_F to characterize the choice of fermion spinor normalization,

$$u^\dagger(\mathbf{p}, r)u(\mathbf{p}, s) = N_F 2E_\mathbf{p}\delta_{rs}. \tag{3.4}$$

For uniformity of notation, we also define $N_B \equiv 1$. The constants J_i, K_i, N_i are constrained by the canonical commutation or anticommutation relations to obey

$$\frac{K_i N_i}{J_i^2} = \frac{1}{(2\pi)^3 2E} \qquad (i = B, F). \tag{3.5}$$

Using the above, one can express the single-particle expectation value of the quantum mechanical probability density as

$$\rho_i = \frac{K_i L_i^2}{(2\pi)^3} \qquad (i = B, F). \tag{3.6}$$

The conventions employed in this book, together with the implied normalization for boson or fermion single-particle states, are

$$L_B = L_F = N_B = N_F = 1, \qquad J_B = J_F = K_B = K_F = 2E(2\pi)^3,$$
$$\langle \mathbf{p}', s | \mathbf{p}, r \rangle = 2E_\mathbf{p}\delta_{rs}(2\pi)^3\delta^{(3)}(\mathbf{p}' - \mathbf{p}), \tag{3.7}$$

where r, s are spin labels. This choice, although somewhat unconventional for fermions,[1] has the advantages that bosons and fermions are treated symmetrically throughout the formalism, the zero-mass limit presents no difficulty, and matrix elements are free of cumbersome kinematic factors.

Lifetimes:

From the decay law $N(t) = N(0)e^{-t/\tau}$, the inverse mean life τ^{-1} is seen to be the transition rate per decaying particle, $\Gamma = \tau^{-1} = -\dot{N}/N$. For decay of a particle of energy E_1 into a total of $n - 1$ bosons and/or fermions, the $\mathcal{S}$-matrix amplitude can be written in terms of a reduced (or invariant) amplitude $\mathcal{M}_{fi}$ as

$$\langle f | \mathcal{S} - 1 | i \rangle = -i(2\pi)^4\delta^{(4)}(p_1 - p_2 \cdots - p_n) \prod_{k=1}^{n} \left(\frac{K_k L_k}{J_k}\right) \mathcal{M}_{fi}$$
$$= -i(2\pi)^4\delta^{(4)}(p_1 - p_2 \cdots - p_n) \prod_{k=1}^{n} \left(\frac{\rho_k}{2E_k N_k}\right)^{1/2} \mathcal{M}_{fi}, \tag{3.8}$$

where the index k labels the individual particles as to whether they are bosons or fermions. The inverse lifetime is computed from the squared S-matrix amplitude per spacetime volume VT and incident particle density ρ_1, integrated over final-state phase space. The choice of phase space is already fixed by our analysis. Thus,

---

[1] Another book sharing this convention is [ChL 84].

defining a parameter of choice $A(\mathbf{p})$ for the (momentum) phase space per particle,

$$\text{Phase space per particle} \equiv \int \frac{d^3\mathbf{k}}{A(\mathbf{k})}, \tag{3.9}$$

the application of completeness to Eq. (3.7) yields

$$\langle \mathbf{p}'|\mathbf{p}\rangle = \int \frac{d^3k}{A} \langle \mathbf{p}'|\mathbf{k}\rangle\langle \mathbf{k}|\mathbf{p}\rangle \;\Rightarrow\; A = KL^2 = (2\pi)^3\rho. \tag{3.10}$$

The inverse lifetime (or decay width) is then given by

$$\tau^{-1} = \Gamma = \frac{1}{\rho_1}\frac{1}{\mathcal{Z}} \int \left(\prod_{k=2}^{n} \frac{d^3 p_k}{(2\pi)^3 \rho_k}\right) \frac{|\mathcal{S}-1|_{\text{fi}}^2}{VT}$$

$$= \frac{1}{2E_1 N_1}\frac{1}{\mathcal{Z}} \int \left(\prod_{k=2}^{n} \frac{d^3 p_k}{(2\pi)^3 2E_k N_k}\right) (2\pi)^4 \delta^4(p_1 - \cdots - p_n) \sum_{\text{int}} |\mathcal{M}_{\text{fi}}|^2, \tag{3.11}$$

where $\mathcal{Z} = \prod_j n_j!$ is a statistical factor accounting for the presence of n_j identical particles of type j in the final state, and the sum 'int' is over internal degrees of freedom such as spin and color.

Cross sections:
For the reaction $1 + 2 \rightarrow 3 + \ldots n$, the cross section σ is the transition rate per incident flux. The incident flux f_{inc} can be represented as

$$f_{\text{inc}} = \rho_1 \rho_2 |\mathbf{v}_1 - \mathbf{v}_2| = \frac{\rho_1 \rho_2}{E_1 E_2}[(p_1 \cdot p_2)^2 - m_1^2 m_2^2]^{1/2}, \tag{3.12}$$

and the cross section becomes

$$\sigma = \frac{1}{\mathcal{Z}} \frac{1}{4\left((p_1 \cdot p_2)^2 - m_1^2 m_2^2\right)^{1/2}}$$

$$\times \int \left(\prod_{k=3}^{n} \frac{d^3 p_k}{(2\pi)^3 2E_k N_k}\right) (2\pi)^4 \delta^4(p_1 + p_2 - \cdots - p_n) \sum_{\text{int}} |\mathcal{M}_{\text{fi}}|^2. \tag{3.13}$$

Watson's theorem:
The scattering operator S is unitary, $S^\dagger S = 1$. Thus, the transition operator $\mathcal{T}$, defined by $S = 1 - i\mathcal{T}$, obeys $i(\mathcal{T} - \mathcal{T}^\dagger) = \mathcal{T}^\dagger \mathcal{T}$. With the aid of the relation $\langle f|\mathcal{T}^\dagger|i\rangle = \langle i|\mathcal{T}|f\rangle^*$, we obtain the unitarity constraint for matrix elements,

$$i\left(\mathcal{T}_{\text{fi}} - \mathcal{T}_{\text{if}}^*\right) = \sum_n \mathcal{T}_{\text{nf}}^* \mathcal{T}_{\text{ni}}, \tag{3.14}$$

where $\mathcal{T}_{\mathrm{fi}} \equiv \langle f | \mathcal{T} | i \rangle$. This constraint implies the existence of phase relations between the various intermediate-state amplitudes. For example, consider a weak transition followed by a strong final-state interaction for which there is a unique intermediate state identical to the final state,

$$A \underset{\text{weak}}{\longrightarrow} BC \underset{\text{strong}}{\longrightarrow} BC, \tag{3.15}$$

i.e., $i = A, n = f = BC$. In this circumstance, time-reversal invariance of the hamiltonian implies $\mathcal{T}_{\mathrm{fi}} = \mathcal{T}_{\mathrm{if}}$, so the left-hand side of the unitarity relation reduces to $-2\mathrm{Im}\mathcal{T}_{\mathrm{if}}$ and both sides of Eq. (3.14) are real-valued. Denoting the weak and strong matrix elements as $|T_{\mathrm{w}}| e^{i\delta_{\mathrm{w}}}$ and $|T_{\mathrm{s}}| e^{i\delta_{\mathrm{s}}}$, it then follows that $\delta_{\mathrm{w}} = \delta_{\mathrm{s}}$.

C–4 Field dimension

We consider a limit in which the theory is invariant under the set of scale transformations $x^{\mu} \to \lambda x^{\mu}$ ($\lambda > 0$) of the spacetime coordinates. Associate with each such coordinate transformation a unitary operator $U(\lambda)$ whose effect on a generic quantum field Φ is given by $U(\lambda)\Phi(x)U^{\dagger}(\lambda) = \lambda^{d_{\Phi}} \Phi(\lambda x)$, where d_{Φ} is the *dimension* of the field Φ. From the canonical commutation relation obeyed by a boson field φ or the canonical anticommutation relation obeyed by a fermion field ψ_{α},

$$[\varphi(0, \mathbf{x}), \dot{\varphi}(0)] = i\delta^{3}(\mathbf{x}), \qquad \left\{ \psi_{\alpha}(0, \mathbf{x}), \psi_{\beta}^{\dagger}(0) \right\} = \delta_{\alpha\beta} \delta^{3}(\mathbf{x}), \tag{4.1}$$

it follows that the *canonical* field dimensions are $d_{\varphi} = 1$ and $d_{\psi} = 3/2$. Composites built from products of these fields carry a dimension of their own, e.g., all fermion bilinears $\overline{\psi}\Gamma\psi$ (Γ is a 4×4 matrix) have canonical dimension 3. Unless protected by some kind of algebraic relation, a field dimension will generally be modified from the canonical value by interaction-dependent *anomalous* dimensions. Field dimensions are particularly useful in ordering the terms contained in a short-distance expansion,

$$A(x)B(0) \underset{x \to 0}{\longrightarrow} \sum_{n} c_{n}(x) O_{n}, \tag{4.2}$$

where A, B, O_{n} are local quantum fields. From the scale invariance of the short-distance limit, it follows that $c_{n}(x) \sim x^{d_{O_{n}} - d_{A} - d_{B}}$. Thus, the fields O_{n} of lowest dimension have the most singular coefficient functions.

C–5 Mathematics in d dimensions

Dirac algebra:

The following set of rules, generally referred to as NDR (*naive dimensional regularization*), is the one most commonly used in the literature. We employ a metric

$g_{\mu\nu}$ corresponding to a spacetime of continuous dimension d and maintain certain $d = 4$ properties of the Dirac matrices such as the trace relations of Eq. (2.6). In the following, I_d is a diagonal d-dimensional matrix with Tr $I_d = 4$ and $\epsilon \equiv (4-d)/2$.

$$g_\mu{}^\mu = d$$
$$\{\gamma_\mu, \gamma_\nu\} = 2g_{\mu\nu}I_d$$
$$\gamma_\mu\gamma^\mu = d\, I_d$$
$$\gamma_\mu\, \slashed{p}\, \gamma^\mu = (2\epsilon - 2)\, \slashed{p}$$
$$\gamma_\mu\, \slashed{p}\, \slashed{q}\, \gamma^\mu = 4p \cdot q\, I_d - 2\epsilon\, \slashed{p}\, \slashed{q}$$
$$\gamma_\mu\, \slashed{p}\, \slashed{q}\, \slashed{r}\, \gamma^\mu = -2\, \slashed{r}\, \slashed{q}\, \slashed{p} + 2\epsilon\, \slashed{p}\, \slashed{q}\, \slashed{r}$$
$$\slashed{p}\slashed{q}\slashed{r} + \slashed{r}\slashed{q}\slashed{p} = 2p \cdot q\, \slashed{r} + 2q \cdot r\, \slashed{p} - 2p \cdot r\, \slashed{q}$$
$$\{\gamma_\mu, \gamma_5\} = 0. \tag{5.1}$$

Note that in NDR, γ_5 anticommutes with the gamma matrices. This will suffice for the calculations appearing in this book, but is not valid for all amplitudes (e.g. closed odd-parity fermion loops).

Integrals:
For the following integrals, we define the denominator function

$$\mathcal{D} \equiv m_1^2 x + m_2^2(1 - x) - q^2 x(1 - x) - i\epsilon, \tag{5.2}$$

take $n_1, n_2 \geq 1$, and denote $i\epsilon$ as the infinitesimal Feynman parameter.

$$\int \frac{d^d p}{(2\pi)^d} \frac{1}{\left[(p - q)^2 - m_1^2 + i\epsilon\right]^{n_1} \left[p^2 - m_2^2 + i\epsilon\right]^{n_2}}$$
$$= (-1)^{n_1+n_2} \frac{i}{(4\pi)^{d/2}} \frac{\Gamma(n_1 + n_2 - d/2)}{\Gamma(n_1)\Gamma(n_2)} \int_0^1 dx\, \frac{x^{n_1-1}(1 - x)^{n_2-1}}{\mathcal{D}^{n_1+n_2-d/2}}, \tag{5.3a}$$

$$\int \frac{d^d p}{(2\pi)^d} \frac{p^\mu}{\left[(p - q)^2 - m_1^2 + i\epsilon\right]^{n_1} \left[p^2 - m_2^2 + i\epsilon\right]^{n_2}}$$
$$= (-1)^{n_1+n_2} q^\mu \frac{i}{(4\pi)^{d/2}} \frac{\Gamma(n_1 + n_2 - d/2)}{\Gamma(n_1)\Gamma(n_2)} \int_0^1 dx\, \frac{x^{n_1}(1 - x)^{n_2-1}}{\mathcal{D}^{n_1+n_2-d/2}}, \tag{5.3b}$$

$$\int \frac{d^d p}{(2\pi)^d} \frac{p^\mu p^\nu}{\left[(p - q)^2 - m_1^2 + i\epsilon\right]^{n_1} \left[p^2 - m_2^2 + i\epsilon\right]^{n_2}}$$
$$= \frac{i}{(4\pi)^{d/2}} \frac{(-1)^{n_1+n_2}}{\Gamma(n_1)\Gamma(n_2)} \left[q^\mu q^\nu \Gamma(n_1 + n_2 - d/2) \int_0^1 dx\, \frac{x^{n_1+1}(1 - x)^{n_2-1}}{\mathcal{D}^{n_1+n_2-d/2}} \right.$$
$$\left. - \frac{g^{\mu\nu}}{2} \Gamma(n_1 + n_2 - 1 - d/2) \int_0^1 dx\, \frac{x^{n_1-1}(1 - x)^{n_2-1}}{\mathcal{D}^{n_1+n_2-1-d/2}} \right], \tag{5.3c}$$

$$\int \frac{d^d p}{(2\pi)^d} \frac{p^\mu p^\nu p^\lambda}{\left[(p-q)+i\epsilon\right]^{n_1} \left[p^2 - m_2^2 + i\epsilon\right]^{n_2}}$$

$$= \frac{i}{(4\pi)^{d/2}} \frac{(-1)^{n_1+n_2}}{\Gamma(n_1)\Gamma(n_2)} \left[q^\mu q^\nu q^\lambda \Gamma(n_1 + n_2 - d/2) \int_0^1 dx \, \frac{x^{n_1+2}(1-x)^{n_2-1}}{\mathcal{D}^{n_1+n_2-d/2}} \right.$$

$$\left. - \frac{1}{2} \left(g^{\mu\nu} q^\lambda + g^{\mu\lambda} q^\nu + g^{\nu\lambda} q^\mu \right) \Gamma(n_1 + n_2 - 1 - d/2) \int_0^1 dx \, \frac{x^{n_1}(1-x)^{n_2-1}}{\mathcal{D}^{n_1+n_2-1-d/2}} \right],$$

$$(5.3d)$$

$$\int \frac{d^d p}{(2\pi)^d} \frac{p^\mu p^\nu p^\lambda p^\sigma}{\left[(p-q)^2 - m_1^2 + i\epsilon\right]^{n_1} \left[p^2 - m_2^2 + i\epsilon\right]^{n_2}}$$

$$= \frac{i}{(4\pi)^{d/2}} \frac{(-1)^{n_1+n_2}}{\Gamma(n_1)\Gamma(n_2)} \left[q^\mu q^\nu q^\lambda q^\sigma \Gamma(n_1 + n_2 - d/2) \int_0^1 dx \, \frac{x^{n_1+3}(1-x)^{n_2-1}}{\mathcal{D}^{n_1+n_2-d/2}} \right.$$

$$- \frac{1}{2} \left(g^{\mu\nu} q^\lambda q^\sigma + g^{\mu\lambda} q^\nu q^\sigma + 4\,\text{perm.} \right) \Gamma(n_1 + n_2 - 1 - d/2) \int_0^1 dx \, \frac{x^{n_1+1}(1-x)^{n_2-1}}{\mathcal{D}^{n_1+n_2-1-d/2}}$$

$$\left. + \frac{1}{4} \left(g^{\mu\nu} g^{\lambda\sigma} + g^{\mu\lambda} g^{\nu\sigma} + g^{\mu\sigma} g^{\nu\lambda} \right) \Gamma(n_1 + n_2 - 2 - d/2) \int_0^1 dx \, \frac{x^{n_1-1}(1-x)^{n_2-1}}{\mathcal{D}^{n_1+n_2-2-d/2}} \right].$$

$$(5.3e)$$

Solid angle:

$$\Omega_d = \int_0^\pi d\theta_{d-1} \sin^{d-2}\theta_{d-1} \ldots \int_0^\pi d\theta_2 \sin\theta_2 \int_0^{2\pi} d\theta_1 = \frac{2\pi^{d/2}}{\Gamma(d/2)}.$$

$$\Omega_2 = 2\pi, \quad \Omega_3 = 4\pi, \quad \Omega_4 = 2\pi^2, \ldots. \tag{5.4}$$

Gamma, psi, beta, and hypergeometric functions:

$$\Gamma(z) = \int_0^\infty dt \, e^{-t} \, t^{z-1} \qquad (\text{Re } z > 0),$$

$$\Gamma(z+1) = z\Gamma(z) = z(z-1)\Gamma(z-1) = \cdots = z!,$$

$$\Gamma(-n+\epsilon) = \frac{(-)^n}{n!} \left[\frac{1}{\epsilon} + \psi(n+1) + \mathcal{O}(\epsilon) \right] \qquad (n \text{ integer}),$$

$$d\Gamma(z)/dz = \Gamma(z)\psi(z) \text{ where } \psi(z+1) = \psi(z) + 1/z,$$

$$\psi(1) = -\gamma = -\lim_{n\to\infty} \left(1 + \frac{1}{2} + \cdots + \frac{1}{n} - \ln n \right) \simeq -0.5772,$$

$$d\psi(z+1)/dz \equiv \psi'(z+1) = \psi'(z) - 1/z^2 \text{ with } \psi'(1) = \pi^2/6,$$

$$B(z,w) = \frac{\Gamma(z)\Gamma(w)}{\Gamma(z+w)} = 2 \int_0^\infty dt \, \frac{t^{2z-1}}{(t^2+1)^{z+w}} \qquad (\text{Re } z, \text{ Re } w > 0),$$

$$F(a, b; c; z) = \frac{\Gamma(c)}{\Gamma(b)\Gamma(c-b)} \int_0^1 dt \; t^{b-1}(1-t)^{c-b-1}(1-zt)^{-a}$$

$$(\text{Re } c > \text{Re } b > 0),$$

$$F(a, b; c; 1) = \frac{\Gamma(c)\Gamma(c-a-b)}{\Gamma(c-a)\Gamma(c-b)},$$

$$\frac{dF(a, b; c; z)}{dz} = \frac{ab}{c} F(a+1, b+1; c+1; z). \tag{5.5}$$

References

In the text, papers are typically cited by using the first two letters of the surname of the first author plus the first letter of those of each remaining authors plus the year of publication. If this becomes impractical due to the number of authors, the catch-all *et al.* is used after the first author. Also, the Review of Particle Properties, *Phys. Rev.* **D86**, 010001 (2012) is cited as [RPP 12]. Experimental results given without citation can be found in [RPP 12].

Aad, G. *et al.* (ATLAS collab.) (2011). Search for contact interactions in dimuon events from *pp* collisions at $\sqrt{s} = 7$ TeV with the ATLAS Detector, *Phys. Rev.* **D84**, 011101.

Aad, G. *et al.* (ATLAS collab.) (2012). Measurement of W^+W^- production in *pp* collisions at $\sqrt{s} = 7$ TeV with the ATLAS detector and limits on anomalous *WWZ* and *WWγ* couplings, *Phys. Rev.* **D87**, 112001.

Aad, G. *et al.* (ATLAS collab.) (2013*a*). Search for a light charged Higgs boson in the decay channel $H^+ \to c\bar{s}$ in $t\bar{t}$ events using *pp* collisions at $\sqrt{s} = 7$ TeV with the ATLAS detector, *Eur. Phys. J.* **C73**, 2465.

Aad, G. *et al.* (ATLAS collab.) (2013*b*). Evidence for the spin-0 nature of the Higgs boson using ATLAS data (arXiv:1307.1432 [hep-ex]).

Aaij, R. *et al.* (LHCb collab.) (2012). A model-independent Dalitz plot analysis of $B^\pm \to DK^\pm$ with $D \to K_S^0 h^+ h^-$ ($h = \pi, K$) decays and constraints on the CKM angle γ, *Phys. Lett.* **B718**, 43.

Aaij, R. *et al.* (LHCb collab.) (2013*a*). Observation of D^0–$\overline{D}^0$ oscillations, *Phys. Rev. Lett.* **110**, 101802.

Aaij R. *et al.* (LHCb collab.) (2013*b*). First evidence for the decay $B_s^0 \to \mu^+\mu^-$, *Phys. Rev. Lett.* **110**, 021801.

Aaij R. *et al.* (LHCb collab.) (2013*c*). Measurement of D^0–$\bar{D}^0$ mixing parameters and search for *CP* violation using $D^0 \to K^+\pi^-$ decays (arXiv:1309.6534 [hep-ex]).

Aaij R. *et al.* (LHCb collab.) (2013*d*). Measurements of indirect *CP* asymmetries in $D^0 \to K^-K^+$ and $D^0 \to \pi^-\pi^+$ decays (arXiv:1310.7201[hep-ex]).

Aaltonen, T. *et al.* (CDF and D0 collab.) (2013). Higgs boson studies at the Tevatron, *Phys. Rev.* **D88**, 052014.

Abbott, L. (1982). Introduction to the background field method, *Acta Phys. Pol.* **B13**, 33.

Abe, K. *et al.* (SLD collab.) (2000). A high precision measurement of the left-right Z boson cross-section asymmetry, *Phys. Rev. Lett.* **84**, 5945.

Abe, K. *et al.* (T2K collab.) (2013). Evidence of electron neutrino appearance in a muon neutrino beam, *Phys. Rev.* **D88**, 032002.

Abe, Y. *et al.* (DOUBLE-CHOOZ collab.) (2012). Indication for the disappearance of reactor electron antineutrinos in the Double Chooz experiment, *Phys. Rev. Lett.* **108**, 131801.

Ablikim, M. *et al.* (BESIII collab.) (2013). Observation of a charged charmoniumlike structure in $e^+e^- \to \pi^+\pi^- J/\psi$ at $\sqrt{s} = 4.26$ GeV, *Phys. Rev. Lett.* **110**, 252001.

Abouzaid, E. *et al.* (KTeV collab.) (2008). Final results from the KTeV experiment on the decay $K_L \to \pi^0\gamma\gamma$, *Phys. Rev.* **D77**, 112004.

Ackerman, N. *et al.* (EXO-200 collab.) (2011). Observation of two-neutrino double-beta decay in ^{136}Xe with EXO-200, *Phys. Rev. Lett.* **107**, 212501.

Adam, J. *et al.* (MEG collab.) (2013). New constraint on the existence of the $\mu^+ \to e^+\gamma$ decay (arXiv:1303.0754 [hep-ex]).

Ade, P.A.R. *et al.* (Planck collab.) (2013). Planck 2013 results. XVI. cosmological parameters (arXiv:1303.5076 [astro-ph.CO]).

Ademollo, M. and Gatto, R. (1964). Nonrenormalization theorem for the strangeness-violating vector currents, *Phys. Rev. Lett.* **13**, 264.

Adkins, G., Nappi, C., and Witten, E. (1983). Static properties of nucleons in the Skyrme model, *Nucl. Phys.* **B228**, 552.

Adler, S.L. (1969). Axial-vector vertex in spinor electrodynamics, *Phys. Rev.* **177**, 2426.

Adler, S.L. (1970). Perturbation theory anomalies, in *Lectures on Elementary Particle Physics,* ed. S. Deser, M. Grisaru and H. Pendleton (MIT Press, Cambridge, MA).

Adler, S.L. and Bardeen, W.A. (1969). Absence of higher-order corrections in the anomalous axial-vector divergence equation, *Phys. Rev.* **182**, 1517.

Adler, S.L. and Dashen, R. (1968). *Current Algebras and Applications to Particle Physics* (Benjamin, New York).

Aguilar, A. *et al.* (LSND collab.) (2001). Evidence for neutrino oscillations from the observation of anti-neutrino(electron) appearance in a anti-neutrino(muon) beam, *Phys. Rev.* **D64**, 112007.

Aharmim, B. *et al.* (SNO collab.) (2011). Combined analysis of all three phases of solar neutrino data from the Sudbury Neutrino Observatory, *Prog. Part. Nucl. Phys.* **71**, 150.

Ahn, J.K. *et al.* (RENO collab.) (2012). Observation of reactor electron antineutrino disappearance in the RENO experiment, *Phys. Rev. Lett.* **108**, 191802.

Ahrens, J. *et al.* (2005). Measurement of the π^+ meson polarizabilities via the $\gamma p \to \gamma\pi^+n$ reaction, *Eur. J. Phys.* **A23**, 113.

Aidala, C.A., Bass, S.D., Hasch D., and Mallot, G.K. (2013). The spin structure of the nucleon, *Rev. Mod. Phys.* **85**, 655.

Akhundov, A.A., Arbuzov, A., Riemann, S., and Riemann, T. (2013). Zfitter 1985-2013 (arXiv:1302.1395 [hep-ph]).

Akhundov, A.A., Bardin, D.Yu., and Reimann, T. (1986). Electroweak one-loop corrections to the decay of the neutral vector boson, *Nucl. Phys.* **B276**, 1.

Alarcon, J.M., Camalich, J.M., and Oller, J.A. (2013). Low energy analysis of $\acute{z}N$ scattering and the pion-nucleon sigma term with covariant baryon chiral perturbation theory (arXiv:1301:3067 [hep-ph]).

Alekhin, S., Djouadi, A., and Moch, S. (2012). The top quark and Higgs boson masses and the stability of the electroweak vacuum, *Phys. Lett.* **B716**, 214.

Ali, A., Hambrock, C., and Wang, W. (2012). Tetraquark interpretation of the charged bottomonium-like states $Z_b(10610)$ and $Z_b(10650)$ and implications, *Phys. Rev.* **D85**, 054011.

Altarelli, G. and Maiani, L. (1974). Octet enhancement of nonleptonic weak interactions in asymptotically free gauge theories, *Phys. Lett.* **B52**, 351.

Altarelli, G. and Ross, G.G. (1988). The anomalous gluon contribution to polarized leptoproduction, *Phys. Lett.* **B212**, 391.

Amaldi, U., de Boer W., and Fürstenau, H. (1991). Comparison of grand unified theories with electroweak and strong coupling constants measured at *LEP*, *Phys. Lett.* **260**, 447.

Amhis, Y. *et al.* (Heavy Flavor Averaging Group collab.) (2012). Averages of b-hadron, c-hadron, and tau-lepton properties as of early 2012 (arXiv:1207.1158 [hep-ex]).

Amsler, C. and Close, F.E. (1996). Is $f_0(1500)$ a scalar glueball?, *Phys. Rev.* **D53**, 295.

An, F.P. *et al.* (DAYA-BAY collab.) (2012). Observation of electron-antineutrino disappearance at Daya Bay, *Phys. Rev. Lett.* **108**, 171803.

Anderson, P.W. (1984). *Basic Notions in Condensed Matter Physics* (Benjamin/Cummings, Menlo Park, CA).

Anthony, P.L. *et al.* (SLAC E158 collab.) (2005). Precision measurement of the weak mixing angle in Moller scattering, *Phys. Rev. Lett.* **95**, 081601.

Antipov, Yu.M. *et al.* (1985). Experimental estimation of the sum of pion electric and magnetic polarizabilities, *Z. Phys.* **C26**, 495.

Antognini, A. *et al.* (2013). Proton Structure from the Measurement of $2S - 2P$ Transition Frequencies of Muonic Hydrogen, *Science* **339**, 417.

Antonelli,V., Miramonti, L., Pena-Garay, C., and Serenelli, A. (2012). Solar neutrinos (arXiv:1208.1356 [hep-ph]).

Anzai, C., Kiyo, Y., and Sumino, Y. (2010). Static QCD potential at three-loop order, *Phys. Rev. Lett.* **104**, 112003.

Aoki, K., Hioki, Z., Kawabe, R., Konuma, M., and Muta, T. (1982). Electroweak theory, *Suppl. Prog. Theor. Phys.* **73**, 1.

Aoyama, T., Hayakawa, M., Kinoshita, T., and Nio, M. (2012). Quantum electrodynamics calculation of lepton anomalous magnetic moments: numerical approach to the perturbation theory of QED, *Prog. Theor. and Exptal. Phys.* **2012**, 01A107.

Appelquist, T. and Bernard, C. (1981). Nonlinear sigma model in the loop expansion, *Phys. Rev.* **D23**, 425.

Appelquist, T. and Carrazone, J. (1975). Infrared singularities and massive fields, *Phys. Rev.* **D11**, 2856.

Armstrong, D.S. and McKeown, R.D. (2012). Parity violating electron scattering and the electric and magnetic strange form factors of the nucleon, *Ann. Rev. Nucl. Part. Sci.* **62**, 337.

Asplund, M., Basu, S., Ferguson, J.W., and Serenelli, A. (2009). New solar composition: the problem with solar models revisited, *Astrophys. J.* **705**, L123.

Aydemir, U., Anber, M.M., and Donoghue, J.F. (2012). Self-healing of unitarity in effective field theories and the onset of New Physics, *Phys. Rev.* **D86**, 014025.

Baak, M. *et al.* (Gfitter group) (2012). Updated status of the global electroweak fit and constraints on New Physics, *Eur. Phys. J.* **C72**, 2003.

Bahcall, J.N. (1964). Solar neutrinos I. Theoretical, *Phys. Rev. Lett.* **12**, 300.

Bahcall, J.N. (1990). *Neutrino Astrophysics* (Cambridge University Press, Cambridge).

Bailey, J.A., Bernard, C., DeTar, C., *et al.* (2009). The $B \to \pi \ell \nu$ semileptonic form factor from three-flavor lattice QCD: a model-independent determination of $abs[V_{ub}]$, *Phys. Rev.* **D79**, 054507.

Balachandran, A.P., Nair, V.P., Rajeev, S.G., and Stern, A. (1983). Soliton states in the QCD effective lagrangian, *Phys. Rev.* **D27**, 1153.

Balantekin, A.B. and Haxton, W.C. (2013). Neutrino oscillations, *Prog. Part. Nucl. Phys.* **71**, 150.

Bali, G.S. *et al.* (1993). A comprehensive lattice study of SU(3) glueballs, *Phys. Lett.* **B309**, 378.

Barbieri, R., Frigens, H., Giuliani, F., and Haber, H.E. (1990). Precision measurements in electroweak physics and supersymmetry, *Nucl. Phys.* **B341**, 309.

Barbieri, R., Pomarol, A., Rattazzi, R., and Strumia, A. (2004). Electroweak symmetry breaking after LEP-1 and LEP-2, *Nucl. Phys.* **B703**, 127.

Bardeen, W.A. (1969). Anomalous Ward identities in spinor field theories, *Phys. Rev.* **184**, 1848.

Bardin, D. and Passarino, G. (1999). *The Standard Model in the Making: Precision Study of the Electroweak Interactions* (Oxford University Press, Oxford).

Barger, V. and Phillips, R.J.N. (1987). *Collider Physics* (Addison-Wesley, Redwood City, CA).

Bass, S.D. (2005). The spin structure of the proton, *Rev. Mod. Phys.* **77**, 1257.

Bauer, C.W., Fleming, S., Pirjol, D., and Stewart, I.W. (2001). An effective field theory for collinear and soft gluons: heavy to light decays, *Phys. Rev.* **D63**, 114020.

Bauer, C.W., Lange B.O., and Ovanesyan, G. (2011). On Glauber modes in Soft-Collinear Effective Theory, *JHEP* **1107**, 077.

Beane, S.R., Bedaque, P.F., Parreno, A., and Savage, M.J. (2005). Exploring hyperons and hypernuclei with lattice QCD, *Nucl. Phys.* **A747**, 55.

Becher, T. (2010). Soft-collinear effective theory - Lectures on 'The infrared structure of gauge theories', ETH Zurich. see http://www.becher.itp.unibe.ch/lectures.html.

Becher, T. and Hill, R.J. (2006). Comment on form-factor shape and extraction of $abs[V_{ub}]$ from $B \to \pi \ell \nu$, *Phys. Lett.* **B633**, 61.

Beck, D.H. and Holstein, B.R. (2001). Nucleon structure and parity violating electron scattering, *Int. J. Mod. Phys.* **E10**, 1.

Bell, J.S. and Jackiw, R. (1967). A PCAC puzzle: $\pi^0 \to \gamma \gamma$ in the sigma model, *Nuovo Cim.* **60A**, 47.

Bellini, G. *et al.* (Borexino collab.) (2012a). First evidence of pep solar neutrinos by direct detection in Borexino, *Phys. Rev. Lett.* **108**, 051302.

Bellini, G. *et al.* (Borexino collab.) (2012b). Absence of day–night asymmetry of 862 keV ^{7}Be solar neutrino rate in Borexino and MSW oscillation parameters, *Phys. Lett.* **B707**, 22.

Beneke M. and Braun, V.M. (1995). Naive non-abelianization and resummation of fermion bubble chains, *Phys. Lett.* **B348**, 513.

Beneke, M., Efthymiopoulos, I., *et al.* (2000). Top quark physics, *Geneva 1999, Standard Model physics (and more) at the LHC* (CERN, Geneva), p. 419.

Beneke, M. and Smirnov, V.A. (1998). Asymptotic expansion of Feynman integrals near threshold, *Nucl. Phys.* **B522**, 321.

Bennett, S.C., Cho, D., Masterson, B.P., Roberts, J.L., Tanner, C.E., Wieman, C.E., and Wood, C.S. (1997). Measurement of parity nonconservation and an anapole moment in cesium, *Science* **275**, 1759.

Bennett, S.C. and Wieman, C.E. (1999). Measurement of the 6S → 7S transition polarizability in atomic cesium and an improved test of the Standard Model, *Phys. Rev. Lett.* **82**, 2484 (Errata *ibid.* **82**, 4153 (1999); **83**, 889 (1999)).

Benson, D., Bigi, I.I., Mannel, T., and Uraltsev, N. (2003). Imprecated, yet impeccable: on the theoretical evaluation of $\Gamma(B \rightarrow X_c \ell \nu)$, *Nucl. Phys.* **B665**, 367.

Bergstrom, L. and Hulth, G. (1985). Induced Higgs couplings to neutral bosons in e+e- collisions, *Nucl. Phys.* **B259**, 137 (Erratum *ibid.* **B276**, 744 (1986)).

Berman, S.M. (1958). Radiative corrections to muon and neutron decay, *Phys. Rev.* **112**, 267.

Bernard, C., *et al.* (2009). The $\bar{B} \rightarrow D^* \ell \bar{\nu}$ form factor at zero recoil from three-flavor lattice QCD: a model independent determination of $arg[V_{cb}]$, *Phys. Rev.* **D79**, 014506.

Bertlmann (2000). *Anomalies in Quantum Field Theory* (Oxford University Press, Oxford).

Bethke, S. (2009). The 2009 world average of alpha(s), *Eur. Phys. J.* **C64**, 689.

Bethke, S. *et al.* (2011). Workshop on precision measurements of α_s (arXiv:1110.0016 [hep-ph]).

Bhattacharya, T., *et al.* (2012). Probing novel scalar and tensor interactions from (ultra)cold neutrons to the LHC, *Phys. Rev.* **D85**, 054512.

Bigi, I.I., Khoze, V.A., Uraltsev, N.G., and Sanda, A.I. (1989). The question of *CP* noninvariance as seen through the eyes of neutral beauty, in *CP Violation*, ed. C. Jarlskog (World Scientific, Singapore).

Bigi, I.I. and Sanda, A.I. (1981). Note on the observability of *CP* violation in *B* decays, *Nucl. Phys.* **B193**, 85.

Bigi, I.I. and Sanda, A.I. (2000). *CP violation* (Cambridge University Press, Cambridge).

Bigi, I.I., Shifman, M.A., Uraltsev, N.G., and Vainshtein, A.I. (1993). QCD predictions for lepton spectra in inclusive heavy flavor decays, *Phys. Rev. Lett.* **71**, 496.

Bigi, I.I., Shifman, M.A., Uraltsev, N.G., and Vainshtein, A.I. (1994). The pole mass of the heavy quark. Perturbation theory and beyond, *Phys. Rev.* **D50**, 2234.

Bigi, I.I. and Uraltsev, N.G. (2001). $D^0 - \bar{D}^0$ oscillations as a probe of quark hadron duality, *Nucl. Phys.* **B592**, 92.

Bijnens, J. (1990). K_{l4} decays and the low energy expansion, *Nucl. Phys.* **B337**, 635.

Bijnens, J., Borg, F., and Dhonte, P. (2003). $K \rightarrow 3\pi$ decays in chiral perturbation theory, *Nucl. Phys.* **B648**, 317.

Bijnens, J. and Jemos, I. (2012). A new global fit of the L_i^r at next-to-next-to-leading order in Chiral Perturbation Theory, *Nucl. Phys.* **B854**, 631.

Bijnens, J. and Wise, M.B. (1984). Electromagnetic contribution to ϵ'/ϵ, *Phys. Lett.* **B137**, 245.

Birrell, N.D. and Davies, P.C.W. (1982). *Quantum Fields in Curved Space* (Cambridge University Press, Cambridge).

Bjorken, J.D. (1966). Applications of the chiral U(6) × (6) algebra of current densities, *Phys. Rev.* **148**, 1467.

Blum, T. *et al.* (2011). K to $\pi\pi$ decay amplitudes from lattice QCD, *Phys. Rev.* **D84**, 114503.

Blum, T. *et al.* (2012). The $K \to (\pi\pi)_{I=2}$ decay amplitude from lattice QCD, *Phys. Rev. Lett.* **108**, 141601.

Bobrowski, M., Lenz, A., Riedl, J., and Rohrwild, J. (2010). How large can the SM contribution to CP violation in $D^0 - \bar{D}^0$ mixing be?, *JHEP* **1003**, 009.

Böhm, M., Hollik, W., and Speisberger, H. (1986). On the one-loop renormalization of the electroweak Standard Model, *Fort. Phys.* **34**, 688.

Boito, D. *et al.* (2012). An updated determination of α_s from τ decays, *Phys. Rev.* **D85**, 093015.

Bollini, C.G. and Giambiagi, J.J. (1972). Dimensional renormalization: the number of dimensions as a regularizing parameter, *Nuovo Cim.* **12B**, 20.

Borasoy, B. and Holstein, B.R. (1999). Nonleptonic hyperon decays in chiral perturbation theory, *Eur. Phys. J.* **C6**, 85.

Braaten, E., Narison, S., and Pich, A. (1992). QCD analysis of the tau hadronic width, *Nucl. Phys.* **B373**, 581.

Brambilla, N. *et al.* (2011). Heavy quarkonium: progress, puzzles, and opportunities, *Eur. Phys. J.* **C71**, 1534.

Brambilla, N., Pineda, A., Soto, J., and Vairo, A. (2005). Effective field theories for heavy quarkonium, *Rev. Mod. Phys.* **77**, 1423.

Branco, G.C., Lavoura, L., and Silva, J.P. (1999). *CP Violation* (Oxford University Press, Oxford).

Brod, J. and Gorbahn, M. (2012). Next-to-next-to-leading-order charm-quark contribution to the CP violation parameter ϵ_K and ΔM_K, *Phys. Rev. Lett.* **108**, 121801.

Brod, J., Gorbahn M., and Stamou, E. (2011). Two-loop electroweak corrections for the $K\pi\nu\bar{\nu}$ decays, *Phys. Rev.* **D83**, 034030.

Brodsky, S.J. and Lepage, G.P. (1980). Exclusive processes in perturbative quantum chromodynamics, *Phys. Rev.* **D22**, 2157.

Buchalla, G., Buras, A.J. and Harlander, M.K. (1990). The anatomy of ϵ'/ϵ in the Standard Model, *Nucl. Phys.* **B337**, 313.

Buchalla, G., Buras, A.J. and Lautenbacher, M.E. (1996). Weak decays beyond leading logarithms, *Rev. Mod. Phys.* **68**, 1125.

Buchmüller, W. and Wyler, D. (1986). Effective lagrangian analysis of new interactions and flavor conservation, *Nucl. Phys.* **B268**, 621.

Buras, A.J., Girrbach, J., Guadagnoli, D., and Isidori, G. (2012). On the Standard Model prediction for $BR(B_{s,d} \to \mu + \mu-)$, *Eur. Phys. J.* **C72**, 2172.

Burgers, G. and Jegerlehner, F. (1989). Δr, or the relation between the electroweak couplings and the weak vector boson masses, in *Z Physics at LEP* 1, ed. Altarelli, G., Kleiss, R., and Verzegnassi, C. (CERN 89-08, Geneva).

Burgess, C.P., Godfrey, S., Konig, H., London, D., and Maksymyk, I. (1994). A global fit to extended oblique parameters, *Phys. Lett.* **B326**, 276.

Cabibbo, N. (1963). Unitary symmetry and leptonic decays, *Phys. Rev. Lett.* **10**, 531.

Cabibbo, N., Swallow, E.C., and Winston, R. (2003). Semileptonic hyperon decays, *Ann. Rev. Nucl. Part. Sci.* **53**, 39.

Cahn, R.N., Chanowitz, M.S., and Fleishon, N. (1979). Higgs particle production by $Z \to H\gamma$, *Phys. Lett.* **B82**, 113.

Callen, C.G., Coleman, S., Wess, J., and Zumino, B. (1969). Structure of phenomenological lagrangians II, *Phys. Rev.* **177**, 2247.

Caprini, I., Colangelo G., and Leutwyler, H. (2006). Mass and width of the lowest resonance in *QCD*, *Phys. Rev. Lett.* **96**, 132001.

Caprini, I., Lellouch, L., and Neubert, M. (1998). Dispersive bounds on the shape of $\bar{B} \to D^* \ell \bar{\nu}$ form-factors, *Nucl. Phys.* **B530**, 153.

Carruthers, P. (1966). *Introduction to Unitary Symmetry* (Wiley Interscience, New York).

Casas, J.A., Di Clemente, V., Ibarra A., and Quiros, M. (2000). Massive neutrinos and the Higgs mass window, *Phys. Rev* **D62**, 053005.

Caswell, W.E. and Lepage, G.P. (1986). Effective lagrangian for bound state problems in *QED*, *QCD* and other field theories, *Phys. Lett.* **B167**, 437.

Chanowitz, M.S. and Ellis, J. (1972). Canonical anomalies and broken scale invariance, *Phys. Lett.* **40B**, 397.

Chanowitz, M.S., Furman, M.A., and Hinchliffe, I. (1978). Weak interactions of ultraheavy termions, *Phys. Lett.* **B78**, 285.

Chatrchyan, S. *et al.* (CMS collab.) (2013). On the mass and spin-parity of the Higgs boson candidate via its decays to Z boson pairs, *Phys. Rev. Lett.* **110**, 081803.

Chay, J., Georgi, H., and Grinstein, B. (1990). Lepton energy distributions in heavy meson decays from *QCD*, *Phys. Lett.* **B247**, 399.

Chen, Y. *et al.* (2006). Glueball spectrum and matrix elements on anisotropic lattices, *Phys. Rev.* **D73**, 014516.

Cheng, T.-P. and Dashen, R. (1971). Is $SU(2) \times SU(2)$ a better symmetry than $SU(3)$?, *Phys. Rev. Lett.* **26**, 594.

Cheng, T.-P. and Li, L.-F. (1984). *Gauge Theory of Elementary Particle Physics* (Clarendon Press, Oxford).

Chetyrkin, K.G. (1997). Quark mass anomalous dimension to O (α_s^4), *Phys. Lett.* **404**, 161.

Chetyrkin, K.G., Harlander, R., Seidensticker, T. and Steinhauser, M. (1999). Second order *QCD* corrections to $\Gamma(t \to Wb)$, *Phys. Rev.* **D60**, 114015.

Chetyrkin, K.G., Kniehl, B.A. and Steinhauser, M. (1998). Decoupling relations to $\mathcal{O}(\alpha_s^3)$ and their connection to low-energy theorems, *Nucl. Phys.* **B510**, 6.

Chodos, A., Jaffe, R.L., Johnson, K., Thorn, C.B., and Weisskopf, V.F. (1974). New extended model of hadrons, *Phys. Rev.* **D9**, 3471.

Cirigliano, V., Donoghue, J.F., and Golowich, E. (2000). Dimension eight operators in the weak OPE, *JHEP* **0010**, 048.

Cirigliano, V., Donoghue, J.F., Golowich, E., and Maltman, K. (2001). Determination of $\langle (\pi\pi)_{I=2} / Q(7, 8) / K^0 \rangle$ in the chiral limit, *Phys. Lett.* **B522**, 245.

Cirigliano, V., Ecker, G., Neufeld, A., Pich, A., and Portoles, J. (2012). Kaon decays in the Standard Model, *Rev. Mod. Phys.* **84**, 399.

Cirigliano, V., Golowich, E., and Maltman, K. (2003). *QCD* condensates for the light quark V–A correlator, *Phys. Rev.* **D68**, 054013.

Cirigliano, V. and Rosell, I. (2007). Two-loop effective theory analysis of $\pi(K) \to e\bar{\nu}_e\gamma$ branching ratios, *Phys. Rev. Lett.* **99**, 231801.

Ciuchini, M., Franco, E. Martinelli, G., Reina, L., and Silvestrini, L. (1995). An upgraded analysis of epsilon-prime epsilon at the next-to-leading order, *Z. Phys.* **C68**, 239.

Cohen, A.G., Glashow, S.L., and Ligeti, Z. (2009). Disentangling neutrino oscillations, *Phys. Lett.* **B678**, 191.

Colangelo, G., Gasser, J., and Leutwyler, H. (2001). $\pi\pi$ scattering, *Nucl. Phys.* **B603**, 125.

Colangelo, P. and Khodjamirian, A. (2000). QCD sum rules, a modern perspective, in *At the Frontier of Particle Physics*, ed. Shifman, M. (World Scientific, Singapore).

Coleman, S. (1985). *Aspects of Symmetry: Selected Erice Lectures* (Cambridge University Press, Cambridge).

Coleman, S. and Glashow, S.L. (1964). Departures from the eightfold way, *Phys. Rev.* **B134**, 671.

Coleman, S. and Mandula, J. (1967). All possible symmetries of the S-matrix, *Phys. Rev.* **159**, 1251.

Coleman, S., Wess, J., and Zumino, B. (1969). Structure of phenomenological lagrangians I, *Phys. Rev.* **177**, 2239.

Coleman, S. and Witten, E. (1980). Chiral symmetry breakdown in large-N_c chromodynamics, *Phys. Rev. Lett.* **45**, 100.

Collins, J., Duncan, A., and Joglekar, S. (1977). Trace and dilation anomalies in gauge theories, *Phys. Rev.* **D16**, 438.

Collins, J., Wilczek, F., and Zee, A. (1978). Low energy manifestations of heavy particles: application to the neutral current, *Phys. Rev.* **D18**, 242.

Collins, J.C. (2011). *Foundations of Perturbative QCD* (Cambridge University Press, Cambridge).

Crewther, R. (1972). Nonperturbative evaluation of the anomalies in low energy theorems, *Phys. Rev. Lett.* **28**, 1421.

Crewther, R. (1978). Effects of topological charge in gauge theory, *Acta Phys. Austriaca (Proc. Suppl.)* **19**, 47.

Cronin, J.A. (1967). Phenomenological model of strong and weak interactions in chiral $U(3)$, *Phys. Rev.* **161**, 1483.

Czarnecki, A. and Marciano, W.J. (1996). Electroweak radiative corrections to polarized Moller scattering asymmetries, *Phys. Rev.* **D53**, 1066.

Czarnecki, A. and Marciano, W.J. (2007). Electroweak radiative corrections to muon capture, *Phys. Rev. Lett.* **99**, 032003.

Czarnecki, A. and Melnikov, K. (1997). Two loop QCD corrections to $b \to c$ transitions at zero recoil: analytical results, *Nucl. Phys.* **B505**, 65.

Czarnecki, A. and Melnikov, K. (1999). Two loop QCD corrections to top quark width, *Nucl. Phys.* **B544**, 520.

Dalgic, E., Gray, A., Wingate, M., Davies, C.T.H., Lepage, G.P., and Shigemitsu, J. (2006). B meson semileptonic form-factors from unquenched lattice QCD, *Phys. Rev.* **D73**, 074502.

D'Ambrosio, G. and Espriu, D. (1986). Rare decay modes of the K meson in the chiral lagrangian, *Phys. Lett.* **B175**, 237.

Das, T. *et al.* (1967). Electromagnetic mass difference of pions, *Phys. Rev. Lett.* **18**, 759.

Dashen, R.F. (1969). Chiral $SU(3) \times SU(3)$ as a symmetry of the strong interactions, *Phys. Rev.* **183**, 1245.

Dashen, R.F., Jenkins, E.E., and Manohar, A.V. (1994). The 1/N(c) expansion for baryons, *Phys. Rev.* **D49**, 4713 (Erratum *ibid.* **D51**, 2489 (1995).

Davier, M., Hoecker, A., Malaescu, B., and Zhang, Z. (2011). Reevaluation of the hadronic contributions to the muon g-2 and to alpha(MZ), *Eur. Phys. J.* **C71**, 1515 (Erratum *ibid.* **C72**, 1874 (2012)).

Davis, R. (1964). Solar neutrinos. II: experimental, *Phys. Rev. Lett.* **12**, 303.

De Bruyn, K., Fleischer, R., Knegjens, R., Koppenburg, P., Merk, M., Pellegrino A., and Tuning, N. (2012). Probing New Physics via the $B_s^0 \rightarrow \mu^+\mu^-$ effective lifetime, *Phys. Rev. Lett.* **109**, 041801.

DeGrand, T. and Detar, C.E. (2010). *Lattice Methods for Quantum Chromodynamics* (World Scientific, Hackensack, NJ).

DeGrand, T., Jaffe, R.L., Johnson, K., and Kiskis, J. (1975). Masses and other parameters of light hadrons, *Phys. Rev.* **D12**, 2060.

Degrassi, G., Di Vita, S., Elias-Mir, J., Espinosa, J.R., Giudice, G., Isidori, G., and Strumia, A. (2012). Higgs mass and vacuum stability in the Standard Model at NNLO, *JHEP* **1208**, 098.

Dehnadi, B., Hoang, A.H., Mateu, V., and Zebarjad, S.M. (2011). Charm mass determination from QCD charmonium sum rules at order α_s^3 (arXiv:1102.2264 [hep-ph]).

Denner, A., Nierste, U., and Scharf, R. (1991). A compact expression for the scalar one loop four point function, *Nucl. Phys.* **B367**, 637.

de Putter, R., *et al.* (2012). New neutrino mass bounds from Sloan Digital Sky Survey III data release 8 photometric luminous galaxies, *Astrophys. J* **761**, 12.

De Roeck (2013). Higgs physics at CMS (plenary talk delivered on 6/24/13 at at 26th Intl. Symp. on Lept. Phot. Ints. at High Energies).

De Rujula, A., Georgi, H., and Glashow, S.L. (1975). Hadron masses in a gauge theory, *Phys. Rev.* **D12**, 147.

De Rujula, A., Lusignoli, M., Maiani, L., Petcov, S., and Petronzio, R. (1980). A fresh look at neutrino oscillations, *Nucl. Phys.* **B168**, 54.

Descotes-Genon, S., Hurth, T., Matias, J., and Virto, J. (2013). $B-> K^*\ell\ell$: The New Frontier of New Physics searches in flavor (arXiv:1305.4808 [hep-ph]).

DeWitt, B. (1967). Quantum theory of gravity: II, III, *Phys. Rev.* **162**, 1195, 1239.

DiVecchia, P. and Veneziano, G. (1980). Chiral dynamics in the large N_c limit, *Nucl. Phys.* **B171**, 253.

Djouadi, A. (2008). The anatomy of electro-weak symmetry breaking. I: the Higgs boson in the Standard Model, *Phys. Rept.* **457**, 1.

Djoudi, A., Kühn, J.H., and Zerwas, P.M. (1990). b-jet asymmetries in Z decays, *Zeit. Phys.* **C46**, 411.

Dobado, A. and Pelaez, J.R. (1997). The inverse amplitude method in chiral perturbation theory, *Phys. Rev.* **D56**, 3057.

Donoghue, J.F. (1994). General relativity as an effective field theory: the leading quantum corrections, *Phys. Rev.* **D50**, 3874.

Donoghue, J.F., Golowich, E., and Holstein, B.R. (1986a). Dispersive effects in D0 anti-D0 Mixing, *Phys. Rev.* **D33**, 179.

Donoghue, J.F., Golowich, E., and Holstein, B.R. (1986b). Low-energy weak interactions of quarks, *Phys. Rep.* **131**, 319.

Donoghue, J.F., He, X.G., and Pakvasa, S. (1986). Hyperon decays and *CP* nonconservation, *Phys. Rev.* **D34**, 833.

Donoghue, J.F. and Holstein, B.R. (1989). Pion transitions and models of chiral symmetry, *Phys. Rev.* **D40**, 2378.

Donoghue, J.F. and Johnson, K. (1980). The pion and an improved static bag, *Phys. Rev.* **D21**, 1975.

Donoghue, J.F. and Li, L.F. (1979). Properties of charged Higgs bosons, *Phys. Rev.* **D19**, 945.

Donoghue, J.F. and Nappi, C. (1986). The quark content of the proton, *Phys. Lett.* **B168**, 105.

Donoghue, J.F., Ramirez, C., and Valencia, G. (1989). Spectrum of *QCD* and chiral lagrangians of the strong and weak interaction, *Phys. Rev.* **D39**, 1947.

Dubnicka, S. *et al.* (2010). Quark model description of the tetraquark state *X*(3872) in a relativistic constituent quark model with infrared confinement, *Phys. Rev.* **D81**, 114007.

Dzuba, V.A., Berengut, J.C., Flambaum, V.V., and Roberts, B. (2012). Revisiting parity non-conservation in cesium, *Phys. Rev. Lett.* **109**, 203003.

Eberhardt, O., Herbert, G., Lacker, H., Lenz, A., Menzel, A., Nierste, U., and Wiebusch, M. (2012). Impact of a Higgs boson at a mass of 126 GeV on the Standard Model with three and four fermion generations, *Phys. Rev. Lett.* **109**, 241802.

Ecker, G., Gasser, J., Pich, A., and de Rafael, E. (1989). The role of resonances in chiral perturbation theory, *Nucl. Phys.* **B321**, 311.

Ecker, G., Pich, A., and de Rafael, E. (1988). Radiative kaon decays and *CP* violation in chiral perturbation theory, *Nucl. Phys.* **B303**, 665.

Eichten, E. (1988). Heavy quarks on the lattice, *Nucl. Phys. (Proc. Suppl.)* **4**, 170.

Eichten, E., Gottfried, K., Kinoshita, T., Lane, K.D. and Yan, T.-M. (1980). Charmonium: comparison with experiment, *Phys. Rev.* **D21**, 203.

Eichten, E. and Hill, B. (1990). Static effective field theory: $1/m$ Corrections, *Phys. Lett.* **B243**, 427

El-Khadra, A.X. and Luke, M. (2002). The mass of the *b* quark, *Ann. Rev. Nucl. Part. Sci.* **52**, 201.

Ellis, J.R., Gabathuler, E., and Karliner, M. (1989). The *OZI* rule does not apply to baryons, *Phys. Lett.* **B217**, 173.

Ellis, J.R., Gaillard, M.K., and Nanopoulos, D.V. (1976). A phenomenological profile of the Higgs Boson, *Nucl. Phys.* **B106**, 292.

Ellis, J.R. and You, T. (2012). Global analysis of the Higgs candidate with mass 125 GeV, *JHEP* **1209**, 123.

Ellis, R.K., Stirling, W.J., and Webber B.R. (2003). *QCD and Collider Physics* (Cambridge University Press, Cambridge).

Ellis, R.K. and Zanderighi, G. (2008). Scalar one-loop integrals for *QCD*, *JHEP* **0802**, 002.

Engel, J., Ramsey-Musolf, M.J., and van Kolck, U. (2013). Electric dipole moments of nucleons, nuclei, and atoms: the Standard Model and beyond, *Prog. Part. Nucl. Phys.* **71**, 21.

Epelbaum, E. and Meißner, U.G. (2012). Chiral dynamics of few- and many-nucleon systems, *Ann. Rev. Nucl. Part. Sci.* **62**, 159.

Erler, J. (2000). Global fits to electroweak data using GAPP (arXiv:0005084 [hep-ph]).

Erler, J. and Ramsey-Musolf, M.J. (2005). The weak mixing angle at low energies, *Phys. Rev.* **D72**, 073003.

Erler, J. and Su, S. (2013). The weak neutral current, *Prog. Part. Nucl. Phys.* **71**, 119.

Espinosa, J.R., Grojean, J., Muhlleitner, M., and Trott, M. (2012). First glimpses at Higgs' face, *JHEP* **1212**, 045.

Faddeev, L.D. and Popov, V.N. (1967). Feynman diagrams for the Yang–Mills field, *Phys. Lett.* **25B**, 29.

Falk, A.F., Grossman, Y., Ligeti, Z., and Petrov, A.A. (2002). $SU(3)$ breaking and D^0–anti-D^0 mixing, *Phys. Rev.* **D65**, 054034.

Ferroglia, A. and Sirlin, A. (2012). Radiative corrections in precision electroweak physics: a historical perspective, *Rev. Mod. Phys.* **85**, 1.

Fleischer, J., Jegerlehner, F., Tarasov, O.V., and Veretin, O.L. (1999). Two loop *QCD* corrections of the massive fermion propagator, *Nucl. Phys.* **B539**, 671 (Erratum *ibid.* **B571** (2000) 511).

Fogli, G.L., Lisi, E., Marrone, A., Montanino, D., Palazzo, A., and Rotunno, A.M. (2012). Global analysis of neutrino masses, mixings and phases: entering the era of leptonic *CP* violation searches, *Phys. Rev.* **D86**, 013012.

Forero, D.V., Tortola, M., and Valle, J.W.F. (2012). Global status of neutrino oscillation parameters after Neutrino-2012, *Phys. Rev.* **D86**, 073012.

Forkel, H. (2005). Direct instantons, topological charge screening, and QCD glueball sum rules, *Phys. Rev.* **D71**, 054008.

Friedrich, J. (2012). Studies in pion dynamics at COMPASS, *Proceedings of Science* (CONFINEMENT) X, 120.

Fritzsch, H. and Gell-Mann, M. (1972). Current algebra: quarks and what else?, in *Proc. XVI Int. Conf. on High Energy Physics*, ed. J.D. Jackson and A. Roberts (National Accelerator Laboratory, Batavia, IL.).

Fujikawa, K. (1979). Path integral measure for gauge invariant field theories, *Phys. Rev. Lett.* **42**, 1195.

Fujikawa, K. (1981). Energy momentum tensor in quantum field theory, *Phys. Rev.* **D23**, 2262.

Fujikawa, K. and Suzuki, H. (2004). *Path Integrals and Quantum Anomalies* (Oxford University Press, Oxford).

Gambino, P., Mannel, T., and Uraltsev, N. (2012). $B \to D^*$ zero-recoil formfactor and the Heavy Quark Expansion in *QCD*: a systematic study, *JHEP* **1210**, 169.

Gambino, P. and Sirlin, A. (1994). Relation between $\sin^2 \hat{\theta}_W(M_Z)$ and $\sin^2 \theta_{\text{eff}}^{\text{lep}}$, *Phys. Rev.* **D49**, 1160.

Gando, A. *et al.* (KamLAND collab.) (2011). Constraints on θ_{13} from a three-flavor oscillation analysis of reactor antineutrinos at KamLAND, *Phys. Rev.* **C83**, 052002.

Gando, A. *et al.* (KamLAND-Zen collab.) (2012). Measurement of the double-β decay half-life of ^{136}Xe with the KamLAND-Zen experiment, *Phys. Rev.* **C85**, 045504.

Gaillard, M.K. and Lee, B.W. (1974). $\Delta I = 1/2$ rule for nonleptonic decays in asymptotically free gauge theories, *Phys. Rev. Lett.* **33**, 108.

Gasser, J. (1987). Chiral perturbation theory and effective lagrangians, *Nucl. Phys.* **B279**, 65.

Gasser, J. and Leutwyler, H. (1984). Chiral perturbation theory to one loop, *Ann. Phys. (N.Y.)* **158**, 142.

Gasser, J. and Leutwyler, H. (1985a). Chiral perturbation theory: expansions in the mass of the strange quark, *Nucl. Phys.* **B250**, 465.

Gasser, J. and Leutwyler, H. (1985b). Low energy expansion of meson form factors, *Nucl. Phys.* **B250**, 517.

Gasser, J., Leutwyler, H., and Sainio, M.E. (1991). Sigma term update, *Phys. Lett.* **B253**, 252.

Gasser, J., Sainio, M.E., and Svarc, A. (1988). Nucleons with chiral loops, *Nucl. Phys.* **B307**, 779.

Gattringer, C. and Lang, C.B. (2010). *Quantum Chromodynamics on the Lattice* (Springer [Lect. Notes Phys. 788], Berlin).

Gavela, M.B., Hernandez, P., Orloff, J., and Pene, O. (1994). Standard Model *CP* violation and baryon asymmetry, *Mod. Phys. Lett.* **A9**, 795.

Gell-Mann, M. (1961). The Eightfold Way: a theory of strong interaction symmetry (CalTech Rept. CTSL-20).

Gell-Mann, M. and Levy, M. (1960). The axial vector current in beta decay, *Nuovo Cim.* **16**, 705.

Gell-Mann, M. and Low, F.E. (1954). Quantum electrodynamics at small distances, *Phys. Rev.* **95**, 1300.

Gell-Mann, M., Oakes, R. and Renner, B. (1968). Behavior of current divergences under $SU(3) \times SU(3)$, *Phys. Rev.* **175**, 2195.

Gell-Mann, M., Ramond, P., and Slansky, R. (1979). Complex spinors and unified theories, *A.I.P. Conf. Proc.* **C790927**, 315.

Georgi, H. (1984). *Weak Interactions and Modern Particle Theory* (Benjamin/Cummings, Menlo Park, CA).

Georgi, H. (1990). An effective field theory for heavy quarks at low-energies, *Phys. Lett.* **B240**, 447.

Georgi, H. (1992). *D*–anti-*D* mixing in heavy quark effective field theory, *Phys. Lett.* **B297**, 353.

Georgi, H. and Glashow, S.L. (1974). Unity of all elementary particle forces, *Phys. Rev. Lett.* **32**, 438.

Georgi, H., Grinstein, B. and Wise, M.B. (1990). Λ_b semileptonic form factors for $m_c \neq \infty$, *Phys. Lett.* **B252**, 456.

Georgi, H., Quinn, H.R., and Weinberg, S. (1974). Hierarchy of interactions in unified gauge theories, *Phys. Rev. Lett.* **33**, 451.

Gerstein, I., Jackiw, R, Lee, B.W., and Weinberg, S. (1971). Chiral loops, *Phys. Rev.* **D3**, 2486.

Gilkey, P. (1975). The spectral geometry of a Riemannian manifold, *J. Diff. Geom.* **10**, 601.

Gilman, F.J. and Wise, M.B. (1979). The $\Delta I = 1/2$ rule and violation of CP in the six-quark model, *Phys. Lett.* **B83**, 83.

Giri, A., Grossman, Y., Soffer, A., and Zupan, J. (2003). Determining γ using $B^\pm \to DK^\pm$ with multibody D decays, *Phys. Rev.* **D68**, 054018.

Giunti, C. and Kim, C.W. (2007). *Fundamentals of Neutrino Physics and Astrophysics* (Oxford University Press, Oxford).

Glashow, S.L. (1961). Partial symmetries of weak interactions, *Nucl. Phys.* **22**, 579.

Goity, J.L. (1986). The decays $K_S^0 \to \gamma\gamma$ and $K_L^0 \to \gamma\gamma$ in the chiral approach, *Zeit. Phys.* **C34**, 341.

Goity, J.L., Bernstein, A.M., and Holstein, B.R. (2002). The decay $\pi^0 \to \gamma\gamma$ to next to leading order in chiral perturbation theory, *Phys. Rev.* **D66**, 076014.

Goldberger, M. and Treiman, S.B. (1958). Conserved currents in the theory of the Fermi interaction, *Phys. Rev.* **110**, 1478.

Goldberger, W.D. and Rothstein, I.Z. (2006). An effective field theory of gravity for extended objects, *Phys. Rev.* **D73**, 104029.

Goldstone, J. (1961). Field theories with superconductor solutions, *Nuovo Cim.* **19**, 154.

Goldstone, J., Salam, A., and Weinberg, S. (1962). Broken symmetries, *Phys. Rev.* **127**, 965.

Golowich, E., Haqq, E., and Karl, G. (1983). Are there baryons which contain constituent gluons?, *Phys. Rev.* **D28**, 160.

Golowich, E., Hewett, J., Pakvasa, S., and Petrov, A.A. (2007). Implications of $D^0-\bar{D}^0$ mixing for New Physics, *Phys. Rev.* **D76**, 095009.

Golowich, E. and Holstein, B.R. (1975). Restrictions on the structure of the $\Delta S = 1$ nonleptonic Hamiltonian, *Phys. Rev. Lett.* **35**, 831.

Golowich, E. and Yang, T.C. (1979). Charged Higgs bosons and decays of heavy flavored mesons, *Phys. Lett.* **B80**, 245.

Gonzalez-Garcia, M.C., Maltoni, M., Salvado, J., and Schwetz, T. (2012). Global fit to three neutrino mixing: critical look at present precision, *JHEP* **1212**, 123.

Grevesse, N. and Sauval, A.I. (1998). Standard solar composition, *Space Sci. Rev.* **85**, 161.

Gronau, M. (2000). U spin symmetry in charmless B decays, *Phys. Lett.* **B492**, 297.

Gronau, M. and London, D. (1990). Isospin analysis of CP asymmetries in B decays, *Phys. Rev. Lett.* **65**, 3381.

Gronau M. and Wyler, D. (1991). On determining a weak phase from CP asymmetries in charged B decays, *Phys. Lett.* **B265**, 172.

Gross, D.J. and Wilczek, F. (1973a). Ultraviolet behavior of nonabelian gauge theories, *Phys. Rev. Lett.* **30**, 1343.

Gross, D.J. and Wilczek, F. (1973b). Asymptotically free gauge theories I, *Phys. Rev.* **D8**, 3633.

Grozin, A.G. (2004). *Heavy Quark Effective Theory* (Springer, New York).

Guberina, B., Peccei, R.D., and Rückl, R. (1980). Dimensional regularization techniques and their uses in calculating infrared safe weak decay processes, *Nucl. Phys.* **B171**, 333.

Gunion, J.F., Haber, H.E., Kane, G.L., and Dawson, S. (1990). *The Higgs Hunters Guide* (Addison-Wesley, Menlo Park, CA).

Haag, R. (1958). Quantum field theories with composite particles and asymptotic conditions, *Phys. Rev.* **112**, 669.

Haber, H.E. and O'Neil, E. (2011). Basis-independent methods for the two-Higgs-doublet model III: the CP-conserving limit, custodial symmetry, and the oblique parameters S, T, U, *Phys. Rev.* **D83**, 055017.

Hagelin, J.S. and Littenberg, L. (1989). Rare kaon decays, *Prog. Part. Nucl. Phys.* **23**, 1.

Hardy, J.C. and Towner, I.S. (2009). Superallowed $0^+ \rightarrow 0^+$ nuclear beta decays: a new survey with precision test of the conserved vector current. hypothesis and the standard model, *Phys. Rev.* **C79**, 055502.

Hardy, J.C. and Towner, I.S. (2010). The evaluation of V_{ud} and its impact on the unitarity of the Cabibbo–Kobayashi–Maskawa quark-mixing matrix, *Rpt. Prog. Phys.* **73**, 046301.

Harnett, D. *et al.* (2011). Near maximal mixing of scalar gluonium and quark mesons: a Gaussian sum rule analysis, *Nucl. Phys.* **A850**, 110.

Harrison, P.F., Perkins, D.H., and Scott,W.G., (2002). Tri-bimaximal mixing and the neutrino oscillation data, *Phys. Lett* **B530**, 167.

Hart, A. *et al.* (2006). A lattice study of the masses of singlet 0^{++} mesons, *Phys. Rev.* **D74**, 114504.

Hart, A. and Teper, M. (2002). Glueball spectrum in $O(a)$ improved lattice QCD, *Phys. Rev.* **D65**, 034502.

Haxton, W.C., Robertson, R.G.H., and Serenelli, A.M. (2012). Solar neutrinos: status and prospects (arXiv:1208.5723 [astro-ph SR]).

He, X.G. (1999). $SU(3)$ analysis of annihilation contributions and CP violating relations in $B \rightarrow PP decays$, *Eur. Phys. J.* **C9**, 443.

Heinemeyer, S. *et al.* (The LHC Higgs Cross Section Working Group collab.) (2013). Handbook of LHC Higgs cross sections: 3. Higgs properties (arXiv:1307.1347 [hep-ph]).

Hinshaw, G. *et al.* (WMAP collab.) (2013). Nine-year Wilkinson Microwave Anisotropy Probe (WMAP) observations: cosmological parameter results (arXiv:1212.5226v2 [astro-ph.CO]).

Hoang, A.H., Ligeti, Z., and Manohar, A.V. (1999). B decays in the upsilon expansion, *Phys. Rev.* **D59**, 074017.

Hoang, A.H., Smith, M.C., Stelzer, T., and Willenbrock, S. (1998). Quarkonia and the pole mass, *Phys. Rev.* **D59**, 114014.

Hoang A.H. and Stewart, I.W. (2003). Ultrasoft renormalization in nonrelativistic QCD, *Phys. Rev.* **D67**, 114020.

Hoang A.H. and Teubner, T. (1999). Top quark pair production close to threshold: top mass, width and momentum distribution, *Phys. Rev.* **D60**, 114027.

Höhler, G. (1983). Pion–nucleon scattering, in *Landolt Börnstein New Series 1-9b2*, ed. H. Schopper (Springer, Berlin).

Hollik, W.F.L. (1990). Radiative corrections in the Standard Model, *Fort. Phys.* **38**, 165.

Holstein, B.R. (1989). *Weak Interactions in Nuclei* (Princeton University Press, Princeton, NJ).

Hughes, R.J. (1981). More comments on asymptotic freedom, *Nucl. Phys.* **B186**, 376.

Iizuka, J. (1966). A systematics and phenomenology of meson family, *Prog. Theor. Phys. Suppl.* **37–38**, 21.

Inami, T. and Lim, C.S. (1981). Effects of superheavy quarks and leptons on low energy weak processes $K_L^0 \to \bar{\mu}\mu$, $K^+ \to \pi^+\bar{\nu}\nu$, and $K^0 - \bar{K}^0$, *Prog. Theor. Phys.* **65**, 297.

Ioffe, B.L. (1981). Calculation of baryon masses in QCD, *Nucl. Phys.* **B188**, 317.

Ioffe, B.L., Fadin, V.S., and Lipatov, L.N. (2010). *Quantum Chromodynamics: Perturbative and Nonperturbative Aspects* (Cambridge University Press, Cambridge).

Ioffe, B.L. and Shifman, M.A. (1980). The decay $\psi' \to J/\psi + \pi^0(\eta)$ and quark masses, *Phys. Lett.* **95B**, 99.

Isgur, N. and Karl, G. (1978). P-wave baryons in the quark model, *Phys. Rev.* **D18**, 4187.

Isgur, N. and Wise, M.B. (1989). Weak decays of heavy mesons in the static quark approximation, *Phys. Lett.* **B232**, 113.

Isgur, N. and Wise, M.B. (1990). Weak transition form factors between heavy mesons, *Phys. Lett.* **237**, 527.

Itzykson, C. and Zuber, J.-B. (1980). *Quantum Field Theory* (McGraw-Hill, New York).

Jackiw, R. and Rebbi, C. (1976). Vacuum periodicity in a Yang–Mills quantum theory, *Phys. Rev. Lett.* **37**, 172.

Jacobs, K. (2013). Higgs physics at ATLAS (plenary talk delivered on 6/24/13 at at 26th Intl. Symp. on Lept. Phot. Ints. at High Energies).

Jaffe, R.L. (1977). Perhaps a stable dibaryon, *Phys. Rev. Lett.* **38**, 195.

Jaffe, R.L., Johnson, K., and Ryzak, Z. (1986). Qualitative features of the glueball spectrum, *Ann. Phys. (N.Y.)* **168**, 334.

Jaffe, R.L. and Manohar, A. (1990). The g_1 problem: deep inelastic electron scattering and the spin of the proton, *Nucl. Phys.* **B337**, 509.

Jarlskog, C. (1985). Commutators of the quark mass matrices in the standard electroweak model and a measure of maximal CP violation, *Phys. Rev. Lett.* **55**, 1039.

Jarlskog, C. (1989). Introduction to CP violation, in *CP Violation*, ed. C. Jarlskog (World Scientific, Singapore).

Jenkins, E.E. (1998). Large $N(c)$ baryons, *Ann. Rev. Nucl. Part. Sci.* **48**, 81.

Ji, X.-D. (1994). Chiral odd and spin dependent quark fragmentation functions and their applications, *Phys. Rev.* **D49**, 114.

Ji, X.-D., Tang, J., and Hoodbhoy, P. (1996). Spin structure of the nucleon in the asymptotic limit, *Phys. Rev. Lett.* **76**, 740.

Johnson, K. (1978). A field theory lagrangian for the MIT bag model, *Phys. Lett.* **78B**, 259.

Johnson, K. and Thorn, C.B. (1976). Stringlike solutions of the bag model, *Phys. Rev.* **D13**, 1934.

Kambor, J. and Holstein, B.R. (1994). $K_S \to \gamma\gamma$, $K_L \to \pi^0\gamma\gamma$ and unitarity, *Phys. Rev.* **D49**, 2346.

Kambor, J., Missimer, J., and Wyler, D. (1990). The chiral loop expansion of the nonleptonic weak interactions of mesons, *Nucl. Phys.* **B346**, 17.

Kaymakcalan, Ö., Rajeev, S., and Schecter, J. (1984). Nonabelian anomaly and vector meson decays, *Phys. Rev.* **D30**, 594.

Kayser, B., Kopp, J., Roberston, R.G.H., and Vogel, P. (2010). On a theory of neutrino oscillations with entanglement, *Phys. Rev.* **D82**, 093003.

Keung, W.Y. and Marciano, W.J. (1984). Higgs scalar decays: $H \rightarrow W + X$, *Phys. Rev.* **D30**, 248.

King, S.F. and Luhn, C. (2013). Neutrino mass and mixing with discrete symmetry, *Rept. Prog. Phys.* **76**, 056201.

Kinoshita, T. and Sirlin, A. (1959). Radiative corrections to Fermi interactions, *Phys. Rev.* **113**, 1652.

Klinkhamer, F.R. and Manton, N.S. (1984). A saddle point solution in the Weinberg–Salam theory, *Phys. Rev.* **D30**, 2212.

Kobayashi, M. and Maskawa, T. (1973). *CP* violation in the renormalizable theory of weak interactions, *Prog. Theo. Phys.* **49**, 652.

Kühn, J.H., Steinhauser, M., and Sturm, C. (2007). Heavy quark masses from sum rules in four-loop approximation, *Nucl. Phys.* **B778**, 192.

Kumar, K.S., Mantry, S., Marciano, W.J., and Souder, P.A. (2013). Low energy measurements of the weak mixing angle (arXiv:1302.6263 [hep-ex]).

Kuzmin, V.A., Rubakov, V.A. and Shaposhnikov, M.E. (1985). On the anomalous electroweak baryon number nonconservation in the early universe, *Phys. Lett.* **B155**, 36.

Kwong, W., Quigg, C., and Rosner, J.L. (1987). Heavy-quark systems, *Ann. Rev. Nucl. Part. Sci.* **37**, 325.

Laiho, J., Lunghi, E., and Van de Water, R.S. (2010). Lattice *QCD* inputs to the CKM unitarity triangle analysis, *Phys. Rev.* **D81**, 034503.

Langacker, P. (1981). Grand unified theories and proton decay, *Phys. Rep.* **C72**, 185.

Langacker, P. (2010). *The Standard Model and Beyond* (Taylor and Francis, Boca Raton, FL).

Lee, B.W., Quigg, C., and Thacker, H.B. (1977). Weak interactions at very high energies: the role of the Higgs-boson mass, *Phys. Rev.* **D16**, 1519.

Lee, B.W. and Swift, A.R. (1964). Dynamical basis of the sum rule $2\Xi_-^- = \Lambda_- + \sqrt{3}\Sigma_0^+$, *Phys. Rev.* **B136**, 228.

Lee, T.D. (1973). A theory of spontaneous *T* violation, *Phys. Rev.* **D8**, 1226.

Lee, T.D. (1982). *Particle Physics and Introduction to Field Theory* (Harwood, New York).

Lehmann, H. (1972). Chiral invariance and effective range expansion for pion pion scattering, *Phys. Lett.* **B41**, 529.

Leibrandt, G. (1975). Introduction to the technique of dimensional regularization, *Rev. Mod. Phys.* **47**, 849.

Lenz, A. and Nierste, U. (2007). Theoretical update of $B_s - \bar{B}_s$ mixing, *JHEP* **0706**, 072.

Lenz, A. and Nierste, U. (2011). Numerical updates of lifetimes and mixing parameters of B mesons, (arXiv:1102.4274 [hep-ph]).

Lepage, G.P. (1998). Perturbative improvement for lattice *QCD*: an update, *Nucl. Phys. Proc. Suppl.* **60A**, 267.

Lepage, G.P. and Thacker, B.A. (1988). Effective lagrangians for simulation of heavy quark systems, *Nucl. Phys.(Proc.Suppl.)* **4**, 199.

Leutwyler, H. and Roos, M. (1984). Determination of the elements V_{us} and V_{ud} of the Kobayashi–Maskawa matrix, *Z. Phys.* **C25**, 91.

Le Yaouanc, A., Oliver, L., Pène, O., and Raynal, J.-C. (1985). Quark model of light mesons with dynamically broken chiral symmetry, *Phys. Rev.* **D31**, 137.

Le Yaouanc, A., Oliver, L., Pène, O., and Raynal, J.-C. (1988). *Hadron Transitions in the Quark Model* (Gordon and Breach, New York).

Lipkin, H.J. (1984). The theoretical basis and phenomenology of the *OZI* rule, *Nucl. Phys.* **B244**, 147.

Lipkin, H.J. (2005). Is observed direct *CP* violation in $B_d \to K^+\pi^-$ due to New Physics? Check Standard Model prediction of equal violation in $B_s \to K^-\pi^+$, *Phys. Lett.* **B621**, 126.

Liu Z.Q. *et al.* (Belle collab.) (2013). Study of $e^+e^- \to \pi^+\pi^- J/\psi$ and observation of a charged charmonium-like state at Belle, *Phys. Rev. Lett.* **110**, 252002.

Lucha, W., Melikhov, D., and Simula, S. (2011). OPE, charm-quark mass, and decay constants of D and D_s mesons from *QCD* sum rules, *Phys. Lett.* **B701**, 82.

Luke, M. (1990). Effects of subleading operators in the heavy quark effective theory, *Phys. Lett.* **B252**, 447.

Ma, E. and Maniatis, M. (2010). Symbiotic symmetries of the two-Higgs-doublet model, *Phys. Lett.* **B683**, 33.

Maki, Z., Nakagawa, M., and Sakata, S. (1962). Remarks on the unified model of elementary particles, *Prog. Theor. Phys.* **28**, 870.

Mangano, G., Miele, G., Pastor, S., Pinto, T., Pisanti, O., and Serpico, T. (2005). Relic neutrino decoupling including flavor oscillations, *Nucl. Phys.* **B729**, 221.

Mannel, T. (1994). Operator product expansion for inclusive semileptonic decays in heavy quark effective field theory, *Nucl. Phys.* **B413**, 396.

Mannel, T. (2004). *Effective Field Theories in Flavour Physics* (Springer, New York).

Manohar, A.V. and Mateu, V. (2008). Dispersion relation bounds for $\pi\pi$ scattering, *Phys. Rev.* **D77**, 094019.

Manohar A.V. and Stewart, I.W. (2007). The zero-bin and mode factorization in quantum field theory, *Phys. Rev.* **D76**, 074002.

Manohar, A.V. and Wise, M.B. (1994). Inclusive semileptonic B and polarized Λ_b decays from *QCD*, *Phys. Rev.* **D49**, 1310.

Manohar, A. and Wise, M.B. (2007). *Heavy Quark Physics* (Cambridge University Press, Cambridge).

Marciano, W.J. (1979). Weak mixing angle and grand unified gauge theories, *Phys. Rev.* **D20**, 274.

Marciano, W.J. (1999). Fermi constants and New Physics, *Phys. Rev.* **D60**, 093006.

Marciano, W.J. (2011). Precision electroweak tests of the Standard Model, *J. Phys. Conf. Ser.* **312**, 102002.

Marciano, W.J. and Querjeiro, A. (1986). Bound on the W boson electric dipole moment, *Phys. Rev.* **D33**, 3449.

Marciano, W.J. and Sirlin, A. (1980). Radiative corrections to neutrino-induced neutral-current phenomena in the $SU(2)_L \times U(1)$ theory, *Phys. Rev.* **D22**, 2695.

Marciano, W.J. and Sirlin, A. (1981). Precise $SU(5)$ predictions for $sin^2\theta_w$, m_W and m_Z, *Phys. Rev. Lett.* **46**, 163.

Marshak, R.E., Riazuddin, and Ryan, C.P. (1969). *Theory of Weak Interactions in Particle Physics* (Wiley, New York).

Martin, A. (1981). A simultaneous fit of $b\bar{b}$, $c\bar{c}$, $s\bar{s}$ and $c\bar{s}$ spectra, *Phys. Lett.* **100B**, 511.

Melnikov, K. and Ritbergen, T.V. (2000). The three loop relation between the MS-bar and the pole quark masses, *Phys. Lett.* **B482**, 99.

Mereghetti, E., Hockings, W.H., and van Kolck, U. (2010). The effective chiral lagrangian from the θ term, *Annals Phys.* **325**, 2363.

Mikheev, S.P. and Smirnov, A.Y. (1985). Resonance amplification of oscillations in matter and spectroscopy of solar neutrinos, *Sov. J. Nucl. Phys.* **42**, 913.

Misiak, M. Asatrian, H.M., *et al.* (2007). Estimate of $B \to X_s\gamma$ at $O(\alpha_s^2)$, *Phys. Rev. Lett.* **98**, 022002.

Mohapatra, R.N., Antusch, S. *et al.* (2007). Theory of neutrinos: a White Paper, *Rept. Prog. Phys.* **70**, 1757.

Mohr, P.J., Newell, D.B., and Taylor, B.N. (2012). CODATA recommended values of the fundamental physical constants: 2010, *Rev. Mod. Phys.* **84**, 1527.

Morningstar, C.J. and Peardon, M.J. (1999). The glueball spectrum from an anisotropic lattice study, *Phys. Rev.* **D60**, 034509.

Muller, T. (2012). New results from the top quark (plenary talk delivered 7/10/12 at 36th Intl. Conf. for High Energy Physics [indico.cern.ch/conferenceTimeTable.py?confId= 181298]).

Nambu, Y. and Lurie, D. (1962). Chirality conservation and soft pion production, *Phys. Rev.* **125**, 1429.

Narison, S. (1989). *QCD Spectral Sum Rules* (World Scientific, Singapore).

Neubert, M. (2005). Effective field theory and heavy quark physics, TASI-2004 (hep-ph/0512222).

Noecker, M.C., Masterson, B.P., and Wieman, C.E. (1988). Precision measurement of parity nonconservation in atomic cesium, *Phys. Rev. Lett.* **61**, 310.

Ochs, W. (2013). The status of glueballs, *J. Phys.* **G40**, 043001.

Ohl, T., Ricciardi, G., and Simmons, E.H. (1993). D–anti-D mixing in heavy quark effective field theory: the sequel, *Nucl. Phys.* **B403**, 605.

Okubo, S. (1962). Note on unitary symmetry in strong interactions, *Prog. Theo. Phys.* **27**, 949.

Okubo, S. (1963). φ meson and unitarity symmetry model, *Phys. Lett.* **5**, 165.

Okun, L. (1982). *Leptons and Quarks* (North-Holland, Amsterdam).

Ovrut, B. and Schnitzer, H. (1980). Decoupling theorems for effective field theories, *Phys. Rev.* **D22**, 2518.

Pak, N.K. and Rossi, P. (1985). Gauged Goldstone boson effective action from direct integration of Bardeen anomaly, *Nucl. Phys.* **B250**, 279.

Parke, S.J. (1986). Nonadiabatic level crossing in resonant neutrino oscillation, *Phys. Rev. Lett.* **57**, 1275.

Peccei, R.D. (1989). The strong *CP* problem, in *CP Violation*, ed. C. Jarlskog (World Scientific, Singapore).

Peccei, R.D. and Quinn, H.R. (1977). *CP* conservation in the presence of instantons, *Phys. Rev. Lett.* **38**, 1440.

Pelaez, J.R. (2004). On the nature of light scalar mesons from their large $N(c)$ behavior, *Phys. Rev. Lett.* **92**, 102001.

Peskin, M.E. and Takeuchi, T. (1990). New constraint on a strongly interacting Higgs sector, *Phys. Rev. Lett.* **65**, 964.

Pich, A. (2013). Review of α_s determinations (arXiv:1303.2262 [hep-ph]).

Pineda, A. and Soto, J. (1998). Effective field theory for ultrasoft momenta in *NRQCD* and *NRQED*, *Nucl. Phys. Proc. Suppl.* **64**, 428.

Politzer, H.D. (1973). Reliable perturbative results for strong interactions?, *Phys. Rev. Lett.* **30**, 1346.

Politzer, H.D. (1974). Asymptotic freedom: an approach to strong interactions, *Phys. Rep.* **14C**, 274.

Pontecorvo, B. (1968). Neutrino experiments and the problem of conservation of leptonic charge, *Sov. Phys. JETP* **26**, 984.

Porsev, S.G., Beloy, K., and Derevianko, A. (2009). Precision determination of electroweak coupling from atomic parity violation and implications for particle physics, *Phys. Rev. Lett.* **102**, 181601.

Porto, R.A., Ross A., and Rothstein, I.Z. (2011). Spin induced multipole moments for the gravitational wave flux from binary inspirals to third Post-Newtonian order, *JCAP* **1103**, 009.

Quaresma, M. (2012). Study of the nucleon spin structure by the Drell–Yan process in the COMPASS-II experiment, *Acta. Phys. Polon. Suppl.* **5**, 1163.

Rafael, E. de (1998). An introduction to sum rules in *QCD*: course (arXiv:9802448 [hep-ph]).

Ramond, P. (1989). *Field Theory: A Modern Primer* (Addison-Wesley, Menlo Park, CA).

Reinders, L.J., Rubenstein, H., and Yazaki, S. (1985). Hadron properties from *QCD* sum rules, *Phys. Rep.* **127**, 1.

Richards, C.S. *et al.* (2010). Glueball mass measurments from improved staggered fermion simulations, *Phys. Rev.* **D86**, 034501.

Richardson, J. L. (1979). The heavy quark potential and the Υ, J/Ψ systems, *Phys. Lett.* **82B**, 272.

Riggenbach, C., Gasser, J., Donoghue, J.F., and Holstein, B.R. (1991). Chiral symmetry and the large N_c limit in $K_{\ell 4}$ decays, *Phys. Rev.* **43**, 127.

Ritbergen. T. van, Vermaseren, J.A.M., and Larin, S.A. (1997). The four loop beta function in quantum chromodynamics, *Phys. Lett.* **B400**, 379.

Rothe, H.J. (2012). *Lattice Gauge Theories: An Introduction* (4th Edition) (World Scientific, Singapore).

Rosenzweig, C., Schechter, J., and Trahern, C.G. (1980). Is the effective lagrangian for *QCD* a σ model, *Phys. Rev.* **D21**, 3388.

Roy, S.M. (1971). Exact integral equation for pion pion scattering involving only physical region partial waves, *Phys. Lett.* **B36**, 353.

Sakharov, A.D. (1967). Violation of *CP* invariance, *C* asymmetry, and baryon asymmetry of the universe, *JETP Lett.* **5**, 24.

Sakurai, J.J. (1969). *Currents and Mesons* (University of Chicago Press, Chicago).

Salam, A. (1969). Weak and electromagnetic interactions, in *Elementary Particle Theory; Nobel Symposium No.8*, ed. N. Svartholm (Almqvist and Wiksell, Stockholm).

Schael, S. *et al.* (ALEPH and DELPHI and L3 and OPAL and SLD and LEP Electroweak Working Group and SLD Electroweak Group and SLD Heavy Flavour Group collabs.) (2006). Precision electroweak measurements on the Z resonance, *Phys. Rept.* **427**, 257.

S. Schael *et al.* (ALEPH and DELPHI and L3 and OPAL and LEP Electroweak Working Group collab.) (2013). Electroweak measurements in electron-positron collisions at W-boson-pair energies at LEP (arXiv:1302.3415 [hep-ex]).

Schnitzer, H.J. (1984). The soft pion Skyrmion lagrangian and strong CP violation, *Phys. Lett.* **B139**, 217.

Schulman, L.S. (1981). *Techniques and Applications of Path Integration* (Wiley, New York).

Schwinger, J. (1951). On gauge invariance and vacuum polarization, *Phys. Rev.* **82**, 664.

Schwinger, J. (1954). The theory of quantized fields, *Phys. Rev.* **93**, 615.

Shifman, M.A. (2010). Vacuum structure and QCD sum rules: introduction, *Int. J. Mod. Phys.* **A25**, 226.

Shifman, M.A., Uraltsev, N.G., and Vainshtein, A.I. (1995). Operator product expansion sum rules for heavy flavor transitions and the determination of $abs[V_{cb}]$, *Phys. Rev.* **D51**, 2217.

Shifman, M.A., Vainshtein, A.I., Voloshin, M.B., and Zakharov, V.I. (1979). Low-energy theorems for Higgs boson couplings to photons, *Sov. J. Nucl. Phys.* **30**, 711.

Shifman, M.A., Vainshtein, A., and Zakharov, V. (1977). Nonleptonic decays of K mesons and hyperons, *JETP* **45**, 670.

Shifman, M.A., Vainshtein, A., and Zakharov, V. (1979a). QCD and resonance physics: I, II, III, *Nucl. Phys.* **B147**, 385,488,519.

Shifman, M.A., Vainshtein, A., and Zakharov, V. (1979b). Nonleptonic decays of strange particles, *Nucl. Phys.* **B120**, 316.

Shifman, M.A. and Voloshin, M. (1988). On production of D^* and D mesons in B meson decay, *Sov. J. Nucl. Phys.* **47**, 511.

Shore, G.M. (1981). On the Meissner effect in gauge theories, *Ann. Phys.* **134**, 259.

Shore, G.M. (2008). The $U(1)_A$ anomaly and QCD phenomenology, *Lect. Notes Phys.* **737**, 235.

Sikivie, P., Susskind, L., Voloshin, M., and Zakharov, V. (1980). Isospin breaking in technicolor models, *Nucl. Phys.* **B173**, 189.

Sirlin, A. (1980). Radiative corrections in the $SU(2)_L \times U(1)$ theory: a simple renormalization framework, *Phys. Rev.* **D22**, 971.

Skiba, W. (2010). TASI lectures on effective field theory and precision electroweak measurements (arXiv:1006.2142 [hep-ph]).

Skyrme, T.H.R. (1961). A non-linear field theory, *Proc. R. Soc. Lon.* **A260**, 127.

Skyrme, T.H.R. (1962). A unified field theory of mesons and baryons, *Nucl. Phys.* **31**, 556.

Smirnov, V.A. (2002). Applied asymptotic expansions in momenta and masses, *Springer Tracts Mod. Phys.* **177**, 1.

Smirnov, V.A. (2012). Analytic tools for Feynman integrals, *Springer Tracts Mod. Phys.* **250**, 1.

Spira, M., Djouadi, A., Graudenz, D., and Zerwas, P.M. (1995). Higgs boson production at the LHC, *Nucl. Phys.* **B453**, 17.

Sutherland, D. (1967). Current algebra and some non-strong meson decays, *Nucl. Phys.* **B2**, 433.

't Hooft, G. (1974). A planar diagram theory of the strong interactions, *Nucl. Phys.* **B72**, 461.

't Hooft, G. (1976a). Computation of the quantum effects due to a four-dimensional pseudoparticle, *Phys. Rev.* **D14**, 3432.

't Hooft, G. (1976b). Symmetry breaking through Bell–Jackiw anomalies, *Phys. Rev. Lett.* **37**, 8.

't Hooft, G., Isidori, G., Maiani, L., Polosa, A., and Riquer, V. (2008). A theory of scalar mesons, *Phys. Lett.* **B662**, 424.

't Hooft, G. and Veltman, M. (1972). Regularization and renormalization of gauge fields, *Nucl. Phys.* **B44**, 189.

't Hooft, G. and Veltman, M.J.G. (1979). Scalar one loop integrals, *Nucl. Phys.* **B153**, 365.

van Kolck, U. (2008). Nuclear Physics from *QCD, Proceedings of Science (CONFINEMENT)* **8**, 030.

Veltman, M. (1967). Theoretical aspects of high energy neutrino interactions, *Proc. R. Soc. Lon.* **A301**, 103.

Veltman, M. (1977a). Limit on mass differences in the Weinberg model, *Nucl. Phys.* **B123**, 89.

Veltman, M. (1977b). Second threshold in weak interactions, *Acta Phys. Polonica* **B8**, 475.

Vermaseren, J.A.M., Larin, S.A., and van Ritbergen, T. (1997). The four loop quark mass anomalous dimension and the invariant quark mass, *Phys. Lett.* **B405**, 327.

Vesterinen M. (on behalf of the LHCb collab.) (2013). LHCb semileptonic asymmetry, (arXiv:1306.0092 [hep-ex]).

Vetterli, D. *et al.* (1989). Effects of vacuum polarization in hadron–hadron scattering, *Phys. Rev. Lett.* **62**, 1453.

Webber, D.M. *et al.* (MuLan collab.) (2011). Measurement of the positive muon lifetime and determination of the Fermi constant to part-per-million precision, *Phys. Rev. Lett.* **106**, 041803.

Weinberg, S. (1966). Pion scattering lengths, *Phys. Rev. Lett.* **17**, 616.

Weinberg, S. (1967a). Precise relations between the spectra of vector and axial vector mesons, *Phys. Rev. Lett.* **18**, 507.

Weinberg, S. (1967b). A model for leptons, *Phys. Rev. Lett.* **19**, 1264.

Weinberg, S. (1968). Nonlinear realizations of chiral symmetry, *Phys. Rev.* **166**, 1568.

Weinberg, S. (1973). New approach to the renormalization group, *Phys. Rev.* **D8**, 3497.

Weinberg, S. (1979a). Baryon and lepton nonconserving processes, *Phys. Rev. Lett.* **43**, 1566.

Weinberg, S. (1979b). Phenomenological lagrangians, *Physica* **A96**, 327.

Weinberg, S. (1990). Nuclear forces from chiral lagrangians, *Phys. Lett.* **B 251**, 288.

Weinstein, J. and Isgur, N. (1983). $qq\bar{q}\bar{q}$ system in a potential model, *Phys. Rev.* **D27**, 588.

Wess, J. and Zumino, B. (1971). Consequences of anomalous Ward identities, *Phys. Lett.* **B37**, 95.

Wilkinson, D.T. and Marrs, R.E. (1972). Finite size effects in allowed beta decay, *Nucl. Inst. Meth.* **105**, 505.

Willenbrock, S. (2004). Symmetries of the Standard Model (arXiv:0410370 [hep-ph]).

Wilson, K. (1969). Nonlagrangian models of current algebra, *Phys. Rev.* **179**, 1499.

Wise, M.B. (1991). New symmetries of the strong interaction (Proceedings of the 1991 Lake Louise Winter Institute and Caltech preprint CALT-68-1721).

Witten, E. (1979). Current algebra for the $U_A(1)$ 'Goldstone boson', *Nucl. Phys.* **B156**, 269.

Witten, E. (1983a). Global aspects of current algebra, *Nucl. Phys.* **B223**, 422.

Witten, E. (1983b). Current algebra, baryons and quark confinement, *Nucl. Phys.* **B223**, 433.

Wolfenstein, L. (1978). Neutrino oscillations in matter, *Phys. Rev.* **D17**, 2369.

Wolfenstein, L. (1983). Parametrization of the Kobayashi–Maskawa matrix, *Phys. Rev. Lett.* **51**, 1945.

Wolfenstein, L. (1985). D^0 anti-D^0 mixing, *Phys. Lett.* **B164**, 170.

Yang, C.N. (1950). Selection rules for the dematerialization of a particle into two photons, *Phys. Rev.* **77**, 242.

Yang, C.N. and Mills, R.L. (1954). Conservation of isotopic spin and isotopic gauge theory, *Phys. Rev.* **96**, 191.

Zeller, G.P. *et al.* (NuTeV collab.) (2001). A precise determination of electroweak parameters in neutrino nucleon scattering, *Phys. Rev. Lett.* **88**, 091802 (Erratum *ibid.* **90**, 239902 (2003)).

Zhan, X. *et al.* (2011). High precision measurement of the proton elastic form factor ratio $\mu_p G_E / G_M$ at low Q^2, *Phys. Lett.* **B705**, 59.

Zweig, G. (1965). Fractional charged particles and $SU(6)$, in *Symmetries in Elementary Particle Physics*, ed. A. Zichichi (Academic Press, New York).

Index